UNIVERSAL HELIOPHYSICAL PROCESSES

IAU SYMPOSIUM No. 257

COVER ILLUSTRATION:

Matter in the Solar System is organized by gravitational and magnetic forces. These forces form structures that evolve through Universal Processes, and Heliophysics is the study of these processes in diverse environments. The major plasma environments in the Solar System are illustrated; Top: the Sun and its immediate vicinity with the extreme-ultraviolet emission from the solar disk and coronal mass ejections representing the electromagnetic and mass emission; the Sun dominates the forces in the solar system; Bottom right: the magnetic fields of planets dominate and define the structure of the space environment surrounding them as they interact with the solar wind; Bottom left: the heliosphere extends out to the interstellar boundary and is influenced by plasma and particles from the interstellar medium.

INTERNATIONAL ASTRONOMICAL UNION

UNION ASTRONOMIQUE INTERNATIONALE

International Astronomical Union

UNIVERSAL
HELIOPHYSICAL PROCESSES

PROCEEDINGS OF THE 257th SYMPOSIUM OF
THE INTERNATIONAL ASTRONOMICAL UNION
HELD IN IOANNINA, GREECE
SEPTEMBER 15–19, 2008

Edited by

NATCHIMUTHUKONAR GOPALSWAMY
NASA Goddard Space Flight Center, Greenbelt, MD, USA

and

DAVID F. WEBB
ISR, Boston College, Chestnut Hill, MA, USA

CAMBRIDGE
UNIVERSITY PRESS

CAMBRIDGE UNIVERSITY PRESS
The Edinburgh Building, Cambridge CB2 8RU, United Kingdom
32 Avenue of the Americas, New York, NY 10013-2473, USA
477 Williamstown Road, Port Melbourne, VIC 3207, Australia
Ruiz de Alarcón 13, 28014 Madrid, Spain
Dock House, The Waterfront, Cape Town 8001, South Africa

First published 2009

Printed in the United Kingdom at the University Press, Cambridge

Typeset in System LaTeX 2_ε

A catalogue record for this book is available from the British Library

Library of Congress Cataloguing in Publication data

ISBN 9780521889889 hardback
ISSN 1743–9213

Table of Contents

Session I: Opening Session
Chair: C. Alissandrakis

Session II: Space Weather
Chair: K. Shibata

Session III: Solar Sources of Heliospheric Variability
Chair: J. Davila

Session IV: Solar-Heliospheric Variability: CMEs
Chair: C. Alissandrakis

Session V: Plasma and Radio Emission Processes
Chair: N. Gopalswamy

Session VI: 3-D Reconnection Processes
Chairs: C. Mandrini & B. Vrsnak

Session VII: Energetic Particles in the Heliosphere
Chair: D. Webb

Session VIII: Heliosphere Boundaries, Interfaces and Shocks
Chair: J. D. Richardson

Session IX: Planetary Atmospheres, Ionospheres, Magnetospheres
Chair: J. Davila

Preface

The IAU Symposium No. 257 on "Universal Heliophysical Processes" was held at the Scientific and Technological Park of Epirus (STEP) in Ioannina, Greece during September 15 19, 2008. The symposium was cosponsored by the International Heliophysical Year (IHY) program and the University of Ioannina. The IHY formally commenced in February 2007, marking the fiftieth anniversary of the International Geophysical Year (IGY, 195758). Like the IGY, IHY activities are centered on international scientific collaboration and global studies but at a much greater physical scale (from Geophysics to Heliophysics) that encompasses the entire solar system and its interaction with the local interstellar medium. The focus of Symposium 257 was on the universality of physical processes in the region of space directly influenced by the Sun through its mass and electromagnetic emissions: the heliospace.

The activities of the IHY revolve around four key elements: Science (coordinated investigation programs, or CIPs conducted as campaigns to investigate specific scientific questions), Observatory Development (deployment of small instruments in developing countries in collaboration with the United Nations), Public Outreach (to communicate the beauty, relevance and significance of space science to the general public and students), and the IGY Gold Program (to identify and honor all those scientists who worked for the IGY program). This IAU symposium is one of the activities planned under the science element.

The focus of the symposium was on the universality of physical processes in the region of space directly influenced by the Sun, called heliospace. The symposium featured about 100 presentations by 96 participants (from 26 countries). In order to capture the spirit of the discussions that followed the talks, we have included the questions asked and the answers provided by the presenters at the end of each paper.

The symposium began with a keynote address that retraced the development of heliophysics from the beginning of space exploration during IGY to IHY. The fierce competition between the Soviets and the Americans, the creation of space agencies in various countries, the development of a broad and active scientific community, and the unprecedented international cooperation were recognized as the primary drivers behind the successes. The inaugural session also provided an overview of how the Sun influences the climate on Earth and other planets.

The next two sessions focused on space weather and its origins at the Sun, and solar variability on various time scales from the solar interior to the atmosphere. The session on mass emissions from the Sun focused on coronal mass ejections (CMEs), which constitute the most energetic phenomenon in the heliosphere and are responsible for most of the severe weather in geospace. CMEs also interact with galactic cosmic rays producing the well-known Forbush effect. The next session on radio emission processes in space plasmas included discussion of the production of energetic electrons at the Sun and in Jupiter and Earths magnetospheres.

Two sessions were devoted to magnetic reconnection, which is one of the ubiquitous, universal processes in heliospace starting with the Sun (flares, chromospheric jets), and extending to planetary magnetospheres and the solar wind itself. The session on energetic particles dealt with both particles accelerated within the heliosphere (by CME driven shocks and flare reconnection) and those coming from outside (accelerated by supernova shocks). Voyager spacecraft crossings of the termination shock have provided new data and new challenges to the theorists to explain cosmic ray modulation in the heliosheath and the acceleration mechanism of anomalous cosmic rays. The discussion continued into the related session on the heliospheric boundaries, interfaces and shocks.

The session on planetary atmospheres, ionospheres and magnetospheres highlighted the recent developments in planetary science. Planetary auroras were discussed as an excellent example of universal process involving energetic particles, planetary magnetic fields, and neutral atmospheres of planets. Similarly, turbulence in the solar wind as well as in the vicinity of interplanetary shocks was discussed as a universal process. A peculiar form of MHD turbulence in closed magnetic structures in the corona was also reported; the turbulence spectrum was found to have a magnetically dominated pre-inertial range, where boundaries have a strong influence. Existence of strong turbulence in the vicinity of CME-driven shocks was found to be critical for the production of large solar energetic particle events.

Finally, the session on flows, obstacles and circulation in the heliospace highlighted the knowledge gained from the three dimensional view of the heliosphere provided by the Voyager and Ulysses spacecraft. The out-of-the ecliptic view from Ulysses over the past 18 years has helped characterize solar wind flows, magnetic fields, energetic particles, cosmic rays, radio and plasma waves, and dust up to 80^0 heliolatitude. The third set of Ulysses polar passes revealed a stunning change in the solar wind: its magnetic field, density, temperature and dynamic pressure all were smaller compared to previous solar minima.

The editors take this opportunity to thank Drs. Joachim Schmidt, Seiji Yashiro, Pertti Makela, and Alexander Nindos for their valuable assistance in preparing this proceedings volume. We also thank the following reviewers who assisted in improving the papers: N. Arge, A. Benz, S. Gibson, C. Mandrini, D. Melrose, L. Ofman, M. Potgieter, J. D. Richardson, S. Spangler, O. C. St. Cyr, G. Thejappa, W. T. Thompson, L. van Driel Gestelyi and B. Vrsnak. Please note that many of the papers contain color figures, which are printed here in black and white but can be viewed online in color.

Natchimuthukonar Gopalswamy and David Webb
Greenbelt, Maryland and Boston, Massachusetts, USA, January 15, 2009

THE ORGANIZING COMMITTEE

Scientific

Nat Gopalswamy (chair, USA)
Kazunari Shibata (co-chair, Japan)
Mei Zhang (China)
Jean-Louis Bougeret (France)
P. K. Manoharan (India)
Sarah Gibson (USA)
Alexander Stepanov (Russia)
Marius Potgieter (South Africa)

David F. Webb (co-chair, USA)
Cristina Mandrini (Argentina)
Claus Froehlich (Switzerland)
Arnold Benz (Switzerland)
Gerard Thullier (France)
Lidia van Driel-Geszelyi (Hungary)
Costas Allisandrakis (Greece)
Bojan Vrsnak (Croatia)

Local

Alexander Nindos (chair, Greece)
Vassiliki Tsikoudi (Greece)
Angeliki Fotiadi (Greece)

Costa Allisandrakis (Greece)
Georgia Tsiropoula (Greece)
Seiji Yashiro (USA)

Acknowledgements

This symposium was coordinated through IAU Division II (Sun and Heliosphere) and sponsored and supported by IAU Divisions III (Planetary Systems Sciences) and IV (Stars), and by IAU Commissions 10 (Solar Activity), 12 (Solar Radiation and Structure), 36 (Theory of Stellar Atmospheres) and 49 (Interplanetary Plasma and Heliosphere).

The Local Organizing Committee was under the auspices of the Physics Department, University of Ioannina, Ioannina, Greece.

Funding support by the
International Astronomical Union,
University of Ioannina,
Greek Ministry of National Education and Religious Affairs,
Greek Ministry of Culture,
International Heliophysical Year
and
National Aeronautics and Space Administration
are gratefully acknowledged.

CONFERENCE PHOTOGRAPH

Participants

Vladimir **Abramov-Maximov**, Pulkovo Observatory, Russia — abramov-maximov@mail.ru

Giorgi **Aburjania**, M. Nodia Insitute of Geophysics, Georgia — aburj@mymail.ge

Costas **Alissandrakis**, University of Ioannina, Greece — calissan@cc.uoi.gr

H. M. **Antia**, Tata Institute of Fundamental Research, Mumbai, India — antia@tifr.res.in

C. Nick **Arge**, Kirtland AFB, USA — Nick.Arge@Kirtland.af.mil

Irina **BakuninaB**, NIRFI, Nizhny Novgorod, Russia — rinbak@mail.ru

Juerg **Beer**, EAWAG, Switzerland — beer@eawag.ch

Anatoly **Belov**, IZMIRAN, Moscow, Russia — abelov@izmiran.ru

Francoise **Bely-Dubau**, Observatoire de la Cote d' Azur, Nice, France — f.bely-dubau@orange.fr

Peter **Bochsler**, Physikalisches Institut, University of Bern, Switzerland — bochsler@soho.unibe.ch

Roger-Maurice **Bonnet**, ISSI, Bern, Switzerland — rmbonnet@issibern.ch

Marusja **Buchvarova**, Space Research Institute, BAS, Bulgaria — marusjab@yahoo.com

Jasa **Calogovic**, Hvar Observatory, Croatia — jcalogovic@geof.hr

Jose R. **Cecatto**, INPE/DAS, S. J. Campos, Brazil — jrc@das.inpe.br

Khatuna **Chargazia**, M. Nodia Institute of Geophysics, Georgia — khatuna.chargazia@gmail.com

Ed **Cliver**, AFRL/RVBXS, Hanscom AFB, USA — edward.cliver@hanscom.af.mil

Norma Bock **Crosby**, Belgian Institute for Space Aeronomy, Brussels, Belgium — norma.crosby@oma.be

Sergio **Dasso**, IAFE, Buenos Aires, Argentina — sdasso@iafe.uba.ar

Joe **Davila**, NASA's GSFC, USA — joseph.m.davila@nasa.gov

Aline **DeLucas**, INPE, S. J. Campos, Brazil — delucas@dge.inpe.br

Michaila **Dimitropoulou**, University of Athens, Greece — michaila.dimitropoulou@nsn.com

Svetla **Dimitrova**, Solar-Terrestrial Influences Laboratory, BAS, Bulgaria — svetla_stil@abv.bg

Ivan **Dorotovic**, Slovak Central Observatory, Hurbanovo, Slovak Republic — id@uninova.pt

Cristiana **Dumitrache**, Astron. Inst. of the Romanian Acad. of Sci., Romania — cristiana_d@yahoo.com

Evgenia **Eroshenko**, IZMIRAN, Moscow, Russia — erosh@izmiran.ru

Victor **Fainshtein**, Institute of Solar-Terrestrial Physics, Irkutsk, Russia — vfain@iszf.irk.ru

Stefan E. S. **Ferreira**, North-West University, Potchefstroom, South Africa — Stefan.Ferreira@nwu.ac.za

Len **Fisk**, University of Michigan, USA — lafisk@umich.edu

Robert **Forsyth**, Imperial College, London, UK — r.forsyth@imperial.ac.uk

Alan **Gabriel**, Institut d' Astrophysique Spatiale, Orsay, France — gabriel@ias.fr

Marina **Galand**, Imperial College, London, UK — m.galand@imperial.ac.uk

Ryszarda **Getko**, Astronomical Institute, University of Wroclaw, Poland — GETKO@ASTRO.UNI.WROC.PL

Marina **Gigolashvili**, Georgian Nat. Astrophys. Obs. at Ilia Chavchavadze State Univ., Georgia — marinagig@yahoo.com

T.E. **Girish**, University College,Trivandrum, India — tegirish5@yahoo.co.in

Thejappa **Golla**, NASA's GSFC, Greenbelt, MD 20771, USA — thejappa.golla@nasa.gov

Nat **Gopalswamy**, NASA's GSFC, Greenbelt, MD 20771, USA — nat.gopalswamy@nasa.gov

Gopalan **Gopkumar**, University College, Trivandrum, India — kggopkumar@yahoo.com

Jack **Gosling**, Lab. for Atmospheric & Space Physics, Univ. of Colorado, USA — jack.gosling@lasp.colorado.edu

Joseph **Grebowsky**, NASA's GSFC, Greenbelt, MD 20771, USA — Joseph.M.Grebowsky@nasa.gov

Irina **Grigoryeva**, Pulkovo Observatory, Russia — irina19752004@mail.ru

Siraj **Hasan**, Indian Institute of Astrophysics, Bangalore, India — hasan@iiap.res.in

Russell **Howard**, Naval Research Laboratory, Washington, DC 20375, USA — russ.howard@nrl.navy.mil

Subhon **Ibadov**, Institute of Astrophysics, Dushanbe, Tajikistan — ibadovsu@yandex.ru

Moira **Jardine**, University of St. Andrews, UK — mmj@st-andrews.ac.uk

Homa **Karimabadi**, University of California at San Diego, USA — homa@ece.ucsd.edu

Iraida S. **Kim**, Sternberg State Astronomical Institute, Moscow, Russia — kim@sai.msu.ru

Alejandro **Lara**, Instituto de Geofisica, Universidad Nacional Autonoma de Mexico — alara@geofisica.unam.mx

George **Livadiotis**, University of Athens, Greece — glivad@phys.uoa.gr

Noe **Lugaz**, University of Hawaii, USA — nlugaz@ifa.hawaii.edu

Robert **MacDowall**, NASA's GSFC, Greenbelt, MD 20771, USA — Robert.Macdowall@nasa.gov

Pertti **Makela**, Catholic University of America, Washington DC, USA — pertti.makela@nasa.gov

Olga **Malandraki**, Office of Space Research and Technology, Academy of Athens, Greece — omaland@xan.duth.gr

Francesco **Malara**, Universita della Calabria, Italy — malara@fis.unical.it

Cristina **Mandrini**, IAFE, Buenos Aires, Argentina — mandrini@iafe.uba.ar

Kurt **Marti**, University of California at San Diego, USA — kmarti@ucsd.edu

Victor **Melnikov**, NIRFI, Nizhny Novgorod, Russia — meln@nirfi.sci-nnov.ru

Donald **Melrose**, University of Sydney, Australia — melrose@physics.usyd.edu.au

Grzegorz **Michalek**, Astronomical Observatory, Jagiellonian University, Krakow, Poland — michalek@oa.uj.edu.pl

Guadalupe **Muñoz-Martinez**, Instituto Politecnico Nacional, Mexico — gmunozm@ipn.mx

Noriyuki **Narukage**, Japan Aerospace Exploration Agency (ISAS/JAXA), Japan — narukage@solar.isas.jaxa.jp

Alexander **Nindos**, University of Ioannina, Greece — anindos@cc.uoi.gr

Alix **Nulsen**, University of Sydney, Australia — a.nulsen@physics.usyd.edu.au

Leon **Ofman**, Catholic University of America, USA — Leon.Ofman@nasa.gov

Kristof **Petrovay**, Eotvos University, Budapest. Hungary — k.petrovay@icsip.elte.hu

Teodor **Pinter**, Slovak Central Observatory, Hurbanovo, Slovak Republic — pinter@suh.sk

Nedelia A. **Popescu**, Astron. Inst. of the Romanian Acad. of Sci., Romania — nedeliapopescu@yahoo.com

Panagiota **Preka-Papadema**, University of Athens, Greece — ppreka@phys.uoa.gr

Dmitry **Prosovetsky**, Institute of Solar-Terrestrial Physics, Irkutsk, Russia — proso@iszf.irk.ru

Veronika **Reznikova**, NIRFI, Nizhny Novgorod, Russia — rez-ver@yandex.ru

John **Richardson**, MIT, USA — jdr@space.mit.edu

Ilan **Roth**, University of California at Berkeley, USA — ilan@ssl.berkeley.edu

Vladimir **Salmin**, Siberian Federal University, Krasnoyarsk, Russia — vsalmin@gmail.com

A. **Satya Narayanan**, Indian Institute of Astrophysics, Bangalore, India — satya@iiap.res.in

Brigitte **Schmieder**, Observatory of Paris, France — brigitte.schmieder@obspm.fr

Kazunari **Shibata**, University of Kyoto, Kwasan and Hida Observatories, Japan — shibata@kwasan.kyoto-u.ac.jp

Steven **Spangler**, University of Iowa, USA — steven-spangler@uiowa.edu

Oliver **Sternal**, Christian-Albrechts-Universitaet Kiel, Germany — sternal@physik.uni-kiel.de

Takeru **Suzuki**, University of Tokyo, Japan — stakeru@ea.c.u-tokyo.ac.jp

Toshio **Terasawa**, Tokyo Institute of Technology, Japan — terasawa@phys.titech.ac.jp

Gilberto **Tisnado**, Tecnofil S.A., Lima, Peru — gilberto@tecnofil.com.pe

Yuri **Tsap**, Crimean Astrophysical Observatory, Ukraine — yur_crao@mail.ru

Vasiliki **Tsikoudi**, University of Ioannina, Greece — vtsikoud@cc.uoi.gr

Georgia **Tsiropoula**, National Observatory of Athens, Greece — georgia@space.noa.gr

Ilya **Usoskin**, University of Oulu, Oulu, Finland — ilya.usoskin@oulu.fi

Rami **Vainio**, University of Helsinki, Finland — rami.vainio@helsinki.fi

Luis Eduardo **Vieira**, Max-Planck-Institut fur Sonnensystemforschung — vieira@linmpi.mpg.de

Gangadharan **Vigeesh**, Indian Institute of Astrophysics, Bangalore, India — vigeesh@gmail.com

Loukas **Vlahos**, University of Thessaloniki, Greece — vlahos@astro.auth.gr

Bojan **Vršnak**, Hvar Observatory, Croatia — bvrsnak@gmail.com

David **Webb**, Boston College, USA — david.webb@hanscom.af.mil

Hong **Xie**, Catholic University of America, Washington DC, USA — hong.xie@nasa.gov

Olesya **Yakovchouk**, Inst. of Nuclear Physics, Moscow State Univ., Russia — olesya@dec1.sinp.msu.ru

Seiji **Yashiro**, Interferometrics, Herndon VA 20171, USA — Seiji.Yashiro@nasa.gov

Leonid **Yasnov**, St.Petersburg State University, Russia — Yasnov@pobox.spbu.ru

Gary **Zank**, University of California, Riverside, USA — gary.zank@ucr.edu

Tomislav **Zic**, Hvar Observatory, Croatia — tzic@geof.hr

Session I

Opening Session

Universal Heliophysical Processes
Proceedings IAU Symposium No. 257, 2008
N. Gopalswamy & D.F. Webb, eds.

Next 50 years of space research

R.-M. Bonnet

International Space Science Institute, Hallerstrasse 6, 3012 Bern, Switzerland

Abstract. Forecasting the next 50 years of space research is a dangerous game and a somewhat irresponsible action. Fortunately, the past 50 years have evidenced what remains in the realm of realism and of the feasible and what definitely belongs to the realm of utopia. Nevertheless those who, like me today, take the risk of forecasting such a relatively long time trend are sure of one thing: to be wrong!

1. What have we learnt from the past 50 years?

By observing the sky from above the Earth's atmosphere, we have accessed all the hidden portions of the electromagnetic spectrum: the UV, the X and the gamma rays, the infrared, and the sub-millimetric wavelengths. We have discovered black holes everywhere confirming in an unprecedented way the prediction of Einstein's theory of Relativity. We have started exploiting with an enormous amount of luck and success that nearly inexhaustible gold mine of discoveries, space astronomy, with telescopes always increasing in size, angular resolution and sensitivity. The revolution in knowledge and in our understanding of the Universe which resulted, has a dimension comparable -if not greater- to that opened by Galileo Galilei with the use of the telescope.

In parallel, we have extensively traveled through the Solar System. We have landed on the Moon, on Mars, Venus, Titan, and on asteroids. Soon, thanks to the Rosetta mission, we will land on the nucleus of a comet and plans are that we will explore more asteroids, the icy moons of Jupiter, return to the Moon and possibly also to Titan. We have discovered water everywhere, on the surface of Mars and underground, on Europa and the moons of Saturn. With the "Pioneers" and the "Voyagers" we have reached the limits of the heliosphere and are just starting to explore the virgin territories of deep space.

In the past 50 years, we have accessed the most thinkable physical extremes: extremes of distances, of temperature -from the several million degrees of the solar corona to the near absolute cold of the deep universe-, extremes of vacuum, of density and gravity, and of time.

May be even more essential for us has been the realization that space observations of the Earth represent one of the most promising tools ever invented to serve humanity. The most mediatic picture of the 20th century will remain for a long time to come the picture of the Earth taken by the Apollo Astronauts hanging above the lunar horizon. With no concerns about political barriers and borders, man-made satellites have proven their indispensable role for measuring our globe and its deformations, observing and forecasting the weather and soon the climate, the melting of ice, the rising of the sea level, the depletion of the ozone layer and the anthropogenic and natural hazards that threaten us more and more.

2. How was this possible?

I can identify four essential elements which helped the development of that genuine scientific revolution.

First, the fierce political competition between the Soviets and the United States, as none of them had any intention to let the control of space in the hands of the rival, injected a lot of money and resources in the development of space systems. That space race was enough to set the political framework and justify the enormous investments in technologies and warfare through a greedy aeronautical industry. The strong synergies between the civilian and the military sectors benefited strongly the scientific community, in particular through the use of declassified technologies.

Second, the creation of space agencies, such as NASA in particular, helped orchestrating the appropriate investments and the planning of the necessary technological developments or the use of existing devices after their adaptation to scientific space systems. That was the case for RTGs and for the development of the DSN without which long-distance missions such as the Pioneer 10 and 11, the Voyager 1 and 2, the two Viking missions to Mars, Galileo , Ulysses, SOHO, Cassini and New Horizons would not have been possible.

Conversely, whenever technological developments were not programmed on time, missions were not able to start. This was clearly the case of the NASA comet mission in the late 70's which required solar sail or electric propulsion, or the long awaited solar probe to which I will come back later.

Third, the education and the development of a very broad and active scientific community, able to take part in space experimentation, in the planning of missions, in their development as well as in their operations, secured the human and indispensable basis for an ambitious space program to start and expand.

Fourth, international cooperation! Of course, international cooperation bears in itself a large weight of political interests for the parties involved but it has also allowed the undertaking of a large number of joint space ventures, in particular between the US and Europe but also the Soviet Union-Russia and several other partners including Japan and recently China. In addition to permitting more ambitious endeavors than would be allowed by just using the cooperating partners' own resources, international cooperation is adding an indispensable element of stability among these partners. Without international cooperation, it is highly probable that Ulysses, SOHO and Cassini would not have been able to make it to the launch pad.

3. The realm of utopia

In parallel to these serious achievements, we could witness the development of what I call utopian concepts. For example, in the wake of the Apollo missions to the Moon, G.K. O' Neill, professor of physics at Princeton University, invented the concept of the "space cities", some kinds of gigantic space stations. The seriousness of that concept can be assessed a posteriori, nearly 40 years after Apollo, looking at the intended original goals and at where we stand today.

The original claim of O'Neill was that: *"Careful engineering and cost analysis shows we can build pleasant, self-sufficient dwelling places in space within the next two decades (i.e. before 1994)"*, solving many of Earth's problems. At first, a space colony of 10,000 people would have been in place in 1988 (i.e. two years after the Challenger accident!) and colonies between 200,000 and 20 million people would be able to live in such habitats in 2008, the year when this IAU symposium is held! The model would lead to establishing by 2050 a space population of about 14 billions and decrease that remaining on Earth from a maximum of 16 billions (according to his estimate) to a stable (?) level of 2 billions onward! For comparison with the real world, we are painfully assembling the International Space Station which is on average occupied by no more than a handful of astronauts, and the operation costs of the station are so high that they may lead to its abandonment.

A very modest example of something akin O'Neill's idea but on Earth was the *Biosphere* project in Arizona, an experiment for a sustainable, isolated human outpost. Surely, building a "space city" on Earth looks a priori simpler than doing it in space. Unfortunately, the project has been abandoned because it was judged much too expensive and requiring enormous amounts of power for maintaining adequate living conditions underneath the dome. Survival was indeed the main activity of the crew of 8 who volunteered to participate in the project, as the life-support system was constantly put into question. The fact that the crew emerged from their 2-year closure still speaking to each other and apparently in better health than when they started was considered at that time as an accomplishment!

The second example shows that even space agencies as serious as NASA are not protected against that kind of utopia. In the early 1970's, at the time when I was involved in the original studies of what was then called the Space Transportation System, STS, which included the Space Shuttle developed by NASA and Spacelab developed by ESA, the flight model included 52 shuttle flights per year and Mr. J. Fltecher, the then NASA administrator lost all sense of humor when such a goal was disrespectfully questioned! Utopia had reached the highest levels of the most powerful space agency. And the story does not end here as we witness today with the manned exploration program.

Ten years ago, the French Space agency CNES under the request of the Minister of Research was committed to develop in cooperation with NASA a Mars sample return mission within the existing and already over-committed French space budget. The project, considered unfeasible, was then offered to ESA to become part of its Aurora Exploration program. Recent talks between ESA, CNES and NASA have led to a cost estimation of the mission ranging between 5 and 8 billion of dollars and the mission is qualified as "the most audacious and technologically challenging space mission since the Apollo lunar landing".

This leads me to ask whether the dreams of sending humans to Mars are still realistic today? The farthest distance to Earth Man has ever been is the Moon, at one light-second. The astronauts on the ISS are circling the Earth at a distance smaller than that which separates Washington from New York, or a little more than a thousandth of a light-second, and the nearest star is still 4 light-years away, reachable, if ever, in some 50,000 to 70,000 years with the most rapid rockets we are able to build presently! In the last 50 years, we have not invented a faster rocket than the Semiorka which sent Sputnik-1 into orbit.

I dare guessing that in the next 50 years the deserts of the Red Planet may well be populated by undefined quantities of robots of different nationalities but most probably not by international colonies of human beings. What should Man do on Mars? As the Late Hubert Curien often said: "we will go there not to do science but for pleasure, or for sport" or, I may add, just because if one goes there the others will follow.

4. So, what will the next 50 years be?

The discoveries of the first 50 years have opened new questions on the evolution of the Universe, on its content, on the existence of Dark Matter and of Dark Energy. Not less fascinating is the prospect of detecting other planets similar to the Earth orbiting other stars, and probably not so long in the future, some signs of life on some of them, responding to the anguishing question of our loneliness in the Universe. These new discoveries require new missions to be developed, large telescopes and space interferometers.

In the next 50 years, we will continue observing our Solar System. We will probably land again on the Moon. We will explore Mars extensively, and also the moons of Jupiter and Saturn, we will land on asteroids and comets.

50 years is indeed a long time to come and to forecast but you all know very well that it may not be long enough for making the right decisions or developing the indispensable technologies. The Solar Probe for example was already under study in the mid seventies, and it is still under study – although in a different incarnation. It took nearly 25 years to get the Hubble Space Telescope from the study stage to the launch pad and it is still in operation as of today. Ulysses which was under study in the mid seventies is just ending its operational life some 30 years later.

We are presently taking little risks and unfortunately, we witness very little progress in the development of new technologies while it is technologies and risk-taking which made it possible to accomplish what we have accomplished in the past 50 years.

5. What are we missing in heliospheric physics?

First: the gravity field. It is in principle simple and understood, except that we are still struggling with the so-.called Pioneer anomaly. We need to understand it. We need more probes flying at long distances. Maybe, New Horizons will help!

Second: The magnetic field. It pervades the whole volume and its sources need to be properly monitored and possibly forecasted: the solar source, the planetary sources and the interstellar source. This requires before all, measuring the solar magnetic field and the solar wind throughout the whole volume of the heliosphere. That can be done (for the Sun) through helioseismology, magnetographs and magnetometers, coronagraphs as well as in situ measurements in the Interstellar medium.

The third "field" is solar radiation. The Sun provides energy to the planets and exerts its influence on the interplanetary and on the interstellar atoms and on dust particles through radiation pressure and absorption. It should be monitored precisely and continuously in all parts of the spectrum.

All three domains require different instruments and different types of missions for studying the Sun, the interplanetary medium and the extreme limits of the heliosphere. There have been a relatively large number of such instruments in operation in the past. However, some are now out of operation and others have never been out of the drawing board. We assume that missions like Yohkoh, SOHO, TRACE, Stereo, ACE, HINODE, Wind and magnetospheric missions and their successors will continue to be implemented in the future. They are relatively modest in size and their supporting technologies are also relatively well in hand. The LWS or ILWS program encompassed a lot of these. This program should be resurrected!

However, on top of these, three missions are constantly mentioned as necessary and are regularly coming back again and again to the discussion and planning table of agencies and academies. They are:

(a) The Solar Probe already mentioned.

(b) A new Out-of-Ecliptic mission, the successor of Ulysses, but with imaging capabilities.

(c) The Interstellar Probe.

6. What will make these missions possible?

6.1. *New technologies*

Because the heliosphere is the largest volume of space ever explored and the largest we can envisage to explore, long travel time characterize these missions and determine their duration.

Heliospheric missions are among the most demanding as far as new technologies are concerned: for going faster, for going further, for going higher. None of the three missions mentioned above will be implemented without an early investment in the necessary technologies and the development of new in situ experiments.

• Investments in new techniques of propulsion are an obvious priority. As already said, in the past 50 years our rockets are not more rapid than the Semiorka which was used to launch Sputnik-1. Here, let me express some concerns. Both the Out-of-Ecliptic mission and the Interstellar Probe claim the utilization of solar sails. But I wonder whether solar sailing is not another utopia. It has been with us for many decades but has never been implemented successfully. If solar sailing is so indispensable, then the first priority should be to plan its development and usage in a proper way. If not, there should be another plan. I am not far from thinking that nuclear propulsion would be a more suitable plan! Since 2001, three successive attempts to fly a demonstrative solar sail have failed: the last one, called Nano Sail-D, was launched recently in August.

• Operating satellites at long distances require non-solar power energy sources. A new generation of RTG is indispensable. Consequently and naturally, the development of nuclear energy systems is required for both fast propulsion and energy sources.

• Operating satellites at long distances with high data rates require also the development of high gain communication systems on board as well as on the ground.

6.2. *International cooperation*

It is obvious that the only organization which is presently able to undertake these missions is NASA. But none of these missions would be realized if NASA goes it alone. This is proven by past history. Past history also tells us that international cooperation is a better approach, even though the success of such joint ventures is never fully guaranteed, in particular against changes in political or budgetary priorities. Nevertheless international cooperation is an absolute necessity.

6.3. *Involving new partners*

What will be possible in the next 50 years will be determined by a context profoundly different from the one which so successfully framed the development of space research in the last 50 years. The early dominance of the United States and of Russia followed by the appearance of strong European and Japanese programs, which formed a very strong East-West dipole, will be succeeded by a multipolar configuration with the development of the Indian and Chinese space programs and possibly others. This will most likely induce new alliances on a front much broader than has been witnessed until now. This is a unique chance and a safeguard against the present tendency to erode space science programs as they are -and will be- more and more confronted to changes in national and international priorities.

6.4. *Joint planning and joint road maps*

International cooperation is also a chance. It should be properly orchestrated and the scientific community through the Space Studies Board of the NAS, the Space Committee of the ESF, COSPAR and other national or international organizations holds in its hands the best tools for managing the scenarios of cooperation. As has been proven by the past 50 years, space research is a vector of international cooperation and integration. It helps the dialogue among countries which are not necessarily politically aligned.

Indeed, the next 50 years will see major changes on this planet and in the way of living of its inhabitants: the gradual disappearance of oil as the main source of energy, with all the multiple consequences on the economy that we are already witnessing today. Several other primary resources will also gradually disappear, such as lithium, platinum, copper, tin etc... The unprecedented evolution of the climate and the acceleration of its changes will require drastic and global measures and more financial support from all governments to counteract these rapid trends. This may well be at the expense of other endeavors considered – wrongly may be- to be of lower priorities. Space agencies for sure will be more and more under pressure to undertake missions to planet Earth in priority. This cannot be criticized as our future is at stake and the management of Planet Earth rests more and more on space systems.

6.5. *Re-juvenilization of the scientific community*

Our future also relies on more science and on Education. The scientists who will be in charge of the missions identified above are still at school today or are not even yet born. This is the condition for space research to survive. This requires vision, faith in the role of space science for education and discovery, the reliance on approaches which have already proven to be successful in the past but will be more and more necessary in the next 50 years, as well as the courage of risk-taking.

By exploring the largest volume of space accessible to our machines, heliospheric physicists offer a unique challenge to future generations, and the only possibility to go where nobody has ever been before.

7. Conclusions

Let me now summarize and conclude.

The space era which started 50 years ago will continue to determine our future, our well being and our knowledge of the Universe and of our galactic environment for many years to come. Its development in the next 50 years will have to take into consideration:

• The necessary and undisputable global effort in Earth sciences which may absorb a substantial amount of the funds allocated to space research.

• The new international context which will involve new and potentially powerful partners.

• The necessity of serious technological preparation which should be undertaken with determination, vision and realism.

• The necessity to involve early the new generations of scientists

In this context, heliospheric sciences with their unique characteristics and highly demanding technologies present important challenges to both the community and the agencies. All should rely very strongly on international cooperation. In that respect, initiatives such as the IHY are welcome and offer great potentialities for organizing the international scientific community.

Because NASA is at present the only Agency with the capability of undertaking most of the required missions, there is a unique opportunity there for intelligent leadership. The future administration to be elected next November is offered an opportunity not to miss.

Discussion

USOSKIN: Do you think ground-based observations with low cost and long duration should be continued through the ages?

BONNET: In several if not all domains of space science there is a tight complementarity between ground based and space based measurements. This is clearly the case for solar physics, Earth sciences, astronomy in general. As a matter of principle we should not do in space what can be done from the ground with similar outcomes. So, by all means, it is very important to maintain the complementarity and contemporarity of ground and space based observations.

CROSBY: At present not enough young people are going into degrees such as space sciences and engineering. What should we do about this problem?

BONNET: As just said, there is no future assured to space science without the education and the formation of future generations of scientists and engineers. For that, the prospects of scientific careers must be made much more appealing. Certainly by emphasizing the challenges of space, the potential of discoveries, and truly for heliospheric physics, the unique capability of going where nobody has ever been before. Careers must also be appealing for scientists. Because science is so essential to our future, the salaries of scientists must be raised at a level that they become attractive. How to do this? A challenge is necessary which only scientists can solve. Challenges will not be missing in the 21st century!

Universal Heliophysical Processes
Proceedings IAU Symposium No. 257, 2008
N. Gopalswamy & D.F. Webb, eds.

© 2009 International Astronomical Union
doi:10.1017/S1743921309029032

Universal processes in heliophysics

Joseph M. Davila[1], Nat Gopalswamy[2] and Barbara J. Thompson[3]

[1] Goddard Space Flight Center, Code 670, Greenbelt, MD 20771, USA
email: `Joseph.M.Davila@nasa.gov`

[2] Goddard Space Flight Center, Code 695, Greenbelt, MD 20771, USA
email: `Nat.Gopalswamy@nasa.gov`

[3] Goddard Space Flight Center, Code 671, Greenbelt, MD 20771, USA
email: `Barbara.J.Thompson@nasa.gov`

Abstract. The structure of the Universe is determined primarily by the interplay of gravity which is dominant in condensed objects, and the magnetic force which is dominant in the rarefied medium between condensed objects. Each of these forces orders the matter into a set of characteristic structures each with the ability to store and release energy in response to changes in the external environment. For the most part, the storage and release of energy proceeds through a number of Universal Processes. The coordinated study of these processes in different settings provides a deeper understanding of the underlying physics governing Universal Processes in astrophysics.

Keywords. Sun: general, interplanetary medium, solar system: general

1. Introduction

The solar system provides us with our best opportunity to study astrophysics in detail. In-situ observations provide measurements of particle distribution functions for both thermal and energetic particles. Details of these distributions can provide important information on the origin of the plasma. Planetary probes have provided detailed observations of planets which fifty years ago could only be studied by telescopic observation. Present day remote sensing observations both from the ground and from spacecraft provide images with resolution at a scale that cannot be approached for objects outside the Solar System, e.g. the Sun is the only star where the surface and sub-surface features can be resolved and studied. This wealth of detail allows the observation of processes in sufficient detail to identify the underlying physics involved. The processes provide the road map for understanding more distant objects in the Universe.

2. Physical Origin of the Universal Process Concept

The large scale structure of the universe is controlled mainly by two forces, gravitation and magnetism. Gravitation is the dominant force controlling the evolution of dense matter in the universe, including planets, stars, planetary systems, galaxies, and clusters of galaxies. Magnetism, a second long-range force, is dominant in the rarefied, ionized matter which generally occupies the space between gravitationally condensed regions. The most easily observed example is the plasma environment of the solar system where the magnetic field is responsible for the storage and subsequent release of large quantities of energy in solar flares, coronal mass ejections (CMEs), magnetic storms, and other transient phenomena. In addition the magnetic field of solid bodies like the Earth, Jupiter, Saturn, and even the Sun, dominate and define the structure of the space environment surrounding them, and exhibit their own energetic evolution.

A common property of both gravitationally and magnetically structured objects is their ability to store, release, and transfer energy in response to internal and external drivers. For example thunderstorms generate atmospheric gravity waves that transfer energy into the stratosphere, the magnetosphere of Earth continually adjusts in response to the variable solar wind, and the magnetic field of the Sun is constantly buffeted by convective motions in the photosphere.

To illustrate this point, consider an isolated magnetic loop like that shown in the left panel in Figure 1. The magnetic field of the Sun is largely contained in individual loops which maintain their identity while interacting with the surrounding corona. Typically these loops are potential, i.e. current free, which is the lowest energy state for the structure. Convective motions at the surface of the Sun twist and deform the loop causing it to store energy. When the stored energy reaches a large enough level, rapid energy release can be triggered returning the loop to its potential state, resulting in the production of plasma flows, particle acceleration, waves, or heating. These are illustrated schematically in Figure 2.

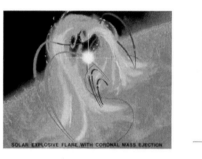 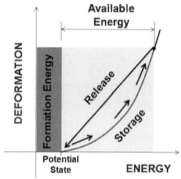

Figure 1. The simplest magnetic structure is a single loop (left panel). On the Sun convective motions shuffle the footpoints of the loop magnetic field generating currents in the corona that store energy (right panel). Eventually this energy is released in other forms.

The evolution of magnetic structures, and planetary atmospheres in the solar system proceeds through a set of Universal Processes (Crooker 2004), e.g. reconnection, particle acceleration, wave generation and propagation, etc. Energy is stored in magnetic structures on the Sun, in Earth's magnetosphere, the magnetospheres of other planets, and in the heliosphere itself. Each of these structures evolves in a similar way. By studying these Universal Processes together, in diverse environments, and in a comparative way, new scientific insights will be gained, and a deeper understanding of the underlying physics can be obtained. These processes are evident in astrophysical settings as well. The study of Universal Processes in heliophysics provides a sound physical basis for the interpretation of astrophysical phenomena.

3. Universal Processes in the IHY

The study of Universal Processes is a primary focus of the International Heliophysical Year (IHY). IHY has three primary objectives, (1) Advancing our understanding of the fundamental heliophysical processes that govern the Sun, Earth and heliosphere; (2) Continuing the tradition of international research and advancing the legacy on the 50th anniversary of the International Geophysical Year; (3) Demonstrating the beauty, relevance and significance of space and Earth science to the world. To accomplish these

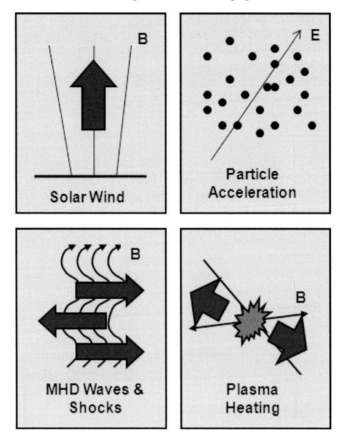

Figure 2. The energy stored in magnetic structures can be released as kinetic energy (upper left), energetic particles (upper right), waves (lower left), or heating (lower right).

objectives, we have identified six goals of IHY, each corresponding to a unique opportunity afforded by IHY:

1. Develop the basic science of heliophysics through cross-disciplinary studies of Universal Processes.

2. Determine the response of terrestrial and planetary magnetospheres and atmospheres to external drivers.

3. Promote research on the Sun-heliosphere system outward to the local interstellar medium - the new frontier.

4. Foster international scientific cooperation in the study of heliophysical phenomena now and in the future.

5. Preserve the history and legacy of the IGY on its 50th Anniversary.

6. Communicate unique IHY results to the scientific community and the general public. IHY is an integrated program of many diverse activities working on an international level to achieve all of the above goals.

4. Examples of Universal Processes

The concept of Universal Processes is perhaps best understood by considering a few examples: (1) Shocks are observed in situ in the interplanetary medium, shocks are believed to play a role in the acceleration of particles in the solar corona, and standing bow shocks and termination shocks separate the major regions in the heliosphere. Shock formation, and particle acceleration are universal processes. (2) Aurorae are observed on Earth, Saturn, and Jupiter, and Jovian auroral "footprints" have been observed on Io, Ganymede and Europa. The formation of aurorae is observed to be the universal response of a magnetized body in the solar wind. Here we consider two examples of Universal Processes in further detail, magnetic reconnection and atmospheric gravity waves.

4.1. *Magnetic Reconnection*

Magnetic reconnection is one of the best examples of a Universal Process in heliophysics. On the Sun solar flares result from reconnection in the corona (e.g., Chen *et al.* 2007). Energy stored in the coronal magnetic field is released as energetic particles, flows, and plasma heating. Figure 3 (after Sui and Holman 2003) shows shows an observation from the RHESSI mission. The flaring region is observed in three different spectral passbands, 6-8, 10-12, and 16-20 keV. Two distinct emission regions are seen. In each region the hottest emission is seen to occur nearest the the center of a line connecting the two emission regions. This emission pattern is consistent with reconnection jets emanating from a central point as indicated in the figure.

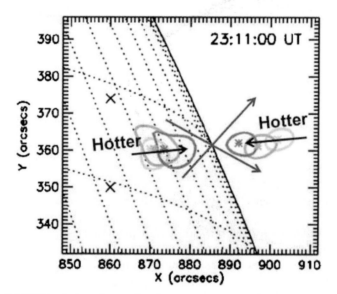

Figure 3. The RHESSI mission observed a flaring region in three passbands, 6–8, 10–12, and 16–20 keV. The spatial positions of the observed regions form a distinct pattern with the hottest to coolest regions symmetrically placed around a central point as would be expected for reconnection (from Sui and Holman 2003). The x-point, arrows, and notations were added for clarity.

Reconnection is also observed in the magnetosphere of Earth (Angelopoulos *et al.* 2007). The THEMIS mission consists of five spacecraft (Figure 4) providing in-situ measurements in the magnetotail supported by extensive simultaneous ground-based observation of auroral activity. With this powerful combination, the THEMIS team was able to demonstrate for the first time that reconnection observed in the tail resulted in substorm activity in the polar region.

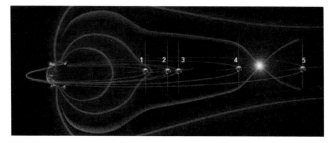

Figure 4. The THEMIS mission consists of five spacecraft orbiting Earth at various distances. This constellation was able to observe magnetic reconnection in the magnetotail while simultaneously viewing substorm activity from the ground.

4.2. *Planetary Gravity Waves*

Universal processes are not confined to the ionized plasma component in the Solar System. The Earth's ionosphere is the boundary between the neutral atmosphere and space (Taylor *et al.* 1987). Energy between these two components is exchanged through this boundary. One process by which this energy exchange takes place is through atmospheric gravity waves. Gravity waves are excited in the neutral atmosphere by thunderstorms, mountain ranges, and other processes. These waves propagate upward from their point of origin steepening as they propagate. Eventually these waves become non-linear and "break" like ocean waves on a beach depositing their energy in the ionosphere-mesosphere, providing a potentially important coupling between the troposphere where climatic effects are evident and the magnetosphere. Figure 5 shows several images of gravity waves in both Earth's and Mars' atmospheres. The potential importance of these processes for energy transfer has only recently been recognized, and much additional work remains to establish the significance of the Universal Process of gravity wave propagation on the structure of Earth, Mars, and other planetary atmospheres.

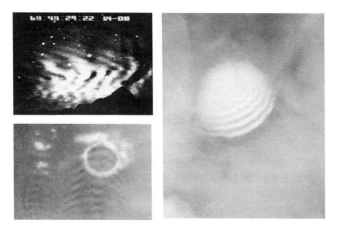

Figure 5. Gravity waves in Earth's atmosphere [upper left panel] transfer energy from the troposphere to the ionosphere-mesosphere (from Taylor *et al.* 1987). Similar waves seen on Mars by the Mars Global Surveyor Mission (lower left and right panels) may couple the lower and upper Martian atmosphere.

5. Summary and Conclusions

The matter in the Solar System is organized primarily by gravitional and magnetic forces. These forces cause the formation of structures that evolve primarily through Universal Processes. Heliophysics, which is the study of Universal Processes in diverse environments, can lead to a deeper physical understanding of magnetic processes like reconnection, particle acceleration, and the generation of waves and other processes in the neutral atmosphere. Heliophysical processes are the foundation for understanding the physical conditions in astrophysical objects, and as such are essential for our understanding of the Universe.

References

Angelopoulos, V., McFadden, J. P., Larson, D., Carlson, C. W., Mende, S. B., Frey, H., Phan, T., Sibeck, D. G., Glassmeier, K.-H., Auster, U., Donovan, E., Mann, I. R., Rae, I. J., Russell, C. T., Runov, A., Zhou, X.-Z., & Kepko, L. 2007,*Sci.*, 321, 931

Chen, P. F., Liu, W. J., & Fang, C. 2007,*Adv. Sp. Res.*, 39, 1421

Crooker, N. U. 2004,*EOS*, 85(37), 351

Sui, L. & Holman G. D. 2003,*Ap. J.*, 596, L251

Taylor, M.J., Hapgood, M.A., & Rothwell, P. 1987, *Planet. Space Sci.*, 35, 413

Discussion

SPANGLER: You stated that the resistivity in heliospheric plasmas is small. Isn't a more accurate statement that collisional resistivity is small? Collisionless processes could enhance the resistivity and associated dissipation.

DAVILA: Yes you are correct. However to get enhanced resistivity the structure must already be evolving through a collisionless process.

BEER: I'm glad that you mentioned the example of atmospheric gravity waves. Universal processes are not reduced to magnetically organized matter but also in gravitationally organized matter.

DAVILA: Absolutely.

GIRISH: What are the early results in the progress of ground-based experiments going on all over the world as a part of IHY started in 2007?

DAVILA: The IHY has approximately 15 instrument projects. These involve generally an instrument donor and an instrument host. Instruments are now operating in South America, Africa and Asia. The project has been well received and is very successful.

CLIVER: The universality of heliospheric processes imposes challenges on our understanding of reconnection, where in the magnetosphere one observes electron skin depth scale of "action", which extrapolates to a few cm in the solar corona. This may require particular conditions to explain the large flaring fluxes of ions and electrons.

DAVILA: There are certainly important differences between reconnection in the magnetosphere and in solar flares. But I believe that the wonderfully detailed in situ observations in the magnetosphere and high resolution images in flares can help us understand reconnection more completely.

Universal Heliophysical Processes
Proceedings IAU Symposium No. 257, 2008
N. Gopalswamy & D.F. Webb, eds.

© 2009 International Astronomical Union
doi:10.1017/S1743921309029044

Composition of matter in the heliosphere

Peter Bochsler[1,2]

[1]Physikalisches Institut, University of Bern,
Sidlerstrasse 5, CH-3012, Bern, Switzerland

[2]Space Science Center, EOS, University of New Hampshire, Durham NH, USA

email: bochsler@soho.unibe.ch

Abstract. The Sun is by far the largest reservoir of matter in the solar system and contains more than 99% of the mass of the solar system. Theories on the formation of the solar system maintain that the gravitational collapse is very efficient and that typically not more than one tenth from the solar nebula is lost during the formation process. Consequently, the Sun can be considered as a representative sample of interstellar matter taken from a well mixed reservoir 4.6 Gy ago, at about 8 kpc from the galactic center. At the same time, the Sun is also a faithful witness of the composition of matter at the beginning of the evolution of the solar system and the formation of planets, asteroids, and comets. Knowledge on the solar composition and a fair account of the related uncertainties is relevant for many fields in astrophysics, planetary sciences, cosmo- and geochemistry. Apart from the basic interest in the chemical evolution of the galaxy and the solar system, compositional studies have also led to many applications in space research, i.e., it has helped to distinguish between different components of diffuse heliospheric matter. The elemental, isotopic, and charge state composition of heliospheric particles (solar wind, interstellar neutrals, pickup ions) has been used for a multitude of applications, such as tracing the source material, constraining parameters for models of the acceleration processes, and of the transport through the interplanetary medium. It is important to realize, that the two mainstream applications, as outlined above – geochemistry and cosmochemistry on one side, and tracing of heliospheric processes on the other side – are not independent of each other. Understanding the physical processes, e.g., of the fractionation of the solar wind, is crucial for the interpretation of compositional data; on the other hand, reliable information on the source composition is the basis for putting constraints on models of the solar wind fractionation.

Keywords. Sun:abundances, solar wind, solar system: general, ISM:abundances

1. Introduction

Diffuse matter in the heliosphere exhibits a wide diversity of compositional features. Solar matter is considered to be the most representative sample with respect to elemental and isotopic composition of the local galactic interstellar medium (ISM) as it existed 4.6 Gy ago at about 8 kpc from the galactic center. The formation of the Sun can be visualized as a crossroads, at which the galactic nucleosynthetic evolution takes its end and the geochemical evolution of matter in the solar system with all its phenotypes (comets, asteroids, planets, planetary atmospheres etc.) takes its origin. Of course, solar matter has not survived in its pristine chemical state as it prevailed in the form of complex molecules, ices, and dust. In the solar core, where hydrogen has been converted to helium, nuclear reactions have also modified the proportions among other light elements and isotopes. In the outer convective zone (OCZ) some light elements and isotopes such as D, ^{3}He, ^{6}Li, and ^{7}Li have either been destroyed or – at least – been modified in their abundances. Furthermore, the composition of the OCZ has also been modified to some extent by secular gravitational settling. Next to the Sun, comets are believed to contain the most pristine material of the solar system. Presumably, some of it is still in its original chemical form as it was imported as dust and ices from the interstellar medium. However, comets do not contain the volatile inventory of the ISM. Whereas the terrestrial planets have accreted most of their material (including some volatiles) in solids, the most abundant elements hydrogen and helium have been accreted onto the Sun and the giant planets in gaseous form. Elemental and isotopic fractionation during the formation of the Sun must have been negligible as the formation process is thought to be very efficient; i.e., only of the order of 10% of the mass has been lost and returned to the ISM during the formation of the solar disk. Hence, to a good approximation the notion of "birth of the solar system", i.e. a short event compared to its lifetime is correct: After a short instant of detachment from the ISM, the solar system has remained a closed system.

Certain types of carbonaceous meteorites contain almost all elements in solar proportions (Anders & Grevesse 1989). The isotopic composition of the non-volatile elements is remarkably similar within this type of meteorites, and it is also comparable with the terrestrial isotopic composition. This indicates a fairly complete homogenization among materials within the inner solar system; an observation which is relevant for inferences on the isotopic composition of these elements also of the Sun – inaccessible to direct measurement. Apart from such inferences it is the solar wind, which provides the most straightforward information on the isotopic composition of solar matter. However, as we will discuss in the following, information on isotopic abundances is biased through fractionation processes during solar wind feeding and acceleration. Experimentally, the isotopic fractionation of the solar wind has been confined by in-situ experiments on moderately refractory elements, which have more than one abundant isotope, and which show little or no variation in meteorites and terrestrial materials, e.g., Mg, Si and Fe (Bochsler *et al.* 1996, Kallenbach *et al.* 1998, Kucharek *et al.* 1998, Ipavich *et al.* 2001).

The purpose of this review is to summarize the current knowledge on the composition of different types of heliospheric matter, and to show how compositional features can be used to trace their sources.

2. Sun and solar wind

The Sun continuously emits charged particles into the surrounding heliosphere. The overwhelming part (95%) of the ions are protons, the rest consists of alpha particles and heavier species – very roughly in proportions similar to solar composition. Figure 1 shows – as an example – the energy distribution of the solar oxygen flux at 1 AU. The solar wind contributes the by far largest fraction of solar particles. In the energy range above typically 10 keV/nucleon the distribution is populated by the so-called suprathermal particles and by solar energetic particles (SEP's). Suprathermal particles are generated by various processes in the corona and in the interplanetary space, i.e., by shock acceleration at corotating interaction regions (CIRs), or interplanetary shocks, created by the propagation of coronal mass ejections (CME's). Solar particles with higher energies are also generated at reconnection sites of merging solar magnetic loops, by the release of magnetic energy in magnetohydrodynamic instabilities in the solar atmosphere, where so-called "impulsive" events produce energetic particles in abundances strongly deviating from normal coronal composition.

2.1. *Solar wind abundances and solar abundances*

2.1.1. *Observations*

As stated previously, a good agreement has generally been found between meteoritic and solar photospheric abundances. Meteorites contain the refractory elements and isotopes approximately in solar proportions, but they lack the volatile complement.

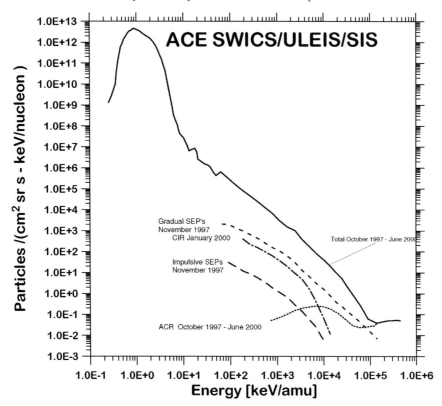

Figure 1. Typical flux of solar oxygen particles as observed with the Advanced Composition Explorer (ACE) from October 1997 through June 2000 at 1 AU. Adapted from Mewaldt *et al.* (2007).

Therefore, comparisons between the photosphere and meteorites rest essentially on the normalization of a few refractory or moderately refractory key elements such as Si and Mg, Fe for which there is on one hand no concern for losses from meteorites (and their parent bodies), and – on the other hand – the photospheric abundances have been determined with fairly high reliability. However, in either case the abundance of these key elements in relation to the most abundant element, hydrogen, is a difficult and sometimes controversial subject. In the case of meteorites – hydrogen is mostly contained in volatile compounds and depleted in all types of meteorites by orders of magnitude – even in the most pristine classes. On the other hand, *absolute* photospheric spectroscopic abundances rest not only on complex calculations in atomic physics but they also depend strongly on assumptions on parameters of model atmospheres. These models have been steadily improved and the agreement between determinations with the intensity and linewidth of different optical lines has become better (cf. Asplund *et al.* 2005). Noble gases (He, Ne, and Ar) have high first ionization potentials and, correspondingly, high excitation potentials, and the determination of their abundances by spectroscopic means poses special difficulties. In many compilations of solar system abundances for volatiles one refers therefore to solar wind abundances or to abundances of solar energetic particles. The noble gases in the solar wind have been determined with the Apollo foil experiments (Geiss *et al.* 1972) and, more recently, with the Genesis sample return mission (Burnett *et al.* 2003, Grimberg *et al.* 2006). Laboratory measurements of Apollo- and Genesis-samples largely agree with each other and have certainly better precisions than long-time averages of in-situ measurements (Bochsler *et al.* 1986, von Steiger *et al.* 2000). To infer solar abundances from solar wind noble gases (and also from SEP's) is somewhat more problematic because one needs information on the fractionation mechanisms between the solar atmosphere and the corona. The most notorious and difficult case is the case of helium, which is depleted roughly by a factor 2 in the solar wind compared to the solar atmosphere. This factor is well known, because from helioseismology the helium abundance in the Sun and in the OCZ is known with very high precision and reliability (e.g., Gough 2007). What is less well established is the mechanism, which causes the depletion of helium in the corona and in the solar wind. If this were well understood, it would be possible to infer how much this mechanism affects oxygen, neon, and other elements, which have been well determined in the solar wind, and it would be possible to create a line of evidence for the abundance of heavy elements in the Sun, which is independent of spectroscopic determinations. In fact such an independent line of evidence is badly needed, since the newly determined low abundances of Asplund *et al.* (2005) have created a problem for helioseismological models. These abundances do not reproduce the observed opacity of solar material. Bahcall *et al.* (2005) and Antia & Basu (2006) have made an attempt to fix this abundance problem by raising somewhat arbitrarily the solar neon abundance by a factor of 3, independent of all other elements. Young (2005) and Bochsler *et al.* (2006) and Bochsler (2007b) have independently criticized this approach on grounds of optical observations of EUV lines in flares and of solar wind determinations, respectively.

Table 1 is an updated compilation of elemental abundances in the solar wind. For a detailed discussion, see Bochsler (2007a), where also a compilation of isotopic abundances is given.

Table 1. Elemental abundances in the solar wind

Element	Interstream	Coronal Hole	References
He	90 ± 30	75 ± 20	[1,2,3]
C	0.68 ± 0.07	0.68 ± 0.07	[3]
N	0.078 ± 0.005	0.114 ± 0.021	[3,4]
O	$\equiv 1$	$\equiv 1$	
Ne	0.14 ± 0.03		[5,6,7,8]
Na	0.0090 ± 0.0015	0.0051 ± 0.0014	[9]
Mg	0.147 ± 0.050	0.106 ± 0.050	[3]
Al	0.0119 ± 0.003	0.0081 ± 0.0004	[10]
Si	0.140 ± 0.050	0.101 ± 0.040	[3,11]
P	0.0014 ± 0.0004		[12]
S	0.050 ± 0.015		[3,13]
Ar	0.0031 ± 0.0008	0.0031 ± 0.0004	[6,7,14,15]
Ca	0.0081 ± 0.0015	0.0053 ± 0.0010	[16,17,18]
Cr	0.0020 ± 0.0003	0.0015 ± 0.0003	[19]
Fe	0.122 ± 0.050	0.088 ± 0.050	[3,20,21,22]
Ni	0.0065 ± 0.0025		[23]

References: [1] Bochsler (1984), [2] Bochsler *et al.* (1986), [3] von Steiger *et al.* (2000), [4] Gloeckler *et al.* (1986), [5] Geiss *et al.* (1972), [6] Geiss *et al.* (2004), [7] Grimberg *et al.* (2008), [8] Heber, [9] Ipavich *et al.* (1999), [10] Bochsler *et al.* (2000), [11] Bochsler (1989), [12] Giammanco *et al.* (2008), [13] Giammanco *et al.* (2007), [14] Cerutti (1974), [15] Weygand *et al.* (2001), [16] Kern *et al.* (1997), [17] Kern (1999), [18] Wurz *et al.* (2003), [19] Paquette *et al.* (2001), [20] Schmid *et al.* (1988), [21] Aellig *et al.* (1999a), [22] Aellig *et al.* (1999b), [23] Karrer *et al.* (2007).

2.1.2. *Theoretical considerations*

How well do the solar wind and the corona reproduce abundances of the OCZ? Understanding the fractionation processes between corona and the solar atmosphere is a necessary ingredient to answer this question. Two main classes of processes have to be considered: The first one is the so-called FIP-Process (FIP = First-Ionization-Potential) (or FIT-First-Ionization-Time), according to which elements can be depleted in the corona depending on their ionization potential. Various models of this mechanism have been proposed, most models contain too many free parameters to make safe predictions and, furthermore, observational data are still of insufficient quality, to eliminate some competing models conclusively. Nevertheless, attempts to establish the importance of this process have been made, by investigating correlations among elements which are expected to be sensitive or not so sensitive to the process. One example is the He/H vs O/H ratio: H/O should be rather insensitive to the FIP-effect, because both elements have the same first ionization potentials and, more importantly, the two elements are expected to be well coupled by resonant charge exchange. Bochsler (2007b) argued that the large variability of H/O, strongly correlated with He/H, speaks against a strong influence of the FIP effect not only on H/O but also on the He/H ratio.

As an alternative, especially in the context of the large He/H variability in the solar wind, collisional effects have regularly been considered as the mechanism responsible for this phenomenon. Geiss *et al.* (1970) have discussed the role of Coulomb friction for accelerating heavy species in the solar wind. Noci & Porri (1983) later pointed out that the non-negligible abundance of helium is important for the momentum balance, and Bürgi & Geiss (1986) elaborated their solar wind models adding He^{++} to H^+ as major particles and found that only the inclusion of helium yields realistic models of the low-speed solar wind, with the characteristic temperature maxima observed in charge state distributions of minor species. Clearly, Coulomb friction is an important agent in carrying heavy species in the low-speed wind. As a consequence, some isotopic fractionation of low-speed wind compared to photospheric abundances must be considered. From the models of Bodmer & Bochsler (2000) one can infer that the strongest effect – of the order of a few tens of percents – has to be expected for the case of the helium isotopes. For medium mass range elements, such as Mg, Si, etc., the effect is of the order of a few percent per mass unit – in the sense of a somewhat elevated abundance of the lighter isotopes – but below the limit for an unambiguous detection by currently operating in-situ experiments. The effect seems, however, detectable with Genesis (Grimberg *et al.* 2008).

At the end of this section also the issue of the variability of the solar wind composition should be briefly addressed. The Genesis mission has confirmed the picture of the Apollo foil experiments: The targets exposed in different regimes: low-speed wind, high-speed wind, and CMEs, show small but detectable variations. Although the distinction of these regimes by the onboard computer of Genesis has not been ideal, the velocity histograms recorded with these exposures look fairly different from each other, and from the in-situ experience one would expect a wider diversity of compositions, especially for elemental ratios with helium involved. For the most reliable in-situ data Figure 2 shows that the helium abundance variability is limited within a range of about a factor of 5 in 99% of the cases. Now, helium is the most likely candidate for strong and variable fractionation since it is special in two respects: It has the highest ionization potential, and He^{++} has the least favorable Coulomb-drag factor of all elements. In contrast, it is difficult to conceive how the Sun could fractionate two mass-neighbors with very similar ionization properties such as Al and Mg, or Al and Si efficiently from each other by a factor of

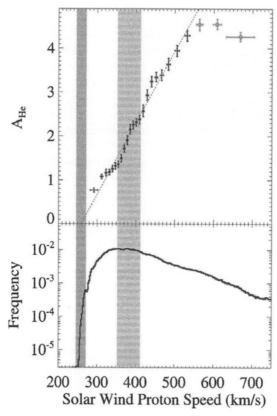

Figure 2. Figure from Kasper *et al.* (2007) (their Figure 4). The lower panel shows a histogram of speeds registered during the entire Wind mission. The heliomagnetic latitudinal dependence of speeds has been removed. The upper panel shows the corresponding helium abundances as observed with the Wind spacecraft. The expansion factor of the magnetic field low in the corona determines the proton speed and the coupling of protons with helium (and minor ions), thus regulating the helium flux at 1 AU.

two or more – even on short time scales, as is sometimes claimed on the basis of in-situ data.

3. Non-solar heliospheric particles

3.1. *Interstellar matter in the heliosphere*

A substantial amount of suprathermal and energetic particles in the heliosphere (see Figure 1) originates from so-called pick-up ions from interstellar matter. These particles are particularly apt for acceleration in shock regions as they are produced with velocity distributions substantially wider than the solar wind. Neutral atoms with high ionization potentials can penetrate, unimpeded by the interplanetary magnetic field, deep into the solar system. Occasionally, as they approach the Sun, they are ionized by solar EUV radiation or by charge exchange with solar wind ions. Upon ionization they begin a cycloidal motion about the ambient magnetic field. As the field moves outwards, pickup ions are swept back to the fringes of the heliosphere. Particularly, pickup ions which are pre-accelerated in the innermost heliosphere, can become energized up to several

MeVs/amu at the heliospheric termination shock, and reenter the heliosphere as the so-called anomalous component of cosmic rays (ACR's, lowermost dotted line in Figure 1).

3.2. *Discovery of the "Outer Source"*

From the historical perspective it is interesting to note that in several cases compositional features and charge state distributions of ions have helped to identify the origin of heliospheric particles. For instance, it has been possible to identify the source of ACR's by the fact that it contained an elemental mix, which differs substantially from what is known as "cosmic" composition. Oxygen and nitrogen were significantly overabundant over C which has a relatively low ionization potential. Fisk *et al.* (1974) proposed that these particles originated from weakly charged pick-up ions from the interstellar medium, which move out to the heliospheric bow-shock, where they are re-accelerated and brought back as ACR's into the inner solar system. Later, also elements with low ionization potentials have been identified in ACR's (Reames 1999 and Cummings *et al.* 2002), and it became clear that these elements could not have been imported from the ISM. This led Schwadron *et al.* (2002) to propose an "Outer Source" of pickup ions for ACRs. Sputtering of interplanetary dust with solar wind ions would release elements with low first ionization potentials and (and consequently, low volatility) out of dust grains such as Si, Mg, and produce the observed enhancement of these elements in the ACR-population (Reames 1999 and Cummings *et al.* 2002).

Similarity of compositional features with known populations may sometimes lead to erroneous interpretations as has recently been discovered by Grimberg *et al.* (2006). An enigmatic component of neon – called "SEP-Neon" – had been found in lunar fines (Wieler *et al.* 1986). Its origin was ascribed to solar energetic particles because it was more deeply implanted in lunar grains than solar wind, and it had an isotopic composition, similar to SEP's. An intriguing problem of this interpretation was its overabundance over several orders of magnitude in relation to the observed fluence of solar wind. Mewaldt *et al.* (2001) had pointed out the possibility that implantation effects could modify the solar wind isotopic composition to the degree that it appeared in a stepwise release process or an etching procedure similar to solar energetic particles. Grimberg *et al.* (2006) and Grimberg *et al.* (2008) now found that this dubious SEP-component was present also in the Genesis target materials despite the fact that the fluence of SEP's during the Genesis exposures was much too small. They confirmed with SRIM-simulations (Ziegler 2004) that indeed – upon implantation into a solid target – the distribution of solar wind particles could be modified in such a way as to produce the observed depth profiles.

3.3. *The "Inner Source"*

Inner Source pickup ions are thought to originate from the interaction of solar wind ions with dust in the inner solar system (Geiss *et al.* 1996). An unsolved puzzle is the overabundance of neon in inner-source pickup ions reported by Gloeckler *et al.* (2000) and by Allegrini *et al.* (2005). Compared to solar wind abundances neon is overrepresented over the neighboring elements C,N,O, and Mg (cf. Figure 4). This is somewhat surprising since one would expect that an interaction of solar wind ions with dust would rather lead to the opposite, i.e. an overrepresentation of dust-constituting elements such as found in the outer source. In general, interplanetary dust seems rather fragile, and furthermore, according to Kehm *et al.* (2006) far from saturated with noble gases. Consequently, sputtering of solids is expected to produce abundant low-FIP pickup ions. A possible solution to this problem might be the relatively large second ionization rates of O, C, and Si compared to Ne. Pickup ions from these elements would undergo second ionization

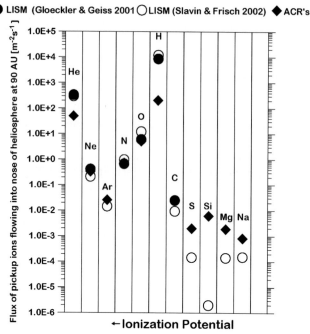

Figure 3. Figure adapted from Cummings *et al.* (2002) (their Figure 10 and Table 4). The anomalous component apparently consists of two different constituents. Both originate from pickup ions, which are generated in the inner solar system and swept out of the heliosphere with the interplanetary magnetic field. At the termination shock pre-accelerated particles are further energized, and some of them re-enter the heliosphere at energies, typically of the order of a few MeV/amu, as the so-called anomalous component of cosmic rays (ACR). Using an elaborated set of ionization rates, the elemental abundances of the LISM derived from Gloeckler & Geiss (2001) (full circles) and from Slavin & Frisch (2002) (empty circles), Cummings *et al.* (2002) succeed to model the abundance pattern of high-FIP elements of ACRs (full diamonds) up to oxygen quite accurately. However, ACR's are overabundant for predominantly dust forming, low-FIP elements (S, Si, Mg, Na), thus requiring another origin of these particles, the so-called "Outer Source" of pickup ions as explained by Schwadron *et al.* (2002).

and become difficult to be detected when they reach the site of observation. This, however, would require a large production or advection of small dust grains in the inner corona (i.e., at heliocentric distances $\leqslant 10 R_\odot$) where photoionization rates are high and/or the electron density is still sufficiently high for collisional ionization (Bochsler *et al.* 2007).

3.4. *Energetic neutrals (ENA)*

With the advent of the technique of surface ionization as a diagnostic tool in space physics, the study of ENA's has become increasingly important. The advantage of neutrals for heliospheric diagnostics is their ability to travel unimpeded through electric and magnetic fields, hence, they can carry information about the source processes, in which they are produced, over large distances across the heliosphere to a remote observer. The backside of the coin is that the detection and analysis of ENA's becomes difficult. For these reasons, compositional analysis of ENAs has still been limited. For a detailed review see Wurz (2000).

Within the inner corona collisional ionization and recombination maintain an equilibrium charge state distribution for a given element. Further out, where the time between collisions with electrons becomes larger than the typical solar wind expansion time, the charge state "freezes" and the static equilibrium can no longer be maintained

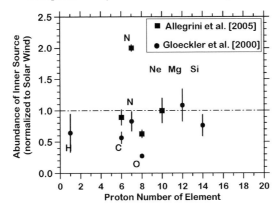

Figure 4. Abundances of Inner Source particles normalized to solar wind abundances and to neon. Full circles are for data from Ulysses/SWICS, published by Gloeckler *et al.* (2000), full squares from the compilation of Allegrini *et al.* (2005) of Ulysses/SWICS data.

(e.g., Owocki *et al.* 1983). Under normal coronal conditions a fraction of the order of 10^{-6} of solar wind particles are expected to flow as neutrals. Assuming a proton flux of $5 \cdot 10^{12}\,\mathrm{m}^{-2}\mathrm{s}^{-1}$, one estimates a neutral hydrogen flux of typically $10^7\,\mathrm{m}^{-2}\mathrm{s}^{-1}$. Collier *et al.* (2001) and Collier *et al.* (2003) reported a fraction of 10^{-3} to 10^{-5} of neutrals in the solar wind flux at 1 AU.

One efficient way of converting solar wind ions into neutrals is backscattering from dust grains. Assuming a typical backscattering and neutralization efficiency of $P_n = 10\%$, and assuming that all backscattered particles are converted into neutrals, one can estimate the probability of a solar wind ion to be backscattered and neutralized, not considering – for the moment – the possibility of re-ionization by charge exchange and photoionization from solar EUV. We assume a heliospheric radial density distribution of dust grains according to n(r) $\propto$ r$^\alpha$. The differential probability of a solar wind ion to hit a dust grain between r and r+dr is then

$$dp = dr \cdot n(r)\sigma = n(R)\sigma \cdot R\left(\frac{r}{R}\right)^{-\alpha}, \qquad (3.1)$$

where R is 1 AU and n(R)σ denotes the integrated cross section density of dust grains at 1 AU. We assume this value to be $5 \cdot 10^{-19}\,\mathrm{m}^2/\mathrm{m}^3$ (Leinert & Grün 1990). Integrated over the range, from where a neutral particle has a chance to survive as neutral at r_0=0.1 AU to the observer at 1 AU, one finds a probability

$$p = P_n \frac{n(R)\sigma R}{\alpha - 1}\left(\left(\frac{r_0}{R}\right)^{1-\alpha} - 1\right). \qquad (3.2)$$

With a typical value of $\alpha = 1.3$ (Leinert & Grün 1990) one obtains then $p = 3 \cdot 10^{-8}$, which is far below the observed lower limit of Collier *et al.* (2003) of 10^{-5}. Again, as in the previous section on the Inner Source, one could seek remedy in increasing the exponent of the density distribution to a probably unrealistic value of $\alpha = 4.7$, thereby concentrating the interplanetary dust towards the Sun. Unfortunately, it has not been possible to identify the composition of this neutral solar wind component. At this time the most likely explanation seems solar wind protons, recombined with electrons in the inner corona, although this process seems to provide flux ratios, which are still at least one magnitude under the observed ones.

4. Conclusions

It has been discovered in recent years that, apart from the solar wind and energetic particles of solar origin, the inner heliosphere is hosting a rich diversity of particles of different sources. Many of them have not yet been identified and explained unambiguously. In all cases the determination of their masses and charge states, together with the investigation of their dynamical properties have provided the essential clues. Heliospheric particles are not just protons and electrons. Understanding their properties and their sources requires also a detailed knowledge of their composition.

5. Acknowledgements

The author gratefully acknowledges helpful discussions with Berndt Klecker, Ansgar Grimberg, and Rainer Wieler. This work has been supported by NASA STEREO contract NAS5-00132.

References

Aellig, M. R., Hefti, S., Grünwaldt, H., *et al.* 1999a, J. Geophys. Res., 104, 24,769

Aellig, M. R., Holweger, H., Bochsler, P., *et al.* 1999b, in Solar Wind Nine, ed. S. R. Habbal, R. Esser, J. V. Hollweg, & P. A. Isenberg (AIP Proceedings 471), 255–258

Allegrini, F., Schwadron, N. A., McComas, D. J., Gloeckler, G., & Geiss, J. 2005, J. Geophys. Res., 110, A05105,doi:10.1029

Anders, E. & Grevesse, N. 1989, Geochim. Cosmochim. Acta, 53, 197

Antia, H. M. & Basu, S. 2006, Astrophys. J., 644, 1292

Asplund, M., Grevesse, N., & Sauval, A. J. 2005, in ASP Conf. Ser. 336: Cosmic Abundances as Records of Stellar Evolution and Nucleosynthesis, ed. T. G. Barnes & F. N. Bash, 25

Bahcall, J. N., Basu, S., & Serenelli, A. M. 2005, Astrophys. J., 631, 1281

Bochsler, P. 1984, Habilitationsschrift, University of Bern

Bochsler, P. 1989, J. Geophys. Res., 94, 2365

Bochsler, P. 2007a, Astron. Astrophys. Rev., 14, 1

Bochsler, P. 2007b, Astron. Astrophys., 471, 315

Bochsler, P., Auchère, F., & Skoug, R. M. 2006, in Proc. SOHO 17 Conference, Taormina (ESA SP 617)

Bochsler, P., Geiss, J., & Kunz, S. 1986, Solar Phys., 103, 177

Bochsler, P., Gonin, M., Sheldon, R. B., *et al.* 1996, in Solar Wind Eight. Proceedings of the Eighth International Solar Wind Conference, ed. D. Winterhalter, J. T. Gosling, S. R. Habbal, W. S. Kurth, & M. Neugebauer, Vol. 382 (Woodbury, N.Y., USA: American Institute of Physics), 199–202

Bochsler, P., Ipavich, F. M., Paquette, J. A., Weygand, J. M., & Wurz, P. 2000, J. Geophys. Res., 105, 12659

Bochsler, P., Möbius, E., & Wimmer-Schweingruber, R. F. 2007, in Proc. Second Solar Orbiter Conference, Athens (ESA SP-641)

Bodmer, R. & Bochsler, P. 2000, J. Geophys. Res., 105, 47

Bürgi, A. & Geiss, J. 1986, Solar Phys., 103, 347

Burnett, D. S., Barraclough, B. L., Bennett, R., *et al.* 2003, Space Sci. Rev., 105, 509

Cerutti, H. 1974, PhD Thesis, University of Bern, Switzerland

Collier, M. R., Moore, T. E., Ogilvie, K., *et al.* 2003, in Proceedings of the Tenth International Solar Wind Conference, ed. M. Velli, R. Bruno, & F. Malara (Melville, New York: AIP Proceedings 679), 790–793

Collier, M. R., Moore, T. E., Ogilvie, K. W., *et al.* 2001, J. Geophys. Res., 106, 24,893

Cummings, A. C., Stone, E. C., & Steenberg, C. D. 2002, Astrophys. J., 578, 194

Fisk, L. A., Kozlovsky, B., & Ramaty, R. 1974, Astrophys. J., 190, L35

Geiss, J., Bühler, F., Cerutti, H., Eberhardt, P., & Filleux, C. 1972, Apollo 16 Prel. Sci. Rep. NASA Special Publication, 315, 14.1

Geiss, J., Bühler, F., Cerutti, H., *et al.* 2004, Space Sci. Rev., 110, 307

Geiss, J., Gloeckler, G., & von Steiger, R. 1996, Space Sci. Rev., 78, 43

Geiss, J., Hirt, P., & Leutwyler, H. 1970, Solar Phys., 12, 458

Giammanco, C., Bochsler, P., Karrer, R., *et al.* 2007, Space Sci. Rev., 130, 329

Giammanco, C., Wurz, P., & Karrer, R. 2008, Astrophys. J., 681, 1703

Gloeckler, G., Fisk, L. A., Geiss, J., Schwadron, N. A., & Zurbuchen, T. H. 2000, J. Geophys. Res., 105, 7459

Gloeckler, G. & Geiss, J. 2001, in Acceleration and transport of energetic particles observed in the Heliosphere, ed. R. A. Mewaldt, J. R. Jokipii, M. A. Lee, E. Moebius, & T. H. Zurbuchen, Vol. 528 (Woodbury, N.Y.: American Institute of Physics), 281–

Gloeckler, G., Ipavich, F. M., Hamilton, D. C., *et al.* 1986, Geophys. Res. Lett., 13, 793

Gough, D. O. 2007, Astron. Nachr./AN, 328, 273

Grimberg, A., Baur, H., Bochsler, P., *et al.* 2006, Science, 314, 1133

Grimberg, A., Baur, H., Bühler, F., Bochsler, P., & Wieler, R. 2008, Geochim. Cosmochim. Acta, 72, 626

Ipavich, F. M., Bochsler, P., Lasley, S. E., Paquette, J. E., & Wurz, P. 1999, EOS Trans. AGU, 80, 256

Ipavich, F. M., Paquette, J. A., Bochsler, P., Lasley, S. E., & Wurz, P. 2001, in Solar and Galactic Composition, ed. R. F. Wimmer-Schweingruber, Vol. CP-598 (Melville, N.Y.: AIP Conf. Proceedings), 121–126

Kallenbach, R., Ipavich, F. M., Kucharek, H., *et al.* 1998, Space Sci. Rev., 85, 357

Karrer, R., Bochsler, P., Giammanco, C., *et al.* 2007, Space Sci. Rev., 130, 317

Kasper, J. C., Stevens, M. L., Lazarus, A. J., Steinberg, J. T., & Ogilvie, K. W. 2007, Astrophys. J., 660, 901

Kehm, K., Flynn, G. J., & Hohenberg, C. M. 2006, Meteoritics and Planetary Sciences, 41, 1199

Kern, O. 1999, PhD Thesis, University of Bern, Switzerland

Kern, O., Wimmer-Schweingruber, R. F., Bochsler, P., Gloeckler, G., & Hamilton, D. C. 1997, in Proceedings of the 31st ESLAB Symp., 'Correlated Phenomena at the Sun, in the Heliosphere and in Geospace', Workshop on Plasma Dynamics and Diagnostics in the Solar Transition Region and Corona (ESA SP-415), 345–348

Kucharek, H., Ipavich, F. M., Kallenbach, R., *et al.* 1998, 103, 26'805

Leinert, C. & Grün, E. 1990, In: Physics of the Inner Heliosphere, ed. R. Schwenn & E. Marsch (Berlin Heidelberg: Springer-Verlag), 207–275

Mewaldt, R. A., Ogliore, R. C., Gloeckler, G., & Mason, G. M. 2001, in Solar and Galactic Composition, ed. R. F. Wimmer-Schweingruber, Vol. CP-598 (Melville, N.Y.: AIP Conf. Proceedings), 393–398

Mewaldt, R. A., Cohen, C. M., Mason, G. M., Haggerty, D. K., & Desai, M. I. 2007, Space Sci. Rev., 130, 323

Noci, G. & Porri, A. 1983, IAGA, Hamburg, paper 4L.04 presented at the 18th General Assembly Meeting

Owocki, S. P., Holzer, T. E., & Hundhausen, A. J. 1983, Astrophys. J., 275, 354

Paquette, J. A., Ipavich, F. M., Lasley, S. E., Bochsler, P., & Wurz, P. 2001, in Solar and Galactic Composition, ed. R. F. Wimmer-Schweingruber, Vol. CP-598 (Melville, N.Y.: AIP Conf. Proceedings), 95–100

Reames, D. V. 1999, Astrophys. J., 518, 473

Schmid, J., Bochsler, P., & Geiss, J. 1988, Astrophys. J., 329, 956

Schwadron, N. A., Combi, M., Huebner, W., & McComas, D. J. 2002, Geophys. Res. Lett., 29, doi:10.1029/2002GL015829

Slavin, J. D. & Frisch, P. C. 2002, Astrophys. J., 565, 364

von Steiger, R., Schwadron, N. A., Hefti, S., *et al.* 2000, J. Geophys. Res., 105, 27,217

Weygand, J. M., Ipavich, F. M., Wurz, P., Paquette, J. A., & Bochsler, P. 2001, in Solar and Galactic Composition, ed. R. F. Wimmer-Schweingruber, Vol. CP-598 (Melville, N.Y.: AIP Conf. Proceedings), 101–106

Wieler, R., Baur, H., & Signer, P. 1986, Geochmim. Cosmochim. Acta, 50, 1997

Wurz, P. 2000, in The outer heliosphere: Beyond the planets, ed. K. Scherer, H. Fichtner, & E. Marsch (Katlenburg-Lindau, Germany: Copernicus Gesellschaft e.V.), 251–288

Wurz, P., Bochsler, P., Paquette, J. A., & Ipavich, F. M. 2003, Astrophys. J., 583, 489
Young, P. R. 2005, Astron. Astrophys., 444, L45
Ziegler, J. 2004, Nucl. Instr. and Methods in Phys. Res. B, 219-220, 1027

Discussion

HOWARD: Do the models for the "inner source" include the "dust free region", in which the dust particles are completely vaporized/ionized?

BOCHSLER: The models we developed assume a simplified radial stationary dust density distribution which ends at 5 solar radii. As long as dust is vaporized and ionized in this region, it produces inner source pick-up ions. Dust which is swept away by the interplanetary magnetic field or by radiation pressure, is less efficient in producing pick-up ions. Hence, it is important to know which of these processes is more important in eliminating dust. The argument in favor of locating the solar-wind-dust interaction very close to the Sun is the apparent overabundance of neon among the singly ionized pick-up ions. Neon has a bigger chance of leaving the corona as a singly ionized species than dust forming elements CNO, Mg, Si which tend to become multiply ionized and are much more difficult to detect at 1 AU than singly ionized species.

FISK: You may wish to note that MESSENGER is providing new observations which appear contradictory to our earlier understanding.

Universal Heliophysical Processes
Proceedings IAU Symposium No. 257, 2008
N. Gopalswamy & D.F. Webb, eds.

Sun and planets from a climate point of view

J. Beer, J. A. Abreu and F. Steinhilber

Swiss Federal Institute of Aquatic Science and Technology, Eawag, 8600 Dübendorf,
Switzerland

Abstract. The Sun plays a dominant role as the gravity centre and the energy source of a planetary system. A simple estimate shows that it is mainly the distance from the Sun that determines the climate of a planet. The solar electromagnetic radiation received by a planet is very unevenly distributed on the dayside of the planet. The climate tries to equilibrate the system by transporting energy through the atmosphere and the oceans provided they exist. These quasi steady state conditions are continuously disturbed by a variety of processes and effects. Potential causes of disturbance on the Sun are the energy generation in the core, the energy transport trough the convection zone, and the energy emission from the photosphere. Well understood are the effects of the orbital parameters responsible for the total amount of solar power received by a planet and its relative distribution on the planet's surface. On a planet, many factors determine how much of the arriving energy enters the climate system and how it is distributed and ultimately reemitted back into space. On Earth, there is growing evidence that in the past solar variability played a significant role in climate change.

Keywords. Sun: activity, (Sun:) solar-terrestrial relations, atmospheric effects

1. Introduction

Climate can be defined as "the prevalent or characteristic meteorological conditions (temperature, pressure, water vapour etc.) of any place or region and their extremes". It is a complex non-linear dynamic system with a large spatial and temporal variability. Since it is impossible to discuss all aspects of climate in a short paper we concentrate here on a few basic considerations which apply to all planets. Basically, the climate can be considered as a machine which is driven by solar energy. Its main purpose is to equally distribute the incoming energy over the planet. Depending on the distance from the Sun, solar luminosity and planetary albedo, a simple estimate of the mean surface temperature of a planet can be made assuming steady state conditions. Then we address the question which processes potentially cause climate variability. Finally we provide some evidence for solar forcing of climate change on Earth.

2. Steady state climate conditions

The energy driving the climate machine on a planet stems to almost 100% from the Sun. By turning every second some 4.2 million tons of mass into energy the Sun generates a power of $2 \cdot 10^{27}$ W which is radiated into space. Fusing hydrogen into helium in the core the Sun is able to maintain this huge power generation for almost 10 billion years. The standard solar model shows that the luminosity is steadily increasing from about 80% of its present values 4 Gyr ago to about 130% in 4 Gyr from now. According to this model the change occurs very smoothly and slowly ($10^{-8}\%$ per year) as shown in Fig. 1.

Other sources of energy on planets are cosmic rays, geothermal energy as a result of radioactive decay and gravitational energy from the time of formation of the solar system,

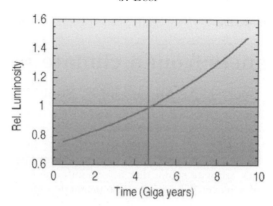

Figure 1. Luminosity of the Sun according to the standard solar model in units relative to the present (Newkirk 1983).

tidal energy from moons, and in some cases gravitational energy released by compression (Jupiter). In the case of the Earth these contributions amount to 10^{10} W, 10^{13} W, 10^{11} W, and 0 W respectively, compared to the 10^{17} W obtained from the Sun. The solar power received by a square meter of a planet depends on the angle of incidence. On Earth it decreases with the cosine of the latitude. This leads to a thermal gradient between low and high latitudes. The climate machine tries to reduce this gradient by transporting energy polewards. Three different mechanisms come into play (Fig. 2): sensible heat flux, latent heat flux and surface heat flux.

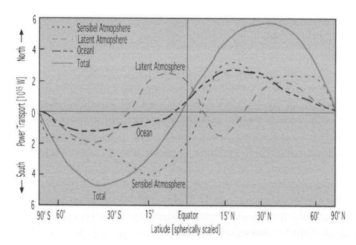

Figure 2. Meridional heat transport on Earth consisting of sensible and latent heat flux through the atmosphere and heat flux through the oceans. The total heat flux is largest between $15°$ and $60°$ latitude for both hemispheres (modified after Bryden and Imawaki 2001).

The sensible heat flux transfers the energy from the planet's surface to the atmosphere by conduction and convection. Then atmospheric circulation transports the energy advectively by moving warm tropical air towards the colder Polar Regions. Latent heat is generated when solid and liquid water is converted into water vapour. This vapour takes part in the atmospheric circulation. When it reaches colder regions the vapour condenses to rain and snow releasing the stored heat again. Finally the surface heat flux consists of warm ocean water flowing polewards in the form of huge streams (e.g. Gulf Stream). Evaporation and sea ice formation generate cold and saline surface water. This dense water descends to greater depth where it flows towards the equator closing the circle. All three processes are affected by a variety of processes such as the distribution of the continents and the Coriolis force. This leads to complex transport patterns. If a planet has no atmosphere and no water the temperature gradient between low and high latitudes can be large. Without this energy transport for example on Earth the Polar Regions would be colder by $25°$ C and the equatorial regions would be warmer by about $15°$ C.

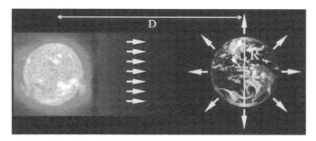

Figure 3. Radiation balance between the Sun and a planet.

Fig. 3 shows a planet with the radius R and the albedo a in the distance D from the Sun with the luminosity L. The power absorbed by a planet is given by the ratio of the planets cross section πR^2 to the power in the distance D from the Sun $4\pi D^2$ corrected for the albedo a (total reflected power):

$$P_{\mathrm{abs}} = (1 - a)\,L\,\frac{\pi R^2}{4\,\pi D^2}. \tag{2.1}$$

If we assume as a first approximation that a planet is a black body and that the incoming solar radiation is equally distributed by the climate machine, the emitted power is given by the law of Stefan-Boltzmann:

$$P_{\mathrm{emi}} = 4\,\pi\,R^2\,\sigma\,T^4. \tag{2.2}$$

In the case of equilibrium, absorption and emission are equal and the temperature T can be calculated:

$$T = \sqrt[4]{\frac{L\,(1-a)}{16\,\pi\,D^2\,\sigma}}. \tag{2.3}$$

Note that the temperature of a planet does not depend on its size. Under the given assumptions it is only determined by the solar luminosity, the albedo, and the distance from the Sun. Fig. 4 shows the dependence of a planet's temperature on the distance for different albedos (upper panel) and luminosities (lower panel). The distance is given in astronomical units (AU) covering the range of the planets from 0.38 AU (Mercury) to 30 AU (Neptune). The luminosity is given in units relative to the present. In the upper panel the luminosity is set to 1 (present value) and in the lower panel an albedo of 0.3 is assumed.

In Table 1 the calculated temperatures for the 8 planets are compared to the measured ones. For each planet a lower value with an albedo of 0.5 and a luminosity of 0.8, an average value with $a = 0.3$ and $L = 1$, and an upper limit with $a = 0.1$ and $L = 1.3$ are given.

Overall there is a reasonable agreement between the estimated and the observed temperatures. The largest discrepancy is observed for Venus. The reason is that Venus has a very dense atmosphere which consists of 96 % of CO_2 with clouds of SO_2 generating the strongest greenhouse effect in the solar system. In the case of Earth the difference between calculated (using the present values $a = 0.3$ and $L = 1$) and measured mean global

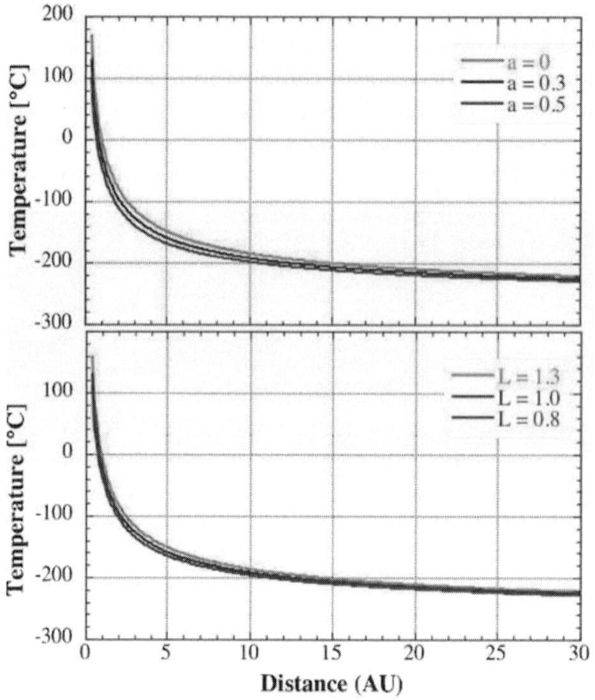

Figure 4. Dependence of planetary temperatures on the distance from the Sun for different values of the albedo a (upper panel) and luminosity L (lower panel).

Table 1. Comparison of the calculated temperatures of the planets for different combinations of albedo and luminosity are compared with the observed temperatures.

Planet	Distance (AU)	Temperature (° C) $a = 0.5$ $L = 0.8$	$a = 0.3$ $L = 1$	$a = 0.1$ $L = 1.2$	Observed
Mercury	0.38	77	130	175	−180 to 420
Venus	0.72	−10	30	66	460
Earth	1	−50	−18	11	15
Mars	1.52	−95	−65	−40	−87 to 5
Jupiter	5.2	−175	−160	−150	−130
Saturn	9.54	−200	−190	−180	−180
Uranus	19.18	−220	−215	−210	−210
Neptune	30.06	−230	−225	−220	−210

temperature is 33° C. This difference is also due to the natural greenhouse effect. It is important to note that the Earth needs the natural greenhouse effect to be habitable, but not necessarily an additional anthropogenic increase. The range of observed temperatures on Mars is very large because Mars has only a very thin atmosphere (0.3 hPa compared to 1000 hPa of Earth) and no liquid water to transport energy. Jupiter is considerably warmer than calculated (−110° C instead of −160° C). Most likely, this difference is due to gravitational compression which provides an additional power at least as large as the solar insolation. In fact, doubling the solar luminosity in formula (2.3) leads to a temperature of about −130° C ($a = 0.3$) in agreement with the observed value at the top of the clouds.

After having discussed the total amount of power received by a planet we address now the question of the distribution of this power on the planet. The distribution is mainly determined by the orbital parameters. Rotation plays a central role in distributing energy. If the axis of rotation is not perpendicular to the ecliptic plane the daily path of the Sun in the sky changes during one orbit giving rise to the seasons on Earth. Finally if the planet orbits around the Sun are elliptical, the distance from the Sun varies continuously and affects accordingly the total power received by the planet. Presently the distance between Sun and Earth varies by 3.4 % during the course of a year causing a variation in the Total Solar Irradiance (TSI) of 87 Wm^{-2}.

3. Climate variability

So far we have assumed that all the conditions determining the climate of a planet are constant. Obviously this is not the case in reality. There are many different sources of variability which ultimately cause deviations from the steady state conditions or, in other words, climate variability. If we stick to our simplified approach by formula (2.3) we already know that the temperature of a planet depends strongly on the luminosity, the distance from the Sun, and the albedo. We discuss now potential changes in these parameters.

3.1. *Luminosity*

The luminosity is the total power emitted by the Sun. Fig. 1 shows that the luminosity increase is very slow and smooth. This is due to the fusion process in the core which is very stable on time scales of millennia. The low luminosity after the formation of the solar system about 4 Gyr ago poses an interesting question called the "faint young sun paradox". According to formula (2.3) a reduction of the luminosity by 25 % leads to a decrease of the mean global temperature on Earth by 18° C. Under the present conditions such a temperature drop would turn the Earth into a "snowball" with a much larger albedo. This would make a return to normal conditions rather impossible. The generally accepted main reason why this did not happen is a higher content of greenhouse gases in the atmosphere at that time. From the derivation of formula (2.3) we obtain:

$$\mathrm{d}T/T = (1/4)\,\mathrm{d}L/L. \tag{3.1}$$

The relative change in temperature is 1/4 of the relative change in L. In other words a change of L of 0.1% as typically measured between solar minimum and maximum during an 11-y Schwabe cycle corresponds to a temperature change of the photosphere of about 1.5 K.

Other potential sources of luminosity changes are energy transport from the core to the solar surface and emission from the photosphere. From the core to about 2/3 of the solar radius the energy is transported radiatively. Then convection becomes more efficient and brings the energy to the photosphere from where it is radiated into space. It is believed that the radiative energy transport is very stable. It is not known to what degree this is also true for the convective transport. However, it cannot be excluded that the magnetic fields generated by the dynamo at the tachocline below the convective zone have some influence on the convection (Kuhn 1988; Kuhn and Libbrecht 1991). The observed changes in the annual mean emission from the photosphere account for only about 0.1 % during an 11-y Schwabe cycle (see Fig. 5) and therefore even very small fluctuations can have comparable effects.

By far, the largest part of the solar power is emitted by the photosphere in the form of electromagnetic radiation. The spectrum resembles that of a blackbody with a temperature of about 5780 K. Only in the UV region of the spectrum there are larger contributions from very high temperatures in the corona, probably induced by reconnections of strong magnetic field lines. The total electromagnetic radiation arriving at the top of the Earth's atmosphere perpendicular to an area of $1\,m^2$ at the distance of $1\,AU$ is called total solar irradiance (TSI). Its spectral distribution is called the solar spectral irradiance (SSI). Direct satellite based monitoring of the TSI over the past 30 years reveals clear variations in phase with the magnetic activity of the 11-y Schwabe cycle (Fig. 5) (Fröhlich and Lean 2004; Fröhlich 2006). The TSI curve is a composite of corrected data from five different instruments as indicated by different colours. There are three different composites based on different data and corrections. Although different in the long-term trend depending on the applied corrections and the used instruments, all composites show consistently lower values for the present solar minimum than for the previous one.

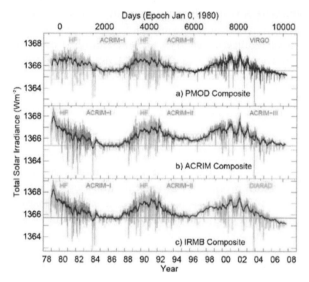

Figure 5. Three composites of the total solar irradiance (TSI) measured by 5 different satellite based radiometers indicated by different colours (The Picard Team, Dewitte, and Schmutz 2006; Lockwood and Fröhlich 2008; Fröhlich 2008; Willson and Mordvinov 2003). Due to composing different instruments and applying different corrections, the long-term trends are slightly different. However, for all composites the last minimum is lower than the previous one.

Simple models describing the TSI as the sum of a constant quiet sun component, a positive component due to bright faculae and the magnetic network, and a negative component composed of the dark sunspots and their penumbra are very successful in explaining all the observed short-term fluctuations (Krivova *et al.* 2003; Solanki and Fligge 2002; Unruh, Solanki, and Fligge 1999; Wenzler *et al.* 2006). However, it is not yet clear whether these models are also applicable to periods of much lower solar activity such as the Maunder minimum when almost no sunspots were observed for about 7 decades. The most recent decline since 2006 raises some serious doubts (Fröhlich 2008). Other potential sources of variability in the solar emission are changes in the solar radius and anisotropic emission. The solar radius is a crucial parameter (Sofia and Li 2005). However, observations did not provide clear evidence for changes in the radius so far (Thuillier, Sofia, and Haberreiter 2005). Clarification is expected from the Picard mission to be launched in 2009 (The Picard Team, Dewitte, and Schmutz 2006). Even without changes in the luminosity anisotropic emission of the total power can lead to changes in the TSI. The fact that sunspots and faculae are more prevalent at lower latitudes clearly points to an anisotropic emission. Whether this is also true for the solar disc free of visible magnetic activity, remains to be checked.

3.2. *Distance*

As we have already mentioned, the distance is the dominant parameter for the temperature of a planet. The first reason is that the solar power decreases with the square of the distance or in other words that the relative change of the temperature is $1/2$ of the relative change of the distance:

$$\mathrm{d}T/T = (1/2)\,\mathrm{d}D/D. \tag{3.2}$$

The second reason is that the distance of the planets changes by almost 2 orders of magnitude from $0.38\,\mathrm{AU}$ (Mercury) to $30\,\mathrm{AU}$ (Neptune).

Since all the planets have elliptical orbits the distance is continuously changing. The eccentricity (fraction of the distance along the semimajor axis at which the focus lies) ranges from 0.0068 of Venus to 0.2056 of Mercury. The situation is further complicated by the fact that the orbital parameters of a planet are disturbed by the gravitational forces of the other planets (mainly Jupiter and Saturn having the largest masses). The mathematical details of these disturbances have been worked out by Milankovic (Milankovic 1930) and more recently by Berger (Berger 1978) and Laskar (Laskar *et al.* 2004). The orbital parameters affected by the other planets are the eccentricity, the obliquity (the tilt angle of the planets axis relative to the ecliptic plane) and the precession of a planet's rotational axis around its mean direction. The calculations for the Earth reveal cyclic variability with characteristic time scales of 100 and $400\,\mathrm{kyr}$ (eccentricity), $40\,\mathrm{kyr}$ (obliquity) and 19–$24\,\mathrm{kyr}$ (precession). While obliquity and precession change only the relative distribution of the solar insolation eccentricity changes affect the total insolation. Since the orbital parameters are very accurately known it is possible to calculate the insolation changes not only for the past several million years but also for the future. As an example Fig. 6 shows the orbital parameters, the insolation deviations in Wm^{-2} for June, December and the season (June minus December) from their corresponding mean values for the past $100\,\mathrm{kyr}$ and the future $20\,\mathrm{kyr}$. Note that, especially at high latitudes these deviations are very large (up to $80\ \mathrm{Wm}^{-2}$) compared to the $2\,\mathrm{Wm}^{-2}$ predicted for a doubling of the atmospheric CO_2 concentration. However, it has to be considered that greenhouse gas forcing acts globally.

Figure 6 shows that the insolation in June was very low at $20\,\mathrm{kyr\,BP}$. This coincides with the last glacial maximum that is followed by a strong warming. The coming $20\,\mathrm{kyr}$ are characterised by relatively small changes in orbital forcing. This relates to the decreasing trend in eccentricity which affects the seasonality.

To see how well the orbital forcing is reflected in paleoclimatic records we compare in Fig. 7 the $\delta^{18}\mathrm{O}$ record from the GRIP ice core drilled in central Greenland ($72°\mathrm{N}$) (Dansgaard *et al.* 1993) with the summer insolation at this latitude. $\delta^{18}\mathrm{O}$ is a measure of the atmospheric temperature when water vapour condenses and snow flakes form. The overall agreement between the two curves is good as far as the long-term trend is concerned. The glacial period is characterised by strong, very rapid changes (so-called Dansgaard-Oeschger events) that are most probably related to abrupt changes in the thermohaline circulation of the ocean. During the Holocene (the past $11\,\mathrm{kyr}$) the climate was comparatively stable and has not clearly followed the insolation curve.

3.3. *Albedo*

The albedo is defined as the ratio of diffusely reflected to incident electromagnetic radiation and, therefore, lies in the interval 0–1. It is difficult to determine the total albedo

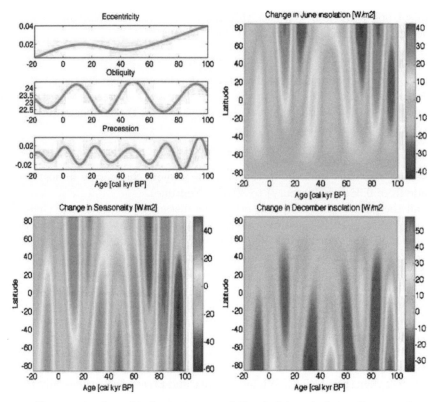

Figure 6. Changes in the orbital parameters of Earth (a) and their effect on the summer (June), the winter (December) and the seasonal (June-December) insolation for the past 100 kyr and the future 20 kyr (-20 kyr BP). Shown are the deviations in Wm^{-2} from the mean values. Note the large changes at high latitudes.

of a planet because it is highly variable ranging from less than 0.1 for water and forests to more than 0.8 for fresh snow. On Earth, the largest contribution comes from the clouds which cover about 50 % of its surface. For the Earth an albedo of 0.3 is usually assumed. Interestingly, the albedo gets much less attention than the TSI although both are equally important as far as solar forcing is concerned. The albedo of clouds plays a central role in the cosmic ray cloud hypothesis put forward by Danish scientists (Svensmark 1998). They claim that the Earth's cloud cover is modulated by the cosmic ray induced ion production in the atmosphere. Later they reduced the effect to low altitude, low latitude clouds (Marsh and Svensmark 2003). This issue is still debated in papers supporting (Usoskin *et al.* 2004) and contradicting (Kernthaler, Toumi, and Haigh 1999; Wagner *et al.* 2001) cosmic ray induced climate change. Other climate relevant effects related to strong atmospheric electrical currents have been proposed by Tinsley (Tinsley 2000). Although there is no doubt that many more different effects take place in the atmosphere there is so far no clear evidence that these processes play a significant role in global climate change (see also contribution by Usoskin in this volume).

3.4. *Aerosols*

Aerosols are liquid droplets or fine solid particles in the atmosphere that influence both directly and indirectly a planet's radiation budget. The direct effect is reflection and scattering of solar radiation back into space, leading to a cooling. As an indirect effect

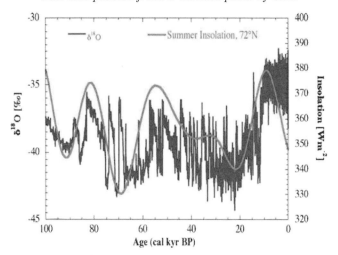

Figure 7. Comparison of the $\delta^{18}O$ record from the GRIP ice core in central Greenland (72°N) (Dansgaard *et al.* 1993) with the corresponding summer insolation for the past 100 kyr.

the aerosols can modify the radiative properties of clouds at low altitudes. The sources of aerosols are amongst others volcanic eruptions, dust from deserts, and anthropogenic activities such as burning of fossil fuel.

3.5. *Greenhouse gases*

Greenhouse gases (H_2O, CO_2, CH_4 and others) let the sunlight with wave lengths in the range of 400–800 nm pass, but absorb the infrared radiation emitted by the Earth. As a result the lower atmosphere gets warmer. As we have already discussed, the Earth's temperature would be about 30° C lower without the natural greenhouse gas concentrations in the atmosphere. Due to the greenhouse effect the emission of the infrared radiation into space does not take place at the Earth's surface but is shifted higher up in the atmosphere where the temperature is about −20° C according to formula (2.3). The greenhouse effect is very important to make the Earth habitable. It was even more crucial at the time when the solar luminosity was considerably lower (faint young sun paradox). However, since the industrialisation mankind has begun to burn large amounts of fossil fuel which is raising the atmospheric CO_2 content to levels unprecedented during the past million years (Luthi *et al.* 2008). A good example for extreme greenhouse gas forcing is Venus. Its atmosphere consists almost entirely (97 %) of CO_2 with clouds containing droplets of sulphuric acid heating the planet to 460° C (Table 1).

4. **Evidence for solar variability and climate change on Earth**

In this chapter we will provide some evidence for solar induced climate change on Earth. In the last decade the number of publications claiming a causal connection between solar variability and climate change was steadily increasing. Nevertheless, there always remains the problem of unequivocal attribution. We still do not know quantitatively how much the total and the spectral solar forcing changed in the past. Furthermore, we do not know how exactly the climate system responds to such forcings. Unquestionably, the climate is a complex non-linear system and even the most advanced general circulation models are far from representing realistically all the complex processes, their couplings, and feedback

effects. A recent overview of the mid- to late Holocene climate change is given by Wanner
et al. 2008. In the following, we present two examples, one from Switzerland and one from
China.

4.1. *The Great Aletsch glacier*

One of the most striking ways to illustrate the ongoing global warming is to compare
old photographs of glaciers with recent ones. The size of a glacier is strongly determined
by the winter precipitation and the summer temperature. It is rather inert and does not
respond immediately to single climatic events but records the climate changes averaged
over the last few decades. By dating trees which were buried during glacier advances it is
possible to reconstruct fluctuations in the length (Denton and Karlén 1973; Holzhauser,
Magny, and Zumbühl 2005; Hormes, Beer, and Schluchter 2006; Joerin, Stocker, and
Schluchter 2006).

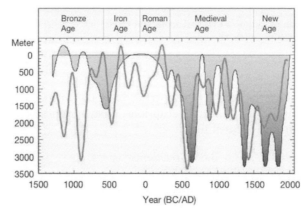

Figure 8. Comparison of the reconstructed length fluctuations of the Great Aletsch glacier
(relative to its present length) in the Swiss Alps (Holzhauser, Magny, and Zumbühl 2005) with
the solar activity record (Steinhilber, Abreu, and Beer 2008; Vonmoos, Beer, and Muscheler
2006). Low solar activity coincides generally with larger extensions of the glacier. Note that the
dating of both the record of glacier length fluctuations and the record of solar activity has some
uncertainties.

Fig. 8 shows the reconstruction of the length fluctuations of the Aletsch glacier, the
largest glacier in the Alps (Holzhauser, Magny, and Zumbühl 2005). The length is given
relative to today. The curve shows that the Aletsch glacier was longer during the little
ice age (about 1350-1850 AD) by almost 3500 m. However, the present situation is not
unique. There were earlier periods when its length was comparable to todays length.
These times coincide with warm epochs such as the Roman and the Medieval Warm
Period. However, taking into account the delayed response the Aletsch glacier reflects not
yet the global warming after 1970 and will therefore continue to melt in the future. The
solar activity record is based on ^{10}Be measurements in polar ice cores. ^{10}Be is produced
in the atmosphere by cosmic ray particles which are magnetically shielded on their way
through the heliosphere depending on the solar activity (Beer, Vonmoos, and Muscheler
2006). This is a good example to illustrate the difficulties we face when we try to attribute
a climate record to solar forcing: the solar activity record is not calibrated in Wm^{-2},
the responses of the climate system in general and of the glacier length in particular are
non-linear, the dating of the length fluctuations and the solar activity have uncertainties,

and there are volcanic and other forcings involved as well. Nevertheless, we believe that the overall agreement points to a significant solar forcing.

4.2. *Speleothems in China*

Speleothems are formed of $CaCO_3$. Ca is dissolved by rainwater percolating through the soil. Investigations of recent deposited $CaCO_3$ shows that the $\delta^{18}O$, the deviation of the $^{18}O/^{16}O$ ratio in $CaCO_3$ from a standard, reflects the amount of precipitation. Using the U/Th dating technique it is possible to determine the time of carbonate formation within a few years for the past 10,000 years.

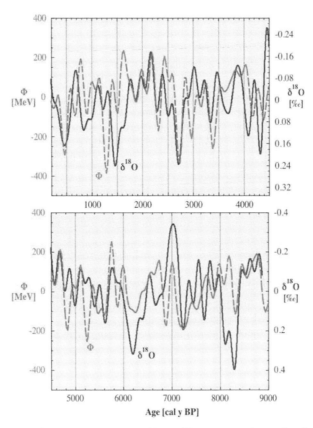

Figure 9. $\delta^{18}O$ record of the Dongge cave in China (Dykoski *et al.* 2005) reflecting the intensity of monsoon (blue line) together with the solar activity record (red line). Both records are low-pass filtered with 200 y.

In Fig. 9 the $\delta^{18}O$ record derived from a stalagmite in the Dongge cave in China (Dykoski *et al.* 2005) is compared with the same solar activity record as in Fig. 8. The Dongge cave is situated near the boundary of the present monsoon. The largest discrepancy occurs at 8200 BP when a large amount of melt water entered the Atlantic reducing the thermohaline circulation (Clark *et al.* 2001). This event is not expected to be reflected in the solar activity record. Beside this event, the overall agreement between the two records also points to a causal relationship. This is corroborated by the fact that spectral analysis reveals common periodicities such as the 208-y de Vries or Suess cycle also known from the analysis of the ^{14}C calibration curve from tree rings (Stuiver and Braziunas 1993).

5. Summary and conclusions

To a large extent the climate of a planet is determined by the radiative energy it receives from the Sun and its distribution. Beside luminosity and albedo the main factor controlling the amount of received energy is the distance between the planet and the Sun. The distribution depends strongly on the existence of an atmosphere and oceans and their properties, but also on the orbital parameters of the planet (rotation, obliquity, precession). When it comes to climate variability potential causes are changes in the energy transport in the convective zone of the Sun, in the emission of the electromagnetic radiation from the photosphere, in the orbital parameters (Milankovic), and in the properties of the planet (albedo, aerosols, greenhouse gases, clouds, vegetation, energy transport etc.). Due to feedback mechanisms, the climatic effect of a change in forcing depends not only on its intensity (Wm^{-2}), but also on its distribution and duration. Predictions are therefore difficult and there are still many open questions. Nevertheless, as far as solar forcing on Earth is concerned there are some facts:

- The Sun is by far the most important source of energy for the climate system.
- The Sun is a variable star. Its irradiance is only well known for the past 30 years - a period of high but relatively stable activity.
- The Sun has the potential for larger TSI variability although it is not yet clear to what extent it was used.
- The climate system was never stable. There were only periods of larger or smaller variability.
- The Holocene atmospheric CO_2 concentration was rather constant before the industrialisation.
- Climate models show that any change in forcing (greenhouse gas, volcanic or solar) leads to responses of the climate system with complex spatial and temporal patterns which makes detection and attribution difficult.
- There is growing evidence that among other forcings the Sun plays a significant role in climate change.

Acknowledgements

We thank Petra Breitenmoser for useful comments. This work was financially supported by the Swiss National Science Foundation.

References

Beer, J., Vonmoos, M., & Muscheler, R.: 2006, Solar Variability Over the Past Several Millennia. *Space Science Reviews* **125**, 79 – 79. doi:**10.1007/s11214-006-9047-4**.

Berger, A. L.: 1978, Long-Term Variations of Daily Insolation and Quaternary Climatic Changes. *Journal of Atmospheric Sciences* **35**, 2367 – 2367.

Bryden, H. L. & Imawaki, S.: 2001, Ocean heat transport. In: Siedler, G., Church, J., Gould, J. (eds.) *Ocean Circulation and Climate*, Academic Press, St. Louis, 474 – 474.

Clark, P. U., Marshall, S. J., Clarke, G. K. C., Hostetler, S. W., Licciardi, J. M., & Teller, J. T.: 2001, Freshwater Forcing of Abrupt Climate Change During the Last Glaciation. *Science* **293**, 287 – 287. doi:**10.1126/science.1062517**.

Dansgaard, W., Johnson, S. J., Clausen, H. B., Dahl-Jensen, D., Gundestrup, N. S., Hammer, C. U., Hvidbjerg, C. S., Steffensen, J. P., Sveinbjörnsdottir, A. E., Jouzel, J., & Bond, G.: 1993, Evidence for general instability of past climnate from a 250-kyr ice-core record. *Nature* **364**, 220 – 220. doi:10.1038/364218a0.

Denton, G. H. & Karlén, W.: 1973, Holocene climatic variations - their pattern and possible cause. *Quaternary Research* **3**, 205 – 205.

Dykoski, C. A., Edwards, R. L., Cheng, H., Yuan, D., Cai, Y., Zhang, M., Lin, Y., Qing, J., An, Z., & Revenaugh, J.: 2005, A high-resolution, absolute-dated Holocene and deglacial Asian monsoon record from Dongge Cave, China. *Earth and Planetary Science Letters* **233**, 86 – 86. doi:10.1016/j.epsl.2005.01.036.

Fröhlich, C.: 2006, Solar Irradiance Variability Since 1978. Revision of the PMOD Composite during Solar Cycle 21. *Space Science Reviews* **125**, 65 – 65. doi:10.1007/s11214-006-9046-5.

Fröhlich, C.: 2008, Total Solar Irradiance Variability: What have we learned about its variability from the record of the last three solar cycles? *Proc. CAWSES Symp., October, 23-27 2007, Kyoto, Japan*.

Fröhlich, C. & Lean, J.: 2004, Solar radiative output and its variability: evidence and mechanisms. *Astronomy and Astrophysicsr* **12**, 320 – 320. doi:10.1007/s00159-004-0024-1.

Holzhauser, H., Magny, M., & Zumbühl, H. J.: 2005, Glacier and lake-level variations in west-central Europe over the last 3500 years. *Holocene* **15**, 801 – 801.

Hormes, A., Beer, J., & Schluchter, C.: 2006, A geochronological approach to understanding the role of solar activity on Holocene glacier length variability in the Swiss Alps. *Geografiska Annaler Series A - Physical Geography* **88**, 294 – 294.

Joerin, U. E., Stocker, T. F., & Schluchter, C.: 2006, Multicentury glacier fluctuations in the Swiss Alps during the Holocene. *Holocene* **16**, 704 – 704.

Kernthaler, S. C., Toumi, R., & Haigh, J. D.: 1999, Some doubts concerning a link between cosmic ray fluxes and global cloudiness. *Geophysical Research Letters* **26**, 866 – 866. doi:10.1029/1999GL900121.

Krivova, N. A., Solanki, S. K., Fligge, M., & Unruh, Y. C.: 2003, Reconstruction of solar irradiance variations in cycle 23: Is solar surface magnetism the cause? *Astronomy and Astrophysics* **399**, 4 – 4. doi:10.1051/0004-6361:20030029.

Kuhn, J. R.: 1988, Helioseismological splitting measurements and the nonspherical solar temperature structure. *Astrophysical Journal Letters* **331**, 134 – 134. doi:10.1086/185251.

Kuhn, J. R. & Libbrecht, K. G.: 1991, Nonfacular solar luminosity variations. *Astrophysical Journal Letters* **381**, 37 – 37. doi:10.1086/186190.

Laskar, J., Robutel, P., Joutel, F., Gastineau, M., Correia, A. C. M. & Levrard, B.: 2004, A long-term numerical solution for the insolation quantities of the Earth. *Astronomy and Astrophysics* **428**, 285 – 285. doi:10.1051/0004-6361:20041335.

Lockwood, M., & Fröhlich, C.: 2008, Recent oppositely directed trends in solar climate forcings and the global mean surface air temperature. II. Different reconstructions of the total solar irradiance variation and dependence on response time scale. *Proceedings of the Royal Society* **464**, 1385 – 1385.

Luthi, D., Le Floch, M., Bereiter, B., Blunier, T., Barnola, U. J. M. Siegenthaler, Raynaud, D., Jouzel, J., Fischer, H., Kawamura, & T. F. K. Stocker: 2008, High-resolution carbon dioxide concentration record 650,000-800,000 years before present. *Nature* **453**, 382 – 382.

Marsh, N. & Svensmark, H.: 2003, Solar Influence on Earth's Climate. *Space Science Reviews* **107**, 325 – 325. doi:10.1023/A:1025573117134.

Milankovic, M.: 1930, Mathematische Klimalehre und atsronomische Theorie der Klimaschwankungen. In: Köppen, W., & Geiger, R. (eds.) *Handbuch der Klimatologie*, Gebrüder Bornträger, Berlin, 176 – 176.

Newkirk, G. Jr.: 1983, Variations in solar luminosity. *Annual review of astronomy and astrophysics* **21**, 467 – 467. doi:10.1146/annurev.aa.21.090183.002241.

Sofia, S. & Li, L. H.: 2005, Mechanisms for global solar variability. *Memorie della Societa Astronomica Italiana* **76**, 768.

Solanki, S. K. & Fligge, M.: 2002, Solar irradiance variations and climate. *Journal of Atmospheric and Solar-Terrestrial Physics* **64**, 685–685.

Steinhilber, F., Abreu, J. A., & Beer, J.: 2008, Solar modulation during the Holocene. *Astrophysics and Space Sciences Transactions* **4**, 6–6.

Stuiver, M. & Braziunas, T. F.: 1993, Sun, Ocean, Climate and Atmospheric $^{14}CO_2$, an evaluation of causal and spectral relationships. *Holocene* **3**, 305–305.

Svensmark, H.: 1998, Influence of Cosmic Rays on Earth's Climate. *Physical Review Letters* **81**, 5030–5030.

The Picard Team, Dewitte S. & Schmutz W.: 2006, Simultaneous measurement of the total solar irradiance and solar diameter by the PICARD mission. *Advances in Space Research* **38**, 1806–1806. doi:**10.1016/j.asr.2006.04.034**.

Thuillier, G., Sofia, S., & Haberreiter, M.: 2005, Past, present and future measurements of the solar diameter. *Advances in Space Research* **35**, 340–340. doi:**10.1016/j.asr.2005.04.021**.

Tinsley, B. A.: 2000, Influence of Solar Wind on the Global Electric Circuit, and Inferred Effects on Cloud Microphysics, Temperature, and Dynamics in the Troposphere. *Space Science Reviews* **94**, 258–258.

Unruh, Y. C., Solanki, S. K., & Fligge, M.: 1999, The spectral dependence of facular contrast and solar irradiance variations. *Astronomy and Astrophysics* **345**, 642–642.

Usoskin, I. G., Marsh, N., Kovaltsov, G. A., Mursula, K., & Gladysheva, O. G.: 2004, Latitudinal dependence of low cloud amount on cosmic ray induced ionization. *Geophysical Research Letters* **31**, 16109. doi:**10.1029/2004GL019507**.

Vonmoos, M., Beer, J., & Muscheler, R.: 2006, Large variations in Holocene solar activity: Constraints from ^{10}Be in the Greenland Ice Core Project ice core. *Journal of Geophysical Research (Space Physics)* **111**(10), 10105. doi:**10.1029/2005JA011500**.

Wagner, G., Livingstone, D. M., Masarik, J., Muscheler, R., & Beer, J.: 2001, Some results relevant to the discussion of a possible link between cosmic rays and the Earth's climate. *Journal of Geophysical Research* **106**, 3388–3388. doi:**10.1029/2000JD900589**.

Wanner, H., Beer, J., Bütikofer, J., Crowley, T. J., Cubasch, U., Flückiger, J., Goosse, H., Grosjean, M., Joos, F., Kaplan, J. O., Küttel, M., Müller, S. A., Prentice, I. C., Solomina, O., Stocker, T. F., Tarasov, P., Wagner, M., & Widmann, M.: 2008, Mid- to Late Holocene climate change: an overview. *Quaternary Science Reviews* **27**, 1828–1828.

Wenzler, T., Solanki, S. K., Krivova, N. A., & Fröhlich, C.: 2006, Reconstruction of solar irradiance variations in cycles 21-23 based on surface magnetic fields. *Astronomy and Astrophysics* **460**, 595–595. doi:**10.1051/0004-6361:20065752**.

Willson, R. C., Mordvinov & A. V.: 2003, Secular total solar irradiance trend during solar cycles 21–23. *Geophysical Research Letters* **30**(5), 1–1.

Discussion

BONNET: Thank you Juerg for an excellent presentation. I am pleased that the topic you discuss brings strong support to my point that space science in the future may face a difficult time when there will be tough competition for funds with Earth sciences: solar influences on the climate are clearly something to be studied more precisely and more continuously in the future. This is not to say that the whole of future heliospheric research should be referring to Sun Earth Climate relations. But there certainly is a connection between heliospheric science and the more immediate concerns of humans on Earth.

BEER: I fully agree with this statement.

BOCHSLER: Is it possible that people crossed the Schnidejoch at all times and climatic periods, but were more negligent during warm periods loosing plants, and other objects more frequently?

BEER: So far the findings clearly point to distinct periods when the snow and ice cover was low enough to cross this pass at almost 3000 m.a.s.l.

Session II

Space Weather

Universal Heliophysical Processes
Proceedings IAU Symposium No. 257, 2008
N. Gopalswamy & D.F. Webb, eds.

© 2009 International Astronomical Union
doi:10.1017/S174392130902907X

Space weather:
science and effects

Norma B. Crosby

Belgian Institute for Space Aeronomy, Ringlaan-3-Avenue Circulaire,
Ringlaan-3-Avenue Circulaire, Belgium
email: norma.crosby@oma.be

Abstract. From the point-of-view of somebody standing outside on a cold winter night looking up at a clear cloudless sky, the space environment seems to be of a peaceful and stable nature. Instead, the opposite is found to be true. In fact the space environment is very dynamic on all spatial and temporal scales, and in some circumstances may have unexpected and hazardous effects on technology and humans both in space and on Earth. In fact the space environment seems to have a weather all of its own – its own "space weather". Our Sun is definitely the driver of our local space weather. Space weather is an interdisciplinary subject covering a vast number of technological, scientific, economic and environmental issues. It is an application-oriented discipline which addresses the needs of "space weather product" users. It can be truly said that space weather affects everybody, either directly or indirectly. The aim of this paper is to give an overview of what space weather encompasses, emphasizing how solar-terrestrial physics is applied to space weather. Examples of "space weather product" users will be given highlighting those products that we as a civilization are most dependent on.

Keywords. Space weather, solar-terrestrial physics, technological and biological effects, forecasting, mitigation

1. Introduction

Since the time of the first space missions back in 1957 engineers and operators have been developing methods to mitigate against space weather induced technological problems; and from the beginning of space research the solar-terrestrial physics community has been investigating the physics behing space weather phenomena. In summary, space environment analysis has always existed. However the term "space weather" linking the engineering and the scientific side of the story is more recent. Space weather is an application-oriented discipline and addresses the needs of "users". It is important to note that basic research in the field of solar-terrestrial physics (STP) is necessary for space weather applications. STP relies on basic research and scientific observations, where the product is "scientific". On the other hand space weather is application oriented and relies on continuous monitoring, where the result becomes a "service product". Space weather is indeed a merging of many topics: space science, engineering, medicine, law, etc..

Space weather is defined by the U.S. National Space Weather Program (NSWP) as Conditions on the Sun and in the solar wind, magnetosphere, ionosphere, and thermosphere that can influence the performance and reliability of space-borne and ground-based technological systems and can endanger human life or health", Wright *et al.* (1993).

A more recent definition of space weather was formulated by members of the COST724 Action, COST724 (2007): "Space weather is the physical and phenomenological state of natural space environments. The associated discipline aims, through observation, monitoring, analysis and modelling, at understanding and predicting the state of the sun, the

interplanetary and planetary environments, and the solar and non-solar driven perturbations that affect them; and also at forecasting and nowcasting the possible impacts on
biological and technological systems."

In more simple terms space weather can be defined as how solar activity may have
unwanted effcts on technological systems and human activity. It is a function of our
location in the solar system, the behaviour of the Sun and the nature of Earth's magnetic
field and atmosphere. Any planet or target in space will have its own local space weather
with its own specific characteristics.

Space weather affects everybody either directly or indirectly. So why has progress
towards organized efforts to improve the practical solutions to space weather problems
become so important lately? There are many good answers to this question: *1*. Present
society on Earth is deeply dependent on reliable space systems and will be more so in the
future (city, rural, and isolated communities) *2*. Technical systems are becoming more
sensitive to the space environment and will continue so in the future due to developments
in technology such as miniaturization *3*. STP science has progressed to a stage where the
possibilities for useful space weather models and predictions are expected soon and in
some cases exist.

Apart from the problems caused to spacecraft by the ultra-high vacuum and extremes
of hot and cold in space, spacecraft also have to survive very hostile environments which
can severely limit space missions as well as pose threats to humans. Missions need to
consider phenomena such as UV, X- and gamma-radiation, energetic charged particles,
plasmas and not to forget neutrals (space debris and meteoroids). Fig. 1 illustrates the
connection between space weather phenomena and the possible induced space weather
effects. This paper offers the reader a short introduction to the fascinating world of space
weather. Various of the space environment phenomena are introduced and followed by the
potential effects which they can have on technological and biological systems. Thereafter
the topic of mitigation is discussed and the paper ends with some final words. For more
detailed information about space weather (general reviews) the reader is referred to the
excellent space weather books that have been written during this decade Bothmer &
Daglis (2007), Carlowicz & Lopez (2002), Daglis (2001), Freeman (2001), Hanslmeier
(2002), Miroshnichenko (2003), Scherer *et al.* (2005), Song *et al.* (2001), Wang & Xu
(2002).

2. The space environment

Interplanetary space, better known as the heliosphere can be thought of as a vast magnetic bubble containing the solar system, the solar wind and the interplanetary magnetic
field, as well as numerous particle populations, and dust. Without the Sun there would
be no life on Earth. However, the Sun is also the origin of phenomena that can influence
our daily existence in a more negative way. The Sun's UV is not the only phenomenon
that we should worry about here on Earth.

Solar flares, coronal mass ejection, solar wind:

The Sun is our closest star and has a characteristic 11 year cycle. At solar maximum
solar activity is largest and at minimum smallest. It is well known that the solar corona is
a very dynamic region which is the seat of many phenomena related to magnetic energy
releases in a large range of sizes and occurring on time scales going from a few seconds
or less to hours. Physical processes leading to magnetic energy releases are analyzed
in individual solar flares using multi-wavelength observations and can last from a few
minutes to a few hours. The energy released in solar flares varies from 10^{28} to 10^{34} ergs,

which is transformed into heating, particle acceleration and mass motions. Solar flares are related to the solar cycle, with their frequency being highest at solar maximum.

Coronographs have allowed us to discover the the world of coronal mass ejections (CMEs). CMEs are huge ejections of plasma in the Sun's outer atmoshere. They are seen as bright features moving outwards through the corona at speeds from 10 to 1000 km/s and correspond to massive expulsions of plasma from the solar atmosphere that cause major transient interplanetary disturbances which have significant terrestrial effects. Halo CMEs reaching the Earth interact with Earth's geomagnetic field and are said to be geoeffective. Magnetic storms are one consequence resulting in numerous space weather signatures including the beautiful northern lights. The shock wave driven by a CME has been found to be an excellent particle accelerator (see next sub-section).

As the solar atmosphere is not gravitationally bound to the Sun, there is a flux of ionized matter that escapes continuously from the Sun. It consists largely of ionized hydrogen, contains a weak magnetic field and is significantly influenced by solar activity. The typical speed of the solar wind is 400 km/s. There exists two regions of the solar wind (high speed and slow speed). The high speed velocity has a velocity that reaches 700 km/s, while the slow speed wind has a velocity of 300 km/s. The interaction of high speed and slow speed winds leads to 3-D corotating interaction regions (CIRs) in the heliosphere. Due to enhnaced field strength and rising wind speed within the compression region CIRs can cause magnetic storms.

High speed solar winds mainly occur at the declining phase of solar maximum and have their origin in coronal holes. They are often connected with enhancements in Earth's outer radiation electron belt. This phenomenon is often observed following two weeks of high

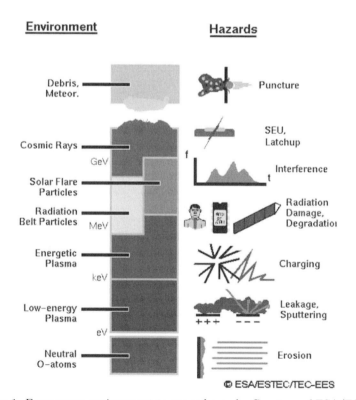

Figure 1. From space environment to space hazards. Courtesy of ESA/ESTEC.

speed winds. Such events have been associated with a space weather effect coined "deep dielectric charging" which will be discussed later.

Energetic particles:

The most energetic particles (energies up to 10^{21} eV) found in our solar system, are those that have origins far outside our solar system, namely galactic cosmic rays (GCRs). They are fully ionized particles and their composition is mostly hydrogen nuclei (protons), 7–10 % Helium and 1 % heavier elements. All GCRs are fully ionized, meaning that they consist of nuclei only. As a first approximation the flux of GCRs in near-Earth space can be considered to be isotropic. During solar maximum the increase in the interplanetary magnetic field strength provides enhanced shielding of the heliosphere against penetrating GCR particles. The result is that the GCR population is most intense during solar minimum. On short-term (a few hours) a rapid decrease in GCR intensity may occur following a CME as the magnetic field of the solar wind sweeps away the particles – this is known as a Forbush decrease.

Solar cosmic rays, also called solar energetic particles (SEPs), are associated with a solar flare and/or the shock wave generated by a CME. SEPs are mainly protons, electrons, and α-particles with small admixtures of 3He-nuclei and heavier ions up to iron. They are sporadic and difficult to predict, lasting from minutes to days with energies from a dozen of keVs to a few GeVs.

Earth's radiation belts are pricipally composed of naturally occuring energetic charged particles trapped in Earth's inner magnetosphere at equatorial distances ranging from approximately 1.2 to 7 Earth radii. The two doughnut-shaped rings (inner and outer belt) consist mainly of protons (tens keV – couple of hundreds of MeV) and electrons (tens keV – several MeV). These particles are trapped in the Earths magnetic field and their motions in the field consist of: *1*. a gyration about field lines, *2*. a bouncing motion between the magnetic mirrors found near the Earths poles and *3*. a drift motion around Earth.

Table 1 presents these energetic particle populations along with their less energetic counterparts. In summary, these particle populations constitute the major radiation

Table 1. Major particle environments in the heliosphere

Particle Populations	Energy Range	Temporal Range	Spatial Range (first order)
Galactic Cosmic Rays	0.1 – 1000 GeV (the 100 to 1000 MeV fluxes constitute the largest contribution)	Continuous (factor 10 variation with solar cycle)	Entire heliosphere
Anomalous Cosmic Rays	< 100 MeV	Continuous	Entire heliosphere
Solar Energetic Particles	keV – GeV	Sporadic (minutes to days)	Source region properties (flare/CME sites and evolution) and bound to CME driven shock
Energetic Storm Particles	keV – (> 10 MeV)	Hours-Day	Bound to shock
Corotating Interaction Regions	keV – MeV	Few days (recurrent)	Bound to CIR shock and compression region
Particles accelerated at planetary bow shocks	keV – MeV	Continuous	Bound to bow shock
Trapped Particle Populations	Tens keV – couple of hundreds of MeV (for protons) Tens keV – several MeV (for electrons)	Variations "minutes-years"	Variations "height-width"

CME: coronal mass ejection   CIR: corotating interaction region environments in the heliosphere (for a detailed overview see Crosby in Bothmer & Daglis (2007)).

Neutrals (particulates):

Particulates include meteoroids, space debris and dust. Meteoroids are particulates in space of natural origin (nearly all of them originate from asteroids or comets). Space debris, on the other hand, is man-made. Dust is a term used for particulates which have a direct relation to a specific solar system body and which are usually found close to the surface of this body (e.g., lunar, Martian or cometary dust). Particulates are not space weather effects, but they are affected by space weather. They can influence spacecraft engineering and operations. The damage caused by collisions is a function of the size, density, and speed distribution of the impacting particles, and also depends on the shielding of the spacecraft.

3. Space weather technological induced effects

Technical users of space weather products include all those technologies that depend on spacecraft and airline reliability, as well as electrical and oil companies. Examples of "technological systems" that encounter space weather induced problems include the electrical networks for power transmission, radio communications, space-borne synthetic aperture radars, global positioning and navigation systems using satellites, geomagnetic surveys, and pipelines (corrosion effects). Satellites monitoring the Earth (e.g., remote sensing) and those used for security purposes may be affected too. It is vital that satellites function for the monitoring of natural catastrophes (e.g., forest fires, floods, avalanches, etc.) or during military operations (e.g., radio communications relying on the ionosphere, localization techniques), so that wrong information is not transmitted, or that system performance is not degraded. The space hazards that a spacecraft may encounter are a function of its orbit, i.e. the various space environment phenomena can affect satellites or spacecraft having different orbits differently.

Auroras are the oldest known manifestation of space weather effects at planet Earth and are due to energetic electrons spiralling down the Earth's magnetic field lines towards the polar regions and striking the upper atmosphere (oxygen atoms or nitrogen molecules). Energy from an atom or molecule excited by a fast electron is released as a photon; different colours of auroral lights are emitted. On the other hand Barlow (1849) was the first person to publish systematic observations of spontaneous electrical currents observed in the wires of an electric telegraph. Since then the effects of space weather on space- and ground-based systems have grown due to our technological developments. The chronological order of the various discoveries of space weather effects is illustrated in Fig. 2.

Radiation effects:

Radiation damage to on-board electronics may be separated into two categories: *1*. Single Event Effects (SEEs), *2*. Cumulative Radiation Damage. SEEs are individual events which occur when a single incident ionizing particle deposits enough energy to cause an effect in a device (e.g. Upset, Latchup, Burnout,). This effect is not only observed on-board spacecraft, but has also been encountered on airplanes. Presently, airplanes fly in average at heights between 9,000 and 12,000 m where one finds both primary radiation (protons, electrons), and secondary radiation (neutrons, mesons and electrons). Neutrons are the most dangerous as they have high penetration power because they are electrically neutral. This is the reason why it is estimated that at such altitudes radiation hazards are comparable to those on satellites situated in LEO (Koskinen, 2001).

Cumulative Radiation Damage is divided into two types: Total Ionizing Dose (TID) and Total Non-Ionizing Dose (also known as "displacement damage dose"). TID occurs when radiation penetrates the constituents of electronic components and the lost energy is stored in the material. Displacement effects are caused by an atom displacement from the normal lattice position to an adjacent position, thus creating structure damage and resulting in recombining centers that deteriorate the electrical characteristics of the material. Energetic particles degrade the semiconductor material and reduce the expected lifetime over which for example solar cells are able to produce energy for the spacecraft.

Electrostatic discharges:

Electrostatic charging occurs on satellites crossing or situated in the near-Earth space environment and is classified into two main categories: Surface Charhing and Internal Charging (also known as "deep dielectric charging"). Surface charging is due to low-energy electrons in the plasma with energies ranging from several keV to several tens of keV and occurs, first of all because of the different mobility characteristics of the charges that constitute the near-Earth plasma. The plasma is energetically neutral. However, electrons are much more mobile (flux of electrons to an "uncharged" surface normally exceeds the flux of ions) and the net effect is that surfaces charge negatively with respect to the local plasma. Deep dielectric charging occurs if electrons have energies higher than 100 keV. These high-energy electrons can penetrate the outer shielding of a satellite and deposit charges in the dielectric materials inside the satellite.

Electrical transients from surface discharging or internal charging can masquerade as "phantom commands" appearing to spacecraft systems as directions from the ground. This can result in loss of control of instruments and power or propulsion systems.

Atmospheric drag:

A denser atmosphere causes more drag. Objects in Low Earth Orbit (LEO) encounter atmospheric drag in the form of gases in the thermosphere (approximately 80–500 km up) or exosphere ($\approx$500 km and up), depending on orbit height. The LEO altitude is usually not less than 300 km because that would be impractical due to the larger atmospheric drag. Emissions from the Sun (including, the highly variable X-ray and ultraviolet output)

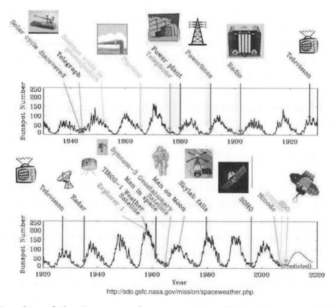

http://sdo.gsfc.nasa.gov/mission/spaceweather.php

Figure 2. Time line of the discovery of space weather effects. Courtesy of NASA/GSFC.

cause the upper atmosphere to heat and expand. If magnetic activity is also triggered in the Earths magnetosphere, intense electric currents flow through the ionosphere and upper atmosphere, and the energy deposited at high latitudes increases the heating and expansion of the atmosphere in these regions markedly.

The most famous example of this effect was Skylab, which burned up in the atmosphere on July 11, 1979 after its orbit deteriorated for 5 years. Some satellites (e.g. Compton Gamma Ray Observatory), have onboard jets to compensate for orbit decay. The Hubble Space Telescope (HST) has no jets or engines of any kind for propulsion, so the only way to restore the altitude is to grab it and move it ["space shuttle during HST servicing missions"]. The drag effect has its beneficial side too and this concerns the problem of space debris. When debris falls towards Earth it burns up in the atmosphere. This is a natural occurring procedure to reduce the number of cosmic debris.

Radio communications:

Earth's ionosphere, created by the ionization of the upper atmosphere by solar EUV and X-ray radiation, plays an important role in radio communications. Spatial scales of inhomogeneities in the ionosphere vary from thousands of kilometres to turbulence with scale sizes of less than a meter. Likewise, the temporal scales vary over many orders of magnitude from many years (solar cycle effects on ionospheric propagation) to hours or even minutes (the scale of weather phenomena). High frequencies (HF) are attenuated in the D layer of the ionosphere and then reflected in the F layer. HF propagation anomalies are due to ionospheric changes resulting from solar flares, proton events and geomagntic storms.

Because of solar flares, the increased intensity of the EUV radiation may result in the increase of the ionization level of the D layer, where the attenuation of the HF waves increases and produces Short Wave Fadeouts (SWF). The attenuation can be so high that it may lead to the interruption of radio transmission. SWFs is a particular ionospheric solar flare effect under the broad category of sudden ionospheric disturbances.

Very High Frequencies (VHF) and Extremely High Frequencies (EHF) are not reflected by the ionosphere. They cross the ionosphere where they are subject to a refraction phenomenon. These frequencies are used to communicate with artificial satellites for command transmission and data reception. At times of more intense ionization and increased concentration and temperature in the ionosphere a phenomenon known as "plasma bubbles" will be created. When radio waves go through such irregularities, the wave intensity and propagation speed are attenuated and changed, respectively. The result is signal fading or data downlink (attenuation of the signal and loss of data).

The ionosphere in polar regions can be badly affected by high energy solar protons (> 10 MeV) arriving hours to days after the solar disturbance (e.g. strong solar flare). Very strong ionisation of the D-region leads to absorption of HF signals similar to an HF Fadeout. This is called a Polar Cap Absorption (PCA) event. PCAs can last for several days following a large solar flare in contrast to a low/mid latitude fadeout which generally lasts just an hour or two.

Ground effects:

Geomagnetically Induced Currents (GICs) are the ground end of the space weather chain. A GIC is defined as a current connected with a geomagnetic variation and a man-made conductor (e.g. power transmission systems, oil and gas pipelines, telecommunication cables, railways). Space weather storms produce intense and rapidly varying currents in the magnetosphere and ionosphere which cause time-dependent magnetic fields seen as geomagnetic disturbances or storms. As expressed by Faradays law of induction, a time variation of the magnetic field is always accompanied by an electric field. The geomagnetic disturbance and the geo-electric field observed at the Earth's surface not only depend on the "primary" space currents, but are also affected by "secondary" currents driven by the electric field within the conducting Earth. In particular for the electric field, the secondary contribution is essential. The horizontal geoelectric field drives ohmic currents (GICs) in technological conductor networks (see Fig.3). Currents induced in power lines flow to ground through substation transformers. Here they cause saturation of the transformer core which can lead to a variety of problems. Increased heating has caused transformers to burn out. Also extra harmonics generated in the transformer produce unwanted relay operations, suddenly tripping out power lines. The stability of the whole system can also be affected as compensators switch out of service. Such a sequence of events led to the Quebec blackout of 13 March 1989.

Biological effects:

High frequency radiation or high-energy particles can knock electrons free from molecules that make up a cell. These molecules with missing electrons are called ions, and their presence disrupts the normal functioning of the cell. Cells that reproduce rapidly (skin, eyes, blood-forming organs) are the most susceptible to damage because they cannot repair themselves easily while replicating. The most severe damage to the cell results when the DeoxyriboNucleic Acid (DNA) is injured. DNA is at the heart of the cell and contains all the instructions for producing new cells. Symptoms of radiation sickness are severe burns that are slow to heal, sterilization, cancer and other damages to organs. Mutations or changes in the DNA can be passed along to offspring.

These above listed biological effects include those associated with current and future astronauts and cosmonauts. However, aircraft crew and passengers onboard airlines are also under risk of radiation, especially on polar routes. In this region of the geomagnetic field, the magnetic field lines are open and energetic particles can reach down to lower altitudes. The Council of the European Union adopted Directive 96/29 Euratom on 13 May 1996. Article 42 of the Directive imposes requirements relating to the assessment

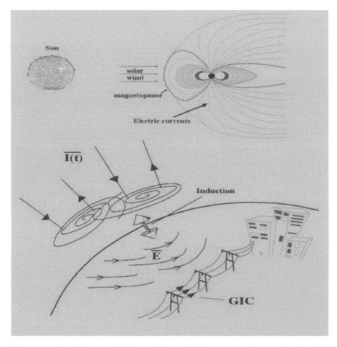

Figure 3. Schematic drawing of the creation of geomagtically induced currents. Courtesy of Finnish Meteorological Institute.

and limitation of air crew members exposure to cosmic radiation and the provision of information on the effect of cosmic radiation.

During the famous Octber–November 2003 solar events all commercial aviation interests were made aware of the radiation storm levels on October 28 to 29, when the Federal Aviation Administration (FAA) issued its first ever advisory suggesting that flights travelling North and South of 35 degrees latitude were subject to excessive radiation doses.

4. Mitigation

Health risks for long duration interplanetary explorative missions and those encountered so far in manned space flight differ significantly in two major features: *1*. "Emergency returns" are ruled out. *2*. The loss of geomagnetic shielding available in low Earth orbit with an associated non-negligible risk for acute early radiation diseases.

There are two approaches in tackling space weather and its unwanted effects. First there is the classical engineering approach which relies on shielding, mitigation for electronics and charging, mitigation in radio communications and medical mitigation. Spacecraft "age" through continual bombardment by energetic particles. The key to radiation protection is the understanding of the space environment and its interaction with shielding. An important issue concerning shielding is the problem of secondary radiation in materials. New forms of shielding materials are imagined and more impetus should be placed on polymer research in regard to the development of resistant light atomic weight shielding. Ofcourse the faster the trip the better, i.e. development of innovative transportation technologies and new propulsion systems as well as orbit optimization, are highly important if not the most important challenges. The ultimate goal is to minimize radiation together with all other health effects and technical hazards by optimizing orbit parameters and shielding. Maris & Crosby (2008) is a review paper concerning the technical effects induced by space weather as well as the techniques used for mitigation.

The second approach in tackling space weather relies on avoiding the phenomena that may cause these effects in the first place, respectively the prevention approach "space weather forecasting". This can only happen if one has a reliable description of the space environment that the spacecraft will encounter? This relies on real-time measurements (space-borne and on Earth) and models of the various environments? In this way one can monitor the space weather and warn potential space weather customers of "bad weather". In the future it will be the users of space weather forecasting centers that must set the requirements for the time scales and precision of the predictions.

5. Final words

Humans and technology (in space and on Earth) are vulnerable to the space weather. Some of the effects can be corrected for and/or mitigated against, others not. Modern society is increasingly becoming dependent on the usage of artificial satellites and will be more so in the future. In summary, scientists, engineers, medical doctors, etc. from many disciplines must learn to work together to optimize our understanding of the space environment and its interaction with our daily lives.

In addition there will be many opportunities in regard to space weather in the near future (Missions to other planets and moons [manned and robotic], colonies on other planets such as Mars, Mining on other planets, moons, asteroids and terra-forming, transportation technology, space tourism, space hotels, emergence of space entrepreneurs). Indeed the field of space weather has a bright future both from the scientific as well as the techniological point-of-view.

References

Bothmer, V. & Daglis, I. A. (eds) 2007, in: *Space Weather: Physics and Effects, Springer Praxis Books*

Carlowicz & Lopez 2002, *Storms from the Sun: The Emerging Science of Space Weather, National Academies Press*

Cost724 2007 *Cost724 Final Report*

Daglis, I. A. 2001 *Space Storms and Space Weather Hazards, Kluwer Academic Publishers*

Freeman, J. W. 2001 *Storms in Space, Cambridge University Press*

Hanslmeier, A. 2002 *Kluwer Academic Publishers*

Koskinen, H., Tankanen, E., Pirjola, R., Pulkkinen, A., Dyer, C., Rodgers, D., Cannon, P., Mandeville, J. C. & Boscher, D. 2001, *Space Weather Effects Catalogue, ESWS-FMI-RP-0001*

Maris, O. & Crosby N. 2008, in: G. Maris & M. D. Popescu (eds.), *Research SignPost Ed. House, India* (In print)

Miroshnichenko, L. 2003, *Radiation Hazard in Space* (Kluwer Academic Publishers)

Scherer, K. *et al.* 2001, in: Scherer, K., Fichtner, H., Heber, B. & Mall, U. (eds.), *Behind a Slogan* (Lecture Notes in Physics)

Song, P. *et al.* (eds) 2001, in: *Space Weather* (American Geophysical Union)

Wang, H. & Xu, R. (eds) 2002, in: *Proceedings of the COSPAR Colloquium on Solar-Terrestrial Magnetic Activity and Space Environment* (Elsevier Science Ltd.)

Wright, Jr., J. M., Lennon, T. J., Corell, R. W., Ostenso, N. A., Huntress, Jr., W. T., Devine, J. F., Crowley, P., & Harrison, J. B. 1995, *The National Space Weather Program, The Strategic Plan., Office of the Federal Coordinator for Meteorological Services and Supporting Research, FCM-P30-1995, Washington DC, USA, August*

Discussion

VLAHOS: Can we make progress with the use of many different codes interacting since we are dealing with a "system" of interacting elements?

CROSBY: There are efforts in trying to connect the different regions (solar atmosphere, magnetosphere, ionosphere/atmosphere) of the Sun-Earth connection. For example, the goal of the Center for Integrated Space Weather Modeling (CISM) is to create a physics-based numerical simulation model that describes the space environment from the Sun to the Earth.

GREBOWSKY: You just briefly discussed radiation belt physics, but it is the prime source of the spacecraft charging effects and spacecraft failures. The one high energy outer belt enhancement mentioned has been attributed actually to an enhanced magnetosphere E-field induced. NASA's RB storm probe mission is geared to solve these problems. Unfortunately there are no ionospheric missions planned.

Universal Heliophysical Processes
Proceedings IAU Symposium No. 257, 2008
N. Gopalswamy & D.F. Webb, eds.

© 2009 International Astronomical Union
doi:10.1017/S1743921309029081

Evolution of several space weather events connected with Forbush decreases

I. Dorotovič[1], K. Kudela[2], M. Lorenc[1], T. Pintér[1] and M. Rybanský[2]

[1]Slovak Central Observatory, P.O. Box 42, SK-94701 Hurbanovo, Slovak Republic
email: `ivan.dorotovic@suh.sk`

[2]Institute of Experimental Physics SAS, Watsonova 47, SK-04353 Košice, Slovak Republic
email: `kkudela@upjs.sk`

Abstract. In our recent paper (Dorotovič *et al.* 2008a) we focused on a study of the Forbush decrease (FD) of January 17–18 and 21–22, 2005. It was shown that the corresponding recovery time can depend on the density of high-energy protons in the CME matter. In this paper we identified several additional events in the period between 1995 and 2007. We found that the majority of FDs studied is accompanied by an abrupt count increase in the proton channel P1 and by a simultaneous decrease in the channel P7 (GOES). However, the analysis of temporal evolution of all FDs did not confirm the hypothesis on different recovery time after FD as a function of the energy distribution of the particles penetrating into radiation belts of the Earth.

Keywords. ISM: cosmic rays, Sun: flares, solar-terrestrial relations, Earth

1. Introduction

More or less regular CR variations of different duration, connected with the length of the solar activity cycle (11-year), the length of the solar rotation (27-days) or related to the rotation of the Earth (24 hours) are interrupted by the sporadic ones – a sudden decrease with slow recovery, FD – Forbush Decrease and an abrupt increase with immediate decrease, GLE – Ground Level Event. Both events are related to the occurrence of solar flares and transient events. Geoeffectivity of some solar events was investigated e.g. by Koskinen & Huttunen (2006), Wang & Wang (2006), Bochníček *et al.* (2007) and Valach *et al.* (2007).

The occurrence of FDs in the period between 1995 and 2007, i.e. in the 23rd cycle of solar activity, was analyzed in this paper. We use, data from the Neutron Monitor (NM) at Lomnický štít and observations from satellites WIND, ACE, SOHO, and GOES for a better description of complex evolution of the selected FDs. The second section describes the input data and methods of processing; results and consequent conclusions are presented in the third section.

2. Data used and their processing

We used the data from the NM at Lomnický štít for the period of 1995 – 2007 as basic input for the analysis. The data from other neutron monitors (Oulu, Haleakala, and South Pole) were used for verification of temporal evolution of CR in certain periods. We defined the following criteria for the selection process:
1) A total decrease in NM counts at Lomnický štít is at least 2% (in normalization of each year to 100%).
2) A decrease occurs at all stations.

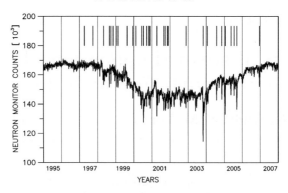

Figure 1. Temporal evolution of cosmic radiation level from NM at Lomnický štít in the period between 1995 and 2007 together with an indication of FDs studied.

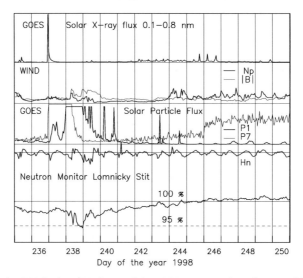

Figure 2. FD on the 238th day (26 August) in 1998: temporal evolution of NM measurements with a resolution of 1 hour (the bottom part of the y-axis), from top: solar X-ray flux (GOES), proton density n_p and interplanetary magnetic field intensity $|B|$ (ACE), proton fluxes P1, P7 and magnetic field component H_n (GOES).

Conditions in the heliosphere between the Earth and the Sun were determined using the data on proton density and magnetic field from the observatories WIND (till 1998), ACE (since 1998), and GOES (proton channels P1: 0.4 – 4.0 MeV and P7: 165 – 500 MeV). The GOES data on solar X-ray flux in the range of 0.1 – 0.8 nm were used for identifying a flare together with its importance.

According to the defined criteria and using the described data we were able to select 34 events in the considered period. The temporal distribution of individual events is shown in Fig. 1. Fig. 2 shows an example of our analysis.

3. Results and discussion

Because the shape of the evolution of individual FDs varies significantly, we decided to define the following types of a FD according to its recovery phase:

A – a classical evolution, i.e. an abrupt decrease and a slow uniform recovery;

B – non-uniform recovery, interrupted by an increase and a decrease;

C – decrease and increase on the same day followed by a second slower decrease and a gradual recovery to the initial level;

D – similar evolution as B, but the interval between the first decrease and increase is longer;

E – decrease to a certain level, then a longer persistent period followed by an abrupt increase to the initial level.

A table with basic properties of studied FDs and detailed description of results can be found in the paper of Dorotovič *et al.* (2008b). We briefly summarize our conclusions as follows:

1) Only in one case out of 34, the origin of FD cannot be connected with a flare in X-ray flux. In many cases (22 of 34) the corresponding flare was followed by a halo type CME.

2) The depth of decrease is maximally 7%, each year being normalized separately.

3) Most FDs are of type A − 14 events; type B − 10 events, type C − 8 events, D and E types − 1 event each.

4) Almost in all cases FD is related to an increase of n_p and $|B|$ at the WIND and ACE satellites. However, there are 7 of 34 cases when this increase follows only after the FD.

5) P1 counts during FD strongly increase, in maximum to 3 − 4 orders of magnitude higher than the quiet value, and P7 decreases. The decrease in P7 is often superimposed by a burst in this energetic range, which usually occurs immediately after the flare or is caused by a flare different from that which caused the FD studied (Fig. 2). Almost each FD is connected with an increase in P1. However, not each increase of P1 causes an FD.

4. Concluding remarks

FD is usually accompanied by an abrupt count increase in the channel P1 and by a simultaneous decrease in the channel P7, often as far as the threshold values of the instrument. This finding can be considered as the main result of the analysis of FD evolution. Then an impression arises that during FD the protons of the primary CR are absorbed on the low-energy protons indicated in the channel P1 and simultaneously also the protons from channel P7.

The analysis of the temporal evolution of FD did not confirm the hypothesis published in the paper by Dorotovič *et al.* (2008a) on different recovery time after a FD as a function of the energy distribution of the particles penetrating into radiation belts of the Earth. There are two reasons: 1) in most cases, the recovery time cannot be determined owing to other variations occurring during the recovery period and 2) the channel P7 is either overloaded due to a previous burst or is at a threshold value.

Acknowledgements

This work was supported by the Slovak Research and Development Agency under contract No. APVV-51-053805 and by VEGA grant project 7063.

References

Bochníček, J., Hejda, P., & Valach, F. 2007, *Stud. Geophys. Geod.*, 51, 439
Dorotovič, I., Kudela, K., Lorenc, M., & Rybanský, M. 2008a, *Solar Phys.*, 250, 339
Dorotovič, I., Kudela, K., Pintér, T., Lorenc, M., & Rybanský, M. 2008b,
    Proc. of the 21st ECRS, 9 - 12 September 2008, Košice, Slovakia, submitted
Koskinen, H. E. J. & Huttunen, K. E. J. 2006, *Space Sci. Rev.*, 124, 169
Valach, F., Hejda, P., & Bochníček, J. 2007, *Stud. Geophys. Geod.*, 51, 551
Wang, R. & Wang, J. 2006, *Adv. Space Res.*, 38, 489

Universal Heliophysical Processes
Proceedings IAU Symposium No. 257, 2008
N. Gopalswamy & D.F. Webb, eds.

Coronal shocks associated with CMEs and flares and their space weather consequences

Marina Laskari[1], Panagiota Preka-Papadema[1], Constantine Caroubalos[2], George Pothitakis[2], Xenophon Moussas[1], Eleftheria Mitsakou[1] and A. Hillaris[1]

[1]Department of Physics, University of Athens, 15784 Athens, Greece

[2]Department of Informatics, University of Athens, 15783 Athens, Greece

Abstract. We study the geoeffectiveness of a sample of complex events; each includes a coronal type II burst, accompanied by a GOES SXR flare and LASCO CME. The radio bursts were recorded by the ARTEMIS-IV radio spectrograph, in the 100-650 MHz range; the GOES SXR flares and SOHO/LASCO CMEs, were obtained from the Solar Geophysical Data (SGD) and the LASCO catalogue respectively. These are compared with changes of solar wind parameters and geomagnetic indices in order to establish a relationship between solar energetic events and their effects on geomagnetic activity.

Keywords. Sun: coronal mass ejections (CMEs), Sun: flares, Sun: radio radiation, (Sun:) solar-terrestrial relations

1. Introduction

The primary sources of geomagnetic phenomena are the solar eruptive events; they initiate the disturbances of solar wind parameters (magnetic field, speed, density and temperature) which in turn drive the Space Weather effects. The latter is characterized by a geomagnetic index such as the *Disturbance storm time* or **Dst** (cf. Gopalswamy *et al.* 2007; the geomagnetic field variations are, on the other hand, quantified, on average, by the **Kp** index. Though the connection of the geomagnetic phenomena with processes on the Sun is well established, the geomagnetic storm effectiveness of CMEs and solar flares is still an open question as published results are not conclusive (for a review cf. Yermolaev *et al.* 2005, also Yermolaev & Yermolaev 2006).

We study the geoeffectiveness of a medium size sample of solar events; each includes a coronal type II burst sometimes extending to an interplanetary type II, accompanied by a GOES SXR flare and SOHO/LASCO CME. The radio bursts were recorded by the ARTEMIS-IV radio spectrograph (Caroubalos *et al.* 2001); the GOES SXR flares and SOHO/LASCO CMEs, were obtained from the Solar Geophysical Data (SGD) and the SOHO/LASCO lists respectively.

We examine the effects on the solar wind parameters and the **Kp**, **Dst** geomagnetic indices variations two to three days after the events recording.

2. Data selection & analysis

The ARTEMIS IV radiospectrograph (Caroubalos *et al.* 2001) observed about 40 type II and/or IV radio bursts which they were published in the form of a catalogue (Caroubalos *et al.* 2004). From this catalogue we have used in our study the same fourteen events studied by Pothitakis *et al.* (these proceedings); here we adopted the same numbering of events. The solar wind parameters (magnetic field (B and Bz component), speed (V),

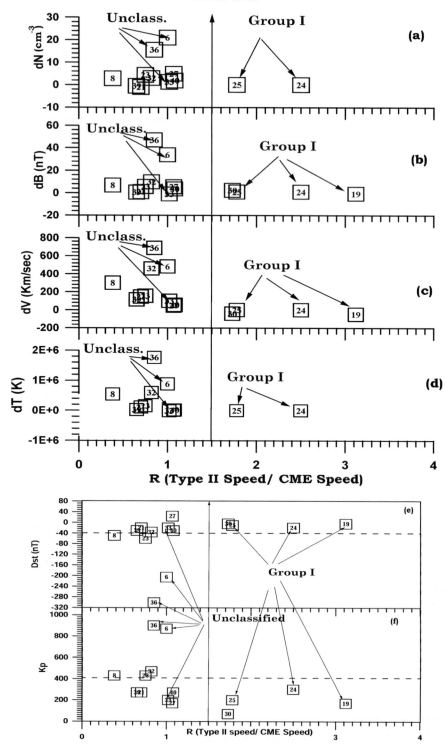

Figure 1. Solar Wind Parameters Variation & Space Weather indices as a function of the ratio of the speed of Type II to the CME speed ($V_r = V_{II}/V_{CME}$): (a)Density (dN) in cm^{-3}, (b) Magnetic Field (dB) in nT, (c) Speed (dV) in Km/sec, (d) Temperature (dT) in K, (e) Dst, (f) Kp. (We have labeled Events of Group I & Unclassified (cf. Table 1 of Pothitakis *et al.* in these Proceedings); The quiet Solar Wind Parameters at 1AU were: B = 5 nT, V = 350 Km/sec, N = 5 cm^{-3}, T = 10^5 K.)

temperature (T) and density (N)) at 1 A.U., and the geomagnetic indices Kp and Dst were obtained from the OMNI data base.

The data set was carefully selected in order to represent solar events within periods of relative inactivity (deduced from the GOES SXR flux 3-4 days prior and after the event) in order to establish an association between the space weather effects and the solar driver.

In Pothitakis *et al.* (cf. their Table 1) the association of the flare–CME–Type II burst parameters was studied; in this report on the other hand, we examine the variation of the geomagnetic indices **Dst** & **Kp** and of the Solar Wind Parameters (dB, dV, dT, dN)) with the ratio of the speed of Type II to the CME speed ($V_r = V_{II}/V_{CME}$); this ratio provides a sort of indication of the CME capability in driving a magnetohydrodynamic shock. The results of this comparison are plotted in Figure 1.

Our examination indicates that a certain class of events may initiate space storms (Events 23, 06, 36, 08 & 32) give **Dst**<-40 or **Kp**>400), the Bastille Event (36) among them.

Comparing this result with the classification of Pothitakis *et al.* we note that the most geoeffective events of our data sample are from Group II (23, 08, 32) plus two of the events which were not classified (36, 06). The events in Group I (24, 25, 19), on the other hand, include fast coronal shocks (1213–1940 km/sec) and CMEs with speeds in the 400–800 km/sec range yet their effect on space weather are small.

Lastly, we note that there is a trend of increased geoeffectiveness and Solar Wind Variations with V_r but only when $V_r < 1.5$.

3. Final remarks

We have studied the geoeffectiveness of a medium size sample of complex events, each including a coronal type II burst, a GOES SXR flare and a SOHO/LASCO CME; the results were compared with the classification of Pothitakis *et al.* (these proceedings) which was based on parameters, related to shock & CME kinetics and radio bursts-flare-CME timing.

Fast coronal shocks do not, always, initiate storms, yet fast shocks in the front of fast CMEs have a higher probability of inciting disturbances in the near earth environment. Furthermore, a trend of increase in geomagnetic effects with R was found when $V_r < 1.5$. This, in fact, excludes fast coronal shocks accompanied by relatively slow CMEs.

References

Caroubalos, C., Maroulis, D., Patavalis, N., Bougeret, J. L., Dumas, G., Perche, C., Alissandrakis, C., Hillaris, A., Moussas, X., Preka-Papadema, P., Kontogeorgos, A., Tsitsipis, P., & Kanelakis, G., 2001, *Exp. Astron.*, 11, 23

Caroubalos, C., Hillaris, A., Bouratzis, C., Alissandrakis, C. E., Preka-Papadema, P., Polygiannakis, J., Tsitsipis, P., Kontogeorgos, A., Moussas, X., Bougeret, J. L., Dumas, G., & Perche, C., 2004, *A&A*, 413, 1125

Gopalswamy, N., Yashiro, S., & Akiyama, S., 2007, *J. Geophys. Res.*, 112(11), 6112

Yermolaev, Y. I., Yermolaev, M. Y., Zastenker, G. N., Zelenyi, L. M., Petrukovich, A. A., & Sauvaud, J. A., 2005, *Planetary and Space Science*, 53, 189

Yermolaev, Y. I. & Yermolaev, M. Y., 2006, *Adv. Sp. Res.*, 37, 1175

Universal Heliophysical Processes
Proceedings IAU Symposium No. 257, 2008
N. Gopalswamy & D.F. Webb, eds.

© 2009 International Astronomical Union
doi:10.1017/S174392130902910X

Possible heliogeophysical effects on human physiological state

Svetla Dimitrova

Solar-Terrestrial Influences Laboratory, Bulgarian Academy of Sciences,
Acad. G. Bonchev Str. Bl. 3 Sofia 1113 Bulgaria
email: svetla_stil@abv.bg

Abstract. A group of 86 healthy volunteers was examined in periods of high solar and geomagnetic activity. In this study hourly Dst-index values and hourly data about intensity of cosmic rays were used. Results revealed statistically significant increments for the mean systolic and diastolic blood pressure, pulse pressure and subjective psycho-physiological complaints of the group with geomagnetic activity increase and cosmic rays intensity decrease.

Keywords. Cosmic rays, geomagnetic activity, blood pressure, subjective complaints

1. Introduction

Space weather is often defined as conditions on the Sun and in the solar wind, magnetosphere, ionosphere and thermosphere that can influence the performance and reliability of space-borne and ground-based technological systems and can endanger human life or health. A variety of physical phenomena are associated with space weather, including cosmic ray (CR) intensity variations and geomagnetic storms. Galactic CRs experience significant variation in response to passing solar wind disturbances such as interplanetary coronal mass ejections (ICMEs) and their accompanying shocks. Arriving at Earth, ICMEs compress the magnetosphere, intensify the magnetosphere currents thus leading to a significant depletion of CR intensity (CRI) and producing geomagnetic storms. Some studies revealed significant effects on myocardial infarctions, brain strokes, and traffic accidents on the days of geomagnetic field (GMF) disturbances accompanied with CRI decreases (Villoresi *et al.* 1995; Ptitsina *et al.* 1998; Dorman 2005).

2. Material and methods

Data were obtained in 86 healthy volunteers in Sofia on working days in autumn and spring in years of high GMA. Systolic, diastolic blood pressure (SBP, DBP) and heart rate (HR) were measured. Pulse pressure (PP) was calculated. Data for some subjective psycho-physiological complaints (SPPC) were gathered also (Dimitrova 2008).

Hourly data about CRI from Rome neutron monitor, were used. Data about GMA, estimated by hourly Dst-index were got from WDC, Kyoto. Fig. 1a,b show hourly CRI and hourly Dst-index variations for the both periods of examinations. Table 1 presents the number of physiological measurements, which were accomplished for the different percents (levels) of CRI decreases. GMA was divided into five levels taking into account Dst-index values (Table 2).

Aalysis of variance (ANOVA), Post-hoc and the method of superimposed epochs were used to study the effect of CRI and GMA up to 3 days before and 3 days after their variations on the physiological parameters.

Table 1. CRI decrease in percents and the number of measurements

CRI, %	3	4	5	6	7	8	9
Meas.	194	715	930	627	253	39	41

Table 2. Dst-index levels and the number of measurements

GMA Level	I Quiet GMA	II Weak storm	III Moderate storm	IV Major storm	V Severe storm
Dst, nT	Dst>-20	$-50<$Dst$\leqslant -20$	$-100<$Dst$\leqslant -50$	$-150<$Dst$\leqslant -100$	Dst$\leqslant -150$
Meas.	1819	544	290	104	42

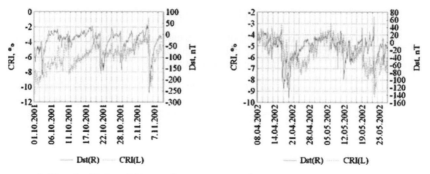

Figure 1. a) Hourly CRI and Dst-index variations during autumn examination period. b) Hourly CRI and Dst-index variations during spring examination period.

Table 3. Significance level p of CRI and GMA variations effects on the physiological parameters; * denotes statistically significant effect.

Factor	SBP	DBP	PP	HR	SPPC
CRI	0.000*	0.000*	0.010*	0.152	0.002*
GMA	0.000*	0.000*	0.003*	0.719	0.000*

3. Results

CRI variations and physiological parameters. ANOVA revealed statistically significant effect for CRI on SBP, DBP, PP and SPPC, Table 3. Fig. 2a shows the mean values of SBP and DBP for the group under different CRI decreases: SBP and DBP increased with the decrease of CRI. The maximal increment for SBP was 10.5% and for DBP 11.4%. Post hoc analyses established that SBP and DBP were significantly higher during CRI decrease with 8% and 9% in comparison to CRI decrease with 3 ÷ 7%.

PP increased also with CRI decrease, having the highest value (10% increment) at 8% CRI decrease. The largest variation for HR was only 2.8%. SPPC increased with CRI decrease and 26.8% of the persons reported SPPC during CRI decrease with 9%.

Statistically significant effect on SBP and DBP (Fig. 2b) from -3rd till +3rd day of different CRI decreases was obtained. It was revealed by Post hoc analyses that arterial blood pressure (ABP) mean values increased significantly from −1st till +3rd day when CRI decreased with 8-9%. ABP was high also on the days before, during and after CRI decrease of 7%.

PP was statistically significantly affected not only on 0 day but also on +1st and +2nd day. PP's mean values of the group were highest from −1st to +3rd day when CRI decreased with 8% and on −3rd and +3rd day of CRI decrease with 9%. SPPC increased statistically significantly on 0, +1st, and +3rd day. Reported subjective complaints in the group were largest from −1st to +3rd day of CRI decrease with 9%.

GMA variations and physiological parameters. GMA effect on SBP, DBP, PP and SPPC was statistically significant, Table 3. GMA increase was followed by an increase of the physiological parameters and the range of changes for SBP and DBP (Fig. 3a), PP and SPPC were respectively 10.7%, 10.4%, 11.2% and 21.3%. Post hoc analyses revealed that the group increased significantly ABP still at moderate storms. The number of the persons who reported SPPC increased significantly still at major storms in comparison with quiet GMA and weak storms.

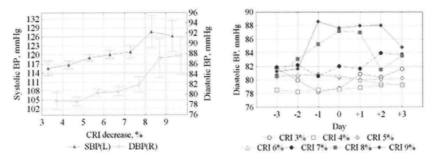

Figure 2. a) CRI effect on SBP and DBP (±95% CI) b) CRI effect on DBP before, during and after CRI decreases.

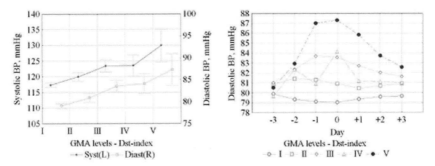

Figure 3. a) GMA effect on SBP and DBP (±95% CI) b) GMA effect on DBP before, during and after geomagnetic storms.

HR of the group increased with GMA increment but with only 1.9%. It was established that SBP and DBP (Fig. 3b) increased statistically significantly from −2nd till +3rd day, PP from −1st till +1st day and SPPC from −1st till 0 day.

4. Conclusions

Both space weather parameters (CRI and Dst-index) were related to statistically significant changes in the human physiological state of the examined group. It was established that SBP, DBP, PP and SPPC of the healthy volunteers increased with CRI decrease and GMA increase and on the days before, during and after their variations.

ABP values of the group were highest from −1st till +3rd day when CRI decreased more than 7% and from −2nd to +3rd of moderate, major and severe geomagnetic storms.Reported SPPC increased the most from −1st till +3rd day of the largest decreases in CRI and from −1st till 0 day of different geomagnetic storms.

The fact that the group increased ABP on average with about 10-11% and almost 1/3 from the persons felt some psycho-physiological discomfort deserves attention from a medical point of view and enhance biological, clinical and social importance of the influences examined.

5. Acknowledgment

This work was partially supported by National Science Fund of Bulgaria under contract NIP L-1530/05. SVIRCO NM is supported by IFSI/INAF-UNIRoma3 Collaboration.

References

Dimitrova, S. 2008, *JASTP*, 70/2-4, 420
Dorman, L. I 2005, *Annales Geophysicae*, 23, 2997
Ptitsyna, N. G., Villoresi, G., Dorman, L. I., Iucci, N., & Tiasto, M. I. 1998, *UFN*, 168 (7), 767
Villoresi, G., Dorman, L. I., Ptitsyna, N. G., Iucci, N., & Tiasto, M. I. 1995, *24th ICRC,* 4, 1106

Session III

Solar Sources of Heliospheric Variability

Universal Heliophysical Processes
Proceedings IAU Symposium No. 257, 2008
N. Gopalswamy & D.F. Webb, eds.

Solar and planetary dynamos: comparison and recent developments

K. Petrovay

Eötvös University, Department of Astronomy
H-1518 Budapest, Pf. 32., Hungary
email: K.Petrovay@astro.elte.hu

Abstract. While obviously having a common root, solar and planetary dynamo theory have taken increasingly divergent routes in the last two or three decades, and there are probably few experts now who can claim to be equally versed in both. Characteristically, even in the fine and comprehensive book "The magnetic Universe" (Rüdiger & Hollerbach 2004), the chapters on planets and on the Sun were written by different authors. Separate reviews written on the two topics include Petrovay (2000), Charbonneau (2005), Choudhuri (2008) on the solar dynamo and Glatzmaier (2002), Stevenson (2003) on the planetary dynamo. In the following I will try to make a systematic comparison between solar and planetary dynamos, presenting analogies and differences, and highlighting some interesting recent results.

Keywords. magnetic fields, MHD, plasmas, turbulence, Sun: magnetic fields, Sun: interior, Earth, planets and satellites: general

1. Approaches to astrophysical dynamos

1.1. *Dimensional analysis: mixing-length vs. magnetostrophic balance*

Faced with a problem like the dynamo, where the governing equations are well known and the source of difficulties is their complexity, it is advisable to start by order of magnitude estimates of the individual terms in the equations. A clear and detailed account of this is given in Starchenko & Jones (2002).

In the case of the Sun, such considerations point to a balance between *buoyancy* and *inertial forces* as the determinant of the resulting flow pattern in the convective zone. Indeed, the order of magnitude equality of these terms is one of the basic formulæ of the mixing length theory of astrophysical convection, as formulated in the 1950s—so this balance is customarily referred to as "mixing length balance" in the dynamo literature.

On the other hand, in rapidly rotating systems the Coriolis term dominates over the inertial term. If the magnetic field is strong enough for Lorentz forces to be comparable to the Coriolis force and the buoyancy, another type of balance known as *magnetostrophic balance* (or MAC balance) sets in. This kind of balance is now generally thought to prevail in the more "mainstream" planetary dynamos, such as those of Earth, Jupiter, Saturn, and possibly Ganymede.

1.2. *Implicit models: Mean field theory*

As the smallest and largest structures present in the strongly turbulent astrophysical dynamos are separated by many orders of magnitude, it is hopeless and perhaps unnecessary to set the full explicit treatment of all scales of motion as a goal. In this sense, all models of astrophysical dynamos are necessarily "mean field models", not resolving scales smaller than a certain level and representing the effect of those scales with some effective diffusivities. Yet the term "mean field theory" is customarily reserved for those models

71

where even the largest scale turbulent motions, thought to be the main contributors to dynamo action, remain unresolved.

Mean field theory has remained the preferred theoretical tool of solar dynamo studies. Even the effect of obviously non-mean-field effects can be included in mean field models in the form of parametrized ad hoc terms. E.g., the emergence of strong magnetic flux loops from the tacholine to the photosphere is now thought by some to be a major contributor to the α-effect (the so-called Babcock-Leighton mechanism for α). The motion of magnetized fluid being highly independent of the rest of the plasma, this may appear to be a case where the mean field description is bound to fail—but the problem is circumvented in mean field dynamo models, e.g. by the introduction of a non-local α-term (Wang *et al.* 1991) or by using different diffusivity values for poloidal and toroidal fields (Chatterjee *et al.* 2004).

1.3. *Explicit models: Numerical simulations*

In contrast to solar dynamo theory, planetary dynamo studies have traditionally focused on models where the large scale, rotationally influenced turbulent motions are explicitly resolved. For a long time such studies were essentially kinematical, prescribing the large scale flows by simple mathematical formulæ that satisfy some more or less well founded basic physical expectations, and solving only the induction equation. Such studies can still significantly contribute to our understanding of dynamos (see e.g., Gubbins 2008 for a recent overview). Yet, from the mid-1990s onwards the rapid increase in computer power has made it possible to develop explicit dynamical models (aka numerical simulations) of the geodynamo, and this has become the main trend in planetary dynamo research.

2. Observational constraints

We are separated from the Earth's outer core by 3000 km of intransparent rock, while the top of Sun's convective zone is directly observable across 1 AU of near-empty space. Although magnetic measurements and seismology in principle allow indirect inferences on conditions in the outer core, these are both limited to relatively large-scale (spherical harmonic degree $l < 13$) magnetic structures, while empirical information on flows in the core is almost completely lacking.

In sharp contrast to this, the brightness of the Sun allows high S/N detection of spectral line profiles, permitting a precise determination of Doppler and Zeeman shifts. This not only provides a wealth of high-resolution observations of flows and magnetic fields at the top of the convective zone: the sensitive detection of waves and oscillations in the solar photosphere also allows a detailed reconstruction of flows and magnetic field patterns in layers lying below the surface.

The different amount of empirical constraints are certainly a major factor in determining the different approaches taken by solar and planetary dynamo studies. The main shortcoming of mean field models impeding progress is the vast number of possibilities available for the choice of parameters and their profiles, which renders the formulation of mean field dynamo models for the planets a rather idle enterprise. The good empirical constraints in the solar case provide an indispensable support by narrowing down the range of admissible models. It is no coincidence that it was the helioseismic determination of the internal rotation profile around 1990 which led to a resurgence in solar mean field dynamo theory.

At the same time, no MHD numerical simulation of the solar convective zone has been able to recover the observed butterfly diagram, and even reproducing the observed differential rotation profile is not trivial (cf. Sect. 6 below). Geodynamo simulations correctly reproduce nearly all the known spatiotemporal variance in the Earth's magnetic field, and recently there are indications that they could provide a successful general scheme to understand the variety of dynamos seen in the Solar System (see Sect. 8).

3. State of matter, stratification

One obvious difference between planetary and stellar dynamos is that while the conducting matter in stars is ionised gas, in planetary dynamos it's conducting liquids. The consequences of this are twofold. On the one hand, in planets incompressibility is often a good approximation, whereas in the Sun there are many scale heights between the top and bottom of the SCZ and the scale height itself varies by several orders of magnitude. This extreme stratification or "stacking" of structures of vastly different scale is one major obstacle in the way of realistic global simulations of the SCZ.

Another consequence of the different state of matter in stellar and planetary dynamos is that molecular transport coefficients or diffusivities (such as viscosity, resistivity or heat diffusivity) are less accurately and reliably known in planetary interiors, given the more complex material structure. Together with the unsatisfactorily constrained thermal state, this means that the exact position, boundaries and even nature of the dynamo shell is in doubt in some planets. In the water giants (Uranus and Neptune) the resistivity of their electrolytic mantles is very uncertain, affecting the extent and depth of the dynamo layer. In Mercury, the unknown thermal state of the planet and the unknown amount of light constituents make the position and thickness of the liquid outer core rather uncertain. (Evidence for a liquid outer core was recently reported by Margot *et al.* 2007.)

4. Energetics and importance of chemistry

The energy source of the convective motions giving rise to the dynamo introduces further divisions into astrophysical dynamos. Jupiter, Saturn and the Sun are luminous enough to drive vigorous convection in their interiors by thermal effects alone. For terrestrial planets the available remnant heat may be only barely sufficient or even insufficient to maintain core convection. In these objects, chemical or compositional driving may be an important contribution to maintaining the convective state of their cores. (See e.g., Stevenson 2003 for numerical estimates.) In the case of the Earth, the most likely candidate is the piling up of light constituents like sulphur at the bottom of the outer core, as iron is freezing out onto the inner core and the light elements are locked out of the solid phase. (Arguments for an analoguous mechanism suggested for Saturn, the "helium rain" now seem to have been weakened by the new results of Stixrude & Jeanloz 2008.)

5. Boundaries and adjacent conducting flows

One important new contribution of numerical simulations was the realization of the crucial role that the choice of boundary conditions play in determining the solutions. Indications for this effect had already been found in kinematic dynamo calculations: e.g. the assumed conductivity of the solid inner core in a geodynamo model affects strongly its capability to maintain a dipole dominated field and the frequency of reversals (Hollerbach & Jones 1993). But the most important effect is due to the thermal boundary conditions. Convection is driven by heat input from below and heat loss from the top of the layer. In terrestrial planets the latter occurs by mantle convection, the efficiency of which is then a fundamental determinant of the behaviour of the dynamo. Indeed, as the timescale of mantle flows is very long compared to those in the dynamo, in addition to the overall mantle convection even the instantaneous convection pattern realized now can have a profound effect on the flow structure in the geodynamo (Glatzmaier 2002). The most widely mentioned possibility why Venus does not seem to support a dynamo is that Venus' mantle convection is less efficient, so its heat flux can be transported conductively throughout the core, and no convective instability arises. The resulting slower cooling of the planet may have the additional result that no inner core has solidified yet, depriving Venus even of the alternative energy source for convection that an inverse molecular weight gradient could provide (Stevenson 2003).

The importance of appropriate boundary conditions is now also recognized in simulations of the solar convective zone. Despite repeated attempts with ever more powerful computers, the hydrodynamical simulations of the new millennium had until very recently not been able to reproduce the observed solar internal rotation profile, the resulting isorotation surfaces being cylindrical rather than conical (e.g., Brun & Toomre 2002). The most recent state-of-the-art simulation (Miesch et al. 2008), however, brought a breakthrough in this respect. The breakthrough is not due to the higher computing power available, but to the introduction of a small ($\sim$0.1 %) pole-equator temperature difference at the bottom boundary. While at first sight ad hoc, the introduction of this difference had the beneficial effect of leading to a realistic differential rotation and a realistic meridional circulation at the same time, raising hopes that this may indeed be the right way out of the dilemma. This is so despite the fact that this solution of the problem may strike one as "æsthetically" less attractive, as the conical isorotation surfaces are attributed to a chance cancellation of two opposing effects, rather than to some deeper physical motive.

A further important factor that may fundamentally affect the behaviour of a dynamo is the presence or absence of conducting stable fluid layers in its vicinity. Such a layer, the solar tachocline, is certainly present below the solar convective zone, and is currently thought to play a major role in the global solar dynamo (Forgács-Dajka & Petrovay 2002; Forgács-Dajka 2003). The strong stratification of the solar convective zone discussed above, has long been expected to give rise to pronounced transport effects or "pumping" mechanisms (see Petrovay 1994 for a comprehensive treatment of these effects). These pumping processes can efficiently remove most of the large-scale magnetic flux from the turbulent region, provided there is an adjacent region with high but finite conductivity that can receive and store this flux. Following earlier simpler numerical experiments, recent MHD numerical simulations have indeed shown the pumping of large scale magnetic flux from the convective zone into the tachocline below, where it forms strong coherent toroidal fields (Browning et al. 2006).

Similar conductive layers adjacent to the turbulent dynamo region may be present in some planetary interiors. Even in the case Earth it has been hypothesized that, as a result of convection, lighter elements ultimately pile up below the the core-mantle boundary, resulting in a stably stratified sublayer within the outer core (Whaler 1980). The effect of such layers on the dynamo has not been studied extensively, but available results indicate that their impact on the observed field may be very important (Schubert et al. 2004).

6. Regime of operation

6.1. *Dimensional and nondimensional parameters*

The number of parameters uniquely determining a dynamo is quite limited. The geometry is a spherical shell between radii r_{in} and r_{out}, with thickness $d = r_{out} - r_{in}$ and relative thickness parameter $x = d/r_{out}$. The shell, rotating with angular velocity Ω, is filled with material of density ρ, characterized by momentum, heat and magnetic diffusivities ν, κ and η, respectively. Finally, convection in the shell is driven by the buoyancy flux F fed in at the bottom of the shell (Olson & Christensen 2006). An internal heating ϵ may be added to the list for cases where volumetric heat loss from secular cooling or radioactive decay is significant.

The number of relevant parameters can be reduced further, realizing that the role of some of the variables is just to set characteristic scales. The length scale is clearly set by d and the mass scale by ρ. For the timescale it has been traditional for certain theoretical considerations to use the resistive time d^2/η in dynamo theory. However, it was recently pointed out by Christensen & Aubert (2006) that for rapid rotators $1/\Omega$ offers a more relevant scaling. Nondimensionalizing all parameters by these scales (and ignoring ϵ) we are left with only 4 nondimensional parameters. Following Olson & Christensen (2006) the effective buoyant Rayleigh number can be defined as $\mathrm{Ra_b} = F/(1-x)d^2\Omega^3$. The nondimensional measures of the diffusivities are the Ekman numbers

$$\mathrm{Ek} = \nu/\Omega d^2 \qquad \mathrm{Ek}_\kappa = \kappa/\Omega d^2 \qquad \mathrm{Ek_m} = \eta/\Omega d^2 \qquad (1)$$

Instead of the three Ekman numbers, one Ekman number and two Prandtl numbers (i.e. diffusivity ratios) are more commonly used:

$$\mathrm{Pr} = \nu/\kappa \qquad \mathrm{Pr_m} = \nu/\eta \qquad (2)$$

The specified parameters then, in principle, determine the solution, i.e. the resulting turbulent flow field and magnetic field. These can be characterized by their respective amplitudes v and B, as well as their typical length scale l—say correlation length, integral scale or similar. (For simplicity a single length scale is assumed for both variables, ignoring anisotropy). In fact, these fiducial scales of the solution may even be estimated on dimensional grounds without actually solving the dynamo equations (Starchenko & Jones 2002). From l and v, turbulent diffusion coefficients can be estimated using the usual *Austausch* recipe $\sim lv$. For the Sun, Earth and gas giants such estimates, summarized in Table 1, are in rough agreement with both observations and simulation results.

As nondimensional measures of v and B we introduce the Rossby and Lorentz numbers:

$$\mathrm{Ro} = v/\Omega d \qquad \mathrm{Lo} = v_A/\Omega d \qquad (3)$$

where $v_A = B/(\rho\mu)^{1/2}$ is the Alfvén speed. It arguably makes more sense to use l instead of d in the above definitions, which leads us to "local" versions of these numbers:

$$\mathrm{Ro_l} = v/\Omega l \qquad \mathrm{Lo_l} = v_A/\Omega l \qquad (4)$$

Were we to adopt the diffusive time scales d^2/ν and d^2/η instead of $1/\Omega$, the nondimensional measures of v and B would be the kinetic and magnetic Reynolds numbers:

$$\mathrm{Re} = dv/\nu \qquad \mathrm{Re_m} = dv/\eta \qquad (5)$$

(Again, using a "local" length parameter l instead of d may be more relevant.)

Table 1. Estimated values of the flow velocity, length scale and turbulent diffusivity

Dynamo	v [m/s]	l [km]	$D_t \sim lv$ [m^2/s]
Sun (deep SCZ):	20	10^5	10^9
Earth:	$3 \cdot 10^{-4}$	100	20
Jupiter	10^{-3}–10^{-2}	10^3	10^3

Let us now consider what the characteristic values of these nondimensional parameters are for astrophysical dynamos and for numerical simulations.

6.2. *Ekman numbers (i.e. diffusivities)*

Ekman numbers are extremely small in astrophysical dynamos, far below the range 10^{-6}–10^{-3} accessible to current numerical simulations. The main obstacle to the further reduction of Ek is that by choosing a realistic Rayleigh number, we fix the input of kinetic and magnetic energy into the system. In a stationary state energy must be dissipated at the same rate as it is fed in. The viscous dissipation rate is $\sim \nu/\lambda^2$ where λ is the smallest resolved scale —so ν cannot be reduced without also increasing the spatial resolution. The situation is similar for resistive (Ohmic) dissipation.

6.3. *Reynolds numbers*

It is interesting to note that the magnetic Reynolds number $\mathrm{Re_m}$ is the only parameter involving a diffusivity whose actual value can be used in current numerical simulations for some planets (including the Earth). Rossby numbers, in contrast, are invariably intractably small in planetary dynamos. The situation in the Sun is the reverse: Rossby numbers are moderate (only slightly below unity in the deep convective zone and actually quite high in shallow layers). Reynolds numbers, however, are all exteremely high in the solar plasma, even in the relatively cool photosphere.

The significance of $\mathrm{Re_m}$ consists in the existence of a critical value $\mathrm{Re_{m,crit}}$ below which no dynamo action is possible. Planetary dynamo simulations have led to the surprising result that $\mathrm{Re_{m,crit}}$ has a universal value of about 40, independently of the other parameters, specifically of the Prandtl numbers (Christensen & Aubert 2006). This is a surprising result as the expectation in turbulence theory, confirmed in numerical simulations of small-scale dynamos, has been that $\mathrm{Re_{m,crit}}$ should be a sensitive function of $\mathrm{Pr_m}$ (Boldyrev & Cattaneo 2004). The solution of this apparent contradiction is not known.

Elementary estimates show that the convective velocity resulting even for very slightly supercritical Rayleigh numbers results in magnetic Reynolds numbers well above $\mathrm{Re_{m,crit}}$. This implies that the conditions for convection and for dynamo action may be virtually identical in astrophysical fluid bodies with a high conductivity (Stevenson 2003). We have already seen that the apparent lack of dynamo action in Venus is attributed to the lack of convection in its core. Planets with very poorly conductive fluid layers, however, may obviously be convecting without supporting a dynamo. A case in point may be the water giants where the dynamo sustaining layer may potentially only extend to a thin sublayer of the convecting water mantle and/or be only slightly supercritical, unable to generate fields strong enough to reach magnetostrophic equilibrium.

6.4. *Rossby number (i.e. velocity amplitude)*

An important issue is how the Rossby and Lorentz numbers (i.e. nondimensional velocity and magnetic field amplitudes) scale with the input parameters of the dynamo problem. The significance of this is twofold. Firstly, as current numerical simulations cannot directly access the parameter range relevant for astrophysical dynamos, such scaling laws can be used to extrapolate their results into the physically interesting domain. Secondly, it is of interest to compare the scaling laws derived from simulation results to those predicted by physical considerations based on the concept of magnetostrophic equilibrium.

For the Rossby number, the scaling law extracted from simulations vs. the law theoretically expected for MAG balance are:

$$\mathrm{Ro} = 0.85\,\mathrm{Ra_b^{0.4}} \qquad \mathrm{vs.} \qquad \mathrm{Ro} \sim \mathrm{Ra_b^{1/2}} \qquad (6)$$

Geometrical factors invoked to explain the slight discrepancy in the values of the exponents do not seem to be capable of explaining it (Christensen & Aubert 2006). However, given the above mentioned doubt that the high values of the diffusivities employed may cast on all simulation results, one may also take the point of view that these two laws are actually in fairly good agreement, and no further explanation is needed. (Note that in this case the theoretical value 0.5 must be deemed the more reliable one.)

6.5. *Lorentz number (i.e. magnetic field strength)*

The scaling of the Lorentz number extracted from dipole-dominated dynamo simulations with no internal heating (Olson & Christensen 2006) viz. the MAC scaling following from the assumption of unit Elsässer number (Starchenko & Jones 2002) are

$$\mathrm{Lo_D} \propto \mathrm{Ra_b}^{1/3} \qquad \mathrm{Lo} \propto \mathrm{Ro_l}^{1/2} \qquad (7)$$

Assuming that the local Rossby number scales similarly to (6), the latter relation may also be turned into a scaling with Ra , but its exponent (0.2–0.25) is even more discrepant from the value yielded by the simulations than in the previous case.

7. Predictive value

In solar dynamo theory there is a widely held belief that predictions for at least the next activity cycle should be possible. Dozens of methods have been proposed for this. The relatively most successful ones are apparently those based on some measure of solar activity or magnetism at the onset of the new cycle. Recently Cameron & Schüssler (2007) convincingly argued that what stands in the background of all such methods is just the well known Waldmeier effect, relating a cycle's amplitude to the rate of rise of activity towards the maximum. As solar cycles are known to overlap by ∼1–2 years, the faster rise of a stronger cycle will result in an earlier epoch for the minimum— so it is natural to expect that at this time, polar magnetic fields and activity indices have not yet decayed to such low values as they reach during minima preceding weak cycles. Indeed, Cameron & Schüssler (2007) demonstrate that a very simple predictor, the value of the sunspot number three years before the minimum performs embarrassingly well (correlation coefficient 0.95) when it comes to predicting the amplitude of the next maximum, but this good performance is fully explained by the effect described above.

This is not to say that it would be pointless to rely on dynamo models to find a physical basis for activity predictions. Firstly, the Waldmeier effect itself is certainly due to some, as yet unclear, aspect of the dynamo. Second, there are grounds to assume that, at least in certain types of dynamos, there really *is* a physical relationship between high-latitude magnetic fields during the minimum and the amplitude of the next maximum (Choudhuri *et al.* 2007). Either way, the persistently low activity during the present solar minimum seems to be a strong indication that cycle 24 will be a rather weak one, in contrast to widely publicized claims based on a certain class of solar dynamo models (Dikpati & Gilman 2006).

Owing to the much longer timescales of planetary dynamos, the possibility for testable temporal predictions is limited. Planetary dynamo models can partly make up for this by the availability of several instances of known planetary dynamos. These inlcude 7 active dynamos (Mercury, Earth, Ganymede and the four giant planets) and one extinct dynamo (Mars), with the remaining planets also providing some important constraints precisely by apparently *not* supporting a dynamo. A testbed of a certain type of modelling approach, then, may be whether it is capable to provide a unified explanatory scheme for

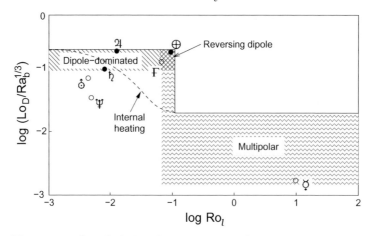

Figure 1. The suggested unified classification scheme of planetary dynamos based on the scalings of Olson & Christensen (2006).

all these systems. Recent results suggest that planetary dynamo simulations may indeed provide such an explanatory scheme.

Collecting and homogenizing the results of hundreds of geodynamo simulations, Olson & Christensen (2006) find an interesting scaling (or rather, "non-scaling") behaviour of the amplitude of the dipolar component of the resulting magnetic field. As mentioned above, considering only dipole-dominated dynamos without internal heating the scaling relationship (7) results for the amplitude of the dipole component. This scaling, however, is not valid for all dynamo models. This is borne out in Figure 1 where the ordinate essentially shows the ratio of the two sides of equation (7). The dipole-dominated cases in this plot will clearly lie along a horizontal line drawn at the ordinate value 0.5, corresponding to the coefficient in equation (7). This is indeed the case for dynamos where the Rossby number is low, i.e. where the driving is relatively weak. However, for Rossby numbers above a critical value of order 0.1, the amplitude of the dipolar component suddenly drops and the solution becomes multipolar. In the case of internally heated models the situation is similar, but the transition between the two regimes is more gradual.

The critical role of the Rossby number in this respect indicates that the underlying cause of the eventual collapse of the dynamo field is the increasing importance of inertial forces in the equation of motion (measured by the Rossby number). This leads to a breakup of the relatively regular columnar structures dominating rapidly rotating convection—the detailed mechanism of this will be discussed further in the next section.

A further interesting finding is that dynamos lying near the top of the Rossby number range of dipolar solutions are generally dipole-dominated, but occasionally undergo excursions and reversals. In a turbulent system this type of behaviour is rather plausible, given that parameters like Ro are expected to fluctuate, and such fluctuations may occasionally take the system into multipolar regime, causing the dipolar field to collapse, and then reform once fluctuations have taken the system back to the dipolar regime.

Olson & Christensen (2006) also attempt to place individual planetary dynamos on the phase plane of Fig. 1, based on known empirical constraints and theoretical considerations. The resulting distribution seems to provide an impressively comprehensive classification system for planetary dynamos. Gas giants lie safely in the dipole-dominated regime, suggesting that their dynamos are in a magnetostrophic state maintaining a non-reversing dipolar field. Earth is found near the limit of the dipole-dominated regime, just where a dipolar dynamo known to be subject to occasional reversals is expected to be.

A somewhat surprising conclusion of this scheme is that the planetary dynamo most similar to Earth's in its behaviour may be Ganymede's, also expected to maintain a reversing dipole. The fact that the dipolar magnetic field amplitude of the water giants is significantly less than relations (7) predict, either be due to a significant amount of internal heating or to the low resistivity in their interiors, which prevents them from reaching magnetostrophic equilibrium. Finally, Fig. 1 would seems to suggest that, despite contrary opinions, Mercury should be expected to sustain a multipolar magnetic field structure.

Some caveats regarding Fig. 1 are in order. All the simulations upon which the figure is based were run with a shell thickness parameter $x = 0.65$, appropriate for the geodynamo. For planets with significantly different shell geometries, especially for those with thin shells (Mercury?; the water giants?), results may prove to be significantly different.

Note also that the parameter on the abscissa of Fig. 1 is actually a rather particular kind of Rossby number, involving one particular local turbulence timescale, defined in a way similar to the the the Taylor microscale. As in current numerical simulations the largest and smallest scales are normally separated by no more than one order of magnitude, this fine distinction between different turbulent scales is not important. But when it comes to application to actual planets, the different turbulent timescales differ by many orders of magnitude, so choosing the right scale is critical. Christensen & Aubert (2006) provide good arguments for the choice of the particular form of Ro_l used in Fig. 1, but it is still somewhat disconcerting that combining large-scale turbulent velocities with a much smaller length scale, the Rossby number loses its widely used physical interpretation as the ratio of rotational and turbulent turnover timescales. In addition, our limited knowledge on the spectrum of magnetostrophic turbulence (Zhang & Schubert 2000, Nataf & Gagnière 2008) makes it hard to derive a reliable value for the Taylor microscale in planetary dynamos: the Kolmogorov spectrum is just a (not too educated) guess.

8. Characteristic flow and field patterns

It is well known that solar activity phenomena appear collectively, in the form of *active regions*. These regions are the solar atmospheric manifestations of large azimuthally oriented magnetic flux loops emerging through the convective zone. Indeed, the development of extremely successful detailed models of this emergence process, first in the thin fluxtube approximation, then in 3D numerical simulations, was probably the most spectacular success story of solar dynamo theory in the last few decades.

In contrast to our familiarity with the basic magnetic structures determining solar activity, nothing about analoguous structures in planetary dynamos had been known until very recently. This situation has spectacularly changed with the development of new visualization and geometrical analysis techniques, which showed that, just like in the Sun, the magnetic field in planetary dynamos is highly intermittent (25 % of the magnetic energy residing in just 1.6 % of the volume), and led to the recognition of a variety of field and flow structures (Aubert *et al.* 2008). The two main classes of such structures are *magnetic cyclones/anticyclones* and *magnetic upwellings*.

Magnetic cyclones and anticyclones are the MHD equivalent of Busse's columnar structures, known to dominate rapidly rotating hydrodynamic convection. The simulations indicate that magnetic anticyclones, in particular, play a key role in generating and maintaining a dipole dominated magnetic field in the upper part of the dynamo layer and above. Near the bottom of the shell the field is invariably multipolar, but the thermal wind-driven upflow in the axis of a magnetic anticyclone amplifies the upward convected fields by stretching, so that at the top of the shell a predominantly dipolar field results.

Magnetic upwellings are buoyancy-driven upflows rising nearly radially through the shell owing to their high velocity. They come in two varieties: polar upwellings, limited to the interior of the tangent cylinder of the inner core, emerge more or less in parallel with the cyclonic/anticyclonic structures, while equatorial upwellings cut through those structures, disrupting their integrity. As a result, equatorial upwellings (distant cousines of the emerging magnetic flux loops in the Sun) are capable of interfering with the mainte-nance of an organized dipolar field by the magnetic anticyclones. As the buoyant driving, and in consequence the turbulent velocity (i.e. the Rossby number) is increased, the num-ber and vigour of upwellings increases, until they can occasionally disrupt the dipolar magnetic field maintained by the magnetic anticyclones. Following such interruptions the dipole may be reformed with a polarity parallel or opposite to its previous polarity: such events correspond to magnetic excursions and reversals, respectively. Finally, with the further increase of Ro the equatorial magnetic upflows permanently discapacitate the anticyclones, and the dominant dipolar field structure collapses. This is the mechanism in the background of the characteristic Ro-dependence of dynamo configurations shown in Fig. 1 and discussed in the previous section. Thus, dynamo studies now seem to have elucidated, at least on a qualitative level, the basic mechanisms underlying the most characteristic phenomena of both solar and planetary dynamos.

Acknowledgements

Support by the Hungarian Science Research Fund (OTKA) under grant no. K67746 and by the EC SOLAIRE Network (MTRN-CT-2006-035484) is gratefully acknowledged.

References

Aubert, J., Aurnou, J., & Wicht, J. 2008, Geophys. J. Intern. 172, 945
Boldyrev, S. & Cattaneo, F. 2004, Phys. Rev. Lett. 92(14), 144501
Browning, M. K., Miesch, M. S., Brun, A. S., & Toomre, J. 2006, ApJ Lett. 648, L157
Brun, A. S. & Toomre, J. 2002, ApJ 570, 865
Cameron, R. & Schüssler, M. 2007, ApJ 659, 801
Charbonneau, P. 2005, Living Rev. Sol. Phys. 2, 2
Chatterjee, P., Nandy, D., & Choudhuri, A. R. 2004, A&A 427, 1019
Choudhuri, A. R. 2008, Adv. Space Res. 41, 868
Choudhuri, A. R., Chatterjee, P., & Jiang, J. 2007, Phys. Rev. Lett. 98(13), 131103
Christensen, U. R. & Aubert, J. 2006, Geophys. J. Intern. 166, 97
Dikpati, M. & Gilman, P. A. 2006, ApJ 649, 498
Forgács-Dajka, E. 2003, A&A 413, 1143
Forgács-Dajka, E. & Petrovay, K. 2002 A&A 389, 629
Glatzmaier, G. 2002, Ann. Rev. Earth Pl. Sci. 30, 237
Gubbins, D. 2008, Geophys. J. Intern. 173, 79
Hollerbach, R. & Jones, C. A. 1993, Nature 365, 541
Margot, J. L., Peale, S. J., Jurgens, R. F., Slade, M. A., & Holin, I. V. 2007, Science 316, 710
Miesch, M. S., Brun, A. S., DeRosa, M. L., & Toomre, J. 2008, ApJ 673, 557
Nataf, H.-C. & Gagnière, N. 2008, ArXiv e-prints, 805
Olson, P. & Christensen, U. R. 2006, Earth Plan. Sci. Lett. 250, 561
Petrovay, K. 1994, in R. J. Rutten, C. J. Schrijver (eds.), Solar Surface Magnetism, NATO ASI
    Series C433, Kluwer, Dordrecht, p. 415
Petrovay, K. 2000, in The Solar Cycle and Terrestrial Climate, ESA Publ. SP-463, p. 3
Rüdiger, G. & Hollerbach, R. 2004, The Magnetic Universe, Wiley-VCH, Weinheim
Schubert, G., Chan, K. H., Liao, X., & Zhang, K. 2004, Icarus 172, 305
Starchenko, S. V. & Jones, C. A. 2002, Icarus 157, 426
Stevenson, D. J. 2003, Earth Pl. Sci. Lett. 208, 1

Stixrude, L. & Jeanloz, R. 2008, Proc. Natl. Acad. Sci. USA 105, 11071
Wang, Y.-M., Sheeley, N. R., & Nash, A. G. 1991, ApJ 383, 431
Whaler, K. A. 1980, Nature 287, 528
Zhang, K. & Schubert 2000, Ann. Rev. Earth Pl. Sci. 32, 409

Discussion

GIRISH: There are studies on the possible effects of planets on solar activity changes or the solar dynamo. From geophysical observations we know Earth's rotation is slowed down over geological time scales by the moon's tidal forces. Can we expect an effect of tidal forces on the geodynamo?

PETROVAY: Tidal forcing/mechanical mixing may indeed have been important in some astrophysical dynamos such as the early lunar dynamo. There has been little effort put into building detailed numerical models for it yet.

Universal Heliophysical Processes
Proceedings IAU Symposium No. 257, 2008
N. Gopalswamy & D.F. Webb, eds.

Solar oscillations

H. M. Antia

Tata Institute of Fundamental Research, Homi Bhabha Road, Mumbai 400005, India
email: antia@tifr.res.in

Abstract. Study of solar oscillations has provided us detailed information about solar structure and dynamics. These in turn provide a test of theories of stellar structure and evolution as well as theories of angular momentum transfer and dynamo. Some of these results about the solar structure and its implication on the recent revision of heavy element abundances are described. Apart from these the solar cycle variations in the rotation rate and its gradients are also discussed.

Keywords. Sun: abundances, Sun: helioseismology, Sun: interior, Sun: oscillations, Sun: rotation

1. Introduction

The Sun oscillates in a set of well defined discrete frequencies, which can be observed at the solar surface. These frequencies are determined by the internal structure and dynamics and hence they contain information about the internal structure and dynamics. During the last solar cycle, detailed observations of solar oscillations have provided a unique opportunity to study the solar structure and dynamics (see Christensen-Dalsgaard 2002 for a review). A detailed seismic study of solar structure led to significant improvements in theoretical solar models and with recent input physics the solar model agreed well with the seismically inferred solar structure (Christensen-Dalsgaard *et al.* 1996; Gough *et al.* 1996). The improvements in solar models include the improvements in input physics like, the equation of state, opacities and nuclear reaction rates as well as the inclusion of diffusion of helium and heavy elements in the solar interior (Christensen-Dalsgaard *et al.* 1993). As a result of these developments it became clear that the discrepancy between the observed flux of solar neutrinos and those calculated in a solar model should be due to neutrino oscillations. In fact, the discrepancy in solar neutrino fluxes was another motivation for extensive tests of solar models, which also contributed to some of these improvements. The neutrino oscillations are now confirmed by measurements from the Sudbury Neutrino Observatory (Ahmad *et al.* 2002). However, this excellent agreement between the standard solar model and seismically inferred structure was spoilt when Asplund *et al.* (2004) found that the Oxygen abundance in the solar photosphere should be reduced by a factor of 1.5. This has led to a crisis in solar models, which would be discussed in Section 2.

Apart from solar structure the frequencies of solar oscillations also give us information about the rotation rate in the solar interior (e.g., Thompson *et al.* 1996; Schou *et al.* 1998). These results established that the differential rotation observed at the solar surface continues through the solar convection zone, while most of the radiative interior has nearly constant rotation rate. There is a sharp transition between these two regions near the base of the convection zone and this transition region has been named tachocline (Spiegel & Zahn 1992). Because of the strong shear in the tachocline region, it is the favoured location for the operation of the solar dynamo. Currently, the major sources of seismic data are (1) the Global Oscillation Network Group (GONG) which is a network of six sites around the world (Hill *et al.* 1996) that is operating since May 1995 and (2) the Michelson Doppler Imager (MDI) instrument (Scherrer *et al.* 1995) on board the SOHO satellite which is operating since May 1996. With the availability of seismic data for the last 13 years it has become possible to study the temporal variations over the solar cycle. In particular, the inferred temporal variations in the solar rotation rate can provide a crucial test of dynamo models.

2. Solar structure and photospheric abundances

Fig. 1 shows the relative differences in sound speed and density between a standard solar model of Christensen-Dalsgaard *et al.* (1996) and the Sun as inferred from seismic data. This solar model was constructed using diffusion of helium and heavy elements below the convection zone and used the heavy element abundances from Grevesse & Noels (1993). It can be seen that there is very good agreement and the difference in sound speed is generally less than 0.1%. The two major regions of discrepancy are near the surface and near the base of the convection zone. There is some difference in the core also, but in that region the errors in inversions are somewhat large and it is not clear if the difference is indeed significant, particularly when systematic errors in inversion are considered. The sharp peak near the base of the convection zone ($r = 0.713R_\odot$) has been attributed to mixing in the tachocline region, and solar models (e.g., Brun *et al.* 1999) which include some mixing in this region do not show this peak. The dip near the surface is most likely to be due to improper estimate of solar radius. If the radial distance, r in the model is scaled by a factor of 1.00018 (i.e., $r/1.00018$) before taking differences then this dip is substantially reduced as seen by the dashed curve in Fig. 1. However, a solar model with a different radius also shows similar dip and hence it is not due to use of incorrect solar radius, but rather because of the uncertainties in treatment of outer layers, the position of solar surface is not correctly estimated in a solar model. The generally adopted value of the solar radius (695.99 Mm) from observations should refer to a layer in the atmosphere about 500 km above the layer with unit optical depth which is normally used as definition of solar surface in a solar model (Brown & Christensen-Dalsgaard 1998). Thus a different value of solar radius should be used in a solar model. This value can be calibrated using frequencies of f-modes (Schou *et al.* 1997; Antia 1998) and gives a value 200 to 300 km less than the standard value. This is about 200 km larger than the value that is expected and this discrepancy is the one which gives the dip in Fig. 1. Of course, the use of the revised value of radius in a solar model will not change Fig. 1 substantially as the dip is due to inadequacy in solar models near the surface and cannot be eliminated by changing the solar radius.

Fig. 1 shows the differences with respect to a solar model with old chemical composition, which has been revised since then. Abundances of most heavy elements are determined spectroscopically and requires a model of solar atmosphere. Traditionally, these solar atmospheric models are 1 dimensional as they assume spherical symmetry. The effect of turbulence is incorporated through ad hoc parameters like micro and macroturbulence. With increase in computing power it has become possible to make limited 3D models of solar atmosphere which attempt to include turbulence over a limited range of length scales in the calculations. Using such models Asplund *et al.* (2004) calculated the abundances of oxygen and found that it needs to be reduced by a factor of almost 1.5. Similarly, abundances of many other elements were also reduced by similar factors. As a result of these reduction the value of Z/X in the Sun reduced from 0.023 (Grevesse & Sauval 1998, henceforth GS98) to 0.0165 (Asplund *et al.* 2005, henceforth AGS05). As a result the opacity in the solar interior is substantially reduced and the structure of

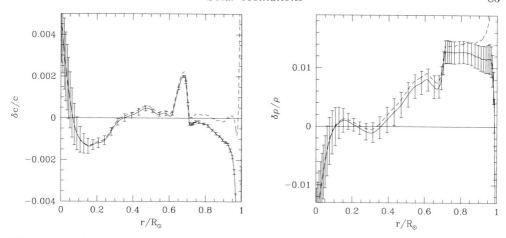

Figure 1. Relative differences in sound speed and density between the standard solar model of Christensen-Dalsgaard *et al.* (1996) and the Sun as inferred from seismic data from GONG. The dashed line shows the difference after scaling the model radius by a factor of 1.00018 before taking the difference.

resulting solar models is quite different from the seismically inferred structure (Bahcall & Pinsonneault 2004; Basu & Antia 2004; Turck-Chiéze *et al.* 2004). Apart from the increased difference in structure variables like the sound speed and density, the depth of the convection zone as well as the helium abundance in the convection zone, reduces substantially below the seismically measured values. Bahcall *et al.* (2006) have done a detailed Monte-Carlo simulations by constructing solar models where various input parameters (including abundances) are randomly varied within the estimated errors to find that solar models with mean heavy element abundances from GS98 are consistent with seismic constraints on the depth of the convection zone and its helium abundance. While models with abundances from AGS05 are not consistent with these constraints. In fact, long before the current revision in Z, low Z solar models were postulated to lower the solar neutrino flux, but such models were ruled out from seismic constraints (e.g., Christensen-Dalsgaard & Gough 1980). Similarly, from a detailed study of the depth of the convection zone, Basu & Antia (1997) concluded that $Z = 0.0245 \pm 0.0008$ assuming that the OPAL opacities were valid. This study did not include the effect of varying mixture of heavy elements, but that effect is found to be small (Basu & Antia 2004).

Numerous attempts have been made to modify the solar models to restore the good agreement between solar models and seismic data. These include increase in opacities (Basu & Antia 2004; Bahcall *et al.* 2004, 2005a), increasing the rate of diffusion of heavy elements below the base of the convection zone (e.g., Basu & Antia 2004; Guzik *et al.* 2005), accretion of low Z material during solar evolution (Castro *et al.* 2007), increasing the abundance of Ne (Antia & Basu 2005; Bahcall *et al.* 2005b). However, none of these attempts have been successful, in the sense that required variations are beyond reasonable estimates and even if a combination of these effects is considered, the resulting solar model doesn't match the seismic structure in full details. For example, the required opacity increase is by 11–25% over the OPAL values (Rogers & Iglesias 1992; Iglesias & Rogers 1996), while recent independent computation of opacities by OP project (Badnell *et al.* 2005) gives a difference of less than 2% near the base of the convection zone. Thus it is unlikely that opacities can be increased by the required amount.

Since the main cause of discrepancy in solar models is the reduction in opacities, Antia & Basu (2005) examined the effect of abundances of different elements on opacities

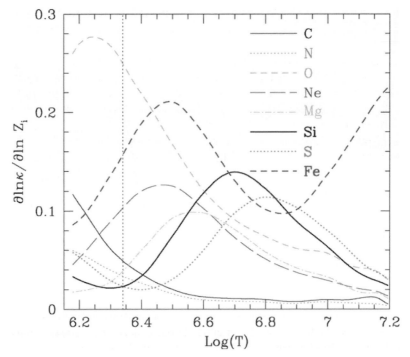

Figure 2. Logarithmic derivative of opacity with respect to abundances of individual heavy elements in a solar model. The OPAL opacity tables are used to calculate the derivatives. The dashed vertical line marks the position of the base of the convection zone.

in the radiative interior. Fig. 2 shows the logarithmic derivatives of opacity with respect to abundances of some of the dominant heavy elements. Apart from oxygen, which plays the most dominant role in opacity near the base of the convection zone, Iron and Neon are also important contributors. Other elements were not found to make significant contributions to opacity in the required region. Of these the Neon abundance in the photosphere cannot be determined directly as Neon doesn't form any line in the photosphere. Thus Neon abundance is determined from coronal lines or solar wind which generally determines the ratio of Ne/O abundances. Hence a reduction in O abundance automatically reduced the Ne abundance. It is well-known that the abundances in the corona/ solar wind are not the same as photospheric abundances as there is the well-known effect of First Ionisation Potential (FIP), the mechanism for which is not understood. Elements with high FIP are known to be depleted in corona or solar wind. Since O and Ne both have relatively high FIP their relative abundance is not expected to be affected by this effect, but that is merely an assumption as the FIP of Ne is almost a factor of 2 higher than that of O and there are no other elements with similar FIP as Ne, whose abundances are independently known for calibrating the FIP effect. Further, it is well known that abundance of helium, which also has high FIP and doesn't form lines in photosphere, was underestimated through similar procedure (Anders & Grevesse 1989). The helium abundance was ultimately determined using seismic data. While the coronal or solar wind measurements give the abundance ratios, for solar models we need absolute value of abundances and it may not be possible to find abundances that are consistent with all abundance ratios. For example, the Ne abundance as determined from Ne/O ratio doesn't agree with that from Ne/Mg ratio after compensating for FIP effect (Feldman & Widing 2003). Considering all these uncertainties, it was suggested that Ne abundance may be increased to compensate for a reduction in oxygen abundance.

It can be easily estimated that in order to compensate for a reduction in oxygen abundance by a factor of 1.5 the neon abundance needs to be increased by a factor of 4 to restore the opacity near the base of the convection zone. Such an increase in neon abundance is clearly unacceptable, and even after that the structure of solar model in the core will be significantly different (Bahcall *et al.* 2005b). However, if the CNO abundances are increased by 1σ of their respective values determined by AGS05, then the required increase in neon abundance is about a factor of 2.5, which is of the same order as the difference between the GS98 and AGS05 values. Soon after this suggestion was made Drake & Testa (2005) measured the Ne/O abundance in nearby stars using Chandra observations to find a value that is a factor of 2.7 higher than that used by AGS05. However, a reanalysis of solar data by Schmelz *et al.* (2005) and Young (2005) found results consistent with AGS05 value and they attributed the higher value found by Drake & Testa to be due to choice of stars with higher activity level. Recently, through a survey of Ne/O coronal abundances in a number of late type stars, Garcia-Alvarez *et al.* (2008) have claimed that Ne/O abundance determined from coronal lines approaches the photospheric value at higher activity levels, thus supporting the higher value of Ne/O determined by Drake & Testa. Many other measurements of Ne abundance in the Sun and other related astrophysical objects have given conflicting results and the issue is not resolved (cf., Basu & Antia 2008).

Since there is considerable uncertainty in spectroscopic determination of heavy element abundances, it is interesting to investigate if it is possible to determine these abundances using seismic techniques similar to those used for determining helium abundance. The helium abundance estimates are obtained from sound speed in the HeII ionisation zone, where the adiabatic index Γ_1 is reduced below its normal value of $5/3$ and the extent of reduction depends on the He abundance. It is often convenient to use the dimensionless gradient of sound speed $W(r) = (1/g)dc^2/dr$, where g is the acceleration due to gravity and c is the sound speed. The function $W(r)$ shows a peak in the HeII ionisation zone which can be calibrated to determine helium abundance (Gough 1984; Basu & Antia 1995). In principle, the same technique can be applied to determination of heavy element abundances, but the main difficulties are that first these abundances (by number) are two orders of magnitude smaller and hence the effect is very small and second the ionisation zones of various elements overlap and it is difficult to isolate the effect of each element. Nevertheless, after detailed study, Antia & Basu (2006) found that it should be possible to determine the total heavy element abundance, Z using this technique and found a value of $Z = 0.0172 \pm 0.002$, which is consistent with GS98 value, but higher than the AGS05 value. This value is also sensitive to the equation of state, but the errorbars in the above estimate also include this effect. This effect is estimated by using various modern equations of state. This provides an independent seismic estimate for Z as other estimates (e.g., Delahaye & Pinsonneault 2006; Chaplin *et al.* 2007) are mainly based on effect of opacity on solar models. Thus all seismic estimates point to higher value of Z that is not consistent with AGS05. The function $W(r)$ can also be used to check if increasing Ne abundance can help to compensate for reduction in O abundance. It may be recalled that this suggestion was based on opacities. It turns out that solar models with AGS05 oxygen abundance are not consistent with observed $W(r)$ even after increasing Ne abundance (Fig. 3). Thus increasing the neon abundance is not likely to solve the problem with solar models in all respects, but if the neon abundance is increased over the GS98 abundances it may help in improving the solar models.

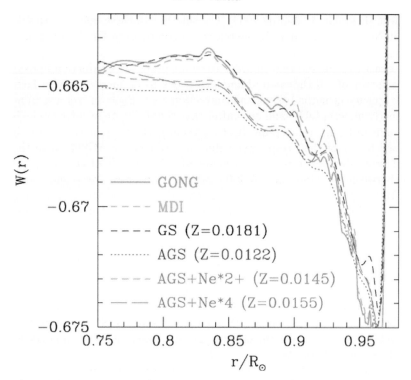

Figure 3. Logarithmic derivative of squared sound speed, $W(r)$ in a few solar models is compared with that inferred from GONG and MDI data. The $Z = 0.0145$ model corresponds to the case where Ne abundance is increased by a factor of 2 and the CNO abundances are increased by 1σ over the AGS05 abundances.

Thus seismic data consistently points to a higher oxygen abundance and if the lower abundances of AGS05 are indeed true, then it will require modifications in almost all input physics, like, opacities, equation of state, diffusion of helium and heavy elements to get the solar models in agreement with seismic data. It may be noted that some recent abundance determinations, e.g., Centeno & Socas-Navarro (2008) using the Ni/O ratio from the blended line and Caffau *et al.* (2008) and Ayres (2008) using an independent 3D atmospheric model also support higher oxygen abundance close to the GS98 value. A part of difference could be due to treatment of non local thermodynamic equilibrium in atmospheric models. More work is clearly required to determine abundances reliably.

3. Rotation in solar interior

The rotational splitting in the frequencies of solar oscillations can be used to infer the rotation rate in the solar interior (e.g., Thompson *et al.* 1996; Schou *et al.* 1998). Such studies give the rotation rate as a function of radius and latitude over most of the solar interior. The reliability degrades as we move towards the core or high latitudes. Further, these studies only give the north–south symmetric component of the rotation rate. North–south asymmetry in the rotation rate in the outer layers can be studied using local helioseismology but we will not consider that in this review. These results have given us the well-known picture of rotation in the solar interior, where the differential rotation continues through the convection zone and near the base of the convection zone, in the

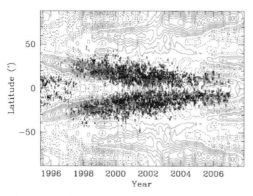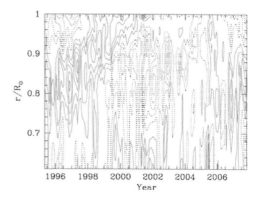

Figure 4. Contours of constant zonal flow velocity, δv_ϕ at $0.98R_\odot$ as a function of time and latitude (Left Panel) and at a latitude of $15°$ as a function of time and radius (Right Panel). The contour spacing is 1 m/s. Solid contours represent positive values, dotted contours show negative value. The zero contour is not shown. In the left panels the points mark the positions of sunspots.

tachocline region, there is a rather sharp transition to nearly uniform rotation in the radiative interior. Apart from this there is also a distinct shear layer near the surface where the rotation rate increases with depth. This shear layer extends to a radius of about $0.95R_\odot$.

With accumulation of GONG and MDI data over the last 13 years it is also possible to study the temporal variations in the rotation rate over the solar cycle. Such studies (e.g., Howe *et al.* 2000; Antia & Basu 2000, 2001; Vorontsov *et al.* 2002) have confirmed the existence of bands of faster and slower than average rotation in the solar interior. This pattern is similar to the torsional oscillations observed at the solar surface (Howard & LaBonte 1986; Ulrich *et al.* 1988) and are referred to as zonal flow. The zonal flow velocity is obtained by subtracting the temporal mean from the rotation rate to get the residual

$$\delta\Omega(r,\theta,t) = \Omega(r,\theta,t) - \langle\Omega(r,\theta,t)\rangle, \tag{1}$$

where θ is the latitude and the angular brackets denote temporal average over the period that the data are available. To account for systematic differences in rotation rate inferred from the GONG and MDI data, the averaging is done separately for GONG and MDI data. The residual $\delta\Omega$ is essentially, the temporally varying component of Ω. The bands of faster or slower than average rotation are found to move towards the equator with time at low latitudes, while at high latitudes they appear to move towards the poles. Further, this pattern penetrates to the base of the convection zone (Vorontsov *et al.* 2002; Basu & Antia 2003). Fig. 4, shows cuts at $r = 0.98R_\odot$ and $\theta = 15°$ in the zonal flow velocity $\delta v_\phi = \delta\Omega r \cos\theta$. It can be seen that at low latitudes, the pattern is rising upwards at a rate of about 1 m s^{-1}. The amplitude of temporal variation in the rotation rate is of the order of a few nHz, which is about 0.5% of the mean rotation rate. While there are significant differences between the temporally averaged rotation rate inferred from GONG and MDI data (Schou *et al.* 2002), these differences essentially cancel when the temporal mean is subtracted while calculating the zonal flow velocity $\delta\Omega$. Thus there is a good agreement between GONG and MDI results for $\delta\Omega$. Below the base of the convection zone, it is difficult to infer any reliable pattern of temporal variation, because of large errors in inversion results. Near the surface, the zonal flow pattern is well correlated to the butterfly diagram representing surface magnetic field (e.g., Snodgrass 1987; Sivaraman *et al.* 2008), as can be seen in Fig. 4, which compares the zonal flow pattern from GONG

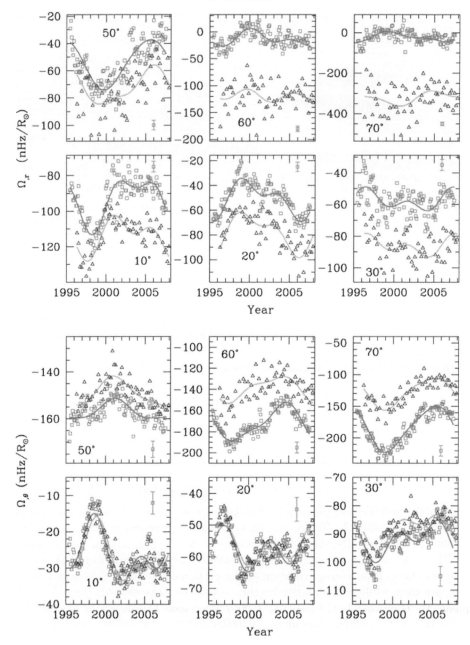

Figure 5. The radial and latitudinal gradients of the rotation rate at $r = 0.95R_\odot$ are shown at a few selected latitudes as a function of time for both GONG (squares) and MDI (triangles) data. For clarity the errorbars are not shown on all points but a sample errorbar is shown in the right corner in each panel.

data with the position of sunspots. The sunspots are generally concentrated in the region around the high latitude edge of the band representing faster than average rotation rate. The zonal flow pattern can also be used to test dynamo models, e.g., Covas *et al.* (2000)

using a mean field dynamo model found variations in rotation rate which qualitatively resemble the observed zonal flow pattern.

The zonal flow pattern in the convection zone is well established (e.g., Howe *et al.* 2006; Antia *et al.* 2008), but for the solar dynamo the gradients of rotation rate are more relevant and hence we need to study these also. The inferred rotation rate can be differentiated to find the gradients, though the errors will be magnified during the process of differentiation. Differentiation of the temporal average of rotation rate, shows that the radial gradient is mainly concentrated in the outer shear layer and in the tachocline, while the latitudinal gradient is of course, confined to the convection zone (Antia *et al.* 2008). It is difficult to determine the radial gradient in the tachocline region reliably, as the thin tachocline region is not adequately resolved in inversion results. Once again there are significant differences in these gradients between the GONG and MDI results, which are largely cancelled while calculating the temporally varying components. Fig. 5 shows the radial ($\Omega_r = \partial\Omega/\partial r$) and latitudinal gradients ($\Omega_\theta = (1/r)\partial\Omega/\partial|\theta|$) at $r = 0.95R_\odot$ at a few selected latitudes. It can be easily seen that temporal variations in these gradients are a sizable fraction of their average values, up to or exceeding 20%. This is much larger than the relative variation of the order of 0.5% in the rotation rate. This substantial variation in the shear should play some role in the solar dynamo.

Temporal variations in these gradients also show bands of higher and lower than average gradients similar to those of zonal flows. These bands are also correlated to the location of sunspots in the butterfly diagram (Antia *et al.* 2008). It is found that the sunspots are predominantly formed in low-latitude regions where the variation of the radial gradient is positive and that of the latitudinal gradient is negative. Since both these gradients are negative, it means that sunspots tend to occur in regions of reduced radial shear but enhanced latitudinal shear.

References

Ahmad, Q. R., *et al.* 2002, *PRL*, 89, 011301

Anders, E. & Grevesse, N. 1989, *Geochimica et Cosmochimica Acta*, 53, 197

Antia, H. M. 1998, *A&A*, 330, 336

Antia, H. M. & Basu, S., 2000, *ApJ*, 541, 442

Antia, H. M. & Basu, S., 2001, *ApJ*, 559, L67

Antia, H. M. & Basu, S. 2005, *ApJ*, 620, L129

Antia, H. M. & Basu, S. 2006, *ApJ*, 644, 1292

Antia, H. M., Basu, S., & Chitre, S. M., 2008, *ApJ*, 681, 680

Asplund, M., Grevesse, N., Sauval, A. J., Allende Prieto, C., & Kiselman, D. 2004, *A&A*, 417, 751 (Erratum in 2005 *A&A*, 435, 339)

Asplund, M., Grevesse, N., & Sauval, A. J. 2005, in *Cosmic Abundances as Records of Stellar Evolution and Nucleosynthesis*, eds., T. G. Barnes, F. N. Bash, ASP Conf. Ser. 336, p. 25 (AGS05)

Ayres, T. R. 2008, *ApJ*, 686, 731

Badnell, N. R., Bautista, M. A., Butler, K., Delahaye, F., Mendoza, C., Palmeri, P., Zeippen, C. J., & Seaton, M. J. 2005, *MNRAS*, 360, 458

Bahcall, J. N. & Pinsonneault, M. H. 2004, *PRL*, 92, 121301

Bahcall, J. N., Serenelli, A. M., & Pinsonneault, M. 2004, *ApJ*, 614, 464

Bahcall, J. N., Basu, S., Pinsonneault, M., & Serenelli, A. M. 2005a, *ApJ*, 618, 1049

Bahcall, J. N., Basu, S., & Serenelli, A. M., 2005b, *ApJ*, 631, 1281

Bahcall, J. N., Serenelli, A. M., & Basu, S. 2006, *ApJS*, 165, 400

Basu, S. & Antia, H. M. 1995, *MNRAS*, 276, 1402

Basu, S. & Antia, H. M. 1997, *MNRAS*, 287, 189

Basu, S. & Antia, H. M., 2003, *ApJ*, 585, 553

Basu, S. & Antia, H. M., 2004, *ApJ*, 606, L85

Basu, S. & Antia, H. M., 2008, *Phys. Rep.*, 457, 217

Brown, T. M. & Christensen-Dalsgaard, J. 1998, *ApJ*, 500, L195

Brun, A. S., Turck-Chièze, S., & Zahn, J.-P. 1999, *ApJ*, 525, 1032

Caffau, E., Ludwig, H.-G., Steffen, M., Ayres, T. R., Bonifacio, P., Cayrel, R., Freytag, B., & Plez, B., 2008, *A&A*, 488, 1031

Castro, M., Vauclair, S., & Richard, O. 2007, *A&A*, 463, 755

Centeno, R. & Socas-Navarro, H. 2008, *ApJ*, 682, L61

Chaplin, W. J., Serenelli, A. M., Basu, S., Elsworth, Y., New, R., & Verner, G. A. 2007, *ApJ*, 670, 872

Christensen-Dalsgaard, J. 2002, *Rev. Mod. Phys.*, 74, 1073

Christensen-Dalsgaard, J. & Gough, D. O. 1980, *Nature* 288, 544

Christensen-Dalsgaard, J., Proffitt, C. R., & Thompson, M. J. 1993, *ApJ*, 403, L75

Christensen-Dalsgaard, J., *et al.*, 1996, *Science* 272, 1286

Covas, E., Tavakol, R., Moss, D., & Tworkowski, A., 2000, *A&A*, 360, L21

Delahaye, F. & Pinsonneault, M. H. 2006, *ApJ*, 649, 529

Drake, J. J. & Testa, P. 2005, *Nature* 436, 525

Feldman, U. & Widing, K. G. 2003, *Space Sci. Rev.*, 107, 665

Garcia-Alvarez, D., Drake, J. J., & Testa, P. 2008, ArXiv:0808.1794

Gough, D. O. 1984, *Mem. Soc. Astron. Ital.*, 55, 13

Gough, D. O., *et al.*, 1996, *Science*, 272, 1296

Grevesse, N. & Noels, A. 1993, in *Origin and Evolution of the Elements*, eds., N. Prantzos, E. Vangioni-Flam, M. Cassè, Cambridge Univ. Press, p. 14

Grevesse, N. & Sauval, A. J. 1998, *Space Sci. Rev.*, 85, 161 (GS98)

Guzik, J. A., Watson, L. S., & Cox, A. N. 2005, *ApJ*, 627, 1049

Hill, F., *et al.*, 1996, *Science* 272, 1292

Howard, R. F. & LaBonte, B. J., 1980, *ApJ*, 239, L33

Howe, R., Christensen-Dalsgaard, J., Hill, F., Komm, R. W., Larsen, R. M., Schou, J., Thompson, M. J., & Toomre, J., 2000, *ApJ*, 533, L163

Howe, R., Rempel, M., Christensen-Dalsgaard, J., Hill, F., Komm, R. W., Larsen, R. M., Schou, J., & Thompson, M. J., 2006, *Astrophys. J.* **649**, 1155

Iglesias, C. A. & Rogers, F. J. 1996, *ApJ*, 464, 943

Rogers, F. J. & Iglesias, C. A. 1992, *ApJS*, 79, 507

Scherrer, P. H., *et al.*, 1995, *Solar Phys.*, 162, 129

Schmelz, J. T., Nasraoui, K., Roames, J. K., Lippner, L. A., & Garst, J. W., 2005, *ApJ*, 634, L197

Schou, J., Kosovichev, A. G., Goode, P. R., & Dziembowski, W. A., 1997, *ApJ*, 489, L197

Schou, J., *et al.*, 1998, *ApJ*, 505, 390

Schou, J., *et al.*, 2002, *ApJ*, 567, 1234

Sivaraman, K. R., Antia, H. M., Chitre, S. M., & Makarova, V. V. 2008, *Solar Phys.*, 251, 149

Snodgrass, H. B., 1987, *Solar Phys.*, 110, 35

Spiegel, E. A. & Zahn, J.-P., 1992, *A&A*, 265, 106

Thompson, M. J., *et al.*, 1996, *Science*, 272, 1300

Turck-Chièze, S., Couvidat, S., Piau, L., Ferguson, J., Lambert, P., Ballot, J., García, R. A., & Nghiem, P., 2004, *PRL*, 93, 211102

Ulrich, R.K., Boyden, J.E., Webster, L., Snodgrass, H.B., Padilla, S.P., Gilman, P., & Shieber, T., 1988, *Solar Phys.*, 117, 291

Vorontsov, S. V., Christensen-Dalsgaard, J., Schou, J., Strakhov, V. N., & Thompson, M. J. 2002, *Science*, 296, 101

Young, P. R. 2005, *A&A*, 444, L45

Discussion

BOCHSLER: I am surprised to see how astrophysicists can spend their time using abundances of high-FIP elements (Ne/O/Ar) as free parameters, despite the fact that these abundances have been reliably determined by many independent methods and their systematics is well understood (e.g. Young 2005, Bochsler 2007 (A&A), Lodden 2008, ApJ.).

ANTIA: Unfortunately, different techniques of measuring elemental abundances give different results and that is why the current controversy has arisen. Hence, we need to check the effect of the range of abundances on other models. If all measurements of solar abundances give the same result this may not be required.

Universal Heliophysical Processes
Proceedings IAU Symposium No. 257, 2008
N. Gopalswamy & D.F. Webb, eds.

A comparison of parameters of 3-minute and 5-minute oscillations in sunspots from synchronous microwave and optical observations

V. E. Abramov-Maximov[1], G. B. Gelfreikh[1], N. I. Kobanov[2] and K. Shibasaki[3]

[1]Central astronomical observatory at Pulkovo, Russian Acad. Sci.,
Pulkovskoe chausse., 65/1, St. Petersburg, 196140, Russia
email: beam@gao.spb.ru

[2]Institute of Solar-terrestrial Physics RAS SB, Lermontov St., 134, Irkutsk, 664033, Russia

[3] Nobeyama Solar Radio Observatory, Minamimaki, Minamisaku, Nagano 384-1305, Japan

Abstract. The observations of 3 and 5 minute oscillations in sunspots present information on propagation of MHD waves in the magnetic tubes of sunspots. We present a comparison of wavelet spectra of radio flux oscillations at $\lambda = 1.76$ cm and oscillations of longitudinal component of the velocity at the chromosphere in sunspot umbra and penumbra in AR 10661 (2004, Aug 18). The radio maps of the Sun obtained with the Nobeyama Radioheliograph were used. The spatial resolution of the radio data was about 10-15 arcsec, and 10 sec cadence was used. On the radio maps sunspot-associated sources were identified and time profiles of their maximum brightness temperatures for each radio source were calculated. Radio data consists of information of oscillations of plasma parameters (in the regions with magnetic field $B = 2000$ G) at the level of the chromosphere-corona transition region. The optical observations were carried out at Sayan observatory. These data included information on longitude component of the magnetic field at the photosphere (line Fe I 6569 Å) and longitudinal component of the velocity at the chromosphere (line $H\alpha$ was used). Comparing the wavelet diagrams covering the same periods of observations at radio and optics showed that some wave trains of time profiles are very similar in both kinds of observations (similar oscillation frequencies and their drifts, variations of amplitudes), however, some significant differences were also registered. The best similarity in optical and radio oscillations was found when the active region (AR) was near the center of the solar disk. The phase shifts between the two kinds of observations reflecting the propagation of MHD waves were also analyzed.

Keywords. Sun: radio radiation, Sun: oscillations, waves, MHD

1. Introduction

The quasi-periodic oscillations are registered practically in all wavelength ranges and in all structures of the solar atmosphere (Bogdan 2000, Fludra 2001, Gelfreikh, Nagovitsyn & Nagovitsyna 2006). Oscillations have periods from seconds to hours or even days. The dominant periods of oscillations for sunspots are 3 minutes and 5 minutes. Most of these oscillations are of unstable nature: both amplitude and frequency are varied with time, often seen as trains of about a dozen of periods. Studing of oscillations can help us to understand such fundamental astrophysical problems as accumulation and release of energy, physics of corona heating and flares origin.

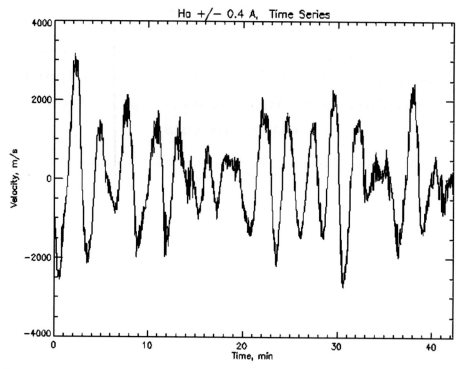

Figure 1. Time-variations of longitudinal component of the chromospheric velocity ($H\alpha$
observations), August 18, 2004, 01:01 - 01:43 UT, AR10661.

The choice of an adequate theory for interpretation of observations includes to essentially complicated problems. For analysis of the three-dimensional structure of the quasi-periodic oscillations it is necessary to use simultaneous observations in a few wavelength ranges. For many years the observations were limited by optical methods representing the phenomena at the level of the photosphere and chromosphere. Modern space techniques made a progress in the study of oscillations at the level of corona and transition region. However, the magnetic tubes of sunspots are not seen on such images like as it is seen from photosphere level. A new progress in such studies was achieved as a result of mapping of the Sun at microwaves. On such maps the sunspots are clearly seen as bright highly polarized small radio sources generated by thermal cyclotron emission of coronal electrons at lower harmonics of its gyro frequency; magnetic field of thousands of Gauss is needed for the effect.

2. Observations

Optical observations were carried out on horizontal solar telescope of the Sayan solar observatory of the Institute of Solar-Terrestrial Physics of Siberian Devision of RAS. The observatory is located at the height of 2000 m above sea level. Variations of the line-of-sight velocity and longitudinal component of the magnetic field (using line FeI 6559 Å) at the photosphere and the longitude velocity component in the chromospheric $H\alpha$ line. The slit of the spectrograph was East-West directed crossing the center of sunspot. The realized space resolution was about one arc sec along the spectrograph slit. The duration of one set of observations was about one hour with cadence of a few seconds. While analyzing the observations we used wavelet spectra with Morle functions of the 6-th order as a base function (Kobanov, Kolobov & Makarchik 2006).

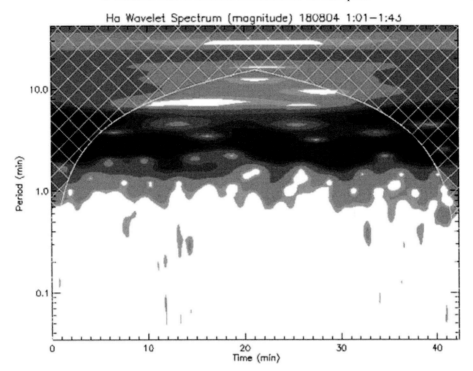

Figure 2. Wavelet spectrum of time-variations of longitudinal component of the chromospheric velocity ($H\alpha$ observations), August 18, 2004, 01:01 - 01:43 UT, AR10661.

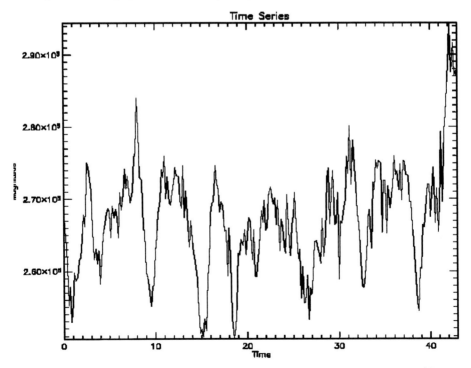

Figure 3. Variations of radio emission of the bright source in the active region AR10661, August 18, 2004, 01:01 - 01:43 UT.

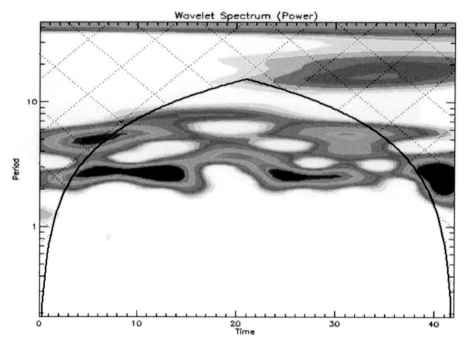

Figure 4. Wavelet spectrum of radio emission of the bright source in the active region
AR10661, August 18, 2004, 01:01 - 01:43 UT.

The time profile of the longitudinal component of the chromospheric velocity is shown
on Figure 1. Wavelet spectrum of time-variations of longitudinal component of the chromospheric velocity is shown on Figure 2.

Data of the Nobeyama Radioheliograph at wavelength 1.76 cm were used to study the
radio sources placed above the same sunspots which were analyzed by optical observations. The radio maps of the whole disk were constructed for the dates of observations
with the time interval of 10 seconds between the maps and 10 sec averaging, both intensity and circular polarization maps were constructed. Having in mind that the sunspot-
associated sources generated thermal gyroresonance emission at the third harmonic of the
electron gyrofrequency are the most sensitive to oscillation processes, we used only the
cases of the presence of such kind of sources, generated for $\lambda = 1.76$ cm in the magnetic
field of $B = 2000$ G at the level of the chromosphere-corona transition region (CCTR).
Identification of the nature of a source is based on its high brightness temperature and
strong polarization. 2D spatial resolution of the radio maps was about of $10 - 15$ arcsec.

The processing of radio data includes:

– imaging NoRH maps (both in total intensity I and circular polarization V) with a
cadence of 10 seconds (averaging 10 seconds),

– interactive extraction of a selected frame from an initial image,

– computation of the new position of the frame corrected for the solar rotation in each
image according to the time of the observation,

– extraction of the frames from all images,

– computations of the time profiles of the total flux as well as the maximum brightness
temperatures over each frame,

– spectral wavelet analysis of time profiles.

Variations of radio emission of the bright source in the AR 10661 (August 18, 2004,
01:01 - 01:43 UT) and wavelet spectrum are shown on Figures 3 and 4.

Detailed comparison of wavelet spectra in radio and optics (Figures 2 and 4) shows two identical trains of the three-minute oscillations with the high level of similarity in the length and periods. At the same time the existence of their shift in time is quite evident. The radio oscillations are seen on 100 ± 20 seconds later than those registered by optical method. This result is in a good agreement with generally accepted idea that 3-minute oscillations demonstrate a MHD wave spreading up in the corona. Its frequency is a result of filtration in the lower underphotospheric region. In our case the velocity of these waves can be found from the observations at the level of the chromosphere and it is equal to 60 $km \cdot sec^{-1}$. The real velocity may be higher. So the delay of 100 sec interpreted as the time of propagation of MHD waves from the chromosphere to corona suggests the height of the CCTR as approximately 6000 km. This value is hardly possible to estimate by other method and height found lies in the reasonable estimations.

3. Conclusions

This work illustrates a possible progress in the new approach in study of quasi-periodic physical processes in the plasma structures of the active regions, especially sunspots, by comparison of parameters of the observed oscillations at different levels of the solar atmosphere. The value of the radio astronomical observations with Nobeyama Radioheliograph is determined by possibility to register the time variations of the parameters of plasma structures definitely inside the magnetic tube of sunspot at the level of the basis of the corona and by full archives of regular daily observations covering more than 15 years with practically any desirable cadence. Comparison of these observations with observations of chromospheric velocity made in $H\alpha$ line resulted in some new estimations of the 3D-dimensional structures and understanding of time evolution of the quiasy-periodic processes (MHD wave including) in magnetic tubes of a sunspots.

Acknowledgement

This work was supported by the Russian Foundation for Basic Research under grants Nos. 06-02-16838, 08-02-91860, 08-02-08744 and also by the Basic Research Program of the Presidium of the Russian Academy of Sciences No 16. V.E. Abramov-Maximov thanks IAU for financial support.

References

Bogdan, T. J. 2000, *Solar Phys.*, 192, p. 373
Fludra, A. 2001, *A&A*, 368, p. 639
Gelfreikh, G., Nagovitsyn, Yu. A, & Nagovitsyna, E. Yu. 2006, *PASJ*, 58, p. 29
Kobanov, N. I., Kolobov, D. Y., & Makarchik, D. V. 2006, *Solar Phys.*, 238, p. 231

Discussion

GOLLA: What is your interpretation of the variation in the amplitude of the waves?

ABRAMOV-MAXIMOV: Probably due to mixing of different periods.

Universal Heliophysical Processes
Proceedings IAU Symposium No. 257, 2008
N. Gopalswamy & D. F. Webb, eds.

© 2009 International Astronomical Union
doi:10.1017/S1743921309029159

The maximum magnetic flux
in an active region

George Livadiotis[a,b,1] and Xenophon Moussas[b,2]

[a] Space Science and Engineering Division, Southwest Research Institute,
San Antonio, TX 78238, US
[b] Department of Astrophysics, Astronomy and Mechanics, Faculty of Physics,
National and Capodistrian University of Athens,
Panepistimiopolis, GR 15784, Zografos, Athens, Greece
[1] email: glivadiotis@swri.edu; glivad@phys.uoa.gr [2] email: xmoussas@phys.uoa.gr

Abstract. The Photometric-Magnetic Dynamical model handles the evolution of an individual sunspot as an autonomous nonlinear, though integrable, dynamical system. The model considers the simultaneous interplay of two different interacted factors: The photometric and magnetic factors, respectively, characterizing the evolution of the sunspot visible area A on the photosphere, and the simultaneous evolution of the sunspot magnetic field strength B. All the possible sunspots are gathered in a specific region of the phase space (A, B). The separatrix of this phase space region determines the upper limit of the values of sunspot area and magnetic strength. Consequently, an upper limit of the magnetic flux in an active region is also determined, found to be $\approx 7.23 \times 10^{23}$ Mx. This value is phenomenologically equal to the magnetic flux concentrated in the totality of the granules of the quite Sun. Hence, the magnetic flux concentrated in an active region cannot exceed the one concentrated in the whole photosphere.

Keywords. Sun: sunspots; activity; magnetic fields; granulation

1. Introduction

During the last sixty years, the phenomenon of the decay phase of sunspots has been a highly controversial issue within the heliophysical community. This involves establishing the exact mathematical expression of the area decay rate of the sunspots. The particular mathematical formula of the decay rate, also called a decay law, is a key point for understanding the specific underlying physical mechanism of sunspot dissolution.

Previous theoretical models, made for describing the sunspot evolution, had the common characteristic of originating in Magnetohydrodynamical equations, constructed after suitable simplifications and modifications. However, each one focuses only on the decay phase of sunspots, resulting to a specific decay law, which indeed, is supported by parts of the observational data.

In such a way, since today three decay laws have prevailed: (i) The linear decay law, $\dot{A} = -w$, where the area decay rate is negative constant ($w > 0$). (ii) The parabolic decay law, $\dot{A} \propto -\sqrt{A}$, exhibiting a slower decay rate than the linear law, because of the positive constant second derivative of area with respect to time, $\ddot{A} > 0$. Finally, (iii) the exponential decay law, $\dot{A} \propto -A$, characterized by a faster decay rate than the linear law (Livadiotis & Moussas (2007), c.f. Introduction). (The dot denotes the derivative with respect to time t.)

On the other hand, the Photometric-Magnetic Dynamical (PhMD) model, introduced by Livadiotis & Moussas (2007), has its own origin, constructed considering several meaningful physical terms. It considers the simultaneous interplay of two different interacted factors: The photometric one, characterizing the evolution of the sunspot visible area A on the photosphere; the magnetic factor, characterizing the simultaneous evolution of the sunspot magnetic field strength B.

In relevance to the previous models, we mention that the PhMD model first of all, describes the whole sunspot lifetime, either the growth or decay phases. It reproduces all the three decay laws: Each decay law is applicable at a different time of the decay phase. Moreover, it predicts the characteristic numerical features of each decay law, as found from the observational data.

In addition, the PhMD model, among others, predicts (i) the distribution of the maximum sunspot areas, (ii) an estimation of the granules mean dimensions, (iii) an upper limit for the maximum values of sunspot area and magnetic strength, and (iv) an upper limit for the maximum value of magnetic flux concentrated in an active region. Here we will discuss the issue of upper limits. We stress that the model is an integrable two-dimensional dynamical system. However, its non-integrable version, that is the perturbed photometric magnetic dynamical model has also been investigated (Livadiotis & Moussas 2008).

2. Construction of the model

The model is formulated by a two-dimensional dynamical system, whereby the two time-dependent functions are the photospheric sunspot area $A(t)$, and its magnetic field strength $B(t)$, characterizing the sunspot as a whole.

$$
\begin{aligned}
\dot{A} &= -a_1(A - A_G) + a_2(B^2 - B_G{}^2) \\
\dot{B} &= -b_{11}A + b_{12} - b_{13}\sqrt{A} + b_2(B - B_G) .
\end{aligned}
\tag{2.1}
$$

As we can see, there are five different additive physical terms, explained below:

The first term, $a_2(B^2 - B_G{}^2)$, yields the creation mechanism of the sunspot. This involves magnetic flux fragments such as those of pores or of other elementary magnetic microstructures, erupting from below the photosphere due to the magnetic buoyancy, while their accumulated smaller flux tubes coalesce into a main core of a larger flux tube. The magnetic buoyancy is a component being proportional to B^2, thus the same holds for the first term, being mentioned.

The next two terms, $-b_{11}A + b_{12}$, $-b_{13}\sqrt{A}$, constitute what we call "collar flow" (Fig. 1a), that is the exchange of inflow and outflow of magnetic flux, respectively (Solanki 2003). Thus, magnetic flux can inflow either through the entire sunspot projected area A, emerging from the solar interior, or through locally specified regions with sufficiently smaller dimensions compared to those of sunspot. On the other hand, magnetic flux can outflow sideways through the area of sunspots, that is a component proportional to the square root of A.

The other two components, $-a_1(A - A_G)$, $+b_2(B - B_G)$, constitute the negative and positive feedbacks, respectively, characterizing the resistance to the area increasing due to the creation mechanism and the magnetic strength decreasing due to the magnetic outflow.

Therefore, the inflow and outflow are combined in order to form the collar flow, which together with the feedbacks, help to stabilized the sunspot. We underline that the model considers that a sunspot does not only maintain its existence, but also evolves during its whole lifetime, by interacting with the simultaneously evolved magnetic field.

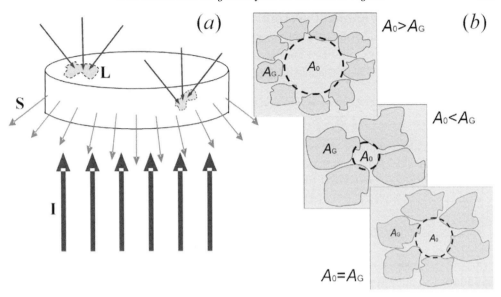

Figure 1. (a) The inflow and outflow, combined, form the well known collar flow, that together with the feedbacks, helps to stabilize the sunspot. The PhMD model considers that a sunspot does not only maintain its existence, but also evolves during its whole lifetime, by interacting with the simultaneously evolved magnetic field. (I: solar interior inflow, L: local inflow, S: sideways outflow.) (b) Scheme of granules arrangement: low density arrangement $A_0 > A_G$; high density arrangement $A_0 < A_G$; initial equilibrium $A_0 = A_G$.

The small characteristic scales A_G, and B_G, are related to the granules, as we shall see further below.

3. Granules characteristic scales

Here we arrive at an interpretation of the area characteristic scales A_G, and B_G. This arises considering the geometric arrangement of granules.

The procedure of magnetic flux tubes coalescence initially takes place in the intermediate space between adjacent granules. Therefore, this is the space, being disposed to the sunspot initial area, A_0.

Now, suppose that granules happens to be distributed in a locally low density arrangement, namely, the initial sunspot area A_0 is larger than the granule area A_G (Fig. 1b). Then, the system shall be spontaneously driven to a higher density arrangement. The enclosed area is compressed, being opposed to the emergence of magnetic elements from below the photosphere, leading to a resistance on the area initial growth, being proportional to the difference between the initial sunspot area A_0 and the granule area A_G.

Obviously, in the opposite case of locally high density granules arrangement, $A_0 < A_G$, the physical system is spontaneously driven to a lower density arrangement, giving advance to the sunspot creation, exhibiting thus, a negative resistance.

The case of neither high nor low density arrangement is thought to express an initial equilibrium, formulated by the initial conditions of zeroth area growth rate and zeroth resistance, that is the initial area being equal to that of the surrounding granules, $A_0 = A_G$. Consequently, we also have that the initial magnetic strength B_0 is equal to the respective scale B_G.

4. Normalized dynamical system

By considering proportionality relations between the functions $A(t)$, $B(t)$, and their respective initial values, A_G, B_G, we can define the new dimensionless variables $x(t) \equiv \frac{A(t)}{A_G}$ and $y(t) \equiv \frac{B(t)}{B_G}$. Then, the system Eq. (2.1), can be rewritten as follows:

$$\dot{A} = a_1[-x + q(y^2 - 1)]$$
$$\dot{B} = b_2[p_1 x - p_2 \sqrt{x} + y + m] . \tag{4.1}$$

The physical meaning of the new parameters is inherited from the old ones, namely,
- $a_1 = b_2$ yields a characteristic time scale $\tau \equiv a_1 t$,
- $q = +\frac{1}{y^2 - 1}(\frac{dx}{d\tau})_C$ is related to the sunspot creation mechanism (C: coalescence),
- $p_1 = +\frac{1}{x}(\frac{dy}{d\tau})_I$ characterizes the inflow of the magnetic flux through the solar interior (I: interior),
- $m = +(\frac{dy}{d\tau})_L$: characterizes the inflow through the small local regions (L: local regions),
- $p_2 = -\frac{1}{\sqrt{x}}(\frac{dy}{d\tau})_S$ characterizes the outflow of the magnetic flux through the sunspot boundary (S: sideways area).

Furthermore, we stress that the dynamical system is conservative, $\vec{\nabla}[\dot{x}(x, y), \dot{y}(x, y)] = 0$, since the condition a_1 being equal to b_2, leads to a closed phase space curve, that is the only phase space curve with physical meaning. Indeed, for $a_1 > b_2$, phase space trajectories converge to a sink, describing a sunspot of infinite lifetime. On the other hand, for $a_1 < b_2$, phase space trajectories diverge to infinity, describing sunspots of infinite dimensions. Both cases are physically unacceptable, and thus, we conclude with the condition $a_1 = b_2$.

Therefore, the Hamiltonian constraint is given by the expression,

$$H(x, y) = \frac{a_1}{3}\left\{qy^3 + 3(1-q-x)y + 2\left[-\frac{3}{4}p_1(x^2 - 1) + p_2(x^{\frac{3}{2}} - 1) - \frac{3}{2}m(x-1) + q\right]\right\}, \tag{4.2}$$

while sunspots found to be described by zeroth constraint value, $H(x, y) = H(x_0 = 1, y_0 = 1) = 0$.

We remark that the normalized expression of the system Eq.(4.1) shall remain invariant if we adopt a common functional type for the positive and negative feedbacks involved in Eq. (2.1), namely, $a_{11}(A - A_G) + a_{12}(B - B_G)$ instead of $a_1(A - A_G)$, and $b_{21}(A - A_G) + b_{22}(B - B_G)$ instead of $b_2(B - B_G)$. However, the normalized dimensionless variables have to be defined by $x(t) \equiv \frac{A(t)}{A_G}$ and $y(t) \equiv \lambda\frac{B(t)}{B_G} - \lambda - 1$, where the new parameter $\lambda \equiv (\frac{a_{12}}{2a_2 B_G} - 1)^{-1}$ is encrypted in the expression of y. (In similar way that the other parameters A_G and B_G are encrypted in the expressions of x and y.) Then, A_G and B_G, λ, have to be respectively taken into account when we reverse the problem and we want to find $A(t)$ and $B(t)$ in terms of $x(t)$ and $y(t)$.

5. Results: The fitting of the model and the predicted decay laws

The fitting of the model to the observational data was derived in three different ways: First, we compared the model predictions with the observational data referring to one single sunspot. In such a way, we selected several compact individual sunspots where the model predicted curve of area $A(t)$ was fitted to the respective photometric light curves, as shown in Figs. 2a,b for the (RGO-UASF/NOAA) AR1485104 and AR1488603, respectively. Second, we dealt with the fitting of the model predicted phase space curves

$[A(t), \dot{A}(t)]$ to the statistically modified forty three respective pairs characterizing the totality of large sunspots $A > 35$ MSH (Millionths of Solar Hemisphere), as shown in Fig. 2c. The statistical analysis, being utilized here, was by Hathaway & Choudhary (2005). They arranged all the decay rates of sunspots into forty three bins of areas, and then, they calculated the mean and the standard deviation of the decay rate of each bin. (This statistical manipulation concerns 24000 values of decay rates from the daily observational data of Royal Greenwich Observatory in 1874-1976, that have been analyzed.) We mentioned that there was another kind of fitting, that is the fitting of the model predicted distribution of the maximum sunspot areas to the respective observational one, found to be a Log-Normal distribution (Martínez Pillet *et al.* 1993). (For further details, see: Livadiotis & Moussas 2007.)

One of the significant results of the fitting was that three model parameters remain constant, having for all the sunspots the same fixed value. Namely, $q = 310 \pm 10, p_1 = (1.85 \pm 0.10) \times 10^{-3}, m = 4.90 \pm 0.03$. On the other hand, there are also variant parameters, such as p_2, a_1, A_G, B_G. However, the only variant parameter, that affects the $[x(t), y(t)]$ trajectories for each sunspot, is the outflow p_2.

In Fig. 2d we depict the evolution of a large sunspot along the trajectories $[A(t), \dot{A}(t)]$. The enlarged panel illustrates the decay phase. We observe the "sub-phase" where the linear decay law is applicable (indicated as L), and also for the parabolic (as P), and the exponential (as E) decay laws. Also, notice the divergence from exponential law (indicated as dE), which has recently been established from observations. (Compare with Fig. 2c.) In addition, we remark that the model predicts the correct numerical features of each of the three decay laws, as found from observations (Livadiotis & Moussas 2007).

We stress the fact that even though the model predicted evolution of sunspot area (photometric evolution) has been suitably adjusted to the observational data, the magnetic evolution has not yet been fitted, since the solar magnetic observational data are

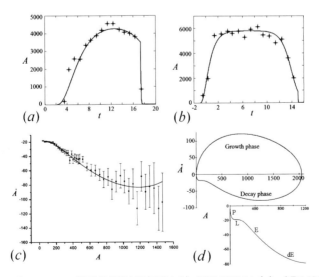

Figure 2. Fitting of sunspots (RGO-USAF/NOAA) AR1485104 (a), AR1488603 (b). (c) Fitting of the model predicted curve $[A(t), \dot{A}(t)]$ (solid line) with the mean decay rates derived from Hathaway & Choudhary (2005). (d) Evolution of a large sunspot along the trajectories $[A(t), \dot{A}(t)]$. (L: linear, P: parabolic, and E: exponential decay laws; dE: divergence from exponential decay law.)

rather poor and inadequate. Nevertheless, the values of B_G and λ have been estimated in the case of large sunspots, that is $B_G \approx 3100$ G and $\lambda \approx 25$. (For details, see: Livadiotis & Moussas 2009.)

6. Phase space trajectories $[x(t), y(t)]$

The various trajectories $[x(t), y(t)]$ can be represented in the phase space diagram as iso-Hamiltonians. Namely, each trajectory is specified by its Hamiltonian constraint value, but all of them are characterized by a fixed value of the outflow p_2. On the other hand, however, trajectories can be represented as iso-outflows. Namely, each trajectory is specified by its outflow p_2 value, but all of them are characterized by a fixed value of the Hamiltonian constraint.

The iso-outflows phase space representation is preferred, since the totality of sunspots are characterized by one single fixed value of Hamiltonian constraint, namely $H(x, y) = 0$, derived from the initial equilibrium condition, described in Sec. 3. Even if we adopt the common functional type of feedbacks, described in Sec. 4, the initial equilibrium condition remains, namely, $A_0 = A_G$, $\dot{A}_0 = 0$, leading to $y_0 = 1$. Then, we readily derive that $x_0 = 1$, $y_0 = 1$, resulting in $H(x, y) = 0$.

This phase space representation is illustrated in Fig. 3, whereby the totality of sunspots can be found in the grey region of the trajectories. This is limited by the peculiar stable point S_1 and by the unstable point U_2. All the trajectories expand, as p_2 decreases. The outflow maximum value, $p_{2,Max} = p_1 + 1 + m \simeq 5.90185$, corresponds to the trivial single point trajectory, that is S_1, while the outflow minimum value, $p_{2,Min} \simeq 0.26284686\ldots$, corresponds to the most expanded trajectory, that passes through U_2.

Hence, the largest value of the maximum normalized area x_{max} corresponds to the outflow minimum value, $p_{2,Min} \simeq 0.26284686\ldots$, and that is the upper limit of the normalized areas, found to be of about $x_{UpLim} = 5700$.

Given the value of the initial area, that is of the area of the initially surrounding granules, we can arrive at the upper limit of the sunspot areas. Therefore, by utilizing the maximum area attained by granules, that is $A_{G,max} = 1.1$ MSH (Foukal 2004), we retrieve the sunspot area upper limit, namely, $A_{UpLim} = A_{G,max} \cdot x_{UpLim}$, hence,

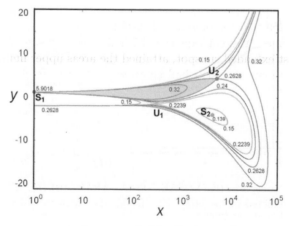

Figure 3. The phase space portrait represented by the iso-outflow curves. ($S_{1,2}$: stable, $U_{1,2}$: unstable fixed points. The Hamiltonian constraint is zero. The outflow p_2 values are indicated.)

$A_{UpLim} = 6200 \pm 400$ MSH. (The errors of the fixed values parameters propagate the error of the upper limit areas.) We note that the peculiar case of the year 1947 is characterized by extremely large sunspots with areas bounded by the estimated upper limit. The light-curve of the largest ever sunspot, that appeared on April 3, 1947, is shown in Fig. 2b.

The unstable point U_2 yields also the upper limit for the normalized magnetic strength y, that is $y_{UpLim} = 4.404$. Given the values of $B_{G,max} \approx 3100$ G and $\lambda_{max} \approx 25$ that correspond to large sunspots, we arrive at the sunspot magnetic strength upper limit, namely, $B_{UpLim} = \frac{B_{G,max}}{\lambda_{max}} \cdot (y_{UpLim} + \lambda_{max} + 1)$, hence, $B_{UpLim} \approx 3770$ G.

Therefore, we finally conclude in the magnetic flux upper limit $\psi_{UpLim} \approx A_{UpLim} \cdot B_{UpLim} \approx 7.23 \times 10^{23}$ Mx. Observations justify this result, since the largest active regions reach a maximum magnetic flux of $\approx 10^{23}$ Mx (Vial 2005, Solanki *et al.* 2006).

7. Discussion and Conclussions

The magnetic flux concentrated in the totality of the granules of the whole area of the quiet Sun (with area $A_{Sun} = 2 \times 10^6$ MSH), having a typical maximum magnetic strength of about $B_G \approx 12$ G (Rabin *et al.* 1991), is equal to $A_{Sun} \cdot B_G \approx 7.2 \times 10^{23}$ Mx (Wang 2004, Fisk 2005). Hence, the upper limit of the magnetic flux in an active region is phenomenologically equal to the magnetic flux concentrated in the totality of the granules of the quite Sun.

Therefore, the magnetic flux concentrated in an active region cannot exceed the magnetic flux concentrated in the photosphere as a whole.

Furthermore, it has to be underlined that the estimated upper limit of the magnetic flux concerns not only the sunspots but also the active regions.

PhMD model concerns the fact of one compact main core of a flux tube, that of an individual sunspot. This is governed by the upper limits of the sunspot photospheric area (photometric factor) and magnetic strength (magnetic factor), as well as by the upper limit of their product, that is of the magnetic flux. Even though, the upper limits of sunspot area and magnetic strength do not indispensably characterize also the host active region, this is thought to hold for the magnetic flux upper limit. Sunspots included in an active region are supposed to be antagonistic in absorbing magnetic flux (inflowing either by solar interior or by local regions), while a dominant sunspot (if there is) shall be the one of outflowing the smaller amount of magnetic flux.

Hence, it is expected that the presence of extremely large sunspots costs the absence of sufficiently smaller ones. As a consequence, it is natural to claim that in the hypothetical presence of the most extended sunspot, attained the areas upper limit, no other sunspot can prevail, as their magnetic flux shall be absorbed by the dominant sunspot. In such a case, all the magnetic flux of the active region has to be transferred to the gigantic sunspot. By means of such consideration we claim that the upper limit for sunspot magnetic flux yields is also an upper limit for active regions.

References

Hathaway, D. H. & Choudhary, D. P. 2005, *NASA Technical Reports Server*, ID: 20050236988

Fisk, L. A. 2005, *ApJ*, 626, 563

Foukal, P. V. 2004, in: *Solar Astrophysics* (Wiley-VCH Verlag GmbH & Co. KGaA,Weinheim), pp. 138-143

Livadiotis, G. & Moussas, X. 2007, *Physica A*, 379, 436

Livadiotis, G. & Moussas, X. 2008, in: G. Contopoulos & P. A. Patsis (eds.), *Chaos in Astronomy* (Springer, Berlin), Ch. 46, p. 455

Livadiotis, G. & Moussas, X. 2009, *Adv. Space Res.*, doi:10.1016/j.asr.2008.09.010, in press

Martínez Pillet, V., Moreno-Insertis, F., & Vázquez, M. 1993, *Astron. Astrophys.*, 274, 521

Rabin, D. M., Devore, C. R., Sheeley, N. R., Harvey, K. L., & Hoeksema, J. T. 1991, in: A. N. Cox, W. C. Livingston & M. S. Matthews (eds.), *Solar Interior and Atmosphere* (The University of Arizona Press), p. 781

Solanki, S. K. 2003, *Astron. Astrophys. Rev.*, 11, 153

Solanki, S. K., Inhester, B., & Schüssler, M. 2006, *Rep. Prog. Phys.*, 69, 563

Vial, J.-C. 2005, *Adv. Space Res.*, 36, 1375

Wang, Y.-M. 2004, *Solar Phys.*, 224, 21

Universal Heliophysical Processes
Proceedings IAU Symposium No. 257, 2008
N. Gopalswamy & D.F. Webb, eds.

© 2009 International Astronomical Union
doi:10.1017/S1743921309029160

The heliospheric magnetic field and the solar wind during the solar cycle

Lennard A. Fisk and Liang Zhao

Dept. of Atmospheric, Oceanic and Space Sciences, University of Michigan,
2455 Hayward St., Ann Arbor, Michigan 48109, USA
email: `lafisk@umich.edu`

Abstract. The heliospheric magnetic field and the solar wind are behaving differently in the current solar minimum, compared to the previous minimum. The radial component of the heliospheric magnetic field, and thus the average value of the component of the solar magnetic field that opens into the heliosphere, the so-called open magnetic flux of the Sun, is lower than it was in the previous solar minimum; in fact, lower than in any previous solar minimum for which there are good spacecraft observations. The mass flux, the ram pressure, and the coronal electron temperature as measured by solar wind charge states are also lower in the current minimum compared to the previous one. This situation provides an opportunity to test some of the concepts for the behavior of the heliospheric magnetic field and the solar wind that have been developed; to improve these theories, and to construct a theory for the solar wind that accounts for the observed behavior throughout the solar cycle, including the current unusual solar minimum.

Keywords. Solar magnetic field, heliospheric magnetic field, coronal processes, solar wind, solar cycle

1. Introduction

Over the last decade we have developed theories for the behavior of the heliospheric magnetic field (Fisk 1996; Zurbuchen *et al.* 1997; Fisk *et al.* 1999a,b; Fisk 1999; Fisk & Schwadron 2001; Fisk 2005; Fisk & Zurbuchen 2006), and for the acceleration of the solar wind (Fisk *et al.* 1998, 1999c; Fisk 2003). During the current solar minimum, the heliospheric magnetic field and the solar wind are behaving in ways that are different from any previous minimum for which we have good spacecraft observations. The current solar minimum thus provides a unique opportunity to test these theories, modify them as necessary, and construct theories that can account for the behavior of the heliospheric magnetic field and the solar wind throughout the solar cycle, including the current unusual minimum.

We begin by reviewing the observations of the heliospheric magnetic field and the solar wind, from the previous solar minimum through the present one, to illustrate the differences between the two successive minima. We use observations from Ulysses, which span the entire time period. We then consider whether our theories for the behavior of the heliospheric magnetic field are consistent with these observations, and suggest an addition to these theories that improves the consistency. We next develop a theory for the solar wind, based on our previous theories for the solar wind, which can account for the behavior of the solar wind mass flux, coronal electron temperatures and thus the ionic charge-states of the solar wind, and for the solar wind flow speed, and which is consistent with observations throughout the solar cycle.

2. Observations

Summarized in Figure 1 are some of the main observations we will deal with:

Radial magnetic field strength. In the top panel is the radial component of the helio-spheric magnetic field, normalized by heliocentric radial distance squared. We will refer to this as the normalized radial component of the heliospheric magnetic field. The measure-ments from Ulysses are from many different latitudes; however, it has been shown from Ulysses that the latitude variations in the heliospheric radial magnetic field are weak, as is to be expected (Smith & Balogh 1995; Balogh & Smith 2001). The magnetic pressure in the outer solar corona, where the magnetic field, dragged outward with the solar wind, is radial, must be constant since there are no latitudinal balancing forces; i.e., the radial magnetic field should be uniform. The radial component of the heliospheric magnetic field is thus a measure of the average value of the component of the solar magnetic field that opens into the heliosphere, the so-called open magnetic flux of the Sun.

Note that the normalized radial component varies over the solar cycle, increasing by a factor of ~ 2 near solar maximum, and it attains its minimum value in solar minimum. Note also that the minimum value of the normalized radial component is lower in the current solar minimum than it was in the previous one.

Solar wind mass flux. In the second panel of Figure 1 is the mass flux of the solar wind. Note that the mass flux roughly tracks the normalized radial component of the heliospheric magnetic field, and it too is lower in the current minimum than in the previous one.

Solar wind ram pressure. In the third panel of Figure 1 is shown the ram pressure of the solar wind. Note that it also roughly tracks the solar wind mass flux in the two successive minima. This indicates that although the mass flux is lower in the current minima, the range of solar wind flow speeds is not different between the two minima.

Solar wind charge state. In the bottom panel of Figure 1 is shown the ratio of O^{7+} to O^{6+} in the solar wind. Note that in the current solar minimum the ratio is lower than in the previous minimum. The charge states of the solar wind are frozen-in in the solar

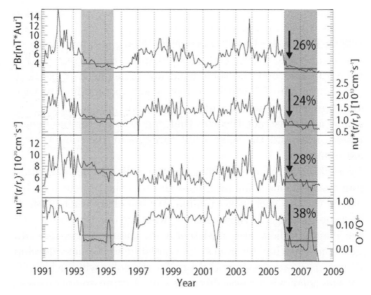

Figure 1. Normalized radial component of the heliospheric magnetic field, solar wind mass flux, solar wind ram pressure, and charge states of O, as observed by Ulysses from 1991 to 2008.

corona, when the density becomes sufficiently low. Thus, in the current solar minimum, the coronal electron temperature is lower than in the previous minimum.

Gloeckler *et al.* (2003) found a remarkable correlation between the coronal electron temperature as determined by solar wind charge states and the solar wind speed. For a period in 1996-97, they found a linear relationship between the square of the solar wind speed and the inverse of the coronal electron temperature, as determined by the solar wind charge states. In Figure 2, we repeat this analysis for the same time period as used by Gloeckler *et al.* (2003), and for a period in 2005-07. We find the exact same linear relationship. It should be noted that the linear relationship is most obvious when there are both fast and slow wind present in the analyzed period. The two time periods chosen are ideal for this. In other time periods, such as the current solar minimum, when the charge states and thus the coronal electron temperature are low, there is a narrower range of solar wind speeds that can be analyzed, and the relationship is less obvious in the data.

We have then several observations to explain: (1) The variation of the normalized radial component of the heliospheric magnetic field, or equivalently the average open magnetic flux of the Sun, during the solar cycle, including the lower value in the current minimum. (2) The correlation of the solar wind mass flux with the normalized radial component of the heliospheric magnetic field. (3) The linear relationship between the square of the solar wind flow speed and the inverse of the coronal electron temperature, and why it is the same during the two different periods shown in Figure 2.

3. The Behavior of the Heliospheric Magnetic Field during the Solar Cycle

The Ulysses mission to date has observed the solar magnetic cycle for ~17 out of the ~22 years required for a full cycle. The picture that has emerged is one of remarkable simplicity: the magnetic field in the heliosphere appears to be organized into two regions of opposite polarity separated by a single current sheet, which appears to persist throughout the solar cycle (Smith *et al.* 2001; Jones *et al.* 2003). In past solar cycles, the average strength of the solar magnetic field that opens into the heliosphere has appeared

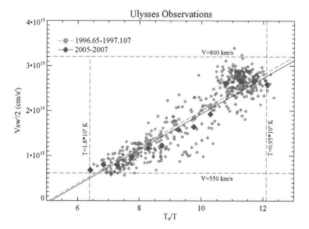

Figure 2. The anticorrelation between the solar wind speed squared and the coronal electron temperature as measured by solar wind charge states (after Gloeckler *et al.* 2003). Two time periods are shown. The first is the same as in the Gloeckler *et al.* (2003) analysis; the second is closer to the current solar minimum.

to be relatively constant, particularly if you compare the field in successive solar minima, and it increases only by a factor of ∼2 at solar maximum (e.g., Wang *et al.* 2000). The current sheet becomes tilted relative to the solar equator as the solar cycle progresses and rotates over. In this simple picture then it is the rotation of the current sheet that accomplishes the field reversal of the Sun, although the more commonly accepted view is that the field reversal occurs by polar field annihilation as in the Babcock (1961) model (e.g., Wang & Sheeley 2003).

The organization and the constancy of the heliospheric magnetic field are related. To eliminate heliospheric magnetic flux, it is necessary that magnetic flux of opposite polarity reconnects, forming an inverted 'U'-shaped loop that is convected out of the heliosphere by the solar wind (Fisk & Schwadron 2001). Such reconnection can only occur at the single current sheet, where magnetic fields of opposite polarity can interact, and within the Alfven radius, which occurs at ∼10 r_{Sun}. Here, a solar loop will also be formed that can return to the Sun, with a net loss of magnetic flux to the heliosphere. While this process is possible, the inverted 'U' loop should be devoid of heat flux, a so-called heat flux dropout (McComas *et al.* 1989, 1992). Such dropouts are rarely observed (Lin & Kahler 1992; Pagel *et al.* 2005), although some controversy remains in the interpretation of the heat-flux dropout data (Pagel *et al.* 2007).

The concept that magnetic flux in the heliosphere cannot be readily eliminated has provided a natural explanation for why the heliospheric magnetic field has appeared to return to the same magnitude at successive solar minima, as documented by Svalgaard & Cliver (2007). There is a background level of magnetic flux always present in the heliosphere. The increases in the heliospheric magnetic field at solar maximum are attributed to an enhanced rate of CMEs. CMEs drag additional magnetic flux into the heliosphere, and if left unabated would result in magnetic flux in the heliosphere that increases without bound (Gosling 1975; McComas 1995). Since this does not occur, it is believed that the magnetic field in the CME reconnects with the background open magnetic flux in a process that has been labeled interchange reconnection (Gosling *et al.* 1995; Fisk & Schwadron 2001; Crooker *et al.* 2002). The CME flux is then converted into open flux, with no net increases in the heliospheric magnetic field. Interchange reconnection can take time to execute, and thus during solar maximum, when the rate of CMEs is high, there is a temporary increase in the heliospheric magnetic field (by a factor ∼2); at solar minimum, the rate of CMEs is lower and the heliospheric field returns to its background level (Owens & Crooker 2006).

As can be seen in Figure 1, unlike our expectations, the normalized radial component of the heliospheric magnetic field, and thus the average open magnetic flux of the Sun, is lower in the current minimum, compared with the previous minimum. The simplest explanation is that we do not know the actual strength of the background, constant level of open flux. In the above argument, we assumed there was no significant contribution of CMEs during solar minimum, and the background level is attained in each minimum. Perhaps that is not the case. If there was still a CME contribution in each previous minimum, but in the current minimum, which is unusually quiet, this contribution is smaller, then we are closer to the background level now than we were previously.

In the remaining sections of this paper, we will consider theories for the solar wind in which open magnetic flux interacts with coronal loops; material is released to provide the mass flux of the solar wind; and the open magnetic flux is displaced resulting in waves and turbulence that heat and accelerate the solar wind. The basis for this theory of the solar wind is the model for the interaction of coronal loops with open flux developed by Fisk (2005). As is illustrated in Figure 3, small loops are emitted through the solar surface. The ends of the loops migrate to the network lanes, where they can reconnect with and thus coalesce with other loops. They can also encounter open flux. If the end of the loop has opposite polarity to the open field line, reconnection occurs. The loop is destroyed and the open field line is displaced. There is a small loop formed at the reconnection site, which subducts back into the photosphere.

Consider what happens to loops that emerge under the current sheet. If the emerging loop is oriented such that it is aligned with the polarity of the open flux, then the end points of the loop never encounter open magnetic flux of opposite polarity. The loop grows without bound. One could even argue that this process provides an origin for all large streamer belt loops.

For our purposes here, there is a natural mechanism whereby small loops could grow without bound around the current sheet, and eventually be continuously emitted as large loops into the heliosphere, even during solar minimum. These large loops would also be subject to interchange reconnection with the background level of open magnetic flux, and the addition of magnetic flux in the heliosphere would not have unlimited growth. However, while these loops are present in the heliosphere, before interchange reconnection, they would raise the level of magnetic flux in the heliosphere above the background level, even at solar minimum.

There is still, however, an electron heat flux problem. Loops are identified in the solar wind by counterstreaming, bi-directional electron fluxes (Gosling *et al.* 1987). There is no particular observational evidence to suggest continuous bi-directional electron fluxes around the current sheet (Zurbuchen & Richardson 2006). We need to remember, however, that bi-directional electrons, as a measure of loops in the solar wind, are mainly observed in large CMEs. Perhaps the signature is not as clear in the loops we are arguing could be continuously emitted around the current sheet. Certainly the geometry of the expansion is different between current sheet loops and large CMEs. The former have particularly long legs compared to the lateral portion of the loop, which must intercept and be influenced by the current sheet. Whether this will affect the bi-directional electron

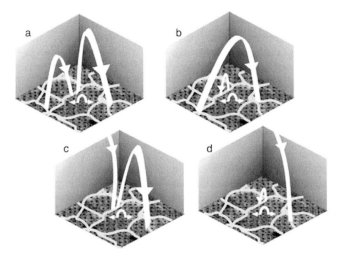

Figure 3. Small loops emerge in the center of supergranules. Each end expands and enters the network lanes, where they move with the random convective motions of the photosphere. In (a) two end points of loops collide and reconnect, coalescing into the large loop shown in (b). In (c), the end point of a loop and an open field line reconnect. In (d), the open file line is displaced to lie over the opposite side of the loop, and the original loop is destroyed. Small loops are created at the reconnection sites, and are assumed to subduct into the photosphere.

fluxes is a problem that could use more theoretical work, and perhaps a closer look at the observations.

Thus, from the observations in Figure 1, and the above arguments, there is a correlation between the normalized radial component of the heliospheric magnetic field and thus the average open magnetic flux, and the rate of emergence of new magnetic flux on the Sun. The correlation is clear at solar maximum, when the emergence of large active regions result in large CMEs. However, there also appears to be a correlation between the level of open flux present in the absence of large CMEs and the rate emergence of small loops on the Sun, which grow under the current sheet and expand into the heliosphere. We would argue that in the current minimum this rate of emergence of small loops is lower, with a resulting lower level of open magnetic flux in the heliosphere, as seen in Figure 1, perhaps even a level that approaches the constant, background level of open magnetic flux.

4. The Behavior of the Solar Wind during the Solar Cycle

There are two basic types of solar wind acceleration theories. The first is the basic Parker theory, which has been elaborated upon and incorporated into many, quite detailed models for the acceleration of the solar wind (e.g., Parker 1958; Isenberg 1991 and references therein; Marsch 1995 and references therein; Hansteen & Leer 1995; Axford & McKenzie 1997; Cranmer *et al.* 2007). In these models you assume there is a deposition of energy and perhaps momentum into the solar corona. This deposition accelerates the solar wind and determines all other flow parameters, such as the solar wind mass flux. In the second type of solar wind acceleration theory, you assume that the mass flux of the solar wind is determined independently of the acceleration (Fisk *et al.* 1998, 1999; Fisk 2003). Matter is released from coronal loops as a result of reconnection with open magnetic flux, and this process determines the mass flux. The temperature and density in the corona then adjust to satisfy two independent constraints, the mass flux and the energy deposition, and result in the required supersonic flow.

In the latter theory, the solar wind is created as a result of the reconnection of open magnetic flux with coronal loops. In coronal holes, there is ample open flux present, which reconnects with the cool, small loops present at the base of the coronal hole. There should also be open flux present outside of coronal holes. Fisk & Zurbuchen (2006) showed that the transport of open magnetic flux resulting from reconnections with coronal loops should result in a uniform, radial component of open flux present in the regions outside of coronal holes. This component of open flux will reconnect with the large, hotter coronal loops on the quiet Sun outside of coronal holes. The mechanism for the origin of the solar wind is the same both inside and outside of coronal holes; the difference is the properties, e.g., the temperature of the coronal loops with which the open flux is reconnecting.

The theory in which the solar wind results from reconnections of open flux with coronal loops has several advantages:

(*a*) It is easy to imagine with this theory, as we shall show, that the mass flux is related to and correlated with the behavior of the open magnetic flux of the Sun, since we concluded in the previous section that the mass flux results from open flux reconnecting with coronal loops.

(*b*) This theory, in which the matter that is released to form the solar wind originates in coronal loops, also explains the composition of the solar wind. As shown by Feldman *et al.* (2005), the composition of coronal loops, the enhancements in elements with low first ionization potential and the electron temperatures and thus charge states, are consistent with those of the solar wind. The composition and coronal electron temperatures of the fast solar wind resemble those of the small, cooler loops at the base of the coronal holes. The composition and coronal electron temperatures of the slow solar wind closely resemble those of large coronal loops on the quiet Sun outside of coronal holes.

(*c*) In this theory it is possible to couple the mass flux of the solar wind with the deposition of energy that accelerates the solar wind. The process of reconnecting open magnetic flux with coronal loops displaces the open magnetic flux in the solar corona. This displacement will produce waves and turbulence in the corona, which when damped deposit energy that heats the solar corona and accelerates the solar wind.

(*d*) This theory is more likely to result in a predictive model for the solar wind, since certain basic flow parameters of the solar wind, e.g., the mass flux and energy deposition can be directly related to observable properties of the Sun, such as the properties of coronal loops.

We should also note that there is now a verified theory for how small loops that emerge on the Sun evolve and interact, through reconnection, with each other and with open magnetic flux. Fisk (2005) developed a relatively simple model for the evolution of coronal loops, based on the transport model of Schrijver *et al.* (1997) for magnetic flux concentrations in the random convective motions of the photosphere. As is illustrated in Figure 3, a small loop emerges through the photosphere. The end points of the loop migrate to the network lane, where each end point behaves independently. If the end points of two loops of opposite polarity encounter each other they reconnect and the two loops coalesce into one. If the end point of a loop encounters an open field line with opposite polarity, it reconnects, destroys the original loop, and displaces the open field line. At the reconnection sites, small loops are formed, which are assumed to subduct back into the photosphere. This theory can be used to determine the interaction rates between loops and open field lines, and thus the transport properties of open magnetic flux on the Sun, since the random displacements of open field lines due to reconnections with loops will cause the open magnetic flux to diffuse along the solar surface.

The theory of Fisk (2005) made a major prediction, which has now been confirmed by two independent sets of observations. The theory predicted that magnetic flux of a single polarity that is reconnecting with small coronal loops will tend to accumulate in regions where the rate of emergence of new magnetic flux is a local minimum. Thus, coronal holes, which are concentrations of open magnetic flux, are predicted to occur in regions where the rate of emergence of new magnetic flux is a local minimum. In Abrahmenko *et al.* (2006) this prediction was confirmed. In a study of 34 coronal holes, the coronal holes were found to occur in regions where the rate of emergence of new magnetic flux is a factor $\sim$2 lower than the surrounding regions. In Hagenaar *et al.* (2008) the more general prediction of the theory was verified. Regions of unipolar magnetic flux, whether in coronal holes or from nearby decaying active regions, tend to occur where the rate of emergence of new magnetic flux is a local minimum.

We thus develop our theory to explain the solar wind observations in Figure 1 & Figure 2 based upon the following concepts: (1) The solar wind mass flux is determined by the release of matter when open magnetic flux reconnects with a coronal loop. (2) The deposition of energy into the corona, which accelerates the solar wind, is the result of the displacement of open magnetic flux due to reconnection of open flux with coronal loops. (3) The evolution and other properties of coronal loops and how they interact are determined by the theory of Fisk (2005).

Solar wind mass flux. In our theory for the solar wind, the mass flux is determined by the amount of material that is released from coronal loops and the frequency with which it is released by reconnection with open magnetic flux. Suppose that the mass that is released from an average loop is M_l and the surface number density of loops is N_l. In

Fisk (2005), the collision frequency, and thus the reconnection frequency, between a loop and an open magnetic field line containing equal amount of magnetic flux as the loop is (Schrijver *et al.* 1997)

$$\frac{1}{\tau_{l,o}} = \frac{3}{4}\frac{\delta h^2}{\delta t}N_o. \tag{4.1}$$

Here, $\delta h^2/2\delta t$ is the diffusion coefficient due to random convective motions in the photosphere with scale size δh, N_0 is the surface number density of open field lines. It follows then that the time-averaged mass flux of the solar wind is given by

$$\rho_{sw}u_{sw}S = M_l N_l \left(\frac{3}{4}\frac{\delta h^2}{\delta t}\right)N_o S_{surf}. \tag{4.2}$$

Here, ρ_{sw} is the mass density of the solar wind, u_{sw} is the mean flow speed of the solar wind, and S is the cross section of a solar wind flux tube. Since the time-averaged mass flux of the solar wind should be constant along the flux tube, the left side of equation (4.2) can be evaluated at any location, including at heliocentric distance, r. The right side of equation (4.2) is evaluated on the solar surface, where S_{surf} is the cross section of the flux tube.

The average open magnetic flux within S_{surf} is $B_o = \phi N_o$, where ϕ is the magnetic flux in an open field line, equal to the magnetic flux in a loop with which the open field line is reconnecting. The total magnetic flux is constant within the flux tube, or $B_o S_{surf} = B_r S_r$, where B_r is the average magnetic field strength normal to the surface S_r at heliocentric distance r. Recall that the radial component of the heliospheric magnetic field is independent of latitude and longitude. We find that $B_r S_r \propto B_r r^2$, the quantity shown in Figure 1, and thus equation (4.2) is consistent with the observed correlation between mass flux and normalized radial magnetic field strength.

In Figure 4, we consider the relationship between the mass flux and the normalized radial magnetic field strength during the same time periods considered in Figure 2, when the Gloeckler *et al.* (2003) anticorrelation between solar wind speed and coronal electron temperature is particularly evident. One-hour values of the mass flux data are binned into ranges of $B_r r^2$ and then averaged within the bin into a single point. This process removes the variations in the mass flux that will be introduced by the other parameters in equation (4.2) besides $B_r r^2$. Clearly, there is a strong linear relationship between mass flux and $B_r r^2$, as predicted in equation (4.2). Note also that there is no non-zero intercept in the linear relationship. This suggests that the entire mass flux is due to the processes described by equation (4.2), as opposed to, e.g., a portion of the mass flux being due only to the acceleration process, as in a standard solar wind model.

The flow speed and charge states of the solar wind. Consider the deposition of energy into the corona. In our theory for the acceleration of the solar wind the source of energy is the random displacements of open field lines due to reconnections with coronal loops. As discussed in Fisk (2003), such random displacements should introduce magnitude variations in the coronal magnetic field comparable to the average magnetic field strength. When these variations are damped there is deposition of energy into the corona comparable to the energy in the average magnetic field of the corona. The characteristic timescale for the deposition should be the characteristic timescale at which open magnetic field lines reconnect with coronal loops, which using equation (4.1), is

$$\frac{1}{\tau_{o,l}} = \frac{3}{4}\frac{\delta h^2}{\delta t}N_l. \tag{4.3}$$

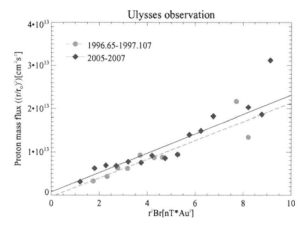

Figure 4. The mass flux of the solar wind versus the bin-averaged, normalized radial component of the heliospheric magnetic field, for the same time periods as shown in Figure 2. The binning technique is described in the text.

In Fisk (2003), the total magnetic energy in a flux tube in the corona is shown to be

$$W_{mag} = \frac{1}{8\pi} \phi N_o S_{surf} \bar{B}_o r_{Sun}. \tag{4.4}$$

Here $\bar{B}_o$ is the average radial component of the open magnetic flux on the solar surface, i.e., $\bar{B}_o = B_r \left(r^2/r_{Sun}^2 \right)$, where r_{Sun} is a solar radius; again, ϕ is the average magnetic flux in an emerging loop or an open magnetic field line. The only assumption in equation (4.4) is that in the solar corona the magnetic field behaves roughly as a potential magnetic field, with zero current.

The conservation of energy equation of the supersonic solar wind is

$$\frac{\rho_{sw} u_{sw}^3 S}{2} = \frac{W_{mag}}{\tau_{o,l}} - \rho_{sw} u_{sw} S \frac{GM_o}{r_{Sun}}. \tag{4.5}$$

Here, GM_o/r_{Sun} is the potential energy per unit mass of the Sun. Thus, substituting in equations (4.2), (4.3) and (4.4) into equation (4.5), we derive that

$$\frac{u_{sw}^2}{2} = \frac{\phi}{M_l} \left(\frac{\bar{B}_o r_{Sun}}{8\pi} \right) - \frac{GM_o}{r_{Sun}}. \tag{4.6}$$

Fisk (2003) noted that the mass contained in loops should be roughly proportional to their temperature, since the density scale height is proportional to the temperature. Thus, $M_l = M_{l,ref} \left(T/T_{ref} \right)$ where $M_{l,ref}$ is the mass of a loop at a reference temperature T_{ref} and T is the temperature of a loop. Further, electron heat flux is large in the corona, and it is readily imaginable that the loops serve as a thermal bath controlling the coronal electron temperature. Thus, the temperature of the loops with which the open flux is reconnecting determines the coronal electron temperature, and the solar wind charge states. Equation (4.6) can then be written as

$$\frac{u_{sw}^2}{2} = \frac{\phi}{M_{l,ref}} \left(\frac{T_{ref}}{T} \right) \left(\frac{\bar{B}_o r_{Sun}}{8\pi} \right) - \frac{GM_o}{r_{Sun}}. \tag{4.7}$$

Equation (4.7) is consistent with the results shown in Figure 2. The intercept on the two linear curves in Figure 2 are $-GM_o/r_{Sun}$, and the linear relationship is proportional to T_{ref}/T. The values of $\bar{B}_o$ differ slightly in the two time frames shown in Figure 2, but the slopes are essentially the same. This suggests that there is a relationship between $M_{l,ref}$ and $\bar{B}_o$.

5. Concluding Remarks

We have examined the behavior of the heliospheric magnetic field and the solar wind during the solar cycle, including the current and previous solar minimum, and used these data to test and refine our previously developed theories:

(a) In previous theories (e.g., Fisk & Schwardon 2001) we assumed that there is a constant background level of open magnetic flux present in the heliosphere, and that the heliospheric field is reduced to this background level in solar minimum. To explain the reduction in open flux below previous levels in the current minimum, we suggest that there is continuous emission of small loops around the current sheet, even in solar minimum, and that there is less such emission during the current, unusually quiet minimum. While this is a possible explanation it is not the only explanation and additional theoretical and observational consideration is required.

(b) In previous theories (e.g., Fisk & Schwardon 2001) we assumed that the mass flux of the solar wind is determined by the release of material from loops. There appears to be strong support for this process, at least in the time periods when the Gloeckler et al. (2003) anticorrelation between solar wind speed and coronal electron temperature holds.

(c) In previous theories (e.g., Fisk 2003) we were able to explain the anticorrelation between the solar wind speed and the coronal electron temperature observed by Gloeckler et al. (2003). These concepts, applied to the solar wind acceleration model developed here, appear to be applicable at least in some different periods throughout the solar cycle.

The relationship for the solar wind speed in equation (4.7) holds some promise for being able to predict the solar wind speed based on observed coronal properties. The main dependence is the temperature of the loops from which the material is released to form the solar wind. In principle, loop temperatures are observable. Although it will be important to be able to correctly map the open magnetic flux from the solar surface into the heliosphere, so that the correct loops, those at the base of the open flux, can be determined. A simple potential field source surface model will not do for this mapping since it cannot determine the correct mapping of open flux into primarily closed field regions. Mapping techniques such as the one developed by Gilbert et al. (2008) may be required.

Acknowledgements

This work was supported in part by the Heliophysics Theory Program, by NASA TR&T grant NNG056M53G, by NASA/JPL contract 1268016, and by NSF grant ATM 0632471.

References

Abramenko, V. I., Fisk, L. A., & Yurhyshyn, V. B. 2006, *ApJ(Letters)*, 641, L65

Axford, W. I. & McKenzie, J. F. 1997, in: J. R. Jokipii, C. P. Sonnet & M. S. Giampapa (eds.), *Cosmic Winds in the Heliosphere* (Tucson: Univ. of Arizona Press), p. 31

Babcock, H. W. 1961, *ApJ*, 133, 572

Balogh, A. & Smith, E. J. 2001, *Space Sci. Revs.*, 97, 147

Cranmer, S. R., van Ballegooijen, A. A., & Edgar, R. J. 2007, *ApJS*, 171, 520

Crooker, N. U., Gosling, J. T., & Kahler, S. W. 2002, *J. Geophys. Res.*, 107, 1028

Feldman, U., Landi, E., & Schwadron, N. A. 2005, *J. Geophys. Res.*, 110(A7), A07109

Feldman, U., Widing, K. G., & Warren, H. P. 1999, *ApJ*, 522, 1133

Fisk, L. A. 1996, *J. Geophys. Res.* 101(A7), 15,547

Fisk, L. A. 2001, *J. Geophys. Res.*, 106(A8), 15,849

Fisk, L. A. 2003, *J. Geophys. Res.*, 108 (A4), p. SSH 7-1

Fisk, L. A. 2005, *ApJ*, 626(1), 563

Fisk, L. A. & Schwadron, N. A. 2001, *ApJ*, 560(1), 425-438

Fisk, L. A. & Zurbuchen, T. H. 2006, *J. Geophys. Res.*, 111, A09115

Fisk, L. A., Schwadron, N. A., & Zurbuchen, T. H. 1998, *Space Sci. Revs.*, 86(1/4), 51-60

Fisk, L. A., Schwadron, N. A., & Zurbuchen, T. H. 1999c, *J. Geophys. Res.*, 104(A9), 19,765

Fisk, L. A., Zurbuchen, T. H., & Schwadron, N. A. 1999a, *ApJ*, 521(2), 868-877

Fisk, L. A., Zurbuchen, T. H., & Schwadron, N. A. 1999b, *Space Sci. Revs.*, 87(1/2), 43-54

Fisk, L. A., Schwadron, N. A., & Zurbuchen, T. H. 1999c, *J. Geophys. Res.*, 104(A9), 19,765

Gilbert, J. A., Zurbuchen, T. H., & Fisk, L. A. 2007, *ApJ*, 663, 583

Gloeckler, G., Zurbuchen, T. H., Geiss, J. 2003, *J. Geophys. Res.*, 108, SSH 8-1

Gosling, J. T. 1975, *Rev. Geophys.*, 13, 1053

Gosling, J. T., Baker, D. N., Bame, S. J., Feldman, W. C., Zwickl, R. D., & Smith, E. J. 1987 *J. Geophys. Res.*, 92, 8519

Gosling, J. T., Birn, J., & Hesse, M. 1995, *Geophys. Res. Lett.*, 22, 869

Hagenaar, H. J., DeRosa, M. L., & Schrijver, C. J. 2008, *ApJ*, in press

Hansteen, V. H., & Leer, E. J. 1995, *J. Geophys. Res.*, 100, 21577

Isenberg, P. A. 1991, in: J. A. Jacobs (ed.), *Geomagnetism* (San Diego: Academic), vol. 4, p. 1

Jones, G. H., Balogh, A., & Smith, E. J. 2003, *Geophys. Res. Lett.*, 30, 2

Marsch, E., von Steiger, R., & Bochsler, P. 1995, *A&A*, 301, 261

Lin, R. P. & Kahler, S. W. 1992, *J. Geophys. Res.*, 97, 8203

McComas, D. J. 1995, *Rev. Geophys.*, 33, 603

McComas, D. J., Gosling, J. T., & Phillips, J. L. 1992, *J. Geophys. Res.*, 97, 171

McComas, D. J., Gosling, J. T., Phillips, J. L., Bame, S. J., Luhmann, J. G., & Smith, E. J. 1989, *J. Geophys. Res.*, 94, 6907

Owens, M. J. & Crooker, N. U. 2006, *J. Geophys. Res.*, 111, A10104

Pagel, C., Crooker, N. U., Larson, D. E., Kahler, S. W., & Owens, M. J. 2005, *J. Geophys. Res.*, 110, A01103

Pagel, C., Gary, S. P., de Koning, C. A., Skoug, R. M., & Steinberg, J. T. 2007, *J. Geophys. Res.*, 112, A04103

Parker, E. N. 1958, *ApJ*, 128, 664

Schrijver, C. J., Title, A. M., van Ballegooijen, A. A., Hagenaar, H. J., & Shine, R. A. 1997, *ApJ*, 487, 424

Smith, E. J. & Balogh, A. 1995 *J. Geophys. Res.*, 105, 27217

Smith, E. J., Balogh, A., Forsyth, R. J., & McComas, D. J. 2001, *Geophys. Res. Lett.*, 28, 4159

Svalgaard, L. & Cliver, E. W. 2007, *ApJ(Letters)*, 661, L203

Wang, Y.-M., Lean, J., & Sheeley, N. R., Jr. 2000, *Geophys. Res. Lett.*, 27, 505

Wang, Y.-M. & Sheeley, N. R., Jr. 2003, *ApJ*, 599, 1404

Zurbuchen, T. H. & Richardson, I. G., 2006, *Space Sci. Revs.*, 123, 31

Zurbuchen, T. H., Schwadron, N. A., & Fisk, L. A. 1997, *J. Geophys. Res.*, 102 (A11), 24, 175

discussion

USOSKIN: What kind of behavior would you expect from the background open flux during the Maunder minimum?

FISK: Disconnection of open flux at the simple current sheet that occurs in the heliosphere appears to be difficult. If there was a simple current sheet with limited disconnection for all cycle back to the Maunder minimum, then the same background level of open flux currently present in the heliosphere should have been present during the Maunder minimum as well.

BOCHSLER: Are you implying that the solar wind would work the same way if there were no helium?

FISK: Our theory for the solar wind assumes the mass flux is determined by the process by which material is released from loops by reconnection with open magnetic flux. There is no special role for helium.

VRŠNAK: Does the formula for $V_{SW}^2 \sim 1/T$ hold also for coronal holes, i.e., does it cover a proper velocity range (say 300–800 km/s)?

FISK: Yes, the temperature of loops under coronal holes is low, and the resulting solar wind speed is high, in our formula. The slow solar wind originates from hotter loops in regions outside of coronal holes. In some ways it is remarkable that a simple linear relationship describes both fast and slow solar wind which have substantially different origins.

Universal Heliophysical Processes
Proceedings IAU Symposium No. 257, 2008
N. Gopalswamy & D.F. Webb, eds.

© 2009 International Astronomical Union
doi:10.1017/S1743921309029172

Processes in the magnetized chromosphere of the Sun

S. S. Hasan

Indian Institute of Astrophysics, Koramangala, Bangalore 560034, India
email: `hasan@iiap.res.in`

Abstract. We review physical processes in magnetized chromospheres on the Sun. In the quiet chromosphere, it is useful to distinguish between the magnetic network on the boundaries of supergranules, where strong magnetic fields are organized in mainly vertical flux tubes and internetwork regions in the cell interiors, which have traditionally been associated with weak magnetic fields. Recent observations from Hinode, however, suggest that there is a significant amount of horizontal magnetic flux in the cell interior with large field strength. Furthermore, processes that heat the magnetic network have not been fully identified. Is the network heated by wave dissipation and if so, what is the nature of these waves? These and other aspects related to the role of spicules will also be highlighted. A critical assessment will be made on the challenges facing theory and observations, particularly in light of the new space experiments and the planned ground facilities.

Keywords. Sun: magnetic – magnetohydrodynamics (MHD), Sun: chromosphere, Sun: oscillations

1. Introduction

It is well known that magnetic fields play an important role in the dynamics of the solar chromosphere. In the chromosphere on the quiet Sun it is useful to distinguish between the *magnetic network* on the boundary of supergranulation cells (Simon & Leighton 1964), where strong magnetic fields are organized in mainly vertical flux tubes, and *internetwork* regions in the cell interiors, where it was earlier believed that the magnetic fields are weaker and dynamically less important. As we shall subsequently see, this picture needs to be modified in the light of new observations.

The main focus of this review is a study of dynamical processes in the magnetized solar chromosphere, which we treat as the region above the photospheric surface (defined as the layer with continuum optical depth unity) with an extension of about 2 Mm. Above the temperature minimum and up to the middle chromosphere (close to a height of about 1.5 Mm), the atmosphere can be effectively regarded as almost isothermal. In the higher layers, the observed properties of the chromosphere are strongly influenced by magnetic fields. A convenient way to parameterize the field strength is in terms of β, defined as the ratio of the gas to magnetic pressure. The $\beta = 1$ (which is not uniform with height) provides a natural separation of the atmosphere into magnetic and non-magnetic (or weakly magnetized) regions. In the lower chromosphere and below, the magnetic field is structured in the form of magnetic flux tubes, which occur at the cell boundaries and constitute the well known magnetic network. These tubes that are mainly vertical and in pressure equilibrium with the outside medium expand upward to conserve magnetic flux. From a low filling factor ($< 1\%$) in the photosphere the tubes spread to 15% in the layers of formation of the emission features in the H and K lines of ionized calcium (at a height of 1 Mm) and to 100% in the so-called magnetic canopy. The remaining quiet Sun outside the network is called the internetwork, sometimes also referred to as cell interior.

The canonical picture of the magnetic network is that it consists of vertical magnetic fields clumped into elements or flux tubes that are located in intergranular lanes, have magnetic field strengths in the kilogauss range, and have diameters of the order of 100 km or less at their footpoints in the photosphere (e.g., Gaizauskas 1985; Zwaan 1987). These magnetic elements can be identified with bright points seen in images taken in the G-band (430.5 nm). High resolution observations show that these flux elements are in a highly dynamical state due to buffeting by convective flows on granular and supergranular scales (e.g., Muller *et al.* 1994; Berger & Title 1996; Nisenson *et al.* 2003). With the availability of new ground-based telescopes at excellent sites and sophisticated image reconstruction techniques it is now possible to examine magnetic elements with an improved resolution of about 0.17″ and investigate their structure and dynamics in unprecedented detail (e.g., Berger *et al.* 2004,; Rouppe van der Voort 2005; Langangen *et al.* 2007). High-quality observations of photospheric magnetic fields are now also being obtained with the Solar Optical Telescope (SOT) on Hinode (e.g., Lites *et al.* 2007, 2008). Magnetoconvection models have been developed to understand the three-dimensional structure and evolution of the magnetic field and its interactions with convective flows (e.g., Vögler *et al.* 2005; Schaffenberger *et al.* 2006; Steiner *et al.* 2008).

The chromospheric network is most clearly seen in filter images taken in the Ca II H & K lines (e.g.,Gaizauskas 1985; Rutten 2007) and in the Ca II IR triplet (Cauzzi *et al.* 2008). In H or K line images the network shows up as a collection of "coarse mottles" or "network grains" that stand out against the darker background. The network grains are continuously bright with intensities that vary slowly in time, in contrast to the "fine mottles" or "cell grains" which are located in the cell interiors and are much more dynamic (e.g., Rutten and Uitenbroek 1991).

Some of the important questions that need to be addressed are: (a) What are the physical processes contributing to the dynamics and heating of the magnetic network and produce the observed enhanced calcium emission? (b) What is the nature of the magnetic field in the internetwork and are these fields dynamically important?and (c) What mechanisms contribute to the fine structure of the chromosphere such as spicules? We shall attempt to shed light on these questions. The plan of this review is as follows: in Sect. 2, we briefly discuss recent observations of the magnetic network, followed in Sect. 3 by a theoretical model for interpreting the chromospheric emission in terms of magneto-acoustic shocks. In Sect. 4, we examine the nature of the internetwork field, particularly in the light of recent observations from Hinode. In Sect. 5, we discuss spicules and their role in the dynamics and energetics of the chromosphere. Finally in Sect. 6, we consider some implications and outstanding issues, particularly for future observational ground and space programmes.

2. Observations of the magnetic network

Recently a large number of high-resolution images of the solar atmosphere have become available from gound-based telescope, thanks largely to adaptive optics and image reconstruction techniques. These support the hypothesis of a network patch consisting of several discrete magnetic elements. A comparison with the Ca II line center image shows that the excess chromospheric emission is localized directly above the photospheric flux tubes, although the bright features seen in Ca II are more diffuse than those seen in the G-band. Time sequences of such images show that the chromospheric network is continually bright (e.g., Tritschler *et al.* 2007). Rutten (2006, 2007) presented reviews and a synthesis of recent high-resolution observations of the solar chromosphere. In the Ca II H & K lines, the network shows up as a collection of "Ca II bright points" or "grains". He

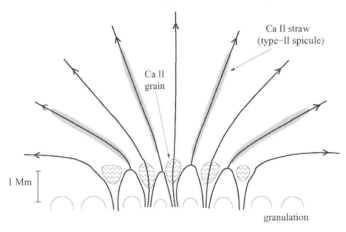

Figure 1. Schematic diagram showing the structure of a magnetic network element on the quiet Sun. The thin half-circles at the bottom of the figure represent the granulation flow field, and the thick curves represent magnetic field lines of flux tubes that are rooted in the intergranular lanes. The Ca II bright grains are thought to be located inside the flux tubes at heights of about 1 Mm above the base of the photosphere. We suggest that the Ca II "straws" (Rutten 2006) may be located at the boundaries between the flux tubes (from Hasan & van Ballegoijen 2008).

also identifies an exciting new phenomenon called "straws" that extend radially outward from network bright points (see Figure 8 in Rutten 2007). These straws are very thin, occur in "hedge rows", and are very short-lived (10-20 s). They appear to be closely related to the so-called "type-II" spicules recently identified in limb observations with SOT on Hinode (De Pontieu 2007a; see Sect. 5 for more details).

Our interpretation of the Ca II observations is summarized in Figure 1, where we show a vertical cross section of a magnetic network element consisting of several discrete flux tubes. We suggest that the Ca II network grains are located inside the magnetic flux tubes, and give rise to the bulk of the Ca II emission from the network element. The grains are thought to be located at heights between 500 km and 1500 km above the photosphere where the flux tubes are no longer "thin" compared to the pressure scale height (about 200 km), but are still well separated from each other. The Ca II straws ("type-II" spicules) have widths of order 100 km, and are located at larger heights (several Mm) where the widths of the flux tubes are much larger than 100 km. Therefore, we suggest the straws are not directly associated with the network grains in the low chromosphere. In Figure 2 we assumed that the straws ("type-II" spicules) are located at the *interfaces* between the flux tubes, as suggested by van Ballegoijen & Nisenson (1998).

It seems unlikely that long-period waves are also responsible for the *heating* of the Ca II network grains in the low chromosphere. Simulations of shock waves with periods $P \sim 200$ s in a plane-parallel, non-magnetic atmosphere have shown that such waves produce large asymmetries in the Ca II H line profiles, and strong variations in the integrated emission. If the grains were heated by such long-period waves, they should exhibit similar strong intensity variations. This is not observed, so the long-period waves observed in network elements cannot be the main source of heating for the Ca II grains. Network grains could possibly be heated by dissipation of waves with shorter periods ($P < 100$ s). Ground-based observations of high-frequency waves in small network elements are affected by seeing, so it is possible that waves with periods $P < 100$ s do exist in network elements but are simply not observable from the ground. This hypothesis could be tested using Ca II H images from Hinode (Kosugi *et al.* 2007), keeping in mind of course that the passband of the Ca II H filter on SOT also includes a significant photospheric

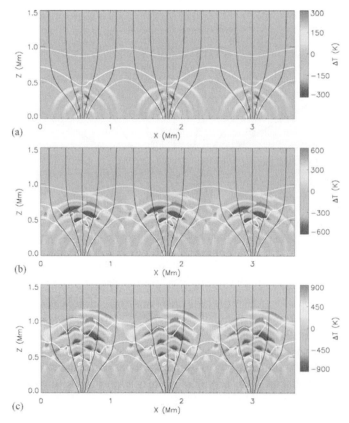

Figure 2. The temperature perturbation, ΔT, (about the initial state) at (a) 75 s, (b) 122 s, and (c) 153 s in a network region consisting of 3 flux tubes. Wave excitation is due to periodic horizontal motion at the lower boundary, with an amplitude of 750 m s^{-1}, and a period of 24 s. The black curves denote the magnetic field lines, and the color scale shows the temperature perturbation. The white curves denote contours of constant β corresponding to $\beta = 0.1$ (upper curve), 1.0 (thick curve) and 10 (lower curve) (from Hasan & van Ballegoiijen 2008).

contribution. Alternatively, network grains could be heated by dissipation of Alfvén waves. At present there is no direct observational evidence for Alfvén waves in flux elements in the photosphere, nor at heights where the Ca II H line is formed.

3. Dynamics of the magnetic network

Earlier work idealized the network in terms of thin flux tubes (e.g., Roberts & Webb 1978) and treated wave propagation in terms of the well known transverse (kink) and longitudinal (sausage) modes (e.g., Spruit 1981). Several investigations have focused on the generation and propagation of transverse and longitudinal wave modes and their dissipation in the chromosphere (e.g., Zhugzhda *et al.* 1995; Ulmschneider 2003 and references therein). Torsional waves have received some attention (e.g., Hollweg *et al.* 1982; Routh, Musielak & Hammer 2007). Hasan and Kalkofen (1999) examined the excitation of transverse and longitudinal waves in magnetic flux tubes by the impact of fast granules on flux tubes, as observed by Muller and Roudier (1992) and Muller *et al.* (1994), and following the investigation by Choudhuri *et al.* (1993), who studied the generation of kink waves by footpoint motion of flux tubes. The observational signature of the modelled process was highly intermittent in radiation emerging in the H and K lines, contrary to observations.

By adding waves that were generated by high-frequency motions due to the turbulence of the medium surrounding flux tubes the energy injection into the gas inside a flux tube became less intermittent, and the time variation of the emergent radiation was in better agreement with the more steady observed intensity from the magnetic network (Hasan *et al.* 2000).

The above studies modelled wave excitation and propagation in terms of the Klein-Gordon equation, motivated by the identification of the power peak near 7 min. in the observed power spectrum (Lites *et al.* 1993) with the cutoff period of kink waves in thin magnetic flux tubes (Kalkofen 1997). This analysis was based on a linear approximation in which the longitudinal and transverse waves are decoupled. However, the motions are expected to become supersonic higher up in the atmosphere. At such heights, nonlinear effects become important, leading to a coupling between the transverse and longitudinal modes. Some progress on this question has been made in one dimension, using the non-linear equations for a thin flux tube by Ulmschneider *et al.* (1991), Huang *et al.* (1995), Zhugzhda *et al.* (1995), and more recently by Hasan *et al.* (2003) and Hasan & Ulmschneider (2004), who examined mode coupling between transverse and longitudinal modes in the magnetic network. By solving the nonlinear, time-dependent MHD equations it was found that significant longitudinal wave generation occurs in the photosphere, typically for Mach numbers as low as 0.2, and that the onset of shock formation occurs at heights of about 600 km above the photospheric base, accompanied by heating (Hasan *et al.* 2003, Huang *et al.* 1995). The efficiency of mode coupling was found to depend on the magnetic field strength in the network and achieved a maximum for field strengths corresponding to $\beta \approx 0.2$, when the kink and tube wave speeds are almost identical. This can have interesting observational implications. Furthermore, even when the two speeds are different, once shock formation occurs, the longitudinal and transverse shocks exhibit strong mode coupling.

The above studies on the magnetic network make use of two important idealizations: they assume that the magnetic flux tubes are thin, an approximation that becomes invalid at about the height of formation of the emission peaks in the cores of the H and K lines; and they neglect the interaction of neighboring flux tubes. Some progress in this direction has been made in recent years by Rosenthal *et al.* (2002) and Bogdan *et al.* (2003), who studied wave propagation in a two-dimensional stratified atmosphere, assuming a potential magnetic field to model the network and internetwork regions on the Sun. They examined the propagation of waves that are excited from a spatially localized source in the photosphere. Their results indicate that there is strong mode coupling between fast and slow waves at the so-called magnetic canopy, which they identify with regions where the magnetic and gas pressures are comparable. Wave propagation in a more realistic configuration consisting of a flux sheet embedded in a field free atmosphere was considered by Hasan *et al.* (2005) and in multiple flux sheets by Hasan & van Ballegoiijen (2008). Other related work has been carried out using 3-D simulations by Vögler *et al.* (2005), Schaffenberger *et al.* (2005) and Carlsson and Bogdan (2006).

Figure 2 shows the wave propagation in a network element idealized in terms of three identical flux tubes (in a 2-D medium) driven by transverse periodic motions at the lower boundary with a velocity amplitude of 750 m s^{-1} and a period of 24 s. The black and white lines respectively denote the magnetic field and contours of constant β for values of $\beta = 0.1$ (upper curve), 1.0 (thick curve) and 10 (lower curve). At the initial epoch (the equilibrium state), the magnetic field strength on the axis of the tubes at the base $z = 0$ is 1000 G, corresponding to a β of about 2.0.

The horizontal motions at the lower boundary produce compressions and decompressions of the gas in the flux tube which generate an acoustic like wave (most effectively

Table 1. Temporal maximum of horizontally averaged vertical component of fluxes (from Vigeesh *et al.* 2008).

Initial Excitation	$F_{A,z}$ $(10^6$ erg cm^{-2} s$^{-1})$			$F_{P,z}$ $(10^6$ erg cm^{-2} s$^{-1})$		
	$z = 100$ km	$z = 500$ km	$z = 1000$ km	$z = 100$ km	$z = 500$ km	$z = 1000$ km
0.75 km s^{-1}, 24s	11.36	1.96	1.33	29.38	1.08	0.14
0.75 km s^{-1}, 120s	35.75	27.70	4.02	134.29	0.79	0.07
0.75 km s^{-1}, 240s	20.90	8.58	3.30	131.84	0.36	0.02
1.50 km s^{-1}, 24s	44.55	7.68	3.34	115.79	4.29	0.57
3.00 km s^{-1}, 24s	168.41	30.40	6.22	434.03	16.90	2.31

at the interface between the tube and ambient medium as shown by Hasan *et al.* 2005) that propagates isotropically with an almost constant sound speed. This can be discerned from the almost constant spacing in the semicircular color pattern. In the central section of the tube, a transverse slow MHD wave is generated that is essentially guided along the field lines. At $t = 75$ s, we find from Figure 2(a) that the wave pattern is confined below the $\beta = 1$ surface ($z \sim 0.5$ Mm). In this region, where the magnetic field can be regarded as weak, the acoustic (fast) mode travels ahead of the (slow) MHD wave (which travels at the Alfvén speed). At the $\beta = 1$ level there is a strong coupling between the two modes as previously demonstrated by Rosenthal *et al.* (2002), Bogdan *et al.* (2003) and Hasan *et al.* (2005). Up to this epoch the waves in the individual tubes are sufficiently well separated from each other and the wave pattern in each tube is qualitatively similar to that in a single tube (Hasan & van Ballegoiijen 2008). However, at $t = 122$ s waves emanating from neighboring tubes interact with each other especially in the ambient medium. However, the wave pattern in any tube is not significantly affected by the presence of its neighbors. Furthermore, the slow magneto-acoustic waves above the $\beta = 1$ surface are confined close to the central regions of the tubes, where they steepen and produce enhanced heating. This heating appears to be dominantly caused by the wave motions generated at the footpoints and *not* by the penetration of acoustic waves from the ambient medium or by waves coming from neighboring tubes (Hasan & van Ballegoiijen 2008).

We now consider the transport of energy in the various wave modes. Following Bogdan *et al.* (2003) we use the linear expression for the energy flux to identify the energy carried by the dominantly acoustic and magnetic (Poynting flux) components. Table 1 (taken from Vigeesh, Hasan & Steiner 2008) shows the temporal maximum of the horizontally averaged vertical components of acoustic and Poynting fluxes at three different heights due to impulsive transverse excitation of the lower boundary. They have considered three different amplitudes for the impusive excitations. As seen in Figure 8 of Vigeesh *et al.* (2008), although the maximum vertical component acoustic fluxes reach values of the order of 10^7 erg cm^{-2} s^{-1} at a height of $z = 1000$ km, depending upon the amplitude of the initial excitation, the average fluxes are an order of magnitude less.

The Poynting fluxes shown in the table represent the maximum value that the fluxes reach in the interval between the start of the simulation until the time when the fast wave reaches the top boundary (around 60 s). Hence these fluxes correspond to the fast mode. The Poynting fluxes associated with the fast mode are relatively lower in magnitude compared to the acoustic fluxes. It should be noted that there is also some Poynting flux associated with the slow mode, since these waves also perturb the magnetic field. But, they are relatively lower than the fluxes that are transported with the fast mode.

Let us now estimate the acoustic energy flux transported into the chromosphere through a single short duration pulse as considered by Vigeesh *et al.* (2008). The maximum values

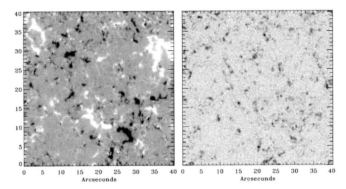

Figure 3. Vertical B_{app}^{L} (left panel) and horizontal B_{app}^{T} apparent flux densities in a quiet-Sun map. The grey scale for B_{app}^{L} saturates at ± 50 Mx cm^{-2} whereas B_{app}^{T} saturates at 200 Mx cm^{-2}. White in the left panel is positive flux density, and dark in the right corresponds to high values of the transverse apparent flux density (from Lites *et al.* 2008).

of the acoustic fluxes at $z = 1000$ km is $\sim 15 \times 10^6$ erg cm^{-2} s^{-1} which is adequate to balance the radiative losses (of the order of 10^7 erg cm^{-2} s^{-1}) in the magnetic network at chromospheric heights. It should be noted that although the fluxes can reach values upto 10^7 erg cm^{-2} s^{-1}, the average values are much less. But multiple excitations can form waves that develop into shocks that follow one another and overtake, increasing the shock strength and thereby increasing the dissipation. Also, in order to be compatible with the observed quasi-steady Ca emission the injection needs to be in the form of sustained multiple short duration pulses as argued by Hasan & van Ballegoiijen (2008). Although long-period acoustic waves (with periods of about 5 min.) have been proposed by De Pontieu *et al.* (2004) as driving spicules, we do not believe that these are responsible for the heating observed in Ca network grains. It should be pointed out that presently there is no observational confirmation for the existence of short-period magneto-acoustic waves in the network. However, SOT on Hinode provides us the possibility to test this hypothesis and ascertain whether such waves can be considered legitimate candidates for heating the magnetized chromophere.

4. Internetwork magnetism

Till recently, it was believed that the magnetic field in the interior of supergranule cells was weak with a mean field strength of a few Gauss. This hypothesis was based on low spatial resolution (greater than 1″) measurements (e.g., Meunier, Solanki & Livingston 1998). However, with increase in resolution, it was found that the magnetic flux in quiet regions increases exponentially with spatial resolution from 1 G for 2″–3″ to around 20 G at 0.5″ (Sanchez Almeida *et al.* 2004) though recent observations from Hinode show that the variation is weak up to 0.3″ resolution (Lites *et al.* 2007). Hanle depolarization signals are consistent with a "hidden" turbulent magnetic field with a typical range of 20-150 G in the internetwork (Trujillo Bueno *et al.* 2004).

A major finding has taken place recently regarding the nature of magnetic fields in the internetwork (IN). New observations from the Hinode Stokes Polarimeter (SP) (with a spatial resolution of 0.3″) reveal the ubiquitous presence of horizontal fields in the range 100-200 G. Figure 3, taken from Lites *et al.* (2008), shows the vertical (left panel) and horizontal (right panel) components of the apparent magnetic flux densities B_{app}^{L} and B_{app}^{T} respectively in a quiet region of the Sun. These observations show that, whereas the vertical magnetic field mainly occurs in the intergranular lanes at the network boundaries, the field in the internetwork regions is dominantly horizontal and well separated from the vertical fields. The horizontal fields, with an average value of at least 55 G, are located preferentially at the edges of bright granules. Lites *et al.* (2008) conjecture that these fields might be important in understanding the "hidden" turbulent flux inferred from the Hanle effect. Recently, Steiner *et al.* (2008) carried out numerical simulations in order to reproduce these observations. They use the line pair of Fe I 630 nm to synthesize Stokes profiles and derive a ratio of 4.3 between the horizontal and vertical components of the magnetic field, which is reasonably close to the observed value of around 5 found by Lites *et al.* (2008). Steiner *et al.* (2008) suggest that the horizontal fields are generated due to flux expulsion.

5. Spicules

Despite their discovery over a century ago, spicules are among the least understood phenomena in the chromosphere. They are jets of gas which can be best observed on the limb in the emission lines of H_α or He I. Spicules are believed to originate in the magnetic network, but the field strength and geometry associated with the spicule channel is not well known. It should be noted that the disc counterpart of spicules are mottles. Recently, some progress has been made in this direction as well as on the mechanism that drives spicules. Trujillo Bueno *et al.* (2005) used the Hanle and Zeeman effects to estimate a field strength of about 10 G at a height of 2000 km above the photosphere and inclined at about 35° to the vertical. Tsiropoula and Tziotziou (2004) independently estimated a field strength of 4.1 G in mottles/spicules based on energy arguments.

From an analysis of high resolution images, De Pontieu *et al.* (2004) inferred a connection between spicules and photospheric p-modes. Although p-mode photospheric oscillations with 5 min. periods are generally evanescent in the upper photosphere, they can leak sufficient energy into the chromosphere. The tunelling becomes particularly effective for waves in inclined magnetic flux tubes because of their higher acoustic cutoff period which can exceed 300 s above the temperature minimum (De Pontieu *et al.* 2004). In the chromosphere, these waves form shocks, which drive spicules similar to the model of Hollweg *et al.* (1982). Using observations from the Solar Optical Telescope on Hinode, De Pontieu *et al.* (2007a) hypothesized that there are at least two species of spicules: "type-I" spicules which are driven by shock waves as discussed above with time scales of 3-7 min. and "type-II" spicules that are much more dynamic and very thin (width ~ 100 km), have lifetimes of 10-150 s, and seem to be rapidly heated to transition region temperatures, sending material though the chromosphere at speeds of 50-150 km s^{-1}. De Pontieu and collaborators suggest that "type-II" spicules may be due to small-scale reconnection events in the chromosphere. De Pontieu *et al.* (2007b) point out that spicules exhibit transverse motions with velocities of 10 to 25 km s^{-1} which may be the signature of Alfvén waves.

An alternative mechanism for spicules based on a model first proposed by Pikelner (1969) is that they are driven by magnetic reconnection in mixed polarity regions at network boundaries (Wilhelm 2000). This model is still schematic and has not been investigated quantitatively. Oscillations with periods of around 50 s have also been observed in spicules, which might be due to kink waves excited by the impact of granules on their footpoints (Kukhianidze *et al.* 2005) or due to Alfvén waves (Hollweg 1982; Kudoh & Shibata 1999; De Pontieu *et al.* 2007b).

6. Implications for future observations

This review has attempted to highlight important new developments concerning processes in the magnetized solar chromosphere as well as to point out some of the outstanding problems. Some implications for future ground and space missions are:

(*a*) Observations as well as simulations reveal that magnetic structures need to be resolved to an accuracy better than 0.1″ on the solar surface. Neither the present missions such as SOHO or TRACE (1″ resolution) and Hinode (150 km resolution) nor those in the near future such as SDO have this capability. It is imperative that the next generation of space missions have instruments that can achieve such an accuracy;

(*b*) Ground-based observations at good sites using adaptive optics can yield a resolution of 1″ or better. However, to simultaneously achieve high, spatial, temporal and spectral resolution, a large aperture ($\geqslant$ 2-m) is required to get a high throughput of photons;

(*c*) Polarimetric measurements with a high sensitivity (a few Gauss) of the magnetic field and its inclination over a large field of view are needed to understand the magnetic topology of the chromosphere;

(*d*) In addition, it is important to examine the distribution of magnetic structures in the network and internetwork regions and determine the magnetic filling factor and also geometry of the magnetic canopy accurately;

(*e*) Spectroscopic measurements in UV using temperature sensitive lines are essential to demarcate the thermal structure of the chromosphere and cross-correlate it with magnetic observations.

References

Berger, T. E., Rouppe van der Voort, L. H. M., Löfdahl, M. G., Carlsson, M., Fossum, A., Hansteen, V., Marthinussen, E., Title, A., & Scharmer, G. 2004, *A&A*, 428, 613

Berger, T. E. & Title, A. M. 1996, *ApJ*, 463, 365

Bogdan, T. J., Carlsson, M., Hansteen, V., McMurry, A., Rosenthal, C. S., Johnson, M., Petty-Powell, S., Zita, E. J., Stein, R. F., McIntosh, S. W., & Nordlund, Å. 2003, *ApJ*, 599, 626

Carlsson, M. & Bogdan, T. J. 2006, *Roy. Soc. Phil. Trans.*, Series A, Vol. 364, p. 395

Cauzzi, G., Reardon, K. P., Uitenbroek, H., Cavallini, F., Falchi, A., Falciani, R., Janssen, K., Timmele, T., Vecchio, A., & Wöger, F. 2008, *A&A*, 480, 515

Choudhuri, A. R., Auffret, H., & Priest, E. R. 1993, *Solar Phys.*, 143, 49

De Pontieu, B., Erdélyi, R., & James, S. P. 2004, *Nature*, 430, 536

De Pontieu, B., McIntosh, S., Hansteen, V. H., Carlsson, M., Schrijver, C. J., Tarbell, T. D., Title, A. M., Shine, R. A., Suematsu, Y., Tsuneta, S., Katsukawa, Y., Ichimoto, K., Shimizu, T., & Nagata, S. 2007a, *PASJ*, 59, S655

De Pontieu, B., McIntosh, S., Carlsson, M., Hansteen, V. H., Tarbell, T. D., Schrijver, C. J., Title, A. M., Shine, R. A., Tsuneta, S., Katsukawa, Y., Ichimoto, K., Suematsu, Y., Shimizu, T., & Nagata, S. 2007b, *Science*, 318, 1574

Gaizauskas, V. 1985, in *Chromospheric Diagnostics and Modeling*, ed. B. W. Lites (National Solar Observatory: Sunspot, NM), 25

Hasan, S. S. & Kalkofen, W. 1999, *ApJ*, 519, 899

Hasan, S. S., Kalkofen, W., & van Ballegooijen, A. A. 2000, *ApJ*, 535, L67

Hasan, S. S., Kalkofen, W., van Ballegooijen, A. A., & Ulmschneider, P. 2003, *ApJ*, 585, 1138

Hasan, S. S., & Ulmschneider, P., 2004, *A&A*, 422, 1085

Hasan, S. S., van Ballegooijen, A. A., Kalkofen, W., & Steiner, O. 2005, *ApJ*, 631, 1270

Hasan, S. S. & van Ballegooijen, A. A. 2008, *ApJ*, 680, 1542

Hollweg, J. V., Jackson, S., & Galloway, D. 1982, *Solar Phys.*, 75, 35

Huang, P., Musielak, Z. E., & Ulmschneider, P. 1995, *A&A*, 297, 579

Kalkofen, W. 1997, ApJ., 486, L145

Kukhianidze, V., Zaqarashvili, T. V., & Khutsishvili, E. 2006, *A & A*, 449, L35

Kosugi, T., Matsuzaki, K., Sakao, T., Shimizu, T., Sone, Y., Tachikawa, S., Hashimoto, T., Minesugi, K., Ohnishi, A., Yamada, T., Tsuneta, S., Hara, H., Ichimoto, K., Suematsu, Y., Shimojo, M., Watanabe, T., Davis, J. M., Hill, L. D., Owens, J. K., Title, A. M., Culhane, J. L., Harra, L., Doschek, G. A., & Golub, L. 2007, *Solar Phys.*, 243, 3

Kudoh, T. & Shibata, K. 1999, *ApJ*, 514, 493

Langangen, Ø., Carlsson, M., & Rouppe van der Voort, L. 2007, *ApJ*, 655, 615

Lites, B. W., Rutten, R. J., & Kalkofen, W. 1993, *ApJ*, 414, 345

Lites, B., Socas-Navarro, H., Kubo, M., Berger, T. E., Frank, Z., Shine, R. A., Tarbell, T. D., Title, A. M., Ichimoto, K., Katsukawa, Y., Tsuneta, S., Suematsu, Y., Shimizu, T., & Nagata, S. 2007, *PASJ*, 59, S571

Lites, B., Kubo, M., Socas-Navarro, H., Berger, T. E., Frank, Z., Shine, R. A., Tarbell, T. D., Title, A. M., Ichimoto, K., Katsukawa, Y., Tsuneta, S., Suematsu, Y., Shimizu, T., & Nagata, S. 2008, *ApJ*, 672, 1237

Meunier, N., Solanki, S. K., & Livingston, W. C. 1998, *A&A*, 331, 771

Muller, R. & Roudier, Th. 1992, *Solar Phys.*, 141, 27

Muller, R., Roudier, Th., Vigneau, J., & Auffret, H. 1994, *A&A*, 283, 232

Nisenson, P., van Ballegooijen, A. A., de Wijn, A. G., & Sütterlin, P. 2003, *ApJ*, 587, 458

Roberts, B. & Webb, A. R. 1978, *Solar Phys.*, 56, 5

Rosenthal, C. S., Bogdan, T. J., Carlsson, M., Dorch, S. B. F., Hansteen, V., McIntosh, S. W., McMurry, A., Nordlund, Å., & Stein, R. F. 2002, *ApJ*, 564, 508

Rouppe van der Voort, L. H. M., Hansteen, V. H., Carlsson, M., *et al.*, 2005, *A&A*, 435, 327

Routh, S., Musielak, Z. E., & Hammer, R., 2007, *Solar Phys.*, 246, 133

Rutten, R. J. & Uitenbroek, H. 1991, *Sol. Phys.*, 134, 15

Rutten, R. J. 2006, in *Solar MHD: Theory and Observations*, Proc. NSO Workshop 23, eds. J. Leibacher, H. Uitenbroek, & R. J. Stein, ASP Conf. Ser. 354 (San Francisco: ASP), p. 276

Rutten, R. J. 2007, in *The Physics of Chromospheric Plasmas*, eds. P. Heinzel, I. Dorotovic, & R. J. Rutten, ASP Conf. Ser. 368 (San Francisco: ASP), p. 27

Sánchez Almeida, J., in *The Solar-B Mission and the Forefront of Solar Physics*, eds. T. Sakurai and T. Sekii, ASP Conf. Ser. 325, (San Francisco: ASP), p. 115

Schaffenberger, W., Wedemeyer-Böhm, S., Steiner, O., & Freytag, B. 2006, in *Solar MHD: Theory and Observations: A High Spatial Resolution Perspective,* eds. J. Leibacher, R. F. Stein, and H. Uitenbroek, ASP Conf. Ser. Vol. 354, (San Francisco: ASP), 345

Simon, G. W. & Leighton, R. B. 1964, *ApJ*, 140, 1120

Spruit, H. C. 1981, *A&A*, 102, 129

Steiner, O., Knölker, M., & Schüssler, M. 1994, in *Solar Surface Magnetism*, eds. R. J. Rutten & C. J. Schrijver, NATO ASI Ser. C-433, (Dordrecht: Kluwer), p. 441

Steiner, O., Rezaei, R., Schaffenberger, W., & Wedemeyer-Böhm, S. 2008, *ApJ*, 680, L85

Tritschler, A., Schmidt, W., Uitenbroek, H., & Wedemeyer-Böhm, S. 2007, *A&A*, 462, 303

Trujillo Bueno, J., Merenda, L., Centeno, R. *et al.* 2005, *ApJ*, 619, L191-L194

Tsiropoula, G., & Tziotziou, K., 2004, *A&A*, 424, 279

Ulmschneider, P., Zähringer, K., & Musielak, Z. E., 1991, *A & A*, 241, 625

Ulmschneider, P. 2003, in *Lectures on Solar Physics,* eds. Antia, H. M., Bhatnagar, A., & Ulmschneider P., Lecture Notes in Physics 619, Springer Verlag, Heidelberg, Berlin, p. 232

van Ballegooijen, A. A., & Nisenson, P. 1998, in *High Resolution Solar Physics: Theory, Observations, and Techniques,* eds. T. R. Rimmele, K. S. Balasubramaniam, & R. R. Radick, ASP Conf. Ser. 183 (San Francisco: ASP), 30

Vigeesh, G., Hasan, S. S., & Steiner, O. 2008, *A&A*, submitted

Vögler, A., Shelyag, S., Schüssler, M., Cattaneo, F., & Emonet, T. 2005, *A&A*, 429, 335

Wilhelm, K. 2000, *A&A*, 360, 351

Zhugzhda, Y. D., Bromm, V., & Ulmschneider, P. 1995, *A&A*, 300, 302

Zwaan, C. 1987, *ARAA*, 25, 83

Discussion

SHIBATA: I have one comment and one question. My comment is about spicule models. You introduced the reconnection model and the Alfven wave model as different models. However, our numerical simulation (Takenchi and Shibata 2001, ApJ 546, L73) shows that Alfven waves are generated by reconnection. So the reconnection model and the Alfven wave model are not necessarily separate models. Furthermore, slow mode magnetoacoustic shocks are generated from nonlinear Alfven waves by mode coupling. Altogether, slow shocks, Alfven waves, and reconnection are all closely related. My question is also about spicules. I am not convinced about the two types of spicules. What is the fraction of type I and type II spicules?

HASAN: The groups that have introduced this classification of two categories of spicules have, to the best of my information, not provided statistics on this fraction. However, I agree with you that it would be important to know this.

WEBB: We have heard that Hinode sees ubiquitous Alfven waves that dissipate sufficient energy to heat the corona. Have you checked energetics of your shock dissipation model to see if it can provide sufficient energy to heat the corona?

HASAN: As I showed, the Poyting flux that corresponds to the energy flux in the magnetic wave in our simulations is an order of magnitude lower than is required for coronal heating. The important question which needs to be addressed is how this energy can be efficiently dissipated in the corona.

Universal Heliophysical Processes
Proceedings IAU Symposium No. 257, 2008
N. Gopalswamy & D.F. Webb, eds.

© 2009 International Astronomical Union
doi:10.1017/S1743921309029184

Magnetic helicity of solar active regions

A. Nindos[1]

[1]Section of Astrogeophysics, Physics Department, University of Ioannina,
Ioannina GR-45110, Greece
email: anindos@cc.uoi.gr

Abstract. Magnetic helicity is a quantity that describes the linkage and twistedness/shear in the magnetic field. It has the unique feature that it is probably the only physical quantity which is approximately conserved even in resistive MHD. This makes magnetic helicity an ideal tool for the exploration of the physics of eruptive events. The concept of magnetic helicity can be used to monitor the whole history of a CME event from the emergence of twisted magnetic flux from the convective zone to the eruption and propagation of the CME into interplanetary space. In this article, I discuss the sources of the magnetic helicity injected into active regions and the role of magnetic helicity in the initiation of solar eruptions.

Keywords. Sun: activity, Sun: magnetic fields, Sun: coronal mass ejections (CMEs), Sun: flares

1. Introduction

Magnetic helicity, H, quantifies the deviation of a magnetic flux tube from its minimum energy state which corresponds to potential magnetic field. In other words it tells us how much a magnetic flux tube is sheared or/and twisted. For an ensemble of magnetic flux tubes, magnetic helicity can be regarded as a measure of the topological complexity of the field giving information about the linkage and twistedness in the field. The "natural" unit of helicity is the square of magnetic flux (Mx^2) and therefore the helicity of a twisted flux tube with N turns and magnetic flux equal to unity is simply N.

It is well established (e.g. see Berger 1984) that magnetic helicity is very well preserved in plasmas with high magnetic Reynolds numbers, even in the presence of dissipative processes such as magnetic reconnection (more accurately, it is approximately conserved on time scales smaller than the global diffusion time scale; see Berger 1984). This property of helicity has important consequences in the evolution of magnetic fields: a stressed magnetic field cannot relax to a potential field. This behavior may have important implications for the initiation of flares and coronal mass ejections (CMEs).

In this article a short review of magnetic helicity of active regions (ARs) is given. For a more detailed review the interested reader is referred to the article by Démoulin (2007). My article is organized as follows. After defining magnetic helicity and its flux, in section 3 I discuss the sources of magnetic helicity that is injected into ARs. The following two sections address problems that are related to its acurate calculation: the computation of maps of helicity flux density (section 4) and the computation of flows that inject helicity into ARs. In section 6, the role of magnetic helicity in the initiation of eruptive events is briefly outlined. Conclusions are presented in section 7.

2. Definitions

2.1. *Magnetic helicity*

For a magnetic field $\mathbf{B}$ fully contained within a volume V (i.e. at any point of its boundary S the normal component $B_n = \mathbf{B} \cdot \hat{\mathbf{n}}$ vanishes), magnetic helicity is defined as

$$H = \int_V \mathbf{A} \cdot \mathbf{B} dV, \tag{2.1}$$

where $\mathbf{A}$ is the magnetic vector potential ($\mathbf{B} = \nabla \times \mathbf{A}$). H is independent of the gauge selection for $\mathbf{A}$ (i.e. independent of the transformation $\mathbf{A} \to \mathbf{A} + \nabla\Phi$, where Φ is any single-valued derivable function of space and time).

In the solar atmosphere magnetic flux passes through S (especially in the photosphere) and therefore the above condition is not satisfied. However, Berger and Field (1984) and Finn and Antonsen (1985) have shown that when $B_n \neq 0$ on S, we can define a gauge-invariant relative magnetic helicity (hereafter refered to as helicity) of $\mathbf{B}$ with respect to the magnetic helicity of a reference field $\mathbf{B}_p$ having the same distribution of normal magnetic flux on the surface S surrounding V:

$$H = \int_V \mathbf{A} \cdot \mathbf{B} dV - \int_V \mathbf{A}_p \cdot \mathbf{B}_p dV, \tag{2.2}$$

where $\mathbf{A}_p$ is the vector potential of $\mathbf{B}_p$. The quantity H does not depend on the common extension of $\mathbf{B}$ and $\mathbf{B}_p$ outside V. Being a potential field it is a convinient choice for $\mathbf{B}_p$. If in addition $\nabla \cdot \mathbf{A}_p = 0$ and $(A_p)_n = 0$ on S then the term $\int_V \mathbf{A}_p \cdot \mathbf{B}_p dV$ vanishes (Berger 1988), so H has the same expression as in the case of the helicity in closed volumes (eq. 2.1).

2.2. Flux of magnetic helicity

Generally, the amount of helicity within V can change either due to helicity flux crossing S or/and due to dissipation within V. Berger (1984) has demonstrated that the helicity dissipation rate is negligible in all processes taking place in the corona, including reconnection and all non-ideal processes. Helicity's dissipation time scale is the global diffusion time scale and consequently it can be regarded as an almost conserved quantity even in resistive MHD.

In the solar atmosphere V is part of the coronal volume, bounded from below by a portion of the photosphere S_p and bounded in the corona by S_c ($S_c = S - S_p$). No data can presently provide $\mathbf{B}$ on any S_c surface. The helicity flux across S_c can only be estimated indirectly by the helicity carried away by CMEs, and estimated in interplanetary space from the associated magnetic clouds. All studies compute the helicity injected at the photospheric level through S_p. Using the gauge $\nabla \cdot \mathbf{A}_p = 0$, and selecting the boundary condition $\mathbf{A}_p \cdot \hat{\mathbf{n}} = 0$ for the vector potential of the potential reference field, Berger & Field (1984) derived the flux of magnetic helicity through a planar surface:

$$\frac{dH}{dt} = 2 \int_{S_p} [(\mathbf{A}_p \cdot \mathbf{B}_t)v_n - (\mathbf{A}_p \cdot \mathbf{v}_t)B_n] dS, \tag{2.3}$$

where B_t and B_n are the tangential and normal components of the photospheric magnetic field and v_t and v_n the tangential and normal compoments of the photospheric plasma velocity.

3. Sources of helicity injected into active regions

The first term of the right-hand side of eq. (2.3) corresponds to the injection of helicity by advection (i.e. emergence of field lines that cross the photosphere) while the second term (also known as shearing term) is the flux of helicity due to motions parallel to S.

Such motions may come either from differential rotation and/or transient photospheric shearing flows.

Differential rotation was the first mechanism that injects helicity into ARs which was studied (DeVore 2000). Even when a single bipole is considered, differential rotation does not provide a monotonous input of magnetic helicity (DeVore 2000). This is because differential rotation rotates both magnetic polarities on themselves and also changes their relative positions, introducing twist and writhe helicity fluxes, respectively. These fluxes always have opposite signs and similar amplitudes, and therefore partially cancel (Démoulin *et al.* 2002a). Démoulin *et al.* (2002b) and Green *et al.* (2002) studied the long-term evolution of the helicity injected by differential rotation into the coronal part of two active regions which were followed from their birth until they decayed. The helicity injection rate from differential rotation was calculated as the sum of the rotation rate of all pairs of elementary fluxes weighted with their magnetic flux. These studies showed that the contribution of differential rotation to the helicity budget of active regions is small.

The total helicity stored into the corona at a given time can be calculated under the force-free field assumption ($\nabla \times \mathbf{B} = \alpha \mathbf{B}$). The best value of α, α_{best}, is determined by comparing the computed field lines with the observed soft X-ray (SXR) or EUV coronal structures. Then the computation of the coronal helicity is relatively straightforward (Berger 1985; see also Georgoulis & LaBonte 2007).

When high-cadence photospheric magnetograms are available, the horizontal velocity appearing in eq. (2.3) can be computed using the local correlation tracking (LCT) technique (November & Simon 1988). Several authors have computed the corresponding helicity injection rate (e.g. Chae 2001; Nindos & Zhang 2002; Moon *et al.* 2002a,b; Nindos *et al.* 2003; Chae *et al.* 2004). Démoulin & Berger (2003) have pointed out that with magnetograms one follows the photospheric intersection of the magnetic flux tubes but not the evolution of the plasma (generally the two velocities are different). Consequently, from the observed magnetic evolution we obtain the flux tube motion and not the plasma motion parallel to the photosphere. If $\mathbf{v}_t$ is the tangential component of the photospheric plasma velocity and v_n the velocity perpendicular to the photosphere, the LCT method detects the velocity of the footpoints of the flux tube which is

$$\mathbf{u} = \mathbf{v}_t - \frac{v_n}{B_n} \mathbf{B}_t. \tag{3.1}$$

The combination of eq. (3.1) and (2.3) shows that the whole helicity flux density can be retrieved within the accuracy of the calculation. Consequently, one may use the quantity $G_A = -2\mathbf{u} \cdot \mathbf{A}_p B_n$ as a proxy to the whole helicity flux density.

The study of the helicity budget of active regions requires knowledge of the helicity carried away from them. It has been recongized (e.g. Low 1996) that CMEs are the primary agents that remove helicity from active regions. The helicity content of a CME can be estimated by the change of coronal helicity of the source region during the event (e.g. Mandrini *et al.* 2005). Inside magnetic clouds H is estimated from in situ measurements of the magnetic field vector. This requires a flux rope model whose parameters are determined by a least square fit to the data because only local measurements are available (e.g. Lepping *et al.* 1990; Daso *et al.* 2003; 2006). In practice, in studies of the long-term evolution of helicity of active regions that are linked to at least one magnetic cloud at 1 AU one assumes that the helicity carried away by each CME is equal to the helicity content in the magnetic cloud. Nindos *et al.* (2003) and Lim *et al.* (2007) were able to partially reconcile the amount of helicity injected into the corona with the helicity carried away by the CMEs in the active regions they studied. However, the uncertainties of these studies are significant primarily due to the large uncertainties in the calculation of the helicity transported away by CMEs.

4. Maps of helicity flux density

The quantity $G_A = -2\mathbf{u} \cdot \mathbf{A}_p B_n$ can be used as a proxy to the helicity flux density (see the discussion in section 3). This proxy has been used extensively in several studies (references are given in section 3). In all these studies G_A maps always appear extremely complex both in space and time, with polarities of both signs present at any time. Pariat et al. (2005) showed that G_A is not a real helicity flux density and that its properties introduce artificial polarities of both signs (see middle column of Fig. 1). For example, G_A is non-zero even in flows that do not inject any magnetic helicity in the field. The spurious signals appear due to the fact that helicity flux densities per unit surface are not physical quantities. Due to the properties of helicity, only helicity flux density per unit of elementary magnetic flux has a physical meaning. But to estimate such quantity using real observations, it is necessary to isolate flux tubes and determine their connectivity, which is actually not possible. Thus any definition of a helicity flux density will only be a proxy of the helicity flux density per unit magnetic flux. Pariat et al. (2005) introduced a new proxy for helicity flux density, G_θ, which does not suffer from G_A's problems. G_θ

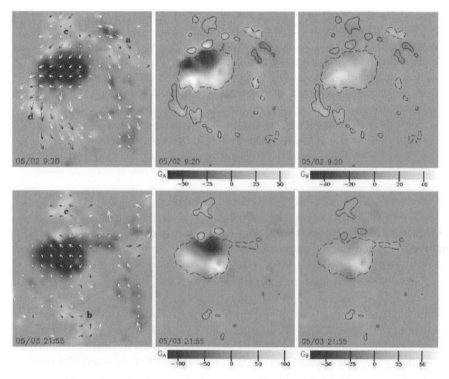

Figure 1. AR 8210 at 09:20 UT on May 2, 1998 (top) and at 21:55 UT on May 3, 1998 (bottom). Left panels: B_n magnetograms with velocity field (arrows). Center panels: G_A maps. Right panels: G_θ maps. G_A and G_θ maps are in units of 10^6 Wb^2 m^{-2}s^{-1} and have ±300 G isocontours of B_n. Note that the scale is not the same for the G_A and the G_θ maps (from Pariat et al. 2006).

implies that the helicity injection rate is the summation of the rotation rate $\frac{d\theta(\mathbf{x}-\mathbf{x}')}{dt}$ of all pairs of elementary fluxes weigthed by their magnetic flux $B_n d^2x$. Therefore it is:

$$G_\theta(\mathbf{x}) = -\frac{B_n}{2\pi} \int_{S_p} \frac{d\theta(\mathbf{x}-\mathbf{x}')}{dt} B_n' d^2x'. \tag{4.1}$$

In order to define the real helicity flux density, the coronal linkage needs to be provided. With it one can represent how all elementary flux tubes move relatively to a given elementary flux tube, and the helicity flux density is defined per elementary flux tube. Using photospheric maps this can be achieved by distributing equally the helicity input between the two footpoints for each elementary flux tube. Then the helicity flux can be rewritten as a flux of magnetic helicity per unit of surface, G_Φ. G_Φ is a field-weighted average of G_θ at both photospheric footpoints, $\mathbf{x}_\pm$, of the photosheric connection:

$$G_\Phi(\mathbf{x}_\pm) = \frac{1}{2}(G_\theta(\mathbf{x}_\pm) + G_\theta(\mathbf{x}_\mp)|B_n(\mathbf{x}_\pm)/B_n(\mathbf{x}_\mp)|). \tag{4.2}$$

While G_Φ provides the true helicity flux density, its practical use is presently limited by our ability to define the coronal linkage for all magnetic polarities. Currently, all we can do is to estimate G_Φ maps for models that resemble certain configurations and evolution patterns (Pariat *et al.* 2006; see below).

Pariat *et al.* (2006; 2007) computed G_A and G_θ maps at several occasions during the evolution of 5 active regions. Unlike the usual G_A maps, most of their G_θ maps showed almost unipolar spatial structures (see Fig. 1) because the nondominant helicity flux densities were significantly suppressed. In a few cases the G_θ maps still contained spurious bipolar signals. With further modelling the computed models of G_Φ were again unipolar. The result of injection of helicity with a coherent sign on the AR scale needs to be checked against statistical studies. If future studies confirm it solar dynamo models will need to explain the formation of twisted flux tubes with either positive or negative helicity but not mixed-sign helicity at the spatial scales resolved by the flow computation methods.

On time scales larger than their transient temporal variations, the time evolution of the total helicity fluxes derived from G_A and G_θ show small differences (see Fig. 2). Theoretically one expects that the helicity flux integrated using G_A and using G_θ should be identical because both definitions are derived from eq. (2.3). The reported small differences may result from the computation of $\mathbf{A}_p$ with a fast Fourier transform of the magnetogram which implies an implicit periodicity of the magnetic flux distribution while with the G_θ computation one assumes that no magnetic flux is present around the magnetogram. Chae (2007) and Jeong & Chae (2007) reported that the integration of G_A typically overestimates the helicity injection about 10-30%. Furthermore, unlike G_A, with G_θ the time evolution of the total flux is determined primarily by the predominant-signed flux while the nondominant-signed flux is roughly stable and probably mostly due to noise (see Fig. 2).

5. Computation of photospheric flows

The discussion in sections 3 and 4 indicates that the computation of photospheric flows is an essential ingredient in any attempt to compute the helicity injected into the coronal part of active regions. The traditionally used LCT method has several limitations that lead to underestimation of the computed helicities (e.g. Démoulin & Berger 2003; Gibson *et al.* 2004). Furthermore, if one uses eq. (3.1) one cannot separate the contribution of the shearing term from the contribution of the advection term to the helicity injected into the corona. Moreover, a note of caution needs to be added regarding the validity of eq. (3.1). In the photosphere there is a sharp stratification of the plasma and also the photosphere is the interface that separates high to low β plasmas. It is a question how a flux tube that is no longer buyoant and has larger radius than the local gravitational scale height will cross this region. Clearly, comparison with MHD simulations are required to check which component(s) of helicity flux will be detected by any method that computes photospheric flows. For this purpose an anelastic MHD simulation was used (Welsch *et al.* 2007; Ravindra *et al.* 2008; Schuck 2008) and the comparisons showed that mostly the shearing term of eq. (2.3) can be determined. However, this simulation did not capture essential features of flux emergence physics and therefore the reported comparisons should be treated with caution.

Alternative approaches have been developed which attempt to compute separately both the shearing and advection term using photospheric vector magnetograms. Kusano *et al.* (2002) proposed a method which uses the vertical component of the induction equation. In fact the velocity of flux tubes cannot be deduced fully from the induction equation and part of the velocity is still computed from the LCT method (Welsch *et al.* 2004). Georgoulis & LaBonte (2005) introduced a minimum structure reconstruction technique to infer the velocity field vector. Their analysis simultaneously determines the field-aligned flows and enforces a unique cross-field solution of the induction equation.

Longcope (2004) introduced a technique (Minimum Energy Fit method; MEF) which demands that the photospheric flow agree with the observed photospheric field evolution according to the induction equation. It selects from all consistent flows, that with the smallest overall flow speed by demanding that it minimize an energy functional. If partial

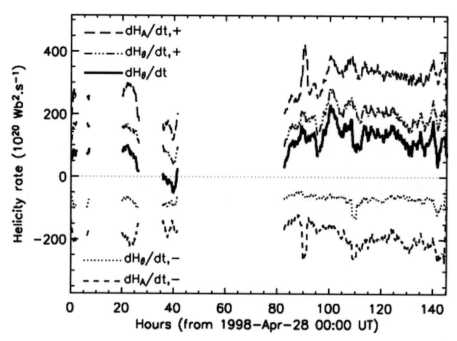

Figure 2. Plots of $(dH_A/dt)_\pm$, $(dH_\theta/dt)_\pm$, and dH_θ/dt as a function of time for AR 8210. The curves have been smoothed on a time interval of 100 min. We do not present the dH_A/dt curve because its differences with respect to the dH_θ/dt curve are too small to be clearly seen (from Pariat *et al.* 2006).

velocity information is available from other measurements, it can be incorporated into the MEF methodology by minimizing the squared difference from that data. Ravindra *et al.* (2008) incorporated velocity information provided by the LCT technique and Doppler velocity measurements. They compared their results with the results of an anelastic MHD simulation (see Fig. 3). The figure shows that LCT largely underestimates the amount of helicity rate while the best performance comes from the MEF method with additional LCT input.

Schuck (2006) developed the Differential Affine Velocity Estimator (DAVE), a method that locally minimizes the square of the continuity equation for the vertical component of the magnetic field subject to an affine velocity profile. Schuck (2008) presented the extension of the DAVE for horizontal magnetic fields, with all plasma components $\mathbf{v}_t$ and v_n described by a local model with linear spatial variations. The new method is called DAVE4VM (DAVE for vector magnetograms) because it requires input from vector magnetogram data.

The above methods, except DAVE4VM, were checked against an anelastic MHD simulation (Welsch *et al.* 2007). The method with the best overall performance was the MEF method. All methods showed weak features that have pointed out by Welsch *et al.* (2007). Schuck (2008) checked the DAVE4VM method against the same anelastic MHD simulation used by Welsch *et al.* (2007) and found that his method could reproduce roughly 95% of the simulation's helicity rates.

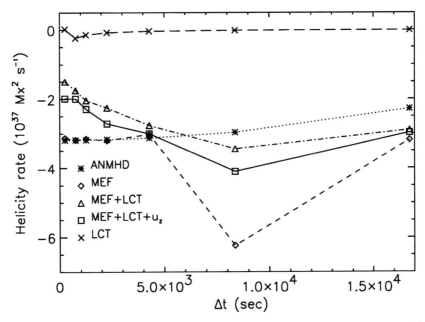

Figure 3. Helicity fluxes obtained by using LCT velocity and various combinations of the MEF algorithm with all velocities plotted as a function of time. The true helicity fluxes (labeled ANMHD) that resulted from an anelastic MHD simulation are also plotted for comparison. The curves labeled MEF, MEF+LCT, MEF+LCT+u_z indicate helicity fluxes computed with MEF without additional data, MEF with additional LCT method data, and MEF with additional LCT method data and additional data for the vertical component of the velocity field provided by the anelastic MHD simulation (from Ravindra *et al.* 2008).

6. Magnetic helicity and CME initiation

A significant fraction of AR's helicity is created by the solar dynamo and then transported into the corona through the photosphere with the emerging magnetic flux. This process together with helicity's property not to be destroyed under reconnection would constantly accumulate helicity into the corona. Furthermore, on the global scale, helicity emerges predominantly negative in the northern hemisphere and predominantly positive in the southern hemisphere (e.g. Pevtsov et al. 1995). And also this hemispheric helicity sign pattern does not change from solar cycle to solar cycle (Pevtsov et al. 2001). Consequently, on the global scale, mutual cancellation of helicity of opposite signs cannot relieve the Sun from excess accumulated helicity. It has been suggested (e.g. Low 1996) that CMEs, as expulsions of twisted magnetic fields, consist the most important process through which accumulated helicity is removed from the corona. Indirect support for this scenario is provided by the work by Zhang et al. (2006) who concluded that there is always a maximum amount of helicity that can be stored in an axisymmetric force-free field outside a sphere.

Low & Zhang (2002) and Zhang & Low (2001; 2003) provided a unified view of CMEs as the last chain of processes that transfer helicity from the convective zone into the interplanetary medium. Their theory exploits Taylor's conjecture that the magnetic field will relax towards a linear force-free field state. A summary of their results is as follows. When new field enters the corona repeated reconnections between the new and pre-existing field take place. This process simplifies the magnetic topology and the dissipated magnetic energy produces flares. The relaxation proceeds according to Taylor's conjecture and results in the formation of a flux rope which contains a significant fraction of the total helicity of the system. The fate of the flux rope is determined by the efficiency of its confinement by its surrounding anchored field. Flux rope ejection occurs when

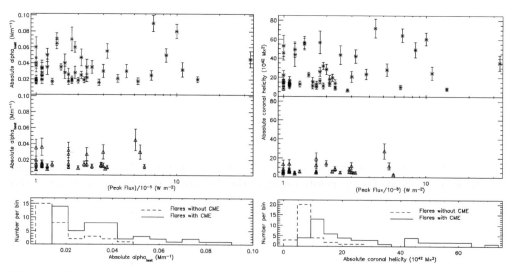

Figure 4. Left column, top: Scatter plot of the pre-flare absolute values of α_{best} as a function of the flare's peak X-ray flux for ARs producing CME-associated flares. Left column, middle: Same as top panel, but for the ARs producing flares without CMEs. Left column, bottom: Histograms of the values of α_{best} appearing in the top and middle panels. The solid line represents the histogram of α_{best} of the ARs which give CME-associated flares while the dashed line is the histogram of α_{best} of the ARs which produce flares that do not have CMEs. Right column: The absolute coronal helicity of the ARs appearing in the left panel. The format is identical to the format of the left column (modified from Nindos & Andrews 2004).

the magnetic energy it contains is sufficient to drive an outward expansion against the confining field.

The above physical view is supported by the work by Nindos & Andrews (2004). They modeled under the linear force-free field approximation the pre-flare coronal field of 78 ARs that produced big flares. Only some 60% of these flares were associated with CMEs. Then from the derived values of α_{best} they computed the corresponding coronal helicities. Their results appear in Fig. 4 and indicate that in a statistical sense both the pre-flare absolute value of α and the corresponding coronal helicity of the ARs producing CME-associated big flares are larger than the absolute value of α and helicity of those that do not have associated CMEs.

There are several other approaches to the initation of CMEs and the role played by magnetic helicity. Amari *et al.* (2003a,b) concluded that the accumulation of helicity is a necessary but not sufficient condition for an eruption to occur. The breakout simulations by Phillips *et al.* (2005) were designed so that no global helicity was injected into the corona. They showed that the eruption occurs at almost the same magnetic energy threshold as in a previous simulation where only positive helicity was injected. In their simulation, the amount of helicity is irrelevant because the negative and positive helicity regions did not reconnect. Contrasting results were found in the simulation by Kusano *et al.* (2004) where the introduction of a reverse helicity is essential for the eruption of a sheared arcade.

7. Conclusions

Magnetic helicity has the unique feature of being conserved even in resistive MHD on time scales less than the global diffusion time scale. This makes helicity probably the only physical quantity which can monitor the entire history of an eruptive event: from the transfer of magnetic field from the convective zone all the way to the eruption and the escape of the CME into interplanetary medium. On the other hand, calculations of helicity are difficult and only relatively recently attempts have been made to measure helicity using solar observations.

Once the importance of helicity was realized, a lot of effort was put on the determination of the sources of the helicity injected into active regions. Theoretical considerations have demonstrated rigorously that shearing motions (either differential rotation or/and transient flows) on the photospheric surface is an inefficient way of providing helicity on the active region scale. However, computations using high-cadence longitudinal magnetograms give the total helicity flux and cannot separate the shearing from the advection term. Furthermore, the computation of velocitites using the LCT method has serious limitations. Attempts for the computation of the shearing and advection term separately have been made using vector magnetograms. But the algorithms that have been developed have not been applied extensively to observations. Even more serious uncertainties are associated with the computation of the helicity carried away by CMEs. All the above problems contribute to the discrepancies concerning the helicity budget of ARs. At this point, these uncertainties have been cleared up only partially and much work needs to be done on this issue.

The role of helicity in the initiation of solar eruptions is a theoretical subject of intense debate. There is a general consensus that for the CME initiation, helicity must be accumulated into the pre-eruption topology. However, it seems that other parameters are also important, for example the location with respect to the pre-existing field where helicity is injected, the efficiency of the reconnection process(es) and how efficiently the helicity-charged stucture is confined by the overlying magnetic field.

References

Amari, T., Luciani, J. F., Aly, J. J., Mikic, Z., & Linker, J. 2003a, *ApJ*, 585, 1073

Amari, T., Luciani, J. F., Aly, J. J., Mikic, Z., & Linker, J. 2003b, *ApJ*, 595, 1231

Berger, M. A. 1984 *Geophys. Astrophys. Fluid Dyn.*, 30, 79

Berger, M. A. 1985 *ApJS*, m59, 433

Berger, M. A. 1988 *A&A*, 201, 355

Berger, M. A., & Field, G. B. 1984 *J. Fluid Mech.*, 147, 133

Chae, J. 2001 *ApJ* (Letters), 560, L95

Chae, J. 2007 *Adv. Sp. Res.*, 39, 1700

Chae, J., Moon, Y.-J., & Park, Y.-D. 2004 *Solar Phys.*, 223, 39

Dasso, S., Mandrini, C. H., Démoulin, P., & Farrugia, C. J. 2003 *JGR*, 108, 1362

Dasso, S., Mandrini, C. H., Démoulin, P., & Luoni, M. L. 2006 *A&A*, 455, 349

Démoulin, P. 2007 *Adv. Sp. Res.*, 39, 1674

Démoulin, P. & Berger, M. A. 2003 *Solar Phys.* 215, 203

Démoulin, P., Mandrini, C. H., van Driel-Gesztelyi, L., Lopez-Fuentes, M. C, & Aulanier, G. 2002a *Solar Phys.* 207, 87

Démoulin, P., Mandrini, C. H., van Driel-Gesztelyi, L., et al. 2002b *A&A*, 382, 650

DeVore, R. C. 2000 *ApJ*, 539, 944

Finn, J. H. & Antonsen, T. M. J. 1985 *Comments Plasma Phys. Contr. Fus.*, 9, 111

Georgoulis, M. K. & LaBonte, B. J. 2006 *ApJ*, 636, 475

Georgoulis, M. K. & LaBonte, B. J. 2007 *ApJ*, 671, 1034

Gibson, S. E., Fan, Y., Mandrini, C. H., Fisher, G., & Démoulin, P. 2004 *ApJ*, 617, 600

Green, L. M., Lopez-Fuentes, M. C., Mandrini, C. H., et al. 2002 *Solar Phys.*, 208, 43

Jeong, H. & Chae, J. 2007 *ApJ*, 671, 1022

Kusano, K., Maeshiro, T., Yokoyama, T., & Sakurai, T. 2002 *ApJ*, 577, 501

Kusano, K., Maeshiro, T., Yokoyama, T., & Sakurai, T. 2004 *ApJ*, 610, 537

Lepping, R. P., Burlaga, L. F., & Jones, J. A. 1990 *JGR*, 95, 11957

Lim, E.-K., Jeong, H., Chae, J., & Moon, Y.-J. 2007 *ApJ*, 656, 1167

Longcope, D. W. 2004 *ApJ*, 612, 1181

Low, B. C. 1996 *Solar Phys.*, 167, 217

Low, B. C. & Zhang, M. 2002 *ApJ* (Letters), 564, L53

Mandrini, C. H., Pohjolainen, S., Dasso, S., et al. 2005 *A&A*, 434, 725

Moon, Y.-J., Chae, J., Choe, G. S., et al. 2002a *ApJ*, 574, 1066

Moon, Y.-J., Chae, J., Wang, H., Choe, G. S., & Park, Y. D. 2002b *ApJ*, 580, 528

Nindos, A. & Zhang, H. 2002 *ApJ* (Letters), 573, L133

Nindos, A., Zhang, J., & Zhang, H. 2003 *ApJ*, 594, 1033

Nindos, A. & Andrews, M. D. 2004 *ApJ* (Letters), 616, L175

November, L. J. & Simon, G. W. 1988 *ApJ*, 333, 427

Pariat, E., Démoulin, P., & Berger, M. A. 2005 *A&A*, 439, 1191

Pariat, E., Démoulin, P., & Nindos, A. 2007 *Adv. Sp. Res.*, 39, 1706

Pariat, E., Nindos, A., Démoulin, P., & Berger, M.A. 2006 *A&A*, 452, 623

Pevtsov, A. A., Canfield, R. C., & Metcalf, T. R. 1995 *ApJ* (Letters), 440, L109

Pevtsov, A. A., Canfield, R. C., & Latushko, S. M. 2001 *ApJ* (Letters), 549, L261

Phillips, A. D., MacNeice, P. J., & Antiochos, S. K. 2005 *ApJ* (Letters), 624, L129

Schuck, P. W. 2006 *ApJ*, 646, 1358

Schuck, P. W. 2008 *ApJ*, 683, 1134

Ravindra, B., Longcope, D. W., & Abbett, W. P. 2008 *ApJ*, 677, 751

Welsch, B. T., Abbett, W. P., DeRosa, M. L., et al. 2007 *ApJ*, 670, 1434

Welsch, B. T., Fisher, G. H., Abbett, W. P., & Regnier, S. 2004 *ApJ*, 610, 1148

Zhang, M. & Low, B. C. 2001 *ApJ*, 561, 406

Zhang, M. & Low, B. C. 2003 *ApJ*, 584, 479

Zhang, M., Flyer, N., & Low, B. C. 2006 *ApJ*, 644, 575

Discussion

DAVILA: Pevtsov *et al.* (2003) report a good temporal correlation between flux emergence and helicity increase in active regions, indicating a subsurface origin of helicity. What is the basis of the doubts you expressed in this regard?

NINDOS: There is no doubt, theoretically, that flux emergence is the most efficient agent. But what do our techniques really measure? In this question, Pevtsov's work simply handles mean parameters.

GIRISH: I want to know whether your model of helicity and magnetic flux carried by CMEs into the corona is valid for high latitude CMEs observed during sunspot maximum outside the active region belt on the Sun?

NINDOS: Yes it is, because "ARs" even when they contain no spots or plages still have helicity.

Universal Heliophysical Processes
Proceedings IAU Symposium No. 257, 2008
N. Gopalswamy & D.F. Webb, eds.

© 2009 International Astronomical Union
doi:10.1017/S1743921309029196

Understanding structures at the base of the solar corona – polar plumes

A. H. Gabriel[1], F. Bely-Dubau[2], E. Tison[1] and L. Abbo[3]

[1] Institut d'Astrophysique Spatial, Université Paris Sud-11,
91405 Orsay Cedex, France
email: `gabriel@ias.fr`

[2] Observatoire de la Côte d'Azur, BP4229, 06304 Nice Cedex, France

[3] Osservatorio Astronomico di Torino, Pino Torinese 10025 – Italy

Abstract. Recent work on coronal polar plumes (Gabriel *et al.* 2003, 2005) has aimed at determining the outflow velocity in plume and interplume regions, using the Doppler dimming technique on oxygen VI observations by SUMER and UVCS on SOHO. By comparing observations of SOHO/EIT with plume modelling, we show that the major part of plumes is the result of chance alignments along the line-of-sight of small enhancements in intensity. This confirms the so-called *curtain* model. These plumes can be attributed to reconnection activity along the boundaries of supergranule cells. A second population of plumes has a lower abundance and arises from surface bright points having a particular magnetic configuration. New observations using the Hinode/EIS spectrometer are in progress, with the aim of providing further insight for this model.

Keywords. Sun: corona, solar wind

The plumes that we are addressing here are those seen extending out to 0.5 $R_\odot$ or more beyond the solar limb, when looking in spectral regions emitting at temperatures around 1 MK or rather less. They are best observed within polar coronal holes and during a solar minimum configuration. They are well observed from SOHO by SUMER and EIT in the oxygen VI 1032 Å and the Fe IX 171 Å band, respectively. Here we concentrate on the EIT observations, since these exist in long synoptic series of images. Existing measurements of physical properties (Wilhelm *et al.* 1998; Wilhelm 2006) show the temperature to be around 0.8 MK and the density of order 10^8 cm^{-3}. Gabriel *et al.* (2003), Gabriel *et al.* (2005) and Teriaca *et al.* (2003) used Doppler dimming techniques (Noci, Kohl & Withbroe 1987) to measure the outflow velocity. Velocities were found to be higher than in the background corona up to 1.5 $R_\odot$, then falling below coronal values at greater heights. We believe that there are two different kind of plume phenomena:

• *Beam plumes*, quasi cylindrical structures rooted in bright magnetic structures on the disk.

• *Curtain plumes*, faint sheet-like structures visible only when seen edge-on.
Our contribution aims to present evidence for this distinction. Many other workers in the field have assumed that plumes are always of the *beam plume* type.

The most convincing evidence that plumes are not always (even, not usually) of the beam type comes from simply looking at images of a sufficient number of plumes. Fig. 1 shows four successive images of the south solar pole at 7 day intervals, equivalent to 90° rotation between each image. Almost all of the plumes, irrespective of the viewing direction, appear to start *at* the solar limb. On the other hand, in the plane of the sky, plumes appear to be evenly distributed over angles of ±20° from the solar axis. If these were beam plumes, we would therefore expect to see about half of them originating on

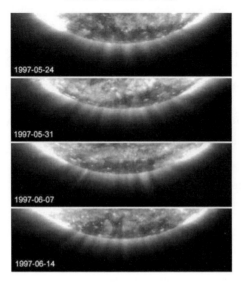

Figure 1. Four EIT images of the south polar coronal hole in Fe IX radiation, taken at 7 day intervals.

the visible solar disk. This is not the case. All of the plumes appear to have their origin *behind* the solar disk, which clearly cannot be the case. We find, and will demonstrate in what follows, that the majority of these plumes are curtain plumes, in which the observed brightness arises through a substantial line-of-sight integration, involving path-lengths of

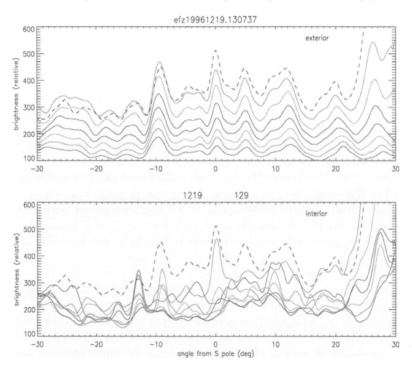

Figure 2. A series of limb scans of a polar coronal hole. The curves in the top panel are outside the limb and the curves in the bottom panel inside. The black dashed curve is shown on both plots and it is at a radius of 1.01 $R_\odot$, close to the radius of maximum brightness. The other curves are stepping outwards (top panel) and inwards (bottom panel) in intervals of 0.01 $R_\odot$.

the order 0.5 $R_\odot$ or greater. Some beam plumes are also present, but are in the minority and usually fainter.

Because of concerns about the misleading impression by images having a limited dynamic range, as well as by problems of visual perception, we further develop this argument in a more quantitative form. Fig. 2 shows the results for the brightness structure of one of these limb images. The curves in the upper panel show clearly the plumes, decreasing in brightness with radial distance. The curves in the lower panel show how this above-limb plume structure decreases rapidly as we pass onto the disk. An obvious plume does rarely maintain its brightness profile for more than 2 further steps or 0.02 $R_\odot$ (equivalent to an absolute height of 0.99 $R_\odot$), giving way to a different structure arising from brightness patterns on the disk. It should be remembered that a real beam plume, on the front side of the Sun is expected to *increase* in brightness as we pass from the limb on to the disk.

To extend our quantitative evaluation, we produce simulated models of what would be expected from the two proposed plume geometries, beam plumes and curtain plumes. The plume material is known to be optically thin, and to fall off in height with a scale-height not too different from the quiet coronal hole corona. The calculation is therefore

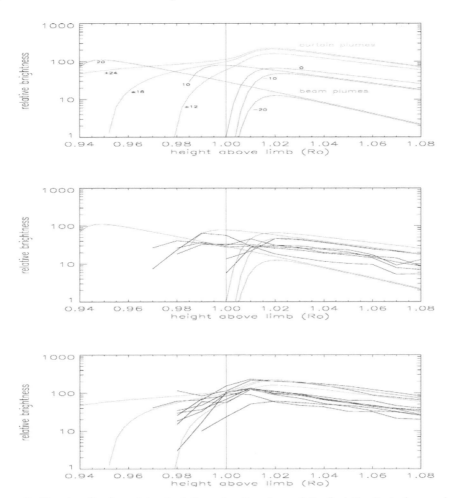

Figure 3. Showing (top) model calculations for the plume fall-off at the limb: beam plumes in red and curtain plumes in green, (middle) beam models with superposed observations, and (bottom) for curtain plumes.

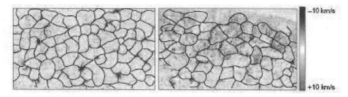

Figure 4. Dopplergram of Ne VIII in a coronal hole showing high outflow velocity at the boundaries of supergranule cells (Hasslet *et al.* 1999).

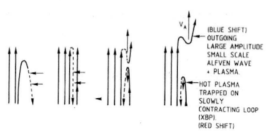

Figure 5. Schematic for the acceleration of fast wind from coronal hole regions (Axford & McKenzie 1992).

straightforward. The only complication is that the 171 Å emission will be absorbed by continuum in the cooler denser chromospheric layers. This effect becomes important when the tangential optical path passes close to the limb. It has the effect of smoothing out some of the curious sharp peaks that otherwise occur in the region $R_\odot = 1.0$ to $R_\odot = 1.01$. An effort has been made to include this in the model. Any uncertainties in this contribution can be shown to have a negligible effect on the interpretation of the data/modelling comparison. The two plume models are shown in red and in green in Fig. 3a. It can be seen that the curtain plumes are predicted to drop significantly in brightness as we cross inside the limb, by a factor extending from 3 to more than 10, depending on the angular extent assumed for the curtain in the line-of-sight, shown here for $\pm 12°$, $\pm 18°$ and $\pm 24°$. On the other hand, the beam plumes in red are predicted to continue to increase with decreasing radius, until they reach either their foot points on the front of the Sun, or the limb if behind the Sun. The two sets of curves are in principle mutually exclusive, and should allow discrimination between the observations. These curves, with brightness shown on a log scale, are in arbitrary units and can be slid vertically.

Since the simulations concern only the plumes and not the background corona, the observations need to be treated in a similar way. The observations used are those from all significant plumes seen in all four views shown in Fig. 1. These are first presented in the quantitative format of Fig. 2. Then for each trace, a series of Gaussian profiles has been fitted, in order to separate the plume from the corona and to correct for some blending of overlapping plumes. This fitting is extended onto the disk, only as far as we are confident for the continuity and identity of the plume (down to around 0.97 or 0.98 $R_\odot$). The resulting radial brightness traces have been separated into the two categories indicated by the two models, and plotted in black in Figs 3b and c, overlaid with their respective model calculations. Their classification as beam or curtain is usually clear, with only one or two uncertain cases due to noise in the observations or in the Gaussian fitting.

1. Discussion and conclusions

Limitations of space preclude a detailed discussion of the distribution of the two plume types. Here we state simply that the curtain plumes are normally much more abundant and they are always present when the solar conditions (viewing angle B, solar minimum) are optimised. The beam plumes are more variable and appear to be associated with small magnetic bright regions within the coronal hole. In the case of the 4 orthogonal views sampled here, it is the fourth on 14 Jun 1997 which contributes most of the beam plumes, corresponding with the increased bright-point activity seen in this image.

The ubiquitous nature of curtain plumes leads us to seek ubiquitous structures as their possible origin. Therefore we consider the supergranular network and its manifestations in the corona as the likely source. Linear structures associated with the cell boundaries, appear in chance alignment as the Sun rotates, leading to the observed plumes, with lifetimes of the order 1 or 2 days, due to the rotation.

It is worth recalling the study by Hassler *et al.* (1999), in which they showed a correlation between the boundaries of supergranular cells and high outflows in the fast solar wind (see Fig. 4). Their wind measurements used Ne VIII lines from SUMER, which are more likely typical of the upper transition region. The availability today of the EIS spectrometer on Hinode allows us to extend their measurements to true coronal temperatures. These studies are currently in progress. It is likely that curtain plumes are associated with the reconnection of emerging closed field loops with the dominant-polarity field in funnels at the boundaries of the cells, as first proposed by Axford & McKenzie 1992. If this is true, they would have a direct association with the heating of the corona and the acceleration of the fast solar wind.

SOHO is a collaborative programme of ESA and NASA. Part of this work was carried out within the framework of the Coronal Plume study team (Leader, Klaus Wilhelm) of the International Space Science Institute (ISSI), Bern, Switzerland.

References

Axford, W. I. & McKenzie, J. F. 1992, in: E. Marsch and R. Schwenn (eds.), *Solar Wind Seven*, Pergamon Press, P. 1

Gabriel, A. H., Bely-Dubau, F., & Lemaire, P. 2003, *ApJ*, 589, 623

Gabriel, A. H., Abbo, L.,Bely-Dubau, F. Llebaria, A., & Antonucci, E. 2005, *ApJ*, 635, L185

Hassler, D. M., Dammasch, I. E., Lemaire, P., Brekke, P., Curdt,W., Mason, H., Vial, J.-C., & Wilhelm, K. 1999, *Science*, 283, 810

Noci, G., Kohl, J. L., & Withbroe G. L. 1987, *ApJ*, 315, 706

Teriaca, L., Poletto, G., Romoli, M., & Biesecker, D. A. 2003, *ApJ*, 588, 566

Wilhelm, K., Marsch, E., Dwivedi, B., Hassler, D. M., Lemaire, P. H., Gabriel, A. H., & Huber, M. C. E. 1998, *ApJ*, 500, 1023

Wilhelm, K. 2006, *A&A*, 455, 697

Discussion

BOCHSLER: Is there an extra free parameter describing the relative orientation of curtain plumes to the line of sight? How are these orientation angles distributed?

GABRIEL: The curtain model is based on a volume emission intensity only 1 to 3 % higher than the basic corona. Thus the brightness only reaches the observed level of 30 % to 50 % of the background when the line of sight reaches levels of 0.5 R or higher. Only a narrow angle of 3 degrees around zero enables the plume to become detectable.

Universal Heliophysical Processes
Proceedings IAU Symposium No. 257, 2008
N. Gopalswamy & D.F. Webb, eds.

© 2009 International Astronomical Union
doi:10.1017/S1743921309029202

Three-dimensional MHD modeling of waves in active region loops

Leon Ofman† and Małgorzata Selwa

Catholic University of America,
NASA Goddard Space Flight Center, Code 671
Washington, DC 20064, USA

Abstract. Observations show that MHD waves are one of the most important universal processes in the heliosphere. These waves are likely to play an important role in energy transfer in the heliosphere, and they can be used as a diagnostic tool of the properties of the local magneto-fluid environment. Recent observations by TRACE and Hinode satellites provide ample evidence of oscillations in coronal active region loops. The oscillations were interpreted as fast (kink), slow, and Alfvén modes, and the properties of the waves were used for coronal seismology. However, due to the complex interactions of the various modes in the inhomogeneous active region plasma, and due to nonlinearity, idealized linear theory is inadequate to properly describe the waves. To overcome this theoretical shortcoming we developed 3D MHD models of waves in active region loops. We investigated the effects of 3D active region magnetic and density structure on the oscillations and the wave dissipation, and we investigated the oscillation of individual loops. Some loops were constructed to contain several threads and twist. Here, we present the results of our models, and show how they can be used to understand better the properties of the waves, and of the active regions.

Keywords. Sun: activity, Sun: corona, Sun: flares, Sun: magnetic fields, MHD, waves

1. Introduction

MHD waves were recently observed in detail in coronal loops (Aschwanden *et al.* 1999; Nakariakov *et al.* 1999; Verwichte *et al.* 2004; Ofman & Wang 2008), and throughout the heliosphere. These waves have been studied for decades as a possible source of energy for coronal heating and solar wind acceleration. The properties of the observed waves and the geometry of the loops can be used for coronal seismology, i.e., the determination of the physical parameters of loops, such as the magnetic field, density and temperature that are difficult to measure with other methods. We present the results of recent 3D MHD models of waves in active region loops motivated by recent observations. Waves in dipole and constant-α force-free field with gravitationally stratified density were studied by Ofman and Thompson (2002). High density coronal loop was introduced in the above configuration by McLaughlin and Ofman (2008). Waves in more realistic active region initialized with potential magnetic field extrapolated from observed photospheric field were studied by Ofman (2007). We show how the effects of active region morphology, twist, and flow influence the oscillation of the loops.

---

† Visiting Associate Professor, Tel Aviv University

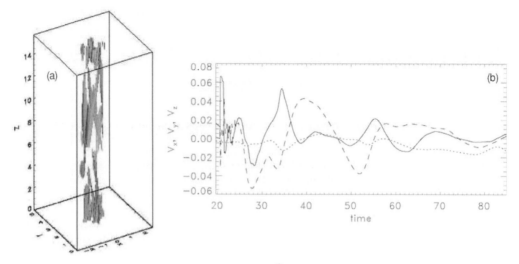

Figure 1. The isosurface of the current density j^2 in the twisted loop, formed as a result of the twist and the impact of the wave.

2. MHD model

We describe solar plasma with the normalized three-dimensional nonlinear resistive MHD equations with gravity included for the dipole case. For simplicity we study the isothermal case. Detailed description of the code can be found Ofman & Thompson (2002).

2.1. *Twisted Loops*

The four-threaded loop is initialized with

$$\rho_0(x, y, z) = \rho_{\min} + (\rho_{\max} - \rho_{\min}) \sum_{i=1}^{4} e^{-\{[(x-x_i)^2 + (y-y_i)^2]/r_0^2\}^2}, \qquad (2.1)$$

where $\rho_{\min} = 0.2$ is the minimal normalized density outside the loops, $\rho_{\max} = 1$ is the maximal normalized density of the loops, $r_0 = 0.25$ is the radius of the loops, and (x_i, y_i), where $i = 1, 2, 3, 4$ are the locations of the axes of the four threads. The twist is introduced by applying circular flow in the $x - y$ plane at $z = 0$ for quarter rotation. Line-tied boundary conditions are used at the top and the bottom of the simulation box, and open boundary conditions on the sides. We launch a velocity pulse that impacts the four-threaded loops and initiates the oscillations.

2.2. *Loops in Dipole Field*

As the initial configuration we take a potential dipole magnetic field and assume a gravitationally stratified equilibrium density (Ofman & Thompson 2002). Following McLaughlin & Ofman (2008) we include a denser loop in our system, however we construct it in such a way that it follows fieldlines and has a smooth density profile:

$$\varrho_i = \varrho_e d \exp\left[-\left([(y-y_0)^2 + (z-z_0)^2]/w\right)^p\right]. \qquad (2.2)$$

At the photospheric boundary we keep all the variables fixed while open boundary conditions are implemented for all the variables at the other five planes. Note, that due to the rapid drop in the magnetic field density, the plasma β increases with height, and it is about 2 at the apex of the loop.

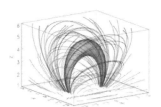 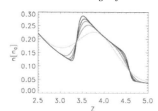 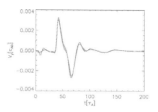

Figure 2. Left panel: Initial 3D magnetic field configuration. Middle panel: Different density profiles (along x = y = 0 cut) of the loop: p = 6 (violet line), p = 4 (blue line), p = 3 (green line), p = 2 (red line) and p = 1 (yellow line). Right panel: Time signatures of transversal component of velocity. Colors correspond to middle panel notation.

3. Numerical Results

3.1. *Twisted Loops*

We find that the twisted loop exhibits transverse oscillations due to the impact of the velocity pulse. The oscillations are more complex than a kink mode in a straight cylinder, and all 3 components of the velocity exhibit damped wave motions. Filamented currents form in the loop due to the combined effects of the twist and the waves (see Figure 1a). The damping rate of the twisted loop is faster, and the phase speed is larger than the kink speed of the parallel-threaded loop case (see Figure 1b).

3.2. *Loops in Dipole Field*

We start our studies by perturbing the loop with a velocity pulse from the side boundary plane. We vary the steepness parameter of the density ratio: $p = 1, 2, 3, 4, 6$. predicted for a straight cylinder that resonance layer width should affect the damping time of the oscillation through the resonant layer width and the density ratio between the loop and surrounding corona.

From the middle panel of Fig. 2 we see that the width of inhomogeneous layer changes by the factor of 5 between steepness parameters $p = 1$ and $p = 6$ for a fixed density ratio. According to Ruderman & Roberts' (2002) theory we should expect the difference in normalized damping times up to 20%. However, we find that resonant absorption does not affect the damping in curved loops (right panel of Fig. 2), because by the time it starts to act the oscillations are damped due to wave leakage enhanced by the large β near loop apex, and the loop curvature.

Next we model the observation of slow standing waves reported by Wang *et al.* (2003a,b). We excite the slow standing wave by a velocity pulse (duration of the pulse ≪ wave-period) from the bottom boundary plane centered in one of loop's footpoints. As the effect of such a pulse launched in the footpoint we observe the initial brightening (left panel of Fig. 3) which is consistent with slow wave oscillation observations with SXT (Fig. 8 from Wang *et al.* 2003b). This kind of excitation produces two pulses launched in both footpoints of the loop simultaneously due to the fast wave coupling (Selwa *et al.* 2007). The time signatures of mass density and velocity at the apex shown in right panel of Fig. 3) match the observational ones (quarter wave-period shift and initial antiphase stage; compare with Fig. 3 from Wang *et al.* 2003a). However, the damping rate is more rapid in our model due to the large value of β and the curvature which lead to increased leakage of the slow wave.

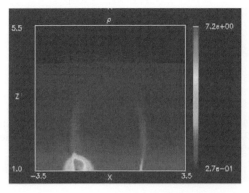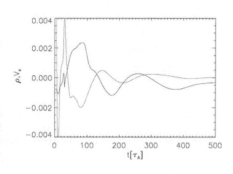

Figure 3. *Left*: Initial brightening due to the pulse in one of the loop's footpoint. *Right*: Time signatures of mass density and velocity at the apex.

4. Conclusions

Motivated by recent observations we model the transverse oscillations of a twisted four-threaded loop. We find that the oscillations of the loop are more complex than the kink mode, with larger phase speed and damping rate than in parallel-thread loop.

For the loop in the dipole field we observe no difference in damping rates due to steepness of density profile across the loop since the dominant damping mechanism is leakage of the fast mode wave due to large plasma β in the loop. We find that for a curved loop footpoint excitation is an efficient mechanism of triggering the slow standing mode: the main observational features (excitation within 1 wave-period and initial footpoint brightening) are reproduced. The results of our study are useful for the development of coronal seismology methods.

Acknowledgement

The authors thank support by NASA grants NNG06GI55G and NNX08AV88G.

References

Aschwanden, M. J., Fletcher, L., Schrijver, C. J., & Alexander, D. 1999, *ApJ*, 520, 880
McLaughlin, J. A. & Ofman, L. 2008, *ApJ*, 682, 1338
Nakariakov, V. M., Ofman, L., DeLuca, E., Roberts, B., & Davila, J. M. 1999, *Science*, 285, 862
Ofman, L. 2007, *ApJ*, 655, 1134
Ofman, L. & Thompson, B. J. 2002, *ApJ*, 574, 440
Ofman, L. & Wang, T. J. 2008, *Astron. Astrophys.*, 482, L9
Ruderman, M. S. & Roberts, B. 2002, *ApJ*, 577, 475
Selwa, M., Ofman, L., & Murawski, K. 2007, *ApJL*, 668, L83
Verwichte, E., Nakariakov, V. M., Ofman, L., & Deluca, E. E. 2004, *Sol. Phys.*, 223, 77
Wang, T. J., Solanki, S. K., Curdt, W., Innes, D. E., Dammasch, I. E., & Kliem, B. 2003a, *Astron. Astrophys.*, 406, 1105
Wang, T. J., Solanki, S. K., Innes, D. E., Curdt, W., & Marsch, E. 2003b, *Astron. Astrophys.*, 402, L17

Universal Heliophysical Processes
Proceedings IAU Symposium No. 257, 2008
N. Gopalswamy & D.F. Webb, eds.

© 2009 International Astronomical Union
doi:10.1017/S1743921309029214

Long period oscillations
of microwave emission of solar active regions:
observations with NoRH and SSRT

I. A. Bakunina[1], V. E. Abramov-Maximov[2], S. V. Lesovoy[3],
K. Shibasaki[4], A. A. Solov'ev[2] and Yu. V. Tikhomirov[1]

[1]Radiophysical Research Institute, B. Pecherskaya st., 25, Nizhny Novgorod, 603950, Russia
email: `rinbak@mail.ru`

[2]Central astronomical observatory at Pulkovo, Russian Acad. Sci.,
Pulkovskoe chaussee., 65/1, St. Petersburg, 196140, Russia
email: `beam@gao.spb.ru`

[3]Institute of Solar-Terrestrial Physics RAS SB, Lermontov St., 134, Irkutsk, 664033, Russia

[4]Nobeyama Solar Radio Observatory, Minamimaki, Minamisaku, Nagano 384-1305, Japan

Abstract. In this work we present the first results of study and comparison of the parameters of quasi-periodic long-term oscillations of microwave emission of large (>0.7 arcmin) sunspots as a result of simultaneous observations with two radioheliographs – NoRH (17 GHz) and Siberian Solar Radio Telescope (SSRT) (5.7 GHz) with 1 minute cadence. Radioheliographs have been working with quite large time overlap (about 5 hours) and have the high spatial resolution: 10 arcsec (NoRH) and 20 arcsec (SSRT). We have found that quasi-periodic long-term oscillations are surely observed at both frequencies with the periods in the range of 20–150 min. We detected common periods for common time of observations with two radioheliographs and interpret this as the consequence of the vertical-radial quasi-periodic displacements of sunspot as a whole structure.

Keywords. high angular resolution, oscillations, magnetic fields, sunspots

---

Quasi-periodic long-term oscillations of the solar flux radio emission with periods about 1 hour have being studied since 1970-th (Kobrin, Pahomov & Prokof'eva 1976, Durasova, Kobrin & Yudin 1986). New radio telescopes, such as Nobeyama Radiohelio-graph (NoRH), opened new possibilities for studying oscillations of solar radio emission from sources above sunspots with high spatial resolution (Gelfreikh, Nagovitsyn & Nagovitsyna 2006). Quasi-periodic long-term oscillations of sunspots registered with the optical methods (Zeeman and Doppler effects) have periods from 40 to 200 minutes and have global character and physical origin sharply different from the well-known 3-5 minute oscillations in sunspots (Efremov, Parfinenko & Solov'ev 2007, Efremov, Parfinenko & Solov'ev 2008, Solov'ev & Kirichek 2006, Solov'ev & Kirichek 2008, Kshevetskii & Solov'ev 2008). Whereas the latter are MHD-waves trapped inside the magnetic flux tubes of the sunspots, the low-frequency oscillations are concerned with quasi-periodic displacements of the whole sunspot as a well localized and stable formation. These are the oscillations of sunspot itself, but not the oscillations of some elements into the sunspot's magnetic tubes (Solov'ev & Kirichek 2008, Kshevetskii & Solov'ev 2008): sunspot oscillates as a whole, keeping its own structure (umbra - penumbra) but changing geometrical sizes, strength and vertical gradient of the magnetic field. These oscillations are possible because sunspots are turned to be relatively "shallow" formations. The depth of their so called lower magnetic boundary is only 3-5 thousands of kilometers. This theoretically predicted general property of sunspots (Solov'ev & Kirichek 2008, Solov'ev 1984a,

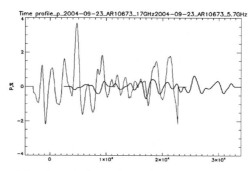

Figure 1. Time profile of the degree of circular polarization P at 17 GHz (NoRH) (thin line, time of observation from - 0.3×10^4 sec till 2.3×10^4 sec) and 5,7 GHz (SSRT) (thick line, time of observation from 0.3×10^4 sec till 3.3×10^4 sec). Vertical axis - P, %, the circular polarization degree, smoothed (9 minutes) and trend component is subtracted; horizontal axis - time of observation, sec

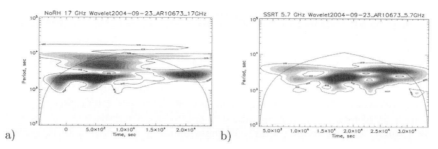

Figure 2. Wavelet spectra ("filter" with moving average 50 min) for the degree of circular polarization P: a) at 17 GHz (time of observation from -0.3×10^4 sec till 2.3×10^4 sec), b) at 5,7 GHz (time of observation: from 0.3×10^4 sec till 3.3×10^4 sec). Vertical axes - period of oscillations, sec; horizontal axes - time of observation, sec

Solov'ev 1984b, Nagovitsyn 1997) is surely confirmed by the data of local helioseismology (Zhao 2001, Kosovichev 2006).

We present the first results of study and comparison of the parameters of quasi-periodic long-term oscillations of microwave emission of large (>0.7 arcmin) sunspots in the non-flare bipolar and unipolar active regions (AR) as a result of simultaneous observations with two radioheliographs – NoRH (17 GHz) and Siberian Solar Radio Telescope (SSRT) (5.7 GHz) with 1 minute cadence for revealing common periods at both frequencies during of two radio-heliographs work time overlap (about 5 hours). These common periods in the microwave range may be considered as a consequence of eigen oscillations of a sunspot (assuming hyrocyclotron emission at both frequencies): as the Alfven times are considerably less than a period of these oscillations, magnetosphere above a sunspot re-forms fastly, and we observe the system's passing through continuous series of the equilibrium states in the microwave range.

On the radio maps sunspot-associated sources were identified and time profiles of their maximum brightness temperatures and circular polarization degree for each radio source were calculated. We studied 11 sunspots-associated radio-sources (20 days of observations) (2001-2006 y.y.) using smoothing procedure (moving average) for both SSRT and NoRH temporal data raws and calculating deviation of original signal from average signal because of nonstationary data and day's changing of the antenna beam and also we used the interpolation method for nonequidistant temporal data raws of SSRT. We calculated the degree of the circular polarization $p = TVmax/TI$ (where $TVmax$ – maximum of the brightness temperature of Stock's parameter V above sunspot, TI – brightness temperature of Stock's parameter I) because it is the most reliable parameter due to nonstability of SSRT data. Wavelet spectra (wavelet Morle–6) (Torrence & Compo 1998) and FFT spectra and cross-correlation function as a function of temporal delays were calculated for parameter P.

One should notice that microwave emission at two different frequencies comes from two different heights of sunspot's magnitosphere, so we can't expect inphase oscillations of P but moreover – antiphase oscillations are more probable because of different sign of the temperature gradient at the heights in the upper chromosphere and lower corona where third hyrolevel at 17 GHz and second and third hyrolevels at 5.7 GHz are located. One example of temporal data raws of P on both frequencies, and wavelet spectra of the microwave emission above the leading sunspot of AR 10673 NOAA, 23 of September 2004, is demonstrated on figures 1 and 2.

We can see from fig. 2 a) and b) that during common time of observations (from 0.3×10^4 sec till 2.3×10^4 sec) there are three common periods: about 33, 40 and 66 minutes. Period of about 40 minutes is most pronounced. Coefficient of cross-correlation function of two spectra is -0.7 under time delay about 0-10 minutes.

Conclusions:

Using wavelet, FFT and cross-correlation analyses we found out that:

1) long period oscillations are surely observed at both frequencies and have wave trains character;

2) periods of the long-term oscillations at both 17 and 5.7 GHz are surely observed in the range of 20 – 150 min, but we should point out at the existence of more long oscillations;

3) we detected common periods (for example – fig. 2 a) and b)) and quite similar spectra for common time of observations with two radioheliographs that may be considered as a consequence of eigen oscillations of a sunspot.

This work is supported by RFBR grants 06-02-39029, 06-02-16295, 06-02-16838, 07-02-01066, 06-02-16981, 08-02-10002 and also by the Basic Research Program of the Presidium of the Russian Academy of Sciences No 16. I.A. Bakunina thanks IAU for financial support.

References

Durasova, M. S., Kobrin, M. M., & Yudin, O. I. 1971, *Nature*, 229, p. 83

Efremov, V. I., Parfinenko, L. D., & Solov'ev, A. A. 2007, *Astron. Reports*, 51, p. 401

Efremov, V. I., Parfinenko, L. D., & Solov'ev, A. A. 2008, *J. Opt. Technol.*, 75, p. 144

Gelfreikh, G., Nagovitsyn, Yu. A, & Nagovitsyna, E. Yu. 2006, *PASJ*, 58, p. 29

Kshevetskii, S. P. & Solov'ev, A. A. 2008, *Astron. Rep.*, 52, p. 772

Kosovichev, A. G. 2006, *Adv. Sp. Res.*, 38, p. 876

Kobrin, M. M., Pahomov, V. V., & Prokof'eva, N. A. 1976, *Solar Phys.*, 50, p. 113

Nagovitsyn, Yu. A. 1997, *PAZ*, 23, p. 859

Solov'ev, A. A. 1984, *Solnechnye Dannye*, p. 73

Solov'ev, A. A. 1984, *Soviet Astron.*, 28, p. 447

Solov'ev, A. A. & Kirichek, E. A. 2006, in: V. Bothmer & A. A. Hady (eds.), *Solar Activity and its Magnetic Origin*, Proc. IAU Symposium No. 233 (Cairo, Egipt), p. 523

Solov'ev, A. A. & Kirichek, E. A. 2006, *Astrophysical Bulletin*, 63, p. 169

Torrence, C. & Compo, G. P. 1998, *Bull. Amer. Meteor. Soc.*, 79, p. 61

Zhao, J., Kosovichev, A. G., & Duval, T. L. 2001, *ApJ*, 557, p. 384

Universal Heliophysical Processes
Proceedings IAU Symposium No. 257, 2008
N. Gopalswamy & D.F. Webb, eds.

ⓒ 2009 International Astronomical Union
doi:10.1017/S1743921309029226

Enhanced Rieger type periodicities' detection in X-ray solar flares and statistical validation of Rossby waves' existence

Michaila Dimitropoulou[1], Xenophon Moussas[2] and Dafni Strintzi[3]

[1]University of Athens, Department of Physics, GR-15483, Athens, Greece,
email: michaila.dimitropoulou@nsn.com
[2]University of Athens, Department of Physics, GR-15483, Athens, Greece,
email: xmoussas@phys.uoa.gr
[3]National Technical University of Athens, GR-15773, Athens, Greece,
email: dafni_strintzi@yahoo.com

Abstract. The known Rieger Periodicity (ranging in literature from 150 up to 160 days) is obvious in numerous solar indices. Many sub-harmonic periodicities have also been observed ($128-, 102-, 78-$, and $51 - days$) in flare, sunspot, radio bursts, neutrino flux and flow data, coined as Rieger Type Periodicities (RTPs). Several attempts are focused to the discovery of their source, as well as the explanation of some intrinsic attributes that they present, such as their connection to extremely active flares, their temporal intermittency as well as their tendency to occur near solar maxima. In this paper, we link the X-ray flare observations made on Geosynchronous Operational Environmental Satellites (GOES) to an existing theoretical model (Lou 2000), suggesting that the mechanism behind the Rieger Type Periodicities is the Rossby Type Waves. The enhanced data analysis methods used in this article (Scargle-Lomb periodogram and Weighted Wavelet Z-Transform) provide the proper resolution needed to argue that RTPs are present also in less energetic flares, contrary to what has been inferred from observations so far.

Keywords. Sun flares, Rieger-Type periodicities, Rossby-type waves

1. Introduction

In 1984 Rieger (Rieger *et al.* 1984) was the first to reveal a 158-day periodicity in the sun, while studying $\gamma-ray$ flare data from Solar Maximum Mission (SMM) in solar cycle 21 (C21). Approximately the same periodicity was also discovered in X-ray flares data taken from the Geosynchronous Operational Environmental Satellites (GOES) for the same solar cycle (Rieger *et al.* 1984). In the context of these attempts, it is noticeable that apart from the Rieger Periodicity itself, numerous other "relative" periodicities were discovered, such as 128-, 102-, 78-, and 51-day periodicities (Bai & Sturrock 1991,Bai 1992). The relevance of these periodicities to the classic Rieger one is that all of them are approximately integer multiples of a principle 25-day periodicity. This remarkable attribute led to coining these periodicities as "Rieger Type Periodicities" (RTPs). A series of additional conclusions came to the surface over the years of observational studies, the most important of which being that Rieger Periodicity was mainly sought in highly explosive flares. Especially when it comes to X-ray flare data from GOES — which is going to be the data source for this study as well – the existence of Rieger periodicity was mainly investigated in flares of class $\geqslant M$ (Rieger *et al.* 1984, Landscheidt 1987 , Bai 2003). Consequently, Rieger periodicity was connected to highly energetic flares, which are presumably triggered by the emergence of photospheric magnetic flux with the same period (Ballester *et al.* 2002, Ballester *et al.* 1999).

Other than their existence, their intermittency and their tendency to occur near solar maxima, there is presently no solid ground for RTPs' theoretical explanation, although numerous models have been proposed over the years. Lou (Lou 2000) attempted to eliminate this distance from the recorded observations. His suggestion was that such periodicities are linked to large-scale equatorially trapped Rossby-type waves. For typical solar parameters, Lou's theoretical model results to periodicities which are very close to the observed ones. More specifically, if sheer Rossby waves are assumed, then the families of periods, which beget from the dispersion relation of these waves in the sun, are:

$$P_r \cong P_{\odot} \left(\frac{|m|}{2} + \frac{(2n+1)\Omega_{\odot} R_{\odot}}{|m|(gD)^{1/2}} \right), \tag{1.1}$$

where $P_{\odot} = \frac{2\pi}{\Omega_{\odot}} \sim 25.1$ days is the solar sidereal rotation period ($\Omega_{\odot}$ being the solar rotation frequency),$n \geqslant 0$ is an integer indicating the number of the considered nodes, m is an integer related to the wavenumber $k_x = \frac{m}{R_{\odot}}$ with $m \geqslant 1$, $R_{\odot} = 7 \times 10^{10} cm$ is the solar radius, $g = 2.7 \times 10^4 cm/s^2$ is the solar surface gravity acceleration and $D = 500 km$ is the average thickness of the photosphere. It is noticeable that for $m = 4, 6, 8, 10, 12$ the $P_r = 53.4, 77.4, 102.000, 126.780, 151.667$ day-periods are produced very consistently by Lou's model in agreement with the observations (Bai & Sturrock 1991, Bai 1992), whereas the odd-valued modes are not so frequently observed.

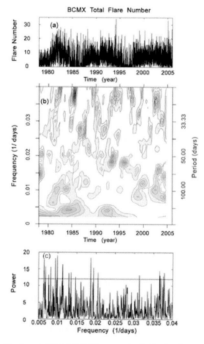

Figure 1. Plot of: (a) Time Series of B,C,M,X summed daily flare numbers, (b) WWZ Wavelet Analysis of B,C,M,X summed daily flare numbers and (c) Scargle-Lomb periodogram for the same data. The dash-dotted line marks the 0.005 level of significance.

2. Motivation and scope

Yet, even Lous recent model leaves several questions unanswered. First and foremost, is it really so, that the RTPs are indeed related to already established active regions, leading mainly to extreme flares? Does the periodic emergence of magnetic flux indeed prefer already formed active regions, thus building complex magnetic configurations which are prompt to produce highly energetic flares? In other words, would it be also possible to clearly observe RTPs in flares of class M, or would RTPs be less prominent in this case? In addition, if we attribute RTPs to Rossby waves according to Lous model, then why observations favor waves with even m values, thus excluding other periodicities which are theoretically possible? The scope of this work is to provide answers to the above-mentioned open questions.

3. Data and methods

In this work, we apply the ScargleLomb periodogram and we test for the first time Wavelet Weighted Z-Transform (WWZ) on X-ray flare data, derived from GOES since 1978. First, the sum of flares per day is considered. This sum can be calculated either over all classes of flares (B, C, M, X) or over the most energetic ones (M, X). Similar data analysis has been attempted several times in the past, like by Bai in 2003 (Bai 2003). This work only considers a prolonged time-period reaching 2006 and adopts daily instead of weekly flare event sums. The novelty presented in this work is the analysis of flares, taken as intensities with time-resolution of 1 min and not as sums of flare events per day. This is possible through the enhanced WWZ method, which is able to handle consistently events which are not regularly distributed in time. The reason for selecting such an analysis is that it can indeed classify flares in a physically continuous

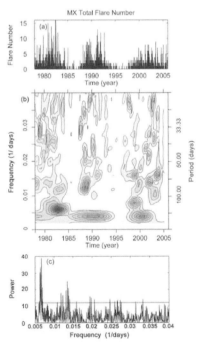

Figure 2. Plot of: (a) Time Series of M,X summed daily flare numbers, (b) WWZ Wavelet Analysis of M,X summed daily flare numbers and (c) Scargle-Lomb periodogram for the same data. The dash-dotted line marks the 0.005 level of significance.

way, without manufacturing artificial sums. In fact, this kind of methodology is applied both for all classes of flares (B, C, M, X) and exclusively for the less energetic ones (B and C), in order to investigate whether RTPs are prominent also in the latter case.

4. Conclusions

In this work, flare data have been analyzed with high resolution methods and were compared with up-to-date theoretical models, targeting RTPs. Starting from Lous theoretical model (Lou 2000), we investigated whether even m periodicities are indeed strangely favored in the Sun. We have proved that when extensive data series are used, odd m periodicities are also frequent and considerably significant. This conclusion places the theoretically foreseen odd m periodicities in the position of normal subharmonics of the 25-d period, exactly as even ones, also from the observational point of view.

As far as the analysis of the X-ray flare data is concerned, the concept of applying statistics on the intensity of the events instead of their summed daily number is tested using the powerful tool of WWZ for the first time, along with the well-established ScargleLomb periodogram. Such methods allow the exploitation of the minute resolution offered by GOES. This enhanced analysis shows that RTPs are present also for weak flare events, thus decoupling these periodicities from magnetic field complexity and extremely active regions. Yet, the occurrence of RTPs in less ener- getic flares needs further theoretical justification. RTPs are observable both/either in sunspot areas and/or in sunspot groups depending on the solar cycle examined (Massi 2007). The former observation is linked with increased magnetic complexity, therefore it would fit closely to RTPs connected with energetic flares (i.e.Mand X class).

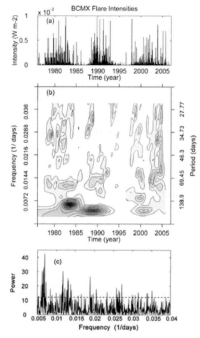

Figure 3. Plot of: (a) Time Series of B,C,M,X flare intensities (minute resolution), (b) WWZ Wavelet Analysis of B,C,M,X flare intensities (minute resolution) and (c) Scargle-Lomb periodogram for the same data. The dash-dotted line marks the 0.005 level of significance.

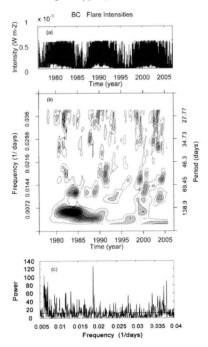

Figure 4. Plot of: (a) Time Series of B,C flare intensities (minute resolution), (b) WWZ Wavelet Analysis of B,C flare intensities (minute resolution) and (c) Scargle-Lomb periodogram for the same data. The dash-dotted line marks the 0.005 level of significance.

Even so, weak looploop interactions resulting from such highly complex magnetic configurations could indeed beget secondary less energetic flares (i.e. B and C class). The latter observation could fit to RTPs connected with any class of flares. The physical mechanism ambiguity of RTP occurrence in less energetic flares could be resolved with future simultaneous analysis on sunspot areas and groups.

The results begetting from Weighted Wavelet Z-Transforms also prove a consistent temporal localization of RTPs among the various analysed flare data series. The Rieger periodicities revealed in this work both in flare number as well as intensity analysis coincide very nicely in time. The temporal occurrence of Rieger periodicities is also consistently cross-checked with previous results (Rieger *et al.* 1984, Ballester *et al.* 1999, Ballester *et al.* 2002) coming from classic wavelet analysis on different solar indices. The prominence of Rieger periodicity in C21, C23 and its absence from C22 is also confirmed within the scope of this work.

References

Bai, T. 1992, *ApJ*, 388, L69

Bai, T. 2003, *ApJ*, 591, 406

Bai, T. & Sturrock, P., 1991, *Nature*, 350, 141

Ballester, J. L., Oliver, R., & Baudin, F. 1999, *ApJ*, 522, L153

Ballester, J. L., Oliver, R., & Carbonell, M. 2002, *ApJ*, 566, 505

Landscheidt, T. 1987, *Solar Phys.*, 107, 195

Lou, Y. Q. 2000, *ApJ*, 540, 1102

Massi, M. 2007, *MemSAI*, 78, 247

Rieger, E., Share, G. H., Forrest, D. J., Kanbach, G., Reppin, C., & Chupp, E. L. 1984, *Nature*, 312, 623

Universal Heliophysical Processes
Proceedings IAU Symposium No. 257, 2008
N. Gopalswamy & D.F. Webb, eds.

© 2009 International Astronomical Union
doi:10.1017/S1743921309029238

A study of spicules from space observations

Ioannis Kontogiannis[1,2], Georgia Tsiropoula[1] and Kostas Tziotziou[1]

[1]Institute for Space Applications and Remote Sensing, National Observatory of Athens,
Lofos Koufos GR-15236 Palea Penteli, Greece
email: [jkonto;georgia;kostas]@space.noa.gr
[2]Section of Astrophysics, Astronomy & Mechanics, Physics Department, University of Athens,
Panepistimiopolis Zografou, GR-15784 Athens, Greece

Abstract. We have studied spicules observed at the northern solar limb by using simultaneous high resolution image sequences. The images were obtained by Hinode/SOT (in the Ca II H passband) and TRACE (in the 1600 Å passband) during a coordinated campaign. Both data sets were reduced and then carefully co-aligned in order to compare the observed patterns in this highly dynamic region of the Sun. The identification of individual structures in both spectral bands allows us to trace their spatial and temporal behaviour. Persistent intensity variations at certain locations, indicate that at least some spicules have a recurrent behavior. Using wavelet analysis we investigate oscillatory phenomena along the axis of off-limb spicules and we construct 2-D maps of the solar limb with the observed oscillations.

Keywords. sun:chromosphere, sun:oscillations

1. Observations, data reduction and analysis

Simultaneous time series observations of the northern solar limb were obtained by TRACE and Hinode/SOT on October 15, 2007. All necessary corrections were carried out, such as dark current, flat-field corrections and spike removal. The images of each data set were carefully co-aligned using cross-correlation between consecutive images, achieving sub-pixel offsets. The Ca II H data set was rebinned to the spatial resolution of TRACE (0.5 arcsec/pixel) and then the two data sets were cross-aligned in order to identify and compare similar structures, if any. The cadence of the observations is 53 s for TRACE and 60 s for SOT and they cover about 1 hour.

The small-scale temporal variation of every pixel on each image was limited by smoothing the time series over 5 consecutive exposures. Then, for every row its minimum value was subtracted from each pixel. Although this last step introduces errors in the form of horizontal lines due to the fact that we did not take into account the curvature of the limb, it helps to increase the contrast of the off-limb structures. To further improve the visibility of the fine-scale off-limb structures we applied the MADMAX operator (Koutchmy & Koutchmy 1988) on all images. This helps discriminate individual spicules (Fig. 1).

2. Morphology and general remarks

The network is visible in both passbands with almost one to one spatial correspondence, while off-limb there are only coarse similarities (Fig. 1). Most spicules are concentrated in groups (bushes), are relatively inclined and show excessive spatio-temporal variations, due to plasma motions and/or ionization. The Ca II H filter has a FWHM of 2.2 Å, while TRACE's 1600 Å passband is very wide (275 Å) and perhaps this is the reason

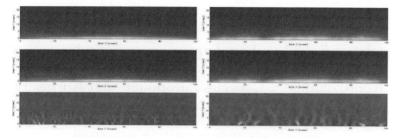

Figure 1. SOT Ca II H *(left)* and TRACE 1600 Å *(right)* filtergrams. *Top to bottom*: the original, enhanced (see text) and after the application of the MADMAX operator images.

why spicules in TRACE appear more diffuse. Thus, emission in both passbands comes from plasma with a wide range of temperatures. Despite that, it seems that some spicules appear at almost the same position when comparing almost co-temporal images. It is very hard, however, to follow their temporal evolution in the two passbands simultaneously. Enhanced images (Fig. 1) show that several spicules attain heights greater than 15″. The shapes of these structures are sometimes irregular and complicated, probably due to superposition effects. Most of them appear and fade within less than 5 frames (i.e. 5 min). Superposition effects, as well as the low cadence of the present observations (1 min) make difficult the study of the evolution of these short-lived structures.

3. Wavelet analysis along individual spicules

In Figure 2 we present two different spicules, each one observed in a different passband. As the same structures do not appear for the same duration in the two passbands a simultaneous analysis was not possible. We performed a wavelet analysis (Torrence & Compo 1998) at every height along their central axes (Fig. 3).

In the Ca II H spicule, periods of 180 s and 300 s were detected very close to the limb. In the TRACE spicule a 300 s period was detected around 3″-4″ above the limb.

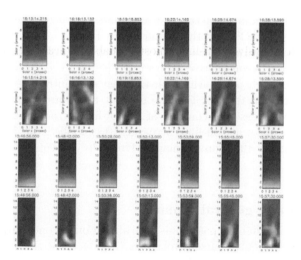

Figure 2. *First and second row*: A spicule in the Ca II H passband. The images in the second row are enhanced using the MADMAX operator. *Third and fourth row*: Same for a spicule at 1600 Å.

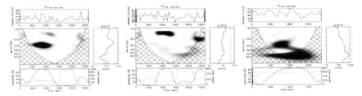

Figure 3. Wavelet analysis of a Ca II H spicule, at 0″.5 (left) and at 1″ (middle) and a 1600 Å spicule, at 3″ above the limb.

Figure 4. 2-D off-limb period maps for Hinode/SOT Ca II H at first and second columns and TRACE 1600 Å passbands, at third and fourth columns.

4. 2-D period maps of oscillatory phenomena

We averaged over a 1″.5 x 1″.5. area of the images and performed wavelet analysis for every "new" pixel above the limb. The periods corresponding to the peaks of the global wavelet spectrum were determined and their probability from the randomization method (Tziotziou *et al.* 2004) was calculated. We considered only periods with probability greater than 80 % to construct 50 s - broad, 2-D period maps (Fig. 4).

Ca II H passband: Periods between 180 s and 320 s are found very close to the limb. In some cases they appear off-limb but no more than 2″ - 3″ above it. They are associated with spicular material but not all spicules show this behaviour. Periods in this range are also found higher than 15″, but are probably due to noise. Periods longer than 400 s are found only off-limb and are probably indicative of spicules' lifetimes.

1600 Å passband: Very few oscillations are detected. Periods up to 350 s are found close to the limb, some of them reaching 5″ high. Almost all of them appear to coincide with periods at the Ca II H line, while the opposite does not happen. Periods longer than 350 s appear almost exclusively off-limb, most probably indicative of spicules' lifetimes.

Acknowledgements

We are grateful to the Hinode and TRACE teams. *Hinode* is a Japanese mission developed and launched by ISAS/JAXA, with NAOJ as domestic partner and NASA and STFC (UK) as international partners. It is operated by these agencies in co-operation with ESA and NSC (Norway).

References

Koutchmy, O. & Koutchmy, S. 1988, *"Optimum Filter and Frame Integration-Application to Granulation Pictures"*, 10th NSO/SPO workshop on *"High Spatial Resolution Solar Observation"*, 1989, 217, O. von der Luhe Ed., 217, 1989

Torrence, C. & Compo, G. P. 1998, *Bull.Amer.Meteor.Soc.*, 79, 61

Tziotziou, K., Tsiropoula, G., & Mein, P. 2004, *A&A*, 423, 1133

Universal Heliophysical Processes
Proceedings IAU Symposium No. 257, 2008
N. Gopalswamy & D.F. Webb, eds.

© 2009 International Astronomical Union
doi:10.1017/S174392130902924X

The mid-term periodicities in sunspot areas

Ryszarda Getko

Astronomical Institute, University of Wroclaw, Wroclaw, Poland
email: getko@astro.uni.wroc.pl

Abstract. The sunspot area fluctuations for the northern and the southern hemispheres of the Sun over the epoch of 12 cycles (12–23) are investigated. Because of the asymmetry of their probability distributions, the positive and the negative fluctuations are considered separately. The auto-correlation analysis of them shows three quasi-periodicities at 10, 17 and 23 solar rotations. The wavelets gives the 10-rotation quasi-periodicity. For the original and the negative fluctuations the correlation coefficient between the wavelet and the auto-correlation results is about 0.9 for 90% of the auto-correlation peaks. For the positive fluctuations it is also 0.9 for 70% of the peaks. For 90% of cycles in both hemispheres the auto-correlation analysis of negative fluctuations shows that two longer periods can be represented as the multiple of the shortest period. For positive fluctuations such dependences are found for more than 50% of cases.

Keywords. Sun: sunspots, methods: data analysis

1. Introduction

In the last decades, the intermediate quasi-periodicities of many solar activity tracers have been discussed. The about 12-rotation periodicity identified Krivova & Solanki (2002) for sunspot data during 1749–2001. It was prominent during times of stronger activity, whereas it diminished and sometimes faded into the background during weak cycles. Getko (2006) found it in both high and low activity periods for the monthly Wolf numbers during cycles 1–22 and for the group sunspot numbers during cycles 5–22. Two longer quasi-periodicities at 17 rotations and at 23 rotations were found in many solar activity parameters from the bottom of the convection zone to the atmosphere. More up-to-date review is by Obridko and Shelting (2007). Here I present results from a statistical study of these periodicities in the sunspot areas during cycles 12–23. It enables one to deduce the mean length of the time period between strong fluctuations.

2. A detailed analysis of quasi-periodicities

I consider the daily sunspot areas for the northern hemisphere (D_l^n), and the southern hemisphere (D_l^s) for solar cycles 12–23 available at the National Geophysical Data Center (NGDC) (http://solarscience.msfc.nasa.gov/greenwch/). For the i-th Carrington rotation I evaluate the mean sunspot area for the northern hemisphere (S_i^n): $S_i^n = \frac{1}{L} \sum_{l=1}^{L} D_l^n$, where L is the number of days for the i-th rotation. I define the fluctuation (F_i^n) of the mean sunspot area (S_i^n) from the smoothed mean sunspot area: $F_i^n = S_i^n - \overline{S_i^n}$ for $i = 1, \ldots, N$, where $\overline{S_i^n} = \frac{1}{13} \sum_{j=i-6}^{i+6} S_i^n$. Each of the time series $\{F_i^n\}$ and $\{F_i^s\}$ contains $N = 1706$ elements. Both have almost the same probability distributions. Fig. 1a shows the histogram of $\{F_i^n\}$ with a fitted Gaussian. The Kolmogorov-Lilliefors and Shapiro-Wilk tests reject the hypothesis of normality for them. Because each distribution has positive skew, the positive and the negative fluctuations are considered separately. For the northern hemisphere they can be defined as follows:

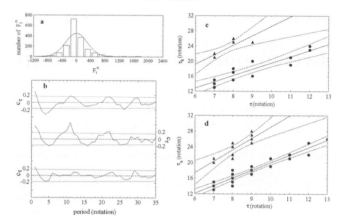

Figure 1. a Histogram of $\{F_i^n\}$ with a fitted Gaussian. **b** *Top:* Auto-correlation function (c_τ) of $\{F_i^n\}$ for cycle 18. *Middle:* Same as for the upper curve, but for $\{F_i^{n-}\}$. *Bottom:* Same as for the upper curve, but for $\{F_i^{n+}\}$. The dotted lines represent two standard errors of each c_τ function. **c** Dependence of τ_2 and τ_3 on τ for positive fluctuations. Lower solid curve represents the regression line for the points (τ, τ_2) (dots). Upper solid curve represents the regression line for the points (τ, τ_3) (triangles). Dashed lines represent the 95 per cent confidence interval for each regression line. **d** Same as for **c**, but for negative fluctuations.

$$F_i^{n+} = \begin{cases} 0 & \text{where} \quad F_i^n \leqslant 0 \\ F_i^n & \text{where} \quad F_i^n > 0 \end{cases} \quad \text{and} \quad F_i^{n-} = \begin{cases} 0 & \text{where} \quad F_i^n > 0 \\ F_i^n & \text{where} \quad F_i^n \leqslant 0 \end{cases} \quad \text{for } i = 1, \dots, N.$$

It is known that for a time series which contains Gaussian white noise and a sinusoidal component a probability distribution is symmetric and the auto-correlation functions (c_τ) of that time series, of its positive fluctuations and of its negative fluctuations should be the same. The functions c_τ of $\{F_i^n\}$, $\{F_i^{n+}\}$ and $\{F_i^{n-}\}$ for one solar cycle are different (Fig. 1b). The functions c_τ of $\{F_i^n\}$ and $\{F_i^{n-}\}$ for cycle 18 have the significant global maxima at $\tau = 11$ rotations and smaller maxima at $\tau_1 = k * \tau$ for $k = 2$ and 3. In 54% of 24 cases (12 cycles in each hemisphere) the functions c_τ of the original fluctuations have significant maxima for $\tau \in [7, 13]$. In 30% cases the maxima for such τ belong to the interval $[1\sigma, 2\sigma]$. For positive fluctuations this contribution is 50% and 46% respectively. For negative fluctuations in 92% cases the maxima at $\tau \in [7, 13]$ are significant. The mean value of all $\tau \in [7, 13]$ for which the maxima are significant is approximately 10 rotations for all three fluctuation groups. I also consider the c_τ maxima for $\tau \in [14, 19]$ and $\tau \in [20, 27]$. For more than 50% of cases the positive fluctuations create the auto-correlation peaks for which the periods $\tau_1 \approx 17$ and $\tau_2 \approx 23$ can be represented as $\tau_k \approx k * \tau$ where $\tau \in [7, 13]$ and $k = 2$ or 3. For each of k the points (τ, τ_k), the regression line (solid) and the 95% confidence interval for each line (dotted) are shown in Fig. 1c. For $k = 2$ the correlation coefficient for 13 points is 0.91, for $k = 3$ it is 0.86 for 8 points. For the negative fluctuations the strong dependence between the considered periodicities was found in $\sim 90\%$ for $k = 2$ and in $\sim 50\%$ for $k = 3$ (Fig. 1d). For $k = 2$ the correlation is 0.95 for 23 points and for $k = 3$ it is 0.91 for 11 points. It is important to add that the c_τ values at $\tau > 27$ are not reliable because of the solar cycle length.

I also applied the Morlet wavelet (Torrence & Compo 1998) to three fluctuation time series for each of 24 cases. Fig. 2 shows the normalized wavelet maps for cycle 18. Black contours denote the 95 per cent significance level for detected peaks. The wavelet map of $\{F_i^n\}$ (top) shows two significant peaks (at $\tau \approx 6$ and 8). They are mainly created by three strong fluctuations at the begining and at the end of the high activity period. During the remaining part of this period a rised power is at $\tau = 11$ rotations. Moreover,

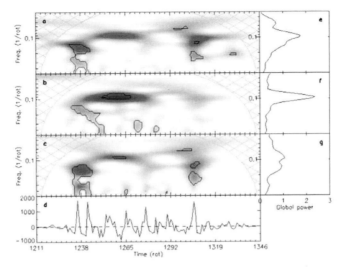

Figure 2. a–c: Wavelet power spectra of **a:** $\{F_i^n\}$, **b:** $\{F_i^{n-}\}$ and **c:** $\{F_i^{n+}\}$ mapping a time-frequency evolution of about 10-rotation periodicity. Top values of wavelet power are denoted by gradual darkening. Black contours denote significance levels of 95 per cent for detected peaks. A cone of influence is marked by the dashed region. **e–g:** Corresponding global wavelet power spectra. **d:** Time series $\{F_i^n\}$ for cycle 18.

the integrated spectrum (right) shows the maximum at $\tau = 11$ and confirms the auto-correlation results. For $\{F_i^{n-}\}$ (middle) the peak at $\tau = 11$ is well detected with 95 per cent level, extends in time during the high activity period and dominates the integrated spectrum. The map of $\{F_i^{n+}\}$ (bottom) is similar to the map of $\{F_i^n\}$, but the peak at $\tau = 11$ is significant. The global spectrum shows two almost the same peaks at $\tau = 11$ and 8. Such an analysis was done for all 24 cases. The auto-correlation and the wavelet results are similar for $\tau \in [7, 13]$ (the correlation between them is 0.9 for 87%, 92% and 72% of the auto-correlation peaks of $\{F_i^n\}$, $\{F_i^{n-}\}$ and $\{F_i^{n+}\}$ respectively).

These results could indicate that the 10-rotation quasi-period is dominant. Getko (2004) showed that large activity complexes were responsible for strong sunspot number fluctuations. Thus, the time between strong fluctuations of toroidal magnetic flux in the tachocline could be on the order of 7–13 rotations. It is also well known that two solar hemispheres show certain hemispheric asymmetries in their solar-cycle features. However, two-sample Kolmogorov-Smirnov test shows that the 10-rotation quasi-periods evaluated for each of 12 solar cycles in each hemispheres do not differ. Two longer quasi-periods at about 17 and 23 rotations could be treated as subharmonics of the 10-rotation quasi-period (Figs. 1c and 1d). This facts could explain a wide range of periodicities in various solar indices at all levels from the tachocline to the Earth.

3. Conclusions

(a) For both hemispheres the probability distributions of fluctuations are similar and have an asymmetry which means that there are more negative than positive fluctuations.

(b) The auto-correlation analysis of the original, the positive and the negative fluctuations prefers three quasi-periods: around 10, 17 and 23 rotations. The wavelet maps show one dominant quasi-period at about 10 rotations.

(c) For 90% of solar cycles in both hemispheres the auto-correlation analysis of negative fluctuations gives peaks for which the period $\tau_2 \approx 17$ rotations can be represented as $\tau_k \approx k * \tau$ where $\tau \in [7, 13]$ and $k = 2$. For $k = 3$ a such dependence are reliable in 50% of considered cases. For positive fluctuations such dependences are found for more than 50% of solar cycles in each hemispheres.

References

Getko, R. 2004, *Solar Phys.* 224, 291
Getko, R. 2006, *Solar Phys.* 238, 187
Krivova, N. A. & Solanki, S. K. 2002, *A & A* 394, 701
Obridko, V. N. & Shelting, B. D. 2007, *Adv. Sp. Res.* 40, 1006
Torrence, C. & Compo, G. P. 1998, *BAAS* 79, 61

Universal Heliophysical Processes
Proceedings IAU Symposium No. 257, 2008
N. Gopalswamy & D.F. Webb, eds.

© 2009 International Astronomical Union
doi:10.1017/S1743921309029251

Solar Differential Rotation of Compact Magnetic Elements and Polarity Reversal of the Sun

Darejan Japaridze, Marina Gigolashvili and Vasili Kukhianidze

Georgian E.Kharadze National Astrophysical Observatory at Ilia Chavchavadze State
University, A. Kazbegi Ave 2a, Tbilisi 0160, Georgia
email: `marinagig@yahoo.com`

Abstract. The differential rotation of the compact elements of the large-scale magnetic fields is studied using Solar Synoptic Charts (1966–1986). It is revealed that compact magnetic elements with the similar polarity of the polar magnetic field of the Sun have a larger rotation rate than the elements with the opposite polarity at all stages in the cycle.

From the comparison of the experimental measuring data of the solar magnetic elements there are received the results: a) The differential rotations of the compact magnetic elements with negative and positive polarities have the similar behavior for the solar 20 and 21 cycles; b) It is established that in the rotation rate of compact magnetic elements there are present some variations at the time of polarity reversal of the Sun.

There is assumed that the physical understanding of the connections of differential rotation of compact magnetic elements and polarity reversal of the Sun depends upon establishing a connection between the temporal variability of spatially resolved solar magnetic elements and polar reversals.

Keywords. Differential rotation, magnetic fields, polarity reversal.

---

1. Introduction

Special attention has always been paid to the study of the solar differential rotation and the solar magnetic field, because they are key to understanding the physical processes in the solar atmosphere. Sunspots have been used as tracers of solar rotation since they were first recognized as features on the Sun (Newton 1924). Other features used are faculae (Newbegin & Newton 1931) and hydrogen filaments (D'Azambuja & D'Azambuja 1948; Japaridze & Gigolashvili 1992; Gigolashvili *et al.* 1995; Gigolashvili *et al.* 2005). Another class of features used for tracking the large-scale solar patterns is neutral lines of the magnetic fields in filtergrams and spectroheliograms (Durrant *et al.* 2002; Gigolashvili *et al.* 2005/2006; Japaridze *et al.* 2006; Japaridze *et al.* 2007). The variations of characteristics of time-varying processes occurring in the solar atmosphere are tightly connected with prolonged large-scale manifestation of solar activity. The interaction between the solar rotation and magnetic fields is indisputably the reason for such activity.

Within the solar interior the surface manifestation of magnetic fields exhibits an unexpected degree of regularity despite such fields being embedded in an extremely turbulent medium. The largest magnetic fields observed at the surface follow episodic patterns of emergence and evolution that collectively form each activity. There is also evidence that smaller-scale magnetic fields also possess an imprinting of such cyclic behavior. Two specific aspects of the coupling between small and large scale structures on the Sun are discussed by DeRosa (2005). McIntosh and coauthors made Carrington maps of H-alpha solar synoptic charts and the results were published in the form of the atlas of stackplots

(McIntosh *et al.* 1991). Large-scale stackplots for the entire range of data for 1966-1986 include the series of plots displaying 10°- zones of solar latitude in the range of 70°. Snodgrass (1992) finds patterns that appear to show features at the same latitude which are moving at different rotation rates. It is also possible to observe the poleward drift of the large-scale unipolar regions and the evolution of the polar cap as well as a variety of other apparent meridional and vertical motions. For compact magnetic elements with negative and positive polarities separately the average values of statistics were calculated for the minimum periods 1964–1966 and 1973–1978, as well as for maximum periods 1967–1972, 1979–1983 for the northern and the southern hemispheres with the 90 % confidence level (Gigolashvili *et al.* 2007). In this paper, we study the differential rotation of compact magnetic elements during solar activity cycles 20 and 21 using the McIntosh's stackplots (McIntosh *et al.* 1991).

2. Data, Method of Treatment and Results

To study the differential rotation of compact magnetic elements for solar activity cycles 20-21 (1966-1986) we used the McIntosh's atlas of synoptic maps. We have chosen only the visually symmetric compact elements with significant angle of a deviation that is possible to be measured at least for 3 days. This is necessary for determination of differential rotation of the features with high accuracy.

For 335 chosen compact magnetic elements, 1675 measurements have been made. In cycles 20 and 21, 990 measurements for 198 features and 685 measurements for 137 features, respectively have been carried out.

We measured the angle between the symmetry axis of a chosen magnetic element and the horizontal line parallel to the horizontal edge chosen among five identical plots. The average slope of long-lived patterns generally varies in a regular way as a function of latitude. Since the frame of reference is the Carrington system of solar longitudes, a vertical pattern in a stackplot represents a pattern rotating at the Carrington synodic rate of 27.2753 days; positive slopes indicate apparent rotation rates slower than the Carrington rate. Negative slopes indicate rotation rates faster than the Carrington rate and these usually occur at latitudes less than 20° (McIntosh, Willock, and Thompson, 1991). The rotation rates for compact magnetic elements were calculated with the help of the empirical formula (Japaridze *et al.*, 2006): $\Omega(\phi) = 1000/(36.664\text{-}\cot\alpha)$, where α is the angle of the slope, ϕ is latitude and $\Omega(\phi)$ is the rotation rate in deg/day. By a method developed by us (Japaridze *et al.*, 2006) measurement of the rotation rates of large-scale features is impossible because of an uncertainty in the determination of the angle of their deviation from the solar central meridian.

As the patterns in McIntosh's stackplots are displayed in both longitude and latitude, we can trace a wealth of various details. Calculated rotation rates of magnetic elements with the positive and negative polarity for low (10°) and middle (60°) latitudinal zones separately for both hemispheres of the Sun are presented on the Figure 1.

The diagrams for every 10°-zone were constructed separately for the northern and southern hemispheres for compact magnetic elements with the positive and negative polarities. In the figure $-CE$ and $+CE$ are rotation rate for the whole cycle of magnetic elements with negative and positive polarities, respectively. Arrows point to the epochs of the polarity reversals of the circumpolar regions of the Sun for northern (big arrows) and southern hemispheres (small arrows). The polarity reversal occurred in 1969, 1971.1, 1974.1 (three-fold polarity reversal) and in 1970.5 for cycle 20 and in 1981.0 and 1981.7 for cycle 21 in the northern and southern hemispheres, respectively (Makarov *et al.*, 1977; Makarov and Sivaraman, 1989a, b).

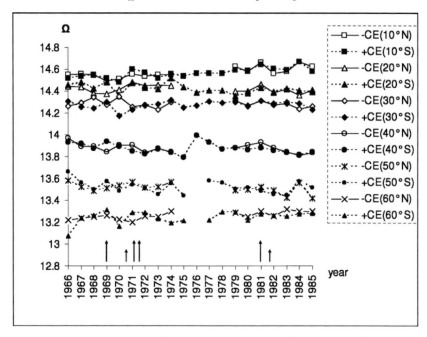

Figure 1. Rotation rate of magnetic elements with the positive and negative polarities for 10°-zones of the northern and southern hemispheres.

From figure 1 we can see that the differential rotations of the compact magnetic elements with negative and positive polarities have the similar behavior for cycles 20 and 21. In the rotation rate of compact magnetic elements some variations are present at the time of the polarity reversal of the Sun.

3. Discussion

It is known that the polar magnetic fields reverse their polarity at maxima of solar cycles. Using homogenous data of hydrogen filaments a quasi-biennial pulse propagation from high latitudes to the equator was found. A pulse drift was observed in the northern hemisphere during 1968–1970, 1979–1981, 1988–1990 and in the southern one during 1969–1971, 1979–1981, 1989–1991. If the polarity reversal is three-fold the residual velocities are great and have secondary peaks at relatively high latitudes. If the polarity reversal is simple, propagation of a quasi-biennial pulse occurs almost simultaneously in both hemispheres and the amplitude of residual velocities is minimal (Gigolashvili *et al.* 2005).

Pulkkinen & Tuominen (1998) found that the solar rotation is continuously changing not only in the course of a cycle but also over a longer time period. They also found strong fluctuations in the equatorial rotation in the course of solar activity.

According to Seeley *et al.* (1987) small-scale compact magnetic elements are parts of larger structures. They have rotation rates different from motions of large-scale structures. Symmetric large-scale magnetic elements (chosen by visual examination) with negative and positive polarity have the same behavior of the rates of differential rotation. During the first polarity reversal of the three-fold changing of circumpolar magnetic field in northern hemisphere the rotation rates of compact magnetic elements with negative and positive polarities change in an anti phase, while all other cases show collective variations of rates.

We investigated the differential rotation of the large-scale magnetic elements during solar activity cycles 20–21. For these cycles the differential rotations of the compact magnetic elements with negative and positive polarities have similar behavior. In the rotation rate of compact magnetic elements some variations are present at the time of polarity reversal of the Sun.

A physical understanding of the connections of differential rotation of compact magnetic elements and polarity reversal of the Sun depends on establishing a connection between the temporal variability of spatially resolved solar magnetic elements and polar reversals.

References

D'Azambuja, M. & D'Azambuja, L. 1948, *Ann. Obs. Paris*, 6, 1

DeRosa, M. L. 2005, *ASP Conference Series*, 346, 337

Durrant, C. J., Turner, J., & Wilson, P. R. 2002, *Solar Phys.*, 211, 103

Gigolashvili, M. Sh., Japaridze, D. R., & Kukhianidze, V. J. 2005, *Solar Phys*, 231, 23

Gigolashvili, M. Sh., Japaridze, D. R., & Kukhianidze, V. J. 2005/2006, *Science without Borders*, 2, 136

Gigolashvili, M. Sh., Japaridze, D. R., Mdzinarishvili, T. G., Chargeishvili, B. B., & Kukhianidze, V. J. 2007, *Advances in Space Research*, 40, 7, 976

Gigolashvili, M. Sh., Japaridze, D. R., Pataraya, A. D., & Zaqarashvili, T. V. 1995, *Solar Phys*, 156, 221

Japaridze, D. R. & Gigolashvili, M. Sh. 1992, *Solar Phys*, 231, 23

Japaridze, D. R., Gigolashvili, M. Sh., & Kukhianidze, V. J. 2006, *Sun and Geosphere*, 1, 31

Japaridze, D. R., Gigolashvili, M. Sh., & Kukhianidze, V. J. 2007, *Advances in Space Research*, 40, 7, 1912

Makarov, V. I. & Sivaraman, K. R. 1989, *Solar Phys*, 123, 367

Makarov, V. I. & Sivaraman, K. R. 1989, *Solar Phys*, 119, 35

Makarov, V. I., Tlatov, A. G., & Callebaut, D. K. 1997, *Solar Phys*, 170, 373

McIntosh, P. S., Willock, E. C., & Thompson, R. J. 1991 *National Geophysical Data Center*, 1

Newbegin, A. M. & Newton, H. W. 1931, *The Observatory*, 54, 20

Newton, H. W. 1924, *MNRAS*, , , 84, 431

Pulkkinen, P. & Tuominen, I. 1998, *Astrophys.*, 338, 748

Snodgrass, H. B. 1992, *ASP Conference Series*, 27, 205

Universal Heliophysical Processes
Proceedings IAU Symposium No. 257, 2008
N. Gopalswamy & D.F. Webb, eds.

© 2009 International Astronomical Union
doi:10.1017/S1743921309029263

Microwave observations with the RATAN-600 radio telescope: detection of the thermal emission sources

Irina Yu. Grigoryeva[1], Larisa K. Kashapova[2], Moisey A. Livshits[3] and Valery N. Borovik[1]

[1]Central Astronomical Observatory at Pulkovo RAS,
196140 Pulkovskoe sh., 65/1, St.Petersburg, Russia. email: `irina19752004@mail.ru`

[2]Institute of Solar-Terrestrial Physics SD RAS, Irkutsk, Russia. email: `lkk@iszf.irk.ru`

[3]Pushkov Institute of Terrestrial Magnetism, Ionosphere and Radio Wave Propagation RAS,
142190 Troitsk, Russia. email: `maliv@mail.ru`

Abstract. We report on two off-limb radio sources of microwave emission which were detected in one-dimensional RATAN-600 solar scans of the post-eruptive loops: on December 2, 2003 (off west limb) and January 25, 2007 (east limb). The microwave spectra showed that the thermal emission was predominant at the early stage of the arcade formation with a small contribution of non-thermal emission. There were no high-energy particles in these events. The microwave spectra of the radio sources associated with the tops of postflare loops show the predominant thermal emission during one hour after the eruption. In case of a small contribution from accelerated particles to the microwave emission, there is a large amount of hot plasma in the region of the loop tops after the eruption.

Keywords. Sun: radio radiation, flares, particle emission, coronal mass ejections (CMEs)

1. Introduction

Systems of post-eruptive arcades could be formed in events of various importance. We suppose that one of the necessary conditions for that is a CME development followed by removal of the main part of the gas mass far into interplanetary space. However, a part of the matter could be held into coronal layers of active regions and take part in the arcade formation. Observations of post-eruptive arcades in various spectral regions were analyzed in several papers (e.g., Feldman *et al.* 1995; Harra-Murnion *et al.* 1998; and Grechnev *et al.* 2006). In this paper we discuss the observations of different cases of post-eruptive arcade formation in events with weak flares (C-class in GOES classification) just at the initial stage. In order to confirm our supposition, it is necessary to analyze the role of thermal and non-thermal processes in the source region of the subsequent arcade formation.

2. Observations

25 January 2007 On January 25, 2007 the solar observations by the radio telescope RATAN-600 (Korol'kov & Parijskij 1979) were made at 07:44 UT, 08:18 UT, 08:52 UT, 9:26 UT (local noon), 10:00 UT, 10:34 UT and 11:08 UT. The limb event consisted of a CME and C6.3 class flare (S08E90). The flare onset was at 06:33 UT and the peak emission occurred at 07:14 UT. At early stages of the post-eruptive arcade formation, the spectrum of microwave emission was practically flat (see Fig. 1d). One may conclude from RHESSI data, that directly after the peak of the flare in the mean HXR spectra the thermal contribution dominated (see Fig. 1c).

2 December 2003 On December 2, 2003 the solar observations were made at 08:02 UT, 09:03 UT (local noon), 10:03 UT and 11:04 UT. This limb event consisted of a filament eruption, CME and C7.2 class flare (S19 W89). The flare onset was earlier than 09:40 UT and the peak emission occurred at 09:48 UT. The nearest moment of radio observations to the beginning of the arcade formation was at 11:04 UT. In X-ray range the source with predominant thermal emission was observed directly above the top of post-eruptive arcade (see Fig. 2a).

3. Discussion

The microwave spectra obtained with the radio telescope RATAN-600 of the radio sources associated with the tops of postflare loops show the predominant thermal emission during one hour after the eruption (see (Fig. 1d and Fig. 2b). In case of a small contribution from accelerated particles to the microwave emission, there is a large amount of

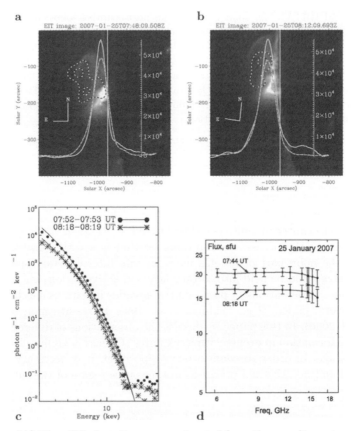

Figure 1. (a and b) The off-limb radio source extracted from the one-dimensional RATAN-600 solar scan (Stokes "I") at 2.03 cm and 5.02 cm at 07:44 UT and at 08:18 UT overlaid on the SOHO/EIT 195 Å solar image at 07:48 UT and 08:12 UT, respectively. Vertical lines show the solar limb. Dashed lines show the overlaid RHESSI image by counters at 90% and 70% of maximum (6–12 keV). Vertical scales show exceeding emission above the quiet Sun's level of the off-limb radio source (antenna temperature in K). (c) Mean HXR photon spectra obtained with RHESSI data. (d) Averaged total flux microwave spectra of the off-limb radio source associated with the post-eruptive arcade during the initial stage formation.

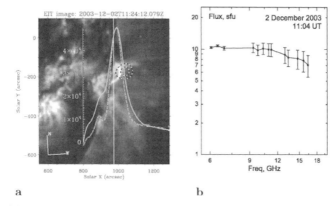

a   b

Figure 2. (a) Data as in Fig. 1a and 1b at 11:04 UT (RATAN data) and at 11:24 UT (SOHO/EIT image, RHESSI data) on December 2, 2003. (b) Data as in Fig. 1d.

hot plasma in the region of the loop tops after the eruption. Simultaneous studies of radio and X-ray radiation provide useful information on the post-eruptive arcade formation. The plasma parameters for both flares are extracted from the RHESSI photon spectrum by fitting the mean spectrum obtained using a standard routine of the RHESSI software (Smith *et al.* 2002). The best fit was obtained for the optically thin bremsstrahlung radiation function (vth). The results confirm the thermal nature of the observed sources and the obtained emission measures and plasma temperatures are: $EM = 6.1 \times 10^{47}$ cm^{-3}, $T = 12.8 \times 10^6$ K and $EM = 2.5 \times 10^{47}$ cm^{-3}, $T = 12.8 \times 10^6$ K for the 25 January 2007 event (at 07:52 UT and 08:18 UT, respectively) and $EM = 1.3 \times 10^{48}$ cm^{-3} and $T = 17.4 \times 10^6$ K for the 2 December 2003 event (at 11:24 UT). Assuming bremsstrahlung as a predominant mechanism of microwave emission and $T = 5 \times 10^6$ K, we estimate from microwave data the emission measure to be $EM = 14.6 \times 10^{48}$ cm^{-3} at 07:44 UT and $EM = 7.5 \times 10^{48}$ cm^{-3} at 08:18 UT on January 25, 2007. These values are higher than those obtained from X-ray data. This may be the evidence of the presence of plasma with different temperatures. It is interesting to study in future the influence of the hot plasma on the post-eruptive processes.

Acknowledgements

The authors thank V. E. Abramov-Maximov for assistance in handing RATAN-600 data and V. F. Melnikov for useful discussion. The research was supported by the RFBR grants 08-02-00872, 06-02-16838. I. Yu. G. thanks LOC IAUS257 for support grant.

References

Feldman, U., Seely, J. F., Doschek G. A., Brown, C. M., Phillips, K. J. H., & Lang, J. 1995, *ApJ*, 446, 860

Grechnev, V. V., Uralov, A. M., Zandanov, V. G., Rudenko, G. V., Borovik, V. N., Grigorieva, I. Y., Slemzin, V. A., Bogachev, S. A., Kuzin, S. V. & Zhitnik, I. A. 2006, *PASJ*, 113, 415.

Harra-Murnion, L. K., Schmieder, B., van Driel-Gesztelyi, L., Sato, J., Plunkett, S. P., Rudawy, P., Rompolt, B., Akioka, M., Sakao, T., & Ichimoto, K. 1995, *AA*, 337, 911

Korol'kov, D. V. & Parijskij, Y. N. 1979, *ST*, 57, 324

Smith, D. M., Lin, R. P., Turin, P., Curtis, D. W., Primbsch, J. H., Campbell, R. D., Abiad, R., Schroeder, P., Cork, C. P., Hull, E. L., Landis, D. A., Madden, N. W., Malone, D., Pehl, R. H., Raudorf, T., Sangsingkeow, P., Boyle, R., Banks, I. S., Shirey, K., & Schwartz, R. 2002, *Solar Phys.*, 210, 33

Universal Heliophysical Processes
Proceedings IAU Symposium No. 257, 2008
N. Gopalswamy & D.F. Webb, eds.

© 2009 International Astronomical Union
doi:10.1017/S1743921309029275

Oscillatory phenomena in a solar network region

Georgia Tsiropoula[1], Kostas Tziotziou[1], Pavol Schwartz[2] and Petr Heinzel[2]

[1]Institute for Space Applications and Remote Sensing, National Observatory of Athens,
Lofos Koufos, 15236 P. Penteli, Greece
email: [georgia; kostas]@space.noa.gr

[2]Academy of Sciences of the Czech Republic, Astronomical Institut,
CZ-25165 Ondřejov, Czech Republic
email: [schwartz; pheinzel]@asu.cas.cz

Abstract. We examine oscillatory phenomena in a solar network region from multi-wavelength, observations obtained by the ground-based Dutch Open Telescope (DOT), and by instruments on the spacecraft Solar and Heliospheric Observatory (SoHO). The observations were obtained during a coordinated observing campaign on October 14, 2005. The temporal variations of the intensities and velocities in two distinct regions of the quiet Sun were investigated: one containing several dark mottles and the other several bright points defining the network boundaries (NB). The aim is to find similarities and/or differences in the oscillatory phenomena observed in these two regions and in different spectral lines formed from the chromosphere to the transition region, as well as propagation characteristics of waves.

Keywords. Sun: chromosphere; transition region; oscillations

1. Introduction

In the chromosphere, the quiet Sun displays a distinct network appearance identical to the photospheric supergranular structure. In the Hα line, especially in its wings, several elongated dark structures, called mottles, outline the cell boundaries. In the transition region, the network stands out as more stable cellular patterning with bright patches identifying the network boundaries (NB) and enclosing dark areas which correspond to the internetwork (IN). This spatial dichotomy is also apparent in the power spectra leading to the suggestion that different physical mechanisms may dominate in each region. Several authors have studied properties of spectral lines and reported that the chromospheric plasma localized in NB oscillates with a dominant period of ∼300 s (Lites *et al.* 1993; Curdt & Heinzel 1998). Cauzzi *et al.* (2000) from the power spectrum computed for the intensity of Hα line center found that for both the NB and IN the power distribution peaks at ∼ 300 s with no enhanced power detectable in the 3 min range in the IN. This is consistent with observations in the Ca II H line by Lites *et al.* (1993) showing a power peak in the 3 min range only for the velocity fluctuations. Tziotziou *et al.* (2004) from a wavelet analysis of Hα observations found intensity and velocity periodicities in dark mottles outlining the NB, and thus overlying the IN, in the range 270–450 s, although a 180 s period was also apparent.

In this work we present observations of a solar network region obtained simultaneously by space-borne and ground-based instruments in several spectral lines. We analyze the fluctuations observed through wavelet and global phase difference analyses in order to search for wave signatures at different heights of the solar atmosphere.

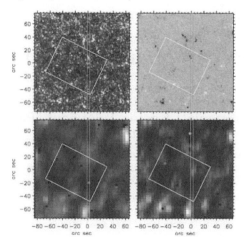

Figure 1. *First row*: C IV TRACE image (*left*) and MDI magnetogram (*right*) obtained at 10:15 UT. *Second row*: Cut-outs of the CDS raster observations of intensities at He I (*left*) and O V (*right*) lines obtained from 10:46 to 11:17 UT and corrected for solar rotation due to the time difference. Solar North is up. The white rectangle inside the images marks the DOT's FOV (pointing to the celestial North), while the two parallel grey lines inside the images mark the location of the sit-and-stare CDS observations.

2. Observations

Co-temporal observations of a quiet region found at the solar disk center were obtained on October 14, 2005 by SoHO (CDS and MDI), the Dutch Open Telescope (DOT) and TRACE. DOT obtained a time sequence in Hα which consists of 26 speckle reconstructed images taken simultaneously at a cadence of 35 s with a pixel size of 0.071″ in 5 wavelengths along the Hα line profile (i.e. at −0.7Å, −0.35Å, line center, 0.35Å and 0.7Å). CDS obtained sit-and-stare observations in several spectral lines among which He I, O V, and Ne VI (effective pixel size of 4Å in the horizontal and 3.36Å in the vertical direction) with 60 exposures and cadence of 49 s. CDS raster scans, as well as MDI magnetograms and TRACE filtergrams were used for the co-alignment of the different data sets (Fig. 1). DOT images were rebinned to CDS spatial pixel sizes both in the X- and Y-direction for intercomparison.

3. Results

In CDS sit-and-stare observations the quiet Sun is dominated by a pattern of bright streaks defining the NB areas, dark streaks defining the IN areas, while mottles regions appear darker than the NB, but brighter than the IN. In the rebinned DOT Hα line center images a bright patch is observed in the place where network bright points were observed and a dark patch in the respective region of dark mottles. A strong positive correlation exists between respective peak intensities of the CDS lines suggesting a strong association between different heights of the solar atmosphere and indicating that we are possibly observing manifestations of the same structural topology at different heights.

3.1. *Wavelet power analysis*

A wavelet analysis is performed in both intensity and Doppler velocity variations, in two regions (NB and mottles' region), for both unfiltered and filtered with a high-pass frequency filter time series of DOT and CDS observations. Hα line-center intensity variations show a most prominent peak at ∼300 s in the mottles' region and an extended

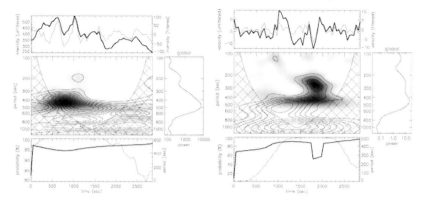

Figure 2. Wavelet analysis at a fixed location of the O V peak intensity (*left*) and velocity (*right*) variations in the mottles' region.

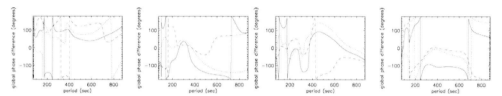

Figure 3. Global phase difference (in degrees) of intensities and velocities as a function of period relative to the Hα line center intensity and Doppler velocity (He I (thick solid line), O V (dotted line), Ne VI (dashed line)) for the mottles' region (*first and third plot, respectively*). The same for the NB region (*second and fourth plot, respectively*).

peak at periods of 300 to 600 s in the NB region. Unfiltered CDS spectra are dominated by a period ∼1000 s, which is subjected to edge effects and has an unclear physical importance. Both CDS intensity and velocity filtered power spectra show several periods with significant power within the cone–of–influence (COI). Most spectra in the mottles' region, as well as in the NB, show dominant oscillation signatures mainly in the 250 s - 400 s range with rather variable probability (Fig. 2, for details see Tsiropoula *et al.* 2009).

3.2. *Global phase difference analysis*

Phase difference obtained with a cross-wavelet transform between the Hα line-center filtered intensity and Doppler velocity time series and the corresponding filtered time series of the CDS lines is used for the study of propagation characteristics of waves at different heights in the solar atmosphere (Fig. 3). Phase difference curves have similar behavior indicating a clear interconnection between processes in the lower and higher solar atmosphere and in both mottles and NB regions. Differences in phase difference for the examined spectral lines suggests formation at slightly different heights. I–I (i.e. between intensities) phase spectra depend on many parameters (i.e. frequency, peculiarities of the line formation) and are difficult to interpret. V–V (i.e. between velocities) phase differences in the NB show vertically propagating waves, since above 200 s the inclination of the phase curve with period is constant for all lines as expected from the theoretical behavior of such waves. For periods above 350 s, the phase difference becomes almost constant indicating no propagation, as expected from the NB cut-off period which is about 350–400 s. Non-constant dependence of phase curves on the period above 200 s in the mottles' region probably indicates: a) non-vertical propagation or b) waves propagating at different inclined mottles along the line-of-sight. The mostly negative phase difference suggests a downward propagation at least for periods of 250–400 s, representing waves refracted from the inclined magnetic field of mottles or waves converted to other wave types, such as fast or slow MHD modes (depending on the field strength).

Acknowledgements

This work was partly supported by a grant PENED 03ED554 [co-financed by EC-European Social Fund (80%) and the Greek Ministry of Development-GSRT (20%)]. P.S. and P.H. are supported by the ESA-PECS project No 98030. Dr E. Khomenko is thanked for very useful discussions concerning the intrepretation of phase differences and waves. The authors are also grateful to the DOT, SoHO and TRACE observers and science planning teams.

References

Cauzzi, G., Falchi, A., & Falciani, R. 2000, *A&A*, 357, 1093
Curdt, W. & Heinzel, P. 1998, *ApJ(Letters)*, 503, L95
Lites, B. W., Rutten, R. J., & Kalkofen, W. 1993, *ApJ*, 414, 345
Tziotziou, K., Tsiropoula, G., & Mein, P. 2004, *A&A*, 423, 1133
Tsiropoula, G., Tziotziou, K., Schwartz, P., & Heinzel, P. 2009, *A&A*, 493, 217

Universal Heliophysical Processes
Proceedings IAU Symposium No. 257, 2008
N. Gopalswamy & D.F. Webb, eds.

© 2009 International Astronomical Union
doi:10.1017/S1743921309029287

Numerical simulation of wave propagation in magnetic network

G. Vigeesh[1], S. S. Hasan[1] and O. Steiner[2]

[1]Indian Institute of Astrophysics,
Bangalore-560034, India
email: `vigeesh@iiap.res.in, hasan@iiap.res.in`

[2]Kiepenheuer-Institut für Sonnenphysik,
79104 Freiburg, Germany
email: `steiner@kis.uni-freiburg.de`

Abstract. We present 2-D numerical simulations of wave propagation in the magnetic network. The network is modelled as consisting of individual magnetic flux sheets located in intergranular lanes. They have a typical horizontal size of about 150 km at the base of the photosphere and expand upward and become uniform. We consider flux sheets of different field strengths. Waves are excited by means of transverse motions at the lower boundary, to simulate the effect of granular buffeting. We look at the magneto-acoustic waves generated within the flux sheet and the acoustic waves generated in the ambient medium due to the excitation. We calculate the wave energy fluxes separating them into contributions from the acoustic and the Poynting part and study the effect of the different field strengths.

Keywords. Sun: magnetic fields – magnetohydrodynamics (MHD) – Sun: photosphere – oscillations

---

The magnetic network in the solar atmosphere consists of intense magnetic field elements located in intergranular lanes. These magnetic elements have field strengths ranging from few hundred gauss to kilogauss (Berger *et al.* 2004). Granular buffeting is believed to excite MHD waves in this medium. Following Hasan *et al.* (2005) and Hasan & Ballegooiijen (2008), we model network magnetic fields as non-potential structures having a horizontal size of about 150 km at the base of the photosphere. Waves are excited in them by means of transverse motions at the lower boundary. The driving generates both fast and slow waves within the flux sheet and acoustic waves in the ambient medium. We consider flux sheets of different field strengths and study the energy transported by the these waves.

The boundary conditions for the magnetic field and the pressure distribution in the physical domain is specified initially. The magnetic field configuration and the density distribution is then iteratively calculated from the magnetohydrostatic equations. We treat two different cases corresponding to field strengths (at $z = 0$) of 800 G (moderate field case) and 1600 G (strong field case), on the axis of the sheet. Waves are excited in this structure by an impulsive transverse motions of the lower boundary with a period of 24 s and an amplitude of 750 m s^{-1}. We solve the MHD equations following the numerical method described in Steiner *et al.* (1994).

The wave energy flux is the energy that is transported by the waves. The linearized wave energy flux is given as (Bray & Loughhead 1974, Bogdan *et al.* 2003),

$$\mathbf{F}_{wave} = \Delta p \mathbf{v} + \frac{1}{4\pi}(\mathbf{B}_0 \cdot \Delta \mathbf{B})\mathbf{v} - \frac{1}{4\pi}(\mathbf{v} \cdot \Delta \mathbf{B})\mathbf{B}_0. \tag{0.1}$$

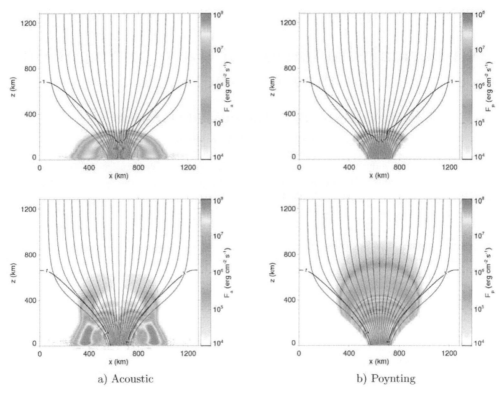

a) Acoustic   b) Poynting

Figure 1. Wave energy fluxes for the case in which the field strength at the axis at $z = 0$ is 800 G (top panels) and 1600 G (bottom panels). The colors show (a) the acoustic flux, and (b) the Poynting flux, at 40 s due to an impulsive horizontal motion at the $z = 0$ boundary with an amplitude of 750 m s^{-1} and a period of $P = 24$ s. The thin black curves are the field lines and the thick black curve represents the contour of $\beta = 1$. The Poynting fluxes are not shown in the ambient medium

The first term is the acoustic flux, and the last two terms are the Poynting flux. The operator Δ gives the perturbations in the variable with respect to the initial time and B_0 refers to the unperturbed magnetic field. Fig. 1. shows the magnitude of acoustic and Poynting fluxes at an elapsed time of 40 s for the two cases under consideration.

We see that the acoustic flux transport is isotropic, while the Poynting fluxes are localized within the flux tube. In the case of moderate field, the excitation at the bottom boundary generates fast (acoustic) and slow (magnetic) waves. These waves undergo mode transmission and conversion (Cally, 2007) as they cross the $\beta = 1$ layer. On the other hand, in the case of the strong field, the fast (magnetic) and slow (acoustic) waves generated within the flux tude do not encounter the $\beta = 1$ layer. There is a significant amount of acoustic emission into the ambient medium in the case of the flux tube with strong field. This is due to the sharp change in the magnetic field strength across the boundary between the flux tube and ambient medium.

To summarize our results, we find that the magnetoacoustic modes generated in the two field configurations are different. The acoustic wave emission from the tube interface depends on the strength of the field. We see a stronger emission in the case of a flux tube with strong field. We conclude that the magnetic field strength significantly effect the wave generation and propagation in the magnetic atmosphere.

Acknowledgements

This work was supported by the German Academic Exchange Service (DAAD), grant D/05/57687, and the Indian Department of Science & Technology (DST), grant DST/INT/DAAD/P146/2006.

References

Berger, T. E., *et al.* 2004, *A&A* 428, 613

Bogdan, T. J. *et al.* 2003, *ApJ* 599, 626

Bray, R. J. & Loughhead, R. E. 1974, The Solar Chrmosphere (London: Chapman & Hall)

Cally, P. S. 2007, *Astronomische Nachrichten* 328, 286

Hasan, S. S., van Ballegooiijen, A. A., Kalkofen W., & Steiner, O. 2005, *ApJ.* 631, 1270

Hasan, S. S. & van Ballegooiijen, A. A. 2008, *ApJ* 680, 1542

Steiner, O., Knölker M., & Schüssler, M. 1994, *NATO ASI Ser. C-433*, (Dordrecht: Kluwer), p. 441

Session IV

Solar-Heliospheric Variability: CMEs

Universal Heliophysical Processes
Proceedings IAU Symposium No. 257, 2008
N. Gopalswamy & D.F. Webb, eds.

Coronal mass ejection: key issues

Richard Harrison

Space Science and Technology Department, Rutherford Appleton Laboratory, Didcot,
Oxfordshire OX11 0QX, United Kingdom, Email: richard.harrison@stfc.ac.uk

Abstract. Coronal Mass Ejections (CMEs) have been addressed by a particularly active research community in recent years. With the advent of the International Heliophysical Year and the new STEREO and Hinode missions, in addition to the on-going SOHO mission, CME research has taken centre stage in a renewed international effort. This review aims to touch on some key observational areas, and their interpretation. First, we consider coronal dimming, which has become synonymous with CME onsets, and stress that recent advances have heralded a move from a perceived association between the two phenomena to a firm, well-defined physical link. What this means for our understanding of CME modeling is discussed. Second, with the new STEREO observations, and noting the on-going SMEI observations, it is important to review the opening field of CME studies in the heliosphere. Finally, we discuss some specific points with regard to EIT-waves and the flare-CME relationship. In the opinion of the author, these issues cover key hot topics which need consideration for significant progress in the field.

Keywords. Sun: corona, Sun: coronal mass ejections, Interplanetary medium

1. Introduction

One of the solar physics 'grand questions' is, how are Coronal Mass Ejections (CMEs) initiated, how do they propagate and influence Solar System bodies? This is not a new question. There have been substantial efforts to address this and to obtain the relevant observations. It is clearly a multi-faceted question addressing the onset of CMEs, the physics of CME propagation into the heliosphere and impacts on bodies such as the Earth. Here we review key issues which are considered by the author to be critical for CME studies at this time. Thus, four quite specific questions are considered, namely:

1. What is the precise relationship between the CME onset and coronal dimming, and what can that tell us about the CME onset?

2. We now have observations of CMEs in the heliosphere; what are these telling us about CME propagation and impacts?

3. What is the relationship between CMEs and EIT waves, and does that relationship help us to understand the onset process?

4. Can we settle the issue of flare/CME asymmetry/symmetry?

2. Coronal dimming – what does it really tell us?

There is no strict definition of coronal dimming, even though it has been a well established topic of interest for some years. Perhaps the only definition that suits all of the reports to date would be: An extreme-UV (EUV) or X-ray intensity depletion of a large region of the corona. However, there are no generally accepted parameters for the degree of depletion, the size of the depletion area, or the EUV/X-ray wavelengths displaying depletion. This is not acceptable; rather loose or variable definitions will not help a proper interpretation of the physics behind these events. However, understanding the dimming may be of critical importance because they have been closely associated with CME onsets.

The dimming phenomenon is not a new discovery. Rust and Hildner (1976) reported a dimming event using Skylab observations. Dimming has been reported and analysed by many researchers using data from SOHO and Yohkoh, e.g. Sterling and Hudson (1997), Gopalswamy and Hanaoka (1998), Zarro *et al.* (1999), Harrison and Lyons (2000) and Harrison *et al.* (2003). Now, with the STEREO and Hinode spacecraft we are seeing the first reports of dimming using these spacecraft.

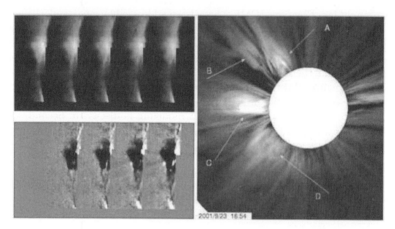

Figure 1. Dimming associated with a CME (Harrison, 2006). The SOHO/LASCO image (right) shows an east limb CME. A sequence of million K Mg IX EUV images (top left) is shown with frames differenced from the first image (bottom left) to reveal dimming on the east limb.

Most papers referring to dimming rely on phenomenological associations. Timing and co-location or alignment with CME activity is discussed but little is said about the physical parameters of the dimming region. So, what is the physical relationship between the dimming and an associated CME? Does the dimming reveal the site of 'lost' mass seen later as (part of) the CME? If so, a determination of the parameters of the dimming plasma, and its history, is key to understanding the CME onset, but this stresses that we need plasma parameters and that requires spectroscopic observations. Thus, we stress the value of the studies utilising spectrometers aboard SOHO and Hinode.

There are studies which clearly indicate, from spectral analysis, that the dimming is due to density depletion (Harrison and Lyons 2000; Harrison *et al.* 2003). These studies also show that the mass loss, again calculated through spectroscopic analysis, is consistent with the associated CME mass, suggesting that we are seeing some or all of the plasma from the dimming as the ascending CME. Similar mass-loss calculations can be made using wide-band EUV/X-ray imager data but only with significant assumptions about the temperature of the plasma. However, it is the author's opinion that the earlier spectroscopic mass calculations do require further confirmation with additional studies using SOHO and Hinode.

Spectroscopic and imaging observations of dimming confirm that the relative timing between the dimming and CME, as well as co-location, show that the dimming events are temporally and spatially associated with the CME activity. These findings allow us to consider physical processes at work but there are two important aspects that we have been missing. These have been addressed recently by Bewsher *et al.* (2008). They produced the first statistical and probability spectroscopic study of the dimming phenomenon, utilising data from almost 200 observational runs using SOHO and utilising an automated dimming identification scheme. Applying an automated scheme means that Bewsher *et al.* (2008) have applied criteria for defining a dimming event. These criteria were selected based on experience from previous observations, and are an attempt to put some constraints on what we believe a dimming event to be. They identified dimming events by the 'depth' of the intensity drop and by the physical size in the image - specifically the minimum depth was twice the statistical error of the intensity measurement and the minimum area was 1.44 arcsec^2 (scale about 5 degrees on the Sun).

Bewsher *et al.*'s first goal was to put the dimming-CME association on a firm footing. There are many reports on individual events, or few events, but no statistical studies to establish the degree of the association. What Bewsher *et al.* (2008) found was that up to 84% of the CMEs in the observation periods could be projected back to dimming regions, and that is the first robust confirmation of the association between CMEs and dimming; these phenomena are clearly related.

The second major point made by Bewsher *et al.* (2008) utilised the spectral capabilities of their dataset. They identified 155 and 146 dimming events in the spectral lines of Mg IX at 368Å and Fe XVI at 360Å, respectively. The abundances of Mg IX and Fe XVI peak at 1 and 2 million K, respectively. In only 96 cases did the code identify the dimming in both lines. This means that for 59 of the million K dimmings, no dimming was seen at 2 million K, and for 50 of the 2 million K dimmings, no dimming was identified at 1 million K. This confirms a point made by Harrison *et al.* (2003); there are significant variations in the degree of dimming between temperatures. Given adequate spectral (temperature) coverage, we can in principle identify all dimming events, but imaging from one wavelength will miss dimming events. There are clearly limitations on dimming identification from imagers alone unless multiple bands are used.

Finally, if dimming is revealing mass loss, can we observe the evacuation process? Harra *et al.* (2007) reported 40 km/s blue-shifted outflows from a dimming region using EIS/Hinode data. This is consistent with Harra and Sterling's (2001) claims to have detected outflows from such a region using CDS/SOHO. Also consistent with this, Harrison and Bewsher (2007) reported CDS/SOHO limb observations of pre-flare diffuse loops ascending from a region as it decayed in intensity, i.e. became a dimming event (see Figure 2).

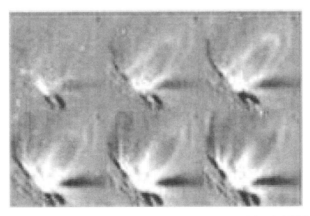

Figure 2. Pre-flare/pre-CME ascending loops detected using SOHO/CDS on 25 July 1999. These rising loops appear to reveal the evacuation of the corona, i.e. the dimming process. These are 4×4 arcmin images in the million K Mg IX line. (Harrison and Bewsher, 2007).

This review of aspects of the dimming observations stresses some key points:

1. The first statistical analysis of a large spectroscopic dataset of dimming has put the CME-onset/dimming association on a firm footing; models must be consistent with this.

2. The same study shows that the degree of dimming varies with temperature between events. This must be taken into account, especially with the use of imager data.

3. Some studies appear to reveal the evacuation process itself as clear mass-outflows or even ascending loops. A key observation for the future is to extend that work.

4. Spectral studies confirm that dimming is due to mass-loss. The consistency between the lost mass and the associated CME mass suggests that the dimming region is the source of at least part of the CME. Again, we do need more observational confirmation of this.

5. There are many consistent observations of relative timing and location that stress the association between the dimming and CME onsets.

3. CMEs in the heliosphere – the potential of new observations

Until recently, observations of CMEs in the heliosphere, including near-Earth space, have been effectively limited to single-point in-situ measurements. With the launch of the SMEI instrument aboard Coriolis, in 2003 (Eyles *et al.* 2003), and the Heliospheric Imagers (HIs) (Harrison *et al.* 2008), launched in 2006 aboard the two STEREO spacecraft, we now have wide-angle imaging of the heliosphere with the ability to detect CMEs out to beyond 1 AU. In particular, the HI instruments provide views from out of the Sun-Earth line, allowing studies of CMEs entering near-Earth space.

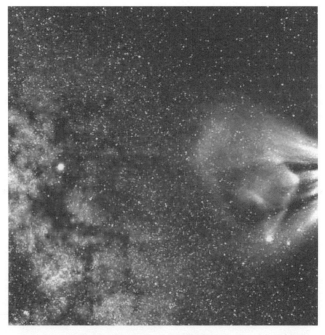

Figure 3. A CME in the heliosphere detected using HI-1 aboard STEREO A on 5 November 2007 (Harrison *et al.* 2009a). The image is 20 degrees across. The horizontal centre-line defines the ecliptic plane. The Sun is 4 degrees off the right hand side. The Milky Way is visible on the left of the frame, with Jupiter at centre-left. Stars down to 12th magnitude can be seen. A CME is clearly seen on the right hand side of the frame.

Figure 3 (from Harrison *et al.* 2009a) shows the quality of the HI instruments; the baffling systems are such that scattered light levels are reduced to 10^{-13} of the solar brightness at worst (Eyles *et al.* 2008). This, combined with the instrument sensitivity allows the imaging of stars down to 12th magnitude. The image shows a range of Solar System and stellar bodies, but the feature of primary interest is the CME on the right hand side of the frame.

The HIs allow imaging from elongations of a few degrees from Sun-centre out to almost 90 degrees. They provide us with the first real chance to study CMEs as they pass through the inner heliosphere and near-Earth space. A number of CMEs are discussed by Harrison *et al.* (2008, 2009a) and we show a few examples in Figure 4. The HI instruments on each spacecraft are identical. Each consists of two telescope systems, one viewing a 20×20 degree field centred on the ecliptic plane from 4 to 24 degrees elongation, and the other occupying a 70×70 degree field, also centred on the ecliptic plane, in this case from 19 to 89 degrees. The instrument concept, fields of view and operation are described by Harrison *et al.* (2008) and Eyles *et al* (2009).

Harrison *et al.* (2009b) have reviewed the CME/ICME relationship using the new HI data and building on early HI results reported by Harrison *et al.* (2008). In a sense, a CME is considered to be a near-Sun eruption imaged using a coronagraph and an Inter-planetary CME (ICME) is a mass ejection in the heliosphere, traditionally only viewed using in-situ measurements. Crooker and Horbury (2006) discussed the connectivity of ICMEs to the Sun, based on interpretations from the in-situ data. It is widely accepted that counter-streaming particle beams in ICMEs indicate that both ends of the ICME are indeed connected to the Sun. On the other hand, uni-directional beams may signal connection at only one end. Logically, then, the lack of beams would appear to signal disconnection at both ends. In this case the ICME has become an isolated plasmoid. The in-situ observations suggest that most ICMEs are connected at both footpoints for a considerable time after the eruption. There is evidence for closed ICMEs even out to Jupiter-like distances. This interpretation is indeed supported by the new HI data which demonstrate the long-duration connectivity of individual CMEs in the heliosphere, at least to Earth-like distances, and stress the fact that there is no evidence for magnetic pinching off of CMEs (Figure 5, left hand panel).

However, McComas (1995) has argued that the heliospheric magnetic flux does not continually build up. Flux must be shed through reconnection somehow. How is this consistent with the lack of observation of closing down magnetic systems behind ascending CMEs? The answer may be in the form of an interchange reconnection process, which has been suggested by Gosling *et al.* (1995) (Figure 5). The idea is that the ascending CME can travel a considerable distance still connected to the Sun, and that days or even weeks after the onset, the legs of the CME, still rooted in the Sun, will interact with adjacent open field lines at low altitude in the corona; reconnection results in the formation of low-lying loops and an outward ascending kink-shaped structure ascends into the heliosphere from the site of one of the original CME footpoints. In Figure 5, this is contrasted with the traditional approach where the ascending CME is magnetically pinched off, resulting in an ascending plasmoid and loops closing down underneath. The interchange mechanism has the attractive feature that the site of the greatest field density, magnetic complexity and field-line motion is the site of reconnection. Harrison *et al.* (2009) argue that although this results in the outward propagation of a kinked field-line configuration, what we might expect to observe with coronagraphs or using HI, would be narrow V-shaped, ascending features or trains of ascending blobs associated with re-connection in the boundaries between closed and open fields; the CME front would be

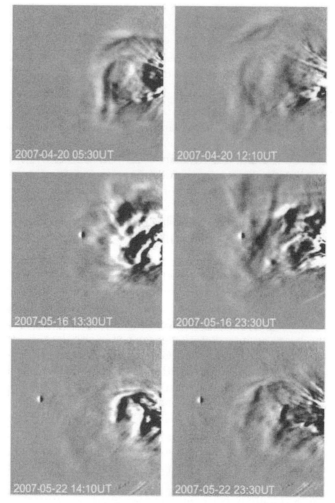

Figure 4. Three CMEs imaged using HI-1, on 20 April, 16 May and 22 May 2007. For each, two frames show their passage through the first 20 degrees elongation from the Sun. These images show that the same basic structure persists.

long gone. Evidence for such phenomena is being sought at this time and could prove to be crucial in understanding the CME process.

Another valuable analysis of the HI data is providing a global view of CME activity in the inner heliosphere. Davies *et al.* (2009) have extended the coronagraph time-altitude plots due to Sheeley *et al.* (1999) to produce a time-elongation display which shows the nature of CME activity from 4 to almost 90 degrees elongation. Figure 6, from Davies *et al.* (2009), shows such a plot which is produced from stacking intensity scans along the ecliptic plane across the HI-1 and HI-2 data. Outward propagating events are revealed by sloping lines whose gradients and shapes are a function of the outward speed and the location of the event with respect to the plane of the sky (Rouillard *et al.* 2008a). Such a plot extends this technique by an order of magnitude from the Sun, over previous studies. The method reveals a plethora of ascending structures in the lowest portions of the image (below 10 degrees elongation - i.e. in the region occupied by the coronagraph fields). From a few tens of degrees we detect far fewer events. Outward propagating features

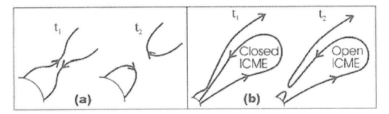

Figure 5. (a) The traditional view of an ascending CME pinching-off, and (b) the interchange reconnection process (from Crooker and Horbury 2006).

may be dissipating or decreasing in intensity as they expand outwards. Observationally, events are best viewed if they are located on the so-called Thomson sphere, defined as the circle or sphere with the Sun-Spacecraft line as the centre-line. This sphere defines the points of 90 degree scatter of the photospheric light off free electrons in the heliosphere to the 'observer' at the STEREO spacecraft (Vourlidas and Howard, 2006). If the forest of ascending structures at the lowest levels of Figure 6 are typical of all longitudes then the effect of the Thomson sphere on interpreting this will not be of great significance.

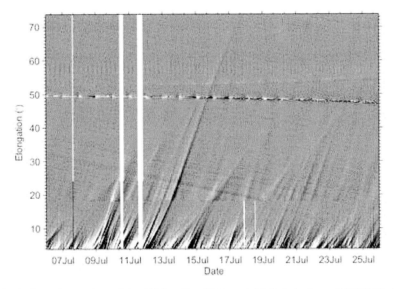

Figure 6. A time-elongation plot utilising the HI-1 and HI-2 data from STEREO A for July 2007 (from Davies *et al.* 2009)

Above 50 degrees elongation we detect just a few ascending structures. Just one event crosses the entire image to the outer edge of the field. The slope and shape of the profile show that this CME ascended at 320 km/s, 48 degrees east of the Spacecraft-Sun line (Davies *et al.* 2009), which is consistent with an association with an active region at E29 (longitude from Earth).

Plots such as these do show a synoptic and global view of ejecta in the inner helio-sphere, demonstrating a simplification of the ascending structure with elongation. This also stresses the value of in-situ measurement near to the Sun to sample pristine solar wind, and this is one of the prime goals of the Solar Orbiter mission.

Whilst considering CMEs in the heliosphere, we must consider their impact on Solar System bodies. A graphic illustration of this has been presented by Vourlidas *et al.* (2007). They show the impact of a CME on comet Encke, which resulted in a complete disconnection of the comet's ion tail, presumably due to the CME magnetic fields interacting with the cometary plasma. This is a demonstration on a small scale of events which could occur with other bodies, including planets.

This short discussion demonstrates the potential for this observational approach. We note that there are on-going studies of planetary impacts of CMEs, as well as other cometary studies and imaging of Co-rotating Interaction Regions (see Rouillard *et al.* 2008a,b; Sheeley 2008a,b).

In conclusion, wide-angle heliospheric imaging is providing us with many new lines of study. Of particular interest is the propagation of CMEs through the heliosphere, and impacts on Solar System bodies. In addition, we identify promising studies of the last phase of CME activity and of global imaging of ejecta in the heliosphere. It is clear that this area will be extremely productive in the coming months and years.

4. Comments on CMEs, EIT-waves and flare/CME symmetry

EIT or coronal waves have become a hot topic. These are rapidly expanding disturbances which propagate around the solar globe, and they are shown to have a close association with CMEs and flares. Biesecker *et al.* (2002) studied 173 EIT-waves and showed an intimate association with CMEs and a less significant flare association. Of interest here is the CME association because it may provide some insight to CME activity. Amongst the models being proposed for this phenomenon, Plunkett *et al.* (2002) suggested that the waves are due to fast-mode MHD waves propagating from CME initiation sites. More recently, Attrill *et al.* (2007) have suggested that the 'waves' are actually due to successive reconnections in the flanks of a CME.

We must address two misconceptions. First, the apparent dimming displayed in differenced EUV images behind the expanding EIT-wave is not the same as the so-called coronal dimming phenomenon. Such coronal dimming is more localised, does not display such rapid expansion, and reveals a much more significant decrease in intensity, though such an event may be encompassed within the area behind an EIT-wave. However, Chen and Fang (2005), for example, suggest that EIT-wave defines a coronal dimming region and refers to the dimming work of Harrison *et al.* (2003). That work was actually concerned with coronal dimming of a large but localised region under a CME and not the shallow dimming found over a large area of the solar globe behind an EIT-wave.

A second, and perhaps more significant, misconception is the association between the legs of the CME and the EIT-wave-front. For example, the Chen and Fang (2005) paper mentioned above discussed a model in which the CME extent (and the coronal dimming) was defined explicitly by the extent of the EIT-wave, i.e. the EIT-wave defined the lateral extent of the ascending CME footprint (and the dimming event). Attrill *et al.* (2007) also state that the coronal wave is the magnetic footprint of a CME; specifically they state that "the diffuse EIT coronal bright fronts are due to driven magnetic reconnections between the skirt of the expanding CME magnetic field and favourably oriented quiet Sun magnetic loops". The EIT-wave may well be due to an MHD wave propagating through the corona, or to magnetic reconnection processes, as suggested by Attrill *et al.* (2007). However, we know from decades of observation that CME legs do not separate significantly with time and certainly do not expand to wrap around the solar disk. Thus, it is imperative that we do not associate intimately the expanding EIT "wave" with the legs of a CME. That is not to say that they cannot be associated at all - they are just not co-located throughout the sequence of events.

Nobody doubts the close association between flares and CMEs. From an analysis of a number of flare events associated with CMEs detected using the Solar Maximum Mission coronagraph, Harrison (1986, 1995) demonstrated asymmetry between flare and associated CME activity. They suggested that a flare associated with a CME could lie anywhere under the CME-span. At the same time, they claimed that the flare and CME onset times were not always coincident – the onset of the CME could appear to precede the associated flare onset. In addition, they pointed out that the CME-spans were an order of magnitude (or more) larger than the scale of associated flares and, in the absence of clear expansion of CME-legs, suggested that CME source regions must be much larger than the flare site. All of this led to the suggestion that the flare and CME are closely associated but that they do not cause one another - they both result from the relaxation of complex magnetic topologies and, as such, can occur in concert (see Harrison 1991).

Many subsequent papers have provided evidence to confirm or deny this scenario. Yashiro *et al.* (2008) examined 496 flare-CME pairs using SOHO data in an essential study, extending the old analyses significantly. Their principal conclusions were:

1. X and M class flares associated with CME onsets are likely to show symmetry within the CME-span, i.e. to lie under the centre of the CME-span;

2. C-class flares showed less symmetry with significant numbers residing near the edge or outside the CME-span.

Yashiro *et al.*'s Figure 4 uses a flare-CME symmetry index used in previous studies and claim that the data show consistency with the so-called CSHKP flare-CME model. This is a combination of models due to Carmichael, Sturrock, Hirayama, Kopp and Pneumann, which calls for reconnection above the flare site with the ascending CME propagating symmetrically above the flare. For this model there is clear symmetry between the CME and associated flare and the onsets of the flare and CME should coincide. Whilst the symmetry displayed by many of the X and M class flare events appears to be consistent with this approach, their results do present an anomaly for several reasons:

1. Although the M and X class events show a high degree of symmetry between flare and CME, there is a tail to the distribution which shows that some M and X class flares occur anywhere under the CME-span.

2. The Yashiro *et al.* (2008) data for the C-class events actually shows the same result as the Harrison *et al.* papers, the flares appear to occur anywhere under the CME-span.

3. We also note that there are flares without obvious CME activity.

The Harrison *et al.* studies were taken from a period lacking in bright flares, so the results of Yashiro *et al.* (2008) are not inconsistent with their results. However, if we wish to adopt the CSHKP approach for the brighter flares, we cannot use that approach for the flares indicated in the three points above. Do we adopt a different flare-CME model for weaker flares and for some of the large flares, not to mention the flare-less CMEs? The author is uncomfortable with the notion that we should devise a flare model that only supports the brighter flares which showed symmetry. We require a model which allows for any symmetry/asymmetry between flare and CME and in the opinion of the author that means adopting the non-cause and effect approach mentioned above.

5. Conclusions

There are key open issues that require clarification or general acceptance, to progress in some areas of CME research. We have addressed coronal dimming and stressed that recent spectroscopic interpretations have far-reaching consequences and must be taken seriously. We have briefly reviewed the opening field of heliospheric imaging; it is clear that we have witnessed rapid progress and there is promise of much to come. There are also some issues relating to EIT-waves and our understanding of how they are associated with CME footprints, and to the flare-CME asymmetry that need careful consideration.

References

Attrill G. D. R., Harra, L. K., van Driel-Gesztelyi, L., & Demoulin, P. 2007, *ApJ* 656, L101

Bewsher, D., Harrison, R. A., & Brown, D. S. 2008, *A&A* 478, 897

Crooker, N. U. & Horbury, T. S. 2006, *Space Sci. Revs* 123, 93

Biesecker D. A., Myers, D. C., Thompson, B. J., Hammer, D. M., & Vourlidas, A., 2002, *Astrophys. J.*, 569, 1009

Chen, P. F. & Fang, C. 2005 in *'Coronal and Stellar Mass Ejections'*, Proc. IAU Symp. 226, 55

Davies J. A., Harrison, R. A., Rouillard, A. P., Sheeley, N. R., Perry, C. H., Bewsher, D., Davis, C. J., Eyles, C. J., Crothers, S. R., & Brown, D. S. 2009, *Geophys. Res. Lett.* 36, L02102

Eyles C. J., Simnett, G. M., Cooke, M. P., Jackson, B. V., Buffington, A., Hick, P. P., Waltham, N. R., King, J. M., Anderson, P. A., Holladay, P. E. 2003 *Solar Phys.* 217, 319

Eyles C. J., Harrison, R. A., Davies, C. J., Waltham, N. R., Shaughnessy, B. M., Mapson-Menard, H.C. A., Bewsher, D., Crothers, S. R., & Davies, J. A. 2009, *Solar Phys.* in press

Gopalswamy N. and Hanaoka, Y. 1998, *ApJ* 498, 179

Gosling, J. T., Birn, J., Hesse, M. 1995, *Geophys. Res. Lett.* 22, 869

Harra, L. K. & Sterling, A. C. 2001, *ApJ* 561, L215

Harra, L. K., Hara, H., Imada, S., Young, P. R., Williams, D., Sterling, A., Korendyke, C., & Attrill, G. 2007, *Publ. Astron. Soc. Japan* 59, S801

Harrison, R. A. 1986,*A&A* 162, 283-291

Harrison R. A. 1991, *Phil. Trans. Roy. Soc. London, A,* 336, 401–412

Harrison R. A. 1995, *A&A* 304, 585-594

Harrison R. A. 2006, in *'Solar eruptions & energetic particles'*, AGU Geophys. Mon. Ser. 165, 73

Harrison, R. A. & Bewsher, D. 2007, *A&A* 461, 1155

Harrison, R. A. & Lyons, M. 2000, *A&A* 358, 1097

Harrison, R. A., Bryans, P., Simnett, G. M., & Lyons, M. 2003, *A&A* 400, 1071

Harrison, R. A., Davis, C. J., Eyles, C. J., & 12 co-authors 2008, *Solar Phys.* 247, 171

Harrison, R. A., Davies, J. A., Rouillard, A. P., Davis, C. J., Eyles, C. J., Bewsher, D., Crothers, S. R., Howard, R. A., Sheeley, N. R., Vourlidas, A., Webb, D. F., Brown, D. S., & Dorrian, G. D. 2009a, *Solar Phys.* in press

Harrison, R. A., Davis, C. J., Bewsher, D., Davies, J. A., Eyles, C. J., & Crothers, S. R. 2009b, *Adv. Space Res.* submitted

McComas D. J. 1995, *Rev. Geophys. Suppl.* 33, 603

Plunkett, S. P., Michels, D. J., Howard, R. A., & 6 co-authors, 2002, *Adv. Sp. Res.* 29, 1473

Rouillard, A., Davies, J. A., Forsyth, R., J., & 9 co-authors, 2008a, *Geophys. Res. Lett.* 35, L10110

Rouillard, A. P., Davies, J. A., Rees, A., & 13 co-authors, 2008b, *J. Geophys. Res.* submitted

Rust, D. M. & Hildner, E. 1976, *Solar Phys.* 48, 381

Sheeley, N. R., Walters, J. H., Wang, Y.-M. & Howard, R. A. 1999, *J. Geophys. Res.* 104, 24,739

Sheeley, N. R., Herbst, A. D., Palatchi, C. A., & 21 co-authors, 2008a, *ApJ* 674, L109

Sheeley, N. R., Herbst, A. D., Palatchi, C. A., & 21 co-authors, 2008b, *ApJ* 675, 853

Sterling, A. C. & Hudson, H. S. 1997, *ApJ* 491, L55

Vourlidas, A. & Howard, R. A. 2006, *ApJ* 642, 1216

Vourlidas, A., Davis, C. J., Eyles, C. J., Crothers, S. R., Harrison, R. A., Howard, R. A., Moses, J. D., & Socker, D. G. 2007, *ApJ* 668, L79

Yashiro, S., Michalek, G., Akiyama, S., Gopalswamy, N., & Howard, R. A. 2008b, *ApJ* 673, 1174

Zarro, D. M., Sterling, A. C., Thompson, B. J., Hudson, H. S., & Nitta, N. 1999, *ApJ* 520, 139

Universal Heliophysical Processes
Proceedings IAU Symposium No. 257, 2008
N. Gopalswamy & D.F. Webb, eds.

Stellar mass ejections

Moira Jardine[1], Jean-Francois Donati[2] and Scott G. Gregory[3]

[1]SUPA, School of Physics and Astronomy, University of St Andrews, North Haugh,
St Andrews, KY16 9SS, UK
email: mmj@st-andrews.ac.uk

[2]LATT, CNRS–UMR 5572, Obs. Midi-Pyrénées, 14 Av. E. Belin, F–31400 Toulouse, France
email: donati@ast.obs-mip.fr

[3]SUPA, School of Physics and Astronomy, University of St Andrews, North Haugh,
St Andrews, KY16 9SS, UK
email: sg64@st-andrews.ac.uk

Abstract. It has been known for some time now that rapidly-rotating solar-like stars possess the stellar equivalent of solar prominences. These may be three orders of magnitude more massive than their solar counterparts, and their ejection from the star may form a significant contribution to the loss of angular momentum and mass in the stellar wind. In addition, their number and distribution provide valuable clues as to the structure of the stellar corona and hence to the nature of magnetic activity in other stars.

Until recently, these "slingshot prominences" had only been observed in mature stars, but their recent detection in an extremely young star suggests that they may be more widespread than previously thought. In this review we will summarise our current understanding of these stellar prominences, their ejection from their stars and their role in elucidating the (sometimes very non-solar) behaviour of stellar magnetic fields.

Keywords. stars:magnetic fields, stars:coronae, stars:imaging, stars:spots

1. Introduction

During this Symposium we have learned a great deal about the Sun and its influence on its environment, but in this review we want to begin by addressing the question "How typical is the Sun as a star?" Stars on the main sequence (i.e. those stars that have settled into the longest phase of their lives, when they are burning hydrogen in their cores) can have very different interior structures depending on their mass, yet magnetic activity is almost ubiquitous among them. High mass stars have a convectively stable (or radiative) outer envelope, so how do they generate their magnetic fields? They can do this in their convective cores, although this raises the question of how to transport the flux to the surface (Charbonneau & MacGregor 2001; Brun *et al.* 2005). They can also generate magnetic fields in the radiative zone, but a very non-solar dynamo process (Spruit 2002; Tout & Pringle 1995; MacDonald & Mullan 2004; Mullan & MacDonald 2005; Maeder & Meynet 2005). Alternatively, the fields may be fossils, left over from the early stages of the formation of the star (Moss 2001; Braithwaite & Spruit 2004; Braithwaite & Nordlund 2006). Very low mass stars also have an internal structure that is very different from that of the Sun in that convection may extend throughout their interiors. In the absence of a tachocline, these stars cannot support a solar-like interface dynamo, yet they, like the high mass stars, exhibit observable magnetic fields. The mechanism by which they generate these magnetic fields has received a great deal of attention recently. While a decade or so ago, it was believed that these stars could only generate small-scale magnetic fields (Durney *et al.* 1993; Cattaneoe 1999), more recent studies have suggested that large

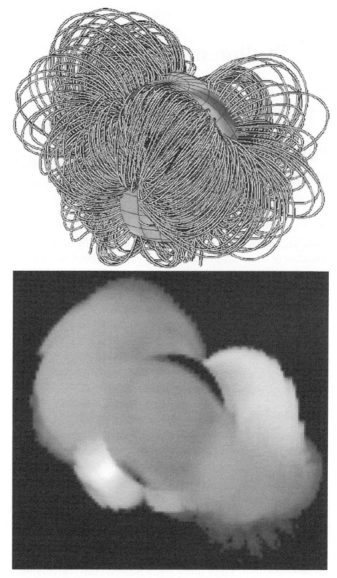

Figure 1. Closed field lines (top) and corresponding X-ray image (bottom) for the rapidly-rotating star LQ Hya. A coronal temperature of 10^6K is assumed.

scale fields may be generated. These models differ, however, in their predictions for the form of this field and the associated latitudinal differential rotation. They predict that the fields should be either axisymmetric with pronounced differential rotation (Dobler *et al.* 2006), non-axisymmetric with minimal differential rotation (Küker & Rüdiger 1997, 1999; Chabrier & Küker 2006) or, in a very recent model, axisymmetric with negligible differential rotation (Browning 2008).

2. Observing stellar prominences and magnetic fields

With this bewildering array of magnetic field geometries, the nature of any prominences that might be confined in and ultimately ejected from these coronae becomes even more

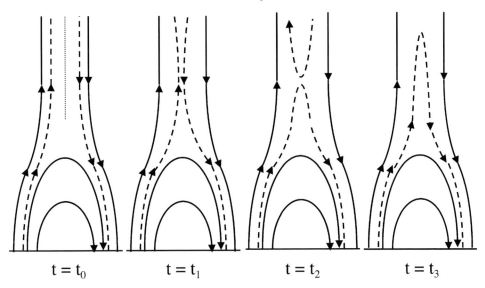

$$t = t_0 \qquad t = t_1 \qquad t = t_2 \qquad t = t_3$$

Figure 2. A schematic diagram of the formation of prominence-bearing loops. Initially, at $t = t_0$ a current sheet is present above the cusp of a helmet streamer. Reconnection in the current sheet at $t = t_1$ produces a closed loop at $t = t_2$. The stellar wind continues to flow until pressure balance is restored, thus increasing the density in the top of this new loop. Increased radiative losses cause the loop to cool and the change in internal pressure forces it to a new equilibrium at $t = t_3$.

interesting, but detecting their presence is not an easy task. They can, however, be observed in rapidly-rotating stars as transient Hα absorption features (Collier Cameron & Robinson 1989b,a; Collier Cameron & Woods 1992; Jeffries 1993; Byrne *et al.* 1996; Eibe 1998; Barnes *et al.* 2000; Donati *et al.* 2000). In many instances these features re-appear on subsequent stellar rotations, often with some change in the time taken to travel through the line profile. These features are interpreted as arising from the presence of clouds of cool, dense gas co-rotating with the star and confined within its outer atmosphere. As many as six may be present in the observable hemisphere. What is most surprising about them is their location, which is inferred from the time taken for the absorption features to travel through the line profile. Values of several stellar radii from the stellar rotation axis are typically found, suggesting that the confinement of these clouds is enforced out to very large distances. Indeed the preferred location of these prominences appears to be at or beyond the equatorial stellar co-rotation radius, where the inward pull of gravity is exactly balanced by the outward pull of centrifugal forces. Beyond this point, the effective gravity (including the centrifugal acceleration) points outwards and the presence of a restraining force, such as the tension in a closed magnetic loop, is required to hold the prominence in place against centrifugal ejection. The presence of these prominences therefore immediately requires that the star have many closed loop systems that extend out for many stellar radii. Maps of the surface brightness distributions of these stars can be obtained by Doppler imaging, while magnetograms are now almost routinely possible with Zeeman-Doppler imaging. These maps typically show a complex distribution of surface spots that is often very different from that of the Sun, with spots and mixed polarity flux elements extending over all latitudes up to the pole (Donati & Collier Cameron 1997; Donati *et al.* 1999; Strassmeier 1996).

From these magnetograms we can extrapolate the coronal magnetic field using a *Potential Field Source Surface* method (Altschuler & Newkirk, Jr. 1969; Jardine *et al.* 1999,

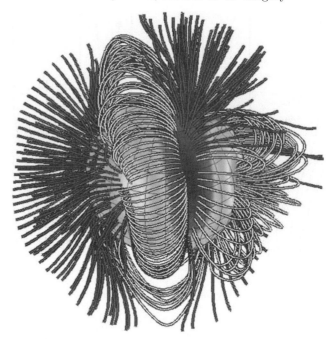

Figure 3. Closed field lines (white) and open field lines (blue) extrapolated from a Zeeman-Doppler image of Tau Sco.

2001, 2002a; McIvor *et al.* 2003), or using non-potential fields (Donati 2001; Hussain *et al.* 2002). By assuming that the gas trapped on these field lines is in isothermal, hydrostatic equilibrium, we can determine the coronal gas pressure, subject to an assumption for the gas pressure at the base of the corona. We assume that it is proportional to the magnetic pressure, i.e. $p_0 \propto B_0^2$, where the constant of proportionality is determined by comparison with X-ray emission measures (Jardine *et al.* 2002b, 2006; Gregory *et al.* 2006a). For an optically thin coronal plasma, this then allows us to produce images of the X-ray emission, as shown in Fig. 1. This immediately highlights one of the greatest puzzles of stellar prominences: that they are confined to a such great distances – several stellar radii - that they may well be outside the extent of the closed, X-ray emitting corona.

One way out of this problem is to confine the prominences in the wind region beyond the closed corona. Jardine & van Ballegooijen (2005) have produced a model for this that predicts a maximum height y_m for the prominence as a function of the co-rotation radius, y_K where

$$\frac{y_m}{R_\star} = \frac{1}{2}\left(-3 + \sqrt{1 + \frac{8GM_\star}{R_\star^3 \omega^2}}\right) \tag{2.1}$$

$$= \frac{1}{2}\left(-3 + \sqrt{1 + 8\left[\frac{y_K}{R_\star} + 1\right]^3}\right). \tag{2.2}$$

Fig. 2 shows the sequence of events that might lead to the formation of one of these "slingshot" prominences. The stellar wind flows along the open field lines that bound a closed field region, forming a helmet streamer. If the current sheet that forms between these oppositely-directed field lines reconnects, then a loop of magnetic field will be formed. The stellar wind will continue to flow for a short time, until pressure balance is

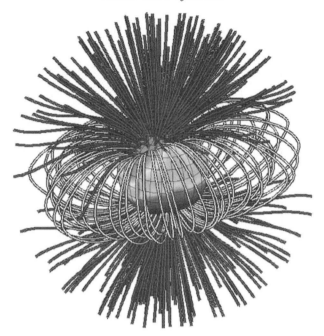

Figure 4. Closed field lines (white) and open field lines (blue) extrapolated from a
Zeeman-Doppler image of V374 Peg.

re-established with a new field configuration. Jardine & van Ballegooijen (2006) showed
that a new, cool equilibrium was possible which could reach out well beyond the co-
rotation radius. The distribution of prominence heights shown in Dunstone *et al.* (2008a,b)
for the ultra-fast rotator Speedy Mic shows prominences forming up to (but not significantly beyond) this maximum height.

3. Stellar magnetic field variation with stellar mass and evolutionary state

But what of the many stars whose internal structure is very different from that of the
Sun? Surface magnetograms are now available for stars of a range of masses. At the high
mass end, Tau Sco is a very interesting example. At 15 $M_\odot$ it has a radiative interior,
and yet as shown in Fig. 3 it displays a complex, strong field (Donati *et al.* 2006b). If
this is a fossil field, it might be expected to be a simple dipole, but the very youth of this
star, at only a million years, may be the reason why the higher-order field components
have not yet decayed away. Interestingly, Tau Sco shows H_α absorption features that are
very similar to prominence signatures in lower mass stars. In this case, however, they
are attributed to a "wind-compressed disk" that forms when sections of the very massive
wind emanating from different parts of the stellar disk collide and cool (Townsend &
Owoki 2005). In contrast, as shown in Fig. 4 the very low mass fully-convective star
V374 Peg has a very simple, dipolar field (Donati *et al.* 2006a). The highly-symmetric
nature of the field and the absence of a measureable differential rotation are consistent
with the recent models of Browning (2008). It is unfortunately not possible at present to
detect any prominences that might be present on these very low mass stars because they
stars are intrinsically too faint. Their detection would, however, be a very clear test of
the magnetic structure, since in a simple dipole any prominences should, by symmetry,
form in the equatorial plane.

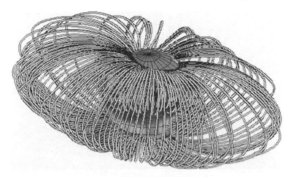

Figure 5. Closed field lines structure extrapolated from Zeeman-Doppler image of BP Tau.

It appears that stars with different internal structures may have dynamos that produce very different types of magnetic field. In particular, the transition from a solar-type interior to one in which the convective zone extends throughout the star appears to be associated with a decrease in field complexity. This transition happens for solar mass stars as they evolve from their very earliest stages when they are fully-convective, through the development of a radiative core as they approach the main sequence. Any associated change in the magnetic field structure is potentially very important, since the magnetic field is believed to channel the flow of material from the accretion disk that surrounds such young stars onto hotspots on the stellar surface. Significant advances have been made in the study of this *magnetospheric accretion* recently, with the advent of large-scale 3D MHD codes. It appears that the structure of the magnetic field can be a crucial factor in determining the nature of the accretion (Rekowski & Brandenburg 2004; Gregory *et al.* 2006b; Long *et al.* 2007).

It is not only the flow of material onto the star that is important, however. The loss of both mass and angular momentum in a wind is also a crucial issue since these young stars should spin up as they contract, but they are observed to have typically only moderate rotation rates. This spin-down may be achieved through the exchange of magnetic torques between the star and the disk (known as *disk-locking*) or through a wind (Königl 1991; Collier Cameron & Campbell 1993; Shu *et al.* 1994; Matt & Pudritz 2005).

Determining the structure of the coronal magnetic field of these young stars is a difficult problem, however, since there are many factors than can influence it. As the stellar magnetic field drags through the disk it will be sheared and may be opened up entirely (Lynden-Bell & Boily 1994). This shearing may act to deposit energy in the corona through reconnection between the magnetic fields of the star and the disk – certainly, some of the very large flares observed in these systems may be attributed to reconnection (Favata *et al.* 2005). These processes will act in addition to the effect of the (possibly evolving) dynamo and surface flows.

Recently, we have successfully acquired Zeeman-Doppler images of two of these very young stars that are still accreting from their disks. One of these, BP Tau is only $0.7M_\odot$ and is believed to be fully convective, while the other, V2129 Oph at $1.4M_\odot$ is believed to have already developed a radiative core (Donati *et al.* 2007, 2008; Jardine *et al.* 2008). As shown in Fig. 5, BP Tau displays a strong (1.2kG) dipolar component to its magnetic field. In contrast, as shown in Fig. 6, the dominant field component in V2129 Oph is the 1.2kG octupole component. The relatively stronger dipole component of BP Tau's field may allow it to carve out a larger inner hole in its disk, relative to the co-rotation radius (Gregory *et al.* 2008).

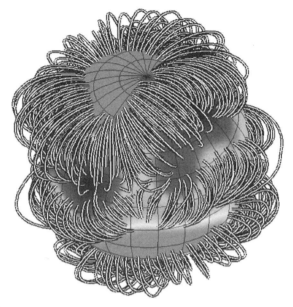

Figure 6. Closed field lines structure extrapolated from Zeeman-Doppler image of V2129 Oph.

4. Prominences in young stars

In young stars that are still accreting it would be impossible to detect prominences, even if they were present, since the H_α line is so strongly affected by the accretion process that variations due to prominences could not be disentangled from those due to accretion. However, in stars that have only recently lost their disks, it is possible and indeed in one such example, TWA6, at least one prominence has been detected (Skelly *et al.* 2008). This is a very interesting example as the star appears to be at the boundary in its evolution between a fully convective state and the development of a radiative core. This star has a heavily-spotted surface with spots extending all the way to the rotation pole. The one prominence detected survived for at least 3 days and was situated at a radius of $4R_\star$ - consistent with the maximum value of $4.8R_\star$ that would be predicted by the Jardine & van Balegooijen (2006) theory.

5. Conclusions

It is clear from solar observations of prominences that they delineate the structure of the magnetic field, but inferring that structure from observations of prominences is not a simple (or even possible) task. This problem is even more challenging in the case of stellar prominences where their large distance from the stellar rotation axis presents a challenge to models of their confinement by the star's magnetic field. The new Zeeman-Doppler maps of stellar surface magnetic fields, however, show that stellar magnetic fields may be very different in stars of different mass and hence internal structure. In particular, the presence of a convectively stable core – and hence of a shear layer or tachocline separating this from the convective outer region – seems to lead to a complex, high-order field. Stars that are fully convective seem to show a much simpler structure. At present, we only have observations of prominences on mature stars with radiative cores. The low mass stars that are fully-convective are too faint to allow the detection of the transient H_α absorption features that are the signature of prominences. The one example we have of prominences forming in a star that is at the boundary between a fully-convective state and the formation of a radiative core is the young star TWA6 which displays a complex field. In a star with a simple dipolar field we might expect prominences to form a torus in the equatorial plane of the star – hence producing no rotational modulation. The detection of prominences in a low mass star would be an interesting test of the field structures detected by Zeeman-Doppler methods, with potentially profound implications for dynamo theories.

References

Altschuler, M. D. & Newkirk, Jr., G. 1969, *Solar Phys.*, 9, 131
Barnes, J., Collier Cameron, A., James, D. J., & Donati, J.-F. 2000, *MNRAS*, 314, 162
Braithwaite, J. & Nordlund, A. 2006, *A&A*, 450, 1077
Braithwaite, J. & Spruit, H. C. 2004, *Nature*, 431, 819
Browning, M. 2008, *ApJ*, in press
Brun, A. S., Browning, M. K., & Toomre, J. 2005, *ApJ*, 629, 461
Byrne, P., Eibe, M., & Rolleston, W. 1996, *A&A*, 311, 651
Cattaneoe, F. 1999, *ApJ*, 515, L39
Chabrier, G. & Küker, M. 2006, *A&A*, 446, 1027
Charbonneau, P. & MacGregor, K. B. 2001, *ApJ*, 559, 1094
Collier Cameron, A. & Campbell, C. G. 1993, *A&A*, 274, 309
Collier Cameron, A. & Robinson, R. D. 1989a, *MNRAS*, 238, 657
Collier Cameron, A. & Robinson, R. D. 1989b, *MNRAS*, 236, 57
Collier Cameron, A. & Woods, J. A. 1992, *MNRAS*, 258, 360
Dobler, W., Stix, M., & Brandenburg, A. 2006, *ApJ*, 638, 336
Donati, J.-F. 2001, *LNP Vol. 573: Astrotomography, Indirect Imaging Methods in Observational Astronomy*, 573, 207
Donati, J.-F. & Collier Cameron, A. 1997, *MNRAS*, 291, 1
Donati, J.-F., Collier Cameron, A., Hussain, G., & Semel, M. 1999, *MNRAS*, 302, 437
Donati, J.-F., Forveille, T., Cameron, A. C., *et al.* 2006a, *Science*, 311, 633
Donati, J.-F., Howarth, I. D., Jardine, M. M., *et al.* 2006b, *MNRAS*, 370, 629
Donati, J.-F., Jardine, M. M., Gregory, S. G., *et al.* 2007, *MNRAS*, 380, 1297
Donati, J.-F., Jardine, M. M., Gregory, S. G., *et al.* 2008, *MNRAS*, 386, 1234
Donati, J.-F., Mengel, M., Carter, B., Cameron, A., & Wichmann, R. 2000, *MNRAS*, 316, 699
Dunstone, N. J., Hussain, G. A. J., Cameron, A. C., *et al.* 2008a, *MNRAS*, 387, 1525
Dunstone, N. J., Hussain, G. A. J., Collier Cameron, A., *et al.* 2008b, *MNRAS*, 387, 481
Durney, B. R., De Young, D. S., & Roxburgh, I. W. 1993, *Solar Phys.*, 145, 207
Eibe, M. T. 1998, *A&A*, 337, 757
Favata, F., Flaccomio, E., Reale, F., *et al.* 2005, *ApJS*, 160, 469
Gregory, S. G., Jardine, M., Cameron, A. C., & Donati, J.-F. 2006a, *MNRAS*, 373, 827
Gregory, S. G., Jardine, M., Simpson, I., & Donati, J.-F. 2006b, *MNRAS*, 371, 999
Gregory, S. G., Matt, S. P., Donati, J.-F., & Jardine, M. 2008, *MNRAS*, in press
Hussain, G. A. J., van Ballegooijen, A. A., Jardine, M., & Collier Cameron, A. 2002, *ApJ*, 575, 1078
Jardine, M., Barnes, J., Donati, J.-F., & Collier Cameron, A. 1999, *MNRAS*, 305, L35
Jardine, M., Collier Cameron, A., & Donati, J.-F. 2002a, *MNRAS*, 333, 339
Jardine, M., Collier Cameron, A., Donati, J.-F., Gregory, S. G., & Wood, K. 2006, *MNRAS*, 367, 917
Jardine, M., Collier Cameron, A., Donati, J.-F., & Pointer, G. 2001, *MNRAS*, 324, 201
Jardine, M., Gregory, S. G., & Donati, J.-F. 2008, *MNRAS*, in press
Jardine, M. & van Ballegooijen, A. A. 2005, *MNRAS*, 361, 1173
Jardine, M., Wood, K., Collier Cameron, A., Donati, J.-F., & Mackay, D. H. 2002b, *MNRAS*, 336, 1364
Jeffries, R. 1993, *MNRAS*, 262, 369

Königl, A. 1991, *ApJ,* 370, L39

Küker, M. & Rüdiger, G. 1997, *A&A,* 328, 253

Küker, M. & Rüdiger, G. 1999, in: *ASP Conf. Ser. 178: Workshop on stellar dynamos,* Vol. 178, 87–96

Long, M., M., R. M., & Lovelace, R. V. E. 2007, *MNRAS,* 374, 436

Lynden-Bell, D. & Boily, C. 1994, *MNRAS,* 267, 146

MacDonald, J. & Mullan, D. J. 2004, *MNRAS,* 348, 702

Maeder, A. & Meynet, G. 2005, *A&A,* 440, 1041

Matt, S. & Pudritz, R. E. 2005, *ApJ,* 632, L135

McIvor, T., Jardine, M., Cameron, A. C., Wood, K., & Donati, J.-F. 2003, *MNRAS,* 345, 601

Moss, D. 2001, in: S. G. Mathys & D. Wickramasinghe (eds.), *ASP Conference Series,* Vol. 248, *Magnetic fields across the Hertzsprung-Russell diagram,* (San Francisco), 305

Mullan, D. J. & MacDonald, J. 2005, *MNRAS,* 356, 1139

Rekowski, B. V. & Brandenburg, A. 2004, *A&A,* 420, 17

Shu, F., Najita, J., Ostriker, E., *et al.* 1994, *ApJ,* 429, 781

Skelly, M. B., Unruh, Y. C., Cameron, A. C., *et al.* 2008, *MNRAS,* 385, 708

Spruit, H. 2002, *A&A,* 381, 923

Strassmeier, K. 1996, in: K. G. Strassmeier & J. L. Linsky, (eds.), *IAU Symposium 176: Stellar Surface Structure,* (Kluwer), 289–298

Tout, C. A. & Pringle, J. E. 1995, *MNRAS,* 272, 528

Townsend, R. H. D. & Owoki, S. P. 2005, *MNRAS,* 357, 251

Discussion

DASSO: Do you see some relationship between the existence of a convection zone in stars and prominences?

JARDINE: We have only observed prominences in stars that have a convective zone. High mass stars which have a radiative interior do show transient H-alpha absorption features. These are interpreted as a "wind compressed disk" — though they show many features in common with prominences.

SPANGLER: There is a well-established empirical relation for the decrease in X-ray luminosity and coronal activity with time for solar-type stars ("the Sun in time"). Are these data useful as constraints on your magnetic field models?

JARDINE: This decrease is largely due to the change in rotation rate as solar-type stars spin down due to the action of a hot, magnetically-channeled wind. We can only observe prominences on rapidly-rotating stars, so it is difficult to follow this whole evolution.

GOPALSWAMY: 1. Do you observe eruptive prominences? 2. Are there prominences in flare stars and what happens to the prominences during flares?

JARDINE: 1. Yes, on Speedy Mic, but it isn't published yet. 2. Yes, there are stellar "Active Prominences" that are associated with flares – but we don't know if they are ejected or simply heated.

GIRISH: I believe that the maximum height of observation of a stellar prominence will depend on the latitude of its origin and on the phase of the stellar activity cycles. Have you looked in the long term observations available for solar prominences to verify this idea?

JARDINE: AB Dor has been observed for over 10 years and there is no evidence for a cyclic change in the magnetic structure, although there is a variation in optical brightness. The nature of magnetic cycles on other stars is as yet an open question.

VRŠNAK: Why our Sun does not have these huge prominences?

JARDINE: It rotates too slowly! Stellar prominences appear to form at or beyond the Keplerian co-rotation radius, which for the Sun is located at 40 R_S.

Universal Heliophysical Processes
Proceedings IAU Symposium No. 257, 2008
N. Gopalswamy & D.F. Webb, eds.

© 2009 International Astronomical Union
doi:10.1017/S1743921309029329

Magnetic flux ropes: Fundamental structures for eruptive phenomena

Tahar Amari[1] and Jean-Jacques Aly[2]

[1] CNRS, Centre de Physique Théorique de l'Ecole Polytechnique,
F-91128 Palaiseau Cedex, France
email: amari@cpht.polytechnique.fr

[2] AIM - Unité Mixte de Recherche CEA - CNRS - Université Paris VII - UMR n⁰ 7158, Centre
d'Etudes de Saclay, F-91191 Gif sur Yvette Cedex, France

Abstract. We consider some general aspects of twisted magnetic flux ropes (TFR), which are thought to play a fundamental role in the structure and dynamics of large scale eruptive events. We first discuss the possibility to show the presence of a TFR in a pre-eruptive configuration by using a model along with observational informations provided by a vector magnetograph. Then we present, in the framework of a generic model in which the coronal field is driven into an evolution by changes imposed at the photospheric level, several mechanisms which may lead to the formation and the disruption of a TFR, including the development of a MHD instability, and we discuss the issues of the energy and helicity contents of an erupting configuration. Finally we report some results of a recent and more ambitious approach to the physics of TFRs in which one tries to describe in a consistent way their rising through the convection zone, their emergence through the photosphere, and their subsequent evolution in the corona.

Keywords. Flux ropes, Eruptive events, MHD

1. Why flux ropes?

Confined eruptive flares, eruptive prominences, coronal mass ejections (CMEs) and interplanetary magnetic clouds (IMCs) are most generally thought to be different observational aspects of a unique phenomenon, the *eruptive event*. Their respective structures exhibit indeed many similarities. In particular, they quite often show directly or indirectly the presence of a twisted magnetic flux rope (TFR).

Let us recall a few basic observational facts concerning these large scale eruptive phenomena to see how TFR are actually involved (a detailed review of the many observations may be found in Gopalswamy *et al.* 2006). A typical CME is constituted of a front, a dark cavity, and a plasmoid which contains about 10^{16} g of material. The latter probably originates from a prominence, i.e., a sheet of relatively cold and dense plasma (compared to the surrounding corona) which may stay in quasi-equilibrium for long periods of time (prominences are highly interesting objects which have been given a great deal of attention by solar physicists; see the review paper by Schmieder in this volume, and Forbes *et al.* 2006). Such a prominence is often seen indeed to rise before the CME and an associated flare, and it may be naturally thought that it gets ejected with the CME at the average speed of 10^3 kms^{-1}. By simply looking at the images provided by observations one may often guess the presence of twist in an eruptive prominence (as in the well known one called "Granddady"). Moreover, the presence in many cases of a TFR has been much supported by Gary & Moore (2004) who made a quantitative study showing indeed that a twisted structure clearly appears during the eruptive phase (see also Gibson *et al.* 2006).

In this paper, we report on some particular issues related to the possible role of TFRs in CMEs. We address in particular the following crucial issues: Is a TFR already present in a pre-eruptive configuration, or does it get created during the eruption; does a TFR containing a prominence just traces out the visible CME phenomenon as a passive entity or does it play a role in the initiation of the ejection itself. Thus this is not a review paper on possible CMEs mechanisms. For detailed up-to-date interesting reviews on that more general topics, we refer the readers to , e.g., Priest & Forbes (2002), Gopalswamy *et al.* (2006), and Mikic & Lee (2006).

2. Flux-Rope and Pre-Eruptive configuration

Let us start with some general remarks on eruptive phenomena. The first and most important one is that they have to be of magnetic origin. This conclusion appears quite inescapable if we just make a comparison between the various possible sources of energy present in the corona: magnetic, thermal, gravitational, kinetic. Only the first one has a sufficient magnitude to power a CME, say. The second point is that, in the preeruptive phase, the low corona appears to be in quasi-equilibrium, the magnetic, pressure, and gravitational forces balancing each other. In fact, owing to the dominance of the magnetic energy, the equilibrium may be considered to a very good approximation as being force-free, with the magnetic pressure being thus balanced by the magnetic tension, while the two other forces just intervene to fix the distribution of the plasma along the field lines (Priest & Forbes 2002). The third point concerns the storage of the energy: It has to be associated with coronal electric currents. In fact the magnetic field **B** can be expressed as the sum of a potential term created by the photospheric currents and of a term created by the coronal currents. The magnetic energy is the sum of the energies of these two fields, with only the second one being liable to get dissipated.

Unfortunately, the magnetic field cannot yet be accessed directly in the corona by observational means. One thus needs to build up models to try to understand the details of the processes leading to an eruptive event. Here, we shall restrict our attention to two classes of models which have a long tradition behind them: The TFR model and the Magnetic Arcade (MA) one, which have both developed into a large variety of submodels (Figure 1). The presence of a TFR as shown on the left figure (Amari *et al.* 2000) is actually a generic feature of a variety of magnetic configurations which have been studied since the middle 80's in the context of solar prominence modelling. The TFR gives indeed to a field the geometric properties needed to ensure the support of cold material, as magnetic dips are obviously present, and a series of models of increasing complexities have been constructed to describe this support, starting with simple pictures in which the prominence is represented by a line current in equilibrium either in a potential field (Anzer &Priest 1985) or in a linear force-free field (Amari &Aly 1989), and ending with the much more sophisticated models reported, e.g., in Aulanier & Demoulin (1998), Titov & Démoulin (1999), and Lionello *et al.* (2002). See Forbes *et al.* (2006) for a more detailed discussion. The arcade model on the right (Antiochos *et al.* 1999) is also generic of a class of models.

To prove or disprove the existence of one type of structure rather than the other in the pre-eruptive configuration is an important issue which may be solved to some extent by using the measurements performed at the photospheric level by a vector magneto-graph, along with a good method to resolve the well-known 180 degrees ambiguity on the transverse field, a problem which has been given recently a great deal of attention

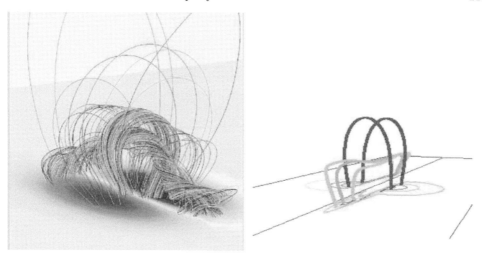

Figure 1. Two generic models which are candidates for describing a pre-eruptive magnetic configuration: On the left, the twisted flux rope from Amari *et al.* (2000), on the right, arcade model from Antiochos *et al.*(1999a)

(Li *et al.* 2007, Metcalf *et al.* 2006). There seems to be two possible ways to proceed: Either one may merely investigate the orientation of the transverse field, or one may reconstruct the coronal field. The first method rests on the following remark, which has been long pointed out in the context of solar prominences: There must be a reversal from a normal to an inverse type configuration – in which the transverse field points from the negative polarity towards the positive one – if a TFR is present down to the photospheric level as it is the case in some MHD models (Amari *et al.* 2003a). Although quite simple in principle, this method has a strong limitation: It does not allow to conclude at the absence of a TFR if no one is revealed – e.g., a TFR could exist at higher coronal altitude above a sheared arcade present at the lower level. In the latter case, it is necessary to appeal to the second and much heavier method, which has been much developed in the last few years (Amari *et al.* 1997, Schrijver *et al.* 2006, Aly & Amari 2007, Wiegelmann 2008) in relation with the availability of several ground based (IVM, SOLIS, ASP, THEMIS, EST) and embarked vector magnetographs HINODE, and the prospect of several new ones (SDO, SOLAR-ORBITER) in the near future. Evidences for TFR have thus been obtained in some pre-eruptive configurations by Bleybel *et al.* (2002), Régnier & Amari (2004) (see Figure 2), Thalmann & Wiegelmann (2008), and Canou *et al.* (2008, in preparation).

3. Flux Rope and the coronal evolution problem

Assuming that configurations with TFR do exist, the question of their formation and evolution immediately arises. For about 35 years, this problem has been mainly studied in the framework of a general model in which the solar corona magnetic field is made to evolve in response to changes occuring at the photospheric level, and this has lead to formulating the *Coronal Evolution Problem*. In the latter one starts from an initial potential or low beta force-free configuration, and makes it evolving by prescribing motions of either one of the following types on the photosphere: Either shearing motions, or flux emergence or submergence, both corresponding to some observations.

When considering a solution to that problem, one has to address the important question of the evolution of two basic quantities: The magnetic energy W – the source powering an

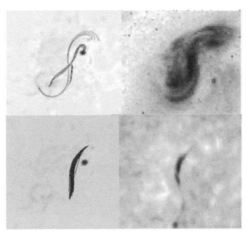

Figure 2. Nonlinear force-free reconstruction of a pre-eruptive configuration from IVM boundary data. Two TFRs may be observed. The first one has a twist exceeding one turn and coincides with the X-ray loops (sigmoid). The second one, located underneath, is less twisted, and its dips (bottom row) support cool material seen in H-alpha line. From Régnier & Amari (2004)

eruptive event –, and the relative magnetic helicity H (Démoulin 2007, Amari *et al.* 2003a, Amari *et al.* 2003b). At least in a nondynamical phase, the evolution of both quantities is essentially controlled by exchanges occurring at the photospheric level (through the Poynting vector for W). A question which thus naturally arises is that one of the possible existence of upper bounds on the amounts of energy and helicity which can be injected into the field by the photospheric motions.

As for the energy we underline two important theorems which do apply to quasi-equilibrium configurations and are thus relevant indeed in a pre-eruptive phase (see Aly & Amari 2007 for a detailed discussion). They do involve two reference fields having the same distribution of photospheric normal component, B_n, as $\mathbf{B}$: The potential field $\mathbf{B}_\pi$, and the open field $\mathbf{B}_\sigma$, the latter having all its lines being open and thus containing current sheets. Theorem 1 states that the energy $W[\mathbf{B}]$ of a force-free field $\mathbf{B}$ is bounded from below by the energy of $\mathbf{B}_\pi$, which justifies the third remark made at the beginning of the previous section. Theorem 2 (Aly 1984) states that $W[\mathbf{B}]$ is bounded from above by a number depending only on B_n, the best possible upper bound – the so-called least upper bound – being conjectured to be equal to the energy of $\mathbf{B}_\sigma$. A weakened form of this guess (Aly 1991, Sturrock 1991), in which one's attention is restricted to configurations having all their lines being connected to the boundary, has been supported by several theoretical arguments and simulations (see however Choe & Cheng 2002). But a few recent numerical examples of axisymmetric solutions exhibiting one or several TFR disconnected from the boundary have been constructed and found to have an energy exceeding $W[\mathbf{B}_\sigma]$ (Flyer *et al.* 2005, Hu *et al.* 2003). This shows that the presence of TFRs can lead to quite large values of the magnetic energy. Note however that no fully 3D configurations with disconnected TFR have yet been obtained.

That the absolute value of the magnetic helicity of a force-free field may be also bounded from above by a number depending only on B_n has been conjectured by Zhang *et al.* (2006), but no proof of that statement has been yet furnished. If true, this would imply immediately that imparting sufficient shear or twist to a force-free field leads in any case to a nonequilibrium process.

4. Flux rope creation and disruption

In the first class of evolution problems which has been considered by solar physicists, the footpoints of the field lines of an initially potential configuration have been imposed shearing motions, and this problem has been treated both analytically and numerically for fields of increasing geometric complexity: Translation invariant, rotation invariant about an axis (axisymmetric), and more recently fully 3D. In the latter case, the profile of the imposed photospheric flow has been found to be of crucial importance. For a flow exhibiting a strong shear localized near a neutral line, one observes the formation of a sheared arcade in equilibrium when the field topology is bipolar, with no disruption occuring (Antiochos *et al.* 1999; in the corresponding 2D situation a plasmoid is ejected when a small resistivity is introduced (Amari *et al.* 1996a)). On the contrary, when the flow leads to a global twisting of the field (which is equivalent to a shearing only near the neutral line), the configuration evolves slowly through a sequence of quasi-equilibria up to a certain twist threshold of about 1 turn. Once the latter is exceeded, the configuration experiences a transition towards a dynamic evolution. A central flux rope is created, which pierces through the overlying field lines and erupts, but without disconnecting from the photosphere. This phenomenon has been called *very fast opening* (Amari *et al.* 1996b) and it has been revisited more recently by Török & Kliem (2003) and Aulanier *et al.* (2005). Several features of this model may be related to observational facts. In particular the existence of strong electric currents localized below the flux rope may explain the well known characteristic sigmoidal structures. The fact that the TFR remains attached to the Sun while expanding in the solar wind and the interplanetary medium may explain why it may look open from the low corona point of view while appearing still closed in the interplanetary medium where it may be possibly identified as a magnetic cloud (Démoulin 2008). Moreover the interaction of this expanding TFR with the overlying field, which leads to the appearance of strong currents at the interface, may be at the origin of EIT waves as recently proposed in Delannée & Amari (2000) and Delannée *et al.* (2008). Finally we point out that the very fast opening phenomenon involves here a partial opening rather than a total one (the open field conjecture is then not challenged). Therefore a full opening is not necessarily implied in a disruption (Amari *et al.* 1996b).

More recently the effects of photospheric flux changes have also been considered. They may mimic the emergence or submergence of flux through the photosphere, and in particular the so-called flux cancellation process (FC). The latter is often observed on the Sun and it has been given a great deal of attention after Martin *et al.* (1985). It has been found for instance to occur in the big X 5.7 "Bastille day" event in 2000 (Kosovichev & Zharkova 2001). Originally proposed as a mechanism leading to the formation of a prominence inside a TFR contained in a 2D equilibrium (van Ballegooijen & Martens 1989), and also to the formation of an erupting plasmoid in an axisymmetric configuration (Forbes & Priest 1995), FC has been studied in 3D (Amari *et al.* 2000, Linker *et al.* 2001) as a possible process leading to the creation and the disruption of a TFR. If one starts from an initial sheared configuration containing a non-zero magnetic helicity, FC leads after a certain threshold to the creation of a TFR in equilibrium, which experiences later on a major global disruption. The key point here is that there is a decrease of the energy of the open field (which depends only on the photospheric distribution of the normal field component) while the energy of the evolving low beta coronal configuration (which is related to the presence of coronal currents) does not change significantly. Thus both these energies become comparable at some critical time, which precludes the existence of a global equilibrium and leads to the disruption (Amari *et al.* 2000). The TFR created by the FC mechanism may possibly explain several observed characteristics such as the presence of a prominence (there are dips), the presence of a sigmoid, and the current sheet/cusp formation below the ejected rope.

Another mechanism associated with a flux change on the boundary has been proposed for explaining the following fact. During the death of an active region due to the dispersion of its flux, large scale eruptive events are nonetheless produced and reformation of filaments from remnants of previous eruptions are observed. Following Leighton (1964) and Wang *et al.* (1991), this dispersion has been modelled by turbulent diffusion (TD) occuring at the photosheric level (Amari *et al.* 1999, Amari *et al.* 2003b). This leads once more to a well defined BVP in which one starts from an initial configuration supposed to represent the remnant of a previous eruptive field which has relaxed to a non-potential configuration and thus has a nonzero helicity. The field is thus made to evolve slowly due to TD, and it is found that in all cases the resulting evolution leads to the formation of a TFR in equilibrium. Depending on the initial helicity contents, either a confined disruption (moderate helicity) or a global one (large enough helicity) is produced eventually. Although this could seem to show that a minimum amount of magnetic helicity be necessary to trigger a CME, say, it should be noted that the total magnetic helicity of the configuration remains unchanged during the evolution. Once more, the results of this model are in agreement with several observational characteristics of eruptive events.

TFR have also been shown to form when the evolution is driven by converging motions applied to an initial configuration with a non-zero helicity. This problem has been first considered in 2D (Priest & Forbes 1990) and more recently in 3D by Amari *et al.* (2003a). By starting from the set of initial configurations previously used in the FC studies, it has been shown that the field evolves through a series of equilibria up to a certain threshold beyond which the topology changes to a TFR-like one. However unlike in the FC or TD mechanisms the TFR is not in equilibrium, and it experiences a disruption. Here TFR formation and disruption appears to be associated.

5. Flux ropes and robustness with respect to magnetic topology

As indicated above a very localized shear applied to a simple topology bipolar configuration does not lead to a disruption. The situation turns out to be quite different, however, if the bipolar configuration is a part of a larger quadrupolar configuration. Taking such a complex topology field as an initial state in the BVP previously solved for the simple bipole, it is found indeed (Antiochos *et al.* 1999) that there is formation in a first stage of a strongly sheared arcade with dips favorable to prominence support. Beyond a critical threshold the field lines above the coronal X-point reconnect with the inner bipolar lines, thus triggering a large scale disruption. For this mechanism to be efficient, it is necessary that the current sheet which forms near the location of the initial X-point be maintained in equilibrium all along the first part of the evolution for otherwise only an insufficient amount of free energy would be stored (note that describing a current sheet is numerically difficult). This interesting mechanism is called the Break out Model (BOM), and it has the merit of showing the role of the magnetic topology in an evolution. It should be noted, however, that this role was also pointed out in earlier 2D studies (Forbes & Isenberg 1991, Isenberg *et al.* 1993).

Some observations have shown that several pre-eruptive configurations had a complex topology (this was the case in particular for the July 14, 1998 flare, as found by Aulanier *et al.* 2000), and one could be tempted to take this fact as an evidence in favor of the only BOM (Lynch *et al.* 2008; see Figure 3). However, that a disruption occurs when the field has a complex topology is not the signature of a particular mechanism, it is just one component of the context in which the BOM may be relevant. In fact, if one

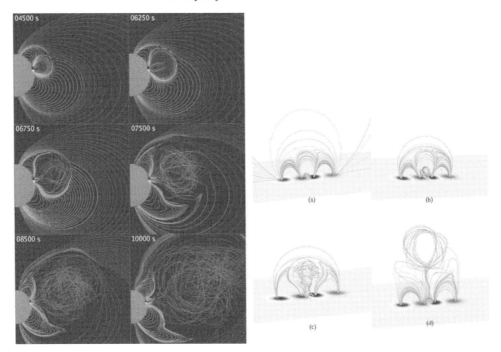

Figure 3. Twisted flux Ropes created in the BOM and in the FC model. In the BOM (left panel, from Lynch *et al.* 2008), which presupposes a complex topology, the TFR is created during the eruptive phase, while in the FC model (right panel, from Amari *et al.* 2007) the TFR is formed prior to the eruption as an equilibrium structure existing independently of the nature (bipolar or multipolar) of the background configuration.

takes a quadrupolar configuration as the one used in the BOM, and submit it to FC, then it is found that a TFR in equilibrium gets formed in a first phase in the inner bipolar part. A disruption is thus suffered by the configuration in a second phase (Amari *et al.* 2007). Compared to the case of a simple dipolar configuration, it is clear that the overlying arcade has weaker confinement properties due to the presence of the X-point, which allows a faster expulsion of the TFR. To conclude this section, we note that the BOM and the FCM share the properties of weakening the confinement, and of producing a TFR. But they involve different processes and different structures in the lower part of the magnetic configuration, the TFR appearing before the eruption as an equilibrium structure in the FCM, while it is created in the BOM by a nonequilibrium process involving a shear transfer by reconnection between two initially disconnected topological cells.

6. Flux ropes prone MHD instabilities

As we have seen previously, a TFR may be produced in an evolving equilibrium which may thus get disrupted at some stage. Basically there are three possibilities to explain this disruption: i) there exists no equilibrium compatible with the photospheric changes as in the FC mechanism; ii) an equilibrium compatible with the photospheric changes may exist but it is too far to be reached (very fast opening mechanism), and iii) an equilibirum exists but it is unstable. We now explore this last possibility in the context of ideal MHD, which may be used here because of the very high conductivity of the low coronal plasma. We first note that simple 2D arcades have never been found to be unstable, and that there is in 3D a known sufficient condition for a force-free equilibrium to be stable: It is that $\alpha L \lesssim 1$ (Aly 1990), where L is the typical length scale of the structure and α the order of magnitude of the force-free function ($\nabla \times \mathbf{B} = \alpha \mathbf{B}$). But no sufficient condition of instability seems to have been established yet.

What about configurations containing TFR? Some informations can be drawn here from the many studies of cylindrical and toroidal configurations which have been conducted up to now in the context of thermonuclear fusion in magnetically confined plasma. For instance, cylindrical and toroidal TFR configurations have long been shown by plasma laboratory physicists to be subject to the kink instability when the poloidal component of the magnetic field becomes of the order of the axial one (Freidberg 1987). Solar physicists have thus looked for the possibility of the development of the kink in a cylindrical coronal TFR exhibiting a twist of about one turn around the axis when the anchoring of the footpoints in two horizontal plates representing the dense photosphere is taken into account (Raadu 1972, Baty & Heyvaerts 1996, Baty 1997). The more realistic case of a toroidal line-tied field has also been studied, mainly by considering the 2D analytical model of Titov & Démoulin (1999), and kink unstable TFR have been obtained in spite of the stabilizing line-tying effect (Török et al. 2004). This mechanism may reproduce some of the observed characteristics of confined disruptions (Török & Kliem 2005, Fan & Gibson 2004, Fan & Gibson 2007). And finally similar conclusions have been suggested by the results of some non-symmetric simulations (Amari & Luciani 1999).

Another property of TFR discovered by studying fusion toroidal configurations is the fact that a poloidal magnetic field exerts a net outward radial force par unit of length on the toroidal current as a simple consequence of flux conservation (Freidberg 1987). In a tokamak there exist some restoring forces due either to the presence of a wall, which induces a restoring pressure build up in the external part, or to an external vertical field $\mathbf{B}_{ext}$. In the latter case, however, a too fast decrease of $\mathbf{B}_{ext}$ with the distance to the axis makes the resulting equilibrium unstable. This is the so-called *torus instability*, and it has been suggested that it could also occur in the solar corona. In that case, the "external" magnetic field is the one of the overlying arcade, and it has been proven indeed that the torus instability may develop for some shape of the inner TFR and some decreasing profile of $\mathbf{B}_{ext}$, with a coronal disruption thus being produced (Kliem & Török 2006).

It is worth noticing that although both the kink and the torus instabilities are interesting exact properties of TFR, their application to the disruption at the origin of eruptive events is not yet completely convincing. In their simple form used up to now, they do develop indeed in a pre-eruptive configuration with a high degree of symmetry. The latter then exhibits the well known phenomenon of *symmetry breaking* once 3D perturbations are allowed. Such a symmetry certainly exists for a laboratory device like a tokamak, but it seems quite difficult to think of a consistent mechanism which could produce a similarly constrained equilibrium in the solar corona. In fact it should be clear from the results above that a low beta symmetric configuration, once twisted or subject to flux changes, evolves quite generally into a non-symmetric state. Moreover it may be noticed that although these models are able to produce interesting quantitative predictions for the acceleration profiles in CMEs, they may disagree with some recent observations (Schrijver et al. 2008).

7. Flux ropes driven from below: Towards a global approach

As stated in Sect. 3, the evolution of the solar corona magnetic field has long been studied as a BVP in which an initial equilibrium is driven into an evolution by prescribing some specific changes in the photospheric parameters. TFRs have thus been shown to be created, e.g., by FC, which can be considered in some sense as mimicking the late phase of TFR emergence. This is clearly why the created TFR does not depend on the overlying configuration, which may or may not have a complex topology. To go a step further towards a realistic model, it is clear that one should add a description of the actual process of emergence of the TFR from the convection zone (CZ) into the corona.

The most ambitious approach to try to deal with such a program consists in setting at once a global MHD model including both the CZ and the corona, which constitutes a huge numerical challenge because of the strong difference in the plasma characteristics in both regions, and the stiff variations in the photospheric transition. In spite of that, this formidable task has yet been faced by several authors (Fan 2001, Archontis *et al.* 2004, Magara & Longcope 2003, Manchester *et al.* 2006, Galsgaard *et al.* 2005, Cheung *et al.* 2007). However, although this global approach has to be taken eventually, it may be useful in a first step to try to gain some simple insights into the complicated physics involved in that problem by following a more modest but complementary approach in which an initial TFR is introduced in the CZ and forced to rise rigidly. This leads once more to solving a coronal BVP as the one driven by photospheric changes. But the imposed changes are now those which generate an electric field associated to a TFR that is vertically crossing the photosphere at a given velocity. In the CZ, the TFR can be taken to be either a torus rising at a uniform speed (Fan & Gibson 2004, Fan & Gibson 2007), or a cylinder deformed by an horizontally varying vertical flow, which makes only the central part to "emerge" (Amari *et al.* 2004). In both cases its confinement in the CZ is insured by the external high pressure field (it has not to be assumed to be force-free). This leads to the emergence of the TFR into the low beta corona, with a nonequilibrium behavior occurring eventually or not depending on the amount of initial twist and on the strength of the coronal confinement. It is worth noticing that this simple model reproduces the characteristic features of the so-called "tongs" which have been observed in the concentration of the photospheric normal component of the field (López Fuentes *et al.* 2000) and were already seen in earlier MHD simulations (Fan 2001).

Although reproducing many observed facts, the approach above has some heavy short-comings. In particular theoretical arguments and numerical simulations show that a TFR rigidly rising in the CZ converts a part of its vertical motion into horizontal one when hitting the photosphere. This of course prevents its rapid emergence, which is further impeded by the accumulation of heavy material in the dips. A possible way to bypass these problems consists in transferring the horizontal component of the magnetic field, which is responsible for the normal component of the electric current, before the vertical rising flow in the CZ becomes too small. That is what is done in the Resistive Layer Model (RLM) (Amari *et al.* 2005) in which the turbulent photospheric transition is represented by a resistive layer. An important feature of that model is that it insures the conservation of the global magnetic helicity. In fact helicity injection through the interface depends only on the tangential component of the electric field and the normal component of the magnetic field, and the latter are in any case always continuous. When one goes from the CZ to the corona, the term in the Ohm's law which is dominant in the determination of the horizontal electric field just changes. Below, it is the effect of the vertical component of the flow (and the horizontal component of the magnetic field) which is important, thus it is the magnetic diffusivity (and the electric current), and finally it is the shear flow, say, (and the vertical component of the magnetic field) which becomes important at the basis of the corona (a remaining vertical flow may of course be added to that horizontal flow). The RLM has been shown to allow indeed the emergence of the TFR into the corona (while no emergence occurs if the resistivity is made to vanish), followed later on by its disruption (Amari *et al.* 2005). The difference with the rigid case considered above is that magnetic helicity does keep increasing but tends to saturate.

8. Conclusion

Magnetic flux ropes are structures which may be easily formed in the solar corona by various mechanisms. They are good candidates to support prominence material which is denser and cooler than the coronal one around. They may lead to either confined disruption or large scale eruptive phenomena such as CMEs and two ribbon flares. Their interest also relies on the fact that they may be subject to various ideal instabilities such as the kink and the torus instabilities. There are several indications that they may be present in the corona prior to some eruptive event. In particular, this has been shown to be true in some cases by using boundary data provided by vector magnetographs and a force-free low corona model, without then making any extra assumptions on their origin. Considering only MHD mechanisms, TFR have been shown either to exist in the pre-eruptive configuration or to be created only during the eruption through reconnection. They may also represent the magnetic structure of interplanetary magnetic clouds. Of course we do not mean that TFRs are the only way of triggering eruptive phenomena, and we acknowledge the fact that many other types of structures may produce such events. Determining if TFR may also come from below is one of the main challenge of current research in solar physics. Answering this question requires the construction of a model allowing to follow TFR from their possible formation in the stable region below the CZ, their rising through the latter, their piercing through the photosphere, and their evolution in the corona. Much help should be also provided in this respect by the arrival of a new generation of vector magnetographs with low noise and high resolution such as those on board of HINODE or SDO or the future ground based EST.

References

Aly, J. J. 1984, *Astrophys. J.*, 283, 349

Aly, J. J. 1990, *Physics of Fluids B*, 2, 1928

Aly, J. J. 1991, *Astrophys. J.*, 375, L61

Aly, J. J. & Amari, T. 2007, *Geophysical and Astrophysical Fluid Dynamics*, 101, 249

Amari, T. & Aly, J. J. 1989, *Astron. Astrophys.*, 208, 261

Amari, T., Luciani, J. F., Aly, J. J., & Tagger, M. 1996, *Astron. Astrophys.*, 306, 913

Amari, T., Luciani, J. F., Aly, J. J., & Tagger, M. 1996, *Astrophys. J.*, 466, L39

Amari, T., Aly, J. J., Luciani, J. F., Boulmezaoud, T. Z., & Mikic, Z. 1997, *Solar Phys.*, 174, 129

Amari, T. & Luciani, J. F. 1999, *Astrophys. J.*, 515, L81

Amari, T., Luciani, J. F., Mikic, Z., & Linker, J. 1999, *Astrophys. J.*, 518, L57

Amari, T., Luciani, J. F., Mikic, Z., & Linker, J. 2000, *Astrophys. J.*, 529, L49

Amari, T., Luciani, J. F., Aly, J. J., Mikic, Z., & Linker, J. 2003, *Astrophys. J.*, 585, 1073

Amari, T., Luciani, J. F., Aly, J. J., Mikic, Z., & Linker, J. 2003, *Astrophys. J.*, 595, 1231

Amari, T., Luciani, J. F., & Aly, J. J. 2004, *Astrophys. J.*, 615, L165

Amari, T., Luciani, J. F., & Aly, J. J. 2005, *Astrophys. J.*, 629, L37

Amari, T., Aly, J. J., Mikic, Z., & Linker, J. 2007, *Astrophys. J.*, 671, L189

Antiochos, S. K., Devore, C. R., & Klimchuk, J. A. 1999, *Bulletin of the American Astronomical Society*, 31, 868

Antiochos, S. K., DeVore, C. R., & Klimchuk, J. A. 1999, *Astrophys. J.*, 510, 485

Anzer, U. & Priest, E. 1985, *Solar Phys.*, 95, 263

Archontis, V., Moreno-Insertis, F., Galsgaard, K., Hood, A., & O'Shea, E. 2004, *Astron. Astrophys.*, 426, 1047

Aulanier, G. & Demoulin, P. 1998, *Astron. Astrophys.*, 329, 1125

Aulanier, G., Démoulin, P., & Grappin, R. 2005, *Astron. Astrophys.*, 430, 1067

Aulanier, G., DeLuca, E. E., Antiochos, S. K., McMullen, R. A., & Golub, L. 2000, *Astrophys. J.* , 540, 1126

Baty, H. & Heyvaerts, J. 1996, *Astron. Astrophys.*, 308, 935

Baty, H. 1997, *Astron. Astrophys.*, 318, 621

Bleybel, A., Amari, T., van Driel-Gesztelyi, L., & Leka, K. D. 2002, *Astron. Astrophys.*, 395, 685

Cheung, M. C. M., Schüssler, M., & Moreno-Insertis, F. 2007, *Astron. Astrophys.*, 467, 703

Choe, G. S. & Cheng, C. Z. 2002, *Astrophys. J.*, 574, L179

Delannée, C. & Amari, T. 2000, *Bulletin of the American Astronomical Society*, 32, 838

Delannée, C., Török, T., Aulanier, G., & Hochedez, J.-F. 2008, *Solar Phys.*, 247, 123

Démoulin, P. 2007, *Advances in Space Research*, 39, 1674

Démoulin, P. 2008, *Ann. Geophys.*, in press

Fan, Y. 2001, *Astrophys. J.*, 554, L111

Fan, Y. & Gibson, S. E. 2004, *Astrophys. J.*, 609, 1123

Fan, Y. & Gibson, S. E. 2007, *Astrophys. J.*, 668, 1232

Flyer, N., Fornberg, B., Thomas, S., & Low, B. C. 2005, *Astrophys. J.*, 631, 1239

Forbes, T. G. & Isenberg, P. A. 1991, *Astrophys. J.*, 373, 294

Forbes, T. G. & Priest, E. R. 1995, *Astrophys. J.*, 446, 377

Forbes, T. G. , Linker,J. A., Chen, J., Cid, C., Chen, J.,Kóta, J., Lee, M. A., Mann, G., Mikić, Z. , Potgieter, M. S., Schmidt, J. M. , Siscoe, G. L., Vainio, R., Antiochos, S. K., & Riley, P. 2006, *Space Science Reviews*, 123, 251

Freidberg, J. P. 1987, *Ideal Magnetohydrodynamics. Plenum Press, New York*

Galsgaard K., Moreno-Insertis F., Archontis V. & Hood A. 2005, *ApJL*, 618, L153

Gary, G. A. & Moore, R. L. 2004, *ApJ*, 611, 545

Gopalswamy, N. , Mikić, Z., Maia, D., Alexander, D., Cremades, H., Kaufmann, P., Tripathi, D., & Wang, Y.-M. 2006, *Space Science Reviews*, 123, 303

Gibson, S. E., Fan, Y., Török, T., & Kliem, B. 2006, *Space Science Reviews*, 124, 131

Hu, Y. Q., Li, G. Q., & Xing, X. Y. 2003, *Journal of Geophysical Research (Space Physics)*, 108, 1072

Isenberg, P. A., Forbes, T. G., & Demoulin, P. 1993, *Astrophys. J.*, 417, 368

Kliem, B. & Török, T. 2006, *Physical Review Letters*, 96, 255002

Kosovichev, A. G. & Zharkova, V. V. 2001, *Astrophys. J.*, 550, L105

Leighton, R. B. 1964, *Astrophys. J.*, 140, 1547

Li, J., Amari, T., & Fan, Y. 2007, *Astrophys. J.*, 654, 675

López Fuentes, M. C., Demoulin, P., Mandrini, C. H., & van Driel-Gesztelyi, L. 2000, *Astrophys. J.*, 544, 540

Lynch, B. J., Antiochos, S. K., DeVore, C. R., Luhmann, J. G., & Zurbuchen, T. H. 2008, *Astrophys. J.*, 683, 1192

Linker, J. A., Lionello, R., Mikić, Z., & Amari, T. 2001, *J. Geophys. Res.*, 106, 25165

Lionello, R., Mikić, Z., Linker, J. A., & Amari, T. 2002, *Astrophys. J.*, 581, 718

Magara, T. & Longcope, D. W. 2003, *Astrophys. J.*, 586, 630

Manchester,W. IV., Gombosi, T., DeZeeuw D., & Fan Y. 2006, *Astrophys. J.*, 610, 161

Martin, S. F., Livi, S. H. B., & Wang, J. 1985, *Australian Journal of Physics*, 38, 929

Metcalf, T. R., *et al.* 2006, *Solar Phys.*, 237, 267

Mikić, Z. & Lee,M. A. 2006, *Space Science Reviews*, 123, 57

Ott, U. 1993, *Nature*, 364, 25

Priest, E. R. & Forbes, T. G. 1990, *Solar Phys.*, 126, 319

Priest, E. R. & Forbes, T. G. 2002, *Astron Astrophys Rev* , 10,313

Raadu, M. A.1972, *Solar Phys.*, 22, 425

Régnier, S. & Amari, T. 2004, *Astron. Astrophys.*, 425, 345

Schrijver, C. J., *et al.* 2006, *Solar Phys.*, 235, 161

Schrijver, C. J., Elmore, C., Kliem, B., Török, T., & Title, A. M. 2008, *Astrophys. J.*, 674, 586

Sturrock, P. A. 1991, *Astrophys. J.*, 380, 655

Thalmann, J. K. & Wiegelmann, T. 2008, *Astron. Astrophys.*, 484, 495

Titov, V. S. & Démoulin, P. 1999, *Astron. Astrophys.*, 351, 707

Török, T. & Kliem, B. 2003, *Astron. Astrophys.*, 406, 1043

Török, T., Kliem, B., & Titov, V. S. 2004, *Astron. Astrophys.*, 413, L27

Török, T. & Kliem, B. 2005, *Astrophys. J.*, 630, L97

van Ballegooijen, A. A. & Martens, P. C. H. 1989, *Astrophys. J.*, 343, 971

Wang, Y.-M., Sheeley, N. R., Jr., & Nash, A. G. 1991, *Astrophys. J.*, 383, 431

Wiegelmann, T. 2008, *Journal of Geophysical Research (Space Physics)*, 113, 3

Zhang, M., Flyer, N., & Low, B. C. 2006, *Astrophys. J.*, 644, 575

Universal Heliophysical Processes
Proceedings IAU Symposium No. 257, 2008
N. Gopalswamy & D.F. Webb, eds.

© 2009 International Astronomical Union
doi:10.1017/S1743921309029330

Solar prominences

Brigitte Schmieder, Guillaume Aulanier and Tibor Török

LESIA, Observatoire de Paris, 5 Place Janssen, Meudon, 92195, France
email: `brigitte.schmieder@obspm.fr`

Abstract. Solar filaments (or prominences) are magnetic structures in the corona. They can be represented by twisted flux ropes in a bipolar magnetic environment. In such models, the dipped field lines of the flux rope carry the filament material and parasitic polarities in the filament channel are responsible for the existence of the lateral feet of prominences.

Very simple laws do exist for the chirality of filaments, the so-called "filament chirality rules": commonly dextral/sinistral filaments corresponding to left- (resp. right) hand magnetic twists are in the North/South hemisphere. Combining these rules with 3D weakly twisted flux tube models, the sign of the magnetic helicity in several filaments were identified. These rules were also applied to the 180° disambiguation of the direction of the photospheric transverse magnetic field around filaments using THEMIS vector magnetograph data (López Ariste *et al.* 2006). Consequently, an unprecedented evidence of horizontal magnetic support in filament feet has been observed, as predicted by former magnetostatic and recent MHD models.

The second part of this review concerns the role of emerging flux in the vicinity of filament channels. It has been suggested that magnetic reconnection between the emerging flux and the pre-existing coronal field can trigger filament eruptions and CMEs. For a particular event, observed with Hinode/XRT, we observe signatures of such a reconnection, but no eruption of the filament. We present a 3D numerical simulation of emerging flux in the vicinity of a flux rope which was performed to reproduce this event and we briefly discuss, based on the simulation results, why the filament did not erupt.

Keywords. Solar prominence, magnetic emerging flux, eruption, 3D MHD model

---

1. Introduction

Solar filaments (or prominences) are cool and dense plasma embedded in the hot corona. The global structure of filaments can sometimes be observed over several solar rotations. Their pressure-scale height is less than 500 km and does not allow static plasma to remain in prominences as high as $10 - 50$ Mm. The plasma in filament fine structures is dynamic (Schmieder *et al.* 1991, Zirker *et al.* 1998, Lin *et al.* 2005). Recent observations of Hinode give the impression of continuous downflows in vertical structures (Figure 1a). However, the velocities do not exceed ≈ 10 kms^{-1}, unless the filament is particularly activated. The role of the dynamics is still unclear. Therefore it is reasonable to assume that filaments are magnetically supported in the corona, the plasma being frozen in magnetic structures. Magnetic flux tubes and arcades have been proposed in theoretical models to support the filament material (e.g. Amari *et al.* 1999, Antiochos *et al.* 1994). Extrapolations of photospheric magnetic fields show that filaments can be indeed modelled by flux tubes with plasma filling the dips. Since then, this "magnetic dip filling" procedure has been applied by various groups to perform model predictions (Aulanier *et al.* 2000) and to analyze real observations with linear magnetohydrostatic, non-linear magnetofrictional and fully MHD models (Aulanier & Schmieder 2002, Lionello *et al.* 2002, van Ballegooijen 2004, Bobra *et al.* 2008). These topologies have also recently been found to be consistent with the evolution of the photospheric vector magnetic field during a filament formation resulting from flux emergence, as observed by

Figure 1. top panel: Hinode prominence observed on 25 April 2007, low panel, Prominence modeling by the dips of field lines extrapolated from an observed photospheric magnetogram of THEMIS (Dudik *et al.* 2008).

Hinode/SOT (Okamoto *et al.* 2008). The barbs and footpoints are mimicked by the dips of field lines which deviate from the main flux tube and are rooted in parasitic polarities, minor polarities of opposite sign as their environment.

Due to their large scales and multi-wavelength manifestations, solar filaments and prominences are key phenomena for the study of high-stressed non-potential current-carrying magnetic fields in the solar corona. They can be used to understand how magnetic helicity slowly accumulates in the Sun's corona, and is then later ejected in the heliosphere in the form of coronal mass ejections, which are known to be the main drivers of extreme space weather. In spite of impressive progress with the Advanced Stokes Polarimeter (ASP) for the 2D measurement of the internal magnetic field in prominences (Casini *et al.* 2003), building a 3D picture still requires the combination of multi-wavelength observations and magnetic models (e.g. Dudik *et al.* 2008), see (Fig. 1b).

In this context, observational laws have been put forward from the chirality of observed features (e.g. chromospheric fibrils, lateral filament feet, overlaying coronal arcades) to derive the direction of the axial magnetic field inside solar filaments. These are the so-called "filament chirality rules". They state that a dominant fraction of filaments located in the northern (resp. southern) solar hemisphere have right- (resp. left-) bearing feet and fibrils, left (resp. right) skewed arcades, and dextral (resp. sinistral) internal axial fields, which point rightward (resp. leftward) as the filament is viewed from the main positive polarity field on the side of the photospheric inversion line above which it is located (Martin *et al.* 1994, Martin 1998).

When combined with the chirality rules, these models predict that dextral (resp. sinistral) filaments correspond to left- (resp. right) hand magnetic twists, hence to negative (resp. positive) magnetic helicities.

Here we report on the use of all these concepts and tools, firstly for the disambiguation between dipped and arcade topologies as observed with THEMIS below a filament foot, and secondly for quantifying the role of relative magnetic helicities between two filaments to predict either their subsequent merging into a longer and potentially less stable structure, or their interaction in the form of a confined flare.

In the second part of the paper we study the conditions to get filament eruption. Most filaments eventually erupt in many cases as part of a coronal mass ejection (CME). Such eruptions are often preceded by detectable changes in the photospheric magnetic field in the vicinity of the filament. Here we focus on emerging flux in the vicinity of filament channels. It has been suggested that magnetic reconnection between the emerging flux and the pre-existing coronal field can trigger filament eruptions and CMEs. For a particular event, observed with Hinode/XRT, we observe signatures of such reconnection, but no eruption of the filament. We present a numerical simulation of this event and we briefly argue why no eruption took place in this case.

2. Role of the magnetic helicity

2.1. *Resolution of the 180 degree of ambiguity and observations of magnetic dips*

A debate is raging in the solar physics community about the magnetic nature of filament feet, which are common underlaying and lateral extensions observed in absorption on the solar disc in Hα and in the EUV. These feet connect filament bodies suspended in the corona to the lower atmospheric layers. Are these feet formed by continuously injected plasma condensation in magnetic arcades, as hinted by some observations and conceptual models (Martin 1998), or do they consist of quasi-static condensations that are maintained against free-fall by the Lorentz force in a low-lying continuous distribution of magnetic hammocks, from the feet ends to up to the filament bodies, as first predicted by linear force-free field and magnetohydrostatic models (Aulanier & Démoulin 1998, Aulanier & Schmieder 2002, Dudik *et al.* 2008)? This debate had been lacking of new discriminators for about ten years, until this issue was recently addressed through new direct measurements of the photospheric magnetic field vector $\vec{B}$ in a filament channel located far from the center of the solar disc, resulting from the PCA-based inversion of high-precision spectropolarimetric observations with the MTR instrument of the THEMIS telescope (López Ariste *et al.* 2006).

A major problem with these measurements is that they still give the direction of the component of the magnetic field vector on the plane of the sky at $\pm 180°$. This fundamental ambiguity does not allow the observations, taken alone, to state whether an arcade or a dip is measured at a given place. So as to solve this paradigm, chirality rules can be applied to the disambiguation of the measured transverse magnetic fields, before deprojecting them to obtain the three components of the magnetic field vector in the reference frame of the solar surface. This procedure was proposed and applied in López Ariste *et al.* (2006), and rephrased by Martin *et al.* (2008). The studied filament was identified to be sinistral, hence with a magnetic field vector globally pointing toward the left, as viewed from the dominant positive magnetic polarity in the photosphere (Fig. 1, top). Interestingly, it was found that, for almost every area analyzed in detail within the observed filament channel, only the sinistral solution that matched the chirality rule on the plane of the sky remained sinistral in the reference frame of the solar surface. Using the chirality-consistent solution to calculate the curvature $B^2/(\vec{B}\cdot\vec{\nabla})\vec{B}$ of the magnetic field at various places within the channel, the first-ever 3D magnetic dip topology was found in the photosphere below a filament foot from observations (Fig. 2, bottom). This is consistent with early linear force-free models for filament feet (Aulanier & Démoulin 1998) and with recently recovered in MHD simulations of prominence formation by twisted magnetic flux tube emergence through the photosphere (Magara 2007).

2.2. Hα/EUV observations and MHD simulations of merging/flaring filament sections

Even from previous papers it was mentioned that filaments can only merge if their chirality is the same (see e.g. Malherbe 1989, Martin et al. 1994, Rust 2001, van Ballegooijen 2004), this condition had never been tested with dedicated observational and theoretical studies until recently.

The recent multi-wavelength analysis of three interacting filament sections F1, F2 and F3 (Fig. 3) observed during several days during a Joint Observing Programme between ground-based instruments in the Canary Islands (the SVST and the MSDP on the VTT) space-borne satellites (here we only refer to TRACE images), was the first dedicated observational study of this issue. Following their evolution over several days, it was shown that F1 and F2 gently merged into a single structure, as observed by a gradual filling in Hα of the gap R1 between both of them. This merging was associated with mild EUV brightenings and with slow Hα Doppler flows at the merging point (Schmieder et al. 2004). While EUV brightenings are a good indicator of magnetic reconnection, the flows revealed that the merging first took place by dynamic exchanges between the two progenitors, until they formed a more stable single long quiet filament. Two days later F2 and F3 produced a confined flare, manifested by the formation of new long EUV post-flare loops, as they got into contact at the point R2 (Deng et al. 2002). In order to

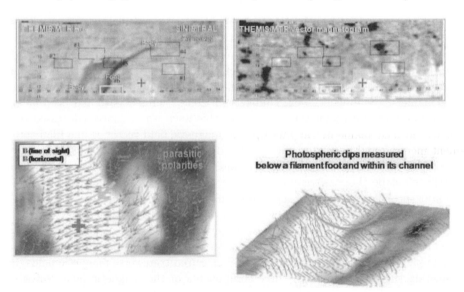

Figure 2. First-ever dentification of a magnetic dip at the footpoint of a filament foot. The vector magnetic field was measured with THEMIS/MTR and the 180° ambiguity was solved using usually observed filament chirality rules. The transverse fields which have an inverse orientation from a − toward a + polarity indicate the presence of magnetic dips above the associated inversion line. (adapted from López Ariste et al. 2006).

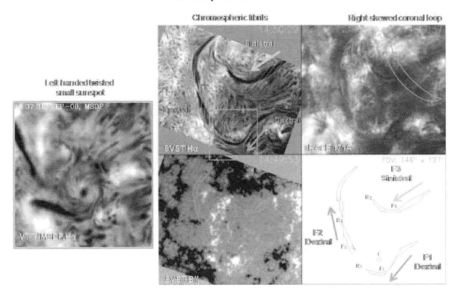

Figure 3. Identification of the chiralities of three interacting filaments F1,2,3, using various observed features. F1 and F2 were observed to merge in the R1 area, whereas a flare took place between F2 and F3 one day later, as they interacted but did not merge (adapted from Deng *et al.* 2002 and Schmieder *et al.* 2004).

address the role of helicity in these two events, Schmieder *et al.* (2004) used the chirality rules for chromospheric fibrils and magnetic field polarity, overlaying coronal arcades and handedness of neighboring sunspots, so as to predict the direction of the axial fields in the filaments. The resulting axial fields are indicated by arrows in the upper-middle panel of Fig. 3. The results of this study were numerous. It was confirmed that when two filaments interact, magnetic reconnection takes place and leads to a sudden flare (resp. a gradual merging) when their helicity signs are of the opposite (resp. the same) sign. It was also shown that magnetic helicity must slowly accumulate prior to filament merging, as seen by the rotation of a small twisted sunspot in the vicinity of the merging point. Finally, it could be deduced that in its early stages, magnetic reconnection accelerates plasma between the previously split filaments, and that it must later result in a change of topology which can sustain quiet and almost non-moving filament material all along the newly formed filament. All these put strong constraints on the MHD modeling of filament merging and flaring.

Numerical MHD simulations of the formation, interaction, and magnetic reconnection between pairs of solar filaments have then been conducted. Line-tied sub-Alfvénic shearing boundary motions were applied to adjacent and initially current-free magnetic bipoles. The simulations were performed in a low-β adiabatic regime, using $500 \times 190 \times 190$ mesh points in a non-uniformed grid, with a Flux Corrected Transport scheme that allows reconnection owing to numerical diffusion at the scale of the mesh. Four possible combinations of chiralities (identical or opposite) and axial magnetic fields (aligned or opposed) between the participating filaments were considered (DeVore *et al.* 2005). It was found that, when the topology of the global flux system comprising the prominences and arcades is bipolar, so that a single polarity inversion line is shared by the two structures, then identical chiralities necessarily imply identical magnetic helicity signs and aligned axial fields. In this case, finite-B slipping magnetic reconnection formed new field lines linking the two initial prominences (see Fig. 4, left). At early times, shear Alfvén waves

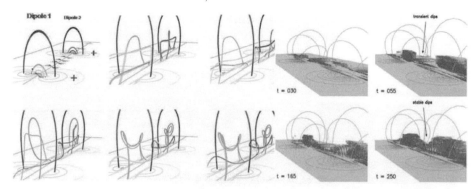

Figure 4. 3D MHD simulation of prominence merging, resulting from finite-B magnetic reconnection between two dipoles that share a common photospheric inversion line, and whose shear result in the same magnetic helicity sign. *(Left:)* Reconnecting field lines. *(Right:)* Resulting distribution of plasma-supporting magnetic dips, simulating the prominence material (adapted from DeVore *et al.* 2005 and Aulanier *et al.* 2006).

propagated through these newly reconnected field lines, which can accelerate plasma condensations from one progenitor to another. As the shear increases, a new distribution of magnetic dips formed and increasingly filled the volume between both progenitors, so that they gradually merged into a single filament. We identified the multistep mechanism, consisting of a complex coupling between photospheric shear, slipping magnetic reconnection in the corona, and formation of quasi bald patches, that is responsible for stable filament merging through dip creation (Aulanier *et al.* 2006). This first model successfully reproduced the observations of filament merging by Schmieder *et al.* (2004). The second model, which also made use of a large-scale bipolar field, but which induced opposite helicities and axial fields between the two prominences, hardly resulted in any magnetic reconnection. The resulting lack of merging is consistent with the observations of Deng *et al.* (2002), although no flare reconnection occured in the model. When the topology instead is quadrupolar, so that a second polarity inversion line crossing the first lies between the prominences, then the converse relation holds between chirality and axial-field alignment. Reconnections that form new linking field lines now occur between prominences with opposite chiralities. They also occur, but only result in footpoint exchanges, between prominences with identical chiralities. These findings do not conflict with the observational rules, since the latter have yet to be derived for non-bipolar filament interactions; they provide new predictions to be tested against future observational campaigns.

3. Flux emergence

The association between emerging flux and filament eruptions has been extensively studied by Feynman and Martin (1995). They found that in 17 out of 22 cases where newly emerging flux in the vicinity of filaments could be observed, the filament erupted, whereas in the remaining 5 cases it did not. The new flux typically started to emerge a few days before the eruption, indicating a slow evolution towards an unstable state before eruption. In 26 out of 31 cases where no emerging flux in the vicinity of the filament was detectable, the filament did not erupt within the period of observation. The authors concluded that filament eruptions are associated with newly emerging flux, but that the latter is not a necessary condition for eruption. In a more recent study, Jing *et al.* (2004) found that 54 out of 80 filament eruptions were associated with flux emergence.

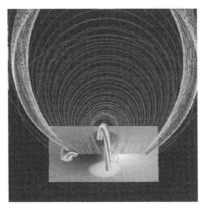

Figure 5. Snapshot from a 3D simulation of flux emergence in the vicinity of a coronal flux rope (from Török 2009). The TD flux rope is visible in the centre. To the left, a small flux rope is emerging and reconnecting with the potential field overlying the TD flux rope.

These results indicate that in about 60-70 % of all cases, newly emerging flux in the vicinity of a filament leads to its eruption. But how can the new flux drive the filament towards eruption? Magnetic reconnection seems to be the key. It was suggested that the emerging flux reconnects with the field overlying the filament and hence destabilizes it (Feynman and Martin 1995) or decreases its tension to a degree that the core flux can not longer be stabilized and erupts (Wang *et al.* 1999). However, as pointed out in Feynman and Martin (1995) and demonstrated in 2.5D numerical simulations (Chen and Shibata 2000), the magnetic orientation of the emerging flux with respect to the pre-existing coronal field must be "favorable" for reconnection to occur. In the next section, we present a recent observation of signatures of such reconnection.

Reconnection is not a necessary condition for eruptions. Eruptive prominences can result directly from instabilities (kink instability, torus instability) leading to loss of equilibrium in the relevant structures (Lin, Forbes and Isenberg 2001, Lin and Forbes 2000, Low 2001, Török and Kliem 2005). In principle disruption occurring in a magnetic structure results from the interaction between the current and the magnetic field in the relevant configuration.

3.1. *The event on 2007 April 24*

Figure 6 shows a recent example of flux emergence close to a filament channel observed by Hinode/XRT on 2007 April 24. The bright X-ray loops indicate that reconnection between the emerging flux and the magnetic field overlying the filament took place. Two main bright systems of loops are visible, one to the left of the filament channel and one arching above the filament, outlining the edge of the filament cavity. We also observed a brightening propagating along this arch in the early phase of the evolution. The filament is not destabilized by the reconnection and does not erupt in this case.

3.2. *3D MHD simulation*

In order to understand the interaction of the emerging flux with the pre-existing coronal field, and to understand why the filament did not erupt, we aimed to reproduce this event in an MHD simulation. The latter will be described in detail elsewhere (Török *et al.* 2009), here we just give a brief summary of the setup and the results.

As initial condition for the simulation we use the analytical model of a bipolar active region by Titov & Démoulin (TD) (1999). The model consists of a force-free, line-tied and twisted coronal flux rope embedded in a potential field arcade. Numerical simulations

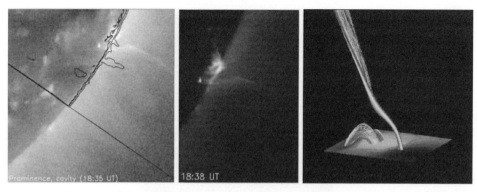

Figure 6. *Left:* Hinode/XRT image (512 x 512 arc sec) overlaid by contours of the H$_\alpha$ prominence observed on 2007 April 25 at 13:38 UT by MSDP. A large cavity is seen above the prominence. *Centre:* The flaring X-ray loops. *Right:* The region of flux emergence in the numerical simulation described in the text. The TD flux rope is located to the right of the zoomed region. Selected newly reconnected magnetic field lines are shown. The field lines qualitatively resemble the shape of the two main X-ray loops visible in the central panel.

(Török *et al.* 2004, Török and Kliem 2007) have demonstrated that the TD flux rope can be subject to the helical kink instability and the torus instability (Kliem and Török 2006). The morphological and kinematic evolution of an erupting filament could be successfully reproduced in another simulation (Török and Kliem 2005, Schrijver *et al.* 2008). For our simulation, we choose the parameters of the TD model such that the flux rope is initially stable with respect to both instabilities.

We then mimic the emergence of new magnetic flux by successively changing the boundary conditions at the bottom plane of the numerical domain (the "photosphere") such that the slow and rigid emergence of another, smaller and uniformly twisted flux rope in the vicinity of the pre-existing rope is modeled (see Figure 5 for a simulation similar to the one described here; see also Fan and Gibson 2004).

As the new flux rope slowly emerges, a magnetic null point is gradually formed slightly above it, within the current sheet which forms at the interface between the two flux ropes, and the emerging rope starts to reconnect with the potential field arcade overlying the TD rope. The amount of flux reconnected depends on the strength and orientation of the magnetic field within the emerging rope with respect to the pre-existing coronal field. We choose the orientation of the rope's field such that it is favorable for reconnection. As the reconnection proceeds, new connectivities are formed (Figure 6, right): field lines rooted in the positive (white) polarity of the emerging flux rope now close down in neighboring regions of negative polarity (black) of the TD model and form low-lying arcades, whereas field lines starting from the negative polarity of the emerging rope reconnect with field lines initially overlying the TD rope. The latter exhibit a kinked shape, just as the bright X-ray loop overlying the filament. This kinked shape corresponds to the field geometry around the 3D null point, which has a typical fan surface and spine field line. Although our simulation does not treat the thermodynamics, it is legitimate to qualitatively compare our simulation with the XRT observations, since newly reconnected field lines are expected to be heated and to brighten in X-ray images. The shapes of the newly reconnected field lines are qualitatively similar to the shape of the bright X-ray features, indicating that the simulation, and its resulting topology, reproduces the magnetic interaction between emerging and pre-existing flux reasonably well.

As the filament in reality, the TD flux rope does not erupt in our simulation. As mentioned earlier, observations indeed show that emerging flux in the vicinity of filaments does not necessarily lead to an eruption. The question arises under which circumstances flux emergence can trigger filament eruptions and CMEs. We think the main factor which decides on the occurrence of an eruption triggered by flux emergence in the vicinity of filaments is how "far" the pre-eruptive coronal configuration is from an unstable state at the time when the reconnection between the emerging flux and the arcade field above the core flux carrying the filament sets in. The reconnection must sufficiently weaken the tension of the overlying field for the core flux to erupt. How effective the reconnection is in this respect will depend on many factors, as for example the field orientation and the spatial distance of the emerging from the core flux. In cases where the core flux is already close to instability, a small amount of new flux emergence might be sufficient to trigger its eruption, whereas in other cases even a large amount of emerging flux might not be able to drive the system towards an unstable state. It seems that the event described above belongs to the latter category. We note that in cases where the new flux emerges just below the filament, it might reconnect directly with the core flux. This could increase the twist of the core flux such that it erupts even if the overlying filed is not weakened significantly. Similar conclusions have been drawn in earlier observational, analytical and numerical studies on the relation between emerging flux and solar eruptions. For a more detailed discussion, we refer the reader to the corresponding papers (e.g. Wang *et al.* 1999, Notoya et al 2007, Chen and Shibata 2000, Lin *et al.* 2001).

4. Summary

Making use of observational chirality rules and theoretical magnetic modeling for filament magnetic fields, we have obtained several important breakthroughs which can now be re-used in future studies. First, a chirality-based disambiguation technique in a THEMIS vector magnetogram of a filament channel have been applied (López Ariste *et al.* 2006); this led to the first observational evidence of magnetic support in filament feet. Second we have analyzed with unprecedented accuracy the interaction between three distinct filaments, for which we identified the sign of magnetic helicity in several independent ways; this led to the observational conjecture that filament mergers must exhibit the same sign of helicity, otherwise their interaction will result in a confined flare (Schmieder *et al.* 2006). Third a parametric study of sheared dipoles with 3D MHD calculations, varying their initial orientation and the direction of their associated shearing motions confirmed the observational conjecture, and extended its results to potential mergers with opposite helicity signs if their progenitor bipoles are oppositely oriented (Aulanier *et al.* 2006). Finally we have presented a new 3D simulation of emerging flux close to a coronal flux rope and discussed that both the "distance" of the flux rope from instability and the "effectivity" of reconnection in driving the rope towards instability will determine whether filament eruptions can be triggered by flux emerging in their vicinity.

Acknowledgements

Financial support by the European Commission through the SOLAIRE network (MTRM-CT-2006-035484) is gratefully acknowledged.

References

Amari, T., Luciani, J. F., Mikic, Z., & Linker, J. 1999, *ApJ*, 518, L60
Antiochos, S. K., DeVore, C. R., & Klimchuk, J. A. 1999, *ApJ*, 510, 495
Aulanier G. & Démoulin, P. 1998, A&A, 329, 1125

Aulanier, G., Srivastava, N., & Martin, S. F. 2000, *ApJ*, 543, 447

Aulanier, G. & Schmieder, B. 2002, *A&A*, 386, 1106

Aulanier, G., DeVore, C. R., & Antiochos, S. K. 2006, *ApJ*, 646, 1349

Bobra, M., van Ballegooijen, A. A., & DeLuca, E. E. 2008, *ApJ*, 672, 1209

Casini, R., López Ariste, A., Tomczyk, S., & Lites, B. W. 2003, *ApJ*, 589, L67

Chen, P. F. & Shibata, K. 2000, *ApJ*, 545, 524

Deng, Y., Schmieder, B., & Engvold, O. 2002, *Sol. Phys.*, 209, 153

DeVore, C. R., Antiochos, S. K., & Aulanier, G. 2005, *ApJ*, 629, 1122

Dudik, J., Aulanier, G., Schmieder, B., Bommier, V., & Roudier, T. 2008, *Sol. Phys.*, 248, 29

Fan, Y. & Gibson, S. E. 2004, *ApJ*, 609, 1123

Feynman, J. & Martin, S. F. 1995, *J. Geophys. Res.*, 100, 3355

Jing, J., Yurchyshyn, V. B., Yang, G., Xu, Y., & Wang, H. 2004, *ApJ*, 614, 1054

Kliem, B. & Török, T. 2006, *Phys. Rev. Lett.*, 96, 255002

Lin, J. 2004, *Solar Physics*, 219, 169

Lin, J. & Forbes, T. G. 2000, *J. Geophys. Res.*, 105, 2375

Lin, J., Forbes, T. G., & Isenberg, P. A. 2001, *J. Geophys. Res.*, 106, 25053

Lin, Y. L., Wiik, J. E., & Engvold, O. 2003, *Solar Phys.*, 216, 109

Lin, Y. L., Wiik, J. E., Engvold, O., et al., 2005, *Solar Phys.*, 227, 283

Low, B. C. 2001, *J. Geophys. Res.* 106, 25141

López Ariste, A., Aulanier, G., Schmieder, B., & Sainz Dalda, A. 2006, *A&A*, 456, 725

Lionello, R., Mikic, Z., Linker, J., & Amari, T. 2002, *ApJ*, 581, 718

Malherbe, J.-M. 1989, in Dynamics and Structure of Quiescent Solar Prominences, Kluwer Ac. Pub., 115

Magara, T. 2007, *PASJ*, 59, L51

Martin, S. F., Bilimoria, N., & Tracadas, P. W. 1994, in Solar Surface Magnetism, Kluwer Ac. Pub., 303

Martin, S. F. 1998, *Sol. Phys.*, 182, 107

Martin, S. F., Lin, Y., & Engvlod, O. 2008, *Sol. Phys.*, 250, 31

Notoya, S. et al., 2007, *ASP Conf. Series*, 369, 381

Okamoto, T. J., Tsuneta, S., Lites, B., et al., 2008, *ApJ*, 673, L215

Rust, D. M. 2001, in Encyclopedia of Astronomy and Astrophysics, http://eaa.iop.org

Schmieder, B., Mein, N., Deng, Y., et al., 2004, *Sol. Phys.*, 223, 119

Schrijver, C. J., Elmore, C., Kliem, B., Török, T., & Title, A. M. 2008, *ApJ*, 674, 586

Titov, V. S. & Démoulin, P. 1999, *A&A*, 351, 707

Török, T. 2009, *in preparation*

Török, T., Kliem, B., & Titov, V. S. 2004, *A&A*, 406, 1043

Török, T. & Kliem, B. 2007, *Astronomische Nachrichten*, 328, 743

Török, T., Schmieder, B., & Aulanier, G. 2009, *in preparation*

van Ballegooijen, A. A. 2004, *ApJ*, 615, 519

Discussion

NINDOS: I have observed a couple of cases where close to a filament, reconnection-favored bipoles emerge. But the filaments didn't erupt. Just for your information.

SCHMIDER: Thank you for your remark. With simulation we can study the conditions of eruptions using a large space of parameters.

Universal Heliophysical Processes
Proceedings IAU Symposium No. 257, 2008
N. Gopalswamy & D.F. Webb, eds.

© 2009 International Astronomical Union
doi:10.1017/S1743921309029342

Statistical relationship between solar flares and coronal mass ejections

Seiji Yashiro[1,2,3] and Nat Gopalswamy[2]

[1]Interferometrics Inc., Herndon, Virginia 20171, USA

[2]NASA Goddard Space Flight Center, Greenbelt, Maryland 20771, USA

[3]The Catholic University of America, Washington, DC 20771, USA
email: Seiji.Yashiro@nasa.gov, Nat.Gopalswamy@nasa.gov

Abstract. We report on the statistical relationships between solar flares and coronal mass ejections (CMEs) observed during 1996-2007 inclusively. We used soft X-ray flares observed by the Geostationary Operational Environmental Satellite (GOES) and CMEs observed by the Large Angle and Spectrometric Coronagraph (LASCO) on board the Solar and Heliospheric Observatory (SOHO) mission. Main results are (1) the CME association rate increases with flare's peak flux, fluence, and duration, (2) the difference between flare and CME onsets shows a Gaussian distribution with the standard deviation $\sigma = 17$ min ($\sigma = 15$ min) for the first (second) order extrapolated CME onset, (3) the most frequent flare site is under the center of the CME span, not near one leg (outer edge) of the CMEs, (4) a good correlation was found between the flare fluence versus the CME kinetic energy. Implications for flare-CME models are discussed.

Keywords. Sun: flares, Sun: coronal mass ejections (CMEs)

1. Introduction

A solar flare is an explosion in the solar atmosphere observed as a sudden flash of electromagnetic radiation emitted from the heated plasma. A coronal mass ejection (CME) is an eruption of magnetized plasma from the Sun into the interplanetary medium. After the discovery of CMEs in 1971, the relationship between them has been examined extensively (see Kahler 1992 for review). Munro *et al.* (1979) reported that ∼40% of CMEs were associated with Hα flares. Andrews (2003) reported that ∼60% of M-class flares were were associated with CMEs. There is no one-to-one flare-CME occurrence (Harrison 1995). However, the similarity between the derivative of the X-ray light curve and CME accelertion profile (Zhang *et al.* 2001, Vršnak *et al.* 2004) suggest a close relation. Both phenomena are thought to be different manifestations of the same process (Harrison 1995, Jing *et al.* 2005).

The Large Angle and Spectrometric Coronagraph (LASCO; Brueckner *et al.* 1995) on board the Solar and Heliospheric Observatory (SOHO) mission has observed more than 13,000 CMEs from 1996, providing a great opportunity to investigate the flare-CME relationship. In this paper we report the statistical relationship between flares and CMEs.

2. Flare-CME association

Yashiro *et al.* (2006) examined the CME association rate as a function of the flare parameters. We repeated that study to include the flare-CME pairs in 2006 and 2007 and restrict the study to limb events. Solar flares have been routinely detected by the

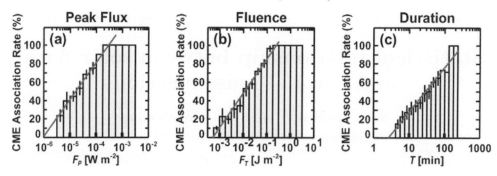

Figure 1. CME association rate as a function of (a) X-ray peak flux F_P, (b) fluence (total flux) F_T and (c) duration T. The plots are similar to Fig. 1 of Yashiro *et al.* 2006, but different data sets were used. See text for details.

X-ray Sensor (XRS) on board the Geostationary Operational Environmental Satellite (GOES) in the two wavelength bands (0.5-4 Å and 1-8 Å). The basic flare parameters are available in the Solar Geophysical Data (SGD) and online Solar Event Reports provided by NOAA. We excluded flares corresponding to SOHO/LASCO downtimes by requiring that at least two LASCO C2 images were obtained between 0 - 2 hours after the flare onset. Yashiro *et al.* (2005) reported that about half of disk CMEs associated with C-class flares and ~16% of disk CMEs associated with M-class flares were invisible to LASCO. In order to examine the flare-CME relationship properly, we used only limbward events with longitudes > 45°. We also eliminated flares at longitudes > 85° because of the possible partial occultation of the X-ray source, resulting in an underestimate of the X-ray flux. There are 6374 flares (above C3 level) listed in SGD, but the locations are not listed for ~2000 of them. For the X- and M-class flares, we identified their locations using solar disk images obtained in X-ray, EUV, Hα, and microwave. For C-class flares, we used only those flares with their locations listed in SGD.

We used the SOHO/LASCO CME Catalog (Yashiro *et al.*, 2004) to find the flare-associated CMEs within a 3-hour time window. However, because the time window analysis by itself could produce false flare-CME pairs, we checked the consistency of the associations by viewing both flare and CME movies available in the Catalog. Eruptive surface signatures, such as filament eruptions and coronal dimmings, helped ascertain the associations. However, in some cases, we could not determine with confidence whether the association was true or not because not all flares had clear eruptive signatures. We abandoned to give a clear true or false answer of the CME association of such events, and left them as ambiguous associations. This way, we classified all the flares into three categories: flares with definite CME association, flares with uncertain CME association, and flares that definitely lacked CMEs.

Figure 1 shows the CME association rate as a function of X-ray peak flux (a), fluence (b), and duration (c). The CME association rate has an error range obtained from the uncertain flare-CME pairs. Assuming that all of the uncertain events were false, the lower limit of the CME association was determined by dividing the number of definitive events by the total number of flares. Similarly we obtained the upper limit by assuming that all uncertain events were true. We used the middle of the lower and upper limits as the representative association rate. This is equivalent to assuming that half of the uncertain events had true CME association.

It is clear that the CME association rate of X-ray flares increased with their peak flux, fluence, and duration. The larger flares tend to be associated with the CMEs, but for even X-class flares about 10% of them lacked CME association (Gopalswamy *et al.*, this volume). Figure 1c is consistent with the well known fact that long duration (or decay) events (LDEs) are well associated with the CMEs, while impulsive flares are not (Sheeley *et al.* 1983). However, even some impulsive flares with duration < 10 min., about 20% of them, have associated CMEs (Kahler *et al.* 1989). There is no critical duration dividing flares into those with and without CMEs.

3. Temporal relationship

The onset time is the most basic parameter to investigate the relation between flares and CMEs. Onset times of CMEs are difficult to identify since they originate below the LASCO C2 field of view (FOV). Exceptions are the CMEs observed by LASCO C1 (e.g. Gopalswamy & Thompson 2000; Zhang *et al.* 2001). These studies showed that the onsets of the flares and CMEs are tightly connected. However, its data are not available for the most of the CMEs, so the CME onset is either estimated from CME-related surface activities (e.g., filament eruptions) or is extrapolated to the solar surface from the CME trajectory in the LASCO C2 and C3 FOVs. We attempt to examine the temporal relationship using the extrapolated CME onset.

The extrapolated CME onset has been used from pre-SOHO era. The extrapolation method and its weakness were described in many literatures (e.g., Harrison & Sime 1989, Harrison *et al.* 1990). We need to make at least two assumptions to obtain the extrapolated onsets. The first is the shape of CME trajectory under the occulting disk. We assume that the unseen trajectory is the same as observed in the LASCO FOVs. In the case of CMEs with constant speed or deceleration motion, the above assumption is apparently incorrect since all CMEs must have an acceleration phase during their initiation. Ignoring the acceleration phase, the extrapolated onsets would be later than the actual onsets. The second necessary assumption is regarding the position where the CME is initiated. We assume that the apex of the CMEs starts from 1 Rs (solar limb). Flare position is a good proxy, but the spatial size of CMEs at the beginning would be larger than that of flares. Ignoring the spatial size of the CMEs, for the limb events, the extrapolated onsets would be earlier than the actual CME onsets.

Figure 2a illustrates how we extrapolate a CME trajectory to obtain the onset. The height-time data are fitted by first-order (dotted line) or second-order (dashed line) polynomials, which correspond to constant speed and constant acceleration (or deceleration), respectively. The extrapolated CME onsets are determined as the time when the fitting lines cross the 1 Rs (horizontal line). T_1 and T_2, respectively indicate the onset time determined from first- and second-order polynomials. The linear lines invariably cross 1 Rs, but the parabolic fits do not necessarily do so (Fig. 2b). Figure 2b is an example showing the parabola does not cross 1 Rs line.

The extrapolated CME onsets may not be accurate due to errors resulting from the assumptions stated above. A good example is the flare-CME event on 2002 April 21 (Fig. 2c). An X1.5 flare started at 00:43 UT and then peaked at 01:51 UT. The flare-associated CME was observed by LASCO at 01:27 UT with a speed of 2393 km/s. The extrapolated CME onset is 01:16 UT for the both linear and parabolic fits (T_1 and T_2). The X1.5 flare started 33 minutes earlier than the associated CME. However, a detailed analysis with TRACE and UVCS observations showed that the eruptive feature associated with the CME started at 00:48 UT at the latest. The difference with the extrapolated CME onset (28 min) should be considered as an error resulting from the assumption that the CME trajectory under the occulting disk is the same as the observed one in the LASCO FOV. Another example is a flare-CME event on 1998 January 25

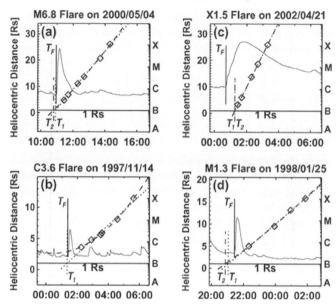

Figure 2. CME Height-time diagram superposed with X-ray light curve to illustrate how we estimate the CME onset. Solid curves are X-ray intensity obtained by GOES satellite (right scale), and T_F indicates flare onset recorded in NOAA/SGD. Diamonds shows heliocentric distance of the CMEs, which are recorded in the CME Catalog. Dotted lines and dashed curves are first- and second-order polynomial fitting, respectively.

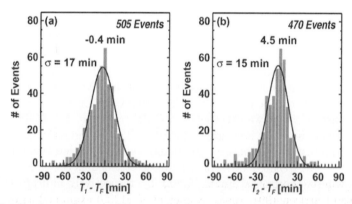

Figure 3. Distributions of onset difference between flares and CMEs. The CME onsets are estimated by (a) first-order and (b) second-order polynomial fitting to the CME height-time data.

(Fig. 2d). An M1.3 flare started at 21:26 UT and the associated CME appeared in C2 FOV at 22:19 UT. In this case, the CME height-time measurements missed the deceleration motion during the CME initiation.

Figure 3 shows the distributions of the difference between the flare and CME onsets. We used only definitive flare-CME pairs for this analysis. The CME onsets (T_1 and T_2) are estimated using linear and parabolic fitting, respectively, as described above. Both distributions are well represented by Gaussians. The parabolic fit must return a better result to the CME trajectory because it has a higher degrees of freedom. The standard deviation is 17 min for linear and 15 min for parabolic fits. Gopalwamy *et al.*

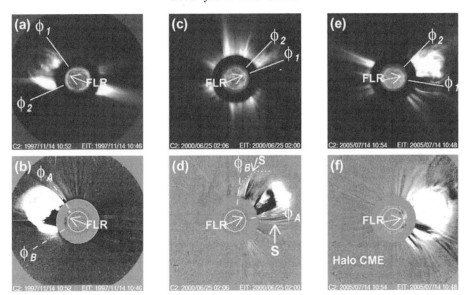

Figure 4. Three CMEs observed by SOHO LASCO to illustrate the measurement of CME span. The top row shows direct images used to measure the main CME body, and the bottom row shows corresponding running difference images used to measure the whole CME. The side edges of the main CME body (the whole CME) are denoted by ϕ_1 and ϕ_2 (ϕ_A and ϕ_B). Arrows point to the position of the flares associated with the CMEs (from Yashiro *et al.* 2008).

(2003) examined the onset times of the prominence eruptions and CMEs and found similar distribution. Since the prominence eruptions and CMEs are parts of the same phenomenon, their onsets should be the same. Therefore it is reaasonable to consider that the differnce between the flare and extrapolated CME onsets results not from the nature of flare-CME relation but from the error of the extrapolated CME onset.

4. Spatial relationship

The spatial relationship between flares and CMEs has been investigated from the 1970s and 1980s with the CME observations obtained by the Solwind and the Solar Maximum Mission (SMM) coronagraphs. Harrison (1986) analyzed 48 flare-CME events observed by SMM and Solwind and reported that many flares occurred near one leg of the associated CMEs (see also Harrison 1991, 1995). Kahler *et al.* (1989) examined 35 events observed by Solwind and reported that flare positions did not peak either at the center or at one leg of the CMEs. However recently Yashiro *et al.* (2008) examined 498 events observed by LASCO from 1996 to 2005 and found that the most probable flare site is at the center of the CME span. Here we extended that study to include the flare-CME pairs in 2006 and 2007.

First we briefly describe the analysis of Yashiro *et al.* For each CME, the angular widths of two structures were measured. One is the main CME body which corresponds to the three-part structure of a CME. The other is the whole CME. Some CMEs possess a faint envelope outside of the main CME body. The envelope structure could be a shock driven by the CME hence not a part of the CME. However, since there is no established way to identify a shock by coronagraph observation itself, we have included the envelope structures as a part of the CME and refer to all the CME features as the whole CME. Figure 4 illustrates how we measured the PAs of the main CME body and the whole

CME. The CME on 1997 November 14 (Figs. 4a and 4b) did not have an envelope, thus the ϕ_1 (ϕ_2) and ϕ_A (ϕ_B) are identical. The angular width is defined as the difference of the two outer edges ($\phi_2 - \phi_1$ for the main CME body and $\phi_B - \phi_A$ for the whole CME). Note that the angular width of a CME would be affected by projection effects if the CME originated away from the solar limb. The measrued CME widths are systematically larger than their true values (See Burkepile *et al.* 2004). The CME on 2000 June 25 had a faint envelope to the north of the main CME body (Figs. 4c and 4d). The northern edge of the envelope denoted by ϕ_B is used for the edge of the whole CME. Since we cannot see an envelope to the south of the CME, the southern edges of the main CME body and whole CME are almost identical. The CME on 2005 July 14 appeared in the C2 FOV at 10:54 UT (Figs. 4e and 4f). The CME had a clear three-part structure with a faint envelope. The envelope covered the occulting disk at 11:54 UT, thus the CME is listed as a halo (Howard *et al.* 1982) in the CME catalog. In this case ϕ_A and ϕ_B cannot be determined.

The distributions of differences between flare PAs and CME CPAs for the main CME body and for the whole CME are shown in Figures 5a and 5e, respectively. Both the distributions are very similar and are well represented by Gaussians. The standard deviation is 17.2° for the difference between the flares and main CME bodies and 17.5° for the difference between the flares and whole CMEs. We separated the events into three groups according to their flare intensity and made the same plots for each group. The second, third, and forth rows in Figure 5 correspond to the events with X-class, M-class, and C-class flares, respectively. The standard deviation is shown in each plot, which ranges from 17.1°-17.7° for the main CME body and 16.8°-21.1° for the whole CME.

In order to investigate the flare position with respect to the main CME body (frontal structure), we normalize the PA differences by the half angular span and show its distribution in Figure 5i. It is clear that most of the flares are located under the span of the main CME body. Figure 5m is the same as the Figure 5i, but for flare locations with respect to the edges of the whole CMEs. Both the distributions are well represented by Gaussians with standard deviations of 0.59 for the main CME body and 0.38 for the whole CME. In both the cases, the peak of the Gaussian is around zero, meaning that the flares frequently occur under the center of the CME span, not near one leg (outer edge) of the CMEs. Then we separated the events into three groups according to their flare intensity. We found that all distributions have a peak around zero, while the width of the distributions is different for different flare levels. The flare-CME events with X-class flares (hereafter X-class events) have a narrower distribution suggesting that many X-class flares lie under the center of the CME span. On the other hand, the C-class events have a broader distribution and a significant number of events occurred outside of the CME span.

5. Energy relationship

Figures 6 and 7 show scatter plots of logarithms of flare and CME parameters. The flare parameters in the X-axis are peak flux (left), fluence (center), and duration (right). The CME parameters in the Y-axis are speed (Fig. 6; top), main body width (Fig. 6; middle), whole CME width (Fig. 6; bottom), mass (Fig. 7; top), and kinetic energy (Fig. 7; bottom). The definition of the main body and whole CME is described in Section 4. The correlation coefficients of the logarithms of these parameters are shown on the plots.

The familiar parameters representing flare and CME properties are peak soft X-ray flux and speed, respectively. The former has been used to determine the rank of solar flares. The scatter plot of these parameters shows that a correlation exists between them,

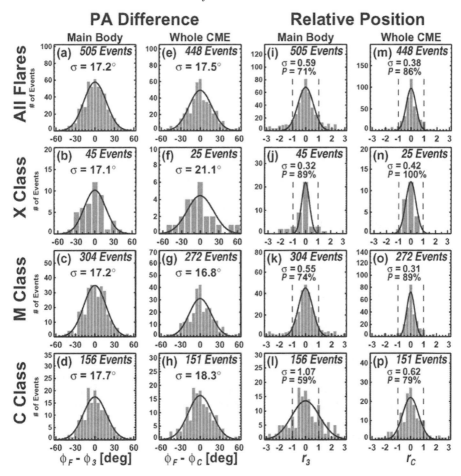

Figure 5. Distribution of flare positions with respect to the CPA of the CME. The first and second colums show the difference of the flare PAs (ϕ_F) and CME CPAs [$\phi_3 = 0.5(\phi_1 + \phi_2)$; $\phi_C = 0.5(\phi_A + \phi_B)$]. The standard deviation obtained by the Gaussian fit is shown in each plot. The third and fourth columns show the distributions of PA difference normalized by the half-CME span [$r_3 = (\phi_F - \phi_3)/0.5(\phi_2 - \phi_1)$; $r_C = (\phi_F - \phi_C)/0.5(\phi_B - \phi_A)$]. The vertical dashed lines mark two side edges of the CMEs. P is the percentage of flares lying inside of the CME span. The second, third, and fourth rows correspond to the event with X-class, M-class, and C-class flares, respectively (from Yashiro *et al.* 2008; Extended data set).

i.e. larger flares are associated with faster CMEs, but the correlation coefficient is not high (0.50). The scatter is wider for weaker flares. For C-class flares, the speed of the associated CMEs ranges from 146 - 1378 km/s (an order of magnitude), while 882 - 3387 km/s (approximately a half order of magnitude) for the largest flares (>X3 level). A better correlation (0.56) is found in the flare fluence (total flux) versus CME speed. Moon *et al.* (2002) reported a correlation coefficient of 0.47 between flare fluence and CME speed observed by GOES and SOHO/LASCO. The coefficient is lower than ours (0.56). Since their data period is from 1996 to 2000, their sample did not have many extreme events. (Out of 13 huge flares (> X3.0 level) in solar cycle 23, ten occurred after Jan 2001). The correlation between flare duration and CME speed is poor (r=0.23). This very weak correlation could be the result of a chain of correlations, since the duration is correlated with the fluence (r=0.68; Veronig *et al.* 2002).

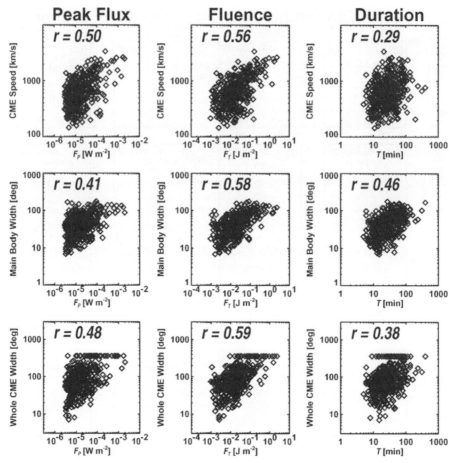

Figure 6. Scatter plots between flare and CME parameters. The X axis is for flare parameters, peak flux (left), fluence (center), and duration (right) and the Y axis is for CME parameters, speed (top), main body width (middle), and whole CME width (bottom). See Figure 4 for the definition of the main body width and whole CME width. The correlation coefficient is shown on the plot.

Since the determination of CME widths is controversial (see Section 4), for each CME we measured angular width of two structures: main body and whole CME. The main body corresponds to the CME three-part structure. The whole CME includes a faint envelope occasionally seen outside of the main body. The evelope could be a shock driven by the CME. If this is the case, the envelope is not a part of the CME. Although we used both widths to investigate the correlation with flare parameters, the scatter plots do not show significant differences between them. Similar to the CME speed, the best correlation is found in the flare fluence versus the CME width. Kahler *et al.* (1989) reported that the correlation coefficient between the CME width and the logarithms of flare duration is 0.67, which is higher than ours (0.46). We should note here that the definitions of the flare duration are different. Kahler *et al.* determined the end of the flare as the time when the X-ray flux returns to the GOES C2 level. Our flare end corresponds to the time when the soft X-ray flux decays to a point halfway between the peak flux and the pre-flare background level. The different definition would explain the different correlation coefficients.

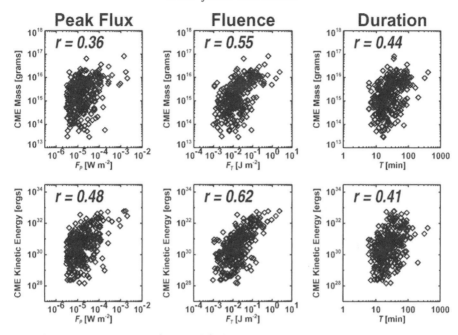

Figure 7. Scatter plots between flare and CME parameters. The X axis is for flare parameters, peak flux (left), fluence (center), and duration (right) and the Y axis is for CME parameters, mass (top) and kinetic energy (bottom).

Hundhausen (1997) reported a correlation between the flare peak flux and CME kinetic energy for 249 flare-CME pairs observed by GOES and SMM. The correlation coefficient of logarithms of these parameters is 0.53. By eliminating the disk events to avoid projection effects, Burkepile *et al.* (2004) obtained a better correlation of 0.74. The former is comparable to ours (0.48) but the latter is significantly higher. For the C-class flares ($10^{-6} \leqslant F_P < 10^{-5} W\ m^{-2}$), the kinetic energy of the associated CMEs ranges over nearly 4 orders of magnitude ($10^{28} - 10^{32}$ ergs) in our plot, while only over 2 orders of magnitude ($10^{29.5} - 10^{31.5}$ ergs) in Burkepile *et al.* plot. It is possible that the higher correlation coefficient obtained by Burkepile *et al.* is due to a smaller number of events (24).

6. Discussion & summary

CME observations by SOHO LASCO over the whole solar cycle enabled us to perform an extensive statistical analysis of flare-CME relationships. We examined the CME associations of flares from 1996 to 2007 and found that the CME association rate clearly increases with the flare's peak flux, fluence, and duration.

Case studies on the flare-CME initiation revealed that the onsets of the flares and CMEs are tightly connected (Zhang *et al.* 2001, Vršnak *et al.* 2004). We attempted to confirm this for a large number of events, but the extrapolated CME onsets are not accurate enough to examine the initiation process. The difference between the flare and CME onsets shows symmetrical Gaussian distributions with standard deviations $\sigma = 17$ min ($\sigma = 15$ min) for the first (second) order extrapolated onset. We conclude that the difference between the flare and extrapolated CME onsets results from the error of the extrapolated CME onset.

Pre-SOHO studies reported that many flares occurred near one leg of the associated CMEs (Harrison *et al.* 1986), or there is no preferred site with respect to the CME (Kahler *et al.* 1989). Yashiro *et al.* (2008) examined the spatial relationship between the flares and CMEs and found that the most probable site is at the center of the CME span. The result is suitable for flare-CME models typified by the CSHKP (Carmichael-Sturrock-Hirayama-Kopp-Pneuman) reconnection model. They also reanalyzed the pre-SOHO data and concluded that the long-term LASCO observation enabled us to obtain the detailed spatial relation between flares and CMEs.

The highest correlation coefficient was found in the scatter plot between the flare fluence and CME kinetic energy (r=0.62). Note that the fluence of a flare, which is obtained by integrating the X-ray flux from its start to end, have been used as a proxy of the flare radiation energy. Their energetic connection is tighter for larger events. For a given C-class flare, the kinetic energy of assocaited CME ranges over 3 orders of magnitude, while under 2 orders of magnitude for a given X-class flare. This would be related to our finding on the spatial relationship: many X-class flares often lie at the center of the associated CME, while C-class flares widely spread to the outside of the CME span. The energy partition between flares and CMEs could be determined by their magnetic configration. The magnetic configuration of the C-class flare-CME events could have large variations, while that of the X-class events might be uniform, e.g. the CSHKP type configuration.

Acknowledgements

SOHO is a project of international cooperation between ESA and NASA. The LASCO data used here are produced by a consortium of the Naval Research Laboratory (USA), Max-Planck-Institut fuer Aeronomie (Germany), Laboratoire d'Astronomie (France), and the University of Birmingham (UK). Part of this effort was supported by NASA (NNX08AD60A).

References

Andrews, M. D. 2003, *Solar Phys.*, 218, 261
Brueckner, G. E., *et al.* 1995, *Solar Phys.*, 162, 357
Burkepile, J. T., *et al.* 2004, *J. of Geophys. Res.*, 109, A03103
Jing, J., *et al.* 2005, *Astrophys. J.*, 620, 1085
Gopalswamy, N. & Thompson, B. J. 2000, *JASTP*, 62, 1457
Gopalswamy, N., *et al.* 2003, *Astrophys. J.*, 586, 562
Harrison, R. A. 1986, *Astron. Astrophys.*, 162, 283
Harrison, R. A. 1995, *Astron. Astrophys.*, 304, 585
Harrison, R. A. & Sime, D. G. 1989, *Astron. Astrophys.*, 208, 274.
Harrison, R. A., *et al.* 1990, *J. of Geophys. Res.*, 95, 917
Howard, R. A., *et al.* 1982, *Astrophys. J.*, 263, L101
Hundhausen, A. J. 1997, AGU Monograph 99, 1
Kahler, S. W. 1992, *ARAA*, 30, 113
Kahler, S. W., *et al.* 1989, *Astrophys. J.*, 344, 1026
Moon, Y.-J., *et al.* 2002, *Astrophys. J.*, 581, 694
Munro, R. H. *et al.* 1979, *Solar Phys.*, 61, 201
Veronig, A., *et al.* 2002, *Astron. Astrophys.*, 382, 1070
Vršnak, *et al.* 2004, *Solar Phys.*, 225, 355
Yashiro, S., *et al.* 2004, *J. of Geophys. Res.*, 109, 7105
Yashiro, S., *et al.* 2005, *J. of Geophys. Res.*, 110, A12S05
Yashiro, S., *et al.* 2006, *Astrophys. J.*, 650, L143
Yashiro, S., *et al.* 2008, *Astrophys. J.*, 673, 1174
Zhang, J., *et al.* 2001, *Astrophys. J.*, 559, 452

Discussion

WEBB: One can use LASCO halo CMEs to study the longitude distribution of flare sources. I have done this for some LASCO halos and found a distribution like you did.

YASHIRO: We have also analyzed the flare location distribution in the longitudinal direction using halo CMEs. The CME launch angle and width were estimated from a cone model. The preliminary analysis shows that may flares located at the center of the CME span, so the result is consistent with yours.

HOWARD: The distribution of "X" class flare site vs CME center looks to be more skewed than symmetric about "0" or co-located. The better correspondence of "X" vs "C" class flares could be a manifestation of the "big flare syndrome" of Gosling in that big events arise from a common instability.

YASHIRO: Not only in spatial relationship but also in energetic relationship, larger events show better correspondence between flares and CMEs. This might mean that the large events occur only in the CSHKP type magnetic configuration, while the small events occur in various configurations.

VRŠNAK: Spatial scatter of C-flares might be due to those events where flare only triggers the eruption of a meta-stable coronal structure (i.e. can be anywhere in the vicinity of the eruptive structure).

YASHIRO: I will check the largely-scattered C-class events with the point of view that you suggested.

MANDRINI: On the disk EUV images slow dimming regions, separated by a large distance. Depending on the location of the active region and orientation of the neutral line with respect to the limb, one can see that the flare may be at different distances from the two dimming regions. This may be responsible for the observed scatter.

YASHIRO: Yes, EIT dimming extent might be used for the proxy of the CME span. It will be nice to examine the flare position with respect to the EIT dimming extent.

Universal Heliophysical Processes
Proceedings IAU Symposium No. 257, 2008
N. Gopalswamy & D.F. Webb, eds.

ⓒ 2009 International Astronomical Union
doi:10.1017/S1743921309029354

Coronal waves in coronal loops during non-flare stage

Dmitry Prosovetsky

The Institute of solar-terrestrial physics, Lermontov st., 126a,
p/o box 291, Irkutsk, 664033, Russia
email: proso@iszf.irk.ru

Abstract. Using SOHO/EIT Fe XII λ195Å observations the new type of oscillations in coronal loops was detected. The oscillation corresponds to wave propagated to outer area of atmosphere of active area. As opposed to most kind of oscillations associated with coronal loops the waves are observed at non-flare stage of active areas evolution. Velocities of the wave propagation were 8-20 km s^{-1} and had quasi-perpendicular direction with magnetic field. Such waves were detected in active areas located on solar disk and loops structures outside solar limb. Investigation of EIT data shows the waves are not result of changes of topology of a magnetic field and loops configuration. The nature and probable sources of waves are discussed.

Keywords. Sun: activity, UV radiation, oscillations

1. Introduction

The oscillations of coronal structures are typical process in solar corona. This phenomenon is observed in wide range of time and spatial scales, any topology of magnetic field. Oscillation is special interesting in solar active regions. There oscillation can be expressed as source of released energy or as their response. In the atmosphere of active regions the oscillations correspond with process of propagation, redistribution and release of energy.

Observations of oscillations on higher atmospheric levels are hard in view of weak emission. Observations become too easy during and after flare events when atmosphere emission increased. For example De Moortel & Brady (2007) detected the higher harmonic loop oscillation after powerful flare. Raouafi *et al.* (2004) investigated shock wave associated with expanding loop system. Such large-scale events as EIT wave connect to flares or CME (e.g. Chertok & Grechnev (2003), Biesecker *et al.* (2002), Wills-Davey *et al.* (2007)).

Quiet events are more difficult for observation. Special observation programs and data processing methods are necessary for their detection. Special program of TRACE observation made possible by De Groof *et al.* (2004) to detect and investigate propagating disturbances in loop. Tomczyk *et al.* (2007) reported about very interesting observation of oscillation of velocities field in quiet system of loop using Coronal Multichannel Polarimeter.

2. Observations

However, today most available data are observations of SOHO/EIT and TRACE. These data are irreplaceable for regular investigations of coronal oscillation. Coronal wave as result this research was detected investigating EIT data.

For demonstration of phenomenon EIT data are chosen EIT data on December 13, 2006. AR 10930 (Fig. 1a) where oscillation seen most clearly was investigated. This active

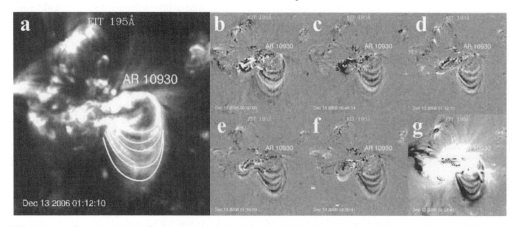

Figure 1. Propagation of wave in the atmosphere of active region NOAA 10930 at Dec 13, 2006. a. Arches system of AR 10930 observed by EIT 195Å. Stable arches marked by curves. b.-g. Allocation of wave train from 00:00:09 to 02:24:41 UT on the difference images.

region produced series of powerful flares and CME events. At December 13 02:14 UT the flare was registered with GOES X-ray class X3.4. At 2:54 UT was detected large coronal mass ejection event in west-south quadrant. All EIT images were rotated to the some equal time before flares. Active region was cut out and frames prior to flare were processed. Unfortunately such frames were only twelve because EIT was "bake out". Before flare we can see the system of stationary arches. Several bright arches were marked (Fig. 1a).

Oscillation was contrasted by special method. The differences between neighboring frames were found for each pixel using formula $I'_{ij}(t'_k) = I_{ij}(t_{k+1}) - I_{ij}(t_k)$. This method of preparation is used usually for allocation fast changing elements of solar images. Fig. 1b-g shows the plasma in arches system is not stationary. In west south part of active region we can see clearly visible waves propagating out from base of active region. There were three or four waves in different time with periodically structure in direction outward from the Sun. In radial direction there is the sequence of waves which starts on height about five thousand kilometers above photosphere. Waves are not seen on EIT images where there are only stationary loops.

The time cadence of used observations could be increased using TRACE observation. Unfortunately the wavetrain aren't seen on TRACE differences sequences. The start of TRACE observation was at 1:47 UT. EIT difference images show the loop appreciably changed between this time and start of flare. Outside flare time on TRACE data is visible only noise. It means the TRACE data are not useful for study quiet events.

The event at December 13 investigation shows the wave is not being detected at levels above EIT arches system. It can be simply explained . First: upper levels of arches system are visible on the quiet sun background. Second, probably the wave's amplitude decreases in low density plasma. On Fig. 2 is shown EIT images and differences sequence at December, 17 2006. Active region 10930 is allocated at west solar limb now. On EIT data there is CME event at 15:30 pm. Before CME magnetic loops were erupted during 30 minutes.

On Fig. 2a there is the stationary loop which stay stable several hours. Wave train with sharp wave front can be found on difference pictures (e.g. Fig. 2b-d) during of 12 hours. Parameters of waves in this day of observation are similar to December, 13.

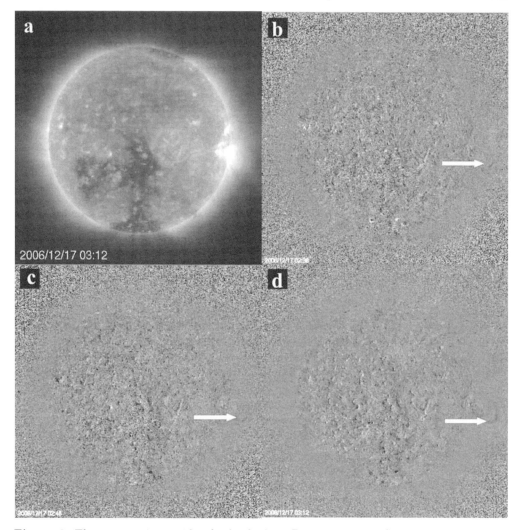

Figure 2. The wave train outside of solar limb at Dec 17, 2006. a. Solar disk according on data of EIT Fe XII 195Åline. b.-d. Allocation of wave front at 02:36 UT (b), 02:48 UT (c) and 03:12 UT (d).

The waves are being detected not so well at as at Dec. 13. It is easy to explain if one takes into account the values of plasma density at coronal level.

3. Discussion

For nature of observed waves explanation, it is important to know the plasma parameters. It is not so simple. But the density can be taken from EIT data. Brosius *et al.* (2002) compared EIT intensity and density taken from Fe XV line data. They found simple dependence $lgN_{e,i} = 8.34 + 0.509lgI_{EIT}$, where $N_{e,i} \geqslant 2.2 \cdot 10^8$. This dependence may help obtain density from EIT intensity range in arches.

In Table 1 is collected some properties of waves events which were observed by SOHO during 2006. Except the wave's properties in Table 1 the information about flares and CME is showed. As we can see the waves velocities was very slow. In Table 1 T and L

Table 1. Some properties of waves.

Date & time, start, end, UT.	Coord., NOAA No	V, km s^{-1}	T, s	L, km	$N_{e,i}$, $\times 10^9$ cm^{-3}	$\Delta N_{e,i}$, $\times 10^9$ cm^{-3}	CME	Flare	GOES Class
2006/12/12 ? -12/13 02:24	S05W23 10930	8-20	750-1900	15000	4.2	1.2	02:54:04	02:40:00	X3.4
2006/12/14 0:59-02:53	S06W46 10930	20	600	12000	3.6	0.8	20:30:04	22:15:00	X1.5
2006/07/06 05:36-07:12	S09E38 10875	15	670	10000	2.5	0.9	08:54:00	08:36:00	M2.5
2006/04/27 9:53-12:44	S08W54 10865	10	1400	14000	3.9	0.8	—	15:52:00	M7.9
2006/04/26 12:12-15:24	S09W35 10898	9	1100	10000	4.7	0.9	—	17:02:00	M1.3
2006/04/06 9:24-12:44	S08W61 10875	18	610	11000	3.1	1.0	—	20:34:00 05:35:00	M1.3 M1.4
2006/12/17 2:34-14:44	**outlimb**	9-15	750-1000	15000	**0.4**	**0.2**	15:30:04	22:15:00	C2.1

are period and wavelength, the plasma density are typical for hot coronal arches and differences between low and high value of density in the waves is labeled as $\Delta N_{e,i}$.

On Fig. 3 some angular and radial properties of waves are shown. From center of arches system base via equal angle depend values of density and velocity. On Fig. 2a for one of waves is shown value of velocity for time of observation. It is seen in some time velocity changed for all angles when wave position corresponded to position of stable arches. In other hand density of waves maxima weakly changes with distance from base of arches system. On Fig. 2b angular dependence for different waves in the same time is shown. Similar properties there are in all of wave's observation.

Thereby the wave always propagates in quasi-perpendicular direction to magnetic field. Properties of wave depend on plasma density and magnetic field value. The value of coronal magnetic field can be found roughly from dipole approximation. For example

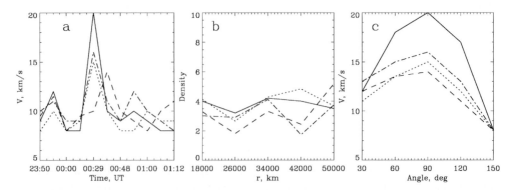

Figure 3. Angular, spacing and time dependence of wave velocities and plasma density. **a.** Angular and time distribution of leading front of wave train at Dec. 13, 2006. Each line corresponds to value of angle with interval of 30 degree between geometric center of loops system and perpendicular to magnetic field **b.** Dependence between distance from photosphere and measured plasma density for the same angles **c.** Angular dependence of velocities for different waves on Dec 13, 2006

at Dec. 13 $B(R) = B_{foot}(1 + \frac{R}{R_D})^{-3} \approx 60G$ where B_{foot} the magnetic field at base of dipole, and R_D the dipole radius. Then the Alfvén speed $v_A = \frac{B}{\sqrt{4\pi\rho}} \approx 2000$ km s^{-1}. The densities were determined, temperature forming Fe XII 195Å line is 1.5 million Kelvin. Unfortunately, magnetic field is known only at photospheric level. In this case sound velocity $c_s^2 = \frac{2\gamma k_B T}{\mu m_p} \approx 180$ km s^{-1} $\leqslant v_A$. Such slow velocity and quasi-perpendicular propagation to magnetic field allow assuming the waves can be acoustic waves with dependence from magnetic field. That is magneto acoustic waves.

The velocity for two fast and slow modes is depending from magnetic field, density and temperature $v_{f,s}^2 = \frac{1}{2}(v_A^2 + c_s^2 \pm \sqrt{v_A^4 + c_s^4 - 2v_A^2 c_s^2 cos2\theta})$. For depended values the velocity of slow magneto acoustic waves there is realistic values.

In conclusion I note such events may transfer the energy to the upper levels. It is necessary to take this event into account for development the models of solar atmosphere. I hope unusual phenomenon will investigated fully henceforward.

4. Acknowledgements

I express sincere gratitude to SOHO/EIT team for data used in this research. The work was supported by RFBR grant No 08-02-08680 and the Program of Presidium RAS No 13.

References

Biesecker, D. A., *et al.* 2002, *ApJ*, 569, 1009

Brosius, J. W., *et al.* 2002, *ApJ*, 574, 453

Cadez, V. M. & Ballester, J. L. 1994, *ApJ*, 292, 669

Chertok, I. M. & Grechnev, V. V. 2003, *Astron. Rep.*, v. 47, 11, 934

De Groof, A., Berghmans, D., van Driel-Gesztelyi, L., & Poedts, S. 2004, *A&A*, 415, 1141

Delaboudiniere, J.-P., Artzner, G. E., Brunaud, J., *et al.* 2004, *Solar Phys.*, 162, 1-2, 291

De Moortel, I. & Brady, C. S. 2007, *ApJ*, v. 664, 2, 1210

O'Shea, E., Srivastava, A. K., Doyle, J. G., & Banerjee, D. 2007, *A&A*, v. 473, 2, L13

Pascoe, D. J., Nakariakov, V. M., & Arber, T. D. 2007, *A&A*, 461, 1149

Raouafi, N.-E., *et al.* 2004, *A&A*, 424, 1039

Tomczyk, S., *et al.* 2007, *Science*, v. 317, 5842, 1192

Wills-Davey, M. J., DeForest, C. E., & Stenflo, J. O. 2007, *ApJ*, 664, 556

Universal Heliophysical Processes
Proceedings IAU Symposium No. 257, 2008
N. Gopalswamy & D.F. Webb, eds.

© 2009 International Astronomical Union
doi:10.1017/S1743921309029366

CMEs 'en rafale' observations and simulations

Cristiana Dumitrache

Astronomical Institute of Romanian Academy
Str. Cutitul de Argint 5,
Bucharest 040557, Romania
email: `crisd@aira.astro.ro`

Abstract. A CME is triggered by the disappearance of a stable equilibrium as a result of the slow evolution of the photospheric magnetic field. This disappearance may be due to a loss of ideal-MHD equilibrium or stability as in the kink mode, or to a loss of resistive-MHD equilibrium as a result of magnetic reconnection. We have obtained CMEs in sequence by a time dependent magnetohydrodynamic computation performed on three solar radii. These successive CMEs resulted from a prominence eruption. Velocities of these CMEs decrease in time, from a CME to another. We present observational evidences for large-scale magnetic reconnections that caused the destabilization of a sigmoid filament. These reconnections covered half of the solar disk and produced CMEs in squall (sequential CMEs).

Keywords. coronal mass ejections, magnetic fields, prominences, instabilities

---

1. Introduction

The slow evolution of the photospheric magnetic fields often result in the destabilization of coronal arcades as filaments and trigger coronal mass ejections. Occurrence of kink instabilities or loss of resistive MHD equilibrium as a consequence of magnetic reconnections rule the equilibrium disappearance and trigger CME.

We have obtained CMEs in 'rafales' (sequence) by a time dependent magnetohydrodynamic computation performed on three solar radii. A prominence destabilization triggered more successive CMEs. We have called this type of ejections "CMEs in squall or en rafales". Velocities of these CMEs decrease in time, from one CME to another, unlike the cannibal mass ejections, where next CME has greater velocity than the precedent, including and devoring it.

We present observational evidences for large-scale magnetic reconnections that caused the destabilization of the filament. These reconnections covered half of the solar disk, as the whole disk Standford magnetograms indicate, and produced CMEs in squall.

2. Numerical simulations

The MHD equations are solved, in two dimensions, with SHASTA method used by Alfven code, developed by Weber (1979), taking into account the gravitation and a complete energy equation. This code was described by Forbes & Priest (1982) and also by Dumitrache (1999). We have considered periodic boundary conditions.

Starting with a current sheet initial configuration and with $\beta = 0.5$ and $Rm = 10^3$, we have performed this numerical experiment on 3 solar radii. No heating has been added ($\sigma = 1$) in the sheet. We obtain a prominences configuration at the Alfven time $t = 0.026$, when the temperature in the sheet is about 5000 K. After that, magnetic reconnections occurred naturally in the sheet and a few CMEs produced in squall. Figures 1 plot

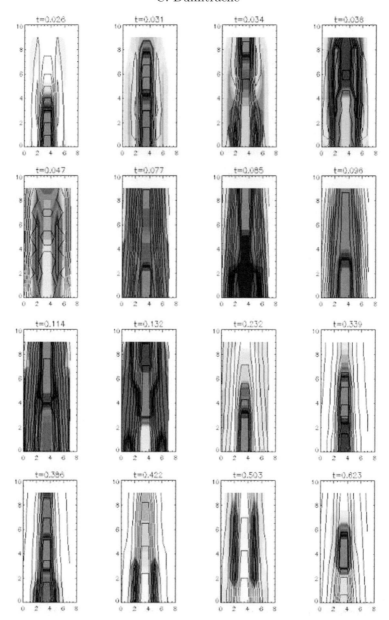

Figure 1. Numerical simulation - CMEs in squal (see the text for explanations).

this prominence and the cartoon of CMEs evolution: the density contours are filled and magnetic field lines are outlined. At $t = 0.031$ the first CME occurred with 1289 km/s. After that one, many other CMEs were triggered and rose with decreasing velocities. The second CME occurred at the Alfven time $t = 0.038$, the third at $t = 0.045$. Later we had more CMEs at $t = 0.085, 0.114, 0.120$ and $t = 0.386$.

After Alfven time $t = 0.623$, a post-CME coronal streamer formed and the maximum velocity increased again at significative values. This phenomenon was described by Dumitrache (2007). Practically, these CMEs took place during 40 minutes.

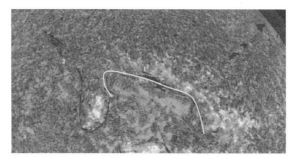

Figure 2. Hα filament superimposed on MDI magnetogram.

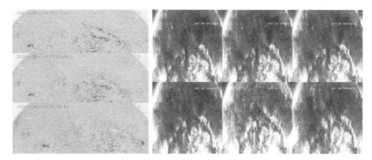

Figure 3. (a) Hα filament in different moments; (b) EIT images of different moments in the filament eruption on 15 August 2001.

3. Observational evidences of sequential CMEs

A large scale magnetic reconnections and shear motions produced CMEs 'en rafale' on 15 August 2001. An important source of these CMEs is a double S filament. This filament is a huge polar feature - no active region exists in its neighborhood. This filament was registered on the solar disk between 11 and 19 August 2001. It appeared as two parallel filaments, but on 13 August a double S-shape linked these two filaments forming a single huge feature. Figure 2 plots the Hα observations superimposed on the MDI magnetogram for that day. We have denoted with (1) the main Western part of the filament, with (2) the Eastern part and with (3) the bend part linking first two ones.

Figure 3a displays images of the filament in Hα for three important moments, on 14 and 15 August 2001. In the picture on top, a two ribbons flare can be observed (indicated by arrows), where the filament denoted by (1) is placed between the ribbons. This flare was observed also in X-ray, by Yohkoh. In the bottom image 3a of the Hα observations both parts of the filament, denoted by (2) and by (3), disappeared after a few CMEs occurred on 15 August. One CME produced on 14 August, issued from the X-ray flare. On 15 August 2001 a few CMEs 'en rafale' produced, having as origin this complex filament: first filament (1) erupted giving a CME at 1:32 UT, but after that, both filaments (1) and (2) erupted violently at 2:54 UT (as seen in C2/LASCO images). The last CME was like a halo, with two branches because of the plasma ejected from both filament parts (2-left and 1-right). A new CME produced later, in the body (1) of the filament. Figure 4 displays few moments observed by C2/LASCO, but more relevant for the CMEs onset are EIT observations. On the EIT movie one can see very well two separated masses erupting from both filaments (1 and 2) and two EIT waves traveling across the Sun, in opposite directions. EIT images of filament changes during CMEs are displayed in Figure 3b. Because of lack of high cadence Hα observations during the time interval of the CMEs onset, we could not say when filaments (2) and (3) disappeared. A new CME occurred

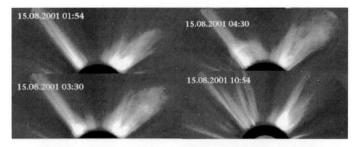

Figure 4. C2/LASCO observations of CMEs in squall.

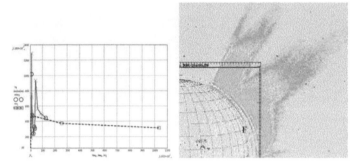

Figure 5. (a)CMEs' velocities plot: simulations and observations; (b)MDI and C2/LASCO image of filament (F) zone on 19 August 2001.

at 10:33 UT. After the last CME, occurred on 15 August 2001, filament (1) continues to exist on the disk but as an S shape (Figure 3a).

According to CDAW Data Center Catalogue, the following data were registered: a CME at 1:31 UT, of 477 km/s velocity; a CME at 2:54 UT, of 370 km/s and a CME at 10:33, of 311 km/s. If we plot these velocities versus time expressed in seconds, we obtain the dashed curve marked with boxes, as in Figure 5a. The continuous line in the same figure represents velocity evolution in time as obtained by our numerical simulations, where the moments of CMEs are marked by circles. We remark that both lines are in continuation. We notice that a new CME occurred on 16 August 2001, and another on 19 August, when the filament (1) reached the visible solar border. After these catastrophic events we should expect that a such huge filament would disappear soon, but surprise: the filament reformed in the same complex S shape (filaments 1+2+3) next solar rotation. This fact indicates us that the magnetic structures supporting the complex filament are strongly rooted in the subphotospheric layers.

The filament arrived at the solar border on 19 August 2001 and figure 5b shows the presence of a coronal streamer above the filament.

4. Discussions

The biggest question is what has caused these successive CMEs and the catastrophic disappearance of the filaments (3) and (2), while filament (1) only changed its shape. Investigating the variations of the tilt angle and differential rotation velocity of all three filaments composing the complex, we notice a sudden increase of the tilt angle for filament (2) after 14 August, while a decrease of these parameters for filament (1) and (3). The differential rotation velocity decreased for filaments (1) and (3) and remained constant for filament (2). We have also computed the shear velocity induced by the differential

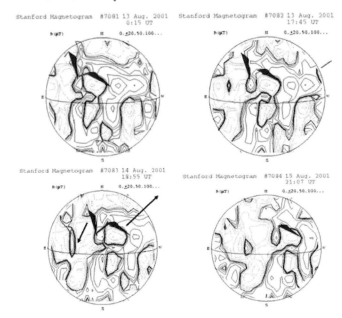

Figure 6. The filament plotted on the Stanford magnetograms

rotation and concluded that shear motions destabilized the filament after 13 August. The images obtained by running difference method applied to Hα observations and also to He observations (Mauna Loa source) indicate height shear motions along the filament channel. This means that the sheared motions were transmitted at all solar atmosphere heights.

We think the key answer is displayed in Figure 6, where Stanford magnetograms are plotted. An important island of positive polarity separated on 13 August and linked again later the same day - it was the day when filament(3) became visible in Hα. On 14 August this magnetic island separated again and also teared in two parts - it was the day when the flaring filament (1) produced a CME. On 15 August both split parts from the central island linked, but each one in another part. It is the day when few successive CMEs occurred and when filaments (2) and (3) disappeared.

We conclude that the filament lost its equilibrium as a result of magnetic reconnections. These reconnections covered half of solar disk and produced CMEs 'en rafales'. After these catastrophic successive coronal mass ejections the double-S filament reformed and was a long-lived entity. The filament lost its stability only temporarily as a consequence of large scale magnetic reconnections and photospheric sheared motions transmitted high into the corona through the solar chromosphere. Since the filament reformed later we believe that the filament magnetic field was strong enough and the filament was very well rooted into the subphotospheric layer, so the magnetic arcades have not been totally affected by large scale magnetic islands reorientation.

References

Dumitrache, C. 2007, *ASP Serie* 370, 180
Dumitrache, C. 1999, *RoAJ* 9, 139
Forbes, T. G. & Priest, E. R. 1982, *Sol.Phys.* 81, 303
Weber, W. J. 1979, *Sol.Phys.* 61, 345

Universal Heliophysical Processes
Proceedings IAU Symposium No. 257, 2008
N. Gopalswamy & D.F. Webb, eds.

The expansion of a coronal mass ejection within LASCO field of view: some regularities

V. G. Fainshtein

Institute of solar-terrestrial physics SB RAS,
P.O. Box 291, 664033, Irkutsk, Russia
email: vfain@iszf.irk.ru

Abstract. Forty five limb CMEs related with eruptive prominences and/or near-to-limb post-eruptive arcades have been tested. It is shown that CMEs can be divided into two groups. The first group includes coronal mass ejections whose "2α" latitude angular sizes apparent in the plane of the sky remain unchanged within measurement accuracy of several degrees. The second one is formed by CMEs that expand "non-radially", namely, their angular sizes increase by the relative value (10-30)% up to the position of the ejection front $R_F = R_{\alpha m}$ and run to the maximal value $2\alpha_m$ at this distance. It has been found that CMEs of the second type are, on the average, wider, faster and have an outer shell brighter and with higher plasma density for long distances. It is shown that on average $R_{\alpha m}$ increases as $2\alpha_m$ rises.

1. Introduction

The apparent angular size (we shall designate it as "2α") of coronal mass ejections (CME) is its geometrical measure and simultaneously reflects important physical properties of CMEs: mass and kinetic energy. Near to the surface of the Sun the angular size of a significant part of CMEs grows as they move. Thus, according to on-the-ground coronagraphs Mark 3 and Mark 4 ($R \leqslant 2.45(2.9)R_0$), such CMEs make up about one third of the total number (Burkepile *et al.* 2007). Here R is the plane of the sky distance from the center of the solar disk, R_0 is the radius of the Sun, the numbers in parentheses refer to Mark 4. A number of researchers have arrived at the conclusion that the angular size of moving CMEs may change even at $R > (2.5 - 3)R_0$ (Eselevich & Filippov 1991, Stockton-Chalk 2002, Yashiro *et al.* 2004) ("non-radial" in terms of Stockton-Chalk 2002). According to Eselevich & Filippov (1990), the angular size of many CMEs recorded in the field of view of the SOLWIND coronagraph, increase by a factor of two or more in the first 2 hours of observation. According to Stockton-Chalk (2002), the angular size of most near-equatorial CMEs observed in the LASCO C3 field of view grow as the CME moves on. In the process, the maximum increase in angular sizes CME was $\approx 5° - 6.5°$, in comparison with radial expansion, in the field of one CME "leg", i.e. was rather small. That CMEs with angular sizes increasing with time (LASCO data) do exist was noted in Yashiro *et al.* (2004). On the other hand, researchers often believe that changes in CME angular sizes in the LASCO field of view are negligible. Thus, there is still no clear idea about CME angular size variations with time at $R > (2.5 - 3)R_0$. This paper relies on LASCO data to examine regularities in the expansion of CMEs related to eruptive prominences and/or post-eruptive arcades on the limb.

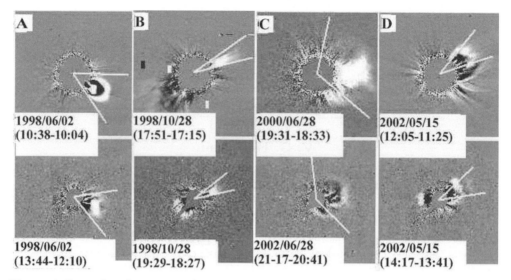

Figure 1. Examples of measuring the angular sizes of CMEs where their boundary determination is hampered by different factors. A and B stands for CMEs with relatively sharp boundaries or with peculiarities at the boundary, which allow the CME angular sizes to be determined within an accuracy of $\approx 3°$. C denotes CMEs with a low brightness gradient at the boundary. D means CMEs for which brightness variations of a previous coronal mass ejection near the CME boundary make it difficult to unumbiguously reveal this boundary.

2. Data and method for analysis

Two types of SOHO/LASCO data were used to define the CME characteristics: difference images of the corona from the http://cdaw.gsfc.nasa.gov/CME_list/daily_movies/ database and calibrated coronal images with image-processing level L1.

In most cases, relatively small time/distance variations can be observed in CME angular sizes 2α. Therefore, it is necessary to provide for the highest accuracy when deriving 2α values. Although CME angular size has been determined by many investigators, and it is almost a routine procedure, finding these sizes within an accuracy of $2° - 5°$ is rather complicated. The causes of limited accuracy in finding 2α include: (1) small brightness gradient in the CME boundary region; (2) brightness variations of the background plasma or of a previous CME near the CME boundary; (3) insignificantly higher CME brightness near the boundary than the surrounding background brightness. In order to measure the CME sizes (within the above accuracy) we use a time sequence of coronal images to determine the CME boundary; two mutually-supplementary methods are used to find the CME sizes. In the first case, we use differential images to take the 2α value as an angle between two rays drawn from the solar disk center in the plane of the sky to CME boundary features. In the second case, CME angular size is determined using latitude scans of the coronal brightness (calibrated data).

Fig. 1 presents examples of using the first method to derive the 2α value both for cases when it is easy to discern the CME boundary, with much precision (Fig. 1A,B), and for problematic CMEs in terms of their boundaries (Fig. 1C-H). In the first case, the accuracy of deriving 2α may be up to $3° - 5°$. In the second case, the accuracy of measuring CME angular sizes may considerably exceed $5°$.

Fig. 2 presents examples of determining CME angular sizes through scanning the brightness of differential calibrated images, at processing level L1. Moreover, in this

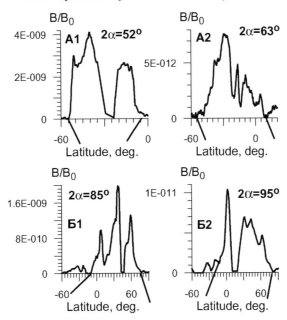

Figure 2. Examples of finding the angular CME sizes using differential scans of calibrated image brightness, at processing level L1. Arrows show CME boundaries. A1-02.06.1998 (10:29-07:02), w-limb, $R = 2.75R_0$; A2-02.06.1998 (13:44-08:47), w-limb, $R = 13R_0$;. B1-07.01.2001 (5:29-3:53), w-limb, $R = 3R_0$; B2-07.01.2001 (9:17-5:17), w-limb, $R = 10R_0$. Differences between the moments of time for which the coronal images were subtracted are parathesised near their respective dates. For comparison, Fig. 2 (A1, A2) shows the scans of brightness for two radii (continuous and dotted lines, respectively).

case the brightness was latitude-smoothed, to diminish the high-frequency (noise) signal component.

It is evident from Fig. 2 that in all the above examples the CME boundaries are quite pronounced as places in which the brightness first reaches zero or a minimum value after a strong decrease in the CME body. The angular size is measured accurate to $\approx 3° - 5°$. CME boundary peculiarity is chosen based on an analysis of the time sequence of coronal brightness scans and, when required, a comparison between the scans and the time sequence of CME images. In this paper, a group of 45 limb CMEs, detected in 1997-2002 and related to near-limb eruptive prominences and/or post-eruptive arcades, were selected for analysis. See the event selection criteria in Fainshtein (2007). The selection of precisely these (limb) CMEs for the analysis is due to the fact that we can determine their true angular sizes. It means that the influence of projective effects should be insignificant.

The EF and PEA characteristics were determined by Sun images in the FeXII λ 195A extreme ultraviolet line (SOHO/EIT). The procedure of the EF (PEA) angle size and values R_1 and R_2 determination is described in the paper Fainshtein (2007). For analysis we also used the $\beta P - A$ angle position of the EF and/or EPA center (within the heliocentric coordinate system). As V1 values we used this velocity values from the "LASCO CME catalog" (http://cdaw.gsfc.nasa.gov/CME_list/), obtained when linearly approximating the $R_1(t)$ points.

3. Results

Fig. 3 shows examples of typical CME angular size variations depending on the position of the front (R_F).

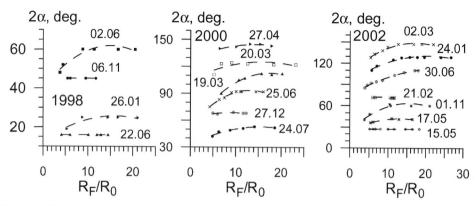

Figure 3. Examples of typical CME angular size variations depending on CME front position (R_F). The numbers near the curves are the CME dates.

The first 2α values were not determined at the moment of CME occurrence, but at the nearest moment when it was possible to determine the CME true angular size. A visual analysis of all dependences $(2\alpha(RF))$ considered indicates that starting from some R_F value (which will be denoted as $R_{\alpha m}$), the CME angular size reaches its maximum $2\alpha_m$ value. Further on, this size reaches saturation and becomes stable, or changes insignificantly within an accuracy of $3°$, or starts decreasing.

It is established that there is a positive correlation between $R_{\alpha m}$ and $2\alpha_m$, and the connection between these parameters becomes stronger for relatively slow coronal mass ejections (Fig. 4).

Connection between $R_{\alpha m}$ and the speed of the CME front V_F (not shown) appears weak and grows for CMEs with rather small angular sizes $R_{\alpha m}$. For the CME velocity V_F we take the velocity resulting from the linear approximation of $R_F(t)$ from `http://lasco-www.nrl.navy.mil/cmelist.html`.

The increase of the CME angular size 2α as the CME moves may be due to different causes. More often, it is related to a faster (in comparison with $2\alpha = const.$) increase of transverse linear CME sizes. In some cases, the non-radial CME expansion is associated

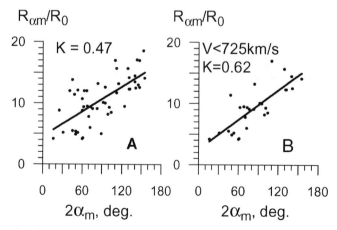

Figure 4. A is the dependence $R_{\alpha m}$ on the maximum CME angular size ($2\alpha_m$). B is the dependence $R_{\alpha m}(2\alpha_m)$ for CMEs with velocities $V_F < 725$ km/s (the median velocity value). K is the correlation coefficient.

with the CME base broadening at the occulter boundary. The third group includes CMEs whose angular sizes increase due to simultaneous action of these two mechanisms. The CME registered on 13 July, 1999 was peculiar. In the course of time, the CME bends and deflects from the radial direction; this results in an apparently strong increase of angular sizes. We used real angular sizes for this CME.

Let us compare some properties of CMEs with radial vs. non-radial expansion. We have selected only those coronal mass ejections into the second group, the increase in whose angular sizes is related to additional expansion due to increased transverse linear sizes. The results of this comparison of the characteristics of the two types of CMEs can be represented using the following figures: the average angular size of radial-expanding CMEs is $64.9°(29.1°)$, while that of non-radially expanding ones is $92.4°(36.8°)$. The respective mean velocity values are 468 (240.9) km/s and 781 (254) km/s. The figures in brackets are the standard deviation.

CMEs of the second type are brighter and denser. Fig. 5 illustrates the higher brightness of the CME outer shell with non-radial expansion. Radial scans of the calibrated brightness of some CMEs are demonstrated here. These scans are drawn along radii near the CME axes. Arrows indicate maximum brightness values in outer CME shells, which we will denote as "CME fronts".

Note that the brightness of the outer part of CMEs with constant angular sizes do not exceed the noise level when the CME front is near the outer boundary of the LASCO C3 field of view ($R = (25 - 30)R_0$). At the same time, the brightness of the outer shell of

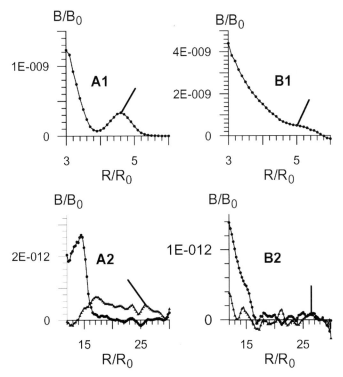

Figure 5. Radial scans of the calibrated brightness for non-radial (A) and radial (B) CMEs. A1: N2, 07.01.01 (05:29-03:53), A2: N3, 07.01.01 (09:17-05:17), circles, (13:41-05:17), triangles. B1: N2, 25.11.97 (20:23-19:30), B2: N3, 25.11.97 (06:42-00:18), circles, (08:42-00:18), triangles. The CME registration dates are accompanied by the differences between the moments for which the coronal images are subtracted from one another.

non-radially expanding CMEs may exceed the noise level by a factor of ≈ 3 to ≈ 10. Fig. 5 proves that, on average, the K-coronal brightness for non-radially expanding CMEs is about 6 times higher than that for radially expanding CMEs; and the angular size of CMEs of the first type is about 1.5 times (see above) as large as the size of the second type. This means that the electron density, and the plasma density, as a whole, in the outer ejection part of CMEs with constant angular size are sometimes lower than in CMEs with non-radial expansion. This suggests that the physical cause of the non-radial CME expansion is increased plasma pressure within the outer CME shell as compared to the background plasma.

In conclusion we shall make one remark. This investigation defined the apparent CME angular size as the size of the angle with its apex at the center of the solar disk, while CME expansion was understood as an increase in this angle. Such a definition of CME angular size is conventional (see (Eselevich & Filippov 1991, Stockton-Chalk 2002, Yashiro *et al.* 2004) and the references therein). At the same time the apex of an angle defining the CME size may be placed, basically, at any other spot. There are studies placing this apex, for example, on the surface of the Sun in the CME axis (at the conditional place of CME emergence). It is easy to ascertain that this results in a changed character of the dependence $2\alpha(R_F)$ in comparison with a case when the apex is located at the center of the solar disk. For example, the angular sizes of CMEs regarded in this work as "radial", will decrease with R_F. At the same time, it is easy to show that in this case the CMEs in question can also be subdivided into two classes. However, the criteria for assigning CMEs to their proper class will be different. A new class will comprise all those CMEs which had been included into the "radial" group of CMEs, while the other class will acquire the "non-radial" CMEs from this study. This question will be dealt with in more detail in the full version of this work.

Acknowledgements. SOHO is a project of international cooperation between ESA and NASA. This work was supported by RF Leading Scientific Schools Support Governmental Grant SS 4741.2006.2 and by P-16 RAS Presidium Fundamental Research Program.

References

Burkepile, J. T., St.Cyr, O. C., Stanger, A. L. Sitongia, L., deToma, G., Gilbert, H. & Darnell, J. A. *The presentation at SOHO20, 27-31 August 2007, Ghent, Belgium*

Eselevich, V. G. & Filippov, M. A. 1991 *Planet. Space Sci.*, V. 737

Fainshtein, V. G. 2007, *Cosmic Research*, 45, 384

Stockton-Chalk, A. 2002 *Proc. "SOLSPA: The Second Solar Cycle and Space Weather Euro-conference", Vico Equencse, Italy, 24-29 September 2001 (ESA SP-477, February 2002)*, 277

Yashiro, S., Gopalswamy, N., Michalek, G., St. Cyr, O. C., Plunkett, S. P., Rich, N. B., & Howard, R. A. 2004 *J. Geophys. Res.* 109, A07105

Discussion

HOWARD: I have not seen "non-radial" expansion in the LASCO data except for CMEs that are well out of the plane-of-the-sky. In this case the effect is due to the projection on the 2D plane.

FAINSHTEIN: I think, for most analyzed CMEs the "non-radial" expansion of the CME is not due to projection onto the plane of the sky. It is a physical effect. But fore some CMEs the "non-radial" expansion possibly may be attributed to projection onto the plane of the sky.

EROSHENKO: Are any of the CMEs full halo or partial halo CMEs?

FAINSHTEIN: No, we have not studied full halo or partial halo CMEs. We have studied "limb" CMEs with axis which is perpendicular to the SunEarth axis.

GIRISH: Have you compared solar wind data for "radial" and "non-radial" CMEs?

FAINSHTEIN: It is hard to do it for CMEs we studied because the CME axis is perpendicular to the Sun-Earth axis. But now such study can be undertaken using data from STEREO.

Universal Heliophysical Processes
Proceedings IAU Symposium No. 257, 2008
N. Gopalswamy & D.F. Webb, eds.

© 2009 International Astronomical Union
doi:10.1017/S174392130902938X

The link between CME-associated dimmings and interplanetary magnetic clouds

Cristina H. Mandrini[1], María S. Nakwacki[1], Gemma Attrill[2,6], Lidia van Driel-Gesztelyi[2,3,4], Sergio Dasso[1,5] and Pascal Démoulin[3]

[1]Instituto de Astronomía y Física del Espacio, CC. 67 Suc. 28, 1428, Buenos Aires, Argentina
email: mandrini@iafe.uba.ar

[2]University College London, Mullard Space Science Laboratory,
Holmbury St.Mary, Dorking, Surrey, RH5 6NT, UK

[3]Observatoire de Paris, LESIA, UMR 8109 (CNRS), 92195 Meudon Principal Cedex, France

[4]Konkoly Observatory of the Hungarian Academy of Sciences, Budapest, Hungary

[5]Departamento de Física, Facultad de Ciencias Exactas y Naturales,
Universidad de Buenos Aires

[6]Harvard-Smithsonian Center for Astrophysics, 60 Garden St., Cambridge, MA 02138, USA

Abstract. Coronal dimmings often develop in the vicinity of erupting magnetic configurations. It has been suggested that they mark the location of the footpoints of ejected flux ropes and, thus, their magnetic flux can be used as a proxy for the ejected flux. If so, this quantity can be compared to the flux in the associated interplanetary magnetic cloud (MC) to find clues about the origin of the ejected flux rope. In the context of this interpretation, we present several events for which we have done a comparative solar-interplanetary analysis. We combine SOHO/Extreme Ultraviolet Imaging Telescope (EIT) data and Michelson Doppler Imager (MDI) magnetic maps to identify and measure the flux in the dimmed regions. We model the associated MCs and compute their magnetic flux using *in situ* observations. We find that the magnetic fluxes in the dimmings and MCs are compatible in some events; though this is not the case for large-scale and intense eruptions that occur in regions that are not isolated from others. We conclude that, in these particular cases, a fraction of the dimmed regions can be formed by reconnection between the erupting field and the surrounding magnetic structures, via a stepping process that can also explain other CME associated events.

Keywords. Sun: coronal mass ejections (CMEs), Sun: magnetic fields, interplanetary medium

1. Introduction

The plasma and magnetic field ejected from the Sun by coronal mass ejections (CMEs) are later observed in the interplanetary (IP) medium as interplanetary coronal mass ejections (ICMEs). When certain characteristics are present (low plasma β, lower proton temperature and stronger magnetic field than in the ambient solar wind, exhibiting a smooth and significant rotation), ICMEs are called magnetic clouds (MCs) (Burlaga *et al.* 1981). Several studies relating qualitatively and/or quantitatively MCs with their solar sources have been published in recent years (see the review by Démoulin 2007).

The two global MHD invariants used to link coronal to IP observations are the magnetic field helicity and flux. Magnetic helicity, at the solar level and in the IP medium, can be derived from observations combined with magnetic field models. At the solar level the ejected helicity can be estimated from the helicity decrease of the coronal field after an ejection, this value in general agrees with the helicity contained in the associated MC (see Mandrini *et al.* 2007 and references therein).

Concerning the ejected magnetic flux, we have to use proper proxies to estimate it. Two solar proxies have been used so far. First, the photospheric flux of the dimming regions forming after CMEs (see Mandrini *et al.* 2007 and references therein) and, second, the one swept out by flare ribbons as they separate, moving away from the magnetic inversion line (see Qiu *et al.* 2007 and references therein). How do these features relate to the ejected flux? What can we learn from the comparison between the estimated ejected flux and the flux in the associated MC when considering both proxies?

In general, coronal dimmings are interpreted as density depletions caused by the eruption of an unstable magnetic configuration. The eruption leads to the expansion of magnetic loops and the evacuation of plasma along them into the IP space (Hudson *et al.* 1996). Two dimmings (primary dimmings) are often seen to form on both sides of the erupting configuration. Two physical models have been proposed. In a first case, dimmings correspond to the footpoints of the ejected flux rope, as suggested by Webb *et al.* (2000). In this context, the flux rope pre-exists in the corona and remains rooted in the dimming regions, as it expands out into the IP. The flux of the dimming regions is comparable to the flux in the axial magnetic field component of the MC in this case. In a second case, the arcade above the flux rope expands significantly before reconnecting (below the flux rope). Dimmings appear at the footpoints of the flux rope and also all along the footpoints of the sheared magnetic arcade. As reconnection proceeds, more flux is progressively added to the erupting flux tube (see next paragraph). In this second case, that corresponds to the theoretical two-dimensional model of Lin & Forbes (2000), the flux in the dimming regions is comparable to the sum of the flux in the axial and azimuthal components of the MC. If the flux rope does not already exist in the corona in this second case, it will be formed during the reconnection process.

The magnetic flux swept out by flare ribbons, a proxy for the reconnected flux, can also be used, as follows (see Qiu *et al.* 2007). In a classical flare model, magnetic reconnection occurs mainly below a pre-existing flux rope (see Lin & Forbes 2000 and references therein). The magnetic flux forming 'post'-flare loops is identical to the flux added to the pre-existing flux rope that will be observed in the IP space as an MC. Then, the flux in the azimuthal component of the MC field will be comparable (if the ejected flux rope was mostly formed by reconnection during the ejection) or larger (when the flux rope was present before ejection) than the flux swept out by flare ribbons.

The above discussion shows that the comparison of coronal and IP associated events can be used to constrain the CME mechanism; in particular, in relation to determining whether the ejected flux rope was pre-existing in the corona, or whether it formed during the eruption. In this paper we present an overview of examples previously analyzed (Section 2). These are: a minor eruptive event on 11 May 1998 (see Mandrini *et al.* 2005, Paper I), a long duration C1.3 class flare and CME (see Attrill *et al.* 2006, Paper II) on 12 May 1997, and the large X17 flare and CME on 28 October 2003 (see Mandrini *et al.* 2007, Paper III). Comparison of these examples allows us to discuss under which circumstances either (i) the magnetic flux in the dimmings or (ii) the magnetic flux swept out by the flare ribbons, may be used as reliable proxies for the ejected flux (Section 3).

2. Overview of three events with different characteristics

The image on the left panel of Fig. 1 shows an elongated feature at disk center. This small, isolated, bipolar and non-numbered active region (AR) showed an eruptive nature during 11 May 1998. The largest event at 08:31 UT was accompanied by the elongation of the sigmoidal AR loops that later disappeared, EUV dimmings and cusp formation. The dimmings were located at both sides of the AR, lying partially over each polarity and

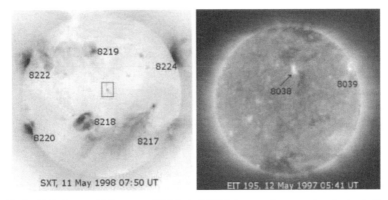

Figure 1. Left: Soft X-ray Telescope (Yohkoh/SXT) image (reversed color) at the maximum extension of the bright loops in the small eruptive AR (within the box) located far from other ARs. Right: EIT image showing the two dimmings on both sides of the isolated AR8038.

extending into the nearby quiet Sun regions. Following the procedure discussed in Paper I, we found that the absolute value of the total net flux in each dimming was $F_{dim} = (13 \pm 2) \times 10^{19}$ Mx, meaning that the flux in these primary dimmings was balanced. Analyzing *in situ* data from the Wind satellite, we were able to identify the resulting tiny MC. We modeled the magnetic data using a cylindrical linear force-free static model (see Paper I) and determined the axial ($F_{z,cloud}$) and azimuthal ($F_{y,cloud}$) fluxes, considering a cloud length constrained by solar and IP observations. These values were: $F_{z,cloud} = 1.3 \times 10^{19}$ Mx and 10×10^{19} Mx $\leqslant F_{y,cloud} \leqslant 20 \times 10^{19}$ Mx.

We proceeded in a similar way for the event shown in the right panel of Fig. 1. AR8038, which was mainly bipolar, was the only region on the solar disk at that time. A long duration C1.3 flare started at 04:42 UT, it was accompanied by a filament eruption, CME, coronal wave and dimmings (see Paper II and references therein). Two main dimmings were present at both sides of the AR lying again partially over the AR polarities and quiet Sun regions. Defining the dimming boundaries as discussed in Paper II, we computed a net flux of $F_{dim} = (21 \pm 7) \times 10^{20}$ Mx for the southern-most dimming at its maximum extension (the southern-most dimming was identified as the sole remaining footpoint of the erupted flux rope, since during the eruption the northern-most edge of the flux rope underwent an interchange reconnection with the north polar coronal hole, transferring its connectivity from the Sun to the interplanetary magnetic field). The MC associated to this eruption was observed by Wind. We modeled the magnetic data using three different cylindrical and static models (see Paper II) and found an axial MC flux $F_{z,cloud} = (4.8 \pm 0.8) \times 10^{20}$ Mx and an azimuthal MC flux $F_{y,cloud} = (17 \pm 8) \times 10^{20}$ Mx, taking a length constrained by solar and IP observations.

One of the largest "Halloween" events was the X17 flare on 28 October 2003 in AR 10486. These events occurred just after solar maximum when the Sun's field was highly complex (Fig. 2, right panel). The X17 flare was accompanied by a filament eruption, CME, coronal EIT and Moreton waves and very extended dimmings (Fig. 2, left panel). Following the method discussed in Paper III, we found for the signed net flux in the numbered dimmings: $F_{dim1} = -4.4 \times 10^{21}$ Mx, $F_{dim2} = 0.7 \times 10^{21}$ Mx, $F_{dim3} = 1.7 \times 10^{21}$ Mx, and $F_{dim4} = -7.0 \times 10^{21}$ Mx. We noticed that none of the dimmed regions lay on the AR main polarities. Though two main dimming regions were located to the west and two to the east of the AR, the net flux was dominantly negative on either side. Therefore, these dimmings were not of the same kind as the ones found in the previous examples. We concluded that primary dimmings were not observed in this case. This

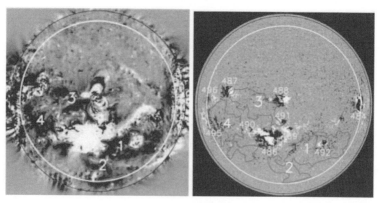

Figure 2. Left: EIT 195 base difference (12:00 UT - 05:00 UT) image showing the largest dimmed regions (1, 2, 3, and 4) around 1 hour after flare onset. The white circle has been drawn at an angular distance of 60^0 from Sun center. MDI magnetic field measurements are done within this limit. Right: Dimmed regions overplotted on an MDI magnetogram at 11:11 UT. All the ARs present on the solar disk are identified (these numbers should start with 10).

happened because the major energy release during the flare was followed by strong chromospheric evaporation and formation of dense and bright loops that make the primary dimmings unobservable. Therefore, the observed distant and extended dimmings must be secondary; these dimmings can be formed by a process illustrated in Fig. 3, as follows. The expanding CME loops reconnect with favorably oriented loops to the west and east. This reconnection process creates two new sets of connectivities: small loops (that become bright due to evaporation) and large-scale loops belonging to the eruptive configuration with displaced footpoints. Dimmings are expected to be associated with these long loops as the plasma expands to fill a larger volume. The MC associated with this event was in strong expansion, so, we modeled its magnetic field using two expansion

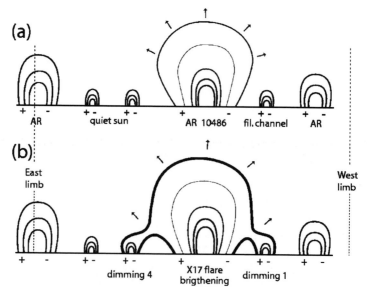

Figure 3. The expanding CME magnetic configuration (a) and its interaction (b) with the surrounding bipoles and ARs resulting in the spread of dimmings (same mechanism proposed by Attrill *et al.* 2007, see text).

models as well as a static one for comparison (see Paper III). We found an axial MC flux 2.8×10^{21} Mx $\leqslant F_{z,cloud} \leqslant 3.1 \times 10^{21}$ Mx and an azimuthal MC flux 11×10^{21} Mx $\leqslant F_{y,cloud} \leqslant 16 \times 10^{21}$ Mx. Following our discussion on the formation of the dimmings for this event via secondary reconnections with the surrounding fields, it is therefore not possible to directly compare the flux in the MC to the flux in the dimmings. Qiu *et al.* (2007) have computed the flux swept out by flare ribbons for this event and found a value of $(18.8 \pm 1.8) \times 10^{21}$ Mx, which is in good agreement with the total MC flux.

3. Discussion and conclusions

In summary, in the three analyzed examples either the magnetic flux in the dimming regions or the one swept out by flare ribbons was found to be compatible primarily with the azimuthal MC flux. This indicates that the ejected rope was mostly formed by magnetic reconnection during the eruption process.

Qiu *et al.* (2007) found a different result concerning the flux in coronal dimming regions. These authors concluded that the flux in dimming regions is comparable in order of magnitude to the axial MC flux. However, in the common studied case, the 28 October 2003 X17 flare and CME, we have shown that the observed extended dimmings are secondary. The formation of these secondary dimmings can be explained via a stepping reconnection process, similar to the mechanism proposed by Attrill *et al.* (2007) for the bright front and diffuse leading edge of coronal waves. Although magnetic reconnection conserves magnetic flux, one cannot be sure how much of the magnetic flux in the secondary dimmings became part of the CME; therefore, these secondary dimming areas do not provide a proper proxy for magnetic flux measurements to be compared with interplanetary data. In fact, a detailed study of each event should be done before comparing the magnetic flux in dimmings to that of the associated MC.

Magnetic flux in coronal dimmings is a reliable proxy for the ejected flux in minor events (i.e. low X-class flares) that occur far from other flux concentrations. In these cases, primary dimmings may be observed. Furthermore, in such cases coronal dimmings can be the only useful proxy since it may not be possible to observe the evolution of flare ribbons or kernels. For large and very intense events, the flux swept out by flare ribbons is a more reliable proxy. We conclude that the most reliable solar proxy for the magnetic flux involved in an ejection depends upon the particular characteristics of the analyzed event.

References

Attrill, G., Harra, L. K., van Driel-Gesztelyi, L., & Démoulin, P. 2007, *ApJ*, 656, L101.

Attrill, G., Nakwacki, M. S., Harra, L. K., van Driel-Gesztelyi, L., Mandrini, C. H., Dasso, S., & Wang, J. 2006, *Solar Phys.* 238, 117.

Burlaga, L., Sittler, E., Mariani, F., & Schwenn, R. 1981, *Jour. Geophys. Res.*, 86, 6673.

Démoulin, P. 2007, *An. Geo.* 207, 87.

Hudson, H. S., Acton, L. W., & Freeland, S. L. 1996, *ApJ* 470, 629.

Lin, J. & Forbes, T. G. 2000, *Jour. Gcophys. Res.* 105, 2375.

Mandrini, C. H., Nakwacki, M. S., Attrill, G., van Driel-Gesztelyi, L., Démoulin, P., Dasso, S., & Elliott, H. 2007, *Solar Phys.*, 244, 25.

Mandrini, C. H., Pohjolainen, S., Dasso, S., Green, L. M., Démoulin, P., van Driel-Gesztelyi, L., Copperwheat, C., & Foley, C. 2005, *A&A*, 434, 725.

Qiu, J., Hu, Q., Howard, T. A., & Yurchyshyn, V. B. 2007, *ApJ*, 659, 758.

Webb, D. F., Lepping, R. P., Burlaga, L. F., DeForest, C. E., Larson, D. E., Martin, S. F., Plunkett, S. P., & Rust, D. M. 2000, *Jour. Geophys. Res.*, 105, 27251.

Discussion

SCHMIEDER: Your cartoon about magnetic reconnection between the expanding magnetic configuration and the nearly magnetic shocks reminds me of the existence of the magnetic carpet with magnetic separatrices. The progression of these reconnections could explain the propagation of waves. On the other hand be careful with the flux measurements of MDI in dimmings.

MANDRINI: Of course this reconnection process can explain the progression of waves. Concerning your question about flux measurements, we have corrected MDI data for underestimation of flux, we have taken into account projection effects and none of the pixels in the dimming had saturated values. Besides all our flux measurements were done above a threshold of 20 Gauss.

VLAHOS: Reconnection is a small scale phenomenon but can have large scale re-arrangements of magnetic topologies. Large scale current sheets do not survive very long (milliseconds). So I prefer the phrase "large magnetic field re-arrangements caused by reconnection". Do you agree?

MANDRINI: Yes, I do.

WEBB: In reply to Vlahos comment: I think we have data on large-scale reconnection in current sheets trailing CMEs. We see ray-like structures following CMEs, evidence of reconnection region moving out, that can last for hours.

MALANDRAKI: Concerning the connectivity issue of the ICME observed in October 2003 you mentioned in the talk, I would like to draw your attention to the paper by Malandraki *et al.* 2005, J. Geophys, Res. Vol. 110, A09506, doi: 101029/2004JA010926, which presents near-relativistic electron observations from the EPAM experiment onboard ACE in the vicinity of and during the passage of the ICME over the s/c. The analysis of bi-directional near-relativistic electron pitch-angle distributions observed during the ICME along with the electron intensity characteristic provide strong evidence that loop-like IMF structures still anchored to the Sun are threading through this ICME. Furthermore, Malandraki *et al.*, 2005 have also used the energetic particle observations by ACE/EPAM to identify more accurately the leading and trailing edge of this ICME compared to these previously identified.

Universal Heliophysical Processes
Proceedings IAU Symposium No. 257, 2008
N. Gopalswamy & D.F. Webb, eds.

© 2009 International Astronomical Union
doi:10.1017/S1743921309029391

The role of aerodynamic drag in dynamics of coronal mass ejections

Bojan Vršnak, Dijana Vrbanec, Jaša Čalogović, and Tomislav Žic

Hvar Observatory, Faculty of Geodesy, Kačićeva 26, HR-10000 Zagreb, Croatia
email: bvrsnak@geof.hr

Abstract. Dynamics of coronal mass ejections (CMEs) is strongly affected by the interaction of the erupting structure with the ambient magnetoplasma: eruptions that are faster than solar wind transfer the momentum and energy to the wind and generally decelerate, whereas slower ones gain the momentum and accelerate. Such a behavior can be expressed in terms of "aerodynamic" drag. We employ a large sample of CMEs to analyze the relationship between kinematics of CMEs and drag-related parameters, such as ambient solar wind speed and the CME mass. Employing coronagraphic observations it is demonstrated that massive CMEs are less affected by the aerodynamic drag than light ones. On the other hand, in situ measurements are used to inspect the role of the solar wind speed and it is shown that the Sun-Earth transit time is more closely related to the wind speed than to take-off speed of CMEs. These findings are interpreted by analyzing solutions of a simple equation of motion based on the standard form for the drag acceleration. The results show that most of the acceleration/deceleration of CMEs on their way through the interplanetary space takes place close to the Sun, where the ambient plasma density is still high. Implications for the space weather forecasting of CME arrival-times are discussed.

Keywords. Sun: coronal mass ejections (CMEs), Sun: corona, (Sun:) solar wind, (Sun:) solar-terrestrial relations, (magnetohydrodynamics:) MHD

1. Introduction

Coronal mass ejections (CMEs) are large-scale solar eruptions during which the magnetic flux of some 10^{23} Wb is launched into the interplanetary space at velocities in the order of 1000 km s^{-1}, carrying along $10^{11} - 10^{13}$ kg of coronal plasma (e.g., Gosling 1990, Webb *et al.* 1994).

After the CME take-off, which is governed by the Lorentz force, in the high corona and interplanetary space the CME dynamics becomes dominated by the aerodynamic drag (Vršnak *et al.* 2004a). Consequently, CMEs faster then solar wind decelerate, whereas slower ones are accelerated, both eventually being adjusted to the solar wind speed (e.g., Lindsay *et al.* 1999, Gopalswamy *et al.* 2001, Manoharan 2006, Vršnak & Žic 2007). In this paper the influence of the aerodynamic drag on the CME kinematics in the high corona and interplanetary space is analyzed, and implications for the space weather forecasting are discussed.

2. Empirical relationships

In Fig. 1a mean accelerations a of CMEs measured in the LASCO (Large Angle and Spectrometric Coronagraph; Brueckner *et al.* 1995) C2/C3 field-of-view, covering $2 - 30\,r_{\odot}$, are presented as a function of their mean plane-of-sky speeds v. The sample includes 3091 CMEs observed in the period from January 1997 to June 2006 (the total number of reported CMEs for this period is 11108, but we excluded events where the estimates of the acceleration and mass are marked as uncertain). The graph reveals a distinct anti-correlation between a and v, i.e., the statistical tendency showing that slow CMEs are on average accelerated, whereas fast ones are decelerated. The intercept of the linear least-squares fit with the abscissa, $v_0 \approx 400$ km s^{-1}, is in the range of an average solar wind speed. Note that there are practically no slow CMEs with $a < 0$. On the other hand, it should be also noted that there are fast CMEs that still accelerate in the considered height range, indicating that the Lorentz force in some events still plays a significant role in the CME dynamics.

The $a(v)$ anti-correlation presented in Fig. 1a indicates that the aerodynamic drag is a dominant force in the majority of events. The aerodynamic drag is usually expressed in the form (Cargill *et al.* 1996, Cargill 2004):

$$a = -\gamma(v - w)|v - w|, \tag{2.1}$$

where v is the CME velocity and w is the ambient solar wind speed. The parameter γ reads:

$$\gamma = c_d \frac{A\rho_w}{m}, \tag{2.2}$$

where c_d is the dimensionless drag coefficient (for details see Cargill 2004), A is the effective CME area perpendicular to the direction of propagation, m is the CME mass, and ρ_w represents the ambient solar wind density.

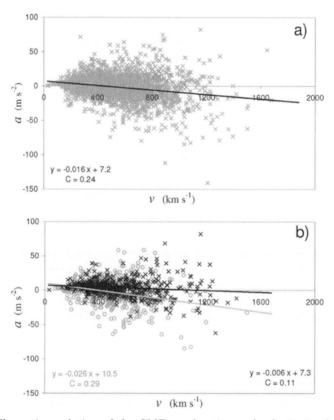

Figure 1. a) The anti-correlation of the CME acceleration and velocity in the LASCO field-of-view. b) The $a(v)$ anti-correlation shown separately for CMEs of low masses (gray) and large masses (black). Linear least square fits are given in the insets, together with the correlation coefficient C.

Thus, we expect that the CME acceleration (after the Lorentz force and the gravity become negligible) should depend on the solar wind speed w and density ρ_w, as well as on the CME speed v, mass m, and size A. Although all of these parameters differ from one event to another, it can be expected that basic relationships should be reflected in the statistical analysis of the kinematical properties of CMEs.

In Fig. 1b we show the $a(v)$ relationship separately for 500 CMEs of the smallest and the largest masses (mean masses are $\overline{m} = 7 \times 10^{10}$ and 7×10^{12} kg, respectively). The low-mass subsample shows a considerably steeper slope k of the $a(v)$ fit (the difference has statistical significance larger than 99%), consistent with the expectation that the effect of drag decreases with increasing mass. Vršnak *et al.* (2008) have shown that the $k(m)$ dependence follows closely the theoretically expected trend $k \propto m^{-1/3}$.

To demonstrate the role of the solar wind speed, we have to involve the interplanetary propagation of CMEs, since measurements of the wind speed are not available near the Sun. In Fig. 2a we first show the the Sun–Earth transit time (TT) as a function of the CME plane-of-sky speed measured in the LASCO C2/C3 field of view. We utilized the sample of 91 events listed by Schwenn *et al.* (2005). Obviously, initially faster CMEs reach the Earth sooner. In Fig. 2b we show a dependence of TT on the solar wind speed w measured at 1 AU (here w represents the mean of the wind speed ahead and behind the CME; for details see Vršnak & Žic 2007). Comparing Fig. 2a with Fig. 2b we find that $TT(w)$ has higher correlation coefficient then $TT(v)$, implying that the solar wind speed plays more important role in determining the transit time than the CME take-off speed

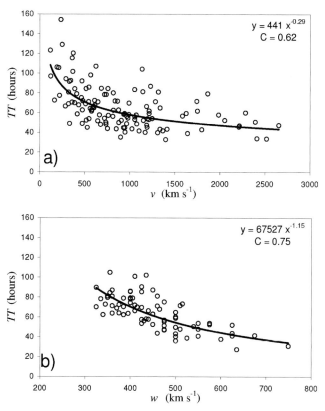

Figure 2. a) Relationship between the Sun-Earth transit time and the CME take-off speed. b) Transit time versus the ambient solar wind speed. Power-law least square fits are given in the insets, together with the correlation coefficient C.

itself. Furthermore, this implies that most of the drag acceleration/deceleration occurs relatively close to the Sun, so that the ejection travels through the interplanetary space at a velocity close to the solar wind speed.

3. Model results and interpretation

Physical background of the empirical results presented in Sect. 2 can be explained employing a simple kinematical model based on Eq. 1. To define the parameter γ (Eq. 2), we assume that the CME dimensions are proportional to the radial distance r. Specifically, in the following we consider a cone-shaped CME of angular (full) width $\phi = 1\,\mathrm{rad}$. For the solar wind density $\rho_w(r)$ we take the empirical density model proposed by Leblanc *et al.* (1998). Given the equation of the continuity (the mass conservation), from that we also get the radial dependence of the solar wind speed $w(r)$. Furthermore, we assume $c_d = 1$ (Cargill 2004). Finally, we assume that Eq. 1 becomes valid beyond the distance r_0. After substituting $a = \mathrm{d}^2r/\mathrm{d}t^2$ and $v = \mathrm{d}r/\mathrm{d}t$ we get a differential equation whose solutions $v(t)$ and $r(t)$ depend on the "initial" CME speed v_0 at r_0. From $v(t)$ and $r(t)$ we also get $v(r)$ and the Sun-Earth transit time TT.

In Fig. 3a we show how the CME velocity decreases with the radial distance for different CME masses. The solar wind speed is normalized to the 1 AU value $w_0 = 400\ \mathrm{km\,s^{-1}}$ and the velocity of the CME at $R_0 \equiv r_0/r_\odot = 10$ is taken to be $v_0 = 1000\ \mathrm{km\,s^{-1}}$. The same situation is considered in In Fig. 3b, but the solar wind speed $w_0 = 600\ \mathrm{km\,s^{-1}}$. Inspecting Figs. 3a and b we see that the speed of low mass CMEs ($m \lesssim 10^{12}$ kg) becomes very close to the solar wind speed already in the LASCO field-of-view ($R < 30$). Given that more than 55% of CMEs has mass $m < 10^{12}$ kg (72% has $m < 2 \times 10^{12}$ kg), this explains why velocities of the majority of LASCO-CMEs are grouped around the solar wind speed (e.g., Yashiro *et al.* 2004). In the same way, this explains why the Sun-Earth

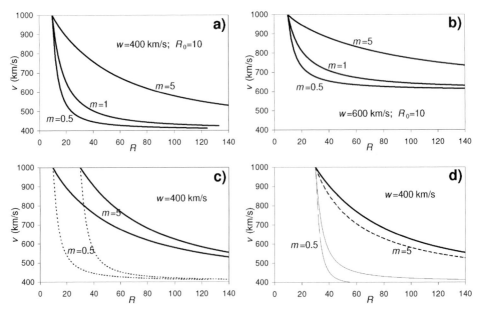

Figure 3. Calculated CME velocity as a function of the radial distance, which is expressed in units of the solar radius $r_\odot$. CME masses m expressed in 10^{12} kg are written by the curves. For details see the main text.

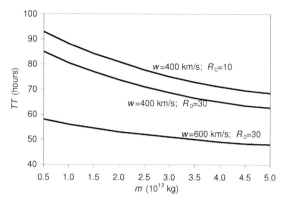

Figure 4. Sun-Earth transit time presented as a function of the CME mass for two solar wind speeds ($w = 400$ and 600 km s^{-1}) and two take-off radial distances $R_0 = 10$ and 30.

transit time is better correlated with the solar wind speed than with the CME take-off velocity. In massive CMEs the adjustment to the solar wind speed lasts longer, which means that the shortest transit times should be expected for fast and massive CMEs, when traveling in fast solar wind, i.e., when being launched from the vicinity of equatorial coronal holes.

Figure 3c shows how the kinematical curves $v(R)$ depend on the height at which the drag becomes effective, i.e., how they depend on the range beyond which the Lorentz force does not compensate the drag anymore. This is illustrated by showing calculations with the initial conditions $v_0 = 1000$ km s^{-1} at $R_0 = 10$ and $R_0 = 30$, respectively. Again we see that the velocity of low-mass CMEs becomes adjusted roughly to the solar wind speed within $\Delta R \approx 20$.

In Fig. 3d we demonstrate how the kinematical curves depend on the solar wind density model. For that purpose we compare the results obtained using the model by Leblanc *et al.* (1998) with outcome for the "hybrid" model proposed by Vršnak *et al.* (2004b). The latter model (results presented by dashed curves) is characterized by a considerably higher density and a steeper density decrease at low heights, so the velocity decrease in the case of light CMEs is extremely fast.

Finally, in Fig. 4 we inspect how the Sun-Earth transit time TT depends on various model parameters. For that purpose we draw three $TT(m)$ curves, calculated employing the density model by Leblanc *et al.* (1998) and taking $v_0 = 1000$ km s^{-1}. We consider solar wind speeds $w_0 = 400$ and 600 km s^{-1}, in combination with $R_0 = 10$ and 30. The graph reveals that differences in TT are larger for low-mass CMEs than for massive ones. Furthermore, one finds out that the difference in solar wind speed is more important than the value of R_0, and that a 50% change of the solar wind speed has larger effect on TT than changing the mass by one order of magnitude.

4. Discussion and conclusion

Comparing empirical relationships with the theoretical results we have demonstrated that the aerodynamic drag is a dominant force that acts on CMEs in the high corona and interplanetary space. Thus, the kinematics of CMEs after the main acceleration stage is determined by the speed and density of the ambient solar wind and by the CME take-off velocity, size, and mass. In the majority of events the CME speed becomes comparable to the solar wind speed close to the Sun, a few tens of solar radii after the Lorentz force becomes negligible. Such a behavior was found also in numerical simulations by Gonzalez-Esparza *et al.* (2003).

Consequently, the solar wind speed is the main parameter that determines the Sun-Earth transit time in most events. This can explain the so-called "Brueckner's 80 h rule" ($TT \approx 80$ h in most events; see Brueckner *et al.* 1998), since it can be presumed that most of CMEs propagate through slow solar wind, and for events with $m < 10^{12}$ kg a typical transit time should be around 80 h (Fig. 4). The shortest transit times ($TT < 1$ day) can be achieved only by massive CMEs of a very high take-off velocity ($v_0 > 2000$ km s^{-1}). Furthermore, a CME has to move through fast solar wind streams and the Lorentz force has to act over large distances to postpone the drag-dominant phase until the solar wind density becomes low.

References

Brueckner, G. E., Howard, R. A., Koomen, M. J., *et al.* 1995, *Solar Phys.*, 162, 357

Brueckner, G. E., Delaboudiniere, J.-P., Howard, R. A., *et al.* 1998, *Geophys. Res. Lett.*, 25, 3019

Cargill, P. J. 2004, *Sol. Phys.*, 221, 135

Cargill, P. J., Chen, J., Spicer, D. S., & Zalesak, S. T. *J. Geophys. Res.*, 101, 4855

Gonzalez-Esparza, J. A., Lara, A., & Perez-Tijerina, E. 2003, *J. Geophys. Res.*, 108, 1039

Gosling, J. T. 1990, in Physics of Magnetic Flux Ropes, *Geophys. Monogr.* Ser. vol. 58, eds. Russell, C. T., Priest, E. R., Lrr, L. C., AGU, Washington, D.C., 343

Gopalswamy, N., Lara, A., Yashiro, S., Kaiser, M. L., & Howard, R. A. 2001 *J. Geophys. Res.*, 106, 29207

Leblanc, Y., Dulk, G. A., & Bougeret, J.-L. 1998, *Solar Phys.*, 183, 165

Lindsay, G. M., Luhmann, J. G., Russell, C. T., and Gosling, J. T. 1999, *J. Geophys. Res.*, 104, 12515

Manoharan, P. K. 2006, *Solar Phys.*, 235, 345

Schwenn, R., dal Lago, A., Huttunen, E., & Gonzalez, W. D. 2005, *Annales Geophysicae*, 23, 1033

Vršnak, B., Ruždjak, D., Sudar, D., & Gopalswamy, N. 2004a, *A&A*, 423, 717

Vršnak, B., Magdalenić, J., & Zlobec, P. 2004b, *A&A*, 413, 753

Vršnak, B. & Žic, T. 2007, *A&A*, 472, 937

Vršnak, B., Vrbanec, D., & Čalogović, J. 2008, *A&A*, in press

Webb, D. F., Forbes, T. G., Aurass, H., *et al.* 1994, *Solar Phys.* 153, 73

Yashiro, S., Gopalswamy, N., Michalek, G., *et al.* 2004, *J. Geophys. Res.*, 109, 7105

Discussion

SPANGLER: I am surprised your drag coefficients were so close to unity. I would have expected that for an obstacle (the CME) moving through an MHD medium, there would be enhanced drag due to radiation of Alfven waves. This effect was responsible for the accelerated orbital decay of the echo-satellite in 1960.

VRSNAK: In fact, the "aerodynamic" drag in the corona and IP space is almost entirely due to the emission of MHD waves since the viscosity is negligible. Numerical simulations by Cargill *et al.* 1995, and later Cargill 2004 show that in the corona and IP space $c_d \approx 1$ except in the case e.g. when CME is of very low density.

IBADOV: If you are not taking into account gravity why CME mass is important for the drag coefficient?

VRSNAK: In principle the drag acceleration depends on the density of the body. Here we used the mass of CME since mass is a measurable quantity (for the density we have to assume the line-of-sight length).

GOPALSWAMY: Your drag essentially is proportional to inverse of CME size scale. Is this correct?

VRSNAK: Yes, that 's right.

SCHMIEDER: If slow and not many CME are accelerated they could loose their identity in the Solar Wind and will be not geoeffective.

VRSNAK: Depends what you mean by the phrase "their identity": they will move with solar wind so we will not detect them in the flow velocity observation. Yet, their magnetic structure will be preserved (that is why slow CMEs can be also geoffective) except in the case of an efficient reconnection which may also "wash-out" their "magnetic identity".

FAINSHTEIN: The force for CME drag must depend on the CME velocity and size. I think that your plot of the acceleration of CME vs mass of CME is a result of this. What do you think about this idea? I assume that CMEs of small size have small masses, and CMEs of large sizes have large masses, on average.

VRSNAK: Basically yes, since CME densities probably do not differ very much, i.e. the mass is primarily determined by the volume.

Universal Heliophysical Processes
Proceedings IAU Symposium No. 257, 2008
N. Gopalswamy & D.F. Webb, eds.

© 2009 International Astronomical Union
doi:10.1017/S1743921309029408

Determining the full halo coronal mass ejection characteristics

V. G. Fainshtein

Institute of solar-terrestrial physics SB RAS,
P.O. Box 291, 664033, Irkutsk, Russia
email: vfain@iszf.irk.ru

Abstract. In this paper we determined the parameters of 45 full halo coronal mass ejections (HCMEs) for various modifications of their cone forms ("ice cream cone models"). We show that the CME determined characteristics depend significantly on the CME chosen form. We show that, regardless of the CME chosen form, the trajectory of practically all the considered HCMEs deviate from the radial direction to the Sun-to-Earth axis at the initial stage of their movement.

1. Introduction

The full halo CME (HCME) was first reported by Howard *et al.* (1982). A lot of HCMEs were detected by LASCO [Yashiro *et al.* (2004)]. Full HCMEs are considered to move from the Sun to the Earth if they are accompanied by activity on the visible disk of the Sun [Webb *et al.* (2000)]. Full HCMEs are responsible for many large geomagnetic storms [Webb *et al.* (2000)]. To time the arrival of such CMEs to the Earth and predictions their geoeffective parameters at $R = 1AE$ it is necessary to determine the true full HCME characteristics near the Sun. In the papers [Zhao *et al.* (2002); Michalek *et al.* (2003); Hie *et al.* (2004); Hue et al (2005); Michalek (2006) and Fainshtein (2006)] some techniques of finding the true full HCME parameters in 3-D space were proposed. In the most techniques CMEs are supposed to have a cone form [Howard *et al.* (1982); Fisher & Munro (1984)]. But, as observations of limb CME show, the form of every CME is best approximated by only one of the cone form three modifications ("ice cream cone models", Fisher & Munro (1984)). In this paper parameters of 45 full HCME for three possible modifications of their cone form are determined using the method [Fainshtein (2006)]. We show that these parameters essentially depend on the CME form chosen modification.

2. Determining the observed full halo CME characteristics

Our analysis showed that there is a positive correlation between the eruptive prominence and/or the limb post-eruptive arcade angular size δ_{P-A} and the 2α angular size of the LASCO C3 CME related to the prominence (arcade), [Fainshtein (2006); Fainshtein (2007)]. The regression line equation for this correlation is $2\alpha = -0.18\,\delta_{P-A}{}^2 + 10.16\delta_{P-A} + 11.3$.

According to Fainshtein (2006), we will assume that this regression line also relates the eruptive filament (EF) angular sizes and/or the post-eruptive arcade (PEA) on the visible disk of the Sun to the EP (PEA) - related full HCME angular size. Then, to find angular size of such CMEs one may use the above regression line, in which δ_{P-A} will mark now the EF (or PEA) angular size on the visible disk of the Sun. To determine other parameters of the full HCMEs we used the relations between the halo CME characteristics obtained within the CME cone model three modifications, Fig. 1. For these models in this

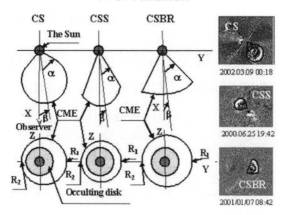

Figure 1. The coronal mass ejection models. A1-C1: the image of the CME moving at β-angle to the Sun-to-Earth axis (X-axis) in the plane X - CME axes. The Y-axis is perpendicular to the X-axis and is located in the plane of the X-axis - CME axes as well as in the plane of the sky. A2-C2: outer-boundary images of a model halo CME in the plane of the sky (plane YZ). To the right of the CME models are the examples of the limb CMEs whose outer boundary form is close to each of model forms.

paper we will use the following designations: CS , CSS, and CSBR models. These models are different in the relation between the cone base size and the size of the structure on which the cone leans. To simplify, we assume the cone base form is a circle, and the cone leans on the part of the sphere in this paper. Using Fig. 1, we may obtain the expressions relating the full HCME angular size with its other parameters. To illustrate this we will give two formulas for the CS model:

$$\sin \beta = \frac{R_1 - R_2}{R_1 + R_2} \sin \alpha \tag{2.1}$$

$$V_{FE} = V_1 \frac{\cos \beta + \sqrt{\sin \alpha^2 - \sin \beta^2}}{\sin \alpha + \sin \beta} \tag{2.2}$$

Here, β is the angle between HCME axis and the Sun - to - Earth axis, V_1 is the velocity of a point with radius R_1, R_1 is the largest, and R_2, respectively, is the smallest positions of the HCME image boundary in the plane of the sky, V_{FE} is the CME front velocity along the Sun-to-Earth axis.

To test the given method of determining the full HCME parameters, 45 coronal mass ejections associated with the eruption of filament (EF) and/or post-eruptive arcade (PEA) have been selected. The EF and PEA characteristics were determined by Sun images in the FeXII λ 195A extreme ultraviolet line (SOHO/EIT). The procedure of the EF (PEA) angle size and values R_1 and R_2 determination is described in the paper Fainshtein (2006). For analysis we also used the β_{P-A} angle position of the EF and/or EPA center (within the heliocentric coordinate system). As V1 values we used the linear fit velocities from the "LASCO CME catalog" ($http : //cdaw.gsfc.nasa.gov/CME\_list/$).

3. Results

(i) It is evident from the our analysis that the CME parameters revealed by using various models differ distinctly. For example, $< \beta > (CS \quad model) = 10.3°$; $< \beta > (CSS \quad model) = 15.4°$; $< \beta > (CSBR \quad model) = 17.05°$, and $< V_{FE}/V_1 >$

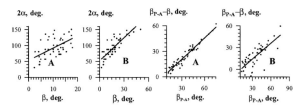

Figure 2. The β_{P-A}-angle dependence of the CME 2α angular size and the $\beta_{P-A} - \beta$ correlation with the β_{P-A} value. A is the CS model, B is the CSBR model.

$(CS \quad model) = 2.3; \; < V_{FE}/V_1 > (CSS \quad model) = 2.05; \; < V_{FE}/V_1 > (CSBR \quad model)$ =1.4. These differences should be considered significant if one applies, e. g., the obtained results to determine the halo CME transit time between the Sun and Earth. Therefore, one has to justify the usage of this or that model in each particular case.

(ii) The β-angle differs significantly from the β_{P-A} angular position of the eruptive filament (of the post-eruptive arcade)related to the HCME. The author believes that this inequality reflects halo CME trajectory deviation from the radial direction towards the Sun-to-Earth axis at the initial stage of their movement. The physical mechanism of such a CME trajectory peculiarity is proposed in the paper Fainshtein (2007).

(iii) With the β_{P-A} (the value of angle position of the CME-related eruptive filament (post-eruptive arcade)) increase, the value of the $(R_1 - R_2)/(R_1 + R_2)$ parameter, that characterizes the halo CME center shift in the plane of the sky relative to the solar disk center along the HCME large axis, also increases.

(iiii) The bulk of the considered full halo CMEs have relatively large angle size with the mean value of $93°$. This value is essentially larger than the "limb" CME mean angular size equal to $\approx 45°$ [Yashiro *et al.* (2004). The similar conclusion was made earlier in the papers [Michalek *et al.* (2003); Fainshtein (2006)]. The CMEs, whose axis are most deviated from the Sun-to-Earth axis, have, on average, larger angle size that the CMEs moving near the Sun-to-Earth axis. In its turn, the $\beta - \beta_{P-A}$ angular difference increases with increasing β_{P-A}, on average (Fig. 2).

Acknowledgements. SOHO is a project of international cooperation between ESA and NASA. This work was supported by RF Leading Scientific Schools Support Governmental Grant SS 4741.2006.2 and by P-16 RAS Presidium Fundamental Research Program.

References

Fainshtein, V. G. 2006, *Geomagnetism and aeronomy*, 46, 384

Fainshtein, V. G. 2007, *Cosmic Research*, 45, 384

Fisher, R. R. & Munro, R. H. 1984, *Astrophys. J.*, 280, 428

Howard, R. A., Michels, D. J., Sheeley, N. R. Jr., & Koomen, M. J. 1982, *Astrophys. J.*, 263, L101.

Michalek, G., Gopolswamy, N., & Yashiro, S. 2003, *Astrophys. J.*, 584, 472

Michalek, G. 2006, *Solar Phys.*, 237, 101

Webb, D. F., Cliver, E. W., Crooker, N. U., St. Cyr, O. C., & Thomson, B. J. 2000, *J. Geophys. Res.*, 105, 7491

Xie, H., Offman, L., & Lawrence, G. 2004, *J. Geophys. Res.*, 109, A03109, doi:10.1029/ 2003JA010226

Hue, X. H., Wang, C. B., & Dou, X. K. 2005, *J. Geophys. Res.*, 110, A08103, doi:10.1029/ 2004A010698

Yashiro, S., Gopalswamy, N., Michalek, G., St. Cyr, O. C., Plunkett, S. P., Rich, N. B., & Howard, R. A. 2004 *J. Geophys. Res.* 109, A07105

Zhao, X. P., Plunkett, S. P., & Liu, W. J. 2002, *J. Geophys. Res.*, 107, SSH (13-1), doi10.1029/2001JA009143

Universal Heliophysical Processes
Proceedings IAU Symposium No. 257, 2008
N. Gopalswamy, & D. F. Webb, eds.

© 2009 International Astronomical Union
doi:10.1017/S174392130902941X

Major solar flares without coronal mass ejections

N. Gopalswamy[1], S. Akiyama[1,2] and S. Yashiro[1,2,3]

[1]NASA Goddard Space Flight Center, Greenbelt, MD 20771, USA
email: `nat.gopalswamy@nasa.gov`

[2]The Catholic University if America, 620 Michigan Ave. NE Washington DC. 20064, USA
email: `sachiko.akiyama@nasa.gov`

[3]Interferometrics Inc., 13454 Sunrise Valley Dr. #240, Herndon, VA 20171, USA
email: `seiji.yashiro@nasa.gov`

Abstract. We examine the source properties of X-class soft X-ray flares that were not associated with coronal mass ejections (CMEs). All the flares were associated with intense microwave bursts implying the production of high energy electrons. However, most (85%) of the flares were not associated with metric type III bursts, even though open field lines existed in all but two of the active regions. The X-class flares seem to be truly confined because there was no material ejection (thermal or nonthermal) away from the flaring region into space.

Keywords. coronal mass ejections (CMEs), flares, particle emission, radio radiation

---

1. Introduction

Magnetic energy release is a ubiquitous phenomenon on the Sun occurring at various spatial and temporal scales. Flares and coronal mass ejections (CMEs) are the two main manifestations of the energy release process. All CMEs are associated with soft X-ray flares (if we include post-eruption arcades in X-ray images), but not all flares are associated with CMEs. The rate at which flares lack CMEs increases as one goes from stronger to weaker flares (Yashiro *et al.* 2005). It is known for a long time that flares with mass motion in H-alpha (also known as eruptive flares) have a high rate of association with CMEs (Munro *et al.* 1979). Flare duration is another important parameter deciding the association with CMEs (Sheeley *et al.* 1983) because eruptive flares are of long duration, while confined flares are impulsive (Kahler *et al.* 1989). Feynman and Hundhausen (1994) reported an X4.0 flare without a CME, although dark and bright surges were associated with the flare. Green *et al.* (2002) reported on the 2000 September 30 X1.2 flare that lacked a CME. Recently, Akiyama *et al.* (2007) compared two active regions, one with flares poorly associated with CMEs and the other with a high rate of association with CMEs. They found that the CME-poor flares were localized near the neutral line between the preceding and following polarity regions. On the other hand, the CME-rich flares were scattered all over the active region. This was also confirmed by Wang and Zhang (2007) using a displacement parameter defined by the surface distance between the flare location and the centroid of magnetic flux distribution in the active region. Nindos and Andrews (2004) found that the coronal helicity is lower in active regions producing compact flares, compared to those with eruptive flares. In this paper, we investigate other aspects of CMEless X-class flares including the escape of nonthermal particles accelerated during the flares. For this purpose, we use the 13 CMEless X-class flares indentified during the interval 1996 to 2005 (inclusive), when the Solar Heliospheric

Observatory (SOHO) mission imaged the corona over a field of view of 2–32 solar radii using the Large Angle and Spectrometric Coronagraph (LASCO).

2. Data Selection

We first collected all the X-class flares reported by the Solar Geophysical Data (SGD), with the start, peak and end times as well as the source locations from H-alpha flares or from the GOES soft X-ray imager. We checked the flare list against the list of CMEs available online, http://cdaw.gsfc.nasa.gov, keeping in mind that the CMEs are expected radially above the flare location (Yashiro *et al.* 2008). The vast majority of X-class flares were associated with CMEs. However, a set of 13 flares did not have corresponding CMEs. For each of these flares, we examined the association of microwave and metric radio bursts (type III and type II) from SGD. We also obtained the source active regions and the heliographic coordinates of the flare locations from SGD. We examined the potential field source surface extrapolation plots available online (http://www.lmsal.com/forecast/, courtesy of M. deRosa & K. Schrijver) to determine if open field lines exist in the active region. Ten of the 13 CMEless flares were listed in Wang and Zhang (2007), although they investigated only 4 flares (one of their flares – the 13:49 UT flare on 2004 July 16 – was associated with an uncataloged CME at position angle 130° and a metric type II burst).

Figure 1 shows one of the CMEless flares – the X1.2 flare of 2005 January 15 at 00:22 UT originating from the northeast quadrant (active region 0720 at N14E08) as seen in EUV images obtained by SOHO's Extreme-ultraviolet Imaging Telescope (EIT). The flare peaked at 00:43 UT and ended at 01:02 UT (duration ∼40 min). The H-alpha flare had an optical importance 1F, but was not eruptive. An intense microwave burst was reported (peak flux ∼3000 SFU at 15.4 GHz) in SGD. The LASCO observations do not show any mass motion (in the direct or the difference image). The radio dynamic spectrum has no observable radio feature during the flare interval (neither type III nor type II bursts). Similar analysis was performed on all the 13 flares and the details are given in Table 1. Specifically, we list the flare start and end times, duration (Dur), X-ray importance (Imp), location, active region (AR) number, H-alpha flare importance (n – no

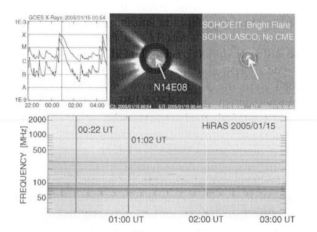

Figure 1. (Top) GOES soft X-ray flare, SOHO/LASCO image with superposed EIT image showing the flare and SOHO/LASCO and EIT difference images. (Bottom) Radio dynamic spectrum from the Hiraiso radio spectrograph (HiRAS) with the start (00:22) and end (01:02) times of the flare marked. The flare of interest is pointed by arrows. Note that there is no radio emission during the flare.

Table 1. X-class flares without CMEs during solar cycle 23 and their properties

#	Flare Start	Peak	Dur	Imp	Location	AR #	Hα	III	μfpk/flux
1	2000/06/06 13:30	13:39	16	X1.1	N18E12	9026[d]	N	N	2.7/560
2	2000/09/30 23:13	23:21	8	X1.2[c]	N07W90	9169	N	N	15.4/2800
3	2001/04/02 10:04	10:14	16	X1.4	N17W60	9393	1B[e]	Y	15.4/1200
4	2001/06/23 04:02	04:08	9	X1.2[c]	N10E23	9511	1B	N	5/100
5[a]	2001/11/25 09:45	09:51	9	X1.1[c]	S16W69	9704[d]	N	N	15.4/130
6	2002/10/31 16:47	16:52	8	X1.2[c]	N29W90	0162	N	N	8.8/3300
7[b]	2004/02/26 01:50	02:03	20	X1.1[c]	N14W15	0564	2N[e]	N	15.4/830
8	2004/07/15 18:15	18:24	13	X1.6	S11E45	0649	N	N	8.8/530
9	2004/07/16 01:41	02:06	29	X1.3	S11E41	0649	N	N	15.4/1900
10	2004/07/16 10:32	10:41	14	X1.1	S10E36	0649	1F[e]	Y	15.4/1200
11	2004/07/17 07:51	07:57	8	X1.0	S11E24	0649	3B[e]	N	5/820
12	2005/01/15 00:22	00:43	40	X1.2	N14E08	0720	1F	N	15.4/3000
13[a]	2005/09/15 08:30	08:38	16	X1.1	S12W14	0808	2N	N	15.4/4100

[a] There were frequent blobs of material along the streamer near the position angle of the flare throughout the day. [b] A small wisp of material was seen close to the north pole (PA ~350) at flare peak. This is probably unrelated to the flare. [c] These flares were isolated; no other X-class flares from these regions. Rest of the regions had other X-class flares with CMEs. [d] These two regions had no open field lines. [e] Listed as eruptive H-alpha flare.

flare reported), type III burst association (N – no, Y – yes), peak frequency of microwave burst (fpk in GHz) and flux in solar flux units (SFU).

3. Results

Several results can be directly extracted from Table 1: (i) The CMEless flares occurred only during the maximum (5 flares) and declining phases (8 flares) of solar cycle 23. (ii) The flares were generally impulsive, with durations ranging from 8 min to 40 min (average value ~16 min). (iii) The X-ray flares were in the low X-class ranging from of X1.0 to X1.6 (average ~X1.2). (iv) The source locations were at all longitudes, some of them being limb events, for which it is easier to detect CMEs. (v) The 13 flares occurred in 10 different active regions, with one region (AR 0649) producing 4 flares. Interestingly, half of the active regions produced other X-class flares (two or more) associated with CMEs. (vi) Only four of the 13 flares were reported to be "eruptive" in H-alpha (SGD). However, the mass motion is likely to be in the horizontal direction. (vii) It is remarkable that only two of the 13 flares were associated with metric type III bursts (very brief). Since there is no spatial information for the type III bursts, we cannot be sure whether the temporal association in the two cases means a spatial association with the flares. (viii) All the flares were associated with intense microwave bursts with peak fluxes varying from 100 SFU to 4100 SFU. The peak emission frequency was around 15.4 GHz for 8 events implying that ~MeV electrons were accelerated during the flare that emit gyrosynchrotron radiation. (ix) All but two active regions had open field lines from one of the polarity patches.

4. Discussion

Microwave emission from flares is caused by high energy (several hundred KeV to MeV) electrons trapped in flare loops. The association of intense microwave bursts during all the 13 flares suggests that nonthermal electrons were produced. However, during 11 out of the 13 (or 85%) flares, no metric type III burst was reported (type III bursts reported in 2 cases, but we cannot say if they were spatially associated with the CMEless flares without spatial information). Only ~25 KeV electrons propagating along open field lines are required to produce type III bursts. The lack of type III bursts may be due to : (i) lack of low energy electrons that produce Langmuir waves, (ii) lack of open field lines in the active region, and (iii) the low energy electrons without access to the open field lines in the active region. It is unlikely that high-energy electrons are produced (for microwave emission) without lower energy electrons (for type III bursts). Potential field extrapolation indicates the presence of open field lines in most of the active regions. Therefore, we can eliminate the possibilities (i) and (ii) and conclude that the accelerated electrons did not have access to the open field lines. If the accelerated electrons are confined to the active region core fields, type III bursts are not expected even if there are open field lines at the edges of active regions. Thus, confined flares lack not only thermal material (CMEs) but also nonthermal material (accelerated electrons) in the anti-Sunward direction. The complete lack of type II bursts during the confined flares has also implications for the source of coronal shocks. The impulsive flares are well-suited to produce blast waves, yet none of the flares was associated with metric type II bursts. This suggests that the presence of CMEs is necessary for type II bursts, supporting the view that coronal shocks are likely to be driven by CMEs.

It is significant that half of the active regions producing CMEless flares also produced two or more X-class flares with CMEs. Some of these were super active regions (AR 9393, 0720, and 0808) that produced a large number of energetic flares and CMEs of great heliospheric consequence (Gopalswamy et al. 2006). It will be worthwhile to investigate the relative positions of the confined and eruptive flares from the same active region to further understand the two types of flares.

5. Conclusions

We identified thirteen X-class flares during cycle 23 that were not associated with CMEs. The flares were generally impulsive (duration ∼16 min) and the X-ray peak flux never exceeded X1.6. The flares showed heating and particle acceleration, but lacked material motion in the vertical direction. While the presence of intense microwave bursts imply copious production of high energy electrons, the lack of type III bursts suggest that the accelerated electrons did not escape from the flare site even though the active regions had open field lines. Many active regions produced both eruptive and confined flares, which needs further investigation to see if confined flares contribute to the eventual occurrence of eruptive flares.

References

Akiyama, S., Yashiro, S., & Gopalswamy N. 2007, *Adv. Space Res.*, 39, 1467
Feynman, J. & Hundhausen, A. J. 1994, *J. Geophys. Res.*, 99, 8451
Gopalswamy, N., Yashiro, S., & Akiyama, S. 2006, in: N. Gopalswamy & A. Battacharyya (eds.), *Solar Influence on the Heliosphere and Earth's Environment: Recent Progress and Prospects* (Mumbai: Quest Publications), p. 79
Green, L. M., Matthews, S. A., van Driel-Gesztelyi, L., Harra, L. K., & Culhane, J. L. 2002, *Solar Phys.*, 205, 325
Kahler, S. W., Sheeley, Jr., N. R., & Liggett, M. 1989, *ApJ*, 344, 1026
Munro, R. H., Gosling, J. T., Hildner, E., MacQueen, R. M., Poland, A. I., & Ross, C. L. 1979, *Solar Phys.*, 61, 201
Nindos, A. & Andrews, M. D. 2004, *ApJ (Letters)*, 616, L175
Sheeley, N., Howard, R. A., Koomen, M. J., & Michels, D. J. 1983, *ApJ*, 272, 349
Wang, Y.-M. & Zhang, J. 2007, *ApJ*, 665, 1428
Yashiro, S., Gopalswamy, N., Akiyama, S., Michalek, G., & Howard, R. A. 2005, *J. Geophys. Res.*, 110, A12S05
Yashiro, S., Michalek, G., Akiyama, S., Gopalswamy, N., & Howard, R. A. 2008, *ApJ*, 673, 1174

Universal Heliophysical Processes
Proceedings IAU Symposium No. 257, 2008
N. Gopalswamy & D.F. Webb, eds.

© 2009 International Astronomical Union
doi:10.1017/S1743921309029421

Dynamics of interplanetary CMEs and associated type II bursts

Alejandro Lara[1] and Andrea I. Borgazzi[1,2]

[1]Instituto de Geofísica
Universidad Nacional Autónoma de México (UNAM), México
email: alara@geofisisca.unam.mx

[2]Divisão de Geofísica Espacial, INPE, Brasil
email: andrea@geofisica.unam.mx

Abstract. Coronal mass ejections (CMEs) are large scale structures of plasma ($\sim 10^{16} g$) and magnetic field expelled from the solar corona to the interplanetary medium. During their travel in the inner heliosphere, these "interplanetary CMEs" (ICMEs), suffer acceleration due to the interaction with the ambient solar wind. Based on hydrodynamic theory, we have developed an analytical model for the ICME transport which reproduce well the observed deceleration of fast ICMEs. In this work we present the results of the model and its application to the CME observed on May 13, 2005 and the associated interplanetary type II burst.

Keywords. Sun: coronal mass ejections (CMEs), Sun: radio radiation, shock waves, solar-terrestrial relations, solar wind

1. Introduction

The transport of interplanetary coronal mass ejections (ICMEs), has been studied for a few years, motivated mainly by the necessity of accurate Sun - Earth travel time predictions. The efforts to explain the ICME behavior may be divided into the following three categories (we include few examples in each case):

• Empirical models (Gopalswamy *et al.* 2000, 2001, 2005; Vršnak 2001; Vršnak *et al.* 2002, 2004, 2007).

• Numerical simulations (Cargill *et al.* 1996; Cargill 2004; Vandas *et al.* 1995; Odstrčil *et al.* 1999a,b; Gonzalez-Esparza *et al.* 2003).

• Theoretical models (Canto *et al.* 2005; Borgazzi *et al.* 2008).

Recently, we have developed an analytical method to explain the dynamics of ICMEs traveling in the ambient solar wind (SW). In this work we present two solutions of the model and the application of this model to the dynamics of the May 13, 2005 ICME and its associated Type II burst.

2. The Model

The forces acting on a body moving through a fluid are generically called "drag forces". Two kinds of drag forces are typically used, depending mainly on the velocity of the body, U (in fact, the selection depends on the Reynolds number): the linear dependence, which we call "laminar", is $F_l = 6\pi \mu R \cdot U$, and the quadratic dependence, which we call "turbulent" is $F_t = \frac{C_d A \rho_{sw} \cdot U^2}{2}$, in this case, ρ_{sw} is the interplanetary medium density; R and A are the ICME radius and cross section; μ is the viscosity and C_d is the drag coefficient, a dimensionless parameter which describes the behavior of the body traveling through any fluid.

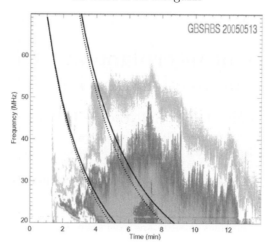

Figure 1. Spectrogram showing the metric type II burst, observed by GBSRBS, overplotted are the fundamental (lower curves) and harmonic (upper curves) solutions of our model. The starting time is 16:41 UT.

Considering separately both forces in the equation of motion, and assuming variations of the ICME radius as $R(x) = x^p$ and SW density as $\rho_{sw} = x^{-2}$ (this approximation is valid from a few tens of $R_\odot$ to beyond 1 AU (Leblanc et $al.$ 1996)), we have the following solutions:

$$-\frac{6\pi\nu\rho_0}{m_{cme}(p-1)}\left[x^{p-1} - x_0^{p-1}\right] = U + U_{sw}\ln\frac{(U - U_{sw})}{(U_0 - U_{sw})} - U_0 \tag{2.1}$$

and

$$-\frac{C_d\pi\rho_0}{2m_{cme}(2p-1)}[x^{(2p-1)} - x_0^{(2p-1)}] \tag{2.2}$$

$$= \frac{U_{sw}}{(U_0 - U_{sw})} - \frac{U_{sw}}{(U - U_{sw})} + \ln\left[\frac{(U - U_{sw})}{(U_0 - U_{sw})}\right]$$

for the laminar and turbulent cases, respectively. Here m_{cme} is the CME mass, ρ_0 is a scaling factor for the density model and corresponds to the density measured at 1 AU; U_0 is the initial CME velocity measured at the initial position x_0; U_{SW} is the SW velocity; and U is the ICME velocity at position x.

3. Type II Bursts

Type II bursts are produced when a disturbing agent is traveling, at relatively moderate speeds, through the solar atmosphere. This agent disturbs the ambient plasma which radiates electromagnetic emission at the plasma frequency. The agent may be a blast wave produced by a flare or a shock wave driven by a CME. At metric wavelengths, both mechanisms are plausible. At hectometric-decimetric and kilometric wavelengths, it is generally accepted that the agent is a shock wave produced by an ICME.

3.1. The May 13, 2005 event

At 17:22 UT on May 13, 2005 a very fast ($v_{cme} \approx 1689$ km/s) halo CME was observed. This CME was associated with a M1.8 flare, localized close to the center of the disk (N12E11), starting at 16:13 UT, peaking at 16:57 UT, and ending at 17:28 UT. At coronal levels, a metric type II burst started at 16:42 UT at 30 and 60 MHz for the fundamental and harmonic emissions, respectively (Figure 1). In the interplanetary medium,

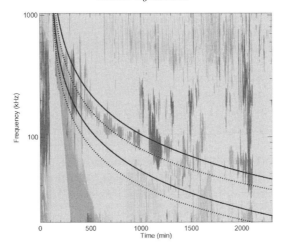

Figure 2. WIND/WAVES spectrogram showing the frequency evolution of the radio emission generated by the ICME driven shock. The starting time is 16:41 UT.

WIND/WAVES experiment detected an interplanetary type II burst, extending in frequency, from 20 kHz to 13.825 MHz and starting $\sim$ 16:00 UT on May 13, 2005 (Figure 2).

At the Earth, a sudden storm commencement (SSC) started at $\sim$ 02:38 UT on May 15, indicating the arriving of the ejecta to 1 AU, followed by a strong (Dst $\approx$ -263 nT) geomagnetic storm. Therefore, the travel time, from May 13 17:22 (first LASCO/C2 observation) to May 15 02:38 (SSC) was 33 hr 16 min.

4. Dynamics of the May 13, 2005 ICME

In order to reproduce the ICME dynamics, indicated by the Type II bursts (which was produced by the ICME driven shock), we have used both, laminar (Eq. 2.1) and turbulent (Eq. 2.2) solutions and changed the parameters accordingly in order to fit both the computed ICME travel time ($\sim$ 33 hr) and the arriving velocity observed at 1 AU ($\sim$ 1097 km/s) for this event. The following parameters correspond to the best fits and the correspondent solutions are over plotted in Figures 1 and 2:

- Laminar regime (solid lines):
  - U_0 = 1689 km/s, p = 0.77, ν = 1.03 $\times$ 10^{21} cm^2/s, m_{CME} = 10^{16} g and U_{SW} = 415 km/s.
  - travel time = 33.8 hr. and ICME velocity at 1 AU = 1169.2 km/s
- Turbulent regime (dotted lines):
  - U_0 = 1689 km/s, p = 0.78, C_d = 3.94 $\times$ 10^4, m_{CME} = 10^{16} g and U_{SW} = 415 km/s.
  - travel time = 33.0 hr. and ICME velocity at 1 AU = 1054.4 km/s

Equations 2.1 and 2.2 give the ICME instant velocity as a function of position. In order to have the ICME position as a function of time, we have divided the interval (1 AU) in 4092 equally spaced sectors, computed the mean velocity in each sector and then, computed the time assuming constant velocity in the sector. Once we had the ICME position as a function of time, we were able to obtain the density by assuming a SW density model ($n \sim 1/x^2$) and therefore, the plasma frequency ($f \sim \sqrt{n}$) as a function of time.

In this way, we can compare our results against the observed Type II behavior. In Figures 1 and 2, we have plotted the computed laminar (continuous line) and turbulent (dotted line) frequency drift. The lower pair of curves represent the fundamental and the upper pair represent the harmonic emissions. We selected the ICME starting time equal to the metric Type II burst starting time.

5. Discussion – Conclusion

The very good fits of both the coronal (Fig. 1) and interplanetary medium (Fig. 2) type II bursts with our model shows that the dynamics in the interplanetary medium of CMEs can be explained by the hydrodynamic theory. This means that the drag force plays a major role in the ICME transport, both linear or quadratic speed dependence are plausible. We note that in this case, and as we are dealing with low density collisionless plasmas, the viscosity and/or drag interactions should be produced by microscopic wave-wave or wave-particle interactions.

At large scales ($\sim$ 1 AU) there are not significant difference between the linear and quadratic dependence of the drag force with the ICME speed. Although, the quadratic model seems to fit better the curvature of the type II frequency drift, the travel time and the ICME velocity at 1 AU.

It is important to note that our approximation has few assumptions, as the radial expansion of the ICME ($R_{ICME}(x) = x^p$) and the density decrease ($\rho(x) = \rho_0/x^2$). The only free parameters are C_d or ν. This fact shows the advantage of using analytical models.

Finally, we quote the values of the drag coefficient, $C_d \approx 4 \times 10^4$ and kinematic viscosity, $\nu \approx 10^{21}$ which seems to reproduce well this event.

Acknowledgements

A. Lara thanks UNAM-PAPIT (IN117309) and CONACyT (49395). A. Borgazzi thanks CNPq-Brazil.

References

Borgazzi, A., Lara, A., Romero-Salazar, L., & Ventura, A. 2008, *Geofísica Internacional*, 47, 301

Canto, J., Gonzalez, R. F., Raga, A. C., de Gouveia Dal Pino, E. M., Lara, A., & Gonzalez-Esparza, A. 2005, *Mon. Not R. Astrom. Soc.*, 357, 572

Cargill, P. J. & Chen, J. 1996, *J. Geophys. Res.*, 101, A3, 4855

Cargill, P. J. 2004, *Sol. Phys.*, 221, 135

Gonzalez-Esparza, A., Lara, A., Perez-Tijerina, E., Santillan, A., & Gopalswamy, N. 2003, *J. Geophys. Res.*, 108, A1, 1039

Gopalswamy, N., Lara, A., Yashiro S., Kaiser, M. L., & Russell H. 2001, *J. Geophys. Res.*, 18, 29.207

Gopalswamy, N., Lara, A., Lepping, R. P., Kaiser, M. L., Berdichevsky, D., & St,. Cyr, O. C. 2000, *Geophys. Res. Lett.*, 27, 145

Gopalswamy, N., Lara, A., Manoharan, P., & Howard, R. 2005, *Adv. Space Res.*, 36, 2289

Leblanc, Y., Dulk, G., & Bougeret, J. 1996, *Solar Phys.*, 183, 165

Odstrčil, D. & Pizzo, V. J. 1999a, *J. Geophys. Res.*, 104, A1, 483

Odstrčil, D. & Pizzo, V. J. 1999b, *J. Geophys. Res.*, 104, A1, 493

Vandas, M., Fisher, S., Dryer, M., Smith, M., & Detman, T. 1995, *J. Geophys. Res.*, 100, A7, 12258

Vršnak, B. 2001, *J. Geophys. Res.*, 106, A11, 25249

Vršnak, B. & Gopalswamy, N. 2002, *J. Geophys. Res.*, 107, A2

Vršnak, B., Ruždjak, D., Sudar, D., & Gopalswamy, N. 2004, *A&A*, 423, 717

Vršnak, B. & Žic, T. 2007, *A&A*, 472, 937

Universal Heliophysical Processes
Proceedings IAU Symposium No. 257, 2008
N. Gopalswamy & D.F. Webb, eds.

© 2009 International Astronomical Union
doi:10.1017/S1743921309029433

The heliosphere mass variations: 1996–2006

T. Pintér[1], I. Dorotovič[1], and M. Rybanský[2]

[1]Slovak Central Observatory, P.O. Box 42, SK-94701 Hurbanovo, Slovak Republic
email: `ivan.dorotovic@suh.sk`

[2]Institute of Experimental Physics SAS, Watsonova 47, SK-04353 Košice, Slovak Republic
email: `milanr@centrum.sk`

Abstract. The variations of the global mass of heliosphere in the 23rd cycle of the solar activity are described. The results are derived from solar corona observations and from 'in situ' measurements made by the space probes SOHO, VOYAGER2, ACE, WIND, and ULYSSES. It has been revealed that though the total mass of corona fluctuates during the solar activity cycle approximately in a ratio of 1 : 3, the specific mass flow (q) in the solar wind does not change in the ecliptic plane. In the polar regions the q decreases during the minimum in a third of the original value and the velocity of expansion is roughly double. These findings are valid for the 23rd solar cycle.

Keywords. Sun: corona, interplanetary medium

1. Introduction

The space controlled by the Sun through its gravitational force is usually called heliosphere. If we take into account that the distance from the nearest stars is approximately 4 LY, then the radius of the heliosphere would be 2 LY, i.e. $9.46\,10^{12}$ km or 63235 AU. However, heliosphere is usually the space with a diameter of $100-200$ AU and the places where the solar and star wind have the same energy density are regarded as the limit of the heliosphere. This space is filled with the matter of the expanding corona solar wind and interplanetary dust as indicated by observations of cometary tails and later confirmed by 'in situ' measurements of space probes. The source of our knowledge about matter distribution in the heliosphere before space missions era was the analysis of observations of the solar corona during total eclipses (K- + F-corona) and the zodiacal light which is an extension of the F-corona. The theoretical basis for such analyses is the knowledge of the mechanisms of light scattering on free electrons (K-corona) and solid dust particles (F-corona and zodiacal light). The second section contains a summary of the results of the analysis and the third section brings information on 'in situ' measurements of proton density and their temperature using space probes. The variations of the heliosphere density in the 23rd cycle of the solar activity (1996–2006) follow from the analysis of the observations of WIND, ACE and SOHO satellites. Conclusions based on this information are presented.

2. The density of the solar corona

According to the generally accepted concepts, the light of corona originates from the scattering on free electrons and dust particles. The distribution can be distinguished according to the absorption lines depth in the spectrum. The scattering on electrons prevails closely to the limb, approximately from $\rho = 2.2$ there prevails the scattering on dust particles (ρ is the distance in solar radii). We can estimate the amount of the ionised matter according to the total brightness of corona during total eclipses. This brightness is conventionally estimated in the range of $1.03 < \rho < 6.0$, whereas the unit is 10^6 of the mean brightness of the solar disc. Fig. 1 brings together the data on total brightness of K-corona during many eclipses. The data are taken from the paper of Rušin & Rybanský (1985) and supplemented by photometric observations carried out in 2001 (Pintér *et al.* 2004) and 2006 (Pintér & Rybanský 2008).

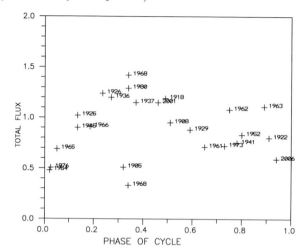

Figure 1. The total brightness of K-corona during total solar eclipses. The value is determined from the observed total brightness of corona in the range of $1.03 < \rho < 6.0$ after subtracting the value of $0.3\,10^6$, pertaining to the F-corona.

Fig. 1 shows that the total brightness and hence the total mass of corona within the indicated limits of the distance fluctuates during the solar activity cycle approximately in a ratio of $1 : 3$. It corresponds to the analysis of observations of K coronameter at Mauna Loa (Hansen *et al.* 1969). Allen (1976) presented a formula for the mean values of the free-electron density. The newer observations of LASCO 2 and LASCO 3 coronagraphs onboard SOHO indicate that short-term variations may be even larger.

3. Density from the in situ measurements

From the 1970s the exploration of the heliosphere has been done by using space probes. The probes HELIOS 1 and 2 approached the Sun to a distance of 0.3 AU, the probe VOYAGER 2 measured some parameters of the heliosphere up to the distance of 84 AU (August 2007). In addition, during their journey, almost all probes determined for the exploration of planets measured the parameters of the surrounding space - heliosphere. However, all these measurements are by far not sufficient for the description of such a huge space. In spite of this, they help to create at least an ordinal idea about its characteristics.

Fig. 2 shows the results of the measurements of proton density as estimated by some space observatories indicated in the Figure. All observations are in the log-log scale concentrating along a line with an inclination of -2.03. If the point at issue was only the expansion with constant velocity, the inclination would be -2. The change of inclination of the approximation line is probably connected with the recombination of protons because the measured kinetic temperature of protons is at a distance of the Earth 10^5 K and the measurements of VOYAGER 2 show that at a distance of more than 80 AU the temperature decreases to a value of about 10^4 K. The amplitude of the measured values is very high, in hourly averages up to $1 : 1000$.

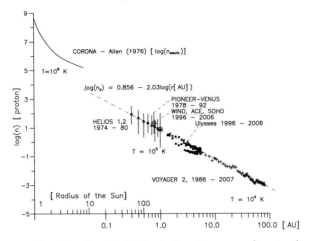

Figure 2. The course of proton density in the heliosphere according to the measurements of various space objects. Details are in the text.

Proton densities, measured at the ULYSSES probe that passes above the solar poles, are approximately half, but the velocity of expansion is roughly double: consequently, the specific mass flow $q = v\,n_p$ is approximately the same as in the ecliptic plane.

4. Conclusions

On the basis of the analysis of evolution of proton densities and velocities measured by sattelites we can assume that:

A) In contrast to solar corona, none of the parameters at WIND and ACE probes shows dependence on the phase of a cycle. The difference in evolution in Fig. 2 cannot be explained by different amount of protons and electrons (by 11% abundance of helium if the number of atoms is $n_p = 0.9\,n_e$). The mass of the corona increases probably only in the closed magnetic structures;

B) The trend of the constant q appears, because the value of v often increases with decreasing n_p;

C) An unexplained increase of v appeared at both probes in 2003;

D) A decrease in both the density and the total flow appears in the cycle minimum at the satellite ULYSSES.

Acknowledgements. We express our sincere thanks and appreciation to the teams of the missions of the HELIOS, PIONIER VENUS, VOYAGER, SOHO, ACE, WIND and ULYSSES space probes for providing a free access to the data. This work was supported by the Slovak Research and Development Agency under contract No. APVV-51-053805 and by VEGA grant project 7063.

References

Allen, C. W. 1976, *Astrophysical Quantities*, 3rd ed., The Athlene Press, University of London
Hansen, R. T., Garcia, Ch.j., Hansen, S. F., & Loomis, H. G. 1969, *Solar Phys.*, 7, 417
Pintér, T., Klocok, L., Minarovjech, M., & Rybanský, M. 2004, *Proceedings of Berzsenyi Dániel College*, Szombathely, 7
Pintér, T. & Rybanský, M. 2008, unpublished, private communication
Rušin, V. & Rybanský, M. 1985, *Bull. Astron. Inst. Czechosl.*, 36, 77

Universal Heliophysical Processes
Proceedings IAU Symposium No. 257, 2008
N. Gopalswamy & D.F. Webb, eds.

© 2009 International Astronomical Union
doi:10.1017/S1743921309029445

Analysis of the ICME on 24 August 2001

Nedelia A. Popescu[1]

[1] Astronomical Institute of Romanian Academy, Bucharest, RO-040557, Romania
email: nedelia@aira.astro.ro

Abstract. The analysis of in-situ data recorded by Ulysses for the ICME on 24 August 2001 is carried out, using different signatures such as He^{++} abundance enhancement, low ion temperature, low velocity, and anomalies of abundance and charge state of heavy ion species.

Keywords. heliosphere, ICME signatures

1. Introduction

In this paper we present detailed description and analysis of the ICME on 24 August 2001 by means of classical identification of ICMEs (He^{++} abundance enhancement, low kinetic temperature, low velocity); plasma dynamics signatures (thermal index $I_{th} > 1$); plasma composition signatures (anomalies of abundance and charge state of heavy ion species, low ion temperature and velocities).

2. 24 August 2001 ICME signatures analysis

The duration of the ICME on 24 August 2001 is 12 hours, between 24 August-13:00 UT, and 25 August-01:00 UT. In order to analyze the characteristic signatures of an ICME we consider the in situ data recorded by Ulysses/SWOOPS and SWICS.

2.1. Classical identification of ICMEs

a) The ICME location is determined by the high helium abundance enhancement, that is greater than 0.08 (Neugebauer & Goldstein 1997). In Fig. 1 solid lines denote the borders of the studied ICME.

b) Both proton and electron temperatures tend to be lower within an ICME, in comparison with the temperature of surrounding solar wind. Richardson *et al.* (1997) proposed as an indicator of an ICME the condition: $T_e/T_p > 2$. This condition is fulfilled by the studied ICME on 24 August 2001 (Fig. 2).

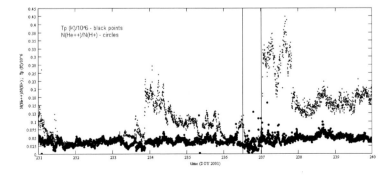

Figure 1. Alpha (He^{++}) abundance distribution and proton temperature distribution.

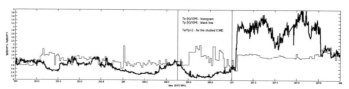

Figure 2. Proton and electron temperatures distributions.

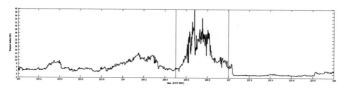

Figure 3. Thermal index I_{th} distribution.

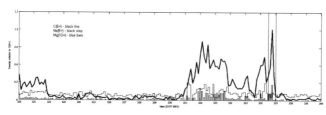

Figure 4. C(6+), Ne(8+), Mg(10+) densities relative to O(6+) distributions.

2.2. Plasma dynamics signature

Neugebauer & Goldstein (1997) defined the thermal index as follows: $I_{th} = (500V_p + 1.75 \times 10^5)/T_p$, where T_p = plasma proton temperature; V_p = plasma proton velocity. If $I_{th} > 1$, plasma seems to be associated with an ICME (Fig. 3).

2.3. Plasma composition signatures

1) *Enhancement of C(6+), Ne(8+), Mg(10+) densities relative to O(6+)* (see Fig. 4).

2) *Criteria of low ion temperature*: In Fig. 5 the temperatures distributions for different heavy ions are presented. In the studied interval all temperatures are low (SWICS data). From Fig. 1 the proton temperature depression can be observed too (SWOOPS data).

3) *Heavy ion species present anomalies of abundance and charge state*:

a) The optimum threshold value for the average Fe charge state is considered $< QFe >=11$ (Lepri & Zurbuchen 2004). Greater values are indicators of the presence of ICMEs. The distributions of Fe average charge state and the Fe/O abundance are presented in Fig. 6 (top panel). In the studied interval, the value of average charge state is between 10.5 and 11 and $Fe/O < 2$.

b) The enhanced charge-state ratios $C(6+)/C(5+)$ and $O(7+)/O(6+)$ can be observed in Fig. 6 (bottom panel). The threshold $O(7+)/O(6+) = 0.8$ corresponds to a freezing-in temperature of $2.05 \times 10^6 K$, this condition is satisfied in the studied case. This result is in good agreement with the velocity distribution (in the studied interval of time the velocity is ~ 550 km/s).

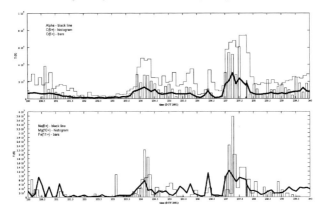

Figure 5. The temperatures distributions for different heavy ions.

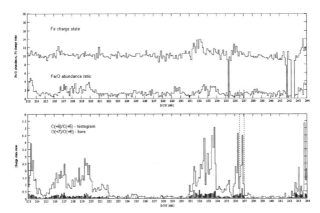

Figure 6. Fe average charge state and the Fe/O abundance distribution (top panel); Charge-state ratios $C(6+)/C(5+)$ and $O(7+)/O(6+)$ (bottom panel)

3. Conclusions

The ICME on 24 August 2001 verifies almost all classical identification conditions, plasma dynamics and composition signatures. These signatures are not necessarily present simultaneously and define exact the same region of the solar wind. Few signatures verify only the threshold values ($N(He^{++})/N(H^{+}) = 0.08$, $O(7+)/O(6+) = 0.8$), but other signatures fulfilled the imposed condition for the detection of an ICME. This analysis has been done in order to follow the interplanetary trajectory of some coronal mass ejections, starting from Ulysses satellite back to the Sun. We suspect this ICME to be the interplanetary counterpart of the 14–15 August 2001 solar event (Oncica, Popescu, & Dumitrache 2008).

References

Lepri, S. T. & Zurbuchen, T. H. 2004, *J. Geophys. Res.*, 105, 29, 231

Neugebauer, M. & Goldstein, R. 1997, in: N. Crooker *et al.* (eds.), *Coronal Mass Ejections*, AGU, p. 245

Oncica, A., Popescu, N., & Dumitrache, C. 2008, in: V.Mioc *et al.* (eds), *Exploring the Solar System and the Universe*, AIP Conference Proceedings, 1043, 323

Richardson, I. G., Farrugia, C. J., & Cane, H. V. 1997, *J. Geophys. Res.*, 102, 4691

Universal Heliophysical Processes
Proceedings IAU Symposium No. 257, 2008
N. Gopalswamy & D.F. Webb, eds.

© 2009 International Astronomical Union
doi:10.1017/S1743921309029457

Relation between coronal type II bursts, associated flares and CMEs

George Pothitakis[1], Panagiota Preka-Papadema[1], Xenophon Moussas[1], Constantine Caroubalos[2], Constantine Alissandrakis[4], Panagiotis Tsitsipis[3], Athanasios Kontogeorgos[3] and Alexander Hillaris[1]

[1]Department of Physics, University of Athens, 15784 Athens, Greece

[2]Department of Informatics, University of Athens, 15783 Athens, Greece

[3]Department of Electronics, Technological Education Institute of Lamia, Lamia, Greece

[4]Department of Physics, University of Ioannina, 45110 Ioannina, Greece

Abstract. We study a sample of complex events; each includes a coronal type II burst, accompanied by a GOES SXR flare and LASCO CME. The radio bursts were recorded by the ARTEMIS-IV radio spectrograph (100-650 MHz range); the GOES SXR flares and SOHO/LASCO CMEs, were obtained from the Solar Geophysical Data (SGD) and the LASCO lists respectively. The radio burst-flare-CME characteristics were compared and two groups of events with similar behavior were isolated. In the first the type II shock exciter appears to be a flare blast wave propagating in the wake of a CME. In the second the type II burst appears CME initiated though it is not always clear if it is driven by the bow or the flanks of the CME or if it is a reconnection shock.

Keywords. Sun: coronal mass ejections (CMEs), Sun: flares, Sun: radio radiation

1. Introduction

The MHD shock radio signatures, in the interplanetary medium and the solar corona, are the kilometric type II bursts and the metric type II radio emissions respectively. Though the former have been, unambiguously, identified with shocks piston-driven by CMEs, the exciter of the latter is somewhat more controversial as it can be either a blast-wave or a CME driven shock. Only the CME driven shocks are expected to propagate into interplanetary space; the blast-waves are damped with distance and, probably, rarely escape the lower corona (cf. Gopalswamy 2006, Pick *et al.* 2006, and references therein). This ambiguity about the exciter of coronal type II bursts has initiated a number of publications (cf. Kahler *et al.* 1984, Claßen & Aurass 2002, etc.) in which the CME, flare and type II parameters are compared. In this report we examine a set of complex events in search of groups with similarities as regards the relationship of type II burst-flare-CME characteristics; each is expected to represent a different shock generation processes.

2. Data Selection and Analysis

The ARTEMIS IV radiospectrograph (Caroubalos *et al.* 2001) observed 40 type II and/or IV radio bursts (1998-2000) which were published in the form of a catalogue (Caroubalos *et al.* 2004). The gross spectral characteristics of these events, and the associated CME and flare parameters were summarized in this catalogue; here we adopted the same numbering of events.

Table 1. Characteristic Parameters of Each Event in the Data Set

Event	$\delta\tau$ (min)	V_r	D (min)	V_{II} (km/sec)	ΔT (min)
Group I					
19	−22	3.12	3	1213	−4
24	−12	2.50	3	1940	−5
25	−32	1.77	1	1477	−4
Average	−22±5.2	2.5 ±0.4	2.33±0.67	1543±212	−4.3±0.33
Group II					
08	2	0.38	11	416	−1
21	−10	0.95	3	430	−1
23	20	0.80	11	806	−1
27	9	1.07	1	375	−1
30	17	1.73	3	737	0
32	17	0.82	4	442	0
39	2	0.65	7	494	−1
40	19	1.12	6	598	−2
Average	9.5±3.8	0.9±0.14	5.8±1.3	537±57	0.9±0.2
Unclassified					
06	−13	1.00	10	940	−5
33	−4	1.02	5	1300	−3
36	22	0.85	17	1430	−3
Overall Average	1.1±4.6	1.27 ±0.20	6.1±1.24	900±132	−2.2±0.5

From the original catalogue we selected fourteen events; (Table 1); they all include flares associated with a type II ARTEMIS-IV metric burst and a SOHO/LASCO CME. The parameters used in our study were:

• The Type II speed, V_{II}, in km/sec. They were calculated from the frequency drift rate of the type II bands, assuming a Newkirk (Newkirk 1961) corona and radial propagation of the MHD shock.

• The ratio of the Type II speed to the CME speed, $V_r = V_{II}/V_{CME}$). V_{CME} was obtained from the on line LASCO lists

• The time interval, $\delta\tau$, between the CME liftoff and the flare onset from the extrapolated value cited in the LASCO lists and the GOES SXR profiles respectively.

• The Type II duration in minutes, D, from the ARTEMIS-IV dynamic spectra.

• The time interval, ΔT, between the Type II launch time and the flare onset. As the emission starts some minutes after the onset of the flare impulsive phase, back-extrapolation of the emission lanes in the ARTEMIS–IV spectra was used to estimate the type II launch time.

In order to quantify similarities between each pair (i,j) of events we computed, a *proximity measure* in the form of *Standard Euclidean Distance*, d_{ij}; the smaller the index d_{ij} between a pair the more its members resemble each other. We define d_{ij} as:

$$d_{ij} = \sqrt{\left[V_{II i} - V_{II j}\right]^2 + \left[V_{r i} - V_{r j}\right]^2 + \left[D_i - D_j\right]^2 + \left[\delta\tau_i - \delta\tau_j\right]^2 + \left[\Delta T_i - \Delta T_j\right]^2}$$

Each coordinate (parameter) in the sum of squares is inversely weighted to the standard deviation of that coordinate. The *proximity measure* (d_{ij}) between pairs of events is used as a criterion for the identification of clusters within our data set. Certain groups of *similar* events emerge thus and are summarized in Table 1:

• Group I (Events 25, 19 & 14). Events in this group exhibit a close time relationship between the type II launch and the flare onset (average ΔT=-4.3±0.3 min); the type IIs are fast (average V_{II}=1543±212 km/sec) while the CME launch precedes the flare by 22±5.2 minutes and the CME speed is almost half the shock speed (average $V_r = V_{II}/V_{CME} = 2.5 \pm 0.4$). This suggests that the type II radio source is located behind the leading edge of the CME and that the associated shock was probably ignited by the flare and was propagating through the transient disturbance at the wake of the CME. This is consistent with the scenario proposed by Wagner & MacQueen (1983) for the 17 April 1980 type II burst and Vršnak *et al.* (2006) for the 3 November 2003 event (cf. also Vršnak & Cliver (2008) for a review). Two of the three events lack interplanetary type II in the WIND/WAVES reports, the only exception being Event 25.

• Group II (Events 21, 30, 39, 32, 40, 27, 08, & 23). Here the type II launch time is well associated with the flare onset (average $\Delta T = -0.90 \pm 0.20$ min) and the CME launch which is about 9.5±3.8 minutes after the flare. The type II speeds (average $V_{II} = 537 \pm 57$ km/sec), on the other hand, are equal or less to the CME speeds (average $V_r = V_{II}/V_{CME} = 0.9 \pm 0.14$); the type II duration (average 5.8±1.3 min) more often than not exceeds the typical values for coronal type IIs. It is expected that the type II are driven by CME bow shocks (when $V_r \approx 1$) or CME flanks, or are reconnection shocks induced by the CME liftoff. Three of the Events (39, 08 & 23) have an interplanetary type II as is often the case with shocks driven by CME front or flanks. For the rest (21, 27, 30, 32 & 40) no interplanetary type II was reported.

The association between the events 33, 06 & 36 remains at present uncertain, however they are characterized by fast CMEs and long duration type II shocks.

3. Discussion & Conclusions

We have studied fourteen complex events, each includes a coronal type II burst a GOES SXR flare and a LASCO CME; certain parameters, related to shock & CME kinetics and radio bursts-flare-CME timing were compared. Uncertainty factors were:

• Projection Effects: They introduce inaccuracies in the CME speed calculation; the errors are minimal in the case of limb CMEs.

• Take off Time: Both CME onset and Type II start are estimated from backward extrapolation neglecting possible acceleration.

• Type II Speed: The calculations rely heavily on the coronal model adopted; this can be resolved in the case of radio images of limb events.

• SXR Flare Onset: Depends on the detection threshold used.

Despite the uncertainties, it was found that most of the events may be grouped together based on their similar behavior; for each group we have conjectured a different shock generation processes, as various mechanisms exist (Vršnak & Cliver (2008)). Further study, with a larger data set of, preferably, limb events, will provide improved results as regards the coronal type II drivers.

References

Caroubalos, C., Hillaris, A., Bouratzis *et al.*, 2004, *A&A*, 413, 1125
Caroubalos, C., Maroulis, D., Patavalis *et al.*, 2001, *Exp. Astron.*, 11, 23
Claßen, H. T. & Aurass, H., 2002, *A&A*, 384, 1098
Gopalswamy, N., 2006, *AGU Geophysical Monograph Series*, 165, 207
Kahler, S., Sheeley, Jr., N. R., Howard *et al.*, 1984, *Solar Phys.*, 93, 133
Newkirk, G. J., 1961, *ApJ*, 133, 983
Pick, M., Forbes, T. G., Mann *et al.*, 2006, *Space Sci. Revs*, 123, 341
Vršnak, B. & Cliver, E. W., 2008, *Solar Phys.*, 142
Vršnak, B., Warmuth, A., Temmer, *et al.*, 2006, *A&A*, 448, 739
Wagner, W. J. & MacQueen, R. M., 1983, *A&A*, 120, 136

Session V

Plasma and Radio Emission Processes

Universal Heliophysical Processes
Proceedings IAU Symposium No. 257, 2008
N. Gopalswamy & D.F. Webb, eds.

© 2009 International Astronomical Union
doi:10.1017/S1743921309029470

Coherent emission

D. B. Melrose

School of Physics, University of Sydney,
NSW 2006, Australia
email: melrose@physics.usyd.edu.au

Abstract. The theory of plasma emission and of electron cyclotron maser emission, and their applications to solar radio bursts and to Jupiter's decametric radioation (DAM) and the Earth's auroral kilometric radiation (AKR) are reviewed, emphasizing the early literature and problems that remain unresolved. It is pointed out that there are quantitative measures of coherence in radio astronomy that have yet to be explored either observationally or theoretically.

Keywords. radiation mechanisms: nonthermal, Sun: radio radiation, interplanetary medium.

1. Introduction

It is 50 years this year since the first papers were published on two radio emission processes that have become central to our understanding of radio emission in the heliosphere: plasma emission (Ginzburg & Zheleznyakov 1958), and electron cyclotron maser emission (Twiss 1958). Ginzburg & Zheleznyakov's work provided a theoretical framework for a qualitatively accepted idea: solar radio bursts (types I, II, III were then known) are due to 'plasma emission' in which Langmuir waves are excited and secondary processes produce escaping radiation near the plasma frequency and its second harmonic. Plasma emission is now well established for other heliospheric radio emissions, notably from shocks and from planetary magnetospheres. In contrast, Twiss' theory (which he applied to type I bursts) was largely ignored; it was about two decades later that electron cyclotron maser emission (ECME) became accepted as the emission process for Jupiter's decametric radiation (DAM), the Earth's auroral kilometric radiation (AKR) and related emissions from the outer planets.

Plasma emission and ECME are two examples of what is called 'coherent' emission. (There is a third astrophysical coherent emission process, pulsar radio emission, which is inadequately understood and is not discussed here.) The characteristic feature used to distinguish coherent from incoherent emission is the high brightness temperature, T_B: any emission that is too intrinsically bright to be explained in terms of an incoherent emission process is assumed to involve a coherent emission process. There are three classes of coherent emission processes (Ginzburg & Zheleznyakov 1975; Melrose 1986a): antenna mechanisms, reactive instabilities and maser instabilities. In an antenna mechanism it is assumed that there exists a bunch of N electrons, localized in both coordinate space, $\mathbf{x}$, and in momentum space, $\mathbf{p}$, such that the bunch radiates like a macro-charge, leading to a power N^2 times the power that an individual electron would radiate. The back-reaction to the emission increases the spread in $\mathbf{x}$, leading to self-suppression. In a reactive instability, there is assumed to be localization in $\mathbf{p}$, but not in $\mathbf{x}$: wave growth is due to a feedback mechanism that leads to self-bunching, causing a phase-coherent wave to grow. The growth tends to increase the spread in $\mathbf{p}$, leading to self-suppression when this spread causes the bandwidth to exceed the growth rate. A maser instability involves an inverted energy population, with growth corresponding to negative absorption; the back

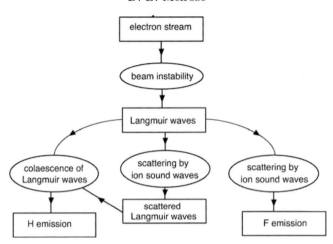

Figure 1. A flow diagram indicating the stages in plasma emission in an updated version on the original theory (Melrose 1970a&b; Zheleznyakov & Zaitsev 1970a&b).

reaction, described quantitatively in terms of quasilinear relaxation, reduces the inverted population, leading to self-suppression.

Which of these coherent emission processes is appropriate in astrophysical and space plasmas? In space plasmas the time available for wave growth is typically very much longer than the growth times and one expects the instabilities to saturate quickly. An antenna mechanism should saturate and evolve into a reactive instability, a reactive instability should saturate and evolve into a maser instability, which should saturate and reach a state of marginal stability. Rapid growth and saturation is necessarily confined to a short time and a small volume, and is expected to produce only highly localized transient bursts of coherent emission. For coherent emission to be observable in radio astronomy, it must come from a sufficiently large volume over a sufficiently long time. In the absence of evidence to the contrary, this argument suggests that only maser instabilities are likely to be relevant and one expects the electron distribution to relax quickly to a marginally stable configuration, minimizing the growth rate of the instability.

After reviews of plasma emission (§2) and ECME (§3), I discuss two aspects of coherent emission that require further development (§4).

2. Plasma emission

The essential idea in plasma emission (notably in type III bursts) is that an electron stream (electron 'beam') generates Langmuir (L) waves, and that these lead to emission of escaping radiation at the fundamental (F) and second harmonic (H) of the plasma frequency, as indicated in the flow diagram Fig. 1. The remarkable contribution of Ginzburg & Zheleznyakov (1958) was that they developed their theory before much of the relevant plasma theory had been properly developed for laboratory plasmas.

2.1. Theory of plasma emission

The first stage in plasma emission involves the beam instability that produces the L-waves. Ginzburg & Zheleznyakov appealed to a reactive version of the beam instability, and this stage was later updated to the slower-growing maser version, sometimes called the bump-in-tail instability. Even this instability grows extremely rapidly compared with the time scale of type III bursts, leading to a dilemma, pointed out by Sturrock (1964). During quasilinear relaxation of the beam, it loses a substantial fraction of its energy, implying that the beam should stop, over a distance that Sturrock estimated in kilometers. This is clearly not the case: type III streams can propagate through the corona, and through the interplanetary medium (IPM) to several AU. One suggested way of overcoming this dilemma involved a recycling in which the faster electrons at the front of the beam lose energy to the L-waves, and the slower electrons at the back of the beam reabsorb them. This effect does indeed seem to occur (Grognard 1983), but the energy loss remains catastrophic unless the efficiency of the recycling is unrealistically close to 100%, whereas one expects the Langmuir waves to evolve due to nonlinear (strong-turbulence) processes (Goldman 1983) before they can be reabsorbed. The actual resolution of Sturrock's dilemma emerged from *in situ* observations of type III burst in the IPM near 1AU.

The first spacecraft to observed type III bursts *in situ* in the IPM initially found no L-waves (Gurnett & Frank 1975; Gurnett & Anderson 1977), trivially resolving the dilemma, but seemingly undermining the theory of plasma emission. Subsequently, it was recognized that L-waves are present, but confined to isolated clumps, implying that the growth of L-waves occurs only in highly localized, transient bursts. The favored explanation is that the beam is in a state of marginal stability. The tendency to growth increases systematically due to faster electrons overtaking slower electrons, tending to increase the positive gradient in momentum space near the front of the beam. This effect is offset by quasilinear relaxation, in the many isolated clumps of L-waves, tending to reduce this gradient. Balancing these two effects leads to marginal stability. Qualitatively, in the marginally stable state, various weak damping processes mostly prevent effective growth, which occurs only in isolated regions where the conditions are particularly favorable for growth. This suggests a 'stochastic growth' model in which many such isolated, localized bursts of growth occur. A specific stochastic growth model, in which the amplification factor in each burst is $\exp G$ with G obeying gaussian statistics, predicts a log-normal statistical distribution of electric fields, and this accounts well for the observed statistical distribution (Robinson, Cairns & Gurnett 1993).

It is now recognized that there is a rich variety of specific processes that can partially convert the energy in Langmuir waves into escaping F and H radiation. In the flow diagram Fig. 1, it is assumed that low-frequency waves, such as ion sound (S) waves, are also excited, and these play two roles. Coalescence and decay processes involving L- and S-waves can produce F emission, $L \pm S \to F$, and backscattered L-waves, $L \pm S \to L'$, which coalesce with the beam-generated waves to produce H emission, $L + L' \to H$.

2.2. *Application to solar radio bursts*

Several problems arose in the early interpretation of the escape of solar radio emission from the corona:

- refraction for F emission implies that only sources within about a degree of the central meridian should be visible;
- modeling of the propagation of type III burst produced consistent results only for the standard coronal density models multiplied by a large factor, typically 30 or 100;
- the radio temperature of the quiet Sun is $\sim 1/10$ of the known temperature;
- F and H emission at a given frequency appear to come from a given height;
- sources at a given frequency have a similar apparent size.

A seemingly plausible interpretation is that scattering off coronal inhomogeneities increases the cone-angle of the escaping radiation and results in a scatter-image with an apparent area much larger than the actual area of the source. However, this violates a fundamental constraint (a Poincaré invariant, or generalized étendue) that the product of the apparent area and the cone angle is a constant for rays propagating through any optical system. This dilemma is resolved by a model in which the radiation is ducted outward in the corona (Duncan 1979) by reflecting off highly collimated overdense structures, called fibers by Bougeret & Steinberg (1977). It is not widely appreciated that the resolution of these problems in the interpretation of the radio data require that the corona be extremely inhomogeneous and highly structured on scales much smaller than can be resolved by present techniques.

Although type III bursts are relatively well understood, there are unresolved problems in identifying the exciting agencies and specific properties of other types (e.g., McLean & Labrum 1985). Type II bursts are perhaps the next best understood, being the most type III-like, but how the shock wave produces type III-like electron streams is still uncertain (Knock & Cairns 2005). Type I emission is less understood; type I emission can include both bursts and a continuum, neither of which are related to flares. No definitive model for type I emission has emerged, despite an enormous amount of detailed data (Elgarøy 1977). It is unclear what the exciting agency is for the type I continuum and indeed for any other broad-band plasma emission such as the flare continuum (e.g., McLean & Labrum 1985).

Observationally, plasma emission is circularly polarized, and various problems arose in connection with the interpretation of the polarization. Variants of the flow diagram Fig. 1, involve including the effect of the magnetic field on the low-frequency waves (replacing the S-waves) and on the F and H emission, which are in the magnetoionic modes. Circular polarization is interpreted as an excess of one magnetoionic wave over the other, and observationally plasma emission favors the o mode over the x mode. An important qualitative result concerning the polarization is that conventional F emission processes produce radiation between the plasma frequency, ω_p, and the cutoff frequency of the x mode, $\omega_x = \Omega_e/2 + (\omega_p^2 + \Omega_e^2/4)^{1/2}$. In this frequency range only the o mode can propagate, and hence F emission should be 100% polarized in the o mode. This is the case for most type I emission, but F emission in type III and type II bursts, although polarized in the o mode, is never 100%. Also the polarization of type I emission decreases as the source approaches the solar limb (Zlobec 1975). This is interpreted as a depolarization due to propagation effects in the inhomogeneous corona. Reflection off duct walls can lead to such depolarization, but only if there are extremely sharp density gradients at the edges of the fibers (Melrose 2006).

3. ECME

The favored interpretation of DAM and AKR is in terms of loss-cone driven ECME.

3.1. *DAM and AKR*

Early interest in DAM (Burke & Franklin 1955) increased dramatically following a report (Bigg 1964) that the radio bursts correlate with the position of the innermost Galilean satellite, Io. An explanation for the Io effect was formulated a few years later (Piddington & Drake 1968), and described as a unipolar inductor by Goldreich & Lynden-Bell (1969). The idea is that the Jovian magnetic field lines that thread Io are frozen into Io and dragged through the corotating Jovian magnetosphere. The relative motion of magnetic field lines sets up an EMF of a few MV that drives a field-aligned current, with the circuit closing across Io and in the Jovian ionosphere. The EMF accelerates electrons up to a few MeV somewhere along the Io flux tube. A major component of DAM is emitted just above the Jovian ionosphere on the Io flux tube. The angular emission pattern of this radiation seems bizarre: confined to the surface of a hollow cone with its apex on the Io flux tube, with an opening angle of the cone $\approx 80°$ and a thickness $\approx 1°$, cf. Fig. 2. DAM

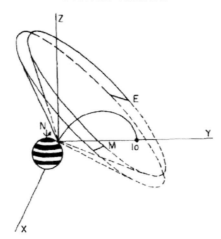

Figure 2. The angular pattern for DAM (Dulk 1967).

is attributed to cyclotron emission by accelerated electrons that mirror and propagate upward with a loss-cone anisotropy. The fly-bys of Jupiter by the Pioneer and Voyager spacecraft largely confirmed the cyclotron interpretation of DAM, notably a pattern of nested arcs about the north Jovian magnetic pole gave strong support to the model of emission on the surface of a hollow cone.

The discovery of AKR in the early 1970s (Gurnett 1974) implied that the Earth is a spectacular radio source below the ionospheric cutoff frequency. The source region for AKR is above the auroral zones and near the last closed field line. The polarization is dominantly x mode. AKR correlates with 'inverted-V' electron precipitation events, with energies ≈ 2 to 10 keV. Data from spacecraft that pass directly through the source show that these electrons are accelerated by a parallel potential gradient, and that the inverted-V events occur in density cavities, cf. Fig. 3. The cavities are localized to regions of strong upward current, that draws electrons downward and exhausts the magnetospheric supply of thermal electrons on the current-carrying flux tube. This leads to 'charge starving', so that the the available EMF localizes and accelerates all the available electrons to several keV.

To explain AKR and DAM one needs to account for the following features.
(1) The emission is near the local cyclotron frequency, with a very high T_B.
(2) The emission is predominantly in the x mode of magnetoionic theory.
(3) The emission pattern for DAM is confined to a thin surface of a wide hollow cone.
(4) AKR correlates with inverted-V precipitating electrons.

3.2. *Theory of ECME*

Electron-cyclotron instability can be classified into four types (Melrose 1986a): ⊥-driven and ∥-driven maser instabilities (Twiss 1958; Schneider 1959), and reactive instabilities due to azimuthal and axial self-bunching (Zhelznyakov 1959; Gaponov 1959). The favored mechanism for ECME in astrophysics is a perpendicular-driven maser, which (like the azimuthal bunching instability) depends on an intrinsically relativistic effect, even for electrons otherwise regarded as nonrelativistic.

The intrinsically relativistic effect in ⊥-driven ECME can be understood by considering the gyroresonance condition

$$\omega - s\Omega_e/\gamma - k_\| v_\| = 0, \tag{3.1}$$

where $s = 0, \pm1, \pm2, \ldots$ is the harmonic number, $\Omega_e = eB/m$ is the electron cyclotron

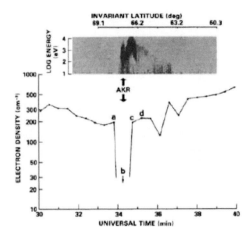

Figure 3. The electron density drops dramatically (to $\omega_p/\Omega_e < 0.2$) as a function of time as the spacecraft (Isis 1) crosses the source region of AKR, indicated by the insert showing the dynamic spectrum of inverted-V electrons (Benson & Calvert 1979).

frequency, and γ is the Lorentz factor. Suppose one considers all the electrons that can resonate at given s, frequency ω, and parallel wavenumber $k_\parallel$. Let the electrons be described by their perpendicular and parallel velocity components, $v_\perp$, $v_\parallel$. The resonant electrons lie on a resonance ellipse (Omidi & Gurnett 1982; Melrose, Rönnmark & Hewitt 1982) in $v_\perp$-$v_\parallel$ space. The absorption coefficient involves an integral around the resonance ellipse, and the sign of the integrand is negative for

$$\left[\frac{s\Omega_e}{v_\perp} \frac{\partial}{\partial p_\perp} + k_\parallel \frac{\partial}{\partial p_\parallel} \right] f(p_\perp, p_\parallel) > 0. \tag{3.2}$$

For a given unstable distribution of electrons, the maximum growth rate corresponds to the specific resonance ellipse that maximizes the positive contribution from (3.2) to the integral around the ellipse.

Suppose one makes the nonrelativistic approximation, $\gamma = 1$, in (3.1); then the resonance ellipse reduces to a line at $v_\parallel = (\omega - s\Omega_e)/k_\parallel$. In this case, the integral along the line $p_\parallel = $ constant may be partially integrated, and one can show that the net contribution to the absorption coefficient from the $p_\perp$-derivative cannot be negative. In this approximation only $\parallel$-driven maser action is possible (Melrose 1976). However, the requirements on the distribution function for effective growth are rather extreme, and not satisfied for the measured inverted-V distribution function.

As emphasized by Twiss (1958), the $\perp$-driven maser involves an intrinsically relativistic effect, even for electrons that one would not regard as relativistic (e.g., a few keV). In the case of perpendicular propagation, $k_\parallel \to 0$, (3.1) with $1/\gamma \approx 1 + v^2/2c^2$ implies that the resonance ellipse becomes a circle centered on the origin. The integral of the $p_\perp$-derivative around this circle can be negative, for example, for a 'shell' distribution with $\partial f/\partial p > 0$, which was the case considered by Twiss. For nearly perpendicular propagation the resonance ellipse is approximately circular, with a center displaced from the origin along the $v_\parallel$-axis by a distance $\propto k_\parallel$. The relativistic effect cannot be ignored when the radius of the resonance circle is comparable with the speed of the electrons that drive the maser. The paradox that one must include the relativistic correction to treat $\perp$-driven ECME, even for nonrelativistic electrons, is resolved by noting that the

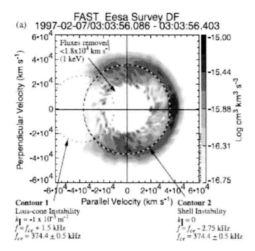

Figure 4. The electron distribution in an inverted-V event; two resonance ellipses that correspond to negative absorption are shown, one inside the loss cone on the left and the other centered on the origin passing through regions with $\partial f/\partial p > 0$ associated with the 'horseshoe' or 'shell' distributions (Ergun *et al.* 2000).

nonrelativistic approximation is formally $c \to \infty$, whereas c is necessarily finite when treating electromagnetic radiation.

Wu & Lee (1979) proposed a loss-cone driven instability for AKR. The maximum growth rate is for the resonance ellipse that lies just inside the loss cone, and samples the region where $\partial f/\partial p_\perp$ has its maximum positive value, as illustrated in Fig. 4. A major success for this theory was that the inverted-V electrons were found to have a loss-cone distribution of the required form. Loss-cone driven ECME leads naturally to narrowband, fundamental x-mode ECME restricted to a narrow surface of a wide hollow cone. This may be understood by noting that the position of the center of the resonance ellipse determines the angle of emission, and that the growth rate is a sensitive function of the position of the center of the ellipse. This property explains the seeming bizarre angular distribution of DAM in Fig. 2. The frequency is determined by the semi-major axis of the ellipse, and a similar argument implies that growth is confined to a narrow bandwidth, centered on the Doppler-shifted cyclotron frequency.

Besides the loss-cone feature, there are other features in the measured distribution function of inverted-V electrons that could produce ECME, specifically a trapped distribution (Louarn *et al.* 1990) and shell and horseshoe features (Bingham & Cairns 2000; Ergun *et al.* 2000). As in Twiss' theory, growth in these cases occurs at $\Omega_e/\gamma < \Omega_e$, and in a cold plasma such radiation is in the z mode, below a stop band between Ω_e and ω_x that precludes escape. The recognition that there is essentially no cold plasma inside the cavity overcomes this difficulty, with warm plasma effects washing out the stop band; the cavity acts like a duct, with AKR reflecting off the sides until it reached a height where ω_x outside the duct is low enough to allow escape (Ergun *et al.* 2000). Due to the absence of cold plasma, trapped, shell or horseshoe distributions are viable alternatives to a loss-cone distribution as the driver of ECME for AKR. However, they cannot account for the angular distribution of DAM, Fig. 2.

3.3. *Stellar applications of ECME*

Loss-cone driven ECME is a candidate for emission in any other astrophysical context where two conditions are satisfied: energetic electrons precipitate in a magnetic bottle,

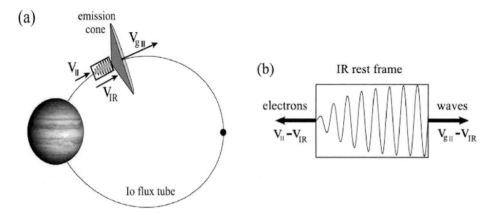

Figure 5. The phase-coherent model for ECME proposed to explain fine structure in DAM: the interaction region is moving at V_{IR}, the wave amplitude is illustrated inside the box, with the waves propagating at the group velocity $V_{g\parallel}$, and the resonant electrons at $V_{\parallel}$ (Willes 2002).

and the ratio ω_p/Ω_e is sufficiently small. One such context is in solar microwave 'spike' bursts (Holman, Eichler & Kundu 1980; Melrose & Dulk 1982). These bursts turn on and off in around 10 ms, and have high brightness temperatures ($\approx 10^{15}$ K). Another application is to the interpretation of radiation from some radio flare stars. There are several classes of flare stars, including M-type (red dwarf) stars, also called UV Ceti variables, which have extremely powerful solar-like flares, and cataclysmic variable stars, which involve mass transfer onto a magnetized star from a binary companion.

A major unsolved problem is how ECME escapes from a stellar corona. Along any prospective escape path, x-mode radiation generated just above ω_x encounters a layer where the local value of Ω_e is equal to $\omega/2$, where second-harmonic cyclotron absorption by thermal electrons is very strong. Four suggested ways of overcoming or avoiding the strong absorption at $2\Omega_e$ have been suggested.

(1) ECME at $s \geqslant 2$, or in the o mode (Melrose, Hewitt & Dulk 1984; Winglee 1985).
(2) When fundamental x mode emission is not allowed, ECME favors fundamental z mode emission, and the escaping radiation results from coalescence of z mode waves to produce higher harmonic x mode emission (Melrose 1991).
(3) There are 'windows' at parallel or perpendicular propagation where some radiation can escape (Robinson 1989).
(4) The radiation can tunnel through the layer: the radiation incident on the absorbing layer from below, with $\omega < 2\Omega_e$, modifies the distribution function of the electrons, allowing re-emission at $\omega > 2\Omega_e$ (McKean, Winglee & Dulk 1989).

One possibility is that ECME generates a large power, with only a tiny fraction of it escaping. An interesting implication is that the absorption of this large power can lead to non-local heating at the absorption layer. These suggested stellar applications of ECME will remain questionable until the escape of the radiation is explained convincingly.

4. Two outstanding problems

Our understanding of coherent emission is far from complete. I comment on two aspects: phase-coherent growth and measures of coherence.

4.1. *Phase-coherent wave growth*

Maser growth applies only if the growth rate is less than the bandwidth of the growing waves. There are examples where fine structures are observed apparently inconsistent with this condition: fine structure in solar radio bursts (Ellis 1969), in S-bursts in DAM (Ellis 1973; Carr & Reyes 1999). In these examples the frequency tends to change with time, somewhat analogous to the narrow-band drifting structures in triggered VLF emissions (Helliwell 1967). It is unclear how such fine structures are related to broader band emission that is consistent with maser theory. The most extreme example of fine structures is in giant pulses from the Crab pulsar (Hankins & Eilek 2007); whereas pulsar emission typically satisfies log-normal statistics, giant pulses do not (Cairns, Johnston & Das 2004). In all three cases, it seems that fine structures involve some distinctive physical process different from other coherent emission. One suggestion, illustrated in Fig. 5, is a modified form of Helliwell's (1967) phenomenological model for VLF emissions. The idea is that a reactive instability grows relatively slowly within an interaction region that is moving along the magnetic field lines, Fig. 5. On setting $\Delta = \omega - s\Omega_e/\gamma - k_{\parallel}v_{\parallel}$ one requires not only $\Delta = 0$, which is the condition (3.1), but also $d\Delta/dt = 0$ along the path of the interaction region to determine how ω changes with t.

4.2. *Measures of coherence*

Is there some systematic way of quantifying the concept of coherence in radio astronomy? The only widely accepted measures of coherence in radio astronomy are either based on a high T_B or on special techniques of high time- and frequency-resolution. However, there are other possible measures that are ignored. Historically, it is interesting that these other measures were initially transferred from radio physics to optics, notably through the Hanbury Brown-Twiss effect (e.g., Hanbury Brown & Twiss 1956), and are now familiar in optics, in terms of photon-counting statistics, but unfamiliar in radio astronomy.

The measures of coherence involve powers of the intensity (e.g., Mandel & Wolf 1995) averaged over a short observation time. The mean value of $\langle I^N \rangle$ for "coherent" radiation is $\langle I^N \rangle = \langle I \rangle^N$, and for "incoherent" (random phase) radiation is $\langle I^N \rangle = N!\langle I \rangle^N$. The quantities $\langle I^N \rangle/\langle I \rangle^N$ are measures of coherence. These measures must depend on the resolution time of the telescope and the coherence time of the radiation, and one expects interesting results only if the resolution time is shorter than the coherence time (which is not known *a priori*). There are additional measures of coherence from the auto- and cross-correlation functions between the Stokes parameters for polarized radiation. (In fact, the stated result for $\langle I^N \rangle$ is for completely polarized radiation, and there is a different value for partially polarized or unpolarized radiation.) At present there is no program for measuring these quantities, for example, by constructing a correlator that automatically determines these higher order correlation functions. There is also no systematic theory predicting what one would expect for various models of coherence.

5. Conclusions

Following the first papers in the field 50 years ago, major progress was made through the 1970s in the theory of plasma emission and its application to solar radio emission, and in ECME and its application to DAM and AKR about a decade later. There is a renewed observational interest in meter-λ radio astronomy (e.g., through LOFAR and MWA) and it is timely to re-appraise some of the unresolved problems left over from this earlier period. Some of these problems are outlined briefly above. There are also general problems relating to coherent emission *per se*. There are examples of fine structures in coherent emission that are inconsistent with maser theory, and appear to involve some intrinsically phase-coherent emission mechanism.

It is pointed out that, in principle, there are measures of coherence available from correlation functions of powers of the intensity and the other Stokes parameters. It is possible to measure these quantities provided that the resolution time of the observations is shorter than the coherence time of the radiation.

Acknowledgement

I thank Mike Wheatland and Matthew Verdon for comments on the manuscript.

References

Benson, R. F. & Calvert, W. 1979, *Geophys. Res. Lett.*, 6, 479

Bigg, E. K. 1964, *Nature*, 203, 1088

Bingham, R. & Cairns, R. A. 2000, *Phys. Plasmas*, 7, 3089

Bougeret, J. L. & Steinberg, J. L. 1977, *A&A*, 6, 777

Burke, B. F. & Franklin, K. L. 1955, *J. Geophys. Res.*, 60, 213

Cairns, I. H., Johnston, S., & Das, P. 2004, *MNRAS*, 353, 270

Carr, T. D. & Reyes, F. 1999, *J. Geophys. Res.*, 104, 25142

Dulk, G. A. 1967, *Icarus*, 7, 173

Duncan, R. A. 1979, *Solar Phys.*, 63, 389

Elgarøy, Ø. 1977 *Solar Noise Storms*, Pergamon Press, Oxford.

Ellis, G. R. A. 1969, *Aust. J. Phys.*, 22, 177

Ellis, G. R. A. 1973, *PASA*, 2, 191

Ergun, R. E., Carlson, C. W., McFadden, J. P., Delory, G. T., Strangeway, R. J., & Pritchett, P. L. 2000, *ApJ*, 538, 456

Gaponov, A. V. 1959, *Izv. VUZ. Radiofiz.*, 2, 450

Ginzburg, V. L. & Zheleznyakov, V. V. 1958, *Soviet Astron.*, 2, 235

Ginzburg, V. L. & Zheleznyakov, V. V. 1975, *ARAA*, 13, 511

Goldman, M. V. 1983, *Solar Phys.*, 89, 403

Goldreich, P. & Lynden-Bell, D. 1969, *ApJ*, 156, 59

Grognard, R. J.-M. 1983, *Solar Phys.* 94, 165

Gurnett, D. A. 1974, *J. Geophys. Res.*, 79, 4277

Gurnett, D. A. & Anderson, R. R. 1977, *J. Geophys. Res.*, 82, 632

Gurnett, D. A. & Frank, L. A. 1975, *Solar Phys.*, 45, 477

Hanbury Brown, R. & Twiss, R. Q. 1956, *Nature*, 178, 1046

Hankins, T. H. & Eilek, J. A. 2007, *ApJ*, 670, 693

Helliwell, R. A. 1967, *J. Geophys. Res.*, 72, 4773

Holman, G. D., Eichler, D., & Kundu, M. R. 1980, in M. R. Kundu, T. E. Gergeley (eds) *Radio physics of the Sun*, D. Reidel (Dordrecht) p. 457

Knock, S. A. & Cairns, I. H. 2005, *J. Geophys. Res.*, 110, A01101

Louarn, P., Roux, A., de Feraudy, H., LeQueau, D., Andre, M., & Matson, L. 1990, *J. Geophys. Res.*, 95, 5983

Mandel, L. & Wolf, E. 1995, *Optical Coherence and Quantum Optics*, Cambridge University Press

McKean, M. E., Winglee, R. M., & Dulk, G. A. 1989, *Solar Phys.*, 122, 53

McLean, D. J. & Labrum, N. R. (eds) 1985, *Solar Radiophysics*, Cambridge University Press

Melrose, D. B. 1970a&b, *Aust. J. Phys.*, 23, 871 & 885

Melrose, D. B. 1976, *ApJ*, 207, 651

Melrose, D. B. 1986a, *Instabilities in space and laboratory plasmas*, Cambridge University Press

Melrose, D. B. 1986b, *J. Geophys. Res.*, 91, 7970

Melrose, D. B. 1991, *ApJ*, 380, 256

Melrose, D. B. 2006, *ApJ*, 637, 1113

Melrose, D. B. & Dulk, G. A. 1982, *ApJ*, 259, 884

Melrose, D. B., Hewitt, R. G., & Dulk, G. A. 1984, *J. Geophys. Res.*, 89, 897

Melrose, D. B., Rönnmark, K. G., & Hewitt, R. G. 1982, *J. Geophys. Res.*, 87, 5140

Omidi, N. & Gurnett, D. A. 1982, *J. Geophys. Res.*, 87, 2377

Piddington, J. H. & Drake, J. F. 1968, *Nature*, 217, 935

Robinson, P. A. 1989, *ApJ*, 341, L99

Robinson, P. A., Cairns, I. H., & Gurnett, D. A. 1993, *ApJ*, 407, 790

Schneider, J. 1959, *Phys. Rev. Lett.*, 2, 504

Sturrock, P. A. 1964, In: W. N. Hess (ed.) *The Physics of Solar Flares*, NASA, p. 357

Twiss, R. Q. 1958, *Aust. J. Phys.*, 11, 564

Willes, A. J. 2002, *J. Geophys. Res.*, 107, 1061

Winglee, R. M. 1985, *J. Geophys. Res.*, 90, 9663

Wu, C. S. & Lee, L. C. 1979, *ApJ*, 230, 621

Zhelznyakov, V. V. 1959, *Izv. VUZ. Radiofiz.*, 2, 14

Zheleznyakov, V. V. & Zaitsev, V. V. 1970a&b *Soviet Astron.*, 14, 47 & 250

Zlobec, P. 1975, *Solar Phys.*, 43, 453

Discussion

SPANGLER: I was intrigued by your proposed test of coherent processes by higher moments of the total power. There is no reason why this could not be measured. N=2 has been routinely measured in interplanetary measurements, and John Armstrong has published measurements, corresponding to N=3.

MELROSE: My point is that it is simple in principle to measure these quantities automatically by an approximately designed correlation. This has not been done, and as yet theoretical interpretations of the measurable quantities $< I^N > / < I >^N$ have not been worked out.

IBADOV: Thanks for your interesting report. Can you briefly give a list of modern problems of Radio Astronomy and a list of modern Radio Telescopes?

MELROSE: It would be presumptuous to shortlist problems in general because there are many different opinions on what is important. From the viewpoint of my paper, the problems that I see as being left unresolved since the 1980s and earlier include the nature of type I emission, the details of the electron acceleration at type II shocks, and the escape of ECME from the solar corona and stellar coronas.

Universal Heliophysical Processes
Proceedings IAU Symposium No. 257, 2008
N. Gopalswamy & D.F. Webb, eds.

© 2009 International Astronomical Union
doi:10.1017/S1743921309029482

Observations of radio spectra at 1–2.5 GHz associated with CME start time

José R. Cecatto[1]

[1] Astrophysics Division, INPE, P.O. Box 515
12227-010, São José dos Campos, Brasil
email: jrc@das.inpe.br

Abstract. We know Coronal Mass Ejections (CME) and flares are the most energetic phenomena happening on the Sun. Until now the information about origin and trigger mechanism of CMEs remains scarce. Also, there is unconclusive information about the association between them and flares although progress has been made in recent years. Multi-spectral observations suggested that the flare energy release occurs in regions from where the decimetric radio emission originates. In this case, investigations of the solar emission in this wavelength range can give us valuable information about these questions. During last solar maximum the Brazilian Solar Spectroscope (BSS) observed the solar radio spectrum (1–2.5 GHz) with high time (100–20 ms) and frequency (50–100 channels) resolutions on a daily (11–19 UT) basis. A survey during the period 1999–2002, shows that a significant fraction (20% −57 events) of CMEs recorded by LASCO has an association with the spectra of radio bursts recorded by BSS. Analysis of the radio spectrum associated to CME shows there is a dominance of continuum and/or pulsation and that the association becomes stronger when we consider the CME acceleration since its origin on the Sun. A statistics of this association between CME dynamics and the characteristics of decimetric radio bursts recorded by BSS is presented. Emphasis is given to observations of the association with CME start time.

Keywords. CME, radio, bursts

1. Introduction

CMEs are among the most energetic phenomena originated on the Sun. They carry enormous kinetic energies (10^{32}–10^{34} erg) out of the Sun due to the high mass ($\sim 10^{-15}$ solar masses) and velocities (hundreds to few thousand km/s). Also, they carry magnetic field in the form of bubbles which become unstable and buoyant, leaving the Sun and propagating out into the interplanetary space. Observed in white light they were first imaged with space-borne coronographs in the early 1970s (Tousey 1973; Gosling *et al.* 1974). Until now several aspects and a controversy in those investigations concerning the relationship between CMEs, solar flares and radio bursts remains. Some authors show that $\leqslant 40\%$ of CME phenomena are associated with flares (Munro *et al.* 1979; Webb & Hundhaunsen 1987; St. Cyr & Webb 1991), while others concentrate on determining why part of the CME are related to flares and which are their common characteristics (Verneta 1997; Svestka 1995; Sheeley *et al.* 1983; Kahler 1994).

Some authors (Hudson *et al.* 2001) reported one case for which the CME occurred 15 minutes after the X-ray/radio emission. At decimetric waves a variant of the pulsation emission have been noticed at the initial phase of CME (Jun Fu, Q. *et al.* 2004). Also, Wang *et al.* (2005) investigated the radio spectra associated to quite strong CME during three days and found the spectra indicate association to type-II, type-III, type-IV, drifting pulsations and fine structures. Yet, fast drift structure and continuum from one radio burst have been found in association to CME by Pohjolainen (2008). However, investigations taking into account for a large sample of data are rare.

Taking this into consideration, this work represents a survey on data from ~4 years period (1999-2002) searching for some association between CME phenomena and BSS bursts spectrum, within the BSS observational window 11–19 UT.

2. Instrumentation, observations and results

The BSS is a digital spectroscope operating daily from 11 to 19 UT since 1998, at INPE, São José dos Campos, Brasil. Jointly with its polar mounted 9 m diameter parabolic antenna, it operates in the frequency range (1-2.5 GHz) with high time (100, 50, 20 ms) and frequency (1–10 MHz) resolutions. Besides, within some limitations it allows us to select observing frequency range, frequency and time resolutions. The data is digitized and recorded in up to 200 frequency channels. Time minimum detectable flux is around 2-3 sfu, for several combinations of the observational parameters (Sawant et al. 2001; Fernandes 1997). Normally, BSS is operated using either 100 or 50 channels within the selected operational bandwidth.

LASCO comprises 3 coronographs which jointly imaged the solar corona since 1.1 until 30 Rs (C1: 1.1–3 Rs, C2: 2–6 Rs, C1: 3.7–30 Rs)(Brueckner et al. 1995). The LASCO (C2, C3) experiment recorded a total of 304 CME phenomena during the period of 1999-2002, 11–19 UT, corresponding to the BSS observing days.

A fraction (20%) of those CMEs has the start time associated with various types of solar events within 5 minutes before the start up to 5 minutes after the end time of the bursts recorded by BSS. Since the CME acceleration - linear, quadratic approximations - is measured its onset time can be calculated. This information is available at CDAW-GSFC(NASA) website. Also, it has to be remarked that the minimum time interval between two consecutive LASCO images is 12 min. Table 1 shows the main characteristics as well as H-α and X-ray activity of all BSS events recorded in association with CME start times.

Two examples of BSS burst spectra recorded in association with CME start time are shown in Figure 1 (left, right). As can be clearly seen from Table 1, the dominant types of radio bursts associated to the CME are pulsations and continuum isolated as well as combined. In several cases the pulsations and continuum emissions are combined together, Figure 2 (left), or with other types of radio burst, e.g. fine structure, as can be seen in Figure 2 (right).

3. Discussions and conclusion

From the sample of all CMEs ($\sim$ 300) observed within the BSS observational window during the period 1999-2002 about 20 % have estimated start time within an interval of 5 minutes before the start time and 5 minutes after the end of associated BSS bursts. Table 1 shows the BSS bursts characteristics as well as H-α and X-ray related flares. From all those 57 associated bursts about 20 % have no H-α flare associated and a little bit more no X-ray flare associated. About those X-ray associated fourteen are class C, twenty two M, and nine X flares. Only half of the cases not associated with H-α flares are simultaneously not associated with X-ray flares. Curiously, a look through the column of active region position (NOAA) shows that statistically the associated Northern hemisphere active regions were observed in about 2/3 of identified active regions while other 1/3 correspond to Southern identified active regions.

Also, a little less than half of CME recorded in association with decimetric radio burst emission is either Halo or Partial Halo CME. The velocity of CME associated ranged from 153 km/s to 2047 km/s with an average 715 km/s. There is a slight dominance of CMEs slower than the average.

Regarding the association between the CME estimated start time with the spectra of radio burst emission observed at decimetric wavelengths the following points need to be remarked. The limited frequency range (1–2.5 GHz) of radio observations. Second, BSS sensitivity does not permit to measure signals lower than 2–3 SFU. Finally, in some cases coronal inhomogeneities can cause absorption of radio waves at decimetric wavelengths. The combination of these factors could explain the relatively low percentage of association between the CME start time and radio bursts emission.

We showed that there is association between a fraction of burst spectra (1–2.5 GHz) recorded by BSS instrument with corresponding CMEs recorded by LASCO C2 and C3 coronographs, mainly concerning about CME start time. This association can give us

Table 1. Characteristics of BSS bursts and flares associated to the onset time of CME.

Date	BSS Begin (UT)	End (UT)	Type	LASCO Vel. (km/s)	Onset time(UT)	H-α Loc.	AR	Begin (UT)	Max (UT)	End (UT)	X-Ray Begin (UT)	End (UT)	Imp.
08/17/99	12:43	18:34	N, FS	776PH	12:44:28	N26E35	8668	B12:47	U12:50	14:05	12:32	13:57	C3
08/17/99	12:43	18:34	N, FS	962PH	15:16:28	N26E35	8668	15:05	15:22	16:34	14:28	17:54	C6
08/30/99	17:50	17:59	G, P	404	17:56:43	S21W58	8673	17:26	17:28	18:13	17:23	18:21	M4
11/27/99	12:09	12:22	G	641	12:14:13	S15W68	8771	12:08	12:12	13:19	12:05	12:16	X1
11/27/99	12:09	12:22	G	235	12:21:55	S15W68	8771	12:08	12:12	13:19	NF*		
04/18/00	14:55	15:01	D T	668	14:53:56	NF*					NF*		
05/02/00	14:43	14:49	CT, FS	1278	14:45:17	N22W68	8971	14:45	14:46	15:05	14:42	14:56	M3
05/19/00	13:12	13:18	IIIi,b	327	13:12:05			13:25	13:30	13:37	NF*		
05/22/00	13:24	13:41	CT	419	13:38:16	NF*					NF*		
06/06/00	15:01	17:17	Pre, CT	929	15:03:42	N20E18	9026	12:06	15:21	18:43	14:58	15:40	X2
06/06/00	15:01	17:17	Pre, CT	1119H	15:21:40	N20E18	9026	12:06	15:21	18:43	14:58	15:40	X2
06/07/00	15:37	15:43	IIIi,b	842H	15:35:42	N23E03	9026	15:04	15:46	18:51	15:34	16:06	X1
06/27/00	13:43	15:02	FB, FS	363	14:37:12	NF*					14:29	14:35	C2
07/04/00	14:58	14:59	D T	562	15:03:45	S22E31	9068	14:58	15:00	15:16	14:57	15:04	C2
07/04/00	15:18	15:18	FS SD	562	15:16:48	N17E49	9070	15:11	15:25	16:22	15:08	15:34	C3
07/06/00	12:26	12:37	CT, P	472	12:31:50	N18E25	9070	12:22	12:23	12:49	12:23	12:49	C4
07/11/00	12:31	12:31	D T	1078H	12:33:29	S18W42	9069	12:31	12:31	12:36	12:12	13:35	X1
07/31/00	15:55	15:55	P, PA	774	15:58:44	NF*					NF*		
09/15/00	14:31	14:37	P	481PH	14:33:13	N13E07	9165	14:31	14:38	15:06	14:29	14:44	M2
09/16/00	13:23	13:34	PA	1056	13:24:54	N14W08	9165	13:08	13:12	13:23	NF*		
11/24/00	14:53	15:20	CT, P	1245H	15:08:32	N22W07	9236	15:01	15:16	15:57	14:51	15:21	X2
04/05/01	16:53	17:21	CT	1390H	16:56:47	S24E50	9415	16:33	17:01	18:49	16:57	18:14	M5
04/06/01	17:18	17:19	P	648	17:20:24	S19E32	9415	17:24	17:30	18:18	17:11	17:50	C5
04/09/01	15:20	16:05	CT, P	1192H	15:32:02	S21W04	9415	15:24	15:34	17:03	15:20	16:00	M8
04/11/01	17:23	17:23	FS nb	1145	17:19:53	S21W27	9415	17:25	17:26	17:28	NF*		
04/25/01	13:42	13:45	FS, OS	856	13:46:51	N18W09	9433	13:44	13:45	14:20	13:39	13:59	M3
04/26/01	11:49	12:01	OS nb	1006H	11:51:38	NF*					11:26	13:19	M8
04/26/01	12:51	13:11	CT, FB, O	844	12:58:31	N17W31	9433	12:11	13:11	14:31	11:26	13:19	M8
06/04/01	15:12	15:14	IV	632	15:11:37	S19E52	9488	15:11	15:17	15:28	15:10	15:23	C1
06/13/01	11:35	11:42	CT, IV	576	11:36:19	S29E66	9502	11:35	11:39	12:18	11:22	11:51	M8
06/13/01	16:22	16:25	IV	276	16:25:57	N20W49	9489	16:21	16:28	16:39	16:20	16:35	C9
08/28/01	16:00	16:02	FS	478PH	16:02:42	N13E68	9601	16:01	16:09	16:14	15:56	16:26	M1
08/28/01	160227	160356	FS	478PH	16:02:42	N13E68	9601	16:01	16:09	16:14	15:56	16:26	M1
08/29/01	18:25	18:27	CT, P, RF	317	18:27:35	N18E58	9600	18:24	U18:39	19:29	18:19	18:50	C3
08/31/01	14:56	14:57	G	310H	14:52:52	N13E29	9601	15:29	15:31	15:52	NF*		
09/03/01	15:52	15:52	CT	196	15:51:07			15:45	17:16	17:37	15:45	17:37	M1
09/03/01	18:22	18:25	CT	1352PH	18:24:11			18:21	18:41	19:10	18:21	19:10	M2
09/11/01	14:17	14:17	D T	791H	14:17:13	N13E35	9615	14:16	14:39	15:30	14:00	15:08	C3
09/18/01	17:05	17:08	PA, SP, CT	376PH	17:00:14	S18E85	9628	17:06	17:06	17:13	17:02	17:09	M1
09/20/01	18:13	18:14	CT, EF	446PH	18:15:39	N09W11	9631	18:15	18:18	18:30	18:12	18:21	M1
09/28/01	14:02	14:08	IV	248	14:07:02	S17W47	9628	13:57	13:59	14:08	NF*		
10/01/01	14:42	14:42	IIIi, n	153	14:42:45	NF*					NF*		
10/19/01	16:22	17:06	P, CT, FB	901H	16:21:18	NF*					16:13	16:43	X2
10/22/01	14:50	15:09	IV	1336H	14:49:53	S21E18	9672	B14:25	15:12	16:02	14:27	15:31	M7
10/25/01	15:06	15:31	CT, LC	1092H	15:01:51	N09E26	9678	14:56	14:56	15:19	14:42	15:28	X1
10/26/01	14:30	14:37	CT, P	350	14:33:08	N07E16	9678	14:30	14:36	14:45	14:28	14:37	M2
11/28/01	16:33	16:35	IV	500H	16:33:37	N04E16	9715	16:32	16:36	16:52	16:26	16:41	M7
12/13/01	14:24	14:24	PA	864H	14:24:54	N16E09	9733	14:24	14:30	15:45	14:20	14:35	X6
04/04/02	15:27	15:27	FS nb	790	15:25:27			15:24	15:32	15:38	15:24	15:38	M6
04/13/02	12:07	12:33	IRF	599	12:07:02	S03E57	9907	12:12	12:13	12:22	12:08	12:20	C3
07/11/02	14:48	14:48	III nb	614	14:48:41	N21E58	10030	14:46	14:48	15:22	14:44	14:57	M6
07/19/02	16:16	16:20	CT	2047H	16:14:01	NF*					NF*		
07/24/02	15:43	15:51	IP	528	15:42:41	S13E49	10039	15:14	15:45	16:59	15:24	16:22	M1
08/28/02	16:40	16:51	OS	447	16:45:45	NF*					16:45	17:09	M1
08/30/02	14:37	14:48	OS	420	14:37:01	N08E75	10095	14:27	14:31	14:36	14:35	14:43	C8

CT - continuum, DT - Dots, FS NB - elementary flare narrow band, FS SD - elementary flare slow drift, FB - fiber, FS - fine structure, G - gradual, IP - intermediate pulsation, IV - intermediate variation, III IS - isolated type III, III NB - narrow band type III, IIIi, b - intermediate and broad band type III, LC - Lace-like, N - Noise, OS - Oscillations, PA - Patch, PRE - Pre-flare, P - Pulsations, RF - Rise and Fall, SP - Split, NF* - No Flare

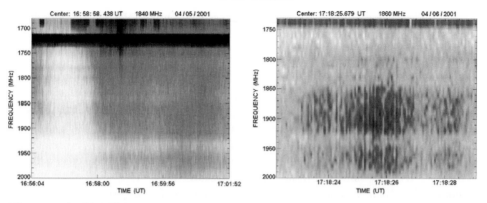

Figure 1. (Left) BSS burst dinamic spectrum associated to CME exhibiting a continuum radio burst emission observed on 04/05/2001 at 16:58 UT. (Right) Pulsation radio emission associated to CME observed on 04/06/2001 at 17:18 UT.

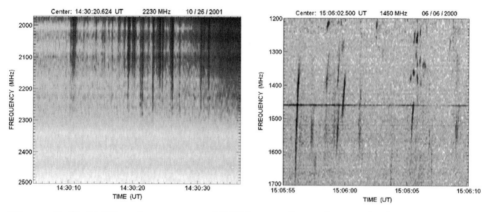

Figure 2. (Left) Dinamic spectrum of BSS bursts observed on 10/26/2001 at 14:30 UT associated to CME. It is clear in the beggining the pulsation emission and its superposition on continuum radio burst emission in the time evolution of the radio burst. (Right) Dinamic spectrum showing a combination of various kinds of radio emission observed by BSS on 06/06/2000 at about 15:06 UT and associated to CME.

important information regarding CMEs origin and trigger mechanism, and also about the associated solar bursts. Further investigation and a deeper analysis of this sample of data as well as statistical analysis of additional data coming from other observatories in other radio bands are required to improve our understanding of the relationship between radio bursts, flares and CME occurrence. Mainly, the investigations have to search for some kind of radio signature associated to CME occurrence. Also, further analysis must be done on those CME phenomena not related to radio bursts and flares.

4. Acknowledgements

We are grateful to the Brazilian Financial Agencies CNPq for the financial support through the grant 475723/2004-0, and FAPESP through the grant 06/55883-0. We would also like to thank the SOHO team for maintaining the database and processing of LASCO and EIT data. SOHO is operated by ESA, CDAW(NASA) scientists whose dedication has made data available to the solar community. Our acknowledgements to the SEC team that maintains The Weekly report and forecast of Solar Geophysical Data. Thanks are also given to the referees for their helpful comments on the manuscript.

References

Brueckner, G. E., Howard, R. A., Koomen, M. J., Korendyke, C. M., Michels, *et al.* 1995, *Solar Phys.*, 162, 357

Fernandes, F. C. R. 1997, in: INPE-6396-TDI/612, , *Tese de Doutorado* (São José dos Campos), p. 178

Gosling, J. T., Hildner, E., MacQueen, R. M., Munro, R. H., Poland, A. I., & Ross, C. L. 1974, *JGR*, 79, 4581

Hudson, H. S., Kosugi, T., Nitta, N. V., & Shimojo, M. 2001, *ApJ*, 561, L211

Jun Fu, Q., Yan, Y. H., Liu, Y. Y., Wang, M., & Wang S. J. 2004, *Chin. J. Astron.Astrophys.*, 4(2), 176

Kahler, S. W. 1994, *ApJ*, 428, 837

Munro, R. H., Gosling, J. T., Hildner, E., MacQueen, R. M., Poland, A. I., & Ross, C. L. 1979, *Solar Phys.*, 61, 201

Pohjolainen, S. 2008, *A&A*, 483, 297

Sawant, H. S., Subramanian, K. R., Faria, C., Fernandes, F. C. R., Sobral, J. H.A., & Cecatto, J. R., *et al.* 2001, *Solar Phys.*, 200, 167

Sheeley Jr., N. R. Howard, R. A., Koomen, M. J., & Michels, D. J. 1983, *ApJ*, 272, 349

St. Cyr, O. C. & Webb, D. F. 1991, *Solar Phys.*, 136, 379

Svestka, Z. 1995, *Private communication*

Tousey, R. 1973, in: Rycroft, M. J. & Kuncorn, S. K. (eds.) *The solar corona-Space Research XIII* (Berlin: Akademie-Verlag), p. 713

Verneta, A. I. 1997, *Solar Phys.*, 170, 357

Verneta, A. I. & Hundhausen, A. J. 1997, *Solar Phys.*, 108, 383

Wang, S. J., Yan, Y., Fu, Q., Liu, Y., & Chen, Z. 2005, in: Dere, K. P., Wang, J., & Yan, Y. (eds.) *Coronal and Stellar Mass Ejections* (Beijing), p. 139

Discussion

MELNIKOV: It is known that dm-continua and pulsations are quite common during solar flares. You say that these types frequently observed in association with CMEs. But possibly this is not a radio signature of CMEs only. Did you study the appearance of dm-continua and pulsations in events without CMEs?

CECATTO: I think there is some reason for a higher relative association between CME onset with continuum and pulsations radio bursts which has to be studied on a physical basis. I think it is not possible to say this is a signature of CME occurrence, yet. Improved statistics and interpretation on a physical basis is required to these observational results. The suggestion to compare with events without CME is welcome.

NINDOS: How do you know that the radio emission comes from the CME and not from the flare?

CECATTO: I did not say "CME radio emission". These investigations are about the radio burst emission associated with the onset of CME on a statistical basis. Also: 1. There are 13 out of 61 cases which have no associated H-alpha flare; 2. I do not believe I can measure the radio emission of a CME with a 9-m-diameter antenna due to the lack of sensitivity.

Universal Heliophysical Processes
Proceedings IAU Symposium No. 257, 2008
N. Gopalswamy & D.F. Webb, eds.

© 2009 International Astronomical Union
doi:10.1017/S1743921309029494

Formation of anisotropic distributions of mildly relativistic electrons in flaring loops

Victor F. Melnikov, Sergey P. Gorbikov and Nikolai P. Pyatakov

Radiophysical Research Institute,
B.Pecherskaya 25, 603950, Nizhny Novgorod, Russia
email: melnikov@nirfi.sci-nnov.ru

Abstract. In this paper we show that different locations of acceleration/injection sites in flaring loops may produce very different types of pitch-angle distributions of accelerated electrons and, as a consequence, different spatial, spectral and polarization properties of the loop microwave emission. It is shown that these properties can be detected using spatially resolved microwave observations of specific flaring loops and be used to choose the most suitable electron acceleration model.

Keywords. Sun: flares, Sun: radio radiation, acceleration of particles

1. Introduction

Today we know quite a wide variety of acceleration mechanisms in solar flares (see for review [Aschwanden 2002, Vlahos 2007]). Among them are: (1) electric DC-field acceleration (in current sheets or in twisted loops); (2) stochastic acceleration (wave turbulence, micro-flares); (3) shock acceleration (propagating MHD shocks; standing MHD shocks in reconnection outflows); (4) betatron acceleration (in collapsing magnetic traps). Their properties are not the same. They may act and inject accelerated electrons in different places inside a flaring loop, for example, a) *near the loop top* after acceleration in the vertical current sheet (so called 'standard model') or in a strong turbulence region, b) *near a footpoint* of a big loop in a double loop configuration, or c) *along a whole loop* if the loop is twisted or contains numerous micro current sheets. Moreover, different acceleration models may produce electrons with different types of pitch-angle distributions (isotropic, with transverse or parallel anisotropy).

Possibly, all of the mentioned mechanisms may operate in solar flares. Only observations can tell us which mechanism is dominant in a specific flare configuration. Analysis of spatially resolved microwave observations of the Nobeyama Radioheliograph has already allowed to discover very interesting and unexpected phenomena. One of them is the presence of the strong optically thin microwave source in the loop top of some single flaring loops (Kundu *et al.* 2001; Melnikov *et al.* 2002a). Later it was found that such events form quite a numerous class of single flaring loops, about 30-50%, most of others are characterized by the brightness peak(s) close to one or two loop footpoints (Martynova *et al.* 2007; Tzatzakis *et al.* 2006). The phenomenon was explained by the enhanced concentration of mildly relativistic electrons in the upper part of microwave flaring loops (Melnikov *et al.* 2002a). Such looptop electron concentration is possible, if electrons have transverse pitch-angle anisotropy, and particle acceleration/injection takes place near the loop top (Melnikov et al. 2006). Another phenomenon is spectral softening of microwave emission near the loop footpoints for disk flares (Yokoyama *et al.* 2002). The discovery was confirmed for other events (Melnikov et al. 2002b, Fleishman *et al.* 2003). It was theoretically explained as the spectral softening of gyrosynchrotron (GS) emission propagating in quasi-parallel direction in the presence of transverse pitch-angle anisotropy of mildly relativistic electrons (Fleishman & Melnikov 2003). Recently, Altyntsev *et al.* (2008) have found ample observational evidences of the existence of parallel to magnetic field pitch-angle anisotropy of energetic electrons in a specific flaring loop. Most interesting evidence of this beam-like anisotropy is the ordinary mode polarization of the emission from optically thin GS microwave source predicted in Fleishman & Melnikov (2003).

The purpose of our paper is to show that the current and future spatially resolved microwave observations are able to provide us with data about the acceleration site and pitch-angle anisotropy of emitting electrons and, therefore, may give us valuable constraints on acceleration models. We focus mostly on the influence of electron distribution dynamics on polarization and spectral properties of GS emission in different parts of a loop.

2. Dynamics of electron distributions

To learn more about the properties of microwave emission and its dynamics in different parts of flaring loops and to study the properties in a more quantitative way then before, we do modelling of the time evolution of the electron spectral, pitch-angle and spatial distributions along a magnetic loop by solving the non-stationary Fokker-Planck equation under different assumptions on the physical conditions in the loop and for different positions of the injection site (loop top, loop legs and feet). We consider the non-stationary Fokker-Planck equation in the form that takes into account Coulomb collisions and magnetic mirroring (Hamilton *et al.* 1990):

$$\frac{\partial f}{\partial t} = -c\beta\mu\frac{\partial f}{\partial s} + c\beta\frac{d\ln B}{ds}\frac{\partial}{\partial\mu}\left[\frac{1-\mu^2}{2}f\right]$$

$$+\frac{c}{\lambda_0}\frac{\partial}{\partial E}\left(\frac{f}{\beta}\right) + \frac{c}{\lambda_0\beta^3\gamma^2}\frac{\partial}{\partial\mu}\left[(1-\mu^2)\frac{\partial f}{\partial\mu}\right] + S, \qquad (2.1)$$

where $f = f(E,\mu,s,t)$ is the electron distribution function of kinetic energy $E = \gamma-1$ (in units of mc^2), pitch-angle cosine $\mu = \cos\alpha$, distance from the flaring loop center s, and time t, $S = S(E,\mu,s,t)$ is the injection rate, $\beta = v/c$, v and c are the electron velocity and speed of light, $\gamma = 1/\sqrt{1-\beta^2}$ is the Lorentz factor, $B = B(s)$ is the magnetic field distribution along the loop, $\lambda_0 = 10^{24}/n(s)\ln\Lambda$, $n(s)$ is the plasma density distribution, $\ln\Lambda$ is the Coulomb logarithm.

In this paper we present the results of our numerical experiments only for two cases using the method developed by Gorbikov and Melnikov (2007). In the first case (Model 1) the source of high energy electrons is located in the magnetic trap center $s = 0$, and in the second one (Model 2) near a trap foot $s = 2.4 \times 10^9$ cm. In both models the trap (loop) is symmetrical and its half-length is 3×10^9 cm and magnetic mirror ratio $B_{max}/B_{min} = 2$, $B_{min} = 200$ G. Plasma density is homogeneous along the loop with $n(s) = 5 \times 10^{10}$ cm^{-3}. The injection function $S(E,\mu,s,t)$ is supposed to be a product of functions dependent only on one variable (energy E, cosine of pitch-angle μ, position s, and time t): $S(E,\mu,s,t) = S_1(E)\,S_2(\mu)\,S_3(s)\,S_4(t)$, where the energy dependence is a power law $S_1(E) = (E/E_{min})^{-\delta}$, $E_{min} = 30$ keV, with the spectral index $\delta = 5$; pitch-angle distribution is isotropic $S_2(\mu) = 1$; time dependence is Gaussian $S_4(t) = exp[-(t-t_m)^2/t_0^2]$, $t_m = 25$ s, $t_0 = 14$ s; spatial distribution is also Gaussian. For

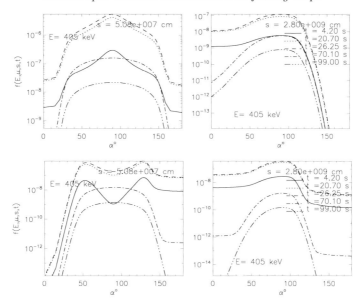

Figure 1. Results of simulations for Model 1 (top panel) and for Model 2 (bottom panel) for electron energy 405 keV and for two positions in the loop: loop top (left panels) and near a footpoint (right panels). The distribution functions over pitch-angle α for the rising phase of injection are shown by solid ($t = 4.2$ s), dotted ($t = 20.7$ s) and dashed ($t = 26.25$ s) lines, and for the decay phase by dot-dashed ($t = 70.1$ s) and dot-dot-dot-dashed ($t = 99$ s) lines.

Model 1: $S_3(s) = exp(-s^2/s_0^2)$, and for Model 2: $S_3(s) = exp[-(s - s_1)^2/s_0^2]$, where $s_0 = 3 \times 10^8$ cm, $s_1 = 2.4 \times 10^9$ cm.

It is known that spectral and polarization properties of GS emission are very sensitive to peculiarities of the electron pitch-angle distribution in a radio source (Fleishman & Melnikov 2003). So, here, in Fig. 1 we present some results of modelling the pitch-angle distributions of mildly relativistic electrons ($E = 405$ keV) in the center and end of the magnetic trap (loop) for Model 1 and Model 2.

Model 1 (injection at the looptop). In the loop center, the distribution remains aniso-tropic perpendicular to magnetic field lines during the injection rise, maximum ($t_m = 25$ s) and decay phases. However, the degree of anisotropy decreases with time, especially in the decay phase. Near a loop footpoint, the electron pitch-angle distribution is clearly asymmetric, showing a considerable amount of electrons with small pitch-angles. In the decay phase, the distribution becomes more and more symmetric with the peak close to $\alpha = 90^o$ (transverse anisotropy increases).

In the case of Model 2 (injection near a footpoint), the pitch-angle distribution and its dynamics near the footpoint are very similar to the ones in Model 1. At the loop center, however, the shape of the distribution and its dynamics are completely different. First of all, we can see two peaks near pitch-angles 50° and 130° that indicates the presence of oblique fluxes (beams) of electrons. Second, the distribution changes dramatically during the decay time getting more isotropic. Obviously, the pitch-angle scattering due to Coulomb collisions and precipitation into the loss-cone play an important role in the mentioned dynamics.

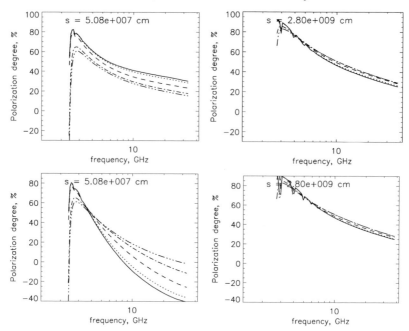

Figure 2. Frequency spectra of polarization degree and its dynamics for Model 1 (top panel) and Model 2 (bottom panel) for two positions in the loop: in the loop top (left plots) and near a footpoint (right plots). The lines meaning is as in Fig. 1.

3. Radio response to the specific electron distributions

In this section we show the influence of electron distribution dynamics on the polarization and spectral properties of microwave GS emission from different parts of a magnetic loop. We do simulations in the frames of assumptions accepted for Model 1 and Model 2 and use the exact formalism described in papers of Ramaty (1969), and Fleishman & Melnikov (2003). The magnetic loop is thin (so that the microwave source is optically thin in the considered frequency range) and located in the plane almost perpendicular to the line of sight ($\theta = 78.5^o$).

Results of our simulations are presented in Figs. 2 and 3. Fig. 2 displays frequency spectra of polarization degree and its dynamics for Model 1 (top panel) and Model 2 (bottom panel) for two positions in the loop: in the loop top (left plots) and near a footpoint (right plots). For both models the polarization spectra of emission from the region near a footpoint are very similar. The polarization is positive (X-mode) at all frequencies and its degree is quite high ($25 - 30\%$) even at the highest frequencies. The time evolution is very weak if present at all.

The polarization spectra from the loop top region are markedly different. They show obvious dynamics. They differ from each other. The most striking differences between Model 1 and Model 2 are the following. First, the polarization degree in Model 2 (isotropic injection near a footpoint) is negative (O-mode) at high frequencies, whereas in Model 1 it is positive both at low and high frequencies. Such unusual phenomenon is explained by the fact that in Model 2 we have an oblique flux (beam) of electrons in the central part of the magnetic trap (Fig.1). The oblique beam of energetic electrons is known to produce O-mode polarized emission in the quasi-transverse direction even in optically thin regime (Fleishman & Melnikov 2003). The second strong difference is the difference in the dynamics of the polarization spectra. In Fig. 2 we can see that in Model 1 the

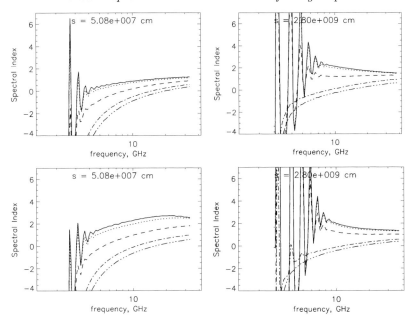

Figure 3. Frequency spectra of the local spectral index $\alpha(f)$. The lines of different styles indicate the moments explained in Fig. 1.

polarization degree decreases with time, by $\approx 10 - 20\%$, whereas in Model 2 it increases considerably, by $\approx 20 - 40\%$, and even may change its sign on the late decay phase of the injection.

Somewhat similar picture of differences between Model 1 and Model 2 is observed for frequency spectra of the local spectral index $\alpha(f)$ and its dynamics (Fig. 3). In Fig. 3 we can see very similar values and dynamics of $\alpha(f)$ near a footpoint (right plots) for both models. At the same time, the values and dynamics of $\alpha(f)$ in the loop top region are noticeably different. For Model 2 the value of $\alpha(f)$ is larger than for Model 1. Moreover, at high frequencies it is even larger than $\alpha(f)$ in the footpoint emission source. The higher values of $\alpha(f)$ in the loop top for Model 2 is definitely associated with the beam-like anisotropy of the energetic electrons in the central part of the magnetic loop (Fig.1). Such anisotropy is known to produce steeper frequency spectra of GS emission in the quasi-transverse direction (Fleishman & Melnikov 2003).

4. Conclusion

The differences in the behavior of polarization and spectra found for two injection models can serve as a diagnostic tool for distinguishing different types of anisotropic distributions in flaring loops. These findings, together with a set of other recent achievements in the theoretical and observational studies, may be developed into a new method of direct diagnostics of acceleration mechanisms and properties of kinetics of high energy electrons in flaring magnetic loops by means of spatially and spectrally resolved microwave observations. It is clear that building new radio instruments such as FASR, CSRH, and modified SSRT, which are able to observe intensity and polarization of microwave emission in a wide frequency range and with high spatial, spectral and temporal resolution is crucially important for solving the key problems in the physics of solar flare particle acceleration.

The work was partly supported by RFBR grants No.06-02-39029, 06-02-16295, 07-02-01066.

References

Altyntsev, A. T., Fleishman, G. D., Huang, G. L., & Melnikov, V. F. 2008 *ApJ*, 677, 1367

Aschwanden, M. J. 2002 *SSRv*, 101, 1

Fleishman G. D. & Melnikov V. F. 2003 *ApJ*, 587, 823

Fleishman, G. D., Gary, D. E., & Nita, G. M. 2003 *ApJ*, 593, 571

Gorbikov, S. P. & Melnikov, V. F. 2007 *Mathematical Modeling*, 19, 112

Hamilton, R. J., Lu, E. T., & Petrosian, V., 1990 *ApJ*, 354, 726

Kundu, M. R., Nindos, A., White, S. M., & Grechnev, V. V. 2001 *ApJ*, 557, 880

Martynova, O. V., Melnikov, V. F., & Reznikova, V. E. 2007 *Proc. of 11th Pulkovo Int. Conf. on Solar Physics, Saint-Peterburg*, 241

Melnikov, V. F., Shibasaki, K., & Reznikova, V. E. 2002 *ApJ*, 580, L185

Melnikov, V. F., Reznikova, V. E., Yokoyama, T., & Shibasaki, K. 2002b *ESA*, SP-506, 339

Melnikov, V. F., Gorbikov, S. P., Reznikova, V. E., & Shibasaki, K. 2006 *Bull. of the RAS Phys.*, 70(10), 1684

Ramaty R. 1969, *ApJ*, 158, 753

Tzatzakis, V., Nindos, A., Alissandrakis, C. E., & Shibasaki, K. 2006 *AIP Conf. Proc.*, 848, 248

Vlahos L. 2007 *Lecture Notes in Physics*, 725, 15

Yokoyama, T., Nakajima, H., Shibasaki, K., Melnikov, V. F., & Stepanov, A. V. 2002 *ApJ*, 576, L87

Discussion

TERASAWA: Is there a reason why effects of anisotropy driven instabilities, such as whistlers are not included in your analysis? They may not be important in the final anisotropy limit considered.

MELNIKOV: This is an important question. Indeed scattering on whistlers is not included in the Fokker-Plank equation in the form we used. However, we did some calculations and found that for the power-law electron energy distribution the level of generated whistler waves is too weak to scatter resonant relativistic electrons effectively. At least, for the small anisotropy limit considered.

SCHMIEDER: What is the expected spatial resolution of the new generation of radio telescopes in Siberia and with FASR?

MELNIKOV: The angular resolution of FASR is expected to be 1 arcsec at 20 GHZ. For the SSRT it will be 10–15 arcsec at the frequency range of 4–9 GHz.

GOPALSWAMY: What is the relation between the microwave loop-top source and the superhot component observed in X-rays?

MELNIKOV: They are coincident. At least for some events we've studied.

Universal Heliophysical Processes
Proceedings IAU Symposium No. 257, 2008
N. Gopalswamy & D.F. Webb, eds.

© 2009 International Astronomical Union
doi:10.1017/S1743921309029500

Localization of type II and type III radio burst sources using multi-spacecraft observations

G. Thejappa[1] and R. J. MacDowall[2]

[1] Department of Astronomy, University of Maryland,
College Park, MD 20742, USA
email: thejappa@astro.umd.edu

[2] NASA, Goddard Space Flight Center
Greenbelt, MD 20771 USA
email: Robert.MacDowall@nasa.gov

Abstract. A method for the localization of the radio burst sources associated with the flare accelerated electron beams and coronal mass ejection (CME) shocks is presented. This method involves the computations of the ray trajectories, time delays, and optical depths in the refracting solar atmosphere. The coordinates of the radiating source can be obtained by comparing the time delays and intensity ratios of the bursts observed by widely separated spacecraft with the computed group delays and intensity ratios at exit points of the rays from the solar atmosphere. This method is applied to a type III radio burst observed by the STEREO spacecraft.

Keywords. Sun, Radio Bursts

1. Introduction

The most intense radio emissions from the the solar corona and interplanetary medium are the type III and type II radio bursts, probably excited by the plasma mechanism at the fundamental and second harmonic of the electron plasma frequency, $f_{pe}(kHz) = 9\sqrt{N_e(cm^{-3})}$, where N_e is the electron density. These bursts are characterized respectively by the fast and slow frequency drifts in the dynamic spectrum. For a given density model, the plasma emission mechanism allows us to convert the observed frequency drifts into the speeds of the outward propagating exciting agencies. For example, the frequency drifts of type III and type II radio bursts yield speeds typical of the flare accelerated electron beams and coronal mass ejection (CME) driven shocks, respectively. The actual association between type III and type II bursts and flare electron beams and CME shocks has been well-established by in situ observations. At kilometric and hectometric wavelengths, the direction finding and triangulation techniques (Reiner *et al.* 1998; Hoang *et al.* 1998) as well as the time difference of the arrival (TDOA) technique (Weber *et al.* 1977) are usually used to obtain the positions of the source observed by two or more spacecraft. Even though the triangulation technique requires the 2 rays from the 2 spacecraft to intersect at a point, such intersection usually does not occur due to refraction and scattering. For example, Hoang *et al.* (1998) have determined the position of a type II burst source using triangulation technique, and found it to be off by several AU from the actual source location due to refraction and scattering. As far as the TDOA technique is concerned, the underlying assumptions are that (1) the effects of propagation on the radio emissions are negligible, and (2) the loci of points that have a fixed difference of distance between two points is a hyperbola (or a hyperboloid) of revolution. Under such circumstances, one can fix the position of the source to a point on such a hyperbola either by assuming that these bursts are excited at the f_{pe} or $2f_{pe}$, or by using observations from an additional spacecraft. However, the radio emissions as they propagate through the inhomogeneous plasma experience time delays due to refraction and scattering. For example, Steinberg *et al.* (1984) have measured the arrival times of the type III radio bursts at the ISEE-3 and Voyager 1 and 2 spacecraft and have found that the radio signals are anomalously delayed up to 500 seconds equivalent to 1 AU of excess path length or even more, when the bursts were detected at two spacecraft. Therefore, unless the measured directions and time delays are corrected for propagation effects, the results obtained by these techniques are very unreliable.

The purpose of this paper is to describe a new method, which can be used to localize the radio burst sources observed by a pair of widely separated spacecraft. This method is based on the time delays and attenuations suffered by the radio emissions due to propagation effects, especially refraction. In section 2, we describe the density and temperature models used in this study, in section 3, we present the ray tracing equations and algorithm to compute the ray trajectories, time delays and optical depths, and in section 4, we apply this method to STEREO observations of a type III radio burst.

2. Density and Temperature Models

For the electron number density of the corona and interplanetary medium, we use the empirical formula derived by Guhathakurta, Holzer & MacQueen (1996) based on the Skylab observations obtained during the declining phase of solar cycle 20 (1973-1976)

$$N_e(r, \theta_{mg}) = N_p(r) + [N_{cs}(r) - N_p(r)]e^{-\theta_{mg}^2/w^2(r)} \text{ cm}^{-3}. \tag{2.1}$$

Here r is the radial distance in units of $R_\odot$, and $\theta_{mg} = \sin^{-1}[-\cos\theta\sin\alpha\sin(\phi - \phi_0) + \sin\theta\cos\alpha]$ is the heliomagnetic latitude of a point from the current sheet, θ and ϕ are the heliographic latitude and longitudes, respectively, $\alpha \simeq 15\,\text{deg}$ is the tilt angle between dipole axis and the rotation axis, and $\phi_0 \simeq 0$ is the angle between the heliomagnetic and heliographic equators. The electron densities at the current sheet $N_{cs}(r)$ and at the poles $N_p(r)$ are given by $N_e(r) = \Sigma_{i=1}^3 c_i r^{-d_i}$, where c_1, c_2 and c_3 are 1.07, 19.94, and 22.10 for the current sheet, and 0.14, 8.02, and 8.12 for the pole, in units of 10^7, and d_1, d_2 and d_3 are 2.8, 8.45, and 16.87, respectively. The half-angular width of the current sheet is $w(r) = \Sigma\gamma_i r^{-\delta_i}$, where γ_1, γ_2, and γ_3 are 16.3, 10.0, and 43.20 degrees, and, δ_1, δ_2, and δ_3 are 0.5, 7.31, and 7.52, respectively. The expression for collision frequency can be written as $\nu = 4.36 N_e T_e^{-3/2}[10.8 + \ln(T_e^{3/2}/f)]$, where T_e is the electron temperature. In this study, we take $T_e = 1.5 \times 10^5$ K. The refractive index can be written as $\mu^2(r) = 1 - \frac{f_{pe}^2(r)}{f^2}$.

3. Ray Tracing Calculations

We use the Cartesian coordinate system with origin at the center of the Sun, and x-axis directed along the line of sight. Haselgrove (1963) has shown that the ray tracing can be performed by using a set of 6 first-order differential equations

$$\frac{d\vec{R}}{dp} = \vec{T} \tag{3.1}$$

$$\frac{d\vec{T}}{dp} = D(\vec{R}) = \frac{1}{2}\frac{\partial\mu^2}{\partial\vec{R}}. \tag{3.2}$$

Here

$$\vec{R} \equiv \begin{pmatrix} x \\ y \\ z \end{pmatrix} \quad \text{and} \quad \vec{T} \equiv \begin{pmatrix} T_x \\ T_y \\ T_z \end{pmatrix}$$

are the position and direction vectors, respectively with $T_x^2 + T_y^2 + T_z^2 = \mu$, and p is the independent variable related to the path length as $dp = \frac{ds}{\mu}$. Using equation (2.1), we can write

$$D(\vec{R}) \equiv \frac{1}{2} \begin{pmatrix} \frac{\partial \mu^2}{\partial x_z} \\ \frac{\partial \mu^2}{\partial y_z} \\ \frac{\partial \mu^2}{\partial z} \end{pmatrix} = \frac{8.90 \times 10^{12}}{f^2} \frac{1}{r^4} N_e(r, \theta_{mg}) \vec{R}. \tag{3.3}$$

The 3rd order Runge-Kutta algorithm, which is used to integrate these ray tracing equations can be written in the vectorial form as

$$R_{n+1} = R_n + \Delta\tau \left[T_n + \frac{1}{6}(A + 2B) \right], \tag{3.4}$$

$$T_{n+1} = T_n + \frac{1}{6}(A + 4B + C), \tag{3.5}$$

$$A = \Delta\tau D(R_n), \tag{3.6}$$

$$B = \Delta\tau D\left(R_n + \frac{\Delta\tau}{2} T_n + \frac{1}{8}\Delta\tau A \right), \tag{3.7}$$

$$C = \Delta\tau D\left(R_n + \Delta\tau T_n + \frac{1}{2}\Delta\tau B \right), \tag{3.8}$$

$$D(R) = \frac{1}{2}\frac{\partial \mu^2}{\partial R}. \tag{3.9}$$

Using this algorithm, one can trace the rays through any medium, i.e., starting from a known point $(\vec{R_0}, \vec{T_0})$, one can generate successively $(\vec{R_1}, \vec{T_1})$, $(\vec{R_2}, \vec{T_2})$......$(\vec{R_n}, \vec{T_n})$. For example, Thejappa, MacDowall & Kaiser (2007) and Thejappa & MacDowall (2008) have successfully applied this algorithm to study the visibility of type III radio bursts and quiet sun radio emissions, respectively. In Fig. 1, we present the computed ray trajectories at 625 kHz, where the smaller circle corresponds to the Sun, and the bigger circle corresponds to the $2f_{pe}$ plasma level. The optical depth τ and the transit time Δt

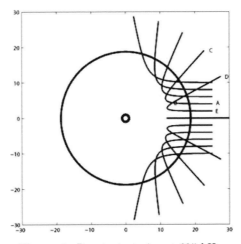

Figure 1. Ray trajectories at 625 kHz.

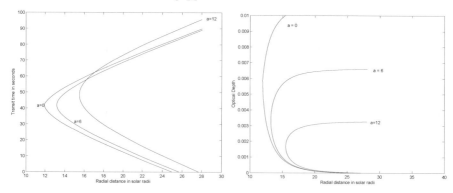

Figure 2. Computed time delays and optical depths.

are computed at each step ΔS as

$$\tau_{i+1} = \tau_i + 2.32 \frac{f_{pe}^2}{f^2} \frac{\nu \Delta S}{\mu_i} \tag{3.10}$$

$$\Delta t_{i+1} = \Delta t_i + 2.32 \frac{\Delta S}{\mu_i}. \tag{3.11}$$

The factor 2.32 is the solar radius $(R_\odot)$ divided by the velocity of light (c). In these two equations, instead of phase velocity c/μ, the group velocity $(c\mu)$ is used to relate Δt to ΔS. In Fig. 2, we present the computed time delays and optical depths against the radial distance, r for various values of the impact distances, a (the distance of the asymptote of the ray from the parallel line through the sun's center).

4. Application

The STEREO mission (Kaiser 2005) consists of two identical spacecraft- one ahead of Earth in its orbit (Stereo 'A'), the other trailing behind (Stereo 'B'). Each spacecraft is equipped with identical instrumentation. For example, the radio receivers (HFR and LFRhi) of the STEREO/Waves experiment (Bougeret et al. 2007) measure the radio wave intensities in the frequency from 40 kHz to 16 MHz, corresponding to source distances of about 1 AU to $1R_\odot$. This experiment is mainly designed to study type II and type III radio bursts. In Fig. 3, we present a typical type II radio burst and a typical type III radio burst detected by both the spacecraft on December 31, 2007. The spacecraft were separated from each other by ~ 42 degrees. The top part of the dynamic spectrum

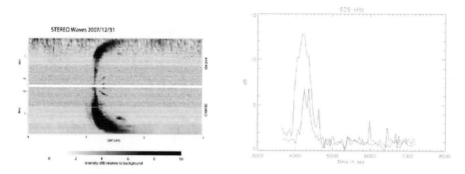

Figure 3. The dynamic Spectrum and the time profile of the type III radio burst observed simultaneously by the two spacecraft of the STEREO mission.

extending from 1 to 16 kHz is obtained by STEREO 'A', and the bottom part extending from 16 MHz to 1 MHz is obtained by STEREO 'B'. The dynamic spectra from both spacecraft clearly show that type III burst extends all the way from 16 MHz to 1 MHz and the slowly drifting type II radio burst contains both fundamental and harmonic bands. In this dynamic spectrum, the LFR frequencies are not shown. Fig. 3 also shows the time profiles of the type III burst (top and bottom profiles correspond to STEREO 'A' and 'B', respectively). From these time profiles, it is clear that the type III emission is delayed almost by ~ 72 seconds at 'B' in comparison with that of 'A'. Here we note that the time delay is estimated using the start time of these bursts since the peak of the emission from 'B' is difficult to establish. Moreover, the burst at 'B' is much weaker than that of 'A' by a factor of ~ 2. The computed ray trajectories, time delays and optical depths in the previous section provide important clues to understanding these observed time delays and intensity ratios. A possible interpretation is that (1) the type III burst is generated by a localized disturbance in the solar atmosphere, and (2) the radiation from the disturbance reaches the widely separated spacecraft by two different paths, namely "direct" path and "echo" path. The "echo" is delayed and attenuated in comparison with the "direct" signal because it has to travel a larger distance with considerably slower group velocity to reach the receiver. For example, these anamolies can be understood using the computed trajectories (Fig. 1). Let us assume that the type III emission is generated by the disturbance located at B. This radiation can reach 'A' by the path BA, and the 'B' by the path BC approximately separated by 40 deg. As seen from these trajectories the "echo" signal which follows the path BC will be more attenuated and will arrive at C later than '"direct"' signal along BA. These time delays and attenuations can be determined from Figure 2. The practical aspect of these anamolies is that by comparing the observed time delays and attenuations with observed quantities, we can find the coordinates of the source (localization). For example, the three dimensional localization of the radio source at one frequency fixes the instantaneous location of the electron beam in the case of a type III burst, and the location of the CME shock in the case of a type II radio burst. By repeating the source localizations at several frequencies, the trajectories of the electron beams and CME shocks can be mapped through interplanetary space. By successfully tracking the radio sources, one can predict the arrival of the electron beams and CME shocks at Earth which is essential for space weather predictions. The detailed procedure and algorithm of localization and tracking of radio sources using the present method will be presented in a separate paper.

Support for T. G. was provided by the Living With a Star Targeted Research and Technology program through NASA Grant 07-LWSTRT-0120. We thank M. L. Kaiser for providing the data and R. A. Hess for his help with data.

References

Bougeret *et al.* 2008, *Space Sci. Rev.*, 136, 487.
Guhathakurta, M., Holzer, T. E., & MacQueen, R. M. 1996, *ApJ*, 458, 817
Haselgrove, J. 1963, *J. Atmos. Terr. Phys.*, 25, 397
Hoang, S., Maksimovic, M., Bougeret, J.-L., Reiner, M. J., & Kaiser, M. L. 1998, *GRL*, 25, 2497
Kaiser, M. L., 2005, *Advances in Space Res.*, 36, 1483
Reiner, M. J., Fainberg, J., Kaiser, M. L., & Stone, R. G. 1998, *JGR*, 103, 1925.
Steinberg, J.-L., Dulk, G. A., Hoang, S., Lecacheux, A., & Aubier, M. G. 1984, *A&A*, 140, 39
Thejappa, G., MacDowall, R. J., & Kaiser, M. L. 2007, *ApJ*, 671, 894
Thejappa, G. & MacDowall, R. J. 2008, *ApJ*, 676, 1338
Weber, R. R., Fitzenreiter, R. J., Novaco, J. C., & Fainberg, J. 1977, *Solar Phys.*, 54, 431

Discussion

ANONYMOUS: Can you measure a real angular size of the cone type III emission propagates in?

THEJAPPA: It is known that type III emission should be beamed if it is excited by plasma mechanism. However, they are visible to widely separated spacecraft. To desire the intrinsic beam pattern, we have to correct the observed visibility for the propagation effects by doing statistical ray tracing.

Universal Heliophysical Processes
Proceedings IAU Symposium No. 257, 2008
N. Gopalswamy & D.F. Webb, eds.

© 2009 International Astronomical Union
doi:10.1017/S1743921309029512

Re-examining the correlation of complex solar type III radio bursts and solar energetic particles

R. J. MacDowall[1], I. G. Richardson[2], R. A. Hess[3], and G. Thejappa[4]

[1]NASA Goddard Space Flight Center, Code 695, Greenbelt, MD 20771, USA
email: robert.macdowall@nasa.gov

[2]Code 661, NASA Goddard Space Flight Center, Greenbelt, MD 20771, and
CRESST and Department of Astronomy, University of Maryland, College Park, MD 20742
email: ian.g.richardson@nasa.gov

[3]NASA Goddard Space Flight Center, Code 695, Greenbelt, MD 20771, USA, and
Wyle Information Systems Group, McLean, VA 22102
email: roger.hess@nasa.gov

[4]NASA Goddard Space Flight Center, Code 695, Greenbelt, MD 20771, USA, and
Department of Astronomy, University of Maryland, College Park, MD 20742
email: thejappa.golla@nasa.gov

Abstract. Interplanetary radio observations provide important information on particle acceleration processes at the Sun and propagation of the accelerated particles in the solar wind. Cane *et al.* (2002) have drawn attention to a class of prominent radio bursts that accompany >20 MeV solar proton events. They call these bursts 'type III-L' because: they are fast drifting (like normal type III bursts associated with electrons accelerated at impulsive solar flares); they are Long-lasting compared to normal type III bursts; they occur Late compared to the onset of the related solar event; and, they commence at Lower frequencies (~100 MHz) than normal type III bursts, suggesting that they originate higher in the corona at ~0.5 R_s above the Sun. We report on an analysis of the correlated radio and SEP events during 1996-2006 using the Wind Waves and near-Earth SEP data sets, and discuss whether the characteristics of the complex type III bursts (at less than 14 MHz) will permit them to serve as proxies for SEP event occurrence and intensity.

Keywords. acceleration of particles, shock waves, Sun: radio radiation, Sun: coronal mass ejections (CMEs), Sun: flares

1. Introduction

Solar radio burst observations provide valuable information about particle acceleration in the corona and inner heliosphere. Type III bursts are radio emissions whose exciter electrons (of energies 1-10 keV) propagate rapidly from a solar flare source outwards along open magnetic field lines. Type II solar bursts result from electrons accelerated at coronal mass ejection (CME)-driven shocks or blast wave shocks in the corona. For both types of solar radio event, the emission mechanism is a plasma process where electron beam energy is converted to a plasma wave intermediary, which is then converted to a propagating electromagnetic wave.

Cane *et al.* (2002) defined a special class of type III bursts, based on examination of radio data from metric to kilometric wavelengths, that were found to accompany >20 MeV solar proton events. These bursts were Long-lasting compared to normal type III bursts; Late compared to the onset of the related solar event; and commenced at Lower frequencies (~100 MHz) than normal type III bursts (suggesting that they

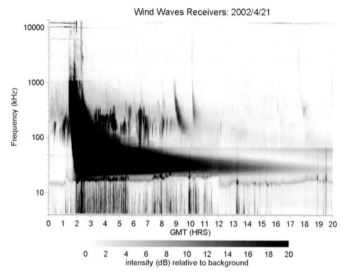

Figure 1. Twenty hours of Wind Waves data (from the RAD2, RAD1, and TNR receivers) on 2002/04/21. The large event can be described as a type III-L burst. Note the long durations at 1 MHz and 100 kHz and the complexity above 1 MHz compared to a typical, intense type III burst (see Figure 2).

originate higher in the corona, at ∼0.5 Rs above the solar surface). Figure 1 presents an example of such a burst, observed by the Wind Waves instrument (Bougeret *et al.* (1995)).

Bursts like that in Figure 1, notable for their long-duration and complexity above 1 MHz and their high intensity, especially at frequencies below 100 kHz, have been studied previously by a number of authors. Cane *et al.* (1981) proposed that the exciter electrons were accelerated by CME-driven shocks; Kahler *et al.* (1986) and Dulk *et al.* (2000) reached similar conclusions. MacDowall *et al.* (1987) and Kundu *et al.* (1990) examined the events in the framework of shock association, rather than shock acceleration. Klein *et al.* (1997) and Reiner *et al.* (2000) pointed to correlations with GHz emissions lower in the corona to argue that the source electrons were not accelerated at shocks. Cane *et al.* (2002) suggested that type III-L bursts were produced by electrons accelerated by reconnection behind fast CMEs.

MacDowall *et al.* (2003) confirmed that bursts with type III-L characteristics are statistically associated with intense SEP (proton) events. They examined durations, intensities, and other characteristics of hectometric type III radio bursts associated with intense SEP events and compared them to several groups of control particle events. They concluded that simple criteria, based on hectometric data alone, can identify the majority (∼ 80%) of type III-L radio bursts associated with > 20 MeV SEP proton events, while excluding almost 100% of the control events. In this paper we examine a concise set of parameters for identifying a type III-L burst and its likely association with an SEP event.

2. Observations

Energetic (>20 MeV) proton data from instruments on various near-Earth spacecraft, including IMP-8 and SOHO, are used to construct a complete list of solar energetic particle (SEP) events during 1996-2006 that are dispersive and above instrumental background

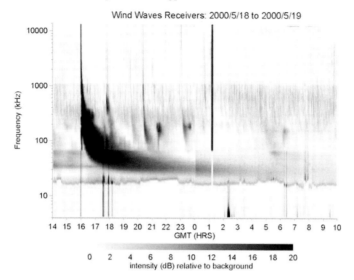

Figure 2. A classical type III burst starting on 2000/05/08. Note the much shorter duration and the simple appearance above 1 MHz. This burst is more intense than the burst in Figure 1 at ∼ 1 MHz, as indicated by the saturation-induced intermodulation seen at 16:00 below ∼ 200 kHz.

levels (a few events may have been missed due to data gaps or high backgrounds from preceding events). We also use data from the Wind Waves receivers, specifically, RAD2 (1-14 MHz), RAD1 (52-1000 kHz), and Thermal Noise Receiver (TNR) (4-64 kHz), as shown in Figure 1, to examine the properties of the radio bursts associated with three groups of proton events. At ∼ 25 MeV during the first 12 hours of an event, these three groups have maximum intensities in well-separated bins defined by (1) 0.0009 - 0.001 protons/ $(cm^2\text{-sec-ster-MeV})$, (2) 0.1 - 0.2 protons/ $(cm^2\text{-sec-ster-MeV})$, and (3) > 1 protons/ $(cm^2\text{-sec-ster-MeV})$. We also selected by visual inspection a control group of intense type III bursts that are not associated with any significant level of SEP activity at or above 25 MeV; an example is shown in Figure 2.

3. Results

In Figure 3, the radio burst durations at 1 MHz (~ 10 R$_s$) determined from RAD1 data are plotted for the 4 groups. Although there is some overlap, the range of burst durations clearly increases with SEP event intensity. Average durations (diamonds) increase from 17 to 25 to 38 minutes for groups (1) to (3). Around a third of the events in group (3) are bursts exceeding 50 minutes, which are not observed for groups (1) or (2). Furthermore, all events in group (3) have durations exceeding ∼ 20 minutes. In contrast, the control bursts clearly have shorter durations ($\leqslant 20$ min) than almost all of the events associated with SEPs, with a mean of ∼ 12 minutes and a smaller spread in duration. Figure 2 shows a similar pattern for the RAD1 durations at 100 kHz except that the durations of the control events now overlap those of the weaker SEP events. Note that the typical durations are longer for all of the groups than at 1 MHz, as is evident for the example bursts in Figures 1 and 2.

Next, we examine how the complexity of the radio burst varies with SEP event intensity. To quantify complexity we use four frequencies in the Waves RAD2 range: 13.825 (RAD2 channel 255), 6.875 (116), 3.475 (48), 1.725 (13) MHz. Radio intensity data for

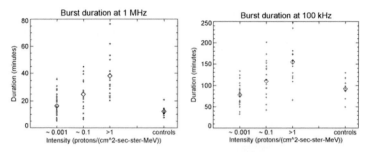

Figure 3. (Left) RAD1 burst durations at 1 MHz for three groups of SEP events with increasing proton intensity at ~ 25 MeV, and a control group of bursts without SEP events. Diamonds show the mean value for each group. (Right) Same format as at left for RAD1 100 kHz data.

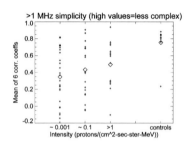

Figure 4. Radio burst simplicity factor as calculated for the 3 groups of SEP events and the control events. See description in text.

pairs of these 4 frequencies (6 pairs in all) are cross correlated and the 6 values are averaged. The result may be called a 'simplicity factor' because the correlation coefficients are larger for simple bursts than for more complicated bursts. The simplicity factor (SF), plotted in Figure 4, has values close to unity for similar burst profiles at the 4 frequencies, with lower values indicating profiles of greater complexity. With one exception, the control events all have simple profiles (SF~ 0.8), whereas bursts associated with SEP events are more complex with a wide range of SF that shows little variation with SEP event intensity. This is consistent with the complexity previously identified as characteristic of type III-L events. Typically, the complex profiles correspond to multiple components that are seen above 1 MHz (see Figure 1). Such components were identified by eye and counted at 2 MHz in the analysis of MacDowall *et al.* (1987).

4. Discussion

The present study of a fraction of the events in the complete SEP event list indicates that duration (~ 100 kHz - 1 MHz) and complexity (1 - 10 MHz) of the radio bursts associated with intense (> 1.0 /(cm^2 s sr MeV) 25 MeV proton events are almost always greater than for the control events. For weaker SEP events, the burst duration generally decreases with proton intensity and there is increasing overlap with the control events, suggesting that it would be difficult to associate all SEP events with a distinct class of type III radio bursts. Nevertheless, long burst duration and complexity at frequencies above 1 MHz appear to reliably select radio events that are associated with intense SEP events. This association may warrant consideration of the radio bursts as a space weather predictor.

There are several outstanding questions which we will discuss in a future work, including: (a) What are the causes of the long duration and complexity of type III-L bursts? (b) Do type III-L bursts suggest that some flare (proton) particles contribute to SEP events? (c) How useful a role can type III-L bursts play in SEP space weather prediction?

Comparison of the complete SEP event list with the Waves radio data will permit us to better investigate the statistics of parameters defining a class of 'type III-L bursts'.

Acknowledgements

The Wind Waves investigation is a collaboration of NASA Goddard Space Flight Center (GSFC), the Observatoire de Paris-Meudon, and the University of Minnesota. Energetic particle data in this study were provided by the GSFC instrument on IMP-8, the ERNE and COSTEP instruments on SOHO, and EPAM instrument on ACE. Observations from these instruments are available from the Space Physics Data Facility (http://spdf.gsfc.nasa.gov/).

References

Bougeret, J. L., *et al.* 1995, *Space Sci. Rev.*, 71, 231

Cane, H. V., Stone, R. G., Fainberg, J., Stewart, R. T., Steinberg, J.-L., & Hoang, S. 1981, *Geophys. Res. Lett.*, 8, 1285

Cane, H. V., Erickson, W. C., & Prestage, N. P. 2002, *J. Geophys. Res.*, 107(A10), 1315, doi:10.1029/2001JA000320

Dulk, G. A., Leblanc Y., Bastian, T. S., & Bougeret, J. L. 2000, *J. Geophys. Res.*, 105, 27,343

Kahler, S. W., Cliver, E. W., & Cane, H. V. 1986, *Adv. Space Res.*, 6, 319

Klein, K.-L., Aurass, H., Soru-Escaut, I., & Kalman, B. 1997, *A&A*, 320, 612 (erratum 322, 1027)

Kundu, M. R., MacDowall, R. J., & Stone, R. G. 1990, *Ap&SS*, 165, 101

MacDowall, R. J., Kundu, M. R., & Stone, R. G. 1987 *Solar Phys.*, 111, 397

MacDowall, R.J., Lara, A., Manoharan, P. K., Nitta, N. V., Rosas, A. M., & Bougeret, J. L. 2003, *Geophys. Res. Lett.*, 30, 12, 8018, doi:10.1029/2002GL016624

Reiner, M. J., Karlicky, J. M., Jiricka, K., Aurass, H., Mann, G., & Kaiser, M. L. 2000, *Astrophys. J.*, 530, 1049

Discussion

SPANGLER: You have plotted your spectra as power level in db above background. This has problems if the background depends on frequency, and in any case, is physically uninterpretable. I would recommend plotting spectra as Janskys or antenna temperature as a function of frequency.

MACDOWALL: We can and often do that; however, given 3 instruments (here) with quite different backgrounds the result may not look the best. Furthermore, given the frequency-dependent backgrounds such events can be seen more easily using relative backgrounds.

ROTH: Fluxes of type III exciting electrons are often related to enhanced MeV heavy ions or He^3. Does it apply to these complex type III events?

MACDOWALL: We haven't looked at this for these events. I recall that, for the 2002 CDAW on SEPs, there did not seem to be a good correlation between heavy ion events and the type III-l events.

GOPALSWAMY: You didn't say much about CMEs. Have you looked at the evolution of magnetic arcades behind the CME to determine how the time scale compares with the radio burst duration?

MACDOWALL: Not yet. As I said, Dr. Cane's hypothesis was that the type III-L bursts electrons are accelerated in reconnection fields behind the CME. We should look at the arcade evolution for time correlation.

Universal Heliophysical Processes
Proceedings IAU Symposium No. 257, 2008
N. Gopalswamy & D.F. Webb, eds.

© 2009 International Astronomical Union
doi:10.1017/S1743921309029524

Explosion of sungrazing comets in the solar atmosphere and solar flares

S. Ibadov[1], F. S. Ibodov[2] and S. S. Grigorian[2]

[1]Institute of Astrophysics, Dushanbe, Tajikistan
email: ibadovsu@yandex.ru

[2]Moscow State University, Moscow, Russia
email: mshtf@sai.msu.ru, grigor@imec.msu.ru

Abstract. Explosive evolution of nuclei of sungrazing comets near the solar surface, which occurs at conditions of intense interaction between the solar atmosphere and falling high-velocity comet nuclei as well as the relation of the phenomenon to the character of solar activity are analytically considered. It is found that, due to aerodynamic fragmentation of the falling body in the solar chromosphere and transversal expansion of the fragmented mass under the action of pressure gradient on the frontal surface, thermalization of the kinetic energy of the body occurs by sharp stopping of the disklike hypervelocity fragmented mass near the solar surface within a relatively very thin subphotospheric layer and has, therefore, an essentially impulsive and strongly explosive character. The specific energy release in the explosion region, erg/g, considerably exceeds the evaporation/sublimation heat of the body so that the process is accompanied by production of a high-temperature plasma. The energetics of such an explosive process corresponds to that of very large solar flares for falling bodies having masses equal to the mass of the nucleus of Comet Halley. Spectral observations of sungrazing comets by SOHO-like telescopes in a wide spectral range, including X rays, with a high time resolution, of the order of 0.1–10 s, are important for revealing solar activity in the form of an impact-generated photospheric flare.

Keywords. comets: general; Sun: flares; explosions

---

1. Introduction

Coronagraphic observations by SOLWIND (Solar Wind), SMM (Solar Maximum Mission) and SOHO (Solar and Heliospheric Observatory) missions indicate the presence of a continuous comet flow passing close to the solar surface or colliding with the Sun (Weissman 1983; Marsden 1989; MacQueen & St. Cyr 1991; Bailey *et al.* 1992; COSPAR 1998). Passages of cometlike bodies, extrasolar comets, near young stars may be responsible for observed changes in stellar spectra, for the origin of the Beta Pictoris like phenomenon, due to evaporation of these bodies (see, e.g., Beust *et al.* 1996, Ibadov *et al.* 2007, and references therein).

At the same time disintegration process of nuclei of sungrazing comets being considered in the framework of traditional sublimation model, i.e., by the action of the solar photospheric thermal radiation, leads to an insignificant decrease in the comet nucleus radii, attaining not more than 20 metres (Weissman 1983; MacQueen & St. Cyr 1991).

We are developing an analytical approach to investigate the evolution of comet nuclei under the conditions of intense interaction between the solar atmosphere and falling nuclei resulting in their aerodynamic fragmentation as well as the relation of the phenomenon to the character of solar activity.

2. Disintegration of comets in the solar atmosphere

The law for velocity variation of fully fragmented comet nuclei with initial radii $R_0 \gtrsim 100$ m in the region close to the endpoint of the deceleration trajectory in the solar atmosphere with the mass density distribution like $\rho_a = \rho_0 \exp(-z/H)$ (Ivanov-Kholodnyi & Nikol'skii 1969) , i.e., at small distances from the solar surface, $z \ll \tilde{z} \ll z_*$, has the following form:

$$V = \tilde{V} \exp\left[-\frac{2b^2}{3C_x C^2}\left(r^2 - \tilde{r}^2\right)\right] = V_0 \exp\left(-\frac{2b^2}{3C_x C^2}r^2\right). \tag{2.1}$$

Here V_0 is the initial orbital velocity of the comet nucleus above the solar photosphere,

$$b = v \exp\left(-\frac{z_*}{H}\right); \quad v = \frac{3C_x \rho_0 H}{4\rho_n R_0 \sin\alpha}; \quad C = \left(\frac{3C_x R_0 \sin\alpha}{8H}\right)^{1/2}; \tag{2.2}$$

$$r = \left[\exp\left(\frac{z_* - z}{H}\right)\right] - 1; \quad \tilde{r} \approx \frac{4C^2}{b}. \tag{2.3}$$

Furthermore, C_x is the coefficient of the aerodynamic drag; $\tilde{r}$, $\tilde{z}$ and $\tilde{V}$ are the characteristic values of r, z and V, which correspond to the value of $R = 2R_0$, i.e., to the time instant when the nucleus is completely fragmented and its transverse radius is equal to the doubled value of the initial radius (Grigoryan et $al.$ 1997; Grigoryan et $al.$ 2000), R_0 and ρ_n are the initial radius and the density of the nucleus respectively; α is the angle between the entry velocity of the nucleus into the atmosphere and the horizon.

From (2.1) and (2.3) it follows that the basic deceleration of the nucleus, the decrease in its velocity from $V_1 = 0.9V_0$ to $V_2 = 0.1V_0$, occurs at $r_2^2(z_2) = 9r_1^2(z_1)$, i.e., in the trajectory segment lying, according to (2.3), in the height range

$$|\Delta z| = |z_2 - z_1| = H \ln\frac{1 + r_1}{1 + r_2} \approx H \ln\frac{r_1}{r_2} \approx 0.7H. \tag{2.4}$$

Using (2.1) we can also obtain an explicit expression for the characteristic value of $r = r_e$ at which the kinetic energy of the fragmented mass falls e times, namely $V = V_0/\sqrt{e}$ at

$$r_e = \frac{\sqrt{3C_x}C}{2b}. \tag{2.5}$$

According to (2.2), (2.3) and (2.5), the height corresponding to the value of $r = r_e$ is

$$z_e = z_* - h\ln(1 + r_e) = H \ln\left(\frac{2b\rho_0 V_0^2}{\sqrt{3C_x}C\sigma_*}\right). \tag{2.6}$$

Assuming $R_0 = 1\,\text{km} = 10^5$ cm, $\sigma_* = 10^4$ dyn/cm^2, $\rho_n = 0.5$ g/cm^3, $C_x = 1$, $\sin\alpha = 0.5$, $H = 1.5 \times 10^7$ cm, from (2.2), (2.3), (2.5) and (2.6) we find $v = 4.5 \times 10^{-5}$, $z_*/H = 8$, $b = 1.3 \times 10^{-8}$, $C = 3.5 \times 10^{-2}$, $\tilde{r} = 4 \times 10^5$, $r_e = 2 \times 10^6$, $R(r_e) = 30R_0$, $z_e = -12H = -1800$ km, $\Delta z = 0.7H = 100$ km. So, aerodynamic fragmentation of a comet nucleus in the solar chromosphere is accompanied by transverse expansion of the fragmented mass and explosion of this high-velocity mass in a relatively very thin subphotosphere sheet: the characteristic timescale of the explosion, thermalization of the kinetic energy of the mass, is of the order of 0.1–1 s.

The specific energy release in the explosion zone, $V^2/2 = 1.8 \times 10^{15}$ erg/g, significantly exceeds the evaporation/sublimation heat of the nucleus material, $E_s = 8 \times 10^{10}$ erg/g, so that the fall of comets onto the Sun will be accompanied by not only evaporation but also production of a plasma with an initial temperature higher than 10^6 K near the solar photosphere (Grigoryan et $al.$ 2000).

The energetics of the process is of the order of 10^{32} erg for a falling mass of the order of 10^{17} g, which corresponds to the mass of the nucleus of Comet Halley 1986 III.

The astrophysical manifestation of the process may be, for instance, an excess of radiation in bright lines of metal atoms during the expansion of the generated high-temperature plasma having a timescale of the order of 10–100 s; a similar process was observed during the collision of Comet Shoemaker–Levy 9 with Jupiter on 16–22 July 1994 (Fortov *et al.* 1996).

3. Conclusion

The passage of comets near the solar surface is accompanied by aerodynamic fragmentation of their nuclei within the solar chromosphere and transverse expansion of the fragmented mass. The sharp stopping of this high-velocity fragmented mass is accompanied by production of a high-temperature plasma near the solar photosphere and by a solar photospheric flare.

The spectral monitoring of solar radiation and sungrazing comets in bright lines of metal atoms and ions involving not only the visual range but also soft X rays with a time resolution of 0.1–10 s is worthwhile.

References

Bailey, M. E., Chambers, J. E., & Hahn, G. 1992, *A&A*, 257, 315

Beust, H., Lagrange, A.-M., Plazy, F., & Mouillet, D. 1996, *A&A*, 310, 181

COSPAR Inform. Bull. 1998, 142, 21

Fortov, V. E., Gnedin, Yu. N., Ivanov, M. F., Ivlev, A. V., & Klumov, B. A. 1996, *Usp. Fiz. Nauk*, 166, 391 [Engl. Transl.: *Phys.-Usp.*, 39, 363]

Grigoryan, S. S., Ibodov, F. S., & Ibadov, S. 1997, *Dokl. Akad. Nauk*, 354, 187 [Engl. Transl.: *Phys.-Dokl.*, 42, 262]

Grigoryan, S. S., Ibadov, S., & Ibodov, F. S. 2000, *Dokl. Akad. Nauk*, 374, 40 [Engl. Transl.: *Phys.-Dokl.*, 45, 463]

Ibadov, S., Ibodov, F. S., & Grigoryan, S. S. 2007, in: *Star-Disk Interaction in Young Stars*, Proc. IAU Symp. No. 243, Grenoble, France, p. VI.4 // www.iaus243.org

Ivanov-Kholodnyi, G. S. & Nikol'skii, G. M. 1969, *The Sun and Ionosphere* (Moscow, Nauka)

MacQueen, R. M. & St. Cyr O. C. 1991, *Icarus*, 91, 96

Marsden, B. G. 1989, *AJ*, 98, 2306

Weissman, P. R. 1983, *Icarus*, 55, 448

Universal Heliophysical Processes
Proceedings IAU Symposium No. 257, 2008
N. Gopalswamy & D.F. Webb, eds.

© 2009 International Astronomical Union
doi:10.1017/S1743921309029536

Dynamics of microwave brightness distribution in the giant 24 August 2002 flare loop

Veronika E. Reznikova[1,2], Victor F. Melnikov[2], Kiyoto Shibasaki[3],
Sergey P. Gorbikov[2], Nikolai P. Pyatakov[2], Irina N. Myagkova[4],
Haisheng Ji[1,5]

[1]Purple Mountain Observatory, Chinese Academy of Sciences, Nanjing, China

[2]Radiophysical Research Institute (NIRFI), Nizhny Novgorod, Russia

[3]Nobeyama Radio Observatory, NAOJ, Nagano, Japan

[4]Lomonosov Moscow State University, Skobeltsyn Institute of Nuclear Physics, Moscow, Russia

[5]Big Bear Solar Observatory, New Jersey Institute of Technology, Big Bear City, USA

Abstract. We have found a similar tendency of the spatial dynamics at 34 GHz for all major temporal sub-peaks of the burst with the re-distribution of the brightness from the footpoints (on the rising phase of each peak) to the upper part of the loop (on the decay phase). Observed dynamics is interpreted by the re-distribution of accelerated electrons number density with their relative enhancement in the loop top. Results of diagnostics show that the ratio of non-thermal electron number density in the loop top and in the footpoint changes 7 times from the peak to decay phase. Model simulations by solving the Fokker-Planck equation allowed to determine an injection type which is able to result in necessary dynamics of energetic electrons.

Keywords. solar flare, microwave radiation, magnetic loop

1. Observed dynamics of brightness distribution

The total flux time profiles of the flare at 17 GHz (thin line) and 35 GHz (thick line) obtained with Nobeyama Polarimeter are shown in Fig. 1. Profiles have multiple emission peaks well separated from each other. Every peak is numbered at the top of Fig. 1.

Fig. 2 represents brightness evolution of 34 GHz emission during the main temporal peak denoted as peak 1 on Fig. 1. On the rising phase of the main peak the southern footpoint (SFP) of the loop is the most bright at 34 GHz and it remains the brightest part of the loop until the maximum. In Figs. 2a, 2b we can also see two other brightness peaks: one near the opposite northern footpoint (NFP) and one near the loop top (LT), but they are much weaker. Only on the decay phase the loop top becomes relatively brighter then the footpoint sources (Fig. 2c) which almost disappear to the moment of the valley (Fig. 2d).

Interestingly, the sources of all major temporal peaks 1-6 are located at the same radio loop and the similar evolution of brightness distribution repeats itself for all these sub-bursts. The absolute brightness temperature of the loop top at the valley times is not reduced compared to the corresponding previous peaks of the flux time profile. Time profiles obtained for different parts of the loop showed that the emission maximum from the loop top is delayed against maxima from the footpoint sources for both frequencies. These delays are more pronounced at 34 GHz than at 17 GHz. Furthermore, the time profile of 34 GHz emission from the loop top is wider and its decay is slower than those from the footpoints. Time profiles of spectral index showed that almost all parts of the

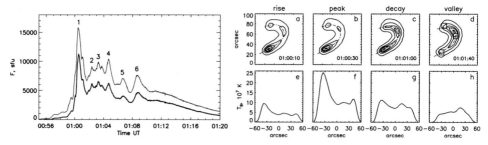

Figure 1. (left) NoRP total flux time profiles at 17 GHz (thin line) and 35 GHz (thick line).

Figure 2. (right) *Top panel*: 34 GHz contour images of the radio source in the rise, peak and decay phases as well as at the end of the decay phase of the main peak just before the start of the second peak. Contours show 0.1, 0.4, 0.6, 0.75, 0.95 levels of the maximum brightness temperature. Dot-dashed line shows the visible flaring loop axis. *Bottom panel*: spatial distributions of radio brightness temperature at 34 GHz along a visible flaring loop axes shown on the top panel at the corresponding moment of time. Abscissa is the distance along the loop, negative values correspond to southern footpoint and zero position to the loop top.

loop (with the exception of SFP) are optically thin at least at 34 GHz since α is negative in the 17–34 GHz frequency range.

2. Discussion

It is important that absolute brightness temperature at 34 GHz in the loop top at the majority of valleys is not reduced compared to corresponding previous peaks of flux densities. This fact indicates the process of the accumulation of accelerated electrons in the upper part of the loop. Furthermore, the delays between microwave emissions from the loop top and footpoints as well as longer decay of the emission from the loop top are also strong evidence of the trapping and accumulation of high energy electrons in the upper part of the flaring loop (see Melnikov, Shibasaki & Reznikova (2002)). Since this event was a limb flare with the viewing angle almost equal for all parts of the loop, the re-distribution of brightness temperature reflects the re-distribution of emitting accelerated electrons along the loop.

Estimation of parameters of nonthermal electrons. To check our assumption we estimated ratios of electron number densities in the loop top and southern footpoint N_{LT}/N_{SFP} for different time moments. The estimations obtained by fitting observed fluxes at 17 GHz and 34 GHz by the spectrum for both sources using exact formulas given by Ramaty (1969), Fleishman & Melnikov (2003) for GS emissivity and absorption coefficient. Electron energy spectral index δ was derived from HXR data obtained by detector SONG aboard of space solar observatory CORONAS-F under the assumption of thick target model. At the maximum of peak 1 δ_{HX}=2.6, at the valley 1 time $\delta_{HX} = 3.1$ for channels 53-150 keV and 150-500 keV. All other parameters were taken from the observations: an ambient plasma density 8×10^{10} cm^{-3} (from GOES/SXT at peak 1); the source depth 3.6×10^8 cm; the viewing angle $84°$.

Estimation showed the ratio N_{LT}/N_{SFP} increases about 7 times from peak 1 time to the end of decay (valley 1) for the mirror ratio 5. Diagnostics of plasma parameters in the loop gave magnetic field strength in the SFP source $B_{SFP} \approx 1000$ G, and in the loop top $B_{LT} \approx 200$ G; number densities of accelerated electrons with energies more than 500 keV in SFP source $N_{SFP} = 3 \times 10^4$ cm^{-3} and $N_{LT} = 9 \times 10^5$ cm^{-3} at the maximum of peak 1 and $N_{SFP} = 3 \times 10^3$ cm^{-3}, $N_{LT} = 6 \times 10^5$ cm^{-3} at the valley time. Thus,

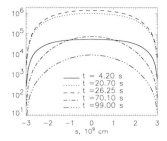

Figure 3. Results of the model simulations using the Fokker-Planck equation. The electron distribution functions $f(E, \mu, s, t)$ for the rising phase of injection are shown by solid and dotted lines, near the injection maximum by dashed line, and for the decay phase by dot-dashed and dot-dot-dot-dashed lines.

diagnostics showed that already at peak 1 time number density in LT is about 30 times higher than in the SFP.

Model calculations using Fokker-Planck equation. On Fig. 3 we present the result of calculation of the time evolution of energetic electron number density along the loop. It is shown for electrons with pitch-angle 85° and energy $E = 405$ keV only for the case when injection is isotropic and continuous along the loop. Magnetic mirror ratio $m = 5$. The flat distribution function at the rising phase should give a radio brightness peak in the footpoint with the strongest magnetic field. Our diagnostics shows that even at the peak time accelerated electron number density is higher in the loop top than in the southern foot point. We can see the similar result in this model, where $N_{LT}/N_{FP} = 7 \div 100$ at the peak time, considering the location of footpoint $s > 2.4 \times 10^9$ cm. At the decay phase the density in the central part of the trap becomes well pronounced. Such dynamics of the particle spatial distribution are caused by pitch-angle scattering due to Coulomb collisions which leads to preferable accumulation of energetic electrons near the center of the trap.

3. Conclusions

For the first time we have found the cyclical dynamics in radio brightness distribution along the loop, synchronized with major temporal peaks: re-distribution of the radio brightness from the footpoints (on the rising phase) to the upper part of the loop (on the decay phase). Since this is the limb event with the loop plane is almost perpendicular to the line of sight, the re-distribution of T_B directly reflects the re-distribution of accelerated electrons number density with they relative enhancement in the loop top. As a result of our diagnostics we have obtained that the ratio N_{LT}/N_{SFP} increases by about 7 times from peak 1 to the end of decay (valley 1) and already at peak 1 the electron number density in the LT is higher than in the SFP. This result of diagnostics is in a good agreement with our calculation of the time evolution of energetic electron number density for the model with isotropic injection which occurs continuous along the loop.

The work was partly supported by RFBR grants No. 06-02-39029, 06-02-16295, 07-02-01066, NSFC grant 10473024 CNSF 10833007, the 973 project with No. 2006CB806302, and Bairen project.

References

Fleishman, G. D., & Melnikov, V. F. 2003, *ApJ*, 584, 1071
Melnikov, V. F., Shibasaki, K., & Reznikova., V. E. 2002, *ApJ*, 480, L185
Ramaty, R. 1969, *ApJ*, 158, 753

Universal Heliophysical Processes
Proceedings IAU Symposium No. 257, 2008
N. Gopalswamy & D.F. Webb, eds.

© 2009 International Astronomical Union
doi:10.1017/S1743921309029548

Magnetic field in active regions of the sun at coronal heights

V. M. Bogod[1] and L. V. Yasnov[2]

[1]Special Astrophysical Observatory, St. Petersburg, Russia
email: vbog_spb@mail.ru

[2]Radiophysical Research Institute, St.-Petersburg State University, Russia
email: Yasnov@pobox.spbu.ru

Abstract. A method is developed for estimation of the vertical structure of the magnetic field in active regions using multi-wave spectral-polarization measurements of radio waves which gives not only the dependence of magnetic field strength on height but also determines two-dimensional form of a magnetic flux tube, emitted in the microwave range of wavelengths.

Keywords. Sun: radio radiation Methods: data analysis

1. Introduction

In Akhmedov *et al.* (1982) it was shown that the magnetic field strength measured for polarized sources above sunspots go down in the transition region by only 20%. Attempts to analyze the structure of magnetic fields using radio data were undertaken many times (Golubchina *et al.* (1981), Mursh & Hurford (1982), Lang & Wilson (1983), Shibasaki (1986), Aschwanden & Bastian (1994), Brosius & White 2006). But the measurements were made with a limited number of wavelengths, which restricted the capabilities of this method.

In this study we use the observational results made with the RATAN-600 radio telescope applying the broadband polarization spectrograph which has a large number of channels (39) from 1.7 cm to 15 cm (Bogod *et al.* 1999). Owing to the use of multi-wave data, it is possibile to study the detailed altitude structure of the magnetic field. On the other hand, the reliability of determination of such a structure increases.

2. Technique of determination of the structure of magnetic field

We use of polarization of radio sources (instead of their intensity). This gives the chance, to a certain degree, to eliminate the interfering influence on the result of adjacent structures of the magnetic field in the active region.

The aim of solar scans processing was to determine the time dependence of the position of a chosen feature in polarized emission of an active region - x_{rad}. Changes in the position of a chosen feature of a spot on the photosphere for a given time interval were corrected using the parallactic angle and the angle of inclination of the Sun's axis in the ecliptic coordinate system.

Using heliolatitude φ of a measured feature of AR, we calculate its location $x_{calc}(h, \lambda, t_i)$ on the solar disk in the coordinate system of the radio scan. We have used the known time dependence of heliolongitude of the sources on the photosphere

$$\lambda(t_i) = (14.35 - 2.77\sin^2(\varphi) - 0.9856)t_i + \lambda, \qquad (2.1)$$

where λ - the constant longitude characterizing the source position. The adequacy of this dependence to the real position of the active region features under study was verified, which showed a high accuracy of calculations limited by the errors of coordinate readout in observational data. Next, we minimized the expression

$$\sum_{i=1}^{N}(x_{calc}(h,\lambda,t_i) - x_{rad}(t_i))^2, \qquad (2.2)$$

where N is the number of data series used (from 2 to 5), and determined the height above photosphere h and λ. Such calculation was made for every wavelength. We believed that the polarized emission of radio sources is determined by cyclotron radiation at the third harmonics of the gyrofrequency.

Because of some uncontrollable errors of the irradiator installation probably occurrence of a systematic error at height definition for the given method. For reduction of this error to the minimum value and for obtaining uniform data the height of the maximum magnetic field magnitude is set equal to the height using model calculations. As it is seen from the subsequent figures, the magnetic field is located low enough in the solar atmosphere. For this reason the usage of model extrapolations does not give essential errors for its definition.

We shall take advantage of one-dimensional dipole approximation of a magnetic field (Takakura 1972)

$$B = \frac{B_o}{(1 + h/d)^3}, \qquad (2.3)$$

where B_o- the magnetic field strength at the level of the photosphere, d - the depth of a dipole under photosphere ($d \approx r$- the radius of a spot), h - the height above photosphere. On SOHO MDI magnetograms we define B_o and r. Knowledge of the magnetic field strength in the lowest point for a given structure (at the highest frequency) B, could help us define its height. The average height from the two adjacent day measurements was calculated. The dispersion in definition h is specified in the figures in the form of vertical line in a point of the maximum magnetic field. Practically in all cases as it is visible from the subsequent figures, the dispersion was insignificant.

3. Magnetic field in active regions at coronal heights

The developed method was applied to observations of various active regions. Let us consider some results. Figure 1 presents the results of measurements of the magnetic field altitude structure for the AR NOAA 0933 observed on 7-8 January 2007. During the same period the AR NOAA 0935 was observed. Magnetic structure of this region for 3-4 January 2007 is represented in Figure 2.

It is shown in Figure 2, that the magnetic flux tube for AR NOAA 0935 is directed upwards with some bends. It can be a projection of the three-dimensional screw structure of the magnetic field to a plane. Such structure is not unique.

4. Discussion

The height of the corona base in an active region for various models is about 2-4 Mm. From the paper of Gary (2001) it follows (in the case of the potential extrapolation of a photospheric magnetic field), that at these heights $B = 400$–700 G, if at the photosphere $B_o = 3000$ G. In our measurements a magnetic field strength in the transition region and the bottom of corona falls no more than by 20%, that is, up to $B \approx 2400$ G, if a field at the photosphere is $B_o = 3000$ G.

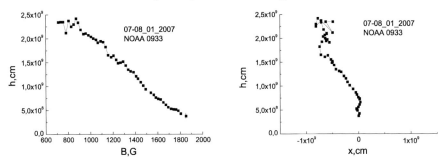

Figure 1. The results of calculations for AR NOAA 0933, observed on 7-8 January 2007. Dependence of the magnetic field with height - on the left and with the structure of the magnetic field - on the right. The value $x = (\lambda(f) - \lambda(f_{max})) \cos(\varphi)/360 \times 43 \cdot 10^{10}$, where f-the frequency, f_{max}- the frequency at which the magnetic field strength is maximum.

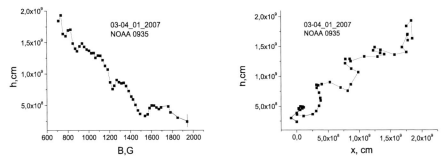

Figure 2. The structure of the magnetic field for AR NOAA 0935, observed on 3-4 January 2007.

Let us give additional arguments in favour of that magnetic fields extend highly in the corona. The loops radiated in soft X-ray, tend to be wider in their tops only by 30% than in their bases (Klimchuk 2000). The loops observed in lines 171 and 195 $\mathring{A}$ have still smaller expansion to tops: 0% for loops without flares and 13 % for flare loops (Watko & Klimchuk 2000). Expansion of loops unequivocally reflects a degree of decrease of the magnetic field strength along a loop. The specified expansion of ultraviolet loops should lead to decrease for the magnetic field in the loop top in comparison with its basis to 0%-27%.

The presence of poorly diverging magnetic flux tubes can be justified theoretically by the presence in them of strong torsion of the magnetic field. In this case the field should be compressed (Zweibel & Boozer (1985), Robertson *et al.* 1992) and tube expansion should decrease and magnetic field strength should increase.

The loop lengths investigated in the paper by Klimchuk (2000) are from $5 \cdot 10^9$cm up to $33 \cdot 10^9$cm. That is, the magnetic field strength in the hottest parts of magnetic flux tubes (radiating in lines 171 and 195 A) should not decrease in their tops, that is, at heights up to $1.5 \cdot 10^{10}$cm, more than for 30%. Our measurements indicate that the magnetic field strength changes with height more considerably but it is essential more slowly than for theoretical computational models. Let us consider an elementary extrapolation of a magnetic field to the corona, using the established fact that the field in transition region should have the value not less than 80-85% of a photosphere field. We shall take advantage as above of one-dimensional dipole approximation of a magnetic field (Takakura (1972)). We shall estimate the value d, proceeding from a vertical gradient of a magnetic field on the photosphere. This gradient equals 0.1 G/km according to Severny

(1965), 0.1-0.2 G/km (Rayrole & Semel (1970)) and 0.3-0.5 G/km (Ioshpa & Obridko (1965)). From (2.3) we have $\acute{B} = 3B_o/d$. Then for $\acute{B} = 0.1$ G/km and $B_o = 3000$ G we obtain $d = 90000$ km, for $\acute{B} = 0.4$ G/km and $B_o = 3000$ G we obtain $d = 22500$ km. Thus the magnetic field strength of 1000 G should be at heights from 10 Mm up to 40 Mm, that corresponds to our measurements.

5. Conclusions

1. A method is developed for the estimation of the vertical structure of the magnetic field in active regions using multi-wave spectral-polarization measurements of radio waves which gives not only the dependence of the magnetic field strength with height but also determines the two-dimentional form of a magnetic flux tube, emitted in a microwave range of wavelengths.

2. Magnetic fields with a strength of about 600 G are located at sufficiently large altitudes in the solar atmosphere (up to 25000 km), which well confirms the ultraviolet and X-ray observations, according to which the divergence of field tubes is small.

3. The topology of a magnetic flux tube emitted at microwaves in some investigated cases can have screw structure.

References

Akhmedov, S. B., Gelfreikh, G. B., Bogod, V. M., & Korzhavin, A. N. 1982, *Solar Phys.*, 79, 41
Golubchina, O. A., Ikhsanova, V. N., Bogod, V. M., & Golubchin, G. S. 1981, *Solnechnye Dannye*, 4, 108 (in Russian)
Mursh, K. A. & Hurford, G. J. 1982, *ARAA*, 20, 497
Lang, K. A. & Wilson, R. F. 1983, *Adv. Sp. Res.*, 11, 2
Shibasaki, K. 1986, *Ap&SS*, 119, 21
Aschwanden, M. J. & Bastian, T. S. 1994, *ApJ*, 426, 425
Brosius, J. W. & White, S. M. 2006, *ApJ*, 641, L69
Bogod, V. M., Garaimov, V. I., Komar, N. P., & Korzhavin, A. N. 1999, *Proc. 9th European Meeting on Solar Physics*, (ESA, SP-448), p. 1253
Takakura, T. 1972, *Solar Phys.*, 26, 151
Gary, G. A. 2001, *Solar Phys.*, 203, 71
Rayrole, J. & Semel, M. 1970, *A&A*, 6, 288
Robertson, J. A., Hood, A. W., & Lothian, R. M. 1992, *Solar Phys.*, 137, 273
Zweibel, E. G. & Boozer, & A. H. 1985, *ApJ*, 295, 642
Klimchuk, J. A. 2000, *Solar Phys.*, 193, 53
Watko, J. A. & Klimchuk, J. A. 2000, *Solar Phys.*, 193, 77
Ioshpa, B. A. & Obridko, V. N. 1965, *Solnechnye dannye*, 5, 62 (in Russian)
Severny, B. 1965, *Izv. Crimea. astr. obs.*, 33, 34 (in Russian)

Universal Heliophysical Processes
Proceedings IAU Symposium No. 257, 2008   © 2009 International Astronomical Union
N. Gopalswamy & D.F. Webb, eds.   doi:10.1017/S174392130902955X

On spatial variations of magnetic field and superthermal electron distribution in cm-radio burst source

Leonid V. Yasnov[1] and Marian Karlický[2]

[1] Radiophysical Research Institute, St.-Petersburg State University
Ul'yanovskaya 1, St.-Petersburg, 198504, Russia

[2] Astronomical Institute, Academy of Sciences of the Czech Republic
251 65 Ondrejov, Czech Republic

Abstract. The paper presents a new method of an estimation of spatial variations of the magnetic field and superthermal electron distribution in solar cm-radio burst sources. The method is based on the analysis of several burst spectra recorded in the different moments of time and on the minimization of the difference between the theoretical and observed radio fluxes. It is found that the measure of the spatial variations of superthermal electron distribution in the radio source is always greater than that for the magnetic field. In most cases this measure has a minimum at the impulsive phase of cm-radio bursts.

Keywords. Sun: radio radiation Methods: data analysis

Many papers have been devoted to a study and interpretation of spectral characteristics of the centimeter radio bursts. These characteristics can be used for a diagnostics of the cm-radio burst sources. An attempt to estimate some physical parameters in the radio source from radio burst spectra has been presented by Böhme *et al.* 1977. The authors concluded that their inversion procedure is ambiguous.

For the X-ray bremsstrahlung and synchrotron emission various inversion methods have been described in the literature (Brown *et al.* 1983; Brown *et al.* 2006; Kontar *et al.* 2004; Prato *et al.* 2006).

In the present paper, for an analysis of the cm-burst radio spectra, a new inversion method is suggested. We propose to reduce an ambiguity of the problem by comparing the radio spectra taken at different times during one specific cm-radio burst.

Let us assume that the radio source has a form of a magnetic loop segment with the squared cross section. The length of this segment along the magnetic field direction (the l coordinate) is L. Δ_0 is the length of the side of the square section at the segment base as well as a location of the segment base at the l-coordinate. Then, for the square side Δ and the magnetic field B in the radio source we can write: $\Delta = \Delta_o(\frac{l}{\Delta_o})^{-\mu}$ and $B = B_o(\frac{l}{\Delta_o})^{2\mu}$, where B_o is the magnetic field at the segment base. Furthermore, let us assume that in the radio source the superthermal electrons are distributed as follows: $n = n_o(\frac{l}{\Delta_o})^{-\nu}$, where n_o is the density of the superthermal electrons at the segment base. The parameters L, Δ_o, B_o, n_o, μ, and ν are unknown parameters and they need to be determined, see the following.

In the centimeter wavelength range the emission is usually produced by the gyrosynchrotron emission mechanism. Therefore all the following analysis is based on this mechanism. In agreement with Dulk (1985), Bastian *et al.* 1998, Nindos *et al.* 2000 we consider the isotropic distribution functions of electrons in the source and thus we can use the emission η and absorption k coefficients derived by Dulk & Marsh (1982).

We suppose that $L \gg \Delta_o$. Because most of the cm-burst sources are located at the top of flare loops (White et al. 2002, Minoshima et al. 2008) and thus the angle between the line of sight and the magnetic field vector θ is much greater than zero, the following integration is made only along the axis of the source (in such approximation the integration along the line of sight is not necessary). Then using the emission and absorption coefficients from the paper by Dulk & Marsh (1982), the radio flux measured in SFU at the specific frequency can be written as

$$F(f) = \frac{1}{10^{-19}R^2} \int_{\Delta_o}^{L} \frac{\left(1 - e^{-k\Delta(l)/\sin(\theta)}\right)\eta\Delta(l)}{k\sin(\theta)} \, dl$$

$$= \frac{C(f)}{10^{-19}R^2}\left[-nca E_{1-nca}(Ldn A(f))Ldn^{nca} - Ldn^{nca} + nca E_{1-nca}(A(f)) + 1\right]$$

where

$$A(f) = 1.28 \times 10^{-10} \frac{n_o\Delta_o}{B_o} e^{-2.308\delta} \sin(\theta)^{0.72\delta - 1.09} \left(\frac{f}{f_{B,o}}\right)^{-0.98\delta - 1.30} \equiv A_o f^{-0.98\delta - 1.30},$$

$$C(f) = 2.42 \times 10^{-13} \frac{B_o^2 \Delta_o^2}{nc - 1} e^{-0.544\delta} \sin(\theta)^{0.07\delta - 1.34} \left(\frac{f}{f_{B,o}}\right)^{0.08\delta + 2.52} \equiv C_o f^{0.08\delta + 2.52},$$

and f is the emission frequency, δ is the energy spectral index of superthermal electrons, $E_n(x)$ is the integral-power function, $nca = \frac{1-nc}{na}$, $Ldn = Ld^{na}$, $nc = (0.16\delta + 2.04)\mu$, $na = (1.96\delta - 0.4)\mu - \nu$, $Ld = \frac{L}{\Delta_o}$, $f_{B,o} = \frac{eB_o}{2\pi m_e c}$, $R = 1.49 \times 10^{13}$ cm is the Sun-Earth distance.

Considering now some real radio observations we need to find the parameters of the theoretical model which give the radio fluxes fitting the observed ones by the best way. For this optimalization procedure, we define the difference of the radio fluxes in the form

$$\sum_{i=1}^{N} (\lg F(f_i) - \lg F^o(f_i))^2,$$

where F and F^o are the theoretical and observed radio fluxes at one specific instant during the radio burst and at the frequency f_i; N is the number of considered frequencies. To find the optimal parameters of the theoretical model, we need to find the minimum of this difference for the auxiliary parameters C_0, A_0. δ, nca, and Ldn. To reduce an ambiguity of the present task, this minimization has be done for at least two radio spectra observed at two different times during one selected radio burst. As explained in the following more spectra represent more information and better results.

There are many papers showing that changes of the magnetic field at the photospheric level during solar flares are negligible (see e.g. Rust 1972). Let us assume the same for the studied radio source (in the cm-range is usually out of the primary energy release) and later we will check the validity of this assumption. Thus the parameters μ, L, and Δ_0 will be taken constant during the whole evolution of the radio burst. Now considering two radio spectra in two different times during the radio burst, i.e. for two sets of the auxiliary parameters (C_0, A_0. δ, nca, and Ldn), which were obtained by the minimization of the foregoing relation, we can express the parameters in the radio source as follows:

$$\mu = \frac{25(nca_i \ln(Ldn_i) - nca_k \ln(Ldn_k))}{nca_i(4\delta_k + 51)\ln(Ldn_i) - nca_k(4\delta_i + 51)\ln(Ldn_k)}$$

$$\nu_i = \frac{(4\delta_i - 4\delta_k + nca_i(49\delta_i - 10))\ln(Ldn_i) + nca_k(10 - 49\delta_i)\ln(Ldn_k)}{nca_i(4\delta_k + 51)\ln(Ldn_i) - nca_k(4\delta_i + 51)\ln(Ldn_k)}$$

Table 1. The parameters of the cm-radio burst of December 24, 1991.

Time(UT)	δ	ν
18:34:00	4.44	3.00 ± 0.05
18:38:02	2.47	1.81 ± 0.08
18:38:14	3.71	2.61 ± 0.04
18:39:26	3.71	2.42 ± 0.17

These relations were derived by algebraic operations of two sets of the auxiliary parameters corresponding to two radio spectra (for their designation we use the indexes i). Let's note that definition of parameters μ, ν does not demand knowledge of the angle θ. The numerical experiments shows that for small amount of frequencies only the parameters μ, δ, and ν are stable (within 15 % of the original values); other parameters differ sometimes several orders of the magnitude.

Let us use the radio spectra which were observed at 18:33 – 18:40 UT, 24 December 1991, at the Sagamore Hill Station, and were analyzed in detail in the paper by Willson (1993). The resulting parameters δ and ν obtained for this burst can be seen in Table 1. Because values of ν are calculated for different pairs of the spectra, and because in each time instant three values of ν can be determined, then an accuracy of the parameter can be estimated. The number of combinations of the spectral pairs is 6 and in all such cases a value of the parameter μ was determined. It gave us the possibility to estimate an accuracy of determination of this parameter as well as estimate the accuracy of the above made assumption about the constant value of the parameter μ during the radio burst. As a result we have $\mu = 0.39 \pm 0.04$, i.e. an inaccuracy is small, which confirms the used assumption and the quality of the determination of these parameters by the proposed procedure.

The impulsive phase of the radio burst takes place at 18:38:02 UT. It can be seen that in the impulsive phase the values of δ and ν sharply decrease. This effect can be connected with an increase of the effectiveness of the acceleration of electrons to higher energies at times of the impulsive burst phase, which corresponds to a decrease of δ and to an increase of the emission volume in this burst phase. Thus, the results are in good agreement with the physics of the burst, and it one again confirms effectiveness of the proposed procedure.

To confirm the found evolution of the parameter ν during the December 24, 1991 radio burst, we used a set of the spectral observations published in Solar Geophysical Data Journal. We used data for bursts recorded on 11 Jan. 68, 14 Feb. 68, 03 Apr. 68, 09 Jul. 68, 21 Aug. 68, 09 Feb. 69, 27 Mar. 69, 17 May 69, 20 May 70, 14 Jul. 70, 12 Aug. 70, 24 Apr. 71 and 07 Aug. 72. From spectral point of view these data are not so good as those for the December 24, 1991 event. However, their number is sufficient for some statistical analysis and thus appropriate to make a statistical verification of this phenomenon. We take into the set of events only the radio bursts with the flux maxima at high-frequencies, i.e. the radio bursts for which reliable parameters δ can be determined. By this way we have selected 68 radio spectra for 13 radio events.

Inaccuracies of observations, minimization procedures and deviations from used assumptions result in the error range of the computed parameters μ and ν_i for all instants during the radio bursts. The error range was estimated from all values of μ and ν_i which were determined for all combinations of spectral pairs for specific instants during the radio bursts.

In this process of the parameter determination, it is possible to verify the validity of the assumption about the constant value of the magnetic field strength during radio bursts. For the radio bursts observed in August 12, 1970 and August 21, 1968 this condition breaks (the value of the relative inaccuracy of μ and ν is large (about 50%), in other cases the relative inaccuracy was about 10%), i.e. the magnetic field structure in the radio source changes during the radio burst. It can be caused by spatial displacements of the sources of the elementary bursts. For other 11 bursts we can say that their magnetic structure is stable. It means that their changes influence the determination of the parameter ν by a negligible way.

Almost all events show a decrease of the parameter ν at the beginning phase of the radio burst. Some exceptions are connected with the following. As we mentioned above, values of ν for the August 12, 1970 event are not trustworthy. For the July 14, 1970 event all data were observed after the maximum burst phase. Only the February 9, 1969 event represents a real exception.

Summarizing the results of all tests and data analysis we can conclude as follows:

1. For all analyzed radio bursts the measure of the spatial variations of the superthermal electron distribution ($\nu = 0.7 - 4.2$) is greater than that for the spatial variations of the magnetic field ($\mu = 0.4 - 0.5$).

2. Almost all events indicate a decrease of the parameter ν at the beginning phase of the radio burst.

Acknowledgements

This research was supported by Grant 06-02-16502 of the RFBR and by Grant IAA300030701 of the Grant Agency of the Academy of Sciences of the Czech Republic.

References

Bastian, T. S., Benz, A. O., & Gary, D. E. 1998, *ARAA*, 36, 131.
Böhme, A., Fürstenberg, F., Hildebrandt, J., Saal, O., Krüger, A., Hoyng, P., & Stevens, G. A. 1977, *Solar Phys.*, 53, 139
Brown, J. C., Craig, I. J. D., & Melrose, D. B. 1983, *Ap&SS*, 92, 105
Brown, J. C., Emslie, A. G., Holman, G. D., Johns-Krull, C. M., Kontar, E. P., Lin, R. P., Massone, A. M., & Piana, M. 2006, *ApJ*, 643, 523
Dulk, G. A. 1985, *ARAA*, 23, 169
Dulk, G. A. & Marsh, K. A. 1982, *ApJ*, 259, 350
Kontar, E. P., Piana, M., Massone, A. M., Emslie, A. G., & Brown, J. C. 2004, *Solar Phys.*, 225, 293
Minoshima, T., Yokoyama, T., & Mitani, N. 2008, *ApJ*, 673, 598
Nindos, A., White, S. M., Kundu, M. R., & Gary, D. E. 2000, *ApJ*, 533, 1053
Prato, M., Piana, M., Brown, J. C., Emslie, A. G., Kontar, E. P., & Massone, A. M. 2006, *Solar Phys.*, 237, 61
Rust, D. M. 1972, *Solar Phys.*, 25, 141
Solar Geophysical Data 1968-1972, NOAA, Boulder, USA
White, S. M., Kundu, M. R., Garaimov, V. I., Yokoyama, T., & Sato, J. 2002, *ApJ*, 576, 505
Wilson, R. F. 1993, *ApJ*, 413, 798

Universal Heliophysical Processes
Proceedings IAU Symposium No. 257, 2008
N. Gopalswamy & D.F. Webb, eds.

© 2009 International Astronomical Union
doi:10.1017/S1743921309029561

Fragmented type II burst emission during CME liftoff

Silja Pohjolainen[1], Jens Pomoell[2] and Rami Vainio[2]

[1]Department of Physics and Astronomy, University of Turku,
Tuorla Observatory, 21500 Piikkiö, Finland
email: `silpoh@utu.fi`

[2]Department of Physics, University of Helsinki,
PO Box 64, 00014 University of Helsinki, Finland
email: `jens.pomoell@helsinki.fi`, `rami.vainio@helsinki.fi`

Abstract. We have performed multiwavelength analysis on an event with a metric type II burst, which appeared first as fragmented emission lanes in the radio dynamic spectrum. The start frequency was unusually high. Since type II bursts are thought to be signatures of propagating shock waves, it is of interest to know how the shocks, and the type II bursts, are formed. This radio event was associated with a flare and a coronal mass ejection (CME), and we investigate their connection. Observations suggested that a propagating shock was formed due to the erupting structures, and the observed radio emission reflects the high densities in active region loops. We then utilised numerical MHD simulations, to study the shock structure induced by an erupting CME, in a model corona including dense loops. Our simulations show that the fragmented part of the type II burst can be formed when a coronal shock driven by a CME passes through a system of dense loops overlying an active region. To produce fragmented emission, the conditions for plasma emission have to be more favourable inside the loop than in the inter-loop area. The obvious hypothesis, consistent with our simulation model, is that the shock strength decreases significantly in the space between the denser loops. Outside the active region, the type II burst dies out when the changing geometry no longer favours the electron shock-acceleration.

Keywords. Sun: coronal mass ejections (CMEs); flares; radio radiation, shock waves, plasmas

1. Introduction

Radio type II bursts are observed in association with flares and coronal mass ejections (CMEs). Metric type II bursts can be observed in dynamic radio spectra as slowly drifting emission lanes, with drift rates approximately at $0.1 - 1.0$ MHz s^{-1} (Nelson & Melrose 1985). The start frequency of metric type II bursts is usually at about $100 - 200$ MHz. The mechanism behind the bursts is generally assumed to be a propagating shock which creates electron beams that excite Langmuir waves, which in turn convert into radio waves at the local plasma frequency and its harmonics.

The exact relationship between solar flares, shocks, and coronal mass ejections is still not well understood, and it is of interest to know how shocks are initiated and under which conditions radio type II bursts can be excited. As shocks can be formed in various ways, it is not evident that all solar radio type II bursts are formed in the same way. In particular, how are untypical, fragmented type II bursts that start at very high frequencies created? Are there differences in shock acceleration or in the surrounding medium that can explain the differences to the "typical" metric type IIs?

2. Analysis

We have analysed in detail one metric type II burst that occurred on 13 May 2001. The burst started at an unusually high frequency and proceeded showing fragmented and curved emission bands, which were visible at the fundamental and second harmonic plasma frequencies (Fig. 1). The first "fragment" showed emission between 500 and 420 MHz, which correspond to densities in the range of $2-3 \times 10^9$ cm^{-3}. The second fragment near 03:02 UT gives densities $1-2 \times 10^9$ cm^{-3} (400–310 MHz), and the third near 03:03 UT gives densities $2-6 \times 10^8$ cm^{-3} (220–130 MHz). These values indicate that the source regions were dense, similar to active region loops. The frequency drifts (fundamental emission) within the fragments were between 1.8 and 4.3 MHz s^{-1}.

The flare was well-observed in X-rays (Yohkoh) and EUV (TRACE). The images show a filament eruption, where most of the material is moving toward the Southeast. The outermost front of the filament moves with a projected speed of about 380 km s^{-1}, but there is also a separate 'blob' that moves more to the Southwest with a projected speed of 450 km s^{-1}. White-light observations from SOHO LASCO revealed a CME front moving toward the South at a speed of 430 km s^{-1}. A soft X-ray loop system was observed to move also southward, at a projected speed of about 650 km s^{-1}. Using the Yohkoh SXT filter ratio method (see, for example, McTiernan *et al.* 1993), we found that the soft X-ray loop densities agreed with the plasma densities of the fragmented metric type II burst (Pohjolainen *et al.* 2008).

3. Simulations

We then utilised numerical MHD simulations to study the shock structure induced by a CME, in a model corona including dense loops. The details of the model are presented in Pomoell *et al.* (2008). We considered three different runs with slightly different parameters. The plot in Fig. 1 shows the radio emission produced by a shock, assuming that the emission is produced immediately in front of the CME leading edge shock. The emission lanes show a frequency drop in accordance with the exponentially decreasing density of the ambient corona. However, when the radio-emitting shock propagates in a dense loop, the frequency is higher, and drops quickly when the shock exits the loop (near $t = 200$ s). Additionally, we indicate the compression ratio of the shock by the size

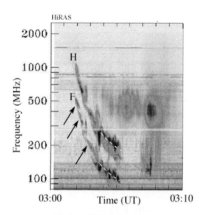 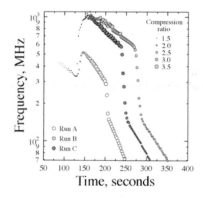

Figure 1. Left: HiRAS dynamic spectrum from the 25-2500 MHz frequency range, at 03:00–03:10 UT on 13 May 2001. 'F' notes emission at the fundamental and 'H' at the second harmonic plasma frequency. Arrows point to the fragmented emission bands and dashed white lines outline the later-appearing "regular" type II burst lanes. Right: Radio track of three different simulation runs. The size of the marker indicates the compression ratio of the shock.

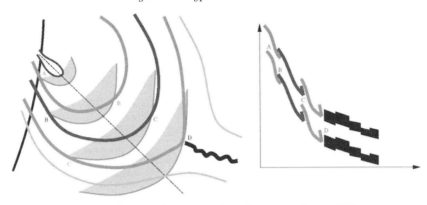

Figure 2. A possible scenario for a fragmented type II burst

of the marker. In all three cases, the shock is strong while it propagates in the loop, with the compression ratio remaining between 3 and 3.5.

The dynamics of the eruption is as follows: As the flux rope starts to rise, a perturbation is formed around the flux rope. Due to the gradient in the Alfvén speed and the increasing speed of the flux rope, the wave steepens to a shock ahead of the flux rope. However, the strength of the shock remains weak in the area below the loop. When the shock reaches the dense loop, it strengthens and slows down quickly due to the low Alfvén speed in the loop. The erupting filament continues to push the loop structure ahead of it, acting as the driver of the shock. Thus, the speed of the shock is roughly that of the displaced loop structure when propagating in the region of low Alfvén speed. As the filament decelerates, the displaced loop and shock escape from the filament. When reaching the region of higher Alfvén speed, the speed of the shock increases with the increasing Alfvén speed, and the shock escapes from the propagating loop structure.

4. Model

A possible scenario for the fragmented type II burst can be the following (Fig. 2): An erupting filament drives a strong shock wave through a system of overlying loops. Among these loops, a few are more favourable sites for plasma emission than others, e.g., because of their higher density or suprathermal electron content. These loops are "lit" in radio as the shock traverses them, producing the fragmented part of the radio burst. The regular part of the burst may then come from the part of the active region above the loop system, e.g., from an overlying current sheet. (See, e.g., Mancuso & Abbo 2004, for a similar geometric model involving the interaction of a shock and a current sheet.) Finally, the burst dies out or even stops abruptly once the shock propagates out of the region completely.

References

Mancuso, S. & Abbo, L. 2004, *A&A*, 415, L17
McTiernan, J.M., Kane, S.R., Loran, J.M., *et al.* 1993, *ApJ*, 416, L91
Nelson, G. J. & Melrose, D. B., 1985, in *Solar Radiophysics*, D. J. McLean and N. R. Labrum (eds.), Cambridge Univ. Press, 333
Pohjolainen, S., Pomoell, J., & Vainio, R. 2008, *A&A*, in press
Pomoell, J., Vainio, R., & Kissmann, R. 2008, *Solar Phys.*, in press

Universal Heliophysical Processes
Proceedings IAU Symposium No. 257, 2008
N. Gopalswamy & D.F. Webb, eds.

© 2009 International Astronomical Union
doi:10.1017/S1743921309029573

A study on the relationship of type III radio bursts CME and solar flares during the active period October-November 2003

Michaella Thanassa[1], Eleftheria Mitsakou[1], Panagiota Preka-Papadema[1], Xenophon Moussas[1], Panagiotis Tsitsipis[2] and Athanasios Kontogeorgos[2]

[1] Department of Physics, University of Athens, 15784 Athens, Greece

[2] Department of Electronics, Technological Education Institute of Lamia, Lamia, Greece

Abstract. Within a period of intense activity (20 October to 5 November 2003), the injection and propagation of near relativistic electrons, resulted in hundreds of type III bursts recorded by the ARTEMISIV radio spectrograph (20–650 MHz). For a number of these type III events association with GOES SXR/Hα flare and/or SOHO/LASCO CME was established. We study the variation of characteristic type III parameters and their relationship with features of the associated flares and/or CMEs.

Keywords. Sun: coronal mass ejections (CMEs), Sun: flares,Sun: radio radiation

1. Introduction

In the period 20 October-5 November 2003 a global complex consisting of three large, remote but connected active regions: AR 0484 (Complex), AR 0486 (Complex) and AR 0488 (Bipolar) produced an abundance of solar energetic phenomena (intense flares, fast CMEs). Several type III bursts were also recorded.

The importance of the study is two fold: Firstly, the electron beams exciting the type III radio emission, stream along the coronal magnetic lines, thus tracing magnetic structures; secondly the origin of the same beams is directly associated with particle acceleration mechanisms in the lower corona.

In this report we expand on the results of a previous study (Mitsakou *et al.* 2006) examining the relationship between the metric solar type III radio bursts and coronal mass ejections (CMEs), and soft X-ray (SXR) flares.

2. Data selection and analysis

Our data set consists of 124 metric type III bursts and groups recorded by the ARTEMIS–IV radio spectrograph (Caroubalos *et al.* 2001, 2006, also Kontogeorgos *et al.* 2006a, 2006b) in the range 20–650 MHz within the active period 20 October-5 November 2003; those were associated with flares (Hα and GOES SXR) from the Solar Geophysical Data (SGD) reports and CMEs from the SOHO/LASCO lists; the association of bursts with GOES/SXR enhancements or flares was established within a time window of about 5 min. From our sample we have eliminated the periods of extreme activity accompanied by type II/IV burst as it was impossible to isolate and analyse the type III bursts in this case.

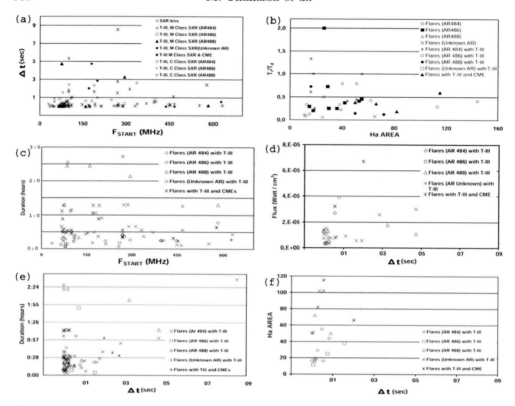

Figure 1. (a) Type III duration Δt as function of start frequency F_{START}, (b) Ratio of SXR rise to decay time (t_r/t_d) as a function of Hα AREA, (c) SXR Duration (D) as a function of start frequency F_{START}, (d) SXR FLUX as a function of Type III duration Δt, (e) SXR Duration (D) as a function of Type III duration Δt, (f) Hα AREA as a function of Type III duration Δt.

From the data sample we deduce a 69% association of type III bursts with recorded flares or SXR enhancements; 9% are also spatially and temporally associated with fast and small-width CME (*jet-CMEs*). A significant fraction (31%) were detected between successive SXR flux maxima (they were labeled as *SXRless* in the text) though not always in the same active region as the SXR flare. The lack of SXR enhancement in *SXRless* type III bursts was probably the result of increased SXR background which prevented detection. As regards event characteristics we have examined:

• The type III starting frequency (F_{START}) depends on the *depth* of the exciter acceleration in the solar corona; it is, usually, less than 300 MHz ($F_{START} < 300$ MHz) with the *SXRless* events having $F_{START} < 150$ MHz. The majority of the *jet-CME* associated bursts have $F_{START} > 200$ MHz. In AR 0484 we have occasionally $F_{START} \approx 500$ MHz (cf. figure 1a).

• The type III (or group) duration (Δt) is characteristic of the group size. We have found that Δt depends on F_{START} in a rather complicated way: For $F_{START} > 350$ MHz we have recorded only isolated bursts (Low Δt), for $F_{START} < 350$ MHz, both groups and isolated bursts appear (cf. figure 1a).

• SXR Flux and Hα flare Area : as we have excluded from our sample the type II/IV periods, most of the Type III-only associated flares are of class C and small (Hα area < 60 and Hα area < 40 in the case of AR 0486). Only the *jet-CMEs* associated flares appear to have Hα area > 50 (cf. figure 1b).

• SXR duration (D), rise time (t_r), and decay time (t_d): Mostly they have $t_r/t_d < 1/2$ (cf. figure 1b & 1c). The flare SXR duration (D) ranges between a few minutes and up to 2 hours, with the majority at approximately 30 minutes. The *jet-CME*-associated flares and the majority of the flares in AR 0488 have durations in excess of 30 minutes.

In Figures 1d,1e & 1f we present the relationship between type III parameters (F_{START} and Δt) with the associated flare characteristics (Hα AREA, SXR Flux and Duration).

3. Final Remarks

The type III bursts and groups, in the absence of type II/IV activity, are, in general associated with medium to small flares since electron acceleration to $\approx 0.3c$ does not require much energy. The start frequency, in the majority of events, is less than 300 MHz with some exceptions of isolated bursts.

The appearance of type III bursts and groups is favored by an increase in SXR flux and Hα Area. Increasing SXR duration (D) on the other hand, does not favor the appearance of bursts of the type III family. In fact as D increases, in the bipolar AR 0488 in particular, the starting frequency and duration of the type IIIs decrease; we note that AR 0488 gives many (50%) LDEs as opposed to the other two ARs.

The two complex active regions (AR 0484 and AR 0486) produced mostly a large number of short duration type IIIs, from the bipolar AR 0488 50% of type IIIs had duration (ΔT) exceeding 2 sec.

References

Caroubalos, C., Alissandrakis, C. E., Hillaris, A., Preka-Papadema, P., Polygiannakis, J., Moussas, X., Tsitsipis, P., Kontogeorgos, A., Petoussis, V., Bouratzis, C., Bougeret, J. L., Dumas, G., & Nindos, A., N. Solomos, ed., *Recent Advances in Astronomy and Astrophysics*, vol. 848 of *American Institute of Physics Conference Series* 2006, pages 864–873

Caroubalos, C., Maroulis, D., Patavalis, N., Bougeret, J. L., Dumas, G., Perche, C., Alissandrakis, C., Hillaris, A., Moussas, X., Preka-Papadema, P., Kontogeorgos, A., Tsitsipis, P., & Kanelakis, G. 2001, *Experimental Astronomy*, 11, 23

Kontogeorgos, A., Tsitsipis, P., Caroubalos, C., Moussas, X., Preka-Papadema, P., Hilaris, A., Petoussis, V., Bouratzis, C., Bougeret, J. L., Alissandrakis, C. E., & Dumas, G. 2006a, *Experimental Astronomy*, 21, 41

Kontogeorgos, A., Tsitsipis, P., Moussas, X., Preka-Papadema, G., Hillaris, A., Caroubalos, C., Alissandrakis, C., Bougeret, J. L., & Dumas, G. 2006b, *Space Sci. Rev.*, 122, 169

Mitsakou, E., Thanasa, M., Preka-Papadema, P., Moussas, X., Hillaris, A., Caroubalos, C., Alissandrakis, C. E., Tsitsipis, P., Kontogeorgos, A., Bougeret, J. L., & Dumas, G., N. Solomos, ed., *Recent Advances in Astronomy and Astrophysics*, vol. 848 of *American Institute of Physics Conference Series*, 2006, pages 234–237

Session VI

3-D Reconnection Processes

Universal Heliophysical Processes
Proceedings IAU Symposium No. 257, 2008
N. Gopalswamy & D.F. Webb, eds.

© 2009 International Astronomical Union
doi:10.1017/S1743921309029597

Magnetic reconnection in the heliosphere: new insights from observations in the solar wind

J. T. Gosling

Laboratory for Atmospheric and Space Physics, University of Colorado
1234 Innovation Drive, Boulder Colorado, USA 80303
email: `jack.gosling@lasp.colorado.edu`

Abstract. Magnetic reconnection plays a central role in the interpretation of a wide variety of observed solar, space, astrophysical, and laboratory plasma phenomena. The relatively recent discovery that reconnection is common at thin current sheets in the solar wind opens up a new laboratory for studying this fundamental plasma process and its after-effects. Here we provide a brief overview of some of the new insights on reconnection derived from observations of reconnection exhaust jets in the solar wind.

Keywords. Magnetic reconnection, solar wind, exhaust jets, plasmas, magnetic fields

1. Introduction

Magnetic reconnection is a physical process that changes magnetic field topology and ultimately converts magnetic field energy to bulk flow energy and plasma heating. It occurs at thin current sheets when the frozen-in field condition of magnetohydrodynamics (MHD) is violated. Original ideas about reconnection had their roots in attempts to explain the sudden release of magnetic energy in solar flares in terms of magnetic neutral points in the solar atmosphere (Giovanelli 1946); however, the concept of reconnection found its first real success in describing the essence of the interaction between the geomagnetic field and the heliospheric magnetic field (HMF) embedded in the solar wind flow (Dungey 1961).

Indirect magnetospheric evidence for reconnection can be found in 1) the association of geomagnetic storms, substorms and erosion of the dayside magnetosphere with southward turnings of the HMF, 2) the sense of magnetospheric convection (generally anti-sunward over the polar caps and sunward at lower latitudes), 3) asymmetric polar cap convection and its dependence on the transverse in-ecliptic component of the HMF, 4) the fact that the polar caps and magnetotail lobes are almost always open to the heliosphere, 5) the hemispheric dependence of polar rain (energetically soft electron precipitation into the polar caps with intensity and spectra comparable to that of the external solar wind electron strahl) on the polarity (toward or away from the Sun) of the HMF, and 6) reconfigurations of the geomagnetic tail in association with geomagnetic activity. Although the above are collectively convincing evidence for reconnection, the first direct evidence came from *in situ* observations of accelerated plasma flows at Earth's magnetopause that were quantitatively consistent with models of the reconnection process (Paschmann *et al.* 1979). Subsequent work has now established beyond any reasonable doubt the fundamental role that reconnection plays in the dynamics of Earth's magnetosphere and, by extension, the dynamics of the magnetospheres of other planets.

We have also come full circle in that scientists now commonly invoke reconnection to explain such diverse solar phenomena as flares, coronal mass ejections (CMEs), post-flare/CME loops, coronal jets, blobs, and down flows and the restructuring of the solar atmosphere in general, as well as coronal heating and impulsive solar energetic particle events. In addition, reconnection is commonly used to explain 1) the mixtures of magnetic field topologies (open, closed, disconnected) often observed within CMEs in the solar wind (ICMEs), 2) the formation of flux rope ICMEs, 3) the rough constancy of open magnetic flux in the heliosphere, and 4) comet tail disconnection events. Indeed, reconnection plays a central role in the interpretation of a wide variety of observed solar, space, astrophysical, and laboratory plasma phenomena (e.g., Priest & Forbes 2000).

2. Magnetic reconnection exhausts in the solar wind

There have been suggestions in the years since reconnection was first "discovered" that the process might also occur in the solar wind far from the Sun, for example at the helio-spheric current sheet, HCS, that separates magnetic fields of opposite magnetic polarity and that generally wraps around the Sun, at the leading edges of ICMEs, or at current sheets formed by solar wind turbulence. Only recently, however, have we learned how to recognize the unambiguous signature of local, quasi-stationary reconnection in the solar wind far from the Sun in the form of Petschek-like exhausts i.e., plasma jets propagating away from a reconnection site and bounded by back-to-back rotational discontinuities or slow mode waves (Petschek 1964; Gosling et al. 2005a).

Figure 1 shows solar wind plasma and magnetic field data encompassing a reconnection exhaust observed at the leading edge of an ICME on 18 February 1999. The exhaust is identified in the data by the roughly Alfvènic accelerated plasma flow (primarily in the n-component) confined to the region where the magnetic field rotated (also primarily in the n-component) and bounded on one side by anticorrelated changes in flow velocity, $\mathbf{V}$, and magnetic field, $\mathbf{B}$, and by correlated changes in $\mathbf{V}$ and $\mathbf{B}$ on the other. This is the characteristic signature by which we identify reconnection exhausts in the solar wind. We note that the field reversal occurred in two distinct, but unequal, steps with the field lingering at an intermediate orientation in between. The total field rotation was about $120°$ and occurred over an interval of about 3.5 minutes, corresponding to a maximum exhaust width of about 1.3×10^5 km. An increased proton temperature, a decreased mag-netic field strength, and a proton density intermediate between the densities on opposite sides also characterized the exhaust, which was embedded within low beta plasma on both sides. Such changes in proton temperature, proton density and field strength are characteristic of many, but certainly not all, reconnection exhausts in the solar wind.

Figure 2 provides a highly idealized planar projection of a slightly asymmetric re-connection exhaust (the oppositely directed exhaust is not shown) convecting with the nearly radial (from the Sun) solar wind flow and a brief explanation of how an exhaust is formed. A large fraction of the field rotation across an exhaust in the solar wind occurs at the two exhaust edges where the plasma entering the exhaust is accelerated. We call these back-to-back rotational discontinuities a bifurcated current sheet since they result from a splitting of a current sheet as an after-effect of the reconnection process.

As illustrated in Figure 3, elevated proton temperatures commonly observed within solar wind reconnection exhausts are associated with interpenetrating (along $\mathbf{B}$) proton beams (not always well resolved) rather than with simple broadened thermal distribu-tions. The counterstreaming proton beams enter into an exhaust from opposite sides as a result of the Alfvènic disturbances that propagate in opposite directions along recon-nected field lines and that mark the edges of an exhaust. The beams thus have a relative

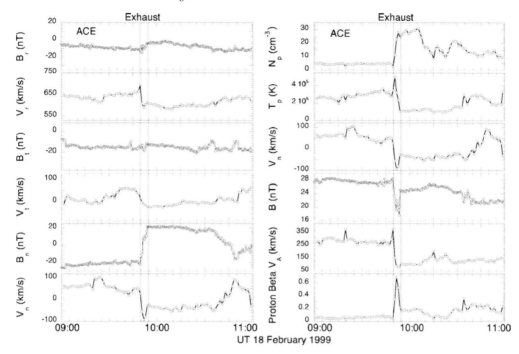

Figure 1. Selected solar wind plasma (at 64-s resolution) and magnetic field (at 16-s resolution) parameters from ACE, then positioned in the solar wind upstream from Earth at (243.2, 0.61, 24.1) R_e in GSE coordinates, surrounding a reconnection exhaust observed on 18 February 1999. The plasma flow velocity and magnetic field components on the left are shown in r, t, n coordinates, where +r is radial out from the Sun, +t points in the direction of solar rotation at constant heliolatitude, and +n completes a right-handed system. Shown from top to bottom on the right are proton number density and temperature, the n-component of the flow velocity, the magnetic field strength, the Alfvèn speed and the proton beta (ratio of gas to field pressure). Vertical lines bracket the reconnection exhaust. The magnetic shear angle across the exhaust was 120°. Adapted from Gosling *et al.* (2007b).

field-aligned speed within an exhaust that is comparable to the sum of the anti-parallel components of the local Alfvèn speeds on opposite sides of a reconnecting current sheet and demonstrate that a magnetic connection exists across an exhaust.

Suprathermal electrons having energies at 1 AU greater than about 70 eV can also be used to demonstrate magnetic connection across an exhaust in reconnection events that occur at the HCS. In the normal solar wind suprathermal electrons are nearly collisionless and can usually be split into two components: 1) a relatively intense focused beam known as the strahl that is directed outward from the Sun along the magnetic field and 2) a roughly isotropic component that we call the halo and that originates largely from scattering out of the strahl beyond 1 AU, the anti-sunward-directed portion of the halo largely resulting from magnetic mirroring of the sunward-directed portion at locations sunward of the observation point. Figure 4a shows the evolution of the suprathermal electron pitch angle distributions, PADs, during a crossing of an antisunward-directed exhaust detected by ACE at the HCS between 03:17:33 and 30:20:29 UT on 17 September 1998. The HCS crossing is identified in Figure 4a by the change in strahl flow polarity, being parallel to **B** (peaking at 0° pitch angle) on one side and antiparallel to **B** (peaking at 180°) on the other. The field was thus directed outward from the Sun prior to the crossing and directed inward afterwards. Notably, however, the strahl disappeared within the exhaust itself where accelerated plasma flow was detected (not shown). There the

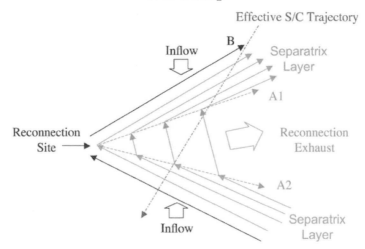

Figure 2. Highly idealized planar projection of a slightly asymmetric reconnection exhaust convecting with the nearly radial (from the Sun) solar wind flow. The sharp field line kink produced by reconnection propagates as a pair of Alfvènic disturbances parallel and antiparallel to a reconnected field line into the plasma on opposite sides of the reconnecting current sheet. As the Alfvènic disturbances propagate they accelerate the plasmas they intercept into the exhaust and away from the reconnection site, thus extracting energy from the reconnecting current sheet. The dashed lines A1 and A2, which pass through the kink pairs on successive reconnected field lines, mark the pair of current sheets (back-to-back rotational discontinuities or slow mode waves) that result from this process and that bound the reconnection exhaust. In practice, the reconnecting fields almost always also have substantial out-of-plane components (parallel to the reconnection X-line). The dash-dot line indicates the projection of an effective spacecraft trajectory through the exhaust. The spacecraft would observe anti-correlated changes in **V** and **B** as it enters the exhaust and correlated change in **V** and **B** as it exits the exhaust since Alfvènic disturbances propagating parallel (antiparallel) to **B** produce anticorrelated (correlated) changes in **V** and **B**, respectively. Adapted from Gosling *et al.* (2005a).

PAD between 0° and 90° was essentially identical to that at those pitch angles after the crossing and the PAD between 90° and 180° was essentially identical to that at those pitch angles prior to the crossing.

As illustrated in Figure 4b, reconnection at the HCS creates closed field lines (connected to the Sun at both ends) sunward of a reconnection site and disconnected (from the Sun) field lines anti-sunward of a reconnection site. In the latter case the evolution of the suprathermal electron PADs from one side of the exhaust to the other should be as illustrated in Figure 4b, exhibiting within the exhaust itself a PAD consisting of suprathermal halo electrons that originally were all sunward-directed on opposite sides of the HCS. The suprathermal electron evolution observed in the 17 September 1998 event (Figure 4a) is consistent with this expectation and thus demonstrates electron interpenetration, magnetic connection across the exhaust, and magnetic disconnection from the Sun. In contrast, one expects to observe closed field lines and counterstreaming strahls within an exhaust sunward of a HCS reconnection site; observations sunward of HCS reconnection sites in other events have confirmed that expectation (Gosling *et al.* 2006).

Multi-spacecraft observations demonstrate that reconnection in the solar wind is commonly a quasi-stationary process that occurs at extended reconnection X-lines (Phan *et al.* 2006; Gosling *et al.* 2007a; Gosling *et al.* 2007d). Estimates of reconnection persistence and spatial extent are limited by the spatial extents and orientations of current sheets present in the solar wind and by available spacecraft separations. The launch of the twin STEREO A and B spacecraft in October 2006 into orbits that increasingly lead

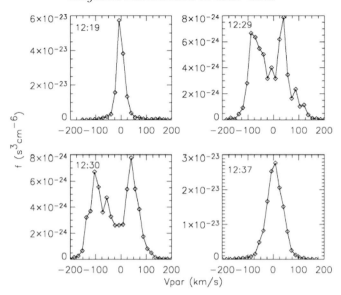

Figure 3. Selected examples of the reduced proton distribution function in the solar wind frame obtained inside (upper right and lower left frames) and outside (upper left and lower right frames) a reconnection exhaust observed on 23 November 1997. Adapted from Gosling *et al.* (2005a).

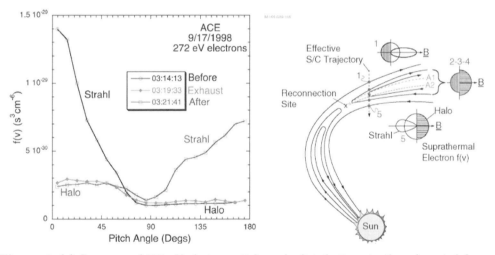

Figure 4. (a) Sequence of 272 eV electron pitch angle distributions in the solar wind frame obtained during a crossing of a reconnection exhaust at the HCS. Adapted from Gosling *et al.* (2005b). (b) Idealized 2-dimensional sketch of reconnection at the HCS illustrating the evolution of the suprathermal electron PADs on the antisunward side of the reconnection site. Adapted from Gosling *et al.* (2005b).

(A) and lag (B) the Earth in its orbit about the Sun opened up a unique opportunity to probe reconnection persistence and the spatial extents of reconnection X-lines in the solar wind.

Figure 5 shows the relative positions of STEREO A and B and 3 other spacecraft (ACE, Wind and Geotail) in the solar wind on 11 March 2007 when all 5 spacecraft observed an extended current sheet. The two STEREO spacecraft were separated by 1215 R_e

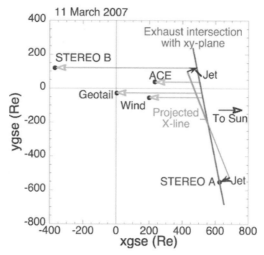

Figure 5. GSE x, y coordinates of STEREO A, ACE, Wind Geotail and STEREO B in the solar wind on 11 March 2007 in Earth radii, Re. All of the spacecraft were nearly in the ecliptic plane. The violet line indicates the intersection of the 11 March 2007 reconnection exhaust with the xy-plane at the time when STEREO A first encountered it. The red line shows the projection of the reconnection X-line onto the xy-plane at that time, the thick (thin) portion corresponding to that part of the X-line lying above (below) the xy-plane. Black arrows at the opposite ends of the X-line projection indicate projections of those portions of the exhaust jets observed by STEREO A and (later) STEREO B, respectively. Blue arrows indicate the motion of the exhaust intersection as the X-line was carried antisunward by the nearly solar wind flow. The lengths of the blue arrows are proportional to the predicted time lags relative to STEREO A for the exhaust encounters at the other spacecraft. Adapted from Gosling *et al.* (2007d).

$(1\ R_e = 6378$ km$)$ at this time, their separation transverse to the radial (from the Sun) direction being $670\ R_e$. Figure 6 shows a time-shifted overlay of plasma and magnetic field data from STEREO A, Wind and STEREO B surrounding a reconnection exhaust observed by all 5 spacecraft on this day. STEREO A observed an increase in flow speed within the exhaust and correlated (anticorrelated) changes in **V** and **B** at the leading (trailing) edge of the exhaust, whereas the other 4 spacecraft observed a decrease in flow speed within the exhaust and, except for STEREO B where 3D flow measurements were not available, anticorrelated (correlated) changes in **V** and **B** at the leading (trailing) edge. Thus STEREO A observed the antisunward-directed exhaust jet from an extended X-line and the other 4 spacecraft observed the oppositely directed exhaust jet, indicating that the reconnection X-line must have crossed the xy-plane somewhere between STEREO A and Wind, as illustrated in Figure 5 (see also Davis *et al.* 2006).

From a minimum variance analysis of the magnetic field data in the vicinity of the exhaust as observed at STEREO A, we find that the exhaust intersection with the xy-plane and the projection of the X-line onto that plane were as shown in Figure 5, with the X-line being tilted relative to the xy-plane by about 7°. Assuming a planar exhaust boundary and a radial solar wind flow of 310 km/s, the predicted delays for the arrival of the exhaust at ACE, Wind, Geotail, and STERO B, respectively were 96, 115, 182 and 309 minutes, in reasonably good accord with observed delays of 105, 126, 184 and 320 minutes. We find that the X-line extended at least $668\ R_e$ $(4.26 \times 10^6$ km$)$ and that reconnection must have persisted for at least 320 minutes. This event thus reveals the tremendous length that a reconnection X-line can have within an extended current sheet and demonstrates how persistent reconnection can be in the solar wind.

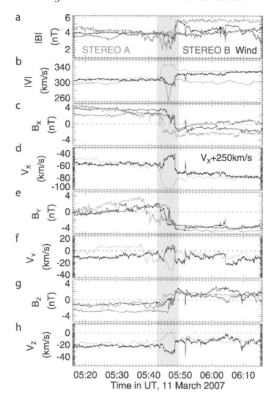

Figure 6. A 1-hr overlay of time-shifted plasma and magnetic field data from STEREO A, Wind, and STEREO B in GSE coordinates. The time shifts for STEREO A and B data relative to the Wind data were +126 and -193 minutes, respectively. Shading indicates a reconnection exhaust observed by all three spacecraft (as well as by ACE and Geotail, not shown). Adapted from Gosling *et al.* (2007d).

Our original examination of 64-s solar wind data from ACE suggested that reconnection occurs only rarely in the solar wind far from the Sun. However, with the development of better techniques for displaying data and by going to higher temporal resolution (3-s) measurements, the situation changed dramatically. We now know that reconnection is relatively common at thin current sheets in the solar wind. Figure 7 provides an overview of solar wind speed and magnetic field variations in March 2006, during the approach to the most recent solar activity minimum. Several high-speed streams of modest amplitude and width were observed during this month; the dominant stream occurred in the March 19-23 interval. Regions of strong magnetic field were present on the leading edges of all the high-speed streams, a result of compression that occurred there. Crossings of the HCS, of which there were at least 4 during the month, can be recognized in Figure 7 as locations where the field azimuthal angle switched from about 135 to about 315° or visa versa. Numerous other current sheets were encountered during March 2006, some of which produced sharp, but often relatively small, changes in the 1-hr averages of the field azimuth and latitude angles. As indicated in Figure 7, we identified at least 46 reconnection exhausts in the Wind 3-s data in this month. All were characterized by correlated changes in **V** and **B** at one edge and by anticorrelated changes in **V** and **B** at the other edge, and thus also by the double-step magnetic field rotations associated with bifurcated current sheets. All but one of the events were observed in relatively low-speed solar wind. A few of the events were observed at times of increasing solar wind speed;

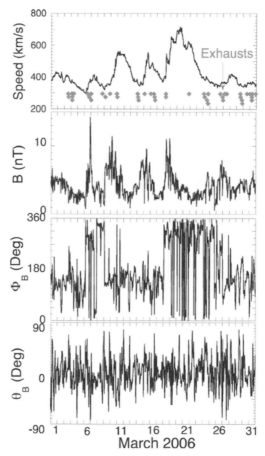

Figure 7. From top to bottom, 1-hr averages of solar wind speed, magnetic field strength, and the azimuth and latitude angles of the heliospheric magnetic field in March 2006. Diamonds placed beneath the speed profile in the top panel indicate times of Wind encounters with reconnection exhausts. Adapted from Gosling *et al.* (2007c).

however, the majority of the events were observed at times of decreasing or nearly constant wind speed, suggesting that reconnection in the solar wind is not typically driven by speed gradients associated with the leading edges of high-speed streams. Although not obvious in the figure, none of the March 2006 exhausts occurred at the HCS and none were associated with ICMEs, since no ICMEs were encountered during the month. Typically, the exhausts were asymmetric; i.e. they occurred at interfaces separating plasmas with somewhat different temperatures, densities, or field strengths. Most of the exhausts occurred in plasma having considerably lower than average plasma beta.

Figure 8 demonstrates that the large majority (89%) of the exhausts observed in March 2006 had local widths less than 4×10^4 km (about 400 ion skin depths), corresponding to exhaust crossing times < about 100 s in a 400 km/s solar wind. Exhausts have been identified down to about the limit of what can be resolved by the Wind 3-s plasma measurement cadence; the narrowest exhaust identified to date in the Wind combined 3-s plasma and magnetic field data had a maximum width of 1×10^3 km (18 ion skin depths). Since that width is still larger than the expected width (about one ion skin depth) of the diffusion region where reconnection actually occurs (Priest & Forbes 2000), there probably are a number of additional exhausts unresolved by 3-s plasma measurements.

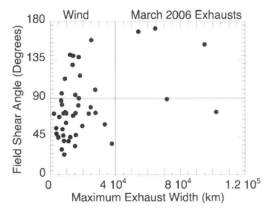

Figure 8. Scatter plot of local magnetic field shear angle versus maximum local exhaust width for the March 2006 reconnection exhausts. Adapted from Gosling *et al.* (2007c).

That expectation is supported by examination of a limited set of current sheets unresolved by a 3-s plasma measurement using 92-ms magnetic field data from Wind that revealed that 10 - 20% of such thin current sheets exhibit the double-step magnetic field rotations characteristic of reconnection exhausts (Gosling & Szabo 2008). In contrast, the broadest exhaust yet identified at 1 AU was 1.85×10^6 km wide and was also characterized by a double-step magnetic field rotation (Gosling *et al.* 2007a).

Perhaps the most remarkable aspect of Figure 8 is that it reveals that the large majority (70%) of the March 2006 exhausts were associated with local field shear angles less than 90°, the smallest field shear angle in this set of events being only 24°. We have recently identified reconnection exhausts associated with local field shear angles as small as 15°. The above clearly indicates that reconnection in the solar wind occurs most often at locations where the so-called guide field component (parallel to the reconnection X-line) considerably exceeds the anti-parallel components. This provides a strong demonstration that, contrary to some perceptions, reconnection does not require or depend on the presence of nearly anti-parallel magnetic fields. The reason for the prevalence of events in the solar wind at current sheets associated with relatively small local field shear angles is simply due to the fact that such current sheets are dominant in the solar wind (see, for example, Figure 7). Further study reveals that our March 2006 results are representative for the solar wind in general near solar activity minimum.

3. Summary

It is now widely appreciated that magnetic reconnection plays a central role in a wide variety of observed solar and space plasma phenomena. The recent recognition that reconnection occurs commonly in the solar wind opens up a new and valuable laboratory for studying the reconnection process and its after-effects. Among other things, work to date has shown that reconnection in the solar wind commonly occurs in a quasi-stationary mode at extended X-lines. It occurs at thin current sheets and produces Petschek-like exhausts of roughly Alfvénic jetting plasma bounded by back-to-back rotational disconti-nuities that bifurcate the reconnecting current sheets. Reconnection most often occurs in the solar wind in the presence of strong guide fields and low plasma beta. It is quite com-mon in the low-speed wind (40-80 events/month are encountered upstream from Earth) and within ICMEs, but is observed less frequently at current sheets in the Alfvènic turbulence characteristic of the high-speed wind from coronal holes (Gosling 2008). Reconnection in the solar wind often appears to be spontaneous and is usually "fast" but not "explosive" - the magnetic energy release occurs over a long interval following reconnection as the Alfvènic disturbances initiated by reconnection propagate into the surrounding solar wind. Finally, although not explicitly demonstrated here, there is as yet no hard evidence to suggest that reconnection in the solar wind ever produces any substantial particle acceleration (Gosling *et al.* 2005c).

The solar wind results complement and extend our understanding of reconnection as derived from *in situ* measurements at Earth's magnetopause and in the geomagnetic tail. Some questions arising from the observations reported here include:

(*a*) Can reconnection be initiated spontaneously or must it be driven by an external flow?

(*b*) What sustains reconnection and what turns it off?

(*c*) What determines whether reconnection is quasi-stationary or transient?

(*d*) Why does reconnection prefer low plasma beta?

(*e*) How do long reconnection X-lines develop?

(*f*) Are slow mode shocks a necessary aspect of fast reconnection in a collisionless plasma?

(*g*) Does reconnection necessarily produce significant particle acceleration?

(*h*) How important is reconnection in dissipating current sheets in general and in the solar wind in particular?

(*i*) Are essentially all thin current sheets eventually disrupted by reconnection?

References

Davis, M. S., Phan, T. D., Gosling, J. T., & Skoug, R. M. 2006. *Geophys. Res. Lett.*, 33, L19102

Dungey, J. W. 1961, *Phys. Rev. Lett.*, 6, 47

Giovanelli, R. G. 1946, *Nature*, 158, 81

Gosling, J. T., 2008, *ApJ (Letters)*, 671, L73

Gosling, J. T. & Szabo, A. 2008, *J. Geophys. Res.*, in press.

Gosling, J. T., Eriksson, S., Blush, L., *et al.* 2007d, *Geophys. Res. Lett.*, 34, L20108

Gosling, J. T., Eriksson, S., McComas, D. J., Phan, T. D., & Skoug, R. M. 2007b, *J. Geophys. Res.*, 112, A08106

Gosling, J. T., Eriksson, S., Phan, T. D., Larson D. E, Skoug, R. M., & McComas, D. J. 2007a, *Geophys. Res. Lett.*, 34, L06102

Gosling, J. T., McComas, D. J., Skoug, R. M., & Smith, C. W. 2006, *Geophys. Res. Lett.*, 33, L17102

Gosling, J. T., Phan, T. D., Lin, R. P., & Szabo, A. 2007c, *Geophys. Res. Lett.*, 34, L15110

Gosling, J. T., Skoug, R. M., McComas, D. J., & Smith. C. W. 2005a, *J. Geophys. Res.*, 110, A01107

Gosling, J. T., Skoug, R. M., Haggerty, D. K., & McComas, D. J., 2005c, *Geophys. Res. Lett.*, 32, L14113

Gosling, J. T., Skoug, R. M., McComas, D. J., & Smith, C. W. 2005b, *Geophys. Res. Lett.*, 32, L05105

Paschmann, G., Sonnerup, B. U. O., Papamastorakis, I. *et al.* 1979, *Nature*, 282, 243

Petschek, H. E. 1964, in: W. Hess (ed), *AAS-NASA Symposium on the Physics of Solar Flares* (NASA Spec. Publ. SP-50), p. 425

Phan, T. D., Gosling, J. T., Davis, M., *et al.* 2006, *Nature*, 439, 175

Priest, E. & Forbes, T. 2000, *Magnetic Reconnection : MHD Theory and Applications* (Cambridge Univ. Press), New York

Discussion

ANTIA: What topological quantum number (like helicity) is changed at magnetic reconnection?

GOSLING: My understanding is that helicity is approximately conserved during reconnection. I am not aware of other quantum number effects.

IBADOV: What do you think about disconnections of plasma tails of comets? Some considerations have been made by Prof. J. Brandt.

GOSLING: Comet tail disconnections may be associated with reconnections in the manner suggested by Brandt and others, but I do not think that sector boundaries are necessarily the sites of the reconnection. There are many other current sheets in the solar wind that might suffice.

VLAHOS: Why do you call the long lived large structures, which can possibly host reconnection, evidence for reconnection?

GOSLING: We identify reconnection exhausts as roughly Alfvenic accelerated plasma flows confined to field reversal regions bounded by correlated changes in velocity and magnetic field on one side and by anti-correlated changes in V and B on the other side. Multi-spacecraft observations reveal that such events can be associated with reconnection persisting for hours at a time at very extended X-lines.

VRŠNAK: Did you check how the inflow/outflow temperature ratio, density ratio, etc depend on the plasma beta and shear angle (and compare that with what is expected from 2.5-Dimensional reconnection models (reconnection in a presence of guiding field).

GOSLING: We have not yet compared such predicted and observed temperature and density ratios.

DASSO: You mentioned several observed and inferred signatures of reconnection in the solar wind, such as a Petschek-like structure of the environment of the current sheet and size < 3 times the ion inertial range of protons. Do you think that these signatures are due to the Hall effect of the reconnection?

GOSLING: We have not yet identified Hall effect signatures in the solar wind reconnection events studied.

Universal Heliophysical Processes
Proceedings IAU Symposium No. 257, 2008
N. Gopalswamy & D.F. Webb, eds.

© 2009 International Astronomical Union
doi:10.1017/S1743921309029603

Magnetic helicity content in solar wind flux ropes

Sergio Dasso

Instituto de Astronomía y Física del Espacio (IAFE), CONICET-UBA
and
Departamento de Física, FCEN-UBA, Buenos Aires, Argentina
email: dasso@df.uba.ar

Abstract. Magnetic helicity (H) is an ideal magnetohydrodynamical (MHD) invariant that quantifies the twist and linkage of magnetic field lines. In magnetofluids with low resistivity, H decays much less than the energy, and it is almost conserved during times shorter than the global diffusion timescale. The extended solar corona (i.e., the heliosphere) is one of the physical scenarios where H is expected to be conserved. The amount of H injected through the photospheric level can be reorganized in the corona, and finally ejected in flux ropes to the interplanetary medium. Thus, coronal mass ejections can appear as magnetic clouds (MCs), which are huge twisted flux tubes that transport large amounts of H through the solar wind. The content of H depends on the global configuration of the structure, then, one of the main difficulties to estimate it from single spacecraft *in situ* observations (one point - multiple times) is that a single spacecraft can only observe a linear (one dimensional) cut of the MC global structure. Another serious difficulty is the intrinsic mixing between its spatial shape and its time evolution that occurs during the observation period. However, using some simple assumptions supported by observations, the global shape of some MCs can be unveiled, and the associated H and magnetic fluxes (F) can be estimated. Different methods to quantify H and F from the analysis of *in situ* observations in MCs are presented in this review. Some of these methods consider a MC in expansion and going through possible magnetic reconnections with its environment. We conclude that H seems to be a 'robust' MHD quantity in MCs, in the sense that variations of H for a given MC deduced using different methods, are typically lower than changes of H when a different cloud is considered. Quantification of H and F lets us constrain models of coronal formation and ejection of flux ropes to the interplanetary medium, as well as of the dynamical evolution of MCs in the solar wind.

Keywords. Magnetohydrodynamics (MHD), Sun: solar wind, Sun: solar-terrestrial relations, Sun: magnetic fields

1. Introduction

Magnetic helicity (H) is a magnetohydrodynamical (MHD) quantity that quantifies the relative twist and linkage between magnetic field lines; H is a conserved quantity (with respect to the typical evolution times of the system) in many media with low dissipation, as the heliosphere.

Twisted magnetic flux tubes (i.e., flux ropes) are ubiquitous in space physics. They are key pieces of magnetic field ($\vec{B}$) configurations and are present in the photosphere of the Sun, in the solar corona, in the solar wind, in different locations of planetary magnetospheres and ionospheres, etc. Thus, magnetic flux ropes can store and transport magnetic energy (E) and, because their magnetic field lines are twisted, also important amounts of magnetic helicity.

1.1. *Coronal mass ejections and magnetic clouds*

Coronal Mass Ejections (CMEs) are massive expulsions of magnetized plasma from the solar atmosphere due to a destabilization of the coronal magnetic configuration that can form flux rope structures (e.g., Gosling *et al.*, 1995). Thus, CMEs remove plasma, energy, and magnetic helicity from the Sun and expel them into interplanetary space. As a consequence of this ejection, CMEs can form confined magnetic structures with both extremes of the magnetic field lines connected to the solar surface, extending far away from the Sun into the solar wind, while the coronal magnetic field is restructured in the low-corona. When they are detected in the interplanetary (IP) medium, they are called interplanetary coronal mass ejections (ICMEs), which are transient structures that perturb the stationary solar wind as they move away from the Sun.

A subset of ICMEs, called magnetic clouds (MCs), are characterized by *in situ* observations of low proton temperature, enhanced magnetic field strength (typically larger than ~ 10 nT), and smooth and large rotation ($\sim 180°$) of the magnetic field vector observed during several hours (Burlaga *et al.*, 1981). The two last characteristics observed in MCs are interpreted as single spacecraft observations of large scale flux ropes traveling in the solar wind (e.g., Burlaga *et al.*, 1981; Bothmer & Schwenn, 1998).

One of the main difficulties determining the global magnetic configuration of a MC from single spacecraft *in situ* observations (one point - multiple times) is that, due to the high speeds of the interplanetary plasma, a single spacecraft can only observe a linear (one dimensional) cut of the MC global structure. Another serious difficulty is the intrinsic mixing between its spatial shape and its time evolution, during the observing period. Despite these difficulties, from the analysis of the observed time profiles of the magnetic field ($\vec{B}$) components, it is possible to infer many of its main features.

The chirality associated with the magnetic configuration of the flux rope (i.e., the sign of H contained in the large scale flux rope) together with its relative orientation with respect to the heliosphere can be estimated in some MCs from the direct analysis of the magnetic field components (Bothmer & Schwenn, 1998).

Coronal mass ejections are frequently associated with filament eruptions. The direction of the MC axis is frequently found to be roughly aligned with that of the disappearing filament (Bothmer & Schwenn, 1994; Bothmer & Schwenn, 1998), preserving their chirality. This result has been also found in some detailed studies of individual cases by Marubashi (1997); Yurchyshyn *et al.* (2001); Ruzmaikin *et al.* (2003); Yurchyshyn *et al.* (2005). However, a few cases presenting a rotated MC axis with respect to its solar counterpart have also been found (e.g., Harra *et al.*, 2007; Foullon *et al.*, 2007).

1.2. *Aim and road map of the paper*

The main aim of this paper is to review the methods to estimate the amount of H contained in magnetic clouds, starting from elemental concepts/definitions and combining them with observations and modeling of flux ropes in the solar wind. We also mention the main difficulties and the main sources of uncertainty associated with these estimations and why these estimations are useful to gain insight in physical mechanisms of heliophysics (Section 1.3).

The definition of H, the reasons of its conservation in space physics, and theoretical expressions for cylindrical flux ropes are presented in Section 2. Then, we present a brief review of MCs (Section 3), different methods to analyze them and compute magnetic flux (Section 4, where an analysis of a case studied is also presented) and H (Section 5). Finally, in Section 6, the conclusions are given.

1.3. *Magnetic helicity in space physics*

When the complexity of a physical system is such that a detailed treatment (taking into account all its degrees of freedom) is not possible, the use of conserved quantities is one of the most useful ways to study it. Even when the system does not have exact conserved quantities, some 'almost' conserved quantities (as e.g., adiabatic invariants) are successfully used to describe its properties.

Magnetic helicity is approximately conserved in the solar atmosphere and the heliosphere (Berger, 1984). Thus, the study of conserved quantities such as magnetic flux (F) and H can help us to understand the physical mechanisms involved in the heliosphere.

Many recent combined quantitative studies of F and H in MCs and their solar sources have been successfully done and have been useful to put some constrains on coronal magnetic configurations and on flux rope formation/eruption models (e.g., Mandrini *et al.*, 2005; Luoni *et al.*, 2005; Attrill *et al.*, 2006; Longcope *et al.*, 2007; Qiu *et al.*, 2007; Harra *et al.*, 2007; Mandrini *et al.*, 2007; Rodriguez *et al.*, 2008; Mostl *et al.*, 2008). Extensions of this kind of analysis and further developments are needed to improve our understanding of this association (see, e.g., the review by Démoulin, 2008).

The solar differential rotation produces an excess of H and it has been suggested that this helicity excess is carried away from the Sun by CMEs (e.g., Ruzmaikin *et al.*, 2003). Thus, the quantification of F and H in MCs is also crucial to improve estimations of the global release of F and H during the solar cycle, putting constrains on solar dynamo models (e.g., Parker 1987; Bieber & Rust, 1995).

Larger amounts of H have been found in the pre-event phase of active regions (ARs) that later produced coronal mass ejections than in ARs having only confined flares (Nindos & Andrews, 2004). However, different models provide different and controversial results on the role of H in the launch of CMEs (e.g., see Démoulin, 2007).

Conservation of H (together with other properties of the plasma) can determine the evolution of a turbulent flow and the best-known consequence of this dynamical evolution is the inverse cascade in 3D-MHD and the evolution toward a force free state (see e.g., Smith, 2003 and references there in). In particular, for spatial scales smaller than the integral/correlation scale of magnetic field fluctuations, which for a heliodistance of one astronomical unit (AU) is $\sim 10^{-2}$ AU (e.g., Matthaeus *et al.*, 2005), the content of H in the interplanetary plasma can be computed from observations of the correlation tensor of magnetic field fluctuations using similar techniques to the ones used by Dasso *et al.* (2005b). The importance of quantifying the content of H in the interplanetary fluctuations is also linked with its influence on the propagation of cosmic rays, for instance modifying the pitch angle scattering coefficient (Bieber *et al.*, 1987).

Thus, H is one of the most (if not the most) important MHD quantity in space physics which quantifies characteristics of magnetic field structures and their consequence for another constituents of the system.

2. Magnetic helicity

For a closed magnetic field configuration inside a volume (Vol), H is defined from the magnetic field and its vector potential ($\vec{A}$, such that $\vec{B} = \vec{\nabla} \times \vec{A}$):

$$H = \int_{Vol} \vec{A} \bullet \vec{B} \ dV \tag{2.1}$$

Elsasser (1956) noticed that for an ideal MHD system, H is a conserved quantity; Moffat (1969) related H with the linking number between two curves, which can be quantified using an integral formula derived by Carl Friedrich Gauss in 1833 (see, e.g., Hirshfeld, 1998), and shown that H can be constructed from the sum of the Gauss linking numbers over every pair of field lines within a volume.

For a non-ideal MHD system (i.e., non-null resistivity, $\eta \neq 0$), magnetic reconnection processes are allowed even for systems with very low η. On the other hand, the dissipation of H is associated with the intensity and the relative alignment between the electric current ($\vec{J}$) and $\vec{B}$: $d_t H \sim -\eta \int \vec{J} \bullet \vec{B} \; dv$.

Thus, and because (i) $\vec{J}$ can increase in current sheets associated with reconnection and (ii) the relative connectivity of magnetic field lines is altered during reconnection processes, in principle it is a valid question 'Is H conserved during reconnection?'.

The answer was given by Berger (1984), who showed that the amount of dissipated H is negligible for transient fast reconnection, as happens in many physical systems in space physics (e.g., the solar corona, the solar wind, planetary magnetospheres).

Part of the answer is based on (i) current sheets associated with reconnection are very small with respect to the bulk of the volume occupied by the fluid (e.g., Morales et al., 2005), (ii) field lines in a plasma form in fact thin flux tubes with internal structure (the twist, which also can add H), and (iii) reconnection between two thin flux tubes is progressive, transforming step by step part of the initial helicity associated with the removed linkage into helicity associated with new additional twist (see, e.g., Section 2.4 of Biskamp, 2000, in particular Figures 2.10 and 2.11). Thus, plasma field lines are not 'naked' but 'dressed with twist' (Biskamp, 2000).

In a turbulent 3D-MHD system there is a net flux of H to larger spatial scales (Frisch et al., 1975; Alexakis et al., 2006) and consequently its dissipation is also inhibited.

The Hall effect can be relevant in space physics, mainly during physical processes occurring at small scales. In particular, it can change the turbulent properties and increase the reconnection rate of a dissipative magnetofluid. However, for a Hall MHD sytem (HMHD) H is also an ideal invariant (Turner, 1986), and the presence of relaxed states with H almost condensed in longest wavelength modes have been also found for weakly dissipative systems (e.g., Servidio et al., 2008).

Because of the gauge freedom of $\vec{A}$, the definition of H given in Eq. (2.1) is physically meaningful only when the magnetic field is fully contained inside the volume V (i.e., the normal component $B_n = \vec{B} \cdot \hat{n}$ vanishes at any point of the surface S surrounding V).

A gauge-independent relative magnetic helicity can be defined even when B_n is different to zero at S (Berger & Field, 1984). The meaning of this relative helicity (H_r) is such that it measures the helicity with respect to the value of H for a given reference field ($\vec{B}_{\rm ref}$), which has the same distribution of B_n on S:

$$H_r(\vec{B}) = H(\vec{B}) - H(\vec{B}_{\rm ref}) = \int_V [\vec{A} \cdot \vec{B} - \vec{A}_{\rm ref} \cdot \vec{B}_{\rm ref}] \, dV . \qquad (2.2)$$

If we choose a different gauge to define the vector potential (i.e., we define $\vec{A}' = \vec{A} + \vec{\nabla}\psi$, where ψ is any scalar function), the new relative helicity H_r' is given by:

$$H_r' = H_r + \int \int_{S(V)} \psi(\vec{B} - \vec{B}_{\rm ref}) \cdot \vec{ds} = H_r . \qquad (2.3)$$

For cylindrical twisted flux tubes, $\vec{B}(\vec{r}) = B_\varphi(r)\hat{\varphi} + B_z(r)\hat{z}$, the reference field can be chosen as $\vec{B}_{\rm ref}(r) = B_z(r)\hat{z}$ with $\vec{A}_{\rm ref}(r) = A_z(R)\hat{z} + A_\varphi(r)\hat{\varphi}$ (so the reference field is chosen with null magnetic helicity since field lines are straight, and $\vec{A} \times \hat{n} = \vec{A}_{\rm ref} \times \hat{n}$ at the surface of the cylinder, [see Dasso et al., 2003]).

Thus, H_r can be expressed independently of $\vec{A}_{\text{ref}}$ and $\vec{B}_{\text{ref}}$ as (Dasso *et al.*, 2005c)

$$H_r = 4\pi L \int_0^R A_\varphi(r)B_\varphi(r)\,r\,dr = 2L \int_0^R B_\varphi(r)F_z(r)\,dr\,, \qquad (2.4)$$

where L is the length along the magnetic tube, R its radius, and $F_z(r)$ the cumulative axial magnetic flux $(F_z(r) = 2\pi \int_0^r B_z(r')\,r'\,dr')$. Thus, as expected, H can be expressed as the sum of the contribution of the azimuthal field 'twisting around' the cumulative axial flux. Because the relative helicity is the one having significant interest for space physics, herein we will refer to H_r just as H.

3. Magnetic clouds

The identification of the MCs/ICMEs boundaries can present some difficulties, and several proxies and techniques can be used (see, e.g., the review by Dasso *et al.*, 2005a and the review by Zurbuchen & Richardson, 2006). Different proxies frequently imply different boundaries, and thus the correct identification of some MCs/ICMEs boundaries is an open problem that needs to be considered (Russel & Shinde, 2005).

In particular, to improve the quantification of magnetic fluxes and helicity, the identification of MC boundaries needs to be made as accurate as possible because (i) these quantities are extensive and thus they critically depend on the size of the object and (ii) because incorrect identification of the boundaries can lead to a wrong interpretation of the real magnetic structure/orientation.

Magnetic clouds can be modeled locally using helical cylindrical geometry as a first approximation (Farrugia *et al.*, 1995). The magnetic field in MCs is relatively well modeled by the so-called Lundquist's model (Lundquist, 1950), which considers a static and axially-symmetric linear force-free magnetic configuration (e.g., Goldstein, 1983), being the cylindrical flux rope in a Taylor state. However, many other different models have been also used to describe MCs, including cylindrical and oblate cross sections, force free and non-force free field configurations, static and expanding magnetic structures, etc. For a review of different models used to describe MCs and comparison of H and F derived from different models see, e.g., Dasso *et al.* (2005a).

To facilitate the understanding of MC properties, we define a system of coordinates linked to the cloud in which $\hat{z}_{\text{cloud}}$ is along the cloud axis (with $B_{z,\text{cloud}} > 0$ at the MC axis). Since the MC moves nearly in the Sun-Earth direction and its speed is much larger than that of the spacecraft (which can be supposed to be at rest during the cloud observing period), we assume a rectilinear spacecraft trajectory in the cloud frame. The trajectory defines a direction $\hat{d}$ (pointing toward the Sun); then, we define $\hat{y}_{\text{cloud}}$ in the direction $\hat{z}_{\text{cloud}} \times \hat{d}$ and $\hat{x}_{\text{cloud}}$ completes the right-handed orthonormal base $(\hat{x}_{\text{cloud}}, \hat{y}_{\text{cloud}}, \hat{z}_{\text{cloud}})$. We also define the impact parameter, p, as the minimum distance from the spacecraft to the cloud axis.

The observed magnetic field in a MC can be expressed in this local frame transforming the observed components (B_x, B_y, B_z) with a rotation matrix to $(B_{x,\text{cloud}}, B_{y,\text{cloud}}, B_{z,\text{cloud}})$. The local system of coordinates is especially useful when p is small compared to the MC radius (R) and for this case the rotation angles can be found using a fitting method (e.g., Dasso *et al.*, 2006) or applying the minimum variance (MV) technique to the normalized time series of the observed magnetic field (e.g., Gulisano *et al.*, 2007). In particular Gulisano *et al.* (2007), from the analysis of a set of cylindrical synthetic MCs, found that the normalized MV technique provides a deviation of the real main MC axis smaller than $10°$ even for p as large as 50% of the MC radius.

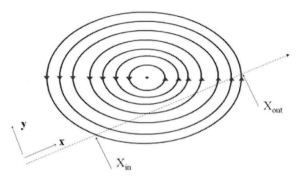

Figure 1. Scheme of the cancellation of F_y (see Eq. 4.3, 2D cut in the plane perpendicular to the axis of symmetry of the flux rope). The trajectory of the spacecraft (projected on this plane) is marked with dashed line, it enters to the flux rope at X_{in} and it exits at X_{out}. F_y will be canceled for a general shape of the flux rope cross section (translation symmetry along the MC axis is the only necessary condition for the cancellation).

4. Reconnection at the MC boundary: progressive pealing of a flux rope

Below we use $\nabla \cdot \vec{B} = 0$ and the local invariance of $\vec{B}$ along the MC axis to define the center and boundaries of twisted flux tubes in the solar wind.

Let us define a plane Π formed by points $\vec{r} = u\,\hat{x}_{cloud} + p_0\,\hat{y}_{cloud} + v\,\hat{z}_{cloud}$, for u and v covering the real numbers and p_0 a fixed value corresponding to the impact parameter associated to the trajectory of the spacecraft that observes the MC. Thus, the magnetic flux of $\vec{B}$ across the plane Π, considering the closed flux rope, is:

$$\int_{flux\ rope} B_{y,cloud}\,dx\,dz = 0\,, \tag{4.1}$$

with x, z being the spatial coordinates in the $\hat{x}_{cloud}$ and $\hat{z}_{cloud}$ directions, respectively.

From the local invariance of $\vec{B}$ along the MC axis ($\hat{z}_{cloud}$), Eq. (4.1) reduces to:

$$\int_{X_{in}}^{X_{out}} B_{y,cloud}\,dx = 0\,, \tag{4.2}$$

being valid for a general shape of the MC cross section, see Figure 1.

If one MC boundary is known, the above flux balance property can be used to find the MC center and the other boundary as follows. We define the cumulative flux

$$F_y(x) = L \int_{X_{in}}^{x} B_{y,cloud}(x')\,dx'\,, \tag{4.3}$$

where X_{in} is the position of the known boundary (e.g., the front of the MC). For $p = 0$ the position where $F_y(x)$ has its absolute extreme corresponds to the x position of the MC center (or to the position where the minimum approach is reached for $p \neq 0$). Then, when $F_y(x)$ goes back to zero at $x = X_{out}$, we have the other boundary. The region from $x = X_{in}$ to $x = X_{out}$ defines the MC flux rope.

The time evolution of the magnetic field while the spacecraft is crossing the MC can affect the correct interpretation of the observations, mixing spatial variation and time evolution. For MCs in strong expansion during the observation time, this expansion effect can have consequences on the correct quantification of F and H using the direct method described above. Expressions to correct for this expansion effect have been presented by Dasso *et al.* (2007).

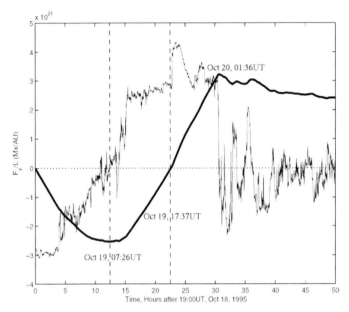

Figure 2. Cumulative flux of $B_{y,cloud}$ (thick solid line) starting from the leading boundary of the cloud. Thin solid line shows $B_{y,cloud}$ as a reference (in arbitrary units). The times of minimum and null flux accumulation are marked with vertical dashed lines. Adapted from Dasso *et al.* (2006)

The cumulative flux F_y for the MC observed on Oct 18, 1995, is shown in Figure 2 (thick solid line), together with the observed $B_{y,cloud}$ component (thin solid line). Vertical dashed lines mark the time at which F_y/L reaches the minimum (Oct 19, 07:26 UT) and the time when the flux cancels (October 19, 17:37 UT). They correspond to the center and the rear boundary of the MC flux rope, respectively. However, different ends for this MC have been chosen in previous studies, e.g., Lepping *et al.* (1997) choose it at 00:00 UT on Oct 20, while Larson *et al.* (1997) choose it at 01:38UT on Oct 20. Other authors have chosen it at other different times. For details of the study of this event, see Dasso *et al.* (2006), and references therein.

We note that the flux balance determines the rear boundary of the flux rope where there is a discontinuity of $B_{y,cloud}$. This is a confirmation of the correct rear boundary location since a current sheet, so a discontinuity of $\vec{B}$, is expected at the boundary of two different magnetic structures.

The analysis of F_y shows that this ICME is not simply formed by a closed flux rope. Some of the MC characteristics, such as the low magnetic variance and the low proton temperature, continue well behind the rear boundary of the flux rope (Lepping *et al.*, 1997). Indeed, the cumulative F_y shows a strong change in the slope at 01:36 UT on Oct 20, expected when the field of the solar wind starts to be integrated in Eq. (4.3). The most plausible physical scenario to create such 'back' magnetic structure is given by a previous (to the observation at 1AU) magnetic reconnection process between the (larger at that moment) original flux rope and the ambient magnetic field. We interpret the lack of flux cancellation in the 'back' as the evidence of a magnetic structure connected with solar wind field lines, which in the past formed the periphery of a larger flux rope (see Figure 6 in Dasso *et al.*, 2006). Thus, the flux rope was progressively peeled before its observation at 1AU.

This 'open' back (or not 'canceled flux') region at the ICME rear has been also recently found in several other events (Dasso et $al.$, 2007; Mostl et $al.$, 2008; Dasso et $al.$, 2008).

5. H in flux ropes

5.1. H from MC models

Theoretical expressions for H can be obtained for the models (see Section 3) used to represent the magnetic structure of MCs. In particular, for a given cylindrical model of the MC, H can be obtained replacing proper expressions for B_φ and A_φ in Eq. (2.4). The free parameters of the models can be obtained from fitting the models to in $situ$ observations inside the MC (see, e.g., Dasso et $al.$, 2003).

From the analysis of 20 well determined MCs using different static models which consider very different distributions of the axial twist of the magnetic field, Gulisano et $al.$ (2005) found that the dispersion of the obtained values of H varying the models considered for a given cloud, was one order of magnitude lower than the dispersion of H for different events. A similar conclusion was reached by Nakwacki et $al.$ (2008a) using dynamical models that permit the radial expansion of the flux rope, and by Nakwacki et $al.$ (2008b) using models that include radial and axial expansions.

5.2. H from a direct method

In this section we present a method to compute F and H directly from the observed magnetic field time series. It requires a transformation of the magnetic data to the cloud frame (Section 3). Three hypotheses are needed: local invariance along the cloud axis, cylindrical symmetry and a moderately low impact parameter.

The position at the minimum approach (similar to the the position of the center of the flux rope when the impact parameter is small) corresponds to the time when F_y is minimum (Section 4, Figs. 1 and 2), and we set the coordinate origin there ($x = 0$). Then, we split the time series of $\vec{B}$ in two subseries for $B_{y,\text{cloud}}$ and $B_{z,\text{cloud}}$. The first subseries corresponds to the in-bound path (the path when the spacecraft is going toward the center of the cloud, $x < 0$) and the second to the out-bound path (when the spacecraft has reached the minimum distance to the cloud axis and is going out the MC, $x > 0$). Differences in the results obtained with these two branches are mainly due to the non-cylindrical symmetry of the flux rope.

Thus, from Eq. (2.4) and for cases when p is small, the magnetic helicity can be estimated as:

$$H \quad \approx \quad 2L \int_0^{X_{out}} B_{y,\text{cloud}}(x') F_z(x') \, dx' , \tag{5.1}$$

For the in-bound path the expressions are the same except that integral limits, $[0, X_{out}]$, are simply replaced by $[X_{in}, 0]$ (with $x < 0$).

If the impact parameter is not zero, the core of the flux rope is not present in the data; so, the fluxes and helicity will be underestimated. However, the relative underestimation for H and F is of the order of $(p/R)^2$ (see Section 5.3 of Dasso et $al.$, 2006).

Estimations of H from models and from this direct method (Dasso et $al.$, 2006), as well as from the in-bound and out-bound branches (Dasso et $al.$, 2005c) are in very good agreement.

6. Conclusions

The quantification of the magnetic helicity (H) and the magnetic flux (F) in magnetic clouds helps us to constrain the formation, ejection, and dynamical evolution of flux ropes in the heliosphere.

We have presented different methods to estimate H and F in MCs which are robust in the sense that for well determined MCs, the obtained values are similar when different methods/techniques are used to compute them. Even more, a very good agreement has been found for the values of H obtained in magnetic clouds (H_{MC}) and the release of H (ΔH_{cor}) from their solar sources during the eruption (e.g., see Dasso *et al.*, 2005d and the review by Démoulin, 2008), a method completely independent and unbiased from the techniques used to estimate H in MCs. In particular a good agreement between H_{MC} and ΔH_{cor} was found from a comparative study of two magnetic clouds with very different sizes, having significantly different values of H: $H_{MC} \sim \Delta H_{cor} \sim 10^{43}$ Mx^2 (Luoni *et al.*, 2005; Dasso *et al.*, 2006) and $H_{MC} \sim \Delta H_{cor} \sim 10^{39}$ Mx^2 (Mandrini *et al.*, 2005).

From the discovery of magnetic clouds more than 25 years ago, very important progress has been made in the understanding of magnetic flux ropes in the solar wind. These astrophysical objects are in fact present in several systems of space physics, as much in our heliosphere as in the broader universe. Some of their main physical processes have been already unveiled; but of course many others are yet waiting to be revealed.

7. Acknowledgments

S.D. thanks the Argentinean grants: UBACyT X425, PICT 03-33370 and PICT-2007-00856 (ANPCyT). S.D. is a member of the Carrera del Investigador Científico, CON-ICET. S.D. thanks P. Démoulin and C.H. Mandrini for useful discussions and improving the manuscript. S.D. thanks the reviewer for improving the manuscript.

References

Alexakis, A., Mininni, P. D., & Pouquet, A. 2006, *Astrophys. J.*, 640, 335

Attrill, G., Nakwacki, M. S., Harra, L. K., van Driel-Gesztelyi, L., Mandrini, C. H., Dasso, S., & Wang, J. 2006, *Solar Phys.*, 238, 117

Berger, M. A. 1984, *Geophysical and Astrophysical Fluid Dynamics*, 30, 79

Berger, M. A. & Field, G. B. 1984, *J. Fluid. Mech.*, 147, 133

Bieber, J. W., Evenson, P. A., & Matthaeus, W. H. 1987, *Astrophys. J.*, 315, 700

Bieber, J. W. & Rust, D. M. 1995, *Astrophys. J.*, 453, 911

Biskamp, D., Magnetic reconnection in plasmas, Cambridge Univ. Press, 2000

Bothmer, V. & Schwenn, R. 1994, *Space Sci. Rev.*, 70, 215

—. 1998, *Annales Geophysicae*, 16, 1

Burlaga, L., Sittler, E., Mariani, F., & Schwenn, R. 1981, *J. Geophys. Res.*, 86, 6673

Dasso, S., Mandrini, C. H., Démoulin, P., & Farrugia, C. J. 2003, *J. Geophys. Res.*, 108, 1362

Dasso, S., Mandrini, C. H., Démoulin, P., Luoni, M. L., & Gulisano, A. M. 2005a, *Adv. Space Res.*, 35, 711

Dasso, S., Milano, L. J., Matthaeus, W. H., & Smith, C. W. 2005b, *Astrophys. J.*, 635, L181

Dasso, S., Gulisano, A. M., Mandrini, C. H., & Démoulin, P. 2005c, *Advances in Space Research*, 35, 2172

Dasso, S., Mandrini, C. H., Luoni, M. L., Gulisano, A. M., Nakwacki, M. S., Pohjolainen, S., van Driel-Gesztelyi, L., & Démoulin, P. 2005d, in ESA Special Publication, Vol. 592, Solar Wind 11/SOHO 16, Connecting Sun and Heliosphere

Dasso, S., Mandrini, C. H., Démoulin, P., & Luoni, M. L. 2006, *Astron. Astrophys.*, 455, 349

Dasso, S., Nakwacki, M., Démoulin, P., & Mandrini, C. H. 2007, *Solar Phys.*, 244, 115

Dasso, S., Mandrini, C. H., Schmieder, B., Cremades, H., Cid, C., Cerrato, Y., Saiz, E., Démoulin, P., Zhukov, A. N., Rodriguez, L., Aran, A., Menvielle, M., & Poedts, S. 2008, *J. Geophys. Res.*, doi:10.1029/2008JA013102, 2009

Démoulin, P. 2007, Advances in Space Research, 39, 1674

—. 2008, *Annales Geophysicae*, 26, 3113

Elsasser, W. M. 1956, *Reviews of Modern Physics*, 28, 135

Farrugia, C. J., Osherovich, V. A., & Burlaga, L. F. 1995, *J. Geophys. Res.*, 100, 12293

Foullon, C., Owen, C. J., Dasso, S., Green, L. M., Dandouras, I., Elliott, H. A., Fazakerley, A. N., Bogdanova, Y. V., & Crooker, N. U. 2007, *Solar Phys.*, 244, 139

Frisch, U., Pouquet, A., Leorat, J., & Mazure, A. 1975, *Journal of Fluid Mechanics*, 68, 769

Goldstein, H. 1983, in Solar Wind Five Conference, 731

Gosling, J. T., Birn, J., & Hesse, M. 1995, *Geophys. Res. Lett.*, 22, 869

Gulisano, A. M., Dasso, S., Mandrini, C. H., & Démoulin, P. 2005, *J. Atmos. Sol. Terr. Phys.*, 67, 1761

—. 2007, *Adv. Space Res.*, 40, 1881

Harra, L. K., Crooker, N. U., Mandrini, C. H., van Driel-Gesztelyi, L., Dasso, S., Wang, J., Elliott, H., Attrill, G., Jackson, B. V., & Bisi, M. M. 2007, *Solar Phys.*, 244, 95

Hirshfeld, A. C. 1998, *Am. J. Phys.*, 66, 1060

Larson, D. E. & *et al.*. 1997, *Geophys. Res. Lett.*, 24, 1911

Lepping, R. P., Burlaga, L. F., Szabo, A., Ogilvie, K. W., Mish, W. H., Vassiliadis, D., Lazarus, A. J., Steinberg, J. T., Farrugia, C. J., Janoo, L., & Mariani, F. 1997, *J. Geophys. Res.*, 102, 14049

Longcope, D., Beveridge, C., Qiu, J., Ravindra, B., Barnes, G., & Dasso, S. 2007, *Solar Phys.*, 244, 45

Lundquist, S. 1950, *Ark. Fys.*, 2, 361

Luoni, M. L., Mandrini, C. H., Dasso, S., van Driel-Gesztelyi, L., & Démoulin, P. 2005, *J. Atmos. Sol. Terr. Phys.*, 67, 1734

Mandrini, C. H., Pohjolainen, S., Dasso, S., Green, L. M., Démoulin, P., van Driel-Gesztelyi, L., Copperwheat, C., & Foley, C. 2005, *Astron. Astrophys.*, 434, 725

Mandrini, C. H., Nakwacki, M. S., Attrill, G., van Driel-Gesztelyi, L., Démoulin, P., Dasso, S., & Elliott, H. 2007, *Solar Phys.*, 244, 25

Marubashi, K. 1997, in Coronal Mass Ejections, Geophysical Monograph 99, 147–156

Matthaeus, W. H., Dasso, S., Weygand, J. M., Milano, L. J., Smith, C. W., & Kivelson, M. G. 2005, *Phys. Rev. Letters*, 95, 231101

Moffatt, H. K. 1969, *Journal of Fluid Mechanics*, 35, 117

Morales, L. F., Dasso, S., & Gómez, D. O. 2005, *J. Geophys. Res.*, 110, 4204

Mostl, C., Miklenic, C., Farrugia, C. J., Temmer, M.,Veronig, A., Galvin, A.B., Vrsnak, B., Biernat, H. K. 2008, *Annales Geophysicae*, 26, 3139

Nakwacki, M. S., Dasso, S., Mandrini, C. H., & Démoulin, P. 2008a, *J. Atmos. Sol. Terr. Phys.*, 70, 1318

Nakwacki, M. S., Dasso, S., Démoulin, P., & Mandrini, C. H. 2008b, *Geof. Int.*, 47, 295

Nindos, A. & Andrews, M. D. 2004, *Astrophys. J. Lett.*, 616, L175

Parker, E. N. 1987, Solar Phys., 110, 11

Qiu, J., Hu, Q., Howard, T. A., & Yurchyshyn, V. B. 2007, *Astrophys. J.*, 659, 758

Rodriguez, L., Zhukov, A. N., Dasso, S., Mandrini, C. H., Cremades, H., Cid, C., Cerrato, Y., Saiz, E., Aran, A., Menvielle, M., Poedts, S., & Schmieder, B. 2008, *Annales Geophysicae*, 26, 213

Russell, C. T. & Shinde, A. A. 2005, *Solar Phys.*, 229, 323

Ruzmaikin, A., Martin, S., & Hu, Q. 2003, *J. Geophys. Res.*, 108, 1096

Servidio, S., Matthaeus, W. H., & Carbone, V. 2008, *Physics of Plasmas*, 15, 042314

Smith, C. W. 2003, *Advances in Space Research*, 32, 1971

Turner, L. 1986, *IEEE Transactions on Plasma Science*, 14, 849

Yurchyshyn, V. B., Wang, H., Goode, P. R., & Deng, Y. 2001, *Astrophys. J.*, 563, 381

Yurchyshyn, V., Hu, Q., & Abramenko, V. 2005, *Space Weather*, 3, 8

Zurbuchen, T. H. & Richardson, I. G. 2006, *Space Science Reviews*, 123, 31

Discussion

GIRISH: Potential field models have been successful in studying closed magnetic structures such as the heliospheric current sheet. But how can potential field models accommodate helicity calculations in the solar corona and solar wind? This is because these models use curl-free magnetic fields.

DASSO: We do not use potential field models to describe the magnetic configuration in the corona or solar wind (in the present work we used linear/non-linear force-free fields, or even non-force-free fields). Because the field configurations we are considering are not contained within closed volume (i.e., B_n can be different to zero at the boundary) we compute their magnetic helicity relative to a reference field configuration, which in the coronal case is generally a potential field.

SCHMIEDER: Has STEREO already observed magnetic clouds?

DASSO: YES, one of the magnetic clouds observed by STEREO presents signatures of a flux rope in one of the spacecraft but not in the other one. This is a result consistent with the interpretation of some ICMEs as flux ropes being observed in their periphery.

Universal Heliophysical Processes
Proceedings IAU Symposium No. 257, 2008
N. Gopalswamy & D.F. Webb, eds.

© 2009 International Astronomical Union
doi:10.1017/S1743921309029615

The August 24, 2002 coronal mass ejection: when a western limb event connects to earth

Noé Lugaz[1], Ilia I. Roussev[1] and Igor V. Sokolov[2]

[1]Institute for Astronomy, University of Hawaii, 2680 Woodlawn Dr., Honolulu, HI, 96822, USA
email: `nlugaz@ifa.hawaii.edu`, `iroussev@ifa.hawaii.edu`

[2]Department of AOSS, University of Michigan, 2455 Hayward St., Ann Arbor, MI 48198
email: `igorosk@umich.edu`

Abstract. We discuss how some coronal mass ejections (CMEs) originating from the western limb of the Sun are associated with space weather effects such as solar energetic particles (SEPs), shocks or geo-effective ejecta at Earth. We focus on the August 24, 2002 coronal mass ejection, a fast (~ 2000 km s^{-1}) eruption originating from W81. Using a three-dimensional magneto-hydrodynamic simulation of this ejection with the Space Weather Modeling Framework (SWMF), we show how a realistic initiation mechanism enables us to study the deflection of the CME in the corona and the heliosphere. Reconnection of the erupting magnetic field with that of neighboring streamers and active regions modify the solar connectivity of the field lines connecting to Earth and can also partly explain the deflection of the eruption during the first tens of minutes. Comparing the results at 1 AU of our simulation with observations by the *ACE* spacecraft, we find that the simulated shock does not reach Earth, but has a maximum angular span of about 120°, and reaches 35° West of Earth in 58 hours. We find no significant deflection of the CME and its associated shock wave in the heliosphere, and we discuss the consequences for the shock angular span.

Keywords. Sun: coronal mass ejections (CMEs), solar-terrestrial relations, acceleration of particles

1. Introduction

1.1. *The Eruption on August 24, 2002*

On August 24, 2002, active region (AR) 10069 was near the western limb of the Sun (W81) when it produced a powerful (X3.1) flare associated with a fast and wide coronal mass ejection (CME). This event has been well studied due to extensive remote observations by SoHO/LASCO and SoHO/UVCS (Raymond *et al.* 2003), in-situ by the *Wind* and *ACE* spacecraft and also its inclusion as one of the Solar Heliospheric Interplanetary Environment (SHINE) campaign events. Based on LASCO observations, this was a wide and fast CME, with an average speed of 1,900 km s^{-1} within the first 20 solar radii. This event has been mostly studied in association with another wide and fast CME on April 21, 2002 for its association with a large Solar Energetic Particle (SEP) event (Tylka *et al.* 2005, 2006). Based on an increase of the iron-to-oxygen ration at large energy, these authors proposed that the shock geometry explain this difference: quasi-perpendicular for the August 24 and quasi-parallel for the April 21 one. The presence or absence of a reflecting boundary at or slightly ahead of Earth associated with a previous eruption has also been recently proposed to explain these differences (Tan *et al.* 2008).

1.2. *Studying Western-Limb Ejections*

Western limb events such as the August 24 and April 21, 2002 CMEs present a number of challenges for space weather prediction. Due to the Parker spiral, the Earth is on average magnetically connected with regions at the solar surface around W55 (for a 400 km s^{-1} background wind). Therefore, SEP events are preferentially associated with western events. Events such as the April 21, 2002 and August 24, 2002 present additional challenges since the SEP arrival time at Earth corresponds to a particle release height of less than 5 $R_\odot$. If one believes that these particles are accelerated by the CME-driven shock wave, there are a number of scenarios to explain the observations. First, the shock must have formed low in the corona, and then, within 5 $R_\odot$ of the solar surface, it either spaned at least 60°, or must have been significantly deflected towards the east, or the magnetic field line connecting Earth to the solar surface significantly diverged from the nominal Parker's spiral. Such differences of up to 30° during SEP events between flare sites and the magnetic footpoint of the Earth on the solar surface have been reported before (Ippolito *et al.* 2005).

Western limb events are often associated with shocks and sometimes ejecta at 1 AU. Both August 24 and April 21, 2002 were associated with a shock wave at 1 AU which transited in about 55 hrs. Among large geomagnetic storms (Dst ⩽ -100 nT) from the past solar cycle (Zhang *et al.* 2008), at least 6 were caused by a shock at Earth associated with an ejection western of W73. This fact, again, seems to imply either a very large span of the shock wave, a large deflection of the CME, or a combination of both.

Until now, it has been hard to study observationally the deflection of a CME in the corona or in the heliosphere due to the paucity of observations, especially in the near-Earth environment. Tripathi *et al.* (2004) reported deflections up to 20° within the first hour after the initiation of an eruption based on a series of LASCO images. Shock span can also be estimated with white-light images, but it is limited by assumptions of symmetry and geometrical effects (Cremades *et al.* 2006). In these two examples, the determination of the CME span and deflection can be only be made for limb CMEs and only in the meridional direction. Additionally, the longitudinal extent of shocks can be estimated from multiple-spacecraft measurements; but until the launch of STEREO, it could only be done with spacecraft at different heliospheric distances, for example by the Helios spacecraft in the 1980s (DeLucas *et al.* 2008). The launches of STEREO and SMEI have also made 3-D tomography of CMEs easier (Jackson & Hick 2002). On the theoretical and numerical sides, previous studies have focused on the deflection of a CME in the heliosphere due to its interaction with the solar wind (Wang *et al.* 2004). Here, based on a 3-D numerical simulation of the August 24, 2002 event, we discuss how, by using a new and realistic model of solar eruptions associated with a realistic model of the coronal magnetic field, we can study these different effects.

2. Simulation Setup

2.1. *Numerical Domains*

In our numerical model, the steady-state solar corona and solar wind are constructed following the methodology of Roussev *et al.* (2003), further described in Roussev *et al.* (2007). The initial condition for the coronal magnetic field is calculated by means of potential field extrapolation, following Altschuler *et al.* (1977), with boundary conditions for the radial magnetic field at the Sun, B_R, provided by full-disk SoHO/MDI observations taken four days before the eruption when the AR was closer to the disk center and better observed. The "solar" boundary in our model is placed at a height of 0.1 $R_\odot$ above the photosphere. The plasma parameters are prescribed in an ad-hoc manner, through a variable polytropic index, to mimic the physical properties of streamers and coronal holes once a steady-state (non-potential) is reached.

The time-dependent MHD equations for a single compressible fluid are solved using the Space Weather Modeling Framework (Tóth *et al.* 2005) using two physical domains: the Solar Corona (SC):, $\{-20 \leqslant x \leqslant 20, -20 \leqslant y \leqslant 20, -20 \leqslant z \leqslant 20\} R_\odot$ and Inner Heliosphere (IH): $\{-220 \leqslant x \leqslant 220, -220 \leqslant y \leqslant 220, -220 \leqslant z \leqslant 220\} R_\odot$. We prescribe the initial grid in such a way as AR 10069 is resolved with cells as small as $4.9 \times 10^{-3} R_\odot$, 4 times finer than the rest of the solar surface. Additionally, the radial direction above the active region (resp. direction of the field lines connecting to Earth) is refined with cells of 0.08 $R_\odot$ up to 8 (resp. 5) $R_\odot$ and 0.16 $R_\odot$ up to 14 (resp. 10) $R_\odot$. The total number of computational cells is of the order of 1.77 and 14.66 millions, with largest meshes of size 1.25 $R_\odot$ and 3.44 $R_\odot$ for SC and IH, respectively.

To the initial magnetic field constrained by SoHO/MDI data, we superimposed newly emerged magnetic flux simulated by a dipolar magnetic field of two point charges. These two charges are initially separated by 5×10^3 km and buried at a depth of 3×10^4 km under the solar surface, and chosen so that the peak value of the radial magnetic field at the solar surface is about 47 Gauss.

2.2. *Solar Eruption Model*

To initiate the eruption, we use the model described in Roussev *et al.* (2007). To summarize, once the steady-state is reached at $t = 0$, the two magnetic charges are moved apart quasi-steadily up to $t = t_S = 20$ min with a speed which is ramped up in $t_S/3$ to 80 km s^{-1}; the charge motion is stopped at $t = t_S$.

In addition to updating the radial component of the dipole field at the boundary, we also impose the accompanying horizontal boundary motions. As the result of moving the charges apart, the magnetic field lines connecting the two spots of the dipole are stretched. With appropriate choice of parameters describing the relative position of the charges, their strength and speed of motion, one can achieve a quasi-steady magnetic field evolution toward a state that is no longer stable. Then, as the result of loss of confinement with the overlying field, the "energized" magnetic field of the dipole erupts, manifesting as a CME.

3. Eruption and Coronal Evolution

3.1. *Loss of Equilibrium*

One of the main results of the work in Roussev *et al.* (2007) was to recognize the importance of the pre-existing magnetic topology in the initiation of the eruption. This is first and foremost because reconnection at the pre-existing null points and quasi-separatrix layers (QSLs) enables the sheared and energized magnetic flux of the dipole to erupt. This work was the continuation and adaptation to realistic background coronal magnetic fields of previous work by Antiochos *et al.* (1999); Galsgaard *et al.* (2005) and Pontin & Galsgaard (2007). There are two main pre-existing topological features important to understand this eruption: a null point between ARs 10067 and 10069 and a QSL between ARs 10066, 10068 and 10069 (the magnetic topology before the shearing phase is illustrated in the left panel of Figure 1). Noteworthy is the fact that there are open field lines originating from AR 10069 and AR 10067 which pass in proximity of the null point.

During the shearing, current builds up along the loops connecting the two magnetic spots of the dipole and the magnetic field lines expand until they reconnect with the overlying field through the QSL. The QSL is disrupted and becomes a current sheet which starts erupting (as illustrated in the right panel of Figure 1). The erupting field lines are now connecting AR 10069 to ARs 10067 and 10066. As some of these field lines expand further, they reconnect though the null point and some of them open up. The

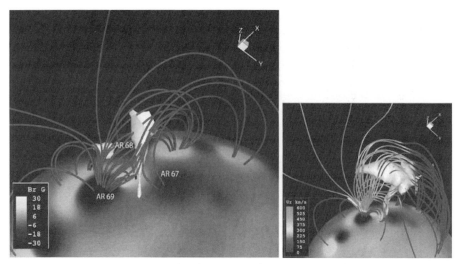

Figure 1. Magnetic topology before (*left*) and after (*right*, $t = 20$min) the shearing phase. The solar surface is color-coded with the radial magnetic field strength and the magnetic field lines with the radial velocity for the right panel. The null point and quasi-separatrix layer (QSL) are visualized as white isosurfaces of plasma beta equal to 0.15 (*left*) and 0.4 (*right*).

main motion of the erupting flux is radially above the initial position of the QSL but the reconnection through the null point between ARs 10069 and 10067 also enables the expansion of the CME towards the east (Earth direction).

3.2. *Magnetic Connection to Earth*

The August 24, 2002 eruption was associated with a large SEP event, even though the magnetic connectivity to Earth is hard to assess due to the position of the AR 81° west of disk center. The timing of the arrival of SEPs at Earth is consistent with particle release at less than 5 $R_\odot$. Additionally, contrary to our model which is not potential, the potential field source surface model (Altschuler *et al.* 1977) shows that there are no open magnetic field lines originating from AR 10069. Our simulation can help explain why a SEP event was indeed observed at Earth. We will not focus on a given field line, but on a set of field lines spanning about 10° at 1 AU including the one connected to Earth on August 24, 2002 01UT. A number of reasons make the determination of the exact footpoint of the field line connected to Earth difficult and compelled us to consider a stack of field lines instead. First, the solar magnetic field is reconstructed from observations on August 21, 2002, 3 days prior to the studied event and the coronal and photospheric magnetic fields have certainly changed in this time span. Second, as noted by Tan *et al.* (2008), the presence of prior ejections in the heliosphere may significantly modify the magnetic connectivity. There was a number of ejections prior to the August 24 ones, although there was no clear ejecta passing Earth at this time according to satellite data. Random walk of the field lines may also account for up to 10° longitudinal variation on the solar surface (Ippolito *et al.* 2005). Last, as can be seen on the left panel of Figure 3, field lines connecting to the vicinity of Earth before the start of the shearing phase have footprints in two very distinct zones on the solar surface through a QSL: one around N20W70 and one around S15W10. The "average" position of the field lines connecting to Earth is about W40, indeed not departing too much from the nominal Parker's spiral. However,

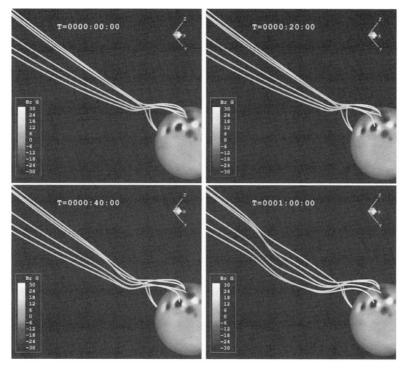

Figure 2. Field lines connecting to the vicinity of Earth and their evolution during the early phase of the eruption. The solar surface is color-coded with the radial magnetic field strength. Note the change of connectivity of one of the field lines at $t = 20$ min which connect AR 10069 to Earth. Note also the propagation of the shock wave along the field lines at later times.

the magnetic topology close to the Sun is such that some of these field lines connect to the proximity of AR 10069.

As described above, we find that, by the end of the shearing phase, the erupting flux has reconnected with open magnetic field lines from AR 10067 through the null point and some erupting field line are now direclty connected to Earth's vicinity. This can be seen in the top right panel of Figure 2. The disruption of these field lines due to the passage of the shock wave can be seen at the later times. We find that the shock wave has been formed by 5 $R_\odot$. It has also been significantly deflected due to reconnection at the northern null point. An approximate visualization of the shock wave can be seen on the right panel of Figure 3 along with the field lines connecting to Earth's vicinity. We should also note that the shock does not become quasi-parallel along Earth-connected field lines until about 90 minutes after the start of the eruption (see also Roussev *et al.* 2008). A similar evolution of the shock angle (from quasi-perpenicular to quasi-parallel) with distance has been previously reported in Manchester *et al.* (2005) for a field line about 37° north of the center of the CME.

4. Heliospheric Evolution and Results at 1 AU

A shock wave associated with the eruption of August 24 was detected at Earth by *ACE* about 58 hours after its start. In our simulation, the shock wave does not extend all the way to Earth. In fact, we find that the shock wave reaches 1 AU approximatively 42 hours after the end of the shearing phase about 60° West of Earth. Its maximum angular

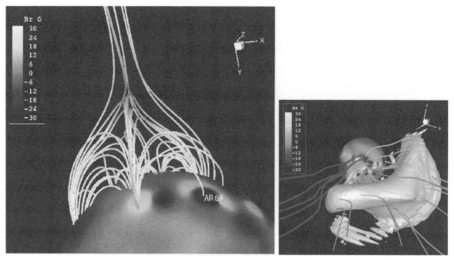

Figure 3. *Left*: Solar footpoints of the magnetic field lines connecting to the vicinity of Earth, which span from S15W05 to N20W75. *Right*: Visualization of the shock wave 1 hour after the start of the shearing phase. Field lines connecting to Earth's vicinity are shown with darker grey and larger radii. The white surface is an isosurface of Aflvénic Mach equal to 1.

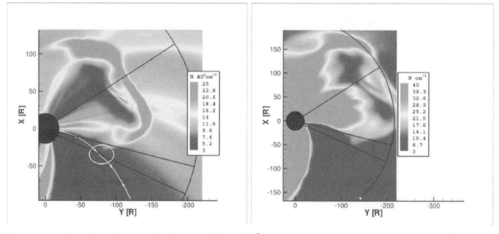

Figure 4. Number density (*left*, scaled by $1/R^2$) in the equatorial plane 29 hours (*left*) and 48 hours (*right*) after the start of the shearing phase. The black circle represent Earth's orbit, Earth's position is shown with a white disk. The three lines are, from top to bottom, the radial directions corresponding to the central position of the eruption, the longitude reached at 58 hours and the maximum longitudinal extent. The field line connecting to Earth is shown in white and the shock position is highlighted with the white ellipse.

span is about 120° and it reaches a maximum of 25° West of Earth. After 58 hours, the simulated shock wave reaches a point about 35° West of Earth. The results at 48 hours are shown on the right panel of Figure 4.

In agreement with previous studies (e.g. Jacobs et al. 2007), we find that there is no significant non-radial expansion of the CME in the heliosphere past the upper corona. Although in our simulation the shock wave does not hit Earth, we find that there is direct magnetic connection between the flank of the shock and Earth during most of the heliospheric evolution of the CME and its associated shock. This is illustrated at time 29 hours in the left panel of Figure 4 and may have important consequences to understand and predict the observed Forbush decrease of ACRs (Eroshenko *et al.* 2008) and the time variation of the SEP event.

5. Discussions and Conclusion

We performed a Sun-to-Earth simulation of the August 24, 2002 CME event with a realistic CME initiation mechanism (Roussev *et al.* 2007). This is one of the first solar-terrestrial simulation involving a CME model more complex than a flux rope or a simple perturbation added onto the solar wind. Using a realistic model is required to study space weather effect such as (i) the formation of the shock wave, (ii) the change of connectivity between the Earth and the solar surface during the eruption, and (iii) the possible deflection of the CME. We find that reconnection of the erupting flux at one null point north-east of the active region results in the deflection of the eruption in the corona (we find no subsequent deflection in the heliosphere) towards the Sun-Earth line. There is an opening of part of the erupting flux due to this reconnection event, and, consequently, a change of magnetic connectivity with Earth. We find that the shock wave has formed by 5 $R_\odot$ and has a sufficient longitudinal extent to accelerate particles along Earth-connected field lines.

Last, we must investigate why, in our simulation, the shock does not reach Earth. First, contrary to what was inferred by Wang *et al.* (2004), we find no consequent deflection of the CME in the heliosphere, even though it is a fast western CME. This might be because the solar wind is not well reproduced. Or it might be because there is in fact no significant deflection of CME in the heliosphere. If we believe that there is no large deflection in the heliosphere, then the shock angular extent must be at least 170° to explain the detection of the shock wave by *ACE*, significantly larger than what is predicted by our model. It is worth noting that the simulated CME is slower in the corona than the observed one by as much as 35%; a faster CME will most likely be associated with a larger shock wave. We will investigate this in future simulations of the same event. In situ observations by STEREO (and Helios) and future polar coronagraphs can valuable information concerning the angular extent of shock wave, and possibly a deflection.

Acknowledgements

The research for this manuscript was supported by NSF grants ATM0639335 and ATM0819653 and NASA grants NNX07AC13G and NNX08AQ16G.

References

Altschuler, M. D., Levine, R. H., Stix, M., & Harvey, J. 1977, *Solar Phys.*, **51**, 345
Antiochos, S. K., DeVore, C. R., & Klimchuk, J. A. 1999, *ApJ*, **510**, 485
Cremades, H., Bothmer, V., & Tripathi, D. 2006, *Adv. Space Res.*, **38**, 461
DeLucas, A. *et al.* 2008, in this volume
Eroshenko, E. *et al.* 2008, in this volume
Galsgaard, K., Moreno-Insertis, F., Archontis, V., & Hood, A. 2005, *ApJ Lett.*, **618**, L153
Ippolito, A., Pommois, P., Zimbardo, G., & Veltri, P. 2005, *A & A*, **438**, 705
Jackson, B. V. & Hick, P. P. 2002, *Solar Phys.*, **211**, 345
Jacobs, C., van der Holst, B., & Poedts, S. 2007, *A & A*, **470**, 359
Manchester, W. B., IV, *et al.* 2005, *ApJ*, **622**, 1225
Pontin, D. I. & Galsgaard, K. 2007, *JGR*, **112**, 3103
Raymond, J. C., *et al.* 2003, *ApJ*, **597**, 1106
Roussev, I. I., *et al.* 2003, *ApJ*, **595**, L57

Roussev, I. I., Lugaz, N., & Sokolov, I. V. 2007, *ApJ Letters*, **668**, L87
Roussev, I. I., Lugaz, N., & Sokolov, I. V. 2008, *AIP Conf.Ser.*, **1039**, 286
Tan, L. C., Reames, D. V., & Ng, C. K. 2008, *ApJ*, **678**, 1471
Tóth, G., *et al.* 2005, *J. Geophys. Res.*, **110**, 12226
Tripathi, D. *et al.* 2004, *Multi-Wavelength Investigations of Solar Activity*, **223**, 401
Tylka, A. J. *et al.* 2005, *ApJ*, **625**, 474
Tylka, A. J. *et al.* 2006, *ApJS*, **164**, 536
Wang, Y., Shen, C., Wang, S., & Ye, P. 2004, *Sol. Phys.*, **222**, 329
Zhang, J., Poomvises, W., & Richardson, I. G. 2008, *Geophys. Res. Lett.*, **35**, 2109

Session VII

Energetic Particles in the Heliosphere

Universal Heliophysical Processes
Proceedings IAU Symposium No. 257, 2008
N. Gopalswamy & D.F. Webb, eds.

© 2009 International Astronomical Union
doi:10.1017/S1743921309029639

History of research on solar energetic particle (SEP) events: the evolving paradigm

Edward W. Cliver[1]

[1] Space Vehicles Directorate, Air Force Research Laboratory
email: afrl.rvb.pa@hanscom.af.mil

Abstract. Forbush initiated research on solar energetic particle (SEP) events in 1946 when he reported ionization chamber observations of the first three ground level events (GLEs). The next key development was the neutron monitor observation of the GLE of 23 February 1956. Meyer, Parker and Simpson attributed this high-energy SEP event to a short time-scale process associated with a solar flare and ascribed the much longer duration of the particle event to scattering in the interplanetary medium. Thus "flare particle" acceleration became the initial paradigm for SEP acceleration at the Sun. A more fully-developed picture was presented by the Australian radio astronomers Wild, Smerd, and Weiss in 1963. They identified two distinct SEP acceleration processes in flares: (1) the first phase accelerated primarily ~ 100 keV electrons that gave rise to fast-drift type III emission as they streamed outward through the solar atmosphere; (2) the second phase was produced by an outward moving (~ 1000 km s^{-1}) magnetohydrodynamic shock, occurring in certain (generally larger) flares. The second phase, manifested by slow-drift metric type II emission, appeared to be required for substantial acceleration of protons and higher-energy electrons. This two-stage (or two-class) picture gained acceptance during the 1980s as composition and charge state measurements strengthened the evidence for two distinct types of particle events which were termed impulsive (attributed to flare-resident acceleration process(es)) and gradual (shock-associated). Reames championed the two-class picture and it is the commonly accepted paradigm today. A key error made in the establishment of this paradigm was revealed in the late 1990s by observations of SEP composition and charge states at higher energies (>10 MeV) than previously available. Specifically, some large and therefore presumably "gradual" SEP events looked "impulsive" at these energies. One group of researchers attributes these unusual events to acceleration of high-energy SEPs by flares and another school favors acceleration of flare seed particles by quasi-perpendicular shocks. A revised SEP classification scheme is proposed to accommodate the new observations and to include ideas on geometry and seed particle composition recently incorporated into models of shock acceleration of SEPs.

Keywords. Sun, Solar Energetic Particles, History

1. Introduction

The nature of particle acceleration at the Sun, particularly at high-energies (>30 MeV), remains a key problem in solar and solar-terrestrial physics. Here I recount the evolution of the paradigm for solar energetic particle (SEP) acceleration, beginning with the initial, implicit, picture of SEP acceleration in flares, through the emergence of the two-class (flare and shock) paradigm from 1963-1993, to challenges in the late 1990s prompted by discordant observations of ion charge states and composition at higher energies (>10 MeV) than previously available, and to recent divergent attempts to account for these discrepancies. The history of paradigm evolution is reviewed in section 2 and in section 3 recent evidence to support shock acceleration of ions at energies >30 MeV is presented. In section 4, a revised SEP classification scheme is proposed to take into account the new SEP observations and theoretical developments on shock acceleration.

2. Historical Review

2.1. *Origins*

SEP research began with a question mark (actually two sets of question marks) indicating curious increases in the cosmic ray intensity observed by ionization chambers on 28 February and 7 March 1942 (Figure 1; Lange & Forbush 1942a). Initially, Lange & Forbush (1942b) interpreted these unusual increases (with off-scale values of ∼7% indicated by arrows in the figure) in terms of the magnetic effects of a ring current on galactic cosmic ray trajectories. When a subsequent event was observed on 25 July 1946, however, Forbush (Figure 2) linked it, as well as the two earlier events, to solar flares. Such increases, caused by protons with energies ∼1 GeV, subsequently came to be referred to as ground level events (GLEs). Forbush's 1946 paper marks the official birth of the field of SEP physics.

2.2. *The GLE of 23 February 1956 and the initial paradigm*

The next essential paper in the field was the pioneering analysis of the GLE on 23 February 1956 GLE by Meyer, Parker, & Simpson (1956); this was the first GLE to be recorded by Simpson-developed neutron monitors. Meyer, Parker, and Simpson deduced a flare particle injection time of 20–30 minutes, an essentially "instantaneous injection" compared to the long duration (∼15 hours) of the particle event above background. They attributed the long isotropic phase of the GLE to interplanetary scattering. Thus the initial (implicit) base level paradigm for particle acceleration at the Sun was that all particles were rapidly accelerated in the flare. Subsequently, the concepts of coronal diffusion (Reid, 1964 and Axford, 1965) and wandering interplanetary field lines (Jokipii & Parker, 1968) were added because of the observed absence of cross-field diffusion in the interplanetary medium.

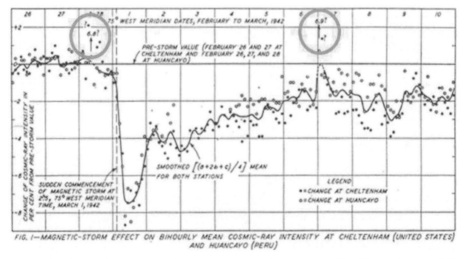

FIG. I—MAGNETIC-STORM EFFECT ON BIHOURLY MEAN COSMIC-RAY INTENSITY AT CHELTENHAM (UNITED STATES) AND HUANCAYO (PERU)

Figure 1. The first indication of solar energetic particle emission. The Cheltenham and Huancayo ionization chamber records revealed two increases (indicated by circled arrows and question marks) that Lange & Forbush (1942a) declined to comment on at the time.

Figure 2. Scott E. Forbush (1904–1984), author of the first paper on SEPs.

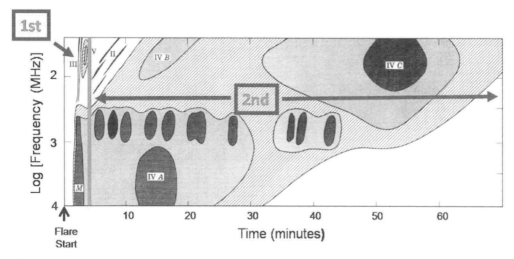

Figure 3. A fully-developed radio burst with the first (impulsive) and second (shock-associated) phases separated by the red vertical line. In the vast majority of flares, only the first phase occurs (after Wild, Smerd, and Weiss, 1963).

2.3. *The two-phase / two-class picture of SEP events*

The earliest evidence for two distinctly different particle acceleration processes on the Sun came from meter wave radio observations (Figure 3). In their *Annual Reviews* paper in 1963, Wild, Smerd, & Weiss outlined a picture that is largely accepted today: SEPs are accelerated both in flares and at coronal/interplanetary shock waves. Type III bursts are the defining signature of the particle acceleration process in solar flares, while coronal shock waves are manifested by slow-drifting type II bursts. The main species accelerated during non-eruptive flares, or the impulsive (first) phase of fully-developed flares, is electrons of energy $\lesssim 100$ keV; a SEP event with higher energy electrons and intense proton emission requires a second-phase of acceleration characterized by a shock.†

Early SEP observations from satellites provided support for this picture. Lin (1970) distinguished between "mixed" SEP events for which both protons and relativistic electrons were observed and low-energy ($\sim$40 keV) "pure electron" events. The mixed events were generally preceded by metric type II solar radio bursts while the pure electron events were characteristically associated with metric type III emission. In 1978 Kahler, Hildner, & Van Hollebeke showed that prompt solar proton events were strongly associated with the then recently discovered coronal mass ejections (CMEs). The standard interpretation of this observation is that fast CMEs drive shock waves that produce type II radio emission and accelerate protons and high-energy electrons.

The underpinnings of the modern two-class picture, which involves $Z \geqslant 2$ ions as well as protons and electrons, can be traced to a remarkable series of papers from 1984-1986. In 1984, Klecker $et~al.$ were the first to show that ^{3}He-rich SEP events and large "normal" SEP events had quite different Fe charge state distributions. While large SEP events typically had low ($\sim$14) Fe charge states, somewhat above that of Fe in the solar wind, the Fe ions in the small ^{3}He-rich events (discovered in 1970 by Hsieh & Simpson) had Fe charge states $\sim$19, suggesting that these events were processed in the high-temperature flare. In a key paper in 1985, Reames, von Rosenvinge, & Lin discovered that ^{3}He ions were an emission of the impulsive phase of flares by associating them with low-energy electrons, the principal particle species of impulsive flares.

Mason $et~al.$ (1986) showed that heavier ions in ^{3}He-rich events had a distinctive abundance pattern, with Fe enriched by a factor of $\sim$10 over that observed in large SEP events. The two types of SEP events inferred by Wild, Smerd, & Weiss (1963) on the basis of solar radio observations and electron and proton spectra were exhibiting distinctions between themselves over a broader range of SEP species and characteristics.

Cane, McGuire, & von Rosenvinge (1986) separated SEP events into two classes on the basis of the durations of their associated flares. Their "impulsive" SEP events had soft X-ray durations $\leqslant 1$ hour, high electron-to-proton (e/p) ratios, and were well-connected, i.e., had associated flares clustered near the $\sim$W60 footpoint of the nominal Parker spiral field line connecting the Sun to Earth. The "gradual" SEP events had flare durations >1 hour, higher proton intensities, and could originate anywhere on the solar disk. What was not appreciated at the time, however, was that many of the impulsive events of Cane, McGuire, and von Rosenvinge were far more energetic than the ^{3}He-rich events of Reames, von Rosenvinge, & Lin (1985).

One other paper from 1984-1986 requires mention here. Breneman & Stone (1985) reported that "the ionic charge-to-mass ratio (Q/M) is the principal organizing factor for the fractionation of . . . SEPs by acceleration and propagation processes and for flare-to-flare variability". It would take $\sim$20 years before this striking result found an explanation within the two-class picture.

In the meantime, the accumulating evidence made a compelling and correct case for two fundamentally different types of SEP events (or acceleration processes on the Sun): (1) "Impulsive" events linked to the flare acceleration process, presumably a resonant wave-particle interaction that resulted in preferential acceleration of electrons, ^{3}He, and high-Z ions; and (2) large "gradual" events attributed to shock acceleration of solar wind

† In 1983, Forrest & Chupp presented evidence from gamma-ray observations that protons and high-energy electrons are also accelerated in the impulsive phase of high energy flares. Subsequently, Ramaty, Murphy, & Dermer (1987) interpreted a delayed "pion-rich" phase of emission in the 3 June 1982 flare in terms of a second phase acceleration process, echoing Wild, Smerd, & Weiss (1963).

Table 1. Two-class Paradigm for SEP Events

	IMPULSIVE	GRADUAL
Particles:	Electron-Rich	Proton-Rich
^{3}He/^{4}He	~1	~0.0005
Fe/O	~1	~0.1
H/He	~10	~100
Q_{Fe}	~20	~14
Duration	Hours	Days
Longitude Cone	$< 30°$	$\sim 180°$
Radio Type	III,V(II)	II,IV
X-Rays	Impulsive	Gradual
Coronograph	—	CME
Solar Wind	—	IP Shock
Events/Year	~1000	~10

particles. The classification scheme is shown in its iconic two-column form in Table 1. Don Reames (1993, 1999) was the primary author of this picture and a driving force in SEP physics for two decades, effectively taking it from a pre-paradigmatic state (in the sense that before ~1990 there was no consensus paradigm or generally-accepted paradigm) to one with a well-defined framework.

As hinted at above, there was an inherent flaw in the scheme, revealed by the fact that "type II" appears in the 'Radio Type' row of both columns in Table 1, when one would think that SEP acceleration by large-scale shocks would only play a role in gradual events. The impulsive class combined two quite different types of events - the ^{3}He-rich events of Reames, von Rosenvinge, & Lin (1985) which characteristically lacked CMEs, type II bursts, and detectable protons (Kahler *et al.*, 1985), and the much more energetic impulsive events from Cane, McGuire, & von Rosenvinge (1986) that were accompanied by each of these phenomena. All that the two types of events had in common was the short duration of the soft X-ray flares. The Cane *et al.* SEP events were selected on the basis of detection of >3 MeV electrons in comparison with the $\lesssim$100 keV electrons typically observed for ^{3}He-rich SEP events.

In an attempt to separate the energetic impulsive SEP events of Cane *et al.* (1986) from the ^{3}He-rich events of Reames, von Rosenvinge, & Lin (1985), Cliver (1996; see also Kallenrode, Cliver, & Wibberenz, 1992) proposed a modification of the two-class picture in which these two types of events were represented by separate columns. The standard two-class paradigm (Table 1) was well-entrenched by 1996, however, and the worry that the higher-energy events of Cane *et al.* were not qualitatively the same as the ^{3}He-rich events was insufficient cause to modify the picture. The case for revision would have to await additional rationale from SEP observations.

2.4. *Recent challenge to the two-class paradigm*

The inherent flaw in the two-class picture was revealed in 1999 with the publication of ACE observations of several large events observed early in cycle 23. Cohen *et al.* (1999) found that events on 6 November 1997, 2 May 1998, 6 May 1998, and 14 November 1998, had elemental compositions at >10 MeV that looked remarkably like those tabulated by Reames for impulsive events at lower energies (Figure 4a). Since all large events were assumed to be gradual, this posed a problem - a problem that was exacerbated when

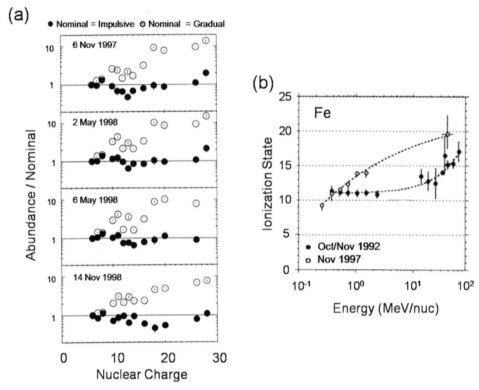

Figure 4. (a) Four of the early large ACE events had impulsive compositions at >10 MeV energies (from Cohen *et al.*, 1999) and (b) at least two of these (one shown) had high Fe charge states (from Mazur *et al.*, 1999).

SAMPEX measurements of charge states at ~40 MeV using the geomagnetic cutoff technique (Mazur *et al.*, 1999) revealed that two of the events (6 November 1997, 14 November 1998) had high Fe charge states (6 November shown in Figure 4b).

Attempts to accommodate the new observations in terms of the two-class picture of SEP events took two disparate paths. The first, after Cane and colleagues (Cane *et al.*, 2002, 2003, 2006), argued that the "impulsive" properties observed in the large "gradual" events were due to flare domination of the gradual events at high (>25 MeV) energies. Alternatively, Tylka and colleagues (Tylka *et al.*, 2005, 2006; Tylka & Lee, 2006) accounted for the enhanced Fe/O ratios and charge states in the events of Cohen *et al.* in terms of quasi-perpendicular shocks operating on a flare suprathermal seed population.

Cane *et al.* (2002) pointed out that strong type III bursts observed with the WAVES instrument on the Wind spacecraft provided a direct link between the flare and the observation of protons at 1 AU. [This argument is problematic because there is evidence that some type IIIs may actually originate in shocks.] For the shock-based solution, the key phenomenon is a coronal shock wave manifested by a type II burst in the decametric-hectometric (DH) range (Gopalswamy *et al.*, 2002; Cliver, Kahler, & Reames, 2004).

Cane *et al.* (2006) noted that events with high Fe/O ratios came from well-connected solar longitudes. This is consistent with a flare source since flares are generally thought to have a smaller "cone of SEP emission" than shocks. These authors linked well-connected events with low Fe/O ratios to strong interplanetary shocks and/or more complex SEP time-intensity profiles.

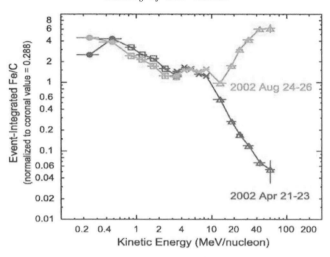

Figure 5. Examples of the two limiting cases of Fe/O variation with energy in large SEP events (from Tylka *et al.*, 2006).

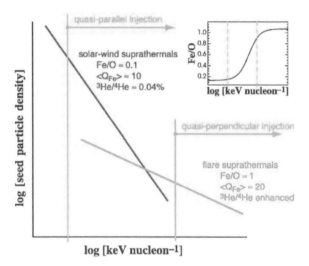

Figure 6. Schematic representation of the seed populations for shock-accelerated SEPs. Because of their higher energies, flare suprathermals (STs) are more accessible to quasi-perpendicular shocks. The inset shows how the Fe/O ratio in the seed population varies with energy (from Tylka *et al.*, 2005).

Tylka *et al.* (2005, 2006) examined the variation of the Fe/O ratio in SEP events with ion energy and noted two extreme possibilities for which Fe/O at ∼40 MeV varied by approximately two orders of magnitude (Figure 5). They noted that similar qualitative behavior was observed *in situ* for shock-associated SEP events observed at 1 AU - an example of a *Universal Heliophysical Process* - and that in that case the difference could be ascribed to shock geometry, with events with increasing Fe/O ratio with energy being linked to quasi-perpendicular shocks. They also noted that quasi-parallel shocks can operate on relatively low energy seed particles such as coronal or solar wind suprathermals while quasi-perpendicular shocks have a higher injection energy requirement, favoring flare suprathermals. Thus, given the seed particle populations shown schematically in Figure 6, quasi-perpendicular shocks are more likely to produce SEP events with enhanced Fe/O while SEP events from quasi-parallel shocks will tend to have low Fe/O ratios. The various combinations of shock geometry and seed particle population variability account for the wide variation observed in SEP composition and charge states for large events. With their shock formulation, Tylka & Lee (2006) were able to reproduce the organization of SEP elemental abundances by charge/mass ratio discovered by Breneman & Stone (1985) and provide the first theoretical explanation for this effect.

3. Evidence in support of shock acceleration of >30 MeV SEPs

In this section I summarize recent work by myself and colleagues relating to the challenge posed by the observations of Cohen *et al.* (1999) and Mazur *et al.* (1999) and the differing explanations suggested by the Cane and Tylka camps.

3.1. *Associations of large SEP events with low-frequency radio bursts*

While metric type II bursts are a quasi-necessary condition for a large SEP event, they are not sufficient. The fact that many metric type IIs are not followed by SEPs at Earth raises a question about the role of type IIs in the cases where there is a linkage. Is there something special about the type IIs linked to high-energy (>20 MeV) SEP events at Earth? To address this question, Cliver, Kahler, & Reames (2004) considered DH type IIs (those observed between 14 - 1 MHz by the Wind/WAVES instrument). They found that from July 1996 - June 2001 metric type IIs with a DH counterpart were ~3.5 times more likely to have >20 MeV SEP association. This suggests simply that stronger shocks, i.e., those capable of persisting to ~3 solar radii, are more likely to produce a detectable >20 MeV SEP event at 1 AU. This correlation of SEP size with an indicator of shock strength provides general support for the shock scenario for large SEP events.

More recently, Cliver & Ling (2009) started with a sample of large favorably-located ~1 MHz (~7 Rs) type III bursts from 1997-2004 and examined their association with large [≥ 1 proton flux unit (pfu)] >30 MeV proton events. They found that to first order such bursts were associated with large >30 MeV SEP events only when they were accompanied by DH type II bursts. This indicates that strong shocks rather than strong flares are required for >30 MeV SEP acceleration at the Sun.

3.2. *Flare time scale and shock geometry*

Flare time-scale is an important parameter in the current two-class picture, with all gradual SEP events having flares with long (>1 hr) soft X-ray durations. Yet well-known intense events such as 3 June 1982 and 6 November 1997 were associated with short duration (<1 hour) flares. Is there a relationship between flare duration, shock geometry and SEP composition? We suggest the following link: Short flare time scale indicates that the eruption originates on compact spatial scales. Upon eruption, the CME undergoes rapid lateral expansion, driving a shock tangential to the solar surface with shock normal perpendicular to the radial magnetic field, i.e., a quasi-perpendicular shock, resulting in enhanced Fe/O and e/p ratios. More gradual eruptions originate on larger spatial scales with a slower rate of energy release and the resultant lateral expansion is less violent, suggesting a reduced role for quasi-perpendicular shocks. In the context of Tylka's shock model, it appears that the short duration flares are preferentially linked to quasi-perpendicular shocks. The cartoon in Figure 7 shows where quasi-parallel and quasi-perpendicular shock acceleration of SEPs might occur in relation to a CME. In this scenario, events like those observed by ACE in 1997-1998 (Cohen *et al.*, 1999) are explained in terms of quasi-perpendicular acceleration and, indeed, in the 6 November

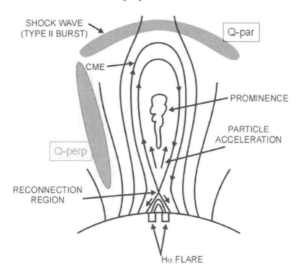

Figure 7. Schematic showing where quasi-parallel and quasi-perpendicular shock acceleration might occur in a solar eruption. In the standard CSHKP picture for eruptive flares (see Hudson & Cliver, 2001), reconnection in the wake of the CME gives rise to a two-ribbon flare. If these flare particles escape the CME, they could become seed particles for the shock.

1997 event, there is evidence (Zhang *et al.*, 2001) for lateral expansion of a CME that could drive such a shock. The incorporation of flare time scale as a parameter in the shock events finds support from the study of Nitta, Cliver, & Tylka (2003) that linked certain events with high Fe/O ratios, including the 6 November 1997 event, to "explosive" eruptions on the Sun, and others with low Fe/O ratios to less abrupt mass ejections.

Cliver (2008b) provides more general evidence that time scale is a crude organizer of Fe/O in large SEP events, with a tendency for shorter flares to be associated with SEP events with higher Fe/O, as well as e/p, ratios. The tendency for enhanced e/p ratios is consistent with the shock-based picture of Tylka and colleagues because on both theoretical (Lee, 2005) and observational (Tsurutani & Lin, 1985) grounds, we would expect quasi-perpendicular shocks (preferentially arising in short-duration flares) to be efficient accelerators of electrons in comparison with quasi-parallel shocks.

4. A proposed revision of the two-class taxonomy of SEP events

The observations of impulsive characteristics in certain large SEP events indicate the need for a modification of the standard two class picture (Cliver, 2008a). Following the lead of Tylka and his co-workers, we propose that the current gradual class be divided into two subclasses on the basis of shock geometry (and seed particles): quasi-perpendicular shocks (operating on flare seed particles) and quasi-parallel shocks (operating on coronal / solar wind suprathermals) (Table 2). This revision reflects the differences between the ^{3}He-rich events [now Flare] of Reames, von Rosenvinge, & Lin (1985) and the energetic impulsive events [Quasi-Perp] of Cane, McGuire, & von Rosenvinge (1986) which were combined in the original two-class picture (Reames, 1993). As a semantic change, we replace the main heading terms "Impulsive" and "Gradual" with "Flare" and "Shock", respectively. It is important to note that the two shock subclasses are extremes of a continuum. The notion of two basic types of SEP events remains intact. SEPs are accelerated in flares and at coronal shock waves. The supporting evidence for altering the classification scheme in this manner (see also Cliver, 2008b) is as follows: (1) The

Table 2. Revised SEP Event Classification[a]

	Flare	Shock Quasi-Perp	Quasi-Par
H Upper Limit[b]	$\sim$3 pr	$\sim$$10^3$ pr	$\sim$$10^4$ pr
e/p[b]	$\sim$$10^2$-$10^4$	$\sim$100	$\sim$50
^{3}He/^{4}He[c]	$\sim$$10^3$-$10^4$	$\sim$$10^1$-$10^2$	$\sim$1
Fe/O[d]	$\sim$8	$\sim$3	<1
Z(>50)/O[e]	$\sim$$10^2$-$10^3$	$\sim$$10^{-1}$-$10^1$	$\sim$$10^{-1}$-$10^1$
Ion Spectra[f]	—	Power-law	Exp. Rollover
QFe[g]	$\sim$20	$\sim$20	$\sim$11
SEP Duration	<1-20 hr	$\sim$1-3 days	$\sim$1-3 days
Longitude Cone[h]	<30-70°	$\sim$100°	$\sim$180°
Seed Particles	N/A	Flare STs	Coronal STs
Radio Type[i]	III	II	II
X-ray Duration	10-60 min	$\sim$1 hr	>1 hr
Coronagraph[j]	*	CME	CME
Solar Wind	—	IP Shock	IP Shock

Notes to Table 2:
[a] See Cliver (1996) for an early attempt to accommodate the quasi-perpendicular shock events.
[b] Cliver & Ling (2007), Cliver (2008b); >10 MeV protons and 0.5 MeV electrons.
[c] Relative to solar wind at $\sim$1 MeV/nuc. The "problem" large SEP events on 6 November 1997, 2 May 1998, and 6 May 1998 had enhanced ^{3}He/^{4}He ratios.
[d] Relative to corona at 5-12 MeV/nuc. For the Flare class, the energy range is 5-12 MeV/nuc (Reames, 1999). For the Shock class, the energy range is 30-40 MeV/nuc (Tylka et al., 2005).
[e] Relative to corona at 5-12 MeV/nuc; Reames and Ng (2004).
[f] For a study of ion spectra in Flare events, see Mason et al. (2002). Spectral shapes for Shock events are for 3-100 MeV/nuc (Tylka et al., 2005).
[g] Flare (<1 MeV/nuc; Klecker et al., 1984); Shock ($\sim$40 MeV/nuc; Mazur et al., 1999).
[h] Lin (1970); Reames, Stone, & Kallenrode (1991); Kallenrode, Cliver, & Wibberenz (1992).
[i] Defining radio type in low frequency range from 14-1 MHz.
[j] The larger flare events can have associated CMEs (Kahler, Reames, & Sheeley, 2001).

similarity between Fe/O variation with energy in quasi-perpendicular solar and inter-planetary shock events (Tylka et al., 2005); 2) New insight on the $\sim$20 year old puzzle of the Breneman & Stone (1985) Q/M fractionation effect provided by Tylka & Lee (2006); (3) Incorporation of flare-time scale into the division of the shock class into two subtypes (it is always a positive sign when a new view of nature encompasses key aspects of the superseded picture); (4) Mounting evidence that strong shocks, rather than strong flares, are a requirement for large >25 MeV SEP events (e.g., Cliver & Ling, 2009); and (5) Occam's razor - if the middle column in Table 2 were attributed to a different type of flare process, the strong (DH) shocks observed in these events would be extraneous.

Since this field began with question marks (Figure 1), it seems appropriate to end this review with some more. Much remains to be done, e.g., detailed mechanisms for both the flare and shock SEP acceleration processes. Sticking points, observations that do not fit neatly into the working hypothesis of Table 2, hold potential for insight and revision. To mention two: (1) the unusual long-duration pion-rich gamma-ray events observed by GRO Comptel (Kanbach, 1993; Ryan, 2000; see Cliver 2006); and (2) flattening spectra for high-energy electron events associated with short-duration flares (Moses et al., 1989). It remains to be seen how or if these phenomena will be incorporated into Table 2.

Acknowledgements

I thank N. Gopalswamy, D. Webb, and K. Shibata for organizing this stimulating meeting and asking me to speak on this topic. Figures 5 and 6 reproduced by permission of the AAS.

References

Axford, W. I. 1965, *Planet. Space Sci.*, 13, 1301
Breneman, H. H. & Stone, E. C. 1985, *ApJ (Lett.)*, 299, L57
Cane, H. V., McGuire, & von Rosenvinge, T. T. 1986, *ApJ*, 301, 448
Cane, H. V., Erickson, W. C., & Prestage, N. P. 2002, *J. Geophys. Res.*, 107(A10), CiteID 1315
Cane, H. V., von Rosenvinge, T. T., *et al.* 2003, *Geophys. Res. Lett.*, 30(12), CiteID 8017
Cane, H. V., Mewaldt, R. A., *et al.* 2006, *J. Geophys. Res.*, 111(A6), CiteID A06S90
Cliver, E. W. 1996, in *High Energy Solar Physics*, eds., R. Ramaty, N. Mandzhavidze, & X.-M. Hua, AIP, Woodbury, NY, vol. 374, p. 45
Cliver, E. W., Kahler, S. W., & Reames, D. V. 2004, *ApJ*, 605, 902
Cliver, E. W. 2006, *ApJ*, 639, 1206
Cliver, E. W. & Ling, A. G. 2007, *ApJ*, 658, 1349
Cliver, E. W. & Ling, A. G. 2009, *ApJ*, 690, 598
Cliver, E. W. 2008a, in *Particle Acceleration and Transport in the Heliosphere and Beyond*, eds., G. Li, Q. Hu, O. Verkhoglyadova, G. Zank, R. Lin, & J. Luhmann, AIP, Melville, NY, vol. 1039, p. 190
Cliver, E. W. 2008b, *Central European Astrophys. Bull.* (in press)
Cohen, C. M. S., *et al.* 1999, *Geophys. Res. Lett.*, 26, 2697
Forbush, S. E. 1946, *Phys. Rev.*, 70, 771
Forrest, D. J. & Chupp, E. L. 1983, *Nature*, 305, 291
Gopalswamy, N., *et al.* 2002, *ApJ (Lett.)*, 572, L103
Hsieh, K. C. & Simpson, J. A. 1970, *ApJ (Lett.)*, 162, L191
Hudson, H. S. & Cliver, E. W. 2001, *J. Geophys. Res.*, 106, 25199
Jokipii, J. R. & Parker, E. N. 1968, *Phys. Rev. Lett.*, 21, 44
Kahler, S. W., Hildner, E., & Van Hollebeke, M. A. I. 1978, *Solar Phys.*, 57, 429
Kahler, S., Reames, D. V., *et al.* 1985, *ApJ*, 290, 742
Kalher, S. W., Reames, D. V., & Sheeley, N. R., Jr. 2001, *ApJ*, 562, 558
Kallenrode, M.-B., Cliver, E. W., & Wibberenz, G. 1992, *ApJ*, 391, 370
Kanbach, G. O., *et al.* 1993, *Aston. Astrophys. Suppl.*, 97, 349
Klecker, B., Hovestadt, D., *et al.* 1984, *ApJ*, 281, 458
Lange, I. & Forbush, S. E. 1942a, *Terr. Mag.*, 47, 185
Lange, I. & Forbush, S. E. 1942b, *Terr. Mag.*, 47, 331
Lee, M. A. 2005, *ApJS*, 158, 38
Lin, R. P. 1970, *Solar Phys.*, 12, 266
Mason, G. M., Reames, D. V., *et al.* 1986, *ApJ*, 303, 849
Mason, G. M., *et al.* 2002, *ApJ*, 574, 1039
Mazur, J. E., Mason, G. M., *et al.* 1999, *Geophys. Res. Lett.*, 26, 173
Meyer, P., Parker, E. N., & Simpson, J. A. 1956, *Phys. Rev.*, 104, 768
Moses, D., Dröge, W., Meyer, P., & Evenson, P. 1989, *ApJ*, 346, 523
Nitta, N. V., Cliver, E. W., & Tylka, A. J. 2003, *ApJ (Lett.)*, 586, L103
Ramaty, R., Murphy, R. J., & Dermer, C. D. 1987, *ApJ (Lett.)*, 216, L41
Reames, D. V., von Rosenvinge, T. T., & Lin, R. P. 1985, *ApJ*, 292, 716
Reames, D. V., Stone, R. G., & Kallenrode, M.-B. 1991, *ApJ*, 380, 287
Reames, D. V. 1993, *Adv. Sp. Res.*, 13(9), 331
Reames, D. V. 1999, *Space Sci. Revs*, 90, 413
Reid, G. C. 1964, *J. Geophys. Res.*, 69, 2659
Ryan, J. M. 2000, *Space Sci. Revs*, 93, 581

Tsurutani, B. T., and Lin, R. P. 1985, *J. Geophys. Res.*, 90, 1

Tylka, A. J., *et al.* 2005, *ApJ*, 625, 474

Tylka, A. J., *et al.* 2006, *ApJS*, 164, 536

Tylka, A. J. & Lee, M. A. 2006, *ApJ*, 646, 1319

Wild, J. P., Smerd, S. F., & Weiss, A. A. 1963, *ARAA*, 1, 291

Zhang, J., Dere, K. P., Howard, R. A., Kundu, M. R., & White, S. M. 2001, *ApJ*, 561, 396

Discussion

SPANGLER: One would think that a CME would generate a quasi-parallel shock (if it expands radially). Are the quasi-perpendicular shocks you mention attributed to fast, transverse expansion of the CME initiation?

CLIVER: Yes, such lateral expansion was observed in the LASCO C1 coronagraph.

FISK: Recent works on suprathermal tails in the solar wind have shown that the tails are enriched in ^{3}He + Fe. How would your paradigm change if this is a seed population for enhanced ^{3}He + Fe in CME-driving solar flare events?

CLIVER: It may improve matters as there is some question that there are sufficient flare seed particles to account for the quasi-perpendicular events.

MELNIKOV: On your last figure you showed a nice correlation between fluxes of mildly relativistic electrons and energetic protons. This means that both populations are accelerated in the same process, and opens an opportunity for diagnostics of proton fluxes using the radio emission generated by those electrons in the corona. For example decameter radio emission would be a good indicator of such particle acceleration and propagation through the high corona. Do you know some studies about this kind of radio diagnostic?

CLIVER: Cane *et al.* (2002) considered the decametric range. But the emission in this frequency range is from decidedly much lower energy electrons.

MACDOWALL: Is it correct that you have not used metric-decametric radio data or 1–10 MHZ type III burst complexity when identifying type III events for your 2009 paper?

CLIVER: Yes, in part because there is at present no objective definition of a complex type III burst. Even then I would expect such bursts to be highly associated with DH type II bursts.

Universal Heliophysical Processes
Proceedings IAU Symposium No. 257, 2008
N. Gopalswamy & D.F. Webb, eds.

© 2009 International Astronomical Union
doi:10.1017/S1743921309029640

Particle acceleration and turbulence transport in heliospheric plasmas

Rami Vainio

Department of Physics, P.O.B. 64, FI-00014 University of Helsinki, Finland
email: `rami.vainio@helsinki.fi`

Abstract. Plasma turbulence at various length scales affects practically all mechanisms proposed to be responsible for particle acceleration in the heliosphere. In this paper, we concentrate on providing a synthesis of some recent efforts to understand particle acceleration in the solar corona and inner heliosphere. Acceleration at coronal and interplanetary shock waves driven by coronal mass ejections (CMEs) is the most viable mechanism for producing large gradual solar energetic particle (SEP) events, whereas particle acceleration in impulsive flares is assumed to be responsible for the generation of smaller impulsive SEP events. Impulsive events show enhanced abundances of ^{3}He and heavy ions over the gradual SEP events. Gradual events often show charge states consistent with acceleration of ions in a dilute plasma at 1–2 MK temperature, while impulsive events have higher charge states. The division of SEP events to gradual and impulsive has been challenged by the discovery of events, which show intensity-vs.-time profiles typical for gradual events but, especially at the highest energies (above 10 MeV/nucl), abundances and charge states more typical of impulsive events. Although a direct flare component cannot be ruled out, we find that particle acceleration at quasi-perpendicular shocks in the low corona also offer a plausible explanation for the hybrid events. By carefully modeling shock acceleration and coronal turbulence and its modification by the accelerated particles, a consistent picture of gradual events thus emerges from the shock acceleration hypothesis.

Keywords. acceleration of particles, instabilities, shock waves, turbulence, waves, Sun: coronal mass ejections (CMEs), Sun: flares, Sun: particle emission

1. Introduction

Plasma turbulence at various length scales affects practically all particle acceleration mechanisms proposed to be responsible for particle acceleration in the heliosphere. Fluctuating electromagnetic fields interact with charged particles transmitting energy and momentum between different particle populations. The turbulent energy itself can act as the source of energy in the acceleration process, like in the stochastic acceleration mechanism. Alternatively, turbulence can transmit the bulk kinetic energy of the system to the accelerated particles, like in the diffusive shock acceleration (DSA) mechanism. Finally, particle confinement in the acceleration region, whatever the mechanism, will be affected by particle transport in turbulent electromagnetic fields. Thus, understanding the properties and evolution of plasma turbulence is key to development of any particle acceleration models in collisionless plasmas.

According to the present paradigm, coronal mass ejection (CME) -driven shocks are the source of solar energetic particles (SEPs) in large gradual SEP events (e.g., Reames 1999). On the other hand, particle acceleration in impulsive flares is assumed to be responsible for the smaller impulsive SEP events, which show enhanced abundances of ^{3}He and heavy ions over the coronal ones (determined from gradual SEP events) (e.g., Reames 1999). Well below MeV/nucleon energies, gradual events show charge states consistent with acceleration of ions from a pool of seed particles at coronal or solar-wind temperature, while impulsive events have higher charge states indicating a higher electron temperature in the source plasma (e.g., Klecker *et al.* 2006). At higher energies, both impulsive and gradual events show increasing charge states. This is consistent with higher temperature of the seed population of particles accelerated to the highest energies and/or proton impact ionization (Kocharov *et al.* 2000) in dense plasma during the particle acceleration process.

The clear-cut division of SEP events to shock-accelerated gradual and flare-accelerated impulsive events has been blurred in recent years by the discovery of hybrid events (e.g., Kocharov & Torsti 2002). It has been found that many events show intensity–time profiles typical for gradual events but, especially at the highest energies (above 10 MeV/nucl), abundances and charge states more typical of impulsive events (Tylka *et al.* 2005). In addition to the apparent interpretation (e.g., Cane *et al.* 2006) that these events are superpositions of flare and shock-accelerated populations with flare acceleration extending to higher energies, such events have been explained by shock acceleration of a seed population containing flare material (Tylka *et al.* 2005; Tylka & Lee 2006). Note that an increase in the ionic charge states as a function of energy is consistent with both scenarios.

In this paper we will concentrate on describing the effects of coronal and heliospheric turbulence on particle acceleration in coronal/interplanetary shocks driven by CMEs. We will show, based on recent empirical and modeling results, that a consistent picture of gradual events emerges from the shock-acceleration hypothesis.

2. Heliospheric turbulence and particle transport

Turbulence in the solar wind plasma and interplanetary magnetic field (IMF) and their relation to heliospheric particle transport has been studied since the sixties (e.g., Jokipii 1966; Coleman 1968; Jokipii & Coleman 1968). The frequency power spectrum of the magnetic fluctuations can be represented as a power law, $P \propto f^{-q}$, over large ranges of frequency. It shows at least three ranges with different values of the spectral index: (i) an energy-containing range at low frequencies (below $\sim 10^{-5}$ Hz at 1 AU), where the power-law spectral index of the fluctuations is about or somewhat below unity; (ii) an inertial range at intermediate frequencies (between $\sim 10^{-5}$ Hz and ~ 1 Hz at 1 AU) where the spectral index is consistent with the Kolmogorov value of 5/3; and (iii) a dissipation range at high frequencies (above ~ 1 Hz at 1 AU), where the spectral index is >2, and variable. For a recent extensive review on solar wind turbulence, see the paper by Bruno & Carbone (2005).

According to the present understanding, the IMF fluctuations can be relatively well described by a two-component model consisting of (i) a two-dimensional (2D) component, in which both the fluctuating magnetic field, $\delta \vec{B}$, and the wave vector, $\vec{k}$, lie in the plane perpendicular to the mean magnetic field; and (ii) a slab component, in which the wave vector is parallel to the mean field (e.g., Matthaeus *et al.* 1990; Bieber *et al.* 1996). The solenoidal condition, $\vec{k} \cdot \delta \vec{B} = 0$, is also fulfilled in both cases. According to observations at 1 AU, the 2D component carries about 80% of the power in inertial range (Bieber *et al.* 1996). Particle transport in such a turbulence has been intensively studied over the last decade (e.g., Bieber *et al.* 1996; Dröge 2003; Bieber *et al.* 2004; Shalchi *et al.* 2004). The main effect of the 2D component is in the perpendicular transport, whereas the slab component seems mainly responsible for the pitch-angle diffusion. In the estimates below, we will simply assume that the slab-mode turbulence is solely responsible for pitch-angle

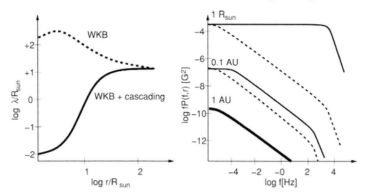

Figure 1. *Left*: A sketch of the the mean free path of 10-MeV protons, backward extrapolated from 1 AU using slab-mode wave intensity obtained by WKB-transported Alfvén waves (dashed curve) and cascading Alfvén waves (solid curve).
Right: A sketch of the unfolding of the cascading Alfvén wave spectrum from 1 AU back to the Sun (solid curves). The dashed curves give the WKB-transported spectra. (At 1 AU, the spectra are assumed to coincide.)

diffusion and, thus, particle diffusion along the mean magnetic field, and that the diffusion coefficient is well approximated by the standard quasi-linear theory at energies of interest (Jokipii 1966; Bieber *et al.* 1996). In this case, the pitch-angle diffusion coefficient can be approximated by

$$D_{\mu\mu} = \tfrac{1}{2}\pi\Omega(1-\mu^2)\frac{|k_{\mathrm{res}}|I_{\mathrm{slab}}(k_{\mathrm{res}})}{B^2}, \qquad (2.1)$$

where $I_{\mathrm{slab}}(k)$ is the intensity of the slab-mode turbulence, $k_{\mathrm{res}} = \Omega/v\mu$ is the cyclotron resonant wavenumber, Ω is the gyrofrequency of the particle, v the particle speed and μ the pitch-angle cosine. The pitch-angle diffusion coefficient gives the scattering mean free path through the well-known expression

$$\lambda = \frac{3v}{8}\int_{-1}^{+1}\frac{(1-\mu^2)^2}{D_{\mu\mu}}\,\mathrm{d}\mu. \qquad (2.2)$$

Observations in the fast solar wind indicate the presence of predominantly outward-propagating Alfvén waves, which seem to be undergoing cascading (Marsch & Tu 1990; see also Bruno & Carbone 2005 and references therein); the spectral index of the waves at low frequencies is about 1, and steepens into about 5/3 at a break point frequency, which moves towards lower values as the distance from the Sun increases. (Note that frequency in spacecraft frame is related to the wavevector approximately as $f = \vec{k}\cdot\vec{V}_{\mathrm{sw}}/2\pi$, where $\vec{V}_{\mathrm{sw}}$ is the solar-wind velocity.) Assuming that the slab-mode turbulence in the inner heliosphere consists of such cascading Alfvén waves, it is possible to estimate the radial evolution of the scattering mean free path of SEPs from the Sun to 1 AU. Vainio *et al.* (2003) and Vainio (2006) used a model of WKB-transported Alfvén waves appended with phenomenological cascading to estimate the mean free path between the Sun and 1 AU. With reasonable values of the interplanetary mean free path ($\sim$0.1 AU) and an assumed f^{-1} spectral form of the waves at the Sun, the backwards extrapolated coronal mean free path of 10-MeV protons is of the order of 0.01 solar radii, which is more than four orders of magnitude below the value obtained by backward extrapolation using only WKB transport (see Fig. 1). Note that the amplitude of the adopted solar f^{-1} spectrum of Alfvén waves is still well below the spectra employed by models of coronal cyclotron heating by high-frequency Alfvén waves (see, e.g., Vainio & Laitinen

2001). Thus, this level of high-frequency fluctuations does not violate any observational constraints concerning coronal Alfvén waves. We will next investigate shock acceleration in an ambient coronal turbulence consistent with the cascading Alfvén wave scenario.

3. Shock acceleration and turbulence

Particle acceleration by CME-driven shocks in turbulent coronal plasma occurs probably via the DSA mechanism (e.g., Bell 1978). In this model, a particle gains energy by repeatedly crossing the shock front and feeling the compression of the flow through its interaction with the small-scale irregularities of the magnetic field. The standard steady-state theory predicts an accelerated particle (species s) distribution at the shock of a power-law form,

$$f_{\text{sh}}^{(s)}(p) = \frac{\sigma \epsilon_s n_{s1}}{4\pi p_{s0}^3} \left(\frac{p}{p_{s0}} \right)^{-\sigma}, \tag{3.1}$$

where p is the particle momentum, $\sigma = 3X/(X-1)$ is the spectral index which depends only on the compression ratio $X = u_{1n}/u_{2n} = \rho_2/\rho_1$ of the shock, n_{s1} is the number density of the s'th species in the upstream region, ϵ_s is fraction of particles (of species s) injected into the acceleration process at the injection momentum p_{s0}, and $u_{1n[2n]}$ and $\rho_{1[2]}$ are the shock-frame plasma flow speed in the shock normal direction and mass density in the upstream [downstream] region. In low-Mach number shocks, the compression ratio X should be replaced by the scattering center compression ratio (Vainio & Schlickeiser 1998, 1999), $X_{\text{sc}} = W_{1n}/W_{2n}$, where the wave speeds $W_n = u_n + \langle v_{\phi,n} \rangle$ contain the effect of the average phase speed, v_ϕ, of the waves with respect to the plasma.

The steady-state assumption is valid only up to energies limited by the available acceleration time. The time scale of particle acceleration in DSA is given by (Drury 1983)

$$\frac{p}{\dot{p}} = \sigma \left(\frac{\kappa_{nn,1}(p)}{u_{1n}^2} + \frac{\kappa_{nn,2}(p)}{X u_{2n}^2} \right), \tag{3.2}$$

where $\kappa_{nn} = \kappa_\parallel \cos^2 \theta_n + \kappa_\perp \sin^2 \theta_n$ is the spatial diffusion coefficient in the shock normal direction, θ_n is the angle between the shock normal and the magnetic field, and $\kappa_{\parallel[\perp]}$ is the spatial diffusion coefficient parallel [perpendicular] to the mean magnetic field. Since shocks amplify turbulence very efficiently, usually the second term in Eq. (3.2) is neglected. Furthermore, in a weakly turbulent plasma with $r_{\text{L}} \ll \lambda_\parallel$, the perpendicular diffusion coefficient is smaller than the parallel one, so if the shock is oblique with, say, $\theta_n < 70°$, we can neglect perpendicular transport and obtain $\kappa_{nn} \simeq \kappa_\parallel \cos^2 \theta_n$ with $\kappa_\parallel = \frac{1}{3} v \lambda$. Thus,

$$\frac{p}{\dot{p}} \simeq \frac{\sigma v \lambda_1(p) \cos^2 \theta_{n,1}}{3 u_{1n}^2}. \tag{3.3}$$

The cut-off momentum in the spectrum, $p_{\text{c}}(t)$, can now be estimated by equating the acceleration time scale with the time available for acceleration, i.e.,

$$\frac{\sigma v_{\text{c}} \lambda_1(p_{\text{c}}) \cos^2 \theta_{n,1}}{3 u_{1n}^2} \simeq \frac{\Delta s_\parallel(t)}{u_{1,n}/\cos \theta_{n,1}} \quad \Rightarrow \quad \lambda_1(p_{\text{c}}) = \frac{3 u_{1,n} \Delta s_\parallel(t)}{\sigma v_{\text{c}} \cos \theta_{n,1}}, \tag{3.4}$$

where $\Delta s_\parallel(t)$ is the distance swept by the shock along a given magnetic field line from the time of first intersection to time t. Here it has been assumed that the shock parameters and λ are not functions of time. If this is not the case, the differential equation (3.3) has to be properly solved, but the form (3.4) still serves as a convenient order-of magnitude approximation for the mean free path required for particle acceleration up to a given cut-off momentum, if the time-dependent parameters are replaced by typical values.

Plugging in the numbers as $\Delta s_\parallel = 1\, R_\odot$, $u_{1,n} = 1000\ \mathrm{km\,s^{-1}}$, $\sigma = 4$ and $\theta_{n,1} = 70°$, and using the coronal mean free path deduced by backward extrapolation from 1 AU produces proton cut-off energies of the order of $E_c \sim 10$ MeV (Vainio 2006). Some tens of MeVs may be obtained by increasing the shock speed and/or the shock normal angle. One should, however, bear in mind, that the neglect of perpendicular transport in the estimates above forbids their use at nearly perpendicular shocks.

If, instead, we take the shock normal angle to be close to $90°$, we might use $\kappa_{nn} \simeq \kappa_\perp$. As an increase in the level of turbulence typically increases perpendicular diffusion, we should now neglect the first term in Eq. (3.2) rather than the second. Assuming that the downstream region is very turbulent, we may use the Bohm limit of the diffusion coefficient, $\kappa_\perp \simeq \kappa_B = \frac{1}{3}\gamma v^2/\Omega_0$ with Ω_0 being the non-relativistic gyrofrequency and γ the Lorentz factor of the particle, and obtain

$$\frac{p}{\dot{p}} \simeq \frac{\sigma\gamma v^2}{3X u_{2n}^2 \Omega_{0,2}} = \frac{\sigma\gamma v^2}{3u_{1n}^2 \Omega_{0,1}} \tag{3.5}$$

where $\Omega_{0,1[0,2]}$ is the non-relativistic gyrofrequency in the upstream [downstream] region of the shock, and $\Omega_{0,2} = X\Omega_{0,1}$ for a perpendicular shock. Thus,

$$dE = v\,dp \simeq (3/\sigma)m\,u_{1n}^2 \Omega_{0,1}\,dt, \tag{3.6}$$

and integrating this from $t = 0$ to the available acceleration time $\tau_\perp$ in the perpendicular shock gives

$$E_c \simeq (3/\sigma)m\,u_{1n}^2 \Omega_{0,1}\tau_\perp. \tag{3.7}$$

Using $\tau_\perp = 100$ s, $u_{1,n} = 1000\ \mathrm{km\,s^{-1}}$, $\sigma = 6$ (i.e., $X = 2$), and $\Omega_{0,1} = 10^4\ \mathrm{s^{-1}}$ (a proton in a 1-gauss field), one obtains $E_c \simeq 5$ GeV for protons. Thus, in a turbulent perpendicular coronal shock particles may be accelerated up to relativistic energies in tens of seconds. Note that the maximum energy per nucleon in this estimate scales like $E_c/A \propto Q/A$, where Q and A are the ionic charge state and the mass number of the particle.

If perpendicular shocks can easily accelerate particles to relativistic energies in the corona, then why do we observe such high energies from the Sun very rarely, during cosmic-ray Ground Level Enhancements (GLEs) only? Of course, the estimate above assumes that a coronal shock may exist in a region of relatively strong perpendicular field for a sufficient amount of time, which is not necessarily easily satisfied in the case of CME-driven shocks. In addition, particles may be able to escape the system along the field lines, if the size of the perpendicular region of the shock is very limited. Our estimate also assumes that particle motion is well approximated by diffusion, at least in the direction perpendicular to the magnetic field. This, however, is not automatically satisfied at all energies of interest. The parallel and perpendicular transport are coupled to each other, and all turbulence scenarios do not lead to efficient diffusion. This is important especially at the lowest energies, where inefficient diffusion transverse to the field may lead to an injection problem: low-energy ions incident on the shock from the upstream region may not be able to return to the shock after being transmitted to the downstream side. Thus, the amount of accelerated particles remains very small unless particles with high initial speeds are available for the shock in the upstream region. Furthermore, while the compression of the perpendicular fields at the shock dramatically decreases the diffusion coefficient in a parallel shock (Vainio & Schlickeiser 1998, 1999), this is not so obvious in a perpendicular shock: there, by compressing the perpendicular field components, the shock compresses the mean field, the possible compressional turbulence components (with $\delta\vec{B} \parallel \vec{B}$), and the transverse turbulent field component in the shock plane, but the field component in the shock normal direction remains uncompressed. This means that at least the field-line random walk in the shock normal direction is actually suppressed. The turbulent component in the shock plane may, however, be strong enough to scatter the particle from one field line to another.

In fact, numerical simulations of particle acceleration in quasi-perpendicular shocks (Giacalone 2005) show that if the upstream region of the shock is turbulent enough, there is no injection problem and perpendicular shocks accelerate particles rapidly into high energies. For an upstream fluctuation amplitude of $(\Delta B/B)_1^2 = 0.1$, Giacalone (2005) obtains a cutoff energy of about $E_c \sim (5 \cdot 10^5)mu_{1n}^2$ for a simulation run with $X = 4$ and $\tau_\perp = 5 \cdot 10^4 \Omega_{0,1}^{-1}$, which is consistent with $\kappa_\perp \sim 0.1 \kappa_B$ rather than Bohm diffusion. Thus, even with a relatively high value of the upstream turbulence amplitude (compared to the cases we are considering for corona) and further compression of the transverse fields at the shock, the transport in the downstream region is not governed by Bohm diffusion. The injection efficiency in Giacalone's (2005) simulation with $(\Delta B/B)_1^2 = 0.1$ is suppressed by an order of magnitude relative to the case with $(\Delta B/B)_1^2 = 1$. Thus, with coronal field fluctuations obtained from our turbulence transport model, i.e., $(\Delta B/B)_1^2 \lesssim 0.001$, the injection problem in quasi-perpendicular shocks may well be a serious one in perpendicular coronal shocks without a preaccelerated seed particle population. Thus, we may infer that particle acceleration at coronal shocks to energies in the GeV range should be limited to rare occasions, as observed.

4. Self-consistent modeling of large gradual SEP events

The geometry of the CMEs implies that most of the time during their propagation through the outer corona they drive oblique or quasi-parallel shocks. However, during large gradual SEP events, shocks seem to be efficiently accelerating particles up to hundreds of MeVs at heights above the low corona. Thus, the turbulence responsible for particle acceleration in these events is not likely to be the rather weak ambient turbulence in the corona.

Already Bell (1978) pointed out that DSA does not have to rely on the ambient turbulence to scatter the particles around the shock. The accelerated particles streaming away from the shock in the upstream plasma frame drive the outward-propagating Alfvén waves unstable. These waves bootstrap the diffusive acceleration process and lead to rapid acceleration at the shock. The idea was applied to traveling interplanetary shocks by Lee (1983), and his steady-state theory has survived, at least in a semi-quantitative sense, the test against observations at 1 AU (Kennel et al. 1986). Lee's steady-state model, appended by the assumption that the acceleration time scale is equal to the dynamic scale of shock propagation, was applied to coronal/interplanetary shock acceleration by Zank et al. (2000) and Rice et al. (2003). Using this acceleration model and an assumption that an ad-hoc fraction of particle flux at the shock will escape from the shock complex towards the upstream region, and following the propagation of the escaping particles in the ambient medium, Li et al. (2003) computed the time-intensity profiles of gradual events at 1 AU. In the coronal shock acceleration problem, however, a self-consistent model of particle escape from the self-generated turbulent trap is needed, before the model can be considered fully adequate. The modeling of these authors, however, convincingly demonstrated that CME-driven shocks can accelerate ions up to hundreds of MeVs and beyond.

The first analytical model to quantitatively address the escape from self-generated waves was the one by Vainio (2003). He considered time dependent excitation of waves and concluded that a relatively large fluence of energetic protons can actually escape from the corona before the waves have grown to substantial amplitudes. In Vainio's (2003) model, small gradual events at MeV energies, and practically all events at relativistic energies have fluences that do not meet the threshold for efficient wave generation. However, Vainio (2003) used an artificially sharpened resonance condition, $k_{\rm res} = \Omega/v$, in his calculations, which meant that high-rigidity particles could not resonate with waves generated by lower-rigidity ones, which is possible if the full quasi-linear resonance condition is employed. Another analytical model that treats the escape from the coronal/interplanetary shock in a consistent manner was developed by Lee (2005). His model is quasi-stationary, but it includes adiabatic focusing in the upstream region, which is able to drive the particles away from the shock allowing them to escape.

More recently, two numerical simulation models combine the effects of wave growth, diffusive acceleration at the shock and focused particle transport in self-generated turbulence. Vainio & Laitinen (2007, 2008) developed a Monte Carlo simulation model tracing individual particles in the upstream region of a shock under the influence of turbulent fields amplified by the particles themselves. The particles were being accelerated by a propagating parallel coronal shock, treated as a boundary condition in the simulation. This is equivalent with the assumption that particle scattering in the downstream region is very efficient, so that the contribution from the residence time in the downstream region to the acceleration time scale can be neglected. The particle scattering rates and wave growth rates in the upstream region were taken from quasi-linear theory but using the sharpened resonance condition, $k_{\rm res} = \Omega/v$. The results indicated that coronal shocks would have no problem in accelerating particles up to hundreds of MeVs in a few minutes even if the injection efficiency of the shock was taken to be rather low to keep the upstream wave intensities in the linear regime. The simulation model agreed with the predictions of the steady-state theory of Bell (1978) in terms of the spectrum of waves and particles at the shock and with the prediction of Vainio (2003) about the fluence of the particle population escaping upstream from the shock before the steady-state wave amplitudes are achieved.

Another simulation model by Ng & Reames (2008) utilizes a different numerical method (finite difference method) and employs many complications of particle and turbulence transport neglected in the model of Vainio & Laitinen (2007). It uses the full resonance condition, includes self-consistent turbulence transmission at the shock (Vainio & Schlickeiser 1999), and also follows particle propagation and wave growth in the downstream region. The main result of previous studies, i.e., efficient acceleration of particles up to hundreds of MeVs in some minutes in parallel CME-driven shocks is recovered also in this model. This implies that the simplified simulation model of Vainio & Laitinen (2007) probably captures the main physical ingredients of the theory.

How would the results of the simulation models, obtained for strictly parallel shocks, change with shock obliquity? Eq. (3.2) indicates that a shock propagating a fixed distance $ds_{\parallel} = u_{n,1}\, dt/\cos\theta_{n,1}$ along the field accelerates particles at a rate $dp/ds_{\parallel} \propto (\lambda\cos\theta_{n,1})^{-1} \propto I(\Omega/v)/\cos\theta_{n,1}$. Theory predicts that the growth rate and, hence, the steady-state intensity of the waves is $I(\Omega/v) \propto f_{\rm sh}^{(\rm p)}(v;\theta_{n,1})\cos\theta_{n,1}$. Thus, assuming a slow rate of change for $\theta_{n,1}$, we can write $dp/ds_{\parallel} \propto f_{\rm sh}^{(\rm p)}(v;\theta_{n,1})$ for the dependence of the acceleration rate on the injection efficiency. There is no explicit dependence of the proton distribution function (3.1) on the shock obliquity, so at least in the quasi-parallel regime (say, $\theta_{n,1} \lesssim 30°$) it should be rather insensitive to $\theta_{n,1}$. In the intermediate obliquity regime (say, $30° \lesssim \theta_{n,1} \lesssim 80°$), simple estimates of the behavior are difficult to make, because the result depends on many details of the shock and the incident seed particle distribution. At nearly perpendicular shocks (say, $\theta_{n,1} \gtrsim 80°$), ion injection is quenched and the number of accelerated particles at a given energy is probably strongly decreasing as a function of the shock normal angle. However, in these shocks we have to include perpendicular transport in the estimate of the acceleration rate, as discussed above.

5. Gradual events with impulsive composition signatures

As discussed above, a class of gradual SEP events shows impulsive-event like composition at high energies. In these events, the iron-to-oxygen abundance ratio first starts to decrease around 1 MeV/nucl, but later increases to values resembling impulsive-flare abundances at tens of MeV/nucl. Tylka et al. (2005) suggested that these compositional signatures were due to diffusive acceleration at quasi-perpendicular coronal shocks of seed populations containing pre-accelerated flare material. Proposing that quasi-perpendicular shocks accelerate particles to higher energies than quasi-parallel ones, and assuming that it would be easier for a coronal shock to inject suprathermal flare ions than quasi-thermal coronal material, Tylka & Lee (2006) developed an analytical model of this scenario. Their model showed extremely good coincidence with the observational results. However, as the model contained several ad-hoc assumptions about the form of the accelerated particle spectra and injection efficiencies for different species and shock obliquities, Sandroos & Vainio (2007) performed test-particle simulations in such a scenario to verify the assumptions. The simulations employed an expanding spherical shock front centered in the low corona sweeping a radial (from the center of the Sun) coronal magnetic field line containing a seed population that was a mixture of low-energy ions with coronal composition and higher-energy ions with impulsive composition. The turbulence spectrum was assumed to be of the form $1/f$ with amplitudes consistent with those extrapolated backwards from the solar wind. The simulation results quantitatively confirmed the results of the model of Tylka & Lee (2006).

The models of Tylka & Lee (2006) and Sandroos & Vainio (2007) are consistent with the assumption that that the mean free path of the accelerated particles is proportional to particle rigidity up to the highest energies in the system. If the upstream turbulence was fully generated by protons accelerated at the shock, this would not be the case. Instead, there would be a low-wavenumber cutoff in the turbulence spectrum at $k_0 \sim m_p \Omega_{p,o}/p_{p,c}$, where $p_{p,c}$ is the cutoff momentum in the proton spectrum. Heavy ions (of species i) resonating with this wavenumber have momenta $p_{i,c} \sim m_i \Omega_{i,c}/k_0 = Q_i p_{p,c}$. For non-relativistic particles, this implies $E_{i,c}/A_i = p_{i,c}^2/(2A_i^2 m_p) = (Q_i/A_i)^2 E_{p,c}$, which becomes an upper limit of the cutoff energy. Recall that shock acceleration with $\kappa \propto vp$ over a finite time yields $E_{i,c}/A_i = (Q_i/A_i)E_{p,c}$, which is the relation adopted by Tylka & Lee (2006) and obtained by Sandroos & Vainio (2007) in their simulation model. Furthermore, as deduced above, self-generated waves have lower intensities in oblique shocks, and the cutoff energies as a function of shock obliquity are most probably decreasing, not increasing as in the case of an external turbulence. Since the model of selective shock acceleration is relying on particles at the highest energies being accelerated by quasi-perpendicular shocks (requiring higher injection energies), it is inconsistent with turbulence being self-generated, or at least with this playing any role in the acceleration process.

If self-generated waves cannot produce the impulsive composition signatures at high energies, we can study the proton fluences of the events to find out if they suggest that wave growth would be important in these events. Tylka et al. (2005) analyzed 30–40 MeV/nucl iron-to-oxygen ratio as a function of >30 MeV proton fluence for 44 gradual

SEP events of the solar cycle 23. A clear organization of the events is evident. All events with enhanced high-energy iron-to-oxygen ratio have small integral proton fluences, below $F \sim 10^7$ cm^{-2} sr^{-1}. We can use the model of Vainio (2003) to estimate the threshold fluence for efficient wave generation in the corona. This requires $E \, dN/dE \sim 10^{33}$ protons to be injected into the flux tube per steradian at the solar surface at the resonant energies. This number translates into a time-integrated net flux per unit logarithmic energy range at 1 AU of $E \, dG_{\mathrm{thr}}/dE \sim 4 \cdot 10^6$ cm^{-2}. This quantity is related to the fluence in a unit logarithmic energy range, $E \, dF/dE$, by $E \, dG/dE = 4\pi\langle\mu\rangle E \, dF/dE$, where $\langle\mu\rangle$ is the average value of the pitch-angle cosine at 1 AU during the event in the considered energy channel. Thus, $E \, dG/dE = 4\pi\alpha\langle\mu\rangle F$, where α is the spectral index of the integral proton fluence assumed to be of form $F \propto E^{-\alpha}$. Noting that the first-order anisotropy $3\langle\mu\rangle \simeq \lambda/L$, where L is the focusing length ($L \sim 1$ AU at 1 AU), we find values of $E \, dG/dE \lesssim \alpha(\lambda/L) \, 4 \cdot 10^7$ cm^{-2} for events with composition anomalies in Tylka's (2005) sample. Note that for $\alpha(\lambda/L) \sim 0.1$, this estimate agrees with the threshold for wave growth, but reasonable values of the mean free path and spectral index may also produce estimates of the time-integrated flux extending up to an order of magnitude above the threshold. However, as the waves produced by 30-MeV protons resonate with iron ions of similar rigidity, the effects of these waves on iron would be most prominent at about an order of magnitude lower energies than the channel considered here. Thus, we may safely state that wave growth is unlikely to have significantly influenced turbulence responsible for iron acceleration at the highest energies in those gradual events showing composition anomalies at the highest energies. This strongly suggests that the shocks accelerating these ions are quasi-perpendicular.

6. Conclusions

We have reviewed some recent modeling efforts to understand particle acceleration in gradual SEP events assuming that they are accelerated by CME-driven shock waves. The following internally consistent picture of the acceleration process emerges:

• Particle acceleration in gradual events can be understood in terms of diffusive shock acceleration in the solar corona and interplanetary medium.

• Large gradual events, with proton fluences exceeding the wave-generation threshold at resonant wavenumbers, can be understood in terms of particle acceleration at shocks propagating through self-generated waves. These events show particle abundances consistent with high-rigidity particles being less effectively accelerated at the shock, i.e., abundances of low-charge-to-mass ratio ions decreasing as a function of energy. This is consistent with the bulk of the acceleration at these energies occurring at the quasi-parallel phase of the shock propagation.

• Smaller gradual events, with proton fluences below the wave-generation threshold, are accelerated in coronal shocks without self-generated turbulence. Extrapolations of turbulence levels from the solar wind to the corona imply that proton acceleration beyond 10 MeV in these events occurs in quasi-perpendicular shocks ($\theta_{n,1} > 70°$ for CME speeds of the order of ~ 1000 km s^{-1}). These events show both decreasing and increasing abundances of low-charge-to-mass ratio ions as a function of energy. As injection in a quasi-perpendicular shock propagating in weak turbulence requires high-velocity ions to be present in the upstream region, we attribute the variations in the high-energy abundance ratios to variations in the seed-particle composition: when iron-rich suprathermal material is present in the ambient plasma, iron-rich composition at the highest energies is obtained.

• GLEs can be understood either in terms of nearly perpendicular shock acceleration, where the acceleration time is governed by Bohm-like diffusion in the downstream region, or in terms of quasi-parallel shock acceleration by self-generated waves in very strong CME-driven shocks with high injection efficiency driving the ambient waves close to non-linear amplitudes. By studying the fluences and time-intensity profiles of the associated >100-MeV proton events it is possible to infer whether wave generation may bootstrap the acceleration to relativistic energies.

More work is still needed to develop simulation models that treat the injection and acceleration process self-consistently in terms of the local shock structure, global and local shock geometry, and waves and instabilities within the shock complex.

References

Bell, A. R. 1978, *MNRAS*, 182, 147

Bieber, J. W., Wanner, W., & Matthaeus, W. H. 1996, *J. Geophys. Res.*, 101, 2511

Bieber, J. W., Matthaeus, W. H., Shalchi, A., & Qin, G. 2004, *Geophys. Res. Lett.*, 31, L10805

Bruno, R. & Carbone, V. 2005, *Living Rev. Solar Phys.*, 2, 4. URL (cited on 3 Oct 2008):
    http://www.livingreviews.org/lrsp-2005-4

Cane, H. V., Mewaldt, R. A., Cohen, C. M. S., & von Rosenvinge, T. T. 2006, *J. Geophys. Res.*,
    111, A06S90

Coleman, P. J. 1968, *ApJ*, 153, 371

Dröge, W. 2003, *ApJ*, 589, 1027

Drury, L. O'C. 1983, *Rep. Prog. Phys.*, 46, 973

Giacalone, J. 2005, *ApJ*, 624, 765

Jokipii, J. R. 1966, *ApJ*, 146, 480

Jokipii, J. R. & Coleman, P. J. 1968, *J. Geophys. Res.*, 73, 5495

Kennel, C. F., Coroniti, F. V., Scarf, F. L., Livesey, W. A., Russell, C. T., & Smith, E. J. 1986,
    J. Geophys. Res., 91, 11 917

Klecker, B., Möbius, E., & Popecki, M. A. 2006, *Space Sci. Rev.*, 124, 289

Kocharov, L. & Torsti, J. 2002, *Sol. Phys.*, 207, 149

Kocharov, L., Kovaltsov, G. A., Torsti, J., & Ostryakov, V. M. 2000, *A&A*, 357, 716

Lee, M. A. 1983, *J. Geophys. Res.*, 88, 6109

Lee, M. A. 2005, *ApJS*, 158, 38

Li, G., Zank, G. P., & Rice, W. K. M. 2003, *J. Geophys. Res.*, 108, 1082

Matthaeus, W. H., Goldstein, M. L., & Roberts, D. A. 1990, *J. Geophys. res.*, 95, 20 673

Marsch, E. & Tu, C.-Y. 1990, *J. Geophys. Res.*, 95, 8211

Ng, C. K. & Reames, D. V. 2008, *ApJL*, 686, L123

Reames, D. V. 1999, *Space Sci. Rev.*, 90, 413

Rice, W. K. M., Zank, G. P., & Li, G. 2003, *J. Geophys. Res.*, 108, 1369

Sandroos, A. & Vainio, R. 2007, *ApJL*, 662, 127

Shalchi, A., Bieber, J. W., Matthaeus, W. H., & Qin, G. 2004, *ApJ*, 616, 617

Tylka, A. J. & Lee, M. A. 2006, *ApJ*, 646, 1319

Tylka, A. J., Cohen, C. M. S., Dietrich, W. F., Lee, M. A., Maclennan, C. G., Mewaldt, R. A.,
    Ng, C. K., & Reames, D. V. 2005, *ApJ*, 625, 474

Vainio, R. 2003, *A&A*, 406, 735

Vainio, R. 2006, in: N. Gopalswamy, R. Mewaldt & J. Torsti (eds.), *Solar Eruptions and Ener-
    getic Particles*, Geophys. Monograph Series 165 (Washington, DC: AGU), p. 253

Vainio, R. & Laitinen, T. 2001, *A&A*, 371, 738

Vainio, R. & Laitinen, T. 2007, *ApJ*, 658, 622

Vainio, R. & Laitinen, T. 2008, *J. Atm. Solar-Terr. Phys.*, 70, 467

Vainio, R. & Schlickeiser, R. 1998, *A&A*, 331, 793

Vainio, R. & Schlickeiser, R. 1999, *A&A*, 343, 303

Vainio, R., Laitinen, T., & Fichtner, H. 2003, *A&A*, 407, 713

Zank, G. P., Rice, W. K. M., & Wu, C. C. 2000, *J. Geophys. Res.*, 105, 25 079

Discussion

VLAHOS: In your model the presence of turbulence before the shock arrival is an essential parameter and in the solar wind this is true but inside the corona we have no idea if turbulence is always there.

VAINIO: For the selective acceleration process we indeed assume that particles are accelerated in ambient turbulence and the values for turbulence power are extrapolated with a transport model from measurements in solar wind. Of course, this is uncertain to some extent and we really don't know if the turbulence is there (in large events, our model does not need almost any ambient turbulence, because the waves are self-generated.)

TSAP: Why did you not consider the drift acceleration mechanism and the mechanism proposed by Sagdeev?

VAINIO: Our models for parallel shocks only employ pitch-angle diffusion, because in such shocks possible drifts do not lead to particle acceleration. In oblique shocks models, we compute particle trajectories in full, so drifts are included (Sagdeev's model, shock surfing, is, however, not included, because we do not employ any cross-shock potential in our shock model. We hope to address this in the future).

Universal Heliophysical Processes
Proceedings IAU Symposium No. 257, 2008
N. Gopalswamy & D.F. Webb, eds.

ⓒ 2009 International Astronomical Union
doi:10.1017/S1743921309029652

Cosmic ray modulation by corotating interaction regions

Jaša Čalogović[1], Bojan Vršnak[1], Manuela Temmer[2] and Astrid M. Veronig[2]

[1] Hvar Observatory, Faculty of Geodesy, Kačićeva 26, HR-10000 Zagreb, Croatia

[2] Institute of Physics, University of Graz, Universitätsplatz 5, A-8010 Graz, Austria
email: jcalogovic@geof.hr

Abstract. We analyzed the relationship between the ground-based modulation of cosmic rays (CR) and corotating interaction regions (CIRs). Daily averaged data from 8 different neutron monitor (NM) stations were used, covering rigidities from $R_c = 0 – 12.91$ GeV. The *in situ* solar wind data were taken from the Advanced Composition Explorer (ACE) database, whereas the coronal hole (CH) areas were derived from the Solar X-Ray Imager onboard GOES-12. For the analysis we have chosen a period in the declining phase of solar cycle 23, covering the period 25 January–5 May 2005. During the CIR periods CR decreased typically from 0.5 % to 2 %. A cross-correlation analysis showed a distinct anti-correlation between the magnetic field and CR, with the correlation coefficient (r) ranging from -0.31 to -0.38 (mean: -0.36) and with the CR time delay of 2 to 3 days. Similar anti-correlations were found for the solar wind density and velocity characterized by the CR time lag of 4 and 1 day, respectively. The relationship was also established between the CR modulation and the area of the CIR-related CH with the CR time lag of 5 days after the central-meridian passage of CH.

Keywords. (ISM:) cosmic rays, (Sun:) solar wind, (Sun:) solar-terrestrial relations

1. Introduction

Corotating interaction regions (CIRs) are regions of compressed plasma formed at the leading edges of corotating high-speed solar wind streams originating in coronal holes (CHs) as they interact with the preceding slow solar wind. CIRs and high speed streams modulate galactic cosmic rays (GCR), thus influencing the flux of cosmic rays (CR) at the Earth (e.g. Richardson 2004). With the aim to improve space weather predictions, we investigated the relationship between CR and CIR.

2. Data and method

The daily averaged count rates from 8 different NM stations were used, supplied by SPIDR website (http://spidr.ngdc.noaa.gov/spidr). NM stations were selected to cover all latitudes and longitudes (3 NM stations in high, 3 in middle and 2 in low latitudes; see Table 1) with the cut-off rigidities ranging from $R_c = 0 – 12.91$ GeV.

Given that high solar activity and CMEs strongly affect CR, we avoided a period with CME activity. A period close to the declining phase of solar cycle 23 was selected, covering 25–125 (25 January until 5 May) days of the year (DOY) 2005. In this period the solar CME activity was particularly low (Vršnak *et al.* 2007a).

The data on proton density n, proton speed v, and magnetic field strength B were taken from the Advanced Composition Explorer (ACE) database (Stone *et al.* 1998). Daily solar coronal hole areas were determined employing the data from Solar X-Ray

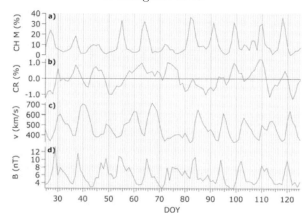

Figure 1. a) Daily measurements of the CH fractional area in the M slice, b) Relative cosmic ray change measured with Climax neutron monitor ($R_c = 3.03$ GeV), c) ACE daily averages of the proton velocity v, and d) magnetic field strength B. The x-axis represents DOY for 2005.

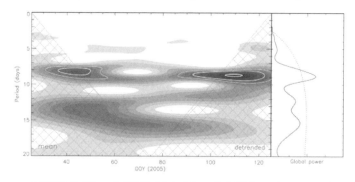

Figure 2. Averaged WPS (left) and GWS (right) for all 8 NM stations together. Analyzed time span is from 25 January 2005 until 5 May 2005 (DOY 25–125). The cone of influence is indicated by cross-hatched regions. With the white contour lines are marked the significance levels of 70 % (thin white line), 80 % (thicker white line) and 90 % (the thickest white line). The dotted line in the GWS represents the 95 % significance level.

Imager (SXI) on board the GOES-12 satellite, where the coronal holes appear as dark features in X-ray images (Hill 2005; Pizzo 2005). Using one SXI level-2 image per day, fractional areas of coronal holes were extracted in three meridional slices in the longitude range: CH E [-40° -20°], CH M [-10° 10°] (Fig. 1a) and CH W [20° 40°], for details see Vršnak *et al.* (2007b).

With an emphasis on daily variations in the CR, the NM data were detrended. The detrended NM data were cross-correlated with the CH areas and solar wind data, allowing for time lags of up to 10 days. The statistical significance was tested using one sided t-test on 5 % level.

To determine the predominant periods in the NM data, as well as their times of appearance, wavelet power spectra (WPS) and global wavelet spectra (GWS) were used (Morlet wavelet; see Temmer *et al.* 2007).

3. Results

In Fig. 1a the CH fractional area for the central meridian slice is presented as a function of DOY 2005. Oscillations of the CH area show the most prominent period of about 9

	CH E	CH M	CH W	v	B	n
NM South Pole	−0.33 (6)	−0.31 (4)	−0.30 (3)	−0.21 (0)	−0.31 (2)	−0.24 (4)
NM Thule	−0.16* (7)	−0.20 (4)	−0.14* (3)	−0.09* (0)	−0.37 (2)	−0.23 (4)
NM Magadan	−0.34 (7)	−0.33 (5)	−0.34 (3)	−0.24 (0)	−0.33 (2)	−0.20 (4)
NM Jungfraujoch	−0.31 (8)	−0.26 (5)	−0.32 (4)	−0.21 (2)	−0.35 (3)	−0.22 (4)
NM Climax	−0.34 (7)	−0.31 (5)	−0.31 (3)	−0.20 (1)	−0.38 (3)	−0.33 (4)
NM Hermanus	−0.27 (7)	−0.27 (5)	−0.27 (3)	−0.11* (1)	−0.38 (3)	−0.32 (4)
NM Haleakala	−0.23 (7)	−0.26 (5)	−0.32 (3)	−0.20 (1)	−0.34 (3)	−0.16* (4)
NM Tibet	−0.34 (7)	−0.39 (5)	−0.41 (3)	−0.27 (1)	−0.38 (3)	−0.18 (4)
average	−0.29 (7.00)	−0.29 (4.75)	−0.30 (3.13)	−0.19 (0.75)	−0.36 (2.63)	−0.24 (4.00)
σ	0.065	0.056	0.078	0.062	0.026	0.063

Table 1. Correlation coefficients between the CR measured at 8 different NM stations and CH areas (three meridional slices in the longitude range: east - CH E, middle - CH M, west - CH W), magnetic field B, proton speed v and density n. The highest correlation coefficient in the 10-days lag period is shown. Numbers in brackets indicate corresponding lag in days. All NM data are detrended. The non-significant values are marked with *.

days (Temmer *et al.* 2007), which can be explained by a "triangular" distribution of large CHs, separated by $\approx 120°$ in longitude. Similar oscillations we find also in the Climax NM data (Fig. 1b), as well as in the solar wind parameters data (proton speed v in Fig. 1c and magnetic field B in Fig. 1d). During the CIR passage the CR decreased typically from 0.5 % to 2 % (Fig. 1b). Figure 2 shows the detected periods in the CR data using a Morlet wavelet, where the most prominent period is also in the range of 9 days (DOY 30 – 125) and slightly weaker periods are in the range of 13.5 days (DOY 20 – 65) as well as 12 and 15 days (DOY 65–115).

The results of a cross-correlation analysis are presented in Table 1 for each NM station, where the highest correlation coefficients (r) within a 10-day lag period are shown. First three columns reveal a distinct anti-correlation (mean $r = -0.29$) between the area of the CIR-related CH and the CR modulation. Depending on the CH slice, CR show time lag of 7, 5, and 3 days for the east, middle and west CH slice, respectively. A significant relationship was also established between CR and solar wind parameters: v, B and n (Table 1). The highest correlations with CR was obtained for the magnetic field B with the CR time delay of 2 to 3 days, followed by proton density n (time lag 4 days) and proton velocity v (time lag 0–1 day).

4. Conclusions

The magnetic field enhancement forming at frontal edge of high-speed streams in the solar wind acts as a shield that reduces GCR flux at 1AU by 0.5 %–2 %. The effect is especially prominent during the declining phase of the solar cycle, when the occurrence rate of equatorial CHs is increased and the CME activity is low. A typical lag between the magnetic field peak and the CR dip is 2–3 days.

References

Hill, S. M. *et al.* 2005, *Solar Phys.*, 226, 255

Pizzo, V. J. *et al.* 2005, *Solar Phys.*, 226, 283

Richardson, I. G. 2004, *Space Sci. Revs*, 111(3), 267

Stone, E. C., Frandsen, A. M., Mewaldt, R. A., Christian, E. R., Margolies, D., Ormes, J. F., & Snow, F. 1998, *Space Sci. Revs*, 86, 1

Temmer, M., Vrsnak, B., & Veronig, M. A. 2007, *Solar Phys.*, 241, 371

Vršnak, B., Temmer, M., & Veronig, M. A. 2007a, *Solar Phys.*, 240, 331

Vršnak, B., Temmer, M., & Veronig, M. A. 2007b, *Solar Phys.*, 240, 315

Universal Heliophysical Processes
Proceedings IAU Symposium No. 257, 2008
N. Gopalswamy & D.F. Webb, eds.

© 2009 International Astronomical Union
doi:10.1017/S1743921309029664

Theory of cosmic ray modulation

Stefan E. S. Ferreira

Unit for Space Physics, North-West University, 2520 Potchefstroom, South Africa
email: `Stefan.Ferreira@nwu.ac.za`

Abstract. This work aims to give a brief overview on the topic of cosmic ray modulation in the heliosphere. The heliosphere, heliospheric magnetic field, transport parameters and the transport equation together with modulation models, which solve this equation in various degree of complexity, are briefly discussed. Results from these models are then presented where first it is shown how cosmic rays are globally distributed in an asymmetrical heliosphere which results from the relative motion between the local interstellar medium and the Sun. Next the focus shifts to low-energy Jovian electrons. The intensities of these electrons, which originate from a point source in the inner heliosphere, exhibit a unique three-dimensional spiral structure where most of the particles are transported along the magnetic field lines. Time-dependent modulation is also discussed where it is shown how drift effects together with propagating diffusion barriers are responsible for modulation over a solar cycle.

Keywords. (ISM:) cosmic rays, (Sun:) solar wind, plasmas, turbulence.

---

1. Introduction

Cosmic rays, when either entering the heliosphere due to their galactic origin, or being produced inside, are subjected to different modulation processes. These include diffusion, convection, energy changes and drifts (see Potgieter 1998; Fichtner 2005, for an overview). The Sun is the source of the heliosphere and as the activity of the Sun periodically changes from maximum to solar minimum conditions (every 11 years) we also observe solar cycle related changes in cosmic ray intensities due to changes in the modulation environment. As these particles travel to Earth they sample the different heliospheric conditions and can therefore provide valuable information about the outer regions of our heliosphere and even distances beyond (Lee & Fichtner 2001).

Apart from a wide range of neutron monitors on Earth (e.g., Moraal *et al.* 2000) which have been measuring cosmic ray intensities over the past few decades, there is also a fleet of spacecraft (like the Helios, Pioneers, Voyagers, Ulysses and PAMELA missions, to name a few) who measure cosmic rays at different energies and positions. Measurements from these missions have increased our knowledge, not only on cosmic ray transport, but also on the details of the background plasma environment in our local astrosphere. These observations have stimulated, to name a few, the development of different state of the art modulation (e.g., Fisk 1971; Kota & Jokipii 1983; le Roux & Potgieter 1995; Hattingh & Burger 1995) and heliospheric (e.g., Baranov & Malama 1995; Pauls & Zank 1996; Washimi & Tanaka 1996; Fahr *et al.* 2000; Pogorelov *et al.* 2006) models, theories on particle transport (e.g., Jokipii 1966; Bieber *et al.* 1994; Burger *et al.* 2000; Lerche & Schlickeiser 2001; Teufel & Schlickeiser 2002), turbulence in the solar wind (e.g., Bieber *et al.* 2004; Matthaeus *et al.* 2003), models of the heliospheric magnetic field (e.g., Parker 1958; Fisk 1996; Burger 2005), detailed theories on acceleration at shocks (e.g., Bell 1978; Blandford & Ostriker 1978) as well as continuous acceleration in the inner heliosheath (e.g., Kallenbach *et al.* 2005; Fisk & Gloeckler 2006) etc.

Furthermore, high-energy galactic cosmic rays may also influence the climate on Earth. Although controversial, Friis-Christensen & Svensmark (1997) and Svensmark (1998) found a correlation between cosmic ray intensity and global cloud coverage on the 11-year time scale of solar activity. There also seems to be a correlation on much longer timescales due to influences by the interstellar environment or spiral arm crossings (Shaviv 2003; Shaviv & Veizer 2003). See e.g. Scherer *et al.* (2006) for more details. Also of importance is the significant danger these high energy particles may have on future human space exploration (Parker 2005).

While the galactic cosmic rays originate outside the solar system due to acceleration at astrophysical shocks, the anomalous cosmic rays (e.g., Fisk *et al.* 1974) are accelerated inside the heliosphere. Voyager measurements (Decker *et al.* 2005; Stone *et al.* 2005) showed that the flux of anomalous cosmic rays in the heliosheath is unexpectedly high compared to expectations before Voyager 1 reached the shock. This might be due to particle acceleration at the flanks of the heliosphere (McComas & Schwadron 2006), dynamic effects (Florinski & Zank 2006) or due to additional continuous acceleration beyond the termination shock (Langner *et al.* 2006; Kallenbach *et al.* 2005; Fisk & Gloeckler 2006; Ferreira *et al.* 2007). Combining this finding with different model results for astrosphere immersed in different interstellar environments Scherer *et al.* (2008) showed that the astrospheric anomalous cosmic ray fluxes of solar-type stars can be a hundred times higher than thought earlier and, consequently, their total contribution to the lower end of the interstellar spectrum can be significant.

The aspects discussed above illustrate the importance of understanding the details of cosmic ray modulation in our heliosphere. With this a detailed understanding of the modulation environment is also necessary. This work will now briefly discuss some different but closely related topics related to cosmic ray transport.

2. The transport equation and modulation models

Cosmic ray modulation in the heliosphere can be described by a transport equation (Parker 1965) (see also work done by Gleeson & Axford 1967; Forman *et al.* 1974; Forman & Jokipii 1978; Webb & Gleeson 1979; Moraal & Potgieter 1982):

$$\frac{\partial f}{\partial t} = -\left(\vec{V} + \langle \vec{v_D} \rangle\right) \cdot \nabla f + \nabla \cdot (\vec{K_S} \cdot \nabla f) \tag{2.1}$$

$$+ \frac{1}{3}(\nabla \cdot \vec{V})\frac{\partial f}{\partial \ln P} + \frac{1}{P^2}\frac{\partial}{\partial P}(P^2 D \frac{\partial f}{\partial P}) + Q.$$

Here t is the time, P is rigidity, Q is any particle sources inside the heliosphere, $\vec{V}$ is the solar wind velocity, $\vec{K_S}$ is the diffusion tensor (discussed below) and $\langle \vec{v_D} \rangle$ the averaged guiding center drift velocity (e.g., Burger *et al.* 2000; Stawicki 2005b) for a near isotropic distribution function f and D a momentum diffusion coefficient. This equation can be solved numerically in various degrees of complexity in so-called modulation models (e.g., Fisk 1971; Kota & Jokipii 1983; le Roux & Potgieter 1990; Hattingh & Burger 1995; Steenberg & Moraal 1996; le Roux & Fichtner 1999; Florinski *et al.* 2003; Langner & Potgieter 2005; Ferreira & Scherer 2006; Zhang 2006) and below some results from these models are presented.

Of primary importance to cosmic ray modulation is the coupling of the transport parameters to the background field and magnetic turbulence (see e.g., Bieber *et al.* 2004; Matthaeus *et al.* 2003; Shalchi *et al.* 2008). Of particular interest is determing the coefficients in $\vec{K_S}$. These can be theoretically calculated from plasma and magnetic field observations using either turbulence theory (e.g., Bieber *et al.* 1994) or the theory of particle wave interaction (e.g., Lerche & Schlickeiser 2001). In both of these approaches the parallel mean free path can be computed using quasi linear theory (e.g., Jokipii 1966; Bieber *et al.* 1994; Burger *et al.* 2000; Teufel & Schlickeiser 2002) or extensions of the latter (e.g., le Roux *et al.* 2005).

For the perpendicular diffusions it was shown via simulations that these may scale as the parallel coefficient (le Roux *et al.* 1999; Giacalone & Jokipii 1999; Qin *et al.* 2002). Recent theoretical advances in perpendicular diffusion include: A nonlinear theory of the perpendicular diffusion of charged particles which include the influence of parallel scattering and dynamical turbulence (Matthaeus *et al.* 2003; Shalchi *et al.* 2004; Shalchi 2006). A nonlinear description for perpendicular particle diffusion in strong electromagnetic fluctuations using the fundamental Newton Lorentz equation (Stawicki 2005a). An equation describing compound and perpendicular diffusion of cosmic rays and random walk of the field lines (Webb *et al.* 2006). See also Minnie *et al.* (2005, 2007) for recent results and incorporation of theories into modulation models.

3. The geometry of the heliosphere and flow profiles of the solar wind plasma inside

The solar wind is the source of the heliosphere. However, interstellar space is not empty and contains matter in the form of the local interstellar medium (LISM). As the heliosphere moves through the LISM it forces the plasma component to flow around it. At some distance the interaction of the LISM and the solar wind causes the supersonic solar wind to decrease to subsonic speeds and a shock is created, called the solar wind termination shock. At larger distances a contact surface is eventually reached, called the heliopause. This separates the solar wind from the interstellar material. Depending on the interstellar conditions there may also be a bow shock around the heliosphere. Concerning the heliospheric geometry and global magnetic field, valuable modeling efforts have been done by Holzer (1989); Suess (1990); Pauls & Zank (1996, 1997); Fahr *et al.* (2000); Scherer & Fahr (2003b,a); Zank & Müller (2003); Izmodenov *et al.* (2005); Borrmann & Fichtner (2005); Pogorelov *et al.* (2006); Opher *et al.* (2006).

The heliosphere, resulting under the relative motion between the LISM and the Sun has an asymmetric structure, for instance the position of the termination shock in the tail region is related to the position in the nose through the relation found by Müller *et al.* (2006) as $r_{TS,tail} = (2.08 \pm 0.04)r_{TS,nose}$ with the position of the heliopause in the nose related to the position of the termination shock as $r_{HP} = (1.39 \pm 0.01)r_{TS,nose}$. In addition, the heliosphere is also more elongated in the poleward directions because of the latitudinal variation of the solar wind momentum flux (McComas *et al.* 2001). This quantity increases by a factor of ~1.5 from the equatorial regions toward the poles, suggesting a poleward elongated termination shock. Nevertheless, the mass flux is reported to be fairly constant with latitude and over a solar cycle (McComas *et al.* 2003), while the dynamic pressure is changing, leading to a more pronounced asymmetry during solar minimum conditions. The effect of a changing solar wind speed over the poles on the heliospheric geometry are illustrated in Fig. 1, which shows the heliosphere in the nose regions in terms of solar wind speed and density for a isotropic wind (solar maximum) and an anisotropic wind (solar minimum). The most important feature shown here, from a cosmic ray modulation point of view, is that as solar activity changes the termination shock moves, especially at the polar and tail regions. See e.g., Scherer & Ferreira (2005) for dynamic modeling results.

In Figure 2 (From Langner & Potgieter 2005) the distribution of cosmic rays in an asymmetrical heliosphere is shown for minimum and moderate solar activity. Compared to a theoretical symmetrical case, the asymmetry of the heliosphere results in insignificant effects for the $A > 0$ (protons drift in from the polar regions), but for the $A < 0$ cycle (protons drift in along the current sheet) the intensities may differ by at least a factor of 1.5 for most of the heliosphere, although in the tail region a factor of 3.5 is evident. As shown in Figure 2 protons are redistributed when the heliosphere is asymmetrically bounded compared to a symmetrical heliosphere (see Figure 7 of Langner & Potgieter 2005). Concerning the inner heliosphere not much evidence can be found in the nose-tail asymmetry when cosmic rays are considered. Also note that the intensities of accelerated particles at the shock differs between the nose and tail region. This is related to the fact that the position of the heliopause in the tail region is increased, causing a smaller population of particles to be accelerated. Also not known is the difference in the continuous acceleration processes between the nose.

4. The heliospheric magnetic field

Embedded in the solar wind is the Sun's magnetic field wounded up in a spiral and transported with the solar wind into space forming the heliospheric magnetic field (Parker 1958). This field determines the passage of charged particles, like cosmic rays in our heliosphere, changing their intensities with time and as a function of energy and position. A very unique signature of this spiral can be observed in low-energy electron intensities. At these energies the Jovian magnetosphere is the dominant source of electrons (Simpson *et al.* 1974; Chenette *et al.* 1974) in the inner heliosphere up to ~10 AU (e.g., Ferreira *et al.* 2001). Shown in Figure 3 is model calculations by Ferreira *et al.* (2001) showing the three-dimensional distribution of Jovian electrons at different polar angles, as indicated. In the left panel, signatures of the Parker spiral is clearly visible in the intensities because of the dominance of parallel diffusion in the inner heliospheric regions.

Concerning the inner heliosphere, deviations from a pure Parker-type field are expected at certain latitudes (Fisk 1996). In a Fisk-type field, magnetic field lines exhibit extensive excursions in heliographic latitude, and this has been cited as a possible explanation for

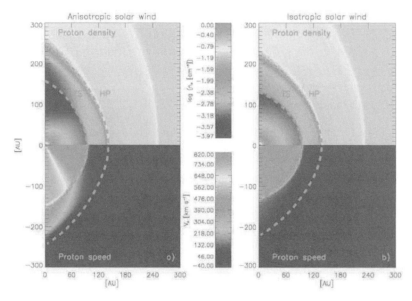

Figure 1. The heliosphere in the nose regions in terms of solar wind speed (top) and density (bottom) for an isotropic wind (solar maximum) and an anisotropic wind (solar minimum)

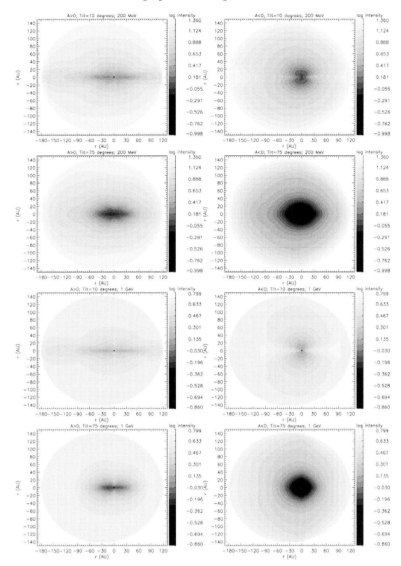

Figure 2. Proton intensity contours in the meridional plane at 200 MeV (top four panels) and 1 GeV (bottom four panels) for an asymmetrical heliosphere for solar minimum (tilt 10 degrees) and moderate solar maximum conditions (tilt 75 degrees), and for the $A > 0$ (left) and $A < 0$ (right) polarity cycles. Note that the legend of the contours corresponds to the exponent of the base 10 on a logarithmic scale and that the scaling differs for the 200 MeV and 1 GeV plots. (from Langner & Potgieter 2005).

recurrent energetic particle events observed by the Ulysses spacecraft at high latitudes (see, e.g., Simpson *et al.* 1995; Zhang 1997; Paizis *et al.* 1999), as well as the smaller than expected cosmic-ray intensities observed at high latitudes (Simpson *et al.* 1996).

The Fisk field and the physics behind it have been discussed in a series of papers (see, e.g., Fisk & Schwadron 2001, and references therein) where it is assumed that the polar coronal hole is symmetric with respect to the solar magnetic axis, and that the magnetic field expands nonradially. The footpoints of the magnetic field lines anchored in the photosphere experience differential rotation. Then, if the magnetic axis of the

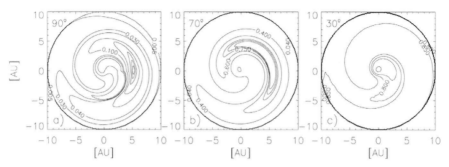

Figure 3. Three-dimensional distribution of Jovian electrons at different polar angles, as indicated.

Sun is assumed to rotate rigidly at the equatorial rate, differential rotation will cause a footpoint to move in heliomagnetic latitude and longitude, thus experiencing different degrees of non-radial expansion. The end result is a field line that moves in heliographic latitude.

Over the last ten years, various attempts to incorporate the Fisk field into numerical modulation models have been reported (Kóta & Jokipii 1999; Burger et al. 2001; Burger & Hitge 2004). A recent overview of various models for the heliospheric magnetic field, including Fisk-type fields, is given by Burger (2005). Recently Burger et al. (2008) presented a Fisk-Parker hybrid field where at high latitudes the field is a mixture of Fisk field and Parker field, and in the equatorial region it is a pure Parker field. They confirmed the result of Burger & Hitge (2004) that a Fisk-type heliospheric magnetic field provides a natural explanation for the observed linear relationship between the amplitude of the recurrent cosmic-ray variations and the global latitude gradient, first reported by Zhang (1997), and showed that this relationship holds for helium, protons, and electrons.

5. Long-term cosmic ray modulation

It was originally shown by Perko & Fisk (1983) and le Roux & Potgieter (1989) that cosmic ray modulation over long periods requires some form of propagating diffusion barriers. This is especially true for solar maximum when step decreases are observed. The largest form of these diffusion barriers are called global merged interaction regions (GMIRs) (Burlaga et al. 1993). Equally important are gradient, curvature and current sheet drifts (Jokipii et al. 1977) as confirmed by comprehensive modeling done by le Roux & Potgieter (1995). These authors showed that it was possible to simulate, to the first-order, a complete 22-year modulation cycle by including a combination of drifts and GMIRs in a time-dependent modulation model. For typical solar minimum conditions, drifts together with changes in the current sheet are responsible modulation while toward solar maximum GMIRs caused the intensities to decrease in a step-like manner and drifts were less important due to the large tilt angles of the heliospheric current sheet.

More recently, Cane et al. (1999) and Wibberenz et al. (2002) argued that the step decreases observed at Earth could not be primarily caused by GMIRs because they occurred before any GMIRs could form beyond 10 AU. Instead they suggested that time-dependent global changes in the heliospheric magnetic field might be responsible for long-term modulation. This was tested by Ferreira & Potgieter (2004) who showed that indeed at neutron monitor energies, solar cycle related changes in the field magnitude alone can explain observations, but for lower energies this is no longer the case and a combination of these approaches were suggested. This can be done by scaling all the transport parameters

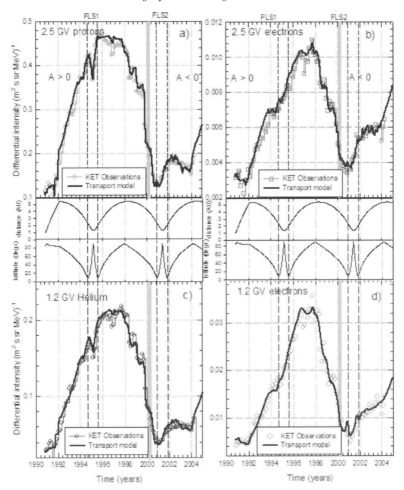

Figure 4. Model results compared to Ulysses/KET observations for protons, electrons and helium at different rigidities as indicated. The position in radial distance and latitude of the Ulysses spacecraft are also shown in the center panels. (from Ndiitwani *et al.* 2005).

with a function depending on the observed magnetic field at Earth and the current sheet tilt angle (Hoeksema 1992). This function results in diffusion coefficients which are roughly a factor of ~ 10 smaller for solar minima compared to solar maxima.

As shown by Ferreira & Potgieter (2004), Ndiitwani *et al.* (2005) and Ferreira & Scherer (2006) this approach incorporated in a numerical modulation model results in very realistic model computations when compared to spacecraft observations at various energies. An example is shown in Figure 4 (from Ndiitwani *et al.* 2005) showing model results compared to Ulysses/KET observations for protons, electrons and helium at different rigidities. Shown here is that the model could simulate the modulation amplitude between solar minimum and maximum correctly as well as step-decreases as they occur.

References

Baranov, V. B. & Malama, Y. G. 1995, *J. Geophys. Res.*, 100, 14755

Bell, A. R. 1978, *MNRAS*, 182, 147

Bieber, J. W., Matthaeus, W. H., Shalchi, A., & Qin, G. 2004, *Geophys. Res. Lett.*, 31, 10805

Bieber, J. W., Matthaeus, W. H., Smith, C. W., Wanner, W., Kallenrode, M.-B., & Wibberenz, G. 1994, *ApJ*, 420, 294

Blandford, R. D. & Ostriker, J. P. 1978, *ApJ*, 221, L29

Borrmann, T. & Fichtner, H. 2005, *Adv. Sp. Res.*, 35, 2091

Burger, R. A. 2005, *Adv. Sp. Res.*, 35, 636

Burger, R. A. & Hitge, M. 2004, *ApJ*, 617, L73

Burger, R. A., Krüger, T. P. J., Hitge, M., & Engelbrecht, N. E. 2008, *ApJ*, 674, 511

Burger, R. A., Potgieter, M. S., & Heber, B. 2000, *J. Geophys. Res.*, 105, 27447

Burger, R. A., van Niekerk, Y., & Potgieter, M. S. 2001, *Space Sci. Revs*, 97, 331

Burlaga, L. F., McDonald, F. B., & Ness, N. F. 1993, *J. Geophys. Res.*, 98, 1

Cane, H. V., Wibberenz, G., Richardson, I. G., & von Rosenvinge, T. T. 1999, *Gephys. Res. Lett.*, 26, 565

Chenette, D. L., Conlon, T. F., & Simpson, J. A. 1974, *J. Geophys. Res.*, 79, 3551

Decker, R. B., Krimigis, S. M., Roelof, E. C., Hill, M. E., Armstrong, T. P., Gloeckler, G., Hamilton, D. C., & Lanzerotti, L. J. 2005, *Science*, 309, 2020

Fahr, H. J., Kausch, T., & Scherer, H. 2000, *A&A*, 357, 268

Ferreira, S. E. S. & Potgieter, M. S. 2004, *ApJ*, 603, 744

Ferreira, S. E. S., Potgieter, M. S., Burger, R. A., Heber, B., Fichtner, H., & Lopate, C. 2001, *J. Geophys. Res.*, 106, 29313

Ferreira, S. E. S., Potgieter, M. S., & Scherer, K. 2007, *J. Geophys. Res.*, 112, 11101

Ferreira, S. E. S. & Scherer, K. 2006, *ApJ*, 642, 1256

Fichtner, H. 2005, *Adv. Sp. Res.*, 35, 512

Fisk, L. A. 1971, *J. Geophys. Res.*, 76, 221

Fisk, L. A. 1996, *J. Geophys. Res.*, 101, 15547

Fisk, L. A. & Gloeckler, G. 2006, *ApJ*, 640, L79

Fisk, L. A., Kozlovsky, B., & Ramaty, R. 1974, *ApJ*, 190, L35

Fisk, L. A. & Schwadron, N. A. 2001, *ApJ*, 560, 425

Florinski, V. & Zank, G. P. 2006, *Geophys. Res. Lett.*, 33, 15110

Florinski, V., Zank, G. P., & Pogorelov, N. V. 2003, *J. Geophys. Res.*, 108, 1

Forman, M. A. & Jokipii, J. R. 1978, *ApSS*, 53, 507

Forman, M. A., Jokipii, J. R., & Owens, A. J. 1974, *ApJ*, 192, 535

Friis-Christensen, E. & Svensmark, H. 1997, *Adv. Sp. Res.*, 20, 913

Giacalone, J. & Jokipii, J. R. 1999, *ApJ*, 520, 204

Gleeson, L. J. & Axford, W. I. 1967, *ApJ*, 149, L115

Hattingh, M. & Burger, R. A. 1995, *Adv. Sp. Res.*, 16, 213

Hoeksema, J. T. 1992, in Solar Wind Seven Colloquium, 191

Holzer, T. E. 1989, *ARAA*, 27, 199

Izmodenov, V., Malama, Y., & Ruderman, M. S. 2005, *A&A*, 429, 1069

Jokipii, J. R. 1966, *ApJ*, 146, 480

Jokipii, J. R., Levy, E. H., & Hubbard, W. B. 1977, *ApJ*, 213, 861

Kallenbach, R., Hilchenbach, M., Chalov, S. V., Le Roux, J. A., & Bamert, K. 2005, *A&A*, 439, 1

Kota, J. & Jokipii, J. R. 1983, *ApJ*, 265, 573

Kóta, J. & Jokipii, J. R. 1999, in Proc. 26th Int. Cosmic Ray Conf., Vol. 7, 9

Langner, U. W. & Potgieter, M. S. 2005, *ApJ*, 630, 1114

Langner, U. W., Potgieter, M. S., Fichtner, H., & Borrmann, T. 2006, *J. Geophys. Res.*, 111, 1106

le Roux, J. A. & Fichtner, H. 1999, *J. Geophys. Res.*, 104, 4709

le Roux, J. A. & Potgieter, M. S. 1989, *Adv. Sp. Res.*, 9, 225

le Roux, J. A. & Potgieter, M. S. 1990, *ApJ*, 361, 275

le Roux, J. A. & Potgieter, M. S. 1995, *ApJ*, 442, 847

le Roux, J. A., Zank, G. P., Li, G., & Webb, G. M. 2005, *ApJ*, 626, 1116

le Roux, J. A., Zank, G. P., & Ptuskin, V. S. 1999, *J. Geophys. Res.*, 104, 24845

Lee, M. A. & Fichtner, H. 2001, in The Outer Heliosphere: *The Next Frontiers*, ed. K. Scherer, H. Fichtner, H. J. Fahr, & E. Marsch, 183

Lerche, I. & Schlickeiser, R. 2001, *A&A*, 378, 279

Matthaeus, W. H., Qin, G., Bieber, J. W., & Zank, G. P. 2003, *ApJ*, 590, L53

McComas, D. J., Elliott, H. A., Schwadron, N. A., Gosling, J. T., Skoug, R. M., & Goldstein, B. E. 2003, *Geophys. Res. Lett.*, 30, 24

McComas, D. J., Goldstein, R., Gosling, J. T., & Skoug, R. M. 2001, *Space Sci. Revs*, 97, 99

McComas, D. J. & Schwadron, N. A. 2006, *Geophys. Res. Lett.*, 33, 4102

Minnie, J., Bieber, J. W., Matthaeus, W. H., & Burger, R. A. 2007, *ApJ*, 663, 1049

Minnie, J., Burger, R. A., Parhi, S., Matthaeus, W. H., & Bieber, J. W. 2005, *Adv. Sp. Res.*, 35, 543

Moraal, H., Belov, A., & Clem, J. M. 2000, *Space Sci. Revs*, 93, 285

Moraal, H. & Potgieter, M. S. 1982, *ApSS*, 84, 519

Müller, H.-R., Frisch, P. C., Florinski, V., & Zank, G. P. 2006, *ApJ*, 647, 1491

Ndiitwani, D. C., Ferreira, S. E. S., Potgieter, M. S., & Heber, B. 2005, *Annales Geophysicae*, 23, 1061

Opher, M., Stone, E. C., & Liewer, P. C. 2006, *ApJ*, 640, L71

Paizis, C., *et al.* 1999, *J. Geophys. Res.*, 104, 28241

Parker, E. N. 1958, *ApJ*, 128, 664

Parker, E. N. 1965, *Planet. Space Sci.*, 13, 9

Parker, E. N. 2005, *Space Weather*, 3, S08004

Pauls, H. L. & Zank, G. P. 1996, *J. Geophys. Res.*, 101, 17081

Pauls, H. L. & Zank, G. P. 1997, *J. Geophys. Res.*, 102, 19779

Perko, J. S. & Fisk, L. A. 1983, *J. Geophys. Res.*, 88, 9033

Pogorelov, N. V., Zank, G. P., & Ogino, T. 2006, *ApJ*, 644, 1299

Potgieter, M. S. 1998, *Space Sci. Revs*, 83, 147

Qin, G., Matthaeus, W. H., & Bieber, J. W. 2002, *Geophys. Res. Lett.*, 29, 7

Scherer, K. & Fahr, H. J. 2003a, *Ann. Geophys.*, 21, 1303

Scherer, K. & Fahr, H. J. 2003b, *Geophys. Res. Lett.*, 30, 17

Scherer, K. & Ferreira, S. E. S. 2005, *ASTRA*, 1, 17

Scherer, K., *et al.* 2006, *Space Sci. Revs*, 127, 327

Scherer, K., Fichtner, H., Ferreira, S. E. S., Büsching, I., & Potgieter, M. S. 2008, *ApJ*, 680, L105

Shalchi, A. 2006, *A&A*, 453, L43

Shalchi, A., Bieber, J. W., & Matthaeus, W. H. 2004, *ApJ*, 604, 675

Shalchi, A., Bieber, J. W., & Matthaeus, W. H. 2008, *A&A*, 483, 371

Shaviv, N. J. 2003, *New Astronomy*, 8, 39

Shaviv, N. J. & Veizer, J. 2003, *GSA Today*, 13, 4

Simpson, J. A., *et al.* 1995, *Science*, 268, 1019

Simpson, J. A., Hamilton, D. C., McKibben, R. B., Mogro-Campero, A., Pyle, K. R., & Tuzzolino, A. J. 1974, *J. Geophys. Res.*, 79, 3522

Simpson, J. A., Zhang, M., & Bame, S. 1996, *ApJ*, 465, L69

Stawicki, O. 2005a, *Adv. Sp. Res.*, 35, 547

Stawicki, O. 2005b, *ApJ*, 624, 178

Steenberg, C. D. & Moraal, H. 1996, *ApJ*, 463, 776

Stone, E. C., Cummings, A. C., McDonald, F. B., Heikkila, B. C., Lal, N., & Webber, W. R. 2005, *Science*, 309, 2017

Suess, S. T. 1990, *Rev. Geophys.*, 28, 97

Svensmark, H. 1998, *Phys. Rev. Lett.*, 81, 5027

Teufel, A. & Schlickeiser, R. 2002, *A&A*, 393, 703

Washimi, H. & Tanaka, T. 1996, *Space Sci. Revs*, 78, 85

Webb, G. M. & Gleeson, L. J. 1979, *ApSS*, 60, 335

Webb, G. M., Zank, G. P., Kaghashvili, E. K., & le Roux, J. A. 2006, *ApJ*, 651, 211

Wibberenz, G., Richardson, I. G., & Cane, H. V. 2002, *J. Geophys. Res.*, 107, 5

Zank, G. P. & Müller, H.-R. 2003, *J. Geophys. Res.*, 108, 7

Zhang, M. 1997, *ApJ*, 488, 841

Zhang, M. 2006, in AIP Conf. Proc. 858: *Physics of the inner heliosheath: 5th Annual IGPP International Astrophysics Conference,* ed. J. Heerikhuisen & *et al.,* 226

Discussion

FISK: I think there is a fundamental problem with your stochastic acceleration mechanism to create the CRs in the heliosheath. You are using a damping mechanism, yet there is inadequate energy in the turbulence in the heliosheath to provide the energy in the ACRs.

FERREIRA: You are certainly correct. We studied stochastic acceleration of ACRs when Voyager1 just crossed the boundary. Recent observations do yield more energy in PUIs than to the magnetic field than expected. Therefore, indeed our model does overestimate the effect and we are busy looking deeper into this.

ANONYMOUS: 1) How strong is the GMIR argument for producing the step decreases, given the fact that a GMIR didn't form during the Halloween 2003 storms. 2) Why don't you consider the difference between high latitude and low latitude CMEs in the $A > 0$ cycle?

FERREIRA: There is evidence that GMIRs do cause the 11-year cycle, at least toward solar maximum. However this is only one aspect, with HMF and current sheet tilt angle also causing time-dependent modulation.

CLIVER: Can you tell us about the current level for cosmic ray intensity? Because the solar polar fields are quite weak at this minimum, is the GCR intensity significantly higher than in previous minima?

FERREIRA: The recent Ulysses/KET data will be published soon. I can show you now one of Bernd Heber's slides.

MELNIKOV: How important is the pitch angle scattering which was not included in the transport equation?

FERREIRA: It is important for second-order effects. We are more interested in global modulation. For work on this concerning cosmic rays see, e.g., Florinski and Le Roux and co-workers.

Universal Heliophysical Processes
Proceedings IAU Symposium No. 257, 2008
N. Gopalswamy & D.F. Webb, eds.

© 2009 International Astronomical Union
doi:10.1017/S1743921309029676

Forbush effects and their connection with solar, interplanetary and geomagnetic phenomena

A. V. Belov

Pushkov Institute of Terrestrial Magnetism, Ionosphere and Radio Wave Propagation
Russian Academy of Sciences (IZMIRAN), 142190 Troitsk,
Moscow region, Russia
email: `abelov@izmiran.ru`

Abstract. Forbush decrease (or, in a broader sense, Forbush effect) - is a storm in cosmic rays, which is a part of heliospheric storm and very often observed simultaneously with a geomagnetic storm. Disturbances in the solar wind, magnetosphere and cosmic rays are closely interrelated and caused by the same active processes on the Sun. Thus, it is natural and useful to investigate them together. Such an investigation in the present work is based on the characteristics of cosmic rays with rigidity of 10 GV. The results are derived using data from the world wide neutron monitor network and are combined with relevant information into a data base on Forbush effects and large interplanetary disturbances.

Keywords. elementary particles, Sun: activity, interplanetary medium, solar-terrestrial relations, coronal mass ejections (CMEs), shock waves.

1. Introduction

A special kind of cosmic ray (CR) variation which we name Forbush decrease (FD) or Forbush effect (FE) was discovered by S. Forbush in 1937 (Forbush 1937, Forbush 1938). Since then hundreds of such events have been observed and discussed, and a large number of papers have been devoted to their study. The present work is not a traditional review of the published literature, such as the fundamental reviews J. Lockwood (Lockwood 1971), and H. Cane (Cane 2000). Much information on the FEs can be found in Dorman's books (eg. Dorman 1963, Dorman 1974) and in other works (eg. Iucci *et al.* 1979, Iucci *et al.* 1986, Cane 1993, Richardson & Cane 2005). The present work should best be considered as a review of FEs observed on Earth during the last five solar activity cycles.

FEs were discovered in ground level measurements of CRs and the data from ground level detectors remain as the main source of the information on these phenomena. This paper is based on the data from the world wide neutron monitor network and contains a discussion of the following themes: FE definition; variety of FEs; time and size distribution of FEs; interrelation of different characteristics of FEs; behavior of the CR anisotropy in the FE; relation of the FE to characteristics of the interplanetary disturbances; relation of FE to geomagnetic activity; dependence of FEs on the solar source characteristics; precursors of FEs.

Since the first years of the space era FEs have been recorded not only on Earth but in space as well - onboard of satellites (Fan *et al.* 1960) and on the remote spacecraft (eg., Lockwood & Webber 1987). These measurements are worth discussing separately. Another theme which is becoming more and more popular although not being discussed here is the influence of Forbush-effects on atmospheric processes and on the climate. It also deserves a separate talk and consideration.

2. Data base on Forbush effects and interplanetary disturbances

Almost all results presented here are obtained by means of databases created in IZMI-RAN during the last 12 years. The first database contains calculations of CR density and anisotropy performed by the global survey method for each hour over 50 years since 1957 (Krymsky *et al.* 1981, Belov *et al.* 2007a). Results for 10 GV rigidity CRs (http://cr20.izmiran.rssi.ru/AnisotropyCR) are combined with the solar wind parameters and geomagnetic activity indexes from OMNI database (http://omniweb.gsfc.nasa.gov). Another database includes information on the interplanetary disturbances and Forbush effects. Selecting events for this database we tried to include all FEs in their broad definition and find a connection with each of them corresponding to a large scale interplanetary disturbance. Into this database are entered: 1. all events followed by the shock, even in the cases when no clear effect in the CR density was observed; 2. all big enough variations in CR density independent of the interplanetary conditions; 3. small events (as a rule, 0.5–1.0%) - if the significant solar wind disturbances were observed at that time. Such a catalogue has some obvious advantages. 1. It is based on physical characteristics (density and anisotropy) of CRs of definite rigidity (10 GV) but not on the data of separate CR detectors. 2. Quantitative characteristics of separate events in this case are much more accurate than in the catalogues constructed on the data from separate stations. 3. The catalogue covers all the events observed in the defined period. Also included are events of small magnitude which may be a consequence of large interplanetary disturbances. 4. Each event (FE) in CRs is associated with certain interplanetary disturbance and possibly with its solar source. Thus, this is a catalogue not only for the FEs but also for the interplanetary disturbances.

At present our database includes about 5900 events covering the period from July 1957 to December 2006 - practically the whole time the world wide neutron monitor network has been operating. Each event is characterized by of the tens of various parameters on the CR variations and relevant phenomena. For the CRs these are the characteristics of charged particles with a rigidity of 10 GV. The main parameter is the FE magnitude A_F which is the maximum variation of CR density during the event.

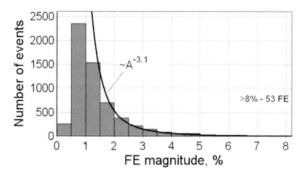

Figure 1. Size distribution of Forbush effect magnitudes.

In Fig. 1 a distribution of the FEs by their magnitude A_F is presented. The maximum of the distribution is located near 1%. The location of maximum is explained by methodological reasons related to the difficulties of selection of the FEs of small magnitude. However, there are probably also physical reasons connected with the characteristic size and other parameters of interplanetary disturbances, first of all, ICMEs. If we ignore the range of small FEs, then, for $A_F > 1.5\%$ the considered distribution is well described by a power law with an index 3.1±0.1. This index is larger than for other solar parameters. For example, the distribution of soft X-ray flare power has an index of 2.19 (Hudson 2007, Belov *et al.* 2007). Once having the full FE distribution by magnitude it is possible to answer the question, which FEs should be considered as large? Comparing the frequency distribution of the FEs with those for geomagnetic storms of different classes (http://www.swpc.noaa.gov/NOAAscales), one can see that FEs with magnitude of >3% correspond to strong geomagnetic storms (with maximum Kp-index ⩾ 7) and such events occur once per 36 days on average. FEs of >12.5% correspond to extreme magnetic storms which occur on average once per three years.

3. What is the Forbush effect?

Forbush effect definition. It is surprising that after a 70-year intensive investigations Forbush effect is not defined conventionally. It is clear that original definition of FE as the CR variation during a geomagnetic storm, found in the old books (e.g., Dorman, 1963), is now out of date. FEs have been already observed many times far from the Earth and other planetary magnetospheres. And even on Earth FE does not always accompany a magnetic storm. And what do we have now apart from the old definition? If we look up the Glossary of solar-terrestrial terms NOAA (http://www.ngdc.noaa.gov), we will read: "Forbush decrease - an abrupt decrease, of at least 10%, of the background galactic cosmic ray intensity as observed by neutron monitors". Unfortunately, this definition is erroneous in almost each word. FD is not necessarily abrupt, very often it occurs gradually. Only very rare FEs reach 10%, the majority of them are much less. FE may be observed not only by neutron monitors but by many other detectors: in the ionization chambers by which they were discovered, in the ground and underground muon telescopes, on various detectors used in space exploration of CRs. It is more often observed in galactic CRs but the FE may be recorded (and is really often observed) in solar CR variations as well. The same concerns the anomalous CR. There is no doubt that this definition does not suit us.

So, what is the FE? FE is often said to be a storm in CRs and this is correct. During the FE we see mostly disturbed galactic CRs (Belov *et al.*, 1997). During the FE the CR often happen to be most modulated, and the magnitude of the largest FEs is higher than 11-year CR variation. It is shown in Fig. 2, where CR variation during the FE and long term CR modulation during the solar cycle for rigidity of 10 GV are plotted on the same magnitude scale but on two different time scales. CR variations in solar cycle 23 did not exceed 19%, whereas in one FE at the end of October 2003 they made about 28%. And finally, during the FE galactic CR flux may be the most anisotropic. FE is a storm in CRs and a manifestation of heliospheric storm. Perhaps, it is reasonable to give the definition of the FE basing on its origin. It may look as follows: "Forbush effect is a result of the influence of coronal mass ejections (CMEs and ICMEs) and/or high speed streams of the solar wind from the coronal holes on the background cosmic rays". Thus, interplanetary disturbances that created the FEs are both of sporadic and recurrent nature. It would be desirable to leave only one class of sources and not consider recurrent phenomena as the FEs. But this is practically impossible because the solar wind disturbance is often a result of interaction of different factors (Crooker & Cliver 1994, Ivanov 1997), both of sporadic (flares, filament disappearances, coronal ejections) and recurrent (coronal holes, streamer structure) origin.

Forbush effect variety. FEs are rather diverse. Among them are large and small, short and long lasting, with fast and gradual fall, with full recovery and without it at all, two-step or not, with simple and complicated time profile, and so on. One reason for

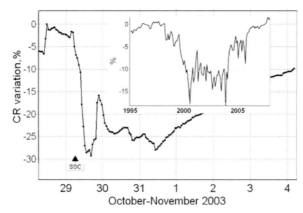

Figure 2. 10 GV cosmic ray variations during the giant Forbush effect in October 2003
(hourly data) and in 23th solar cycle (monthly data).

such a variety is a diversity of solar sources and their combinations. A second reason is
the variety of interplanetary situations arising before and during the event. Therefore
we observe peculiarities in developing of a solar wind disturbance and its interaction
with other heliospheric structures. A third and most important reason is the nature of
the FE observation. FE is usually recorded at one point, mostly on Earth, and such
observation cannot be complete. We should remember that FE occurs within a large
volume occupying a significant part of heliosphere, and the same event at another point
may look quite different, e.g. a considerable decrease of CR density may change into an
increase (Belov *et al.* 1999).

Sporadic and recurrent FEs. The majority of FEs are of sporadic character and caused
by ICMEs. In this case CR decrease is created by expansion of a disturbed solar wind
region that is partly screened from outside by strong and/or transverse magnetic fields.
The earliest phase of FE is observed prior to arrival of the interplanetary disturbance
at Earth as a pre-increase of CR intensity caused by the acceleration of high energy
charged particles on the outer boundary of an ICME, and a pre-decrease due to magnetic
connection between the Earth and the region of FD inside of ICME. When the shock
created by ICME movement, and/or solar ejecta itself arrives at Earth, the main phase of
FE begins (Forbush decrease), when the CR behavior reflects directly magnetic structure
of the propagating disturbance. Finally, at the last stage when a ICME propagates beyond
the Earth orbit we see the phase of FE recovery, showing that an expanded disturbed
region continues to modulate CRs. Solar wind recurrent high speed streams may be the
main reason of many FEs, but these effects are never too large. It seems the limiting
value of FE caused by coronal holes does not exceed 5%. Herewith in the large recurrent
FEs it is also possible to find the effect of CME. We may certainly state that all FEs of
large magnitude and the majority of FEs of middle and small magnitude are caused by
CMEs. However, in the periods when large and effective CMEs are rare the FEs, caused
by coronal holes, dominate. A striking examples of such quiescent periods are the years
of 1995-1996 or 2007-2008 when we did not practically see the sporadic FEs.

4. When do Forbush effects occur?

Fig. 3 demonstrates quasi 11-year cycles in the behavior of the FE magnitude averaged
monthly and yearly. 11-year periodicity appears also in the variations of numbers of different magnitude FEs (Fig. 4). Since all sufficiently large FEs (e.g., >5%) are connected

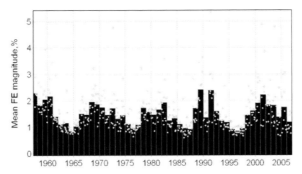

Figure 3. Monthly (points) and yearly (columns) mean FE magnitudes over 1957-2006.

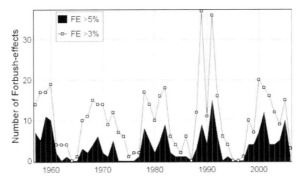

Figure 4. Annual numbers of Forbush effects with magnitudes >3% and >5% in 1957-2006.

with CMEs, by studying the variations in the number of large FEs we can get information about CMEs during those time periods when there were no CME observations.

If we compare the number of different magnitude FEs within different solar cycles, then some unexpected distinctions appear. The maximum sunspot number in cycle 23 is very similar to that in cycle 20. But this similarity goes away if we compare the numbers of FEs in these cycles. In cycle 20 the FEs of >5% magnitude registered was 23, whereas in cycle 23 it was 50. Large FEs appear in cycle 23 more often compared with any other cycle, starting from cycle 19. FEs very often appear in series and each series is usually connected with a burst of solar activity. This peculiarity allows the rate and magnitude of FE to be used for special index calculation (Belov *et al.* 2005) for quantitatively describing the rate of bursts of solar activity and forecasting their probability.

5. Cosmic ray anisotropy in Forbush Effects

If we look at the long term behavior of the first harmonic of the anisotropy in Fig. 5 where all results obtained so far by IZMIRAN group are combined, an opinion may be formed about slow, regular and well ordered variations of anisotropy. Note, that one of the most active researchers of long term variations of anisotropy and their connection with the solar magnetic cycle (which is evident from Fig. 5) was the same Scott Forbush (e.g., Forbush 1973). However, if we blow up any part of this curve one can see that the anisotropy is extremely variable, especially during large FDs.

It is during the FEs that the biggest anisotropy of the galactic CRs is observed. For example, on February 15, 1978 in one of the largest FE both the first and the second

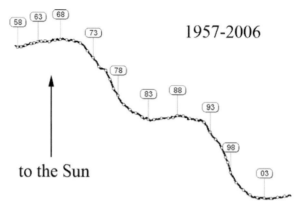

Figure 5. Behavior of the equatorial component of CR (10 GV) anisotropy in 1957-2006.

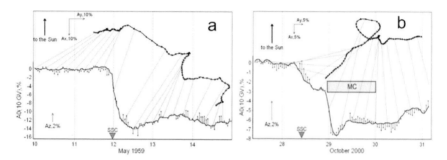

Figure 6. Behavior of the density, north-south and equatorial component of CR (10 GV) anisotropy in two Forbush effects.

harmonics of CR anisotropy reached about 10% (Belov *et al.* 1979) that is 1-2 orders higher than usual values.

Although the anisotropy of CRs changes in magnitude and direction during the whole FE, the fastest and essential variations usually occur near the interplanetary shock (SSC) and close to minimum CR density in the FD. That is clearly seen in the large FE recorded in May 1959 (Fig. 6). The CR anisotropy reflects a more detailed structure of interplanetary disturbance than CR density variations. In particular, the boundaries of magnetic cloud are normally clearly seen in the behavior of anisotropy. The magnetic cloud as a specific part of interplanetary disturbance is a cause for a second step in the two-step FD structure (Barnden 1973, Cane 1993, Wibberenz *et al.* 1997). In the event recorded at the end of October 2000 (Fig. 6) the first decrease of CR density started with the shock arrival near the SSC, another (much deeper) decrease is connected with a magnetic cloud that was passing the Earth from 23UT on October 28 to 00UT on October 30 (http://lepmfi.gsfc.nasa.gov/). In this case during the passage of the cloud the CR anisotropy vector rotated almost a full circle. We should remember also that magnetic clouds can influence the second harmonic of CR anisotropy even more strongly than the first harmonic (Richardson *et al.* 2000).

6. Some relations of Forbush effect characteristics

The FE is one part in a complex of sporadic phenomena characterizing a solar-heliospheric storm (eg. Belov *et al.* 2007c). It is natural to expect at least statistical

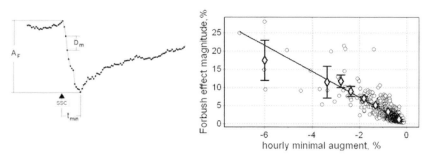

Figure 7. Typical Forbush effect time profile (left part) and relation of FE magnitudes A_F and maximal hourly decrements D_m (right part).

relations between the parameters of FEs and other phenomena, from solar flares to geomagnetic disturbances. Here we start from the inner interrelations among different characteristics of the FE itself.

Inner interrelations. The interrelations exist between parameters of either FE. Thus, maximum CR anisotropy in the FE well correlates with its magnitude A_F. In the right half of Fig. 7 we consider a relation of FE magnitude A_F to the maximum decrement D_m, i.e., minimal hourly augment of CR density variation. One can see that D_m and A_F (depicted in left half of the figure for typical FE time profile) are rather closely related (correlation coefficient is -0.87). It means that a reasonable forecast of maximum depth of FE can be made on the basis of FE evolving in a decrease phase.

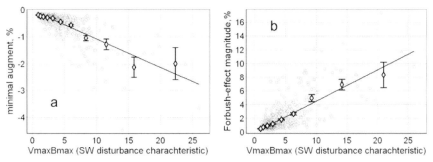

Figure 8. Dependence of Forbush effect maximal hourly decrements D_m (a) and magnitudes A_F (b) on product of maximum solar wind speed and IMF intensity in inteplanetary disturbances.

Relations to characteristic of solar wind disturbances. It is clear that the faster the propagation of interplanetary disturbance and the stronger its magnetic field, the faster will be the decrease of CR density during the main phase of FE. It may be assumed that the rate of a decrease for CR of certain rigidity R will vary in inverse proportion to the time t_r, which is necessary for a disturbance to go a distance equal to gyroradius of the same rigidity particle. Then D_m is proportional to VB, if V and B are solar wind velocity and interplanetary magnetic field (IMF) intensity within the most disturbed part of interplanetary disturbance. The parameter VB has been calculated in our data base for each event as VmaxBmax - the product of maximum values of solar wind velocity and IMF intensity in this disturbance, normalized to 400 km/s and 5 nT respectively. Using this parameter we can verify the above relation. A statistical relation between D_m and VmaxBmax in Fig. 8 is evident enough and close to the linear dependence. The magnitude A_F of FE is also related statistically to the same characteristic VmaxBmax, as obtained by (Belov *et al.* 2001) from the 1977-1998 data. To plot Fig. 8 the results calculated over the period 1964-2006 have been used. A relation of FD depth with the product VB is evident in some theoretical models (e.g., Parker 1963, Krymsky *et al.* 1981, Wibberenz *et al.* 1998).

Relations to geomagnetic activity. Belov *et al.* (2005) calculated mean values of FEs connected with different levels of geomagnetic activity, starting from very low (Kp-index <2-) and up to extreme magnetic storm (Kp = 9) using the data through 1977-1999. In the present work these calculations are made for a longer period (1957-2006) and presented in Fig. 9.

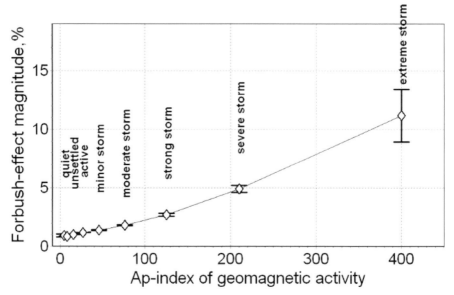

Figure 9. Dependence of mean Forbush effect magnitude on Ap-index of associated geomagnetic acivity.

We see a statistical relation between the FE magnitude and the geomagnetic activity level. Small values of FE (<1%) usually correspond to quiet and unsettled geomagnetic conditions. On the contrary, extremely large magnetic storms are usually followed by giant FEs, for example, the events in August 1972, July 1982 and October 2003. Of 16 magnetic storms with maximum Kp = 9, 13 were followed by the FEs >5%. On the other hand, half of the 10 biggest FEs during 50 years were followed by extreme geomagnetic storms, and in other cases there were severe or strong storms.

However, a relation between FEs and magnetic storms is statistical and is often violated. Two such exceptions are shown in Fig. 10, where the situation of 8 days in November 2002 is presented during which two FEs were recorded. The first one started on November 17, reached very large magnitude (7.4%), but was not followed by a magnetic storm (maximum Kp = 4). The next one, the onset of which is marked by SSC on November 20, was almost unnoticed (its magnitude was 0.8%), but it occurred on the background of a strong magnetic storm (maximum Kp = 7-). Both FEs and geomagnetic storms are created by the same disturbances of the solar wind, but in different ways. In a magnetic storm the key role may be played by a sign of the Bz component of the IMF, which generally doesn't affect the FD magnitude. The same may be said about the density of solar wind plasma. Geomagnetic disturbances are determined by the local characteristics of solar wind flowing around the Earth magnetosphere, whereas CR modulation is a result of the influence of the whole large scale interplanetary disturbance.

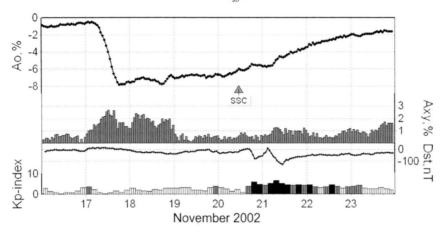

Figure 10. Density variation A0 and equatorial component of anisotropy Axy for 10 GV cosmic rays together with Kp- and Dst- indexes of geomagnetic activity during November 16-23, 2002.

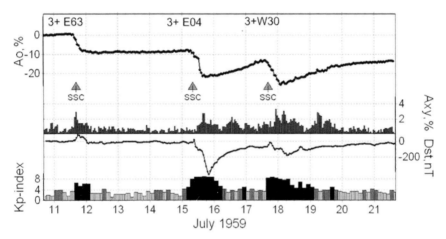

Figure 11. Variations of density and equatorial component of anisotropy Axy for 10 GV cosmic rays together with Kp- and Dst- indexes of geomagnetic activity during July 10-21, 1959. In the upper part optical importances and heliolongitudes are indicated for solar flares associated with the shocks that caused SSCs.

7. Forbush effect connection with solar flares and precursors of FEs

Forbush effects and solar flares. In Fig. 11 the series of three large FEs occurred in July 1959 is presented.

During one week, on July 11, 15 and 17 1959 three powerful shocks arrived at Earth after which strong geomagnetic storms and very deep FDs with magnitudes 10.1, 14.8 and 14.4% occurred. Each of these three events turned out to be associated with a powerful solar flare with optical importance 3+ (Dorman 1963) recorded on July 10, 14 and 16 correspondingly. All three flares generated in one active region at heliolatitudes N15-N26 and are distinguished mainly by heliolongitudes: E64 for the first flare, E04 and W30 - for the second and third. As we understand now, in all three cases large interplanetary disturbances came to Earth with high solar wind velocity, caused by powerful and fast CMEs, whose centers were near the mentioned flares (Yashiro *et al.* 2008). Those three CR events (together with the associated magnetic storms) were connected with the eastern, central and western CMEs, and it is the difference in the relative position of these ejections and Earth that explains the distinctions between those events. The first FE started on July 11, had a prolonged profile and it hardly recovered by the time of the onset of the next FE. It was followed by a short and moderate magnetic storm (maximum Kp = 7-, Dst-index is only -36 nT). The second FE was the biggest in this series. It was characterized by a fast two-step decrease and relatively quick recovery. The geomagnetic storm in that event was not only the biggest of these three but was one of the largest over the history of Dst index observations which fell down to -429 nT on July 15 (Kp index herewith went to a upper limit value 9). The third FE looked similar to the first one in the rate of decrease but had the fastest recovery. It was followed by a severe storm (Kp = 9-, minimum Dst was at -183 nT). In the first case Earth entered a remote western periphery of the eastern ejection and then gradually approached its central part. In the second case it is natural to think that Earth passed through the central part of disturbance including a magnetic cloud. Finally, in the third case the main part of disturbance passed to the west of the Earth and Earth rapidly left the main region of FD.

The dependence of FE characteristics on heliolongitude of solar sources was found long ago and had numerous discussions (eg. Sinno 1961, Iucci et al. 1986, Cane 2000). Recentlty higher efficiency of central CME in the FE creation and east-west asymmetry of such efficiency was demonstated on large experimental material (Belov et al. 2007c). The majority of ejections associated with far from the central meridian flares normally don't create the FE near Earth, but if they do these FEs are very small. However, there are exceptions from this rule, and they are not so rare (eg. Belov et al. 2003a). Usually they are associated with anomalously large and essentially powerful CMEs. Observing CR variations near Earth in these cases we get information about larger variations at the great distances [to the east or west of Earth].

Precursors of Forbush effects. In a similar way we obtain the information from CR variations about remote heliospheric events by observing the precursors of FEs. It was discovered long ago (eg. Fenton et al. 1959, Bloch et al. 1959), that sometimes the changes in CR behavior begin well before the arrival of the interplanetary shock or solar wind disturbance at Earth. Now we know that the effect of approaching shock (precursor) is a complicated combination of pre-increase and pre-decrease in CR variations and assumes specific angle distribution of CR intensity which is difficult to describe by the sum of the first spherical harmonics. However, this effect should result in CR density variations and in the behavior of the first spherical harmonic of CR anisotropy.

Fig. 12 is obtained for all the FEs over 1964-2006 that occurred on the quiet background. Presented results indicate a connection (correlation coefficient is 0.74) between CR anisotropy measured at the last hour before the shock and CR anisotropy inside the FE and far behind the shock. More detailed analysis shows that in some groups of events (especially with western solar sources) the precursors are sufficiently large and become noticeable more than a day before the shock arrives at Earth. This should not be surprising if we remember that the FE is a heliospheric phenomenon which starts well before the interplanetary disturbance arrives at Earth, when the disturbance is being formed near the Sun.

8. Conclusion

After 70 years of study FEs continue to be a hot topic. An array of problems connected with FE await to be solved. A complete FEs theory doesn't exist yet although both interesting and useful attempts of theoretical description have been made (eg. Parker 1963, Krymsky et al. 1981). Still there is no clear understanding of FE contribution in the full heliospheric CR modulation and long term CR variations. For a long time we

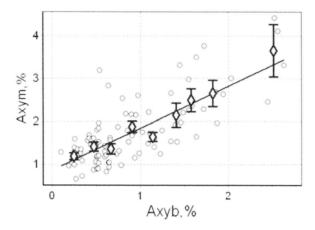

Figure 12. Maximum amplitude Axym of the equatorial component of CR anisotropy during the Forbush decrease versus amplitude Axyb of the same component just before the shock arrival.

have been aware of the benefit of FE observation for space weather tasks, but effective tools for practical use of FE information are not developed so far. FEs originate in the largest area of short term solar activity manifestations observed on Earth and nearby, which distinguishes FEs from geomagnetic and ionospheric disturbances and even solar particle events. That is why the connection of FE characteristics with the parameters of interplanetary disturbances or with geomagnetic activity indices exists but is not very close. Deflections from the average dependencies are informative, they are affected by specific features of the given interplanetary disturbance and its solar sources and provide information on heliospheric processes in interplanetary space remote from the Earth. The ability to reflect large scale processes, that are very often quite remote from the observation point make CR variations a unique tool for the study of solar activity and heliospheric processes. Without FE observation the picture of solar and heliospheric storms would not be complete and clear. Still more unique is the information on the old FEs (eg. 19 solar cycle events) when CMEs were not known, solar wind parameters were not measured and there was not even a tenth part of a customary now solar and heliospheric information. The study of old FEs (complete data of them have existed since 1957) is the most natural and straightforward method to investigate CME and other solar activity manifestations in long time period. Thus, FE data provide information on previous and coming storms. Being very complicated phenomena FEs need a complex approach and accumulation a great amount of various information. Our epoch of large data bases and Internet technologies should make such exploration both easier and more effective.

Acknowledgments. I am thankful to all teams providing continued ground level CR monitoring (http://cr0.izmiran.rssi.ru/ThankYou/). Special gratitude to International Astronomical Union and IAUS257 Organizing Committee for initiating and support this work.

References

Barnden L. R. 1973, *Proc. 13th Internat. Cosmic Ray Conf.*, 2, 1277
Barouch, E. & Burlaga, L. F. 1975, *J. Geophys. Res.*, 80, 449
Belov, A. V., Blokh, Ya. L., Dorman, L. I., Eroshenko, E. A., Guschchina, R. T., Kaminer, N. S., & Libin, I. Ya. 1979, *Proc. 16th Internat. Cosmic Ray Conf.*, 3, 449

Belov, A. V., Dorman, L. I., Eroshenko, E. A., Iucci, N., Villoresi, G., & Yanke, V. G. 1995, *Proc. 24th Internat. Cosmic Ray Conf.*, 4, 912

Belov, A. V., Eroshenko, E. A., & Yanke, V. G. 1997, *Correlated Phenomena at the Sun, in the Helosphere and in Geospace*, 463

Belov, A. V., Eroshenko, E. A., & Yanke, V. G. 1999, *Proc. 25th ICRC*, 6, 431

Belov, A. V., Eroshenko, E. A., Struminsky, A. B., & Yanke, V. G. 2001, *Adv. Space Res.*, 27, 625

Belov, A. V., Buetikofer, R., Eroshenko, E. A., Flueckiger, E. O., Oleneva, V. A., & Yanke V. G. 2003, *Proc. 28th Internat. Cosmic Ray Conf.*, 6, 3581

Belov, A., Bieber, J., Erosh., E. A., Even., P., Pyle, R. & Yanke, V. G. 2003, *AdSR*, 31, 919

Belov, A. V., Buetikofer, R., Eroshenko, E. A., Flueckiger, E. O., Gushchina, R. T., Oleneva, V. A., & Yanke V. G. 2005, *Proc. 29th Internat. Cosmic Ray Conf.*, 1, 375

Belov, A. V., Baisultanova L., Eroshenko E., Mavromichalaki H., Yanke V., Pchelkin V., Plainaki C., & Mariatos G. 2007, *J. Geophys. Res.*, 110, A09S20

Belov, A., Kurt, V., Mavromichalaki, H., & Gerontidou, M. 2007, *Solar Phys.*, 246, 457

Belov, A. V., Eroshenko E., Oleneva V., & Yanke V. 2007, *J. of Atmos. and Solar-Terr. Phys.*

Bloch, Ya. L., Dorman, L. I., & Kaminer, N. S. 1959, *Proc. 6th Internat. Cosmic Ray Conf.*, 4, 77

Cane, H. V. 1993, *J. Geophys. Res.*, 98, 3509

Cane, H. V. 2000, *ISSI Space Science Series, 10, "Cosmic Rays and Earth"*, 41

Crooker, N. U. & Cliver, E. W. 1994, *J. Geophys. Res.*, 99, 23383

Dorman, L. I. 1963, *Cosmic Ray Variations and Space Research*, oscow, N USSR, p.1027

Dorman, L. I. 1974, *Cosmic Ray Variations and Space Explorations*, North-Holland Publ. Co.

Fan, C. Y., Meyer, P., & Simpson, J. A. 1960, *Phys. Rev. Letters*, 4, 421

Fenton, A. G., McCracken, R. G., Rose, D. C. & Wilson, B. G. 1959, *Can. J. Phys.*, 37, 970

Forbush, S. E. 1937, *Phys. Rev.*, 51, 1108

Forbush, S. E. 1938, *Phys. Rev.*, 54, 975

Forbush, S. E. 1973, *J. Geophys. Res.*, 78, 7933

Hudson, H. S. 2007, *ApJ.*, 663, 45

Iucci, N., Parisi M., Storini M., & Villoresi G. 1979, *Nuovo Cimento*, 2C, 1

Iucci, N., Pinter, S., Parisi M., Storini M. & Villoresi G. 1986, *Nuovo Cimento*, 9C, 39

Krymsky G. F. *et al.* 1981, *Cosmic Rays and Solar Wind, Nauka, Novosibirsk*, p. 224

Lockwood, J. A. 1971, *Space Sci. Rev.*, 12, 658

Lockwood, J. A. & Webber, W. R. 1987, *Proc. 26th Int. Cosmic Ray Conf.*, 4, 87

Lockwood, J. A., Webber, W. R., & Debrunner, H. 1991, *J. Geophys. Res.*, 96, 11587

Parker E. N. 1963, *Interplanetary dynamical Processes, John Wiley and Sons, London*

Richardson, I. G., Dvornikov, V. M, Sdobnov, V. E. & Cane, H. V. 2005, *JGR*, 105, 12579

Richardson, I. G. & Cane, H. V. 2005, *Proc. Solar Wind 11*, 755

Sinno, K. 1961, *Ionosphere a. Space Res. Japan*, 15, 276

Wibberenz, G., Cane, H. V., & Richardson, I. G. 1997, *Proc. 25th Int. Cosmic Ray Conf.*, 1, 397

Wibberenz, G., Le Roux J. A., Potgieter M. S., & Bieber J. W. 1998, *Space Sci. Rev.*, 83, 309

Yashiro, S., Michalek, G., Akiyama, S., Gopalswamy, N., & Howard, R. A. 2008, *ApJ*, 673, 1174

Discussion

GOPALSWAMY: What are the differences between the Forbush effects of shock-driving and non-shock CMEs?

BELOV: There are some distinctions between these groups. Forbush effects with shocks are bigger on average than Forbush effects without shocks. CR variations in the first group are more complex than in the second group. However these differences are not so pronounced.

Universal Heliophysical Processes
Proceedings IAU Symposium No. 257, 2008
N. Gopalswamy & D.F. Webb, eds.

© 2009 International Astronomical Union
doi:10.1017/S1743921309029688

Anomalous Forbush effects
from sources far from Sun center

E. Eroshenko[1], Belov A.[1], Mavromichalaki, H.[2], Oleneva, V.[1], Papaioannou, A.[2] and Yanke V.[1]

[1] Pushkov Institute of Terrestrial Magnetism, Ionosphere and Radio Wave Propagation
Russian Academy of Sciences (IZMIRAN), 142190 Troitsk,
Moscow region, Russia
email: abelov@izmiran.ru

[2] Nuclear and Particle Physics Section, Physics Department, University of Athens,
Panepistimiopolis, Zografos GR-15783, Athens, Greece

Abstract. The Forbush effects associated with far western and eastern powerful sources on the Sun that occurred on the background of unsettled and moderate interplanetary and geomagnetic disturbances have been studied by data from neutron monitor networks and relevant measurements of the solar wind parameters. These Forbush effects may be referred to a special sub-class of events, with the characteristics like the event in July 2005, and incorporated by the common conditions: absence of a significant disturbance in the Earth vicinity; absence of a strong geomagnetic storm; slow decrease of cosmic ray intensity during the main phase of the Forbush effect. General features and separate properties in behavior of density and anisotropy of 10 GV cosmic rays for this subclass are investigated.

Keywords. elementary particles, Sun: activity, interplanetary medium, solar-terrestrial relations, coronal mass ejections (CMEs), shock waves.

1. Introduction

We consider the Forbush effect (FE) as a result of influence of interplanetary disturbance on the galactic cosmic rays (CR): it is the response of cosmic rays to the propagating disturbance including precursors (pre-increase and pre-decrease in CR variations), CR intensity decrease as the main phase (FD), and the recovery phase while the Earth exits a disturbance area (Belov *et al.* 2003a, Belov *et al.* 2007). Usually FE is observed simultaneously with the interplanetary magnetic field (IMF) increase. Cosmic ray intensity decrease (FD) during the FE is created in a separate region of interplanetary space where the particle access from outside is difficult. The stronger the IMF created in this special area, the wider is rigidity diapason of CR exposed to this effect, the stronger the CR modulation. As a rule, the deep Forbush decreases correspond to the big increases of the IMF intensity. In Belov *et al.* (2005) we studied the exceptions to this rule where FEs with a magnitude more than 6 % were created on the background of weak disturbances of the IMF. It was found that in a large fraction of the efficient events, the Earth is struck only by a peripheral area of the interplanetary disturbance and the main part misses the Earth, as a rule to the east. However, the event in July 2005 caused by the far western source on the Sun indicated analogous features of higher efficiency of a disturbance modulating influence on the CR (Papaioannou *et al.* 2008). In the present work we continue our study of such anomalous effects. Among all FEs we distinguished a subclass of events characterized by relatively unsettled interplanetary and geomagnetic conditions (IMF < 15 nT, Kp < 6), gradual decrease of the CR intensity on the main phase of FE and slow recovery phase. It turned out that all these events were caused by

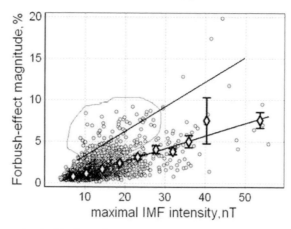

Figure 1. FE magnitude (A_m) vs the maximal IMF intensity B_m. Diamonds are the A_F values averaged by the equal B_m intervals.

far eastern or western solar sources and have very characteristic features in a behavior of CR variations. The goal of our work is to analyze the subclasses of such anomalous FEs associated with far eastern and western solar flares, to study the variations of CR density and anisotropy during these effects and to compare and contrast the CR behavior for these two groups with the aim to use these results for diagnostics of the interplanetary space and near-Earth vicinity.

2. Data and methods

The database on the Forbush effects and interplanetary disturbances, created in IZMI-RAN, has been used for the above study. This database includes as parameters of CR density (A0) and anisotropy (equatorial component of the first harmonic, Axy) and data on the solar wind, interplanetary magnetic field, solar data and geomagnetic activity indices Kp and Dst (Archive SPIDR Data Base, available from http://spidr.ngdc.noaa.gov). The updated measurements on GOES, and OMNI data base have been utilized from the web sites: http://www.ngdc.noaa.gov, http://omniweb.gsfc.nasa.gov, http://sec.ts.astro.it and CME data from http://lasco-www.nrl.navy.mil. The list of sudden storm commence-ments (SSC) has been also used (ftp://ftp.ngdc.noaa.gov) as a proxy for interplane-tary shocks; the time of SSC generally defines the onset of FE. Density and anisotropy for 10 GV CR was derived by the global survey method (Belov *et al.* 2005, Belov *et al.* 2007) over the period 1965-2006 by the hourly data from neutron monitor net-work (http://cr0.izmiran.rssi.ru/common/links.htm) and included into the IZMIRAN database (Asipenka *et al.* 2008). Normally, the stronger the interplanetary disturbance, the bigger is the FE amplitude (A_F) which has in total a linear dependence on B_m (Belov *et al.* 2001). Only the events separated by at least 24 hours have been taken for the analysis and only those where the IMF data provided has at least 80 % coverage.

3. Discussion of the results

The well known relation between IMF intensity and FD magnitude (Belov *et al.*, 2001) is obtained in the present work on the extremely large statistical base (2000 events).

In Fig. 1 one can see a dependence of FE magnitude (A_F) on the maximum IMF intensity B_m. The majority of points are concentrated near the regression line, but there is

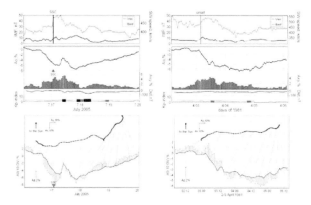

Figure 2. Examples of the FEs in July 2005 and April 1981 caused by the far western solar sources: IMF, solar wind, geomagnetic data (Kp-index and Dst variations), cosmic ray density and anisotropy (A0 and Axy) during the FEs associated with the western solar flares W79 and W52. In the bottom panels vector diagram of CR anisotropy (equatorial component Axy) and density (A0) are presented. Vertical vectors mean north-south component of the CR anisotropy. Thin lines connect equal time points in each 6 hours in vector diagram and density curves.

a group placed significantly higher than this line. A correlation coefficient of 0.67 obtained by the 1987 events shows a relation between FE magnitudes and B_m, so the statistical character of this relation and the possibility to be essentially violated in some cases: large FEs observed under relatively low values of the IMF intensity, and on the contrary, relatively small FEs corresponded sometimes to large IMF intensity. We were interested in the events (circled group) where small and moderate values of IMF corresponded to large A_F. It looks as a result of more effective influence of a disturbance on the CR. A degree of such the efficiency might be estimated as $K_F = A_F/B_m$, %/nT.

This coefficient may be considered as a characteristic of the modulating ability of the associated IMF disturbance and it was one of the main parameters by which a selection of events for analysis was carried out. Our experience shows the events with $K_F > 0.36$ belong to a group of anomalous events when near-Earth measured IMF parameters could not provide the observable Forbush effect. Selection of the FEs by this parameter among events identified with associated solar sources gives the following distribution by the source longitudes within 60° ranges.

type	East (E90-E30)	Centre (E30-W30)	West (W30-W90)
$K_F > 0.36$ anomalous	14	10	9
$K_F < 0.36$ normal	24	62	13

One can see that in normal events the central sources dominate whereas anomalous FEs are mainly caused by extreme eastern or western sources. All anomalous FEs turned out to be of >3 % magnitude and occurred in relatively weakly disturbed interplanetary and geomagnetic conditions (IMF < 15 nT, Kp < 6), with gradual fall of CR intensity in the main phase of Forbush-effect. We compared the characteristics of CR variations during the FEs from western and eastern sources.

In Fig. 2 the examples of FEs in July 2005 (western source at the longitude W79°) and in April 1981 (longitude W52°) are presented. In both cases on the background of relatively quiescent interplanetary and geomagnetic conditions we observe FEs with the

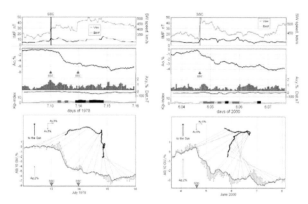

Figure 3. Parameters of the cosmic ray, interplanetary and geomagnetic activity during the FEs associated with far eastern solar flares: in July 1978 (E58) and June 2000 (E60). There designations are the same as in Fig 2.

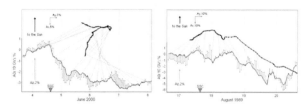

Figure 4. Vector diagrams for equatorial component of CR anisotropy Axy during the FEs associated with eastern (June, 2000, E60) and western (August 1989, W60) solar sources. The designations are the same as in Fig 2.

magnitude of FD >4 % and significant increases of Axy, up to 4.6 % in the main phase (whereas in normal events it doesn't exceed 2 %). The Axy increase started at once with the FE onset thereafter the direction of Axy vector varied very little and remained usually westward during the recovery phase.

In Fig. 3 we give the same presentation of different parameters for the FEs associated with far eastern solar sources. The difference in behavior of the vector Axy in these cases and in western events during the FEs is evident.

The magnitude of FEs caused by the eastern sources turned out to be a little less than for the western ones, but the vector of the equatorial component of CR anisotropy Axy shows quite different behavior by direction (see Fig. 4). For 9 western events the FE averaged magnitude A0 was 4.5 %, maximum value of the equatorial component of anisotropy Axy = 3.5 %, solar wind velocity did not exceed 600 km/s. For 10 eastern events the mean value of A0 = 4.2 %, maximum Axy = 2.1 %.

In Fig. 4 vector diagrams of the anisotropy Axy are presented comparing the eastern and western events. We can note very specific behavior of the CR anisotropy before and during the FEs which is different for eastern and western events. By an order of magnitude the increase of Axy is essentially larger for western FEs, however for 'east' events we observe sharp changes of direction of Axy during the main phase of Forbush-effect. The properties of Axy behavior may help with the identification of a disturbance source causing the FE when there is not sufficient information to make it unambiguously (e.g. Fig. 5). In these plots one can see the characteristic behavior of Axy vector which evidences the FE is caused by a western source.

All the above considered FEs are associated with disturbances originating from sources far from Sun center, and every time the Earth was localized at the edge of the propagating

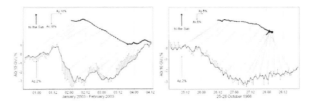

Figure 5. Two examples of events from non-identified sources (February 2003 and October 1966).

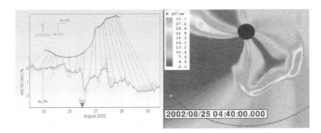

Figure 6. Vector diagram of CR anisotropy behavior and CME model calculations (Lugaz *et al.* 2008) during the August 2002 event.

disturbance, the near-Earth measurements did not show a high perturbation in the Earth's vicinity. The high magnitude of FEs from such remote sources on this quiet background implies great power of a disturbance and tells about modulating over an area wider than the size of the disturbance. How this may occur in the case of a western source may be, in particular, illustrated by Fig. 6 where the picture of propagating disturbance (adopted from Lugaz *et al.* 2008) and behavior of CR anisotropy are presented together.

The FE started on August 26, 2002 after weak shock arrival on the background of moderate interplanetary and geomagnetic activity ($B_m = 15$ nT, Kp $= 5$-, minimum Dst $= -47$ nT). The FE magnitude was not large (about 2%), but the CR anisotropy was rather large for such small effect - 3.3%. It was caused by a CME originating from the western limb (W81) and associated with X3.1/1F solar flare. Fig. 6 demonstrates that the western flank of the CME came to open magnetic field lines connected to the Earth. This effect is one of the reasons why this CME originating from the western limb caused a noticeable FE with large CR anisotropy on the unsettled background.

4. Conclusions

• In the majority of effective events the Earth enters only a periphery of the interplanetary disturbance, the main part of which misses Earth (to the East or to the West).

• Those events are preceded by powerful flares on the Sun, which are generally located far from the center of the solar disk. The CMEs and interplanetary disturbances, originating from the near limb longitudes, appear to be of larger size and more complicated structure than is visible near Earth.

• It was possible to separate small but very definite groups of large FEs which were not followed by strong interplanetary disturbances near Earth, nor by high geomagnetic activity but seemed to be associated with great solar wind disturbances that missed Earth.

• The anomalous eastern FEs have a prolonged descent phase with a later minimum but larger magnitude in CR density than western events. They show a sharp change of the anisotropy direction in the minimum of FE.

• The anomalous western FEs strongly differ from typical FEs caused by western sources which are usually not large and very short. They also differ from the anomalous eastern FEs by a bigger size of CR anisotropy and less variability of its direction.

• These properties may be used for the inner heliosphere as diagnostics and Space Weather predictions.

• In the cases of far sources the CR observations give better information about the real power and size of a disturbance than near-Earth measurements of the solar wind. Moreover, it may be a useful tool in the cases of events where there is not sufficient data for their identification.

Acknowledgement

This work is supported by RFBR Grants 07-02-00915, 07-02-13525, by the Program of Presidium RAS "Neutrino Physics". Authors are thankful for the collections from all stations of world networks providing the monitoring of the CR neutron component http://cr0.izmiran.rssi.ru/ThankYou/main.htm.

References

Asipenka, A. S., Belov, A. V., Eroshenko, E. F., Klepach, E. G., & Yanke, V. G. 2008, *J. Adv. Space Res.*, doi:10.1016/j.asr.2008.09.022

Belov, A. V., Gushchina, R. T., Eroshenko, E. A., Yudakhin, K. F., Yanke, & V. G. 2007, *Geomagnetism and Aeronomy*, 47, 251

Belov, A. V., Gushchina, R. T., Eroshenko, E. A., Ivanus, D., & Yanke, V. G. 2005, *Proc. 29th Internat. Cosmic Ray Conf.*, 2, 239

Belov, A. V., Eroshenko, E. A., Struminsky, A. B., & Yanke, V. G. 2001, *Adv. Sp. Res.*, 27, 625

Belov, A., Bieber, J., Eroshenko, E. A., Evenson, P., Pyle, R., & Yanke, V. G. 2003, *Adv. Sp. Res.*, 31, 919

Belov, A. V., Buetikofer, R., Eroshenko, E. A., Flueckiger, E. O., Oleneva, V. A., & Yanke V. G. 2003, *Proc. 28th Internat. Cosmic Ray Conf.*, 6, 3581

Lugaz, N., Roussev, I. I., Sokolov I. V., & Jacobs C. 2008, *Proc. IAU275 Symposium*, in this volume

Papaioannou, A., Belov, A., Mavromichalaki, H., Eroshenko, E., Oleneva, V. J., & Yanke, V. G. 2008, *J. Adv. Space Res.*, doi:10.1016/j.asr.2008.09.003

Discussion

GOPALSWAMY: In the anomalous FE events, the correlation is different because the maximum magnetic field used is not the real field in the ICME. If you estimate the correct B_{max}, the anomalous FEs will look like the normal FE events.

EROSHENKO: The anomalous FEs look so because we do not have a correct B_{max} (it might be obtained either from the full observations or from any good model). Since we have only near Earth observations, it doesn't reflect the situation near (or inside) the disturbance and we are talking about "anomalous" effect with respect to data in the Earth environment.

Universal Heliophysical Processes
Proceedings IAU Symposium No. 257, 2008
N. Gopalswamy & D.F. Webb, eds.

© 2009 International Astronomical Union
doi:10.1017/S174392130902969X

Bootstrap energization of relativistic electrons in magnetized plasmas

Ilan Roth

Space Sciences, University of California at Berkeley, Berkeley, CA 94720, USA
email: ilan@ssl.berkeley.edu

Abstract. *In situ* and remote observations indicate that relativistic or ultra relativistic electrons are formed at various magnetized configurations. It is suggested that a specific bootstrap mechanism operates in some of these environments. The mechanism applies to (a) relativistic electrons observed on localized field lines in outer radiation belt - through a process initiated at a distant substorm injection; (b) relativistic electrons observed at the interplanetary medium - through a process initiated via coronal injection, at large distances from flares or propagating CME; (c) ultra-relativistic electrons deduced at the galactic jets - through a process initiated via local injection at the small-scale magnetic field. The injected nonisotropic electrons excite whistler waves which boost efficiently the tail of the electron distribution.

Keywords. acceleration of particles, plasmas, jets, radiation mechanisms: nonthermal

1. Introduction

Direct *in situ* heliospheric measurements on board spacecraft and remote solar, heliospheric and astrophysical observations through electromagnetic emissions indicate that energization mechanisms of electron populations to quasi-relativistic, relativistic or ultra-relativistic energies operate at various magnetized plasmas. The in situ measurements relate principally to enhanced fluxes of relativistic electrons (i) in the terrestrial outer radiation belts at L~ 4-10 (denoting equatorial distance of a dipole-like field in units of Earth radius), and (ii) at the interplanetary medium, mainly at heliospheric distances of ~1 AU, capturing electrons of solar origin. The magnetospheric enhancements of relativistic electrons coincide with the recovery phase of geomagnetic storms and possibly with sub-storm injections, while the relation of the relativistic heliospheric electrons to flares or to Coronal Mass Injections (CMEs) shocks still remains debatable. Remote solar and heliospheric electromagnetic observations at various wavelengths reveal the energy spectra of the non-thermal electrons, the locations of the emission processes and together with the in-situ measurements impose constrains on the energization mechanisms. The fluxes of (ultra)relativistic electrons of astrophysical origin are remotely deduced through observation of radiative emissions. Some of the most intense occurrences of these electromagnetic waves are related to jets emanating from accretion discs in radio active galactic nuclei. Although all these observations relate to vastly different magnetized plasmas with various geometries, the question rises regarding a possibility that similar processes operate in these environments.

It is conjectured, therefore, that an analogous physical process at various magnetic configurations may enhance a subset of relativistic electron fluxes in magnetospheric, solar, and astrophysical jet plasmas. The bootstrap mechanism requires existence of a stressed, large-scale magnetic structure, distant injection of seed non-isotropic electrons due to reconnection and energization on closed, inhomogeneous magnetic fields lines.

2. Overview

Electrons form excellent tracers of magnetic field and an important source of electromagnetic radiation: coherent emissions due to collective plasma processes and incoherent single particle emissions due to interaction with plasma or with magnetic field.

a) Terrestrial (planetary) magnetic storms are initiated by an intense, persistent southward interplanetary field which deforms the geomagnetic field, increasing the ring current and decreasing the energetic electron population in the radiation belt (Fig. 1a). In the storm recovery phase the fluxes of relativistic electrons increase often by orders of magnitude above the pre-storm level over few days as a result of Ultra Low Frequency waves diffusion across the actively distorted magnetospheric fields, towards lower L shells (stronger magnetic field), by preserving the first adiabatic invariant and increasing the energy (Baker et al., 1986; Reeves et al., 2003); however, many satellite crossings observe fast ($\sim$1 hour) and large enhancement with a peak at L $\sim$ 4.0, indicating local energization at a confined region of field lines. Various observations (Meredith et al., 2003) correlate this enhancement with the observed substorm injection of sub-relativistic, non isotropic electrons and excitation of magnetospherically reflected whistler waves along the inhomogeneous field lines (Bortnik et al., 2006, Shklyar et al., 2004). The interaction between whistler waves and electrons extends the tail of the distribution into the relativistic domain. Therefore, the observed peaks of relativistic electron fluxes at low L-shells are formed by a bootstrap process initiated via distant injection at large L shells.

b) Solar flares are initiated due to a reconfiguration of the magnetically unstable coronal field, with ensuing injection of electron and ion fluxes towards the chromosphere, exciting emissions at radio, X-ray and γ-ray frequencies. Additionally, magnetic stress release may result in a detachment of a large blob of plasma ($\sim 10^{15}$ g) and its propulsion into the interplanetary medium in the form of Coronal Mass Ejections (CME; Fig 1b). Type III bursts are observed when an electron beam propagates along magnetic field and excites Langmuir waves at the local plasma frequency, which are then partially converted into coherent electromagnetic radiation (Lin, 1985, Wang et al., 2006). The Langmuir waves indicate the local density, allowing one to determine the coronal or interplanetary excitation locations and the beam propagation speed. Type II emissions reflect the location of the propagating CME shock. Precise timing of the observed electron fluxes at $\sim$1 AU (Krucker et al., 1999) showed that one may distinguish between low-energy electrons (< 20 keV) which are injected almost instantaneously with the type III emission and more energetic electrons with a delay of 10-30 minutes. Similarly, mildly relativistic fluxes at 30-350 keV were delayed by up to 40 minutes with respect to the metric type-III, hard X rays and microwave electromagnetic emissions (Haggerty et al., 2003). Long-lasting relativistic electron fluxes, which are observed in conjunction with flares (X rays and type III) and intense CMEs (type II emissions) show an onset of 25 minutes after the type III initialization (Klassen et al., 2005), while Nancay Radioheliograph observations correlate these relativistic enhancements to coronal bursts of 100-s MHz emissions, without connectivity to the flare site and behind the intense CME (Maia et al, 2004; Pick et al., 2005). Hence, these bursts release magnetic energy in the CME evacuated domain in the form of subrelativistic seed electrons, which, in analogy to terrestrial substorms or lightenings, excite whistlers that extend the electron tail into relativistic energies (Roth, 2008). Therefore, the observed delayed relativistic electrons in the interplanetary medium are energized through a bootstrap process initiated via coronal injection, on closed field lines, at large distances from the flare sites or/and CME shock.

c) An active galactic nucleus is a compact region at the center of a galaxy with an unusually high luminosity powered by accretion onto massive black hole; in these accretion discs the conversion of the gravitational to radiation energy can reach the Eddington luminosity limit. Some accretion discs produce highly collimated and fast outflowing jets, whose formation mechanism is not fully understood but believed to result due to acceleration and squeezing of plasma by a twisting magnetic field (Blandford and Payne, 1982; Fendt and Memola, 2001). These relativistic jets extend as far as tens/hundreds of kilo-parsecs from the central black hole (Fig 1c) and are known to provide electrons with a huge relativistic factor up to $\gamma \sim 10^6$. They exhibit obvious observational effects in the radio waveband, where Very Large Array, Hubble Space Telescope and Chandra X-ray Observatory can be used to study the radiation they emit down to sub-parsec scales. Two correlated problems of jet emission include (i) confirmation of the radiation process and (ii) long-duration of its energetic source. The main emission mode is (Lorentz boosted) synchrotron and the observed cutoff (due to the progressive emission softening with increasing energy) is attributed to a cooling process of a single source electron population in a large scale magnetic field (Meisenheimer and Heavens, 1986), with a time scale much shorter than the temporal extent of the jets, which necessitates continuous replenishment of the energetic electrons. Additionally, several recent measurements at a very high resolution of 0.3 arcsec (Jester *et al.*, 2005) indicate flattening of the UV spectrum, requiring either second electron component or a different emission process. An elegant solution for this flattening stipulates that a significant part of the magnetic energy density exists in the form of inhomogeneous, small-scale magnetic fields such that electron trajectory is distorted from a simple gyration through interaction with random small-scale fields, resulting in a "jitter" emission (Medvedev, 2000) and in flattening of the spectrum through diffusive synchrotron radiation (Fleishman, 2006). Simulations of similar configurations indicate formation of these small scale structures (Frederiksen *et al.*, 2004; Nishikawa *et al.*, 2005) far behind the shock. It is conjectured that the adjacent, marginally stable, reconnecting small-scale magnetic field arcs inject a seed of non-isotropic electrons into the closed field lines, analogously to the solar scenario, and through efficient resonant interactions with whistlers the tail of the electron distribution is boosted to ultra relativistic energies. Therefore, the ultra-relativistic jet electrons are energized through a bootstrap process initiated via local injection at the small-scale magnetic field.

3. The Bootstrap model

The above-described configurations and processes feature several similarities, in spite of significant differences in the magnetic geometry and in energization time scales. In all of the configurations there exists directly measurable or indirectly deduced large-scale,

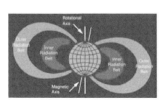  

Figure 1. (a) Sketch of the terrestrial radiation belts. (b) Imaged solar corona with an uplifting CME (LASCO). (c) Jet from Galaxy M87 (Hubble Heritage Project).

deformed field and additional marginally stable field configuration which injects non-isotropic electron populations into closed magnetic fields. These electrons excite whistler waves which propagate and reflect along the magnetic field lines boosting efficiently the energetic tail of the distribution.

3.1. Wave propagation

The nonisotropic distribution of the injected electrons constitutes a source of propagating waves with group velocity directed mainly along the magnetic field. The main excited mode is whistler; its phase/group velocities and amplitude depend on the electron pitch angle distribution and the local plasma parameters.

Oblique whistler eigenmode with wavenumber $\mathbf{k} = (k_\perp, k_\parallel) = (k\ sin\theta, 0, k\ cos\theta)$ and frequency ω, propagating at an angle $\theta = cos^{-1}(k_\parallel/k)$ with respect to the magnetic field is supported by the bulk of injected electrons; neglecting thermal and relativistic effects (justified for the terrestrial and solar injections while requiring correction for more energetic jet injection), the local dispersion relation, with negligible gradients, becomes:

$$(kc/\omega)^2 = [B + (B^2 - 4AC)^{1/2}]/2A \tag{3.1}$$

where $A = \epsilon_1\ sin^2\theta + \epsilon_3\ cos^2\theta$, $B = \epsilon_1\ (\epsilon_3 + A) - \epsilon_2^2\ sin^2\theta$, $C = \epsilon_3\ (\epsilon_1^2 - \epsilon_2^2)$, while ϵ_i denote the components of the dielectric tensor. For parallel propagation ($\theta = 0$), neglecting the ions, Eq (3.1) degenerates into $(kc/\omega)^2 = 1 + \omega_e^2/[(\Omega - \omega)\omega]$, where ω_e and Ω denote the plasma and the nonrelativistic gyrofrequency of the injected electrons, respectively. Whistler ray propagating along an inhomogeneous magnetic field may undergo reflection when its wave normal passes through $\pi/2$ and its longitudinal group velocity $V_{g\parallel}$ reverses sign (e.g., Kimura, 1966, 1985). Hence, this reflecting wave, as observed by numerous satellites, may resonate with bouncing particles numerous times.

3.2. Resonant interaction

Over short interaction times between electrons with gyroradius $\rho = \gamma v_\perp/\Omega$ and gyrofrequency Ω/γ (γ denotes the relativistic factor) and whistler waves, when the plasma and wave parameters change slowly, irreversible changes in energy, adiabatic invariant and pitch angle may take place. Direct integration of the unperturbed trajectories ($z = v_\parallel t, x = \rho sin(\Omega t/\gamma)$) in the propagating wave frame gives

$$cos(k_\parallel z + k_\perp x - \omega t) \sim \Sigma J_n (k_\perp\rho)cos[k_\parallel v_\parallel\ t - (\omega - n\Omega/\gamma)t] \tag{3.2}$$

indicating that as an electron and a whistler propagate along the magnetic field, they may encounter numerous locations where the phase is almost stationary, resulting in the resonance condition (for an integer n)

$$k_\parallel v_\parallel = \omega - n\Omega/\gamma. \tag{3.3}$$

For finite $k_\perp\rho$ all regular $n < 0$ and anomalous $n > 0$ harmonics contribute to changes in energy and pitch angles. Bessel function J_n in (3.2) signifies the increased interaction effectiveness with a higher electron energy. Hence, if the phase angle ζ between the perpendicular electric wave field $E_\perp$ and velocity is hardly modified during the interaction time, while the parallel velocity satisfies (3.3), the electron undergoes an irreversible energy change

$$d\gamma/dt \sim (e/mc^2)E_\perp v_\perp sin\zeta \tag{3.4}$$

Electron dynamics with a monochromatic (3.2) whistler wave may be analyzed trough the normalized relativistic Hamiltonian $H(\mathbf{x}, \mathbf{P})$ with $\mathbf{P} = m\mathbf{v}\gamma + q\mathbf{A}(\mathbf{x})/c$ and (3.2):

$$H = [1 + (\mathbf{p} - \mathbf{A}\ (\mathbf{x}))^2]^{1/2} + \Phi(\mathbf{x}) \tag{3.5}$$

where the canonical momentum P is normalized to mc, the time t to the inverse gy-rofrequency at the reference position Ω_o^{-1}, the spatial coordinates $\mathbf{x}$ to c/Ω_o, and the wavenumber k to Ω_o/c. The background magnetic field is represented by the normalized potential $\mathbf{A_o} = x\eta(\delta z)\mathbf{y}$, with the normalized gyrofrequency $\eta(\delta z) = \Omega(\delta z)/\Omega_o$, where the δ dependence denotes the slowly changing gyrofrequency (mirror force). The wave electric field $[E_x \cos\psi, E_y \sin\psi, E_z \cos\psi]$ is derived from the electrostatic $\Phi = \delta_o \sin\psi$ and the electromagnetic potential $\mathbf{A} = [\delta_1 (k_\parallel/k)\sin\psi, \delta_2 \cos\psi, -\delta_1 (k_\perp/k)\sin\psi]$, with the phase $\psi = [\int \mathbf{k}\,(\delta z)\,\mathbf{x} - \omega t]$. Two canonical transformations cast the equations of motion around a single resonance into (Roth *et al.*, 1999)

$$dZ/dt \sim [(P_\parallel + l\eta/k_\parallel)/\gamma_o - \omega/k_\parallel] + (\partial G_l/\partial P_\parallel)\,cos(k_\parallel Z)/k_\parallel] \tag{3.6}$$

$$dP_\parallel/dt = -(I/\gamma_o)(\partial\eta/\partial z) + G_l k_\parallel sin(k_\parallel Z) \tag{3.7}$$

where $Z = z + l\theta/k_\parallel - \pi/2k_\parallel - (\omega/k_\parallel)t$, $\quad I = I' + lP_\parallel/k_\parallel$, and

$$G_l = [\delta_1 (P_\parallel sin\phi - l\,\eta\,cos\phi/k_\perp)/\gamma_o + \delta_o]\,J_l[k_\perp\sqrt{2I/\eta}] + \delta_2[\sqrt{2I\eta}/\gamma_o]J_l^{\,'}[k_\perp\sqrt{(2I/\eta)}] \tag{3.8}$$

The G_l terms describe higher harmonic coupling while a negligible stationary phase $k_\parallel Z$ (similarly to (3.2)) satisfies the resonance condition (first term of 3.6) and assures small parallel acceleration (3.7); I' is an adiabatic invariant and the irreversible change in action I, $\Delta I = l\Delta P_\parallel/k_\parallel$, which is the major contribution to energy diffusion, is determined by the phase (ζ in Eq 3.4). Very intense nonlinear whistlers, as observed recently on board Stereo satellite (Cattell *et al.*, 2008) may additionally enhance the initial energization (Omura *et al.*, 2008).

4. Implications

There may exist an interesting similarity in energization processes of electrons to (ultra) relativistic energies in a variety of vastly different magnetized plasmas. A common thread connecting these processes includes a strongly distorted large-scale magnetic field and a distant electron seed injection. In the discussed examples the large-scale magnetic fields exhibit the following distorted configurations: (a) terrestrial (planetary) magnetospheric field with a northward polarity is distorted due to a persistent southward interplanetary magnetic field; (b) solar coronal magnetic field is radically distorted when a large blob of plasma is detached from the corona and is propelled into the interplanetary medium as CME, preserving its magnetic connection to the solar surface; (c) strongly distorted magnetic field around accretion disc is frozen in the collimated jet plasma over vast spatial distances. In order for the proposed bootstrap mechanism to operate, an additional marginally stable, stressed magnetic field configuration is required: (i) pinching of the distant terrestrial magnetotail field and thinning of the supporting current results in a substorm, with a directly observed injection of nonisotropic, sub-relativistic electrons into closed terrestrial field lines; (ii) coronal field behind the propagating CME releases its tension via (indirectly deduced through 100s MHz emissions) injection of nonisotropic, sub-relativistic electrons into the closed coronal field lines; (iii) small-scale, inhomogeneous field arcs, required for consistency with observed synchrotron radiation emitted by the relativistic electrons in the galactic jets, reconnect and inject nonisotropic electrons into these closed field lines.

In all cases the injected electrons excite coherent whistler waves which propagate and reflect along the closed field lines, interacting efficiently with the tail of the electron population and boosting its energy to relativistic or ultra relativistic energies. Hence, the bootstrap acceleration mechanism imposes several observational predictions: (a) the local enhancement of relativistic fluxes in the magnetospheric radiation belts will not be produced without bursty substorm injection, (b) the formation of the (delayed) heliospheric relativistic electrons will not occur without bursty coronal injection behind the intense propagating CME (which later opens the venue for the energetic electrons to the interplanetary medium); (c) the re-acceleration of the jet electrons would not occur via the bootstrap mechanism without formation of small-scale inhomogeneous magnetic fields. The energization time scale decreases dramatically with higher injected energy opening availability of additional resonant sites, and with a more anisotropic seed population resulting in higher whistler wave amplitudes.

References

Baker, D, et al., 1986, *Jour. Geoph. Res.*, **91**, 4265

Blandford, R. D. & D. G. Payne, 1982 *MNRAS*, **199** , 883

Bortnik, J., U. S. Inan, & T. F. Bell, 2006, *Geoph. Res. Lett.*, **33**, 3102

Cattell, C. et al., 2008, *Geoph. Res. Lett*, **35**, L01105

Fendt, C. & E. Memola, 2001, *Astronomy & Astrophysics*, **365**, 631.

Fleishman, G. D., 2006, *Ap. Jour.*, 638, 348.

Frederiksen, J. T. et al., 2004, *Astrophys. J.*, **608**, L13

Haggerty, D. K. et al., 2003, *Adv, Space. Res.*, **32**, 2673

Jester, S. et al, 2004, *Astronomy & Astrophysics*, **431**, 477.

Kimura, I., 1966, *Radio Sci.*, **1**, 269

Kimura, I., 1985, *Space Sci. Rev.*, **42**, 449

Klassen et al., 2005, *Jour. Geoph. Res.*, **110**, AO9S04

Krucker, S. et al., 1999, *Astrophys. J.*, **519**, 864

Lin, R. P., *Sol. Phys.*, **100**, 519, 1985.

Maia, D. J. F. & M. Pick, *Astrophys. J.*, **609**, 1082, 2004.

Medvedev, M. V., *Astrophys. J.*, **540**, 704, 2000.

Meisenheimer, K. & A. F. Heavens,, 1986, *Nature*, **323**, 419

Meredith, N. et al, *Geoph. Res. Lett.*, **30**, 1871, 2003.

Nishikawa, 2005, et al., *Astrophys. J.*, **622**, 927, 2005.

Omura, Y., Y. Katoh, & D. Summers, *Jour. Geoph. Res.*, **113**, AO4223, 2008.

Pick, M. & D. J. F. Maia, *Adv, Space Res*, **35**, 1876, 2005.

Reeves , G. D., et al., *Geoph. Res. Lett.*, **30**, 1529, 2003.

Roth, I., M. Temerin, & M. K. Hudson, *Annales Geophysicae*, **17**, 631, 1999.

Roth, I, *Jour. Atm. Sol. Terr. Phys.*, 70, 490, 2008.

Shklyar, D. R., J. Chum, & F. Jiricek, *Ann. Geoph.* , **22**, 3589, 2004.

Wang, L., R. P. Lin, S. Krucker, & J. T. Gosling, *Geoph. Res. Lett.*, **33**, L03106, 2006.

Discussion

THEJAPPA: In your picture, you show that type II emission is coming from quasiparallel shock and other emissions are coming from quasiperpendicular part of the shock in the downstream. How do you explain the electron acceleration by a quasiparallel shock and downstream near the flanks of such a shock?

ROTH: The type II low electron energy emissions are due to the propagating shock but the relativistic electrons are formed due to interaction with whistler waves, which are excited by the non-isotropic electrons, as observed by the NRH emissions, in resemblance to the magnetospheric observations.

SPANGLER: You mentioned the necessity of strong small-scale magnetic fields in extragalactic radio sources. Radio astronomical observations of such sources have given us much information on the magnetic field and energetic particles. The level that the synchrotron radiation is almost always polarized, and often highly polarized, places constraints on the amplitude and/or isotropy of small scale magnetic fluctuations.

ROTH: Several recent observations (Jester, 2005, and reference therein) eliminate the possibility of various electron sources, and the only possibility to explain the observed spectra asserts that small scale inhomogeneous magnetic field are crucial, and that their integrated energy density is of the order of the macroscopic magnetic energy density (Fleishman, 2005). The small scale features are very inhomogeneous. Polarization issue is not resolved yet experimentally.

Universal Heliophysical Processes
Proceedings IAU Symposium No. 257, 2008
N. Gopalswamy & D.F. Webb, eds.

© 2009 International Astronomical Union
doi:10.1017/S1743921309029706

History of the solar environment

Kurt Marti[1] and Bernard Lavielle[2]

[1]University of California at San Diego, Dept. of Chemistry (0317),
9500 Gilman Dr., La Jolla, CA 92093-0317

[2]University of Bordeaux, CNRS, Laboratoire de Chimie Nucléaire Analytique
et Bio-environnementale (CNAB), Domaine Le Haut Vigneau - BP 120,
33175 Gradignan cedex, France

Abstract. Galactic cosmic rays (GCR) provide information on the solar neighborhood during the sun's motion in the galaxy. There is now considerable evidence for GCR acceleration by shock waves of supernova in active star-forming regions (OB associations) in the galactic spiral arms. During times of passage into star-forming regions increases in the GCR-flux are expected. Recent data from the Spitzer Space Telescope (SST) are shedding light on the structure of the Milky Way and of its star-forming-regions in spiral arms. Records of flux variations may be found in solar system detectors, and iron meteorites with GCR-exposure times of several hundred million years have long been considered to be potential detectors (Voshage, 1962). Variable concentration ratios of GCR-produced stable and radioactive nuclides, with varying half-lives and therefore integration times, were reported by Lavielle *et al.* (1999), indicating a recent 38% GCR-flux increase. Potential flux recorders consisting of different pairs of nuclides can measure average fluxes over different time scales (Lavielle *et al.*, 2007; Mathew and Marti, 2008). Specific characteristics of two pairs of recorders (^{81}Kr-Kr and ^{129}I-^{129}Xe) are the properties of self-correction for GCR-shielding (flux variability within meteorites of varying sizes). The ^{81}Kr-Kr method (Marti, 1967) is based on Kr isotope ratios, while stable ^{129}Xe is the decay product of the radionuclide ^{129}I, which is produced by secondary neutron reactions on Te in troilites of iron meteorites. The two chronometers provide records of the average GCR flux over 1 and 100 million year time scales, respectively.

Keywords. Galactic cosmic ray (GCR) flux, Starforming regions in galaxy, Chronometers for GCR flux calibration, Iron meteorites as flux recorders

1. Introduction

Galactic cosmic rays (GCR) provide information on the energy density and on the discrete sources in the "local" region of the galaxy. The energies of GCR (typically 100 MeV to 10 GeV) probably derive from supernova (SN) explosions, which occur once every 30 to 60a in the galaxy (e.g. Axford, 1981). In order to maintain the currently observed intensity of GCR over millions of years, only a few percent of the SN energy has to be used for the GCR acceleration. There is considerable evidence that this acceleration is accomplished in the shock waves as they propagate through the surrounding interstellar gas. This GCR origin is not only suggested from direct observational evidence, but is based on the recognition that the predominant types of SN in spiral galaxies like our own are those which originate from massive stars, predominantly in spiral arms where most massive stars are born and die (Dragicevich *et al.* 1999). The recent Spitzer (SST, NASA, 2008) data provide a considerably different structure of the spiral arms in our own galaxy. High contrasts in the non-thermal radio emission are observed between the spiral arms and the disks of spiral galaxies. Whenever our Sun was located in a region of the galactic spiral arms which show star formation regions (OB associations), an increased GCR flux has to be expected.

The GCR flux and the path lengths of GCR reaching the solar system are also affected by the interstellar medium (ISM). Begelman and Rees (1976) calculated that while crossing moderately dense ISM clouds with densities of 10^2 to 10^3 atoms cm^{-3}, the bow shock of the heliosphere will be pushed inward further than 1 AU. As a consequence, the GCR slowing down effect by the heliosphere will cease to work and the flux of lower energy particles will be significantly increased. It has also been speculated (e.g. Shaviv 2002, 2003) that variable GCR fluxes may be connected to the Earth's cloud cover and possibly the appearance of ice ages.

The position of our Sun with regard to the star formation regions (the original OB associations have lost their O-members, because these have already ended their cycles as SNe), has been investigated in several surveys. The SST measurements of the spiral structure, and in particular the number of arms, has changed and these data suggest that the picture of two major arms is more favorable than for four. They appear to govern the flux systematics for the Sun's orbit in the Milky Way. Direct evidence for recent close-by events is not strong, but Knie et $al.$ (2004) report ^{60}Fe in a deep-sea manganese crust as evidence for a supernova 2.8 Ma ago. A comprehensive census of the stellar content of the OB associations within 1 kps from the Sun was reported by De Zeeuw et $al.$ (1999). This is a project which studies the formation, structure, and evolution of nearby young stellar groups and related star-forming regions. The OB associations are unbound moving groups which can be detected kinematically because of their small internal velocity dispersion.

In general, the intrinsic flux of cosmic rays reaching the outskirts of the solar system is proportional to the star formation rate in the solar system's vicinity. Although there is a lag of several million years between the birth and death of the massive stars which is ultimately responsible for cosmic ray acceleration, this lag may be small when compared with the relevant time scale of GCR flux variations during the last $\sim$ 100 Ma. In order to investigate the effects of the spatial and temporal variations and of clusterings of cosmic-ray sources, Higdon and Lingenfelter (2003) developed a model to calculate the age and path length distributions of cosmic rays reaching the solar system by summation of the diffusive contributions of known discrete sources. This model calculation includes the galactic star-forming regions and the OB stars lost subsequently as SN. In principle, this allows the separation of effects due to spatial and temporal clusterings in the solar neighborhood on the local GCR flux variations.

Direct measurements appear possible and require measurements of GCR flux changes and production rate calibrations, based on the systematics of GCR-produced nuclides in iron meteorites. A variability may be inferred from differences in calculated average-fluxes based on cosmic-ray-produced radionuclides of 0.2 to 16 Ma half-lives, as well as associated stable decay product nuclides, which integrate over the time of exposure of these natural monitors. We address some of these efforts and recent experimental progress. In analyzing such GCR flux monitors data, one can first show that the use of constant production rates leads to contradictions in CRE ages, and then evaluate flux models which yield consistent CRE ages and select those which are consistent with dynamic models for the clustering of cosmic ray sources and inferred temporal variations.

2. Development of GCR Flux Monitors

It has long been recognized that iron meteorites are excellent fossil detectors of cosmic rays since some of them were exposed for periods in excess of 1 Ga. A substantial database of cosmic-ray-produced nuclides already exists, including the data from the ^{40}K/^{41}K method, first developed by Voshage (1962). Several studies indicated systematic differences between ages obtained by radioactive and stable nuclide pairs, such as ^{36}Cl/^{36}Ar, ^{39}Ar/^{38}Ar, and ^{10}Be/^{21}Ne when compared to the ^{40}K/^{41}K results. Although several authors interpreted this as evidence for variability in the cosmic ray flux, the data from several iron meteorites also indicated complex exposure histories. Complexities were also found in spallation records from different locations within the same meteorites, specifically for large recovered masses. In many cases the evidence for complex exposure histories in iron meteorites was difficult to assess, but cosmic-ray records in iron meteorites must take into account evidence for multiple breakups in the evaluation of possible variations in cosmic-ray intensity. A combined study of complex exposure histories and of the constancy of galactic cosmic rays was carried out by Lavielle *et al.* (1999). The authors' goal was to recalibrate the ^{40}K/^{41}K ages for a large number of iron meteorites which were inferred to have experienced simple exposure histories.

In this flux calibration the authors assumed a time-independent cosmic ray flux for a 0.5 Ga period. Their results show that average fluxes based on the ^{40}K halflife (1.26 Ga) disagree in a systematic way with calibrations based on radionuclides of about million year half-lives (^{81}Kr, ^{36}Cl, ^{10}Be, ^{53}Mn). Their calculated average production rates of ^{36}Cl or of ^{36}Ar over the calibration interval (0.15-0.70 Ga) were lower by 28%, when compared to production rates commonly used for the recent cosmic-ray flux. These authors concluded that a recent cosmic ray flux increase offered a most straightforward explanation, but they cautioned that uncertainties in spallation systematics for product nuclei close to a doubly magic mass number (40) should not be ignored. They further concluded that the magnitude of the resulting shifts in exposure ages depend on adopted models for flux changes. The recent GCR flux increase inferred by Lavielle *et al.* (1999) compared to average fluxes during the time interval 150 to 700 Ma ago did not provide specific information about the timing of the increase, nor the possibility of a cyclic variation. The geometries and crossings into spiral arms of our galaxy need to be reevaluated based on the new SST data.

3. Shielding-correcting nuclide pairs

1. Flux monitors for the past 1 Ma

Two chronometers are useful for the determination of the recent GCR flux: ^{81}Kr - ^{83}Kr$_c$ ($t_{1/2} = 0.23$ Ma), and ^{36}Cl - ^{36}Ar ($t_{1/2} = 0.30$ Ma).

These chronometers are ^{81}Kr $-$ ^{83}Kr$_c$ ($\lambda_{81} = 3.24$ x 10^{-5} a^{-1}) with ^{81}Kr(t) = (P_{81}/λ_{81}) $(1 - \exp(-\lambda_{81} t))$ and a measured ratio $[^{83}$Kr$_c]$ / $[^{81}$Kr] = (P_{83}/P_{81}) λ_{81} $T_{CRE}/(1 - \exp(-\lambda_{81} T_{CRE}))$ and the chronometer ^{36}Cl $-$ ^{36}Ar$_c$ ($\lambda_{36} = 2.46 \times 10^{-5}$ a^{-1}) with ^{36}Cl(t) = (P (^{36}Cl)/λ_{36}) $(1 - \exp(-\lambda_{36} t))$ and measured $[^{36}$Ar$_c]/[^{36}$Cl] = $P_{36}(^{36}$Ar)/$P_{36}(^{36}$Cl) λ_{36} $T_{CRE}/(1 - \exp(-\lambda_{36} T_{CRE}))$ which in both cases self-correct for GCR-shielding variations (assuming a single exposure geometry).

They are expected to give identical CRE ages of iron meteorites, and can be used for cross-calibrations, as long as the recent flux was constant. The first chronometer (Marti, 1967) has been considered to be one of the most reliable methods for cosmic-ray-exposure dating. Concentrations of the radionuclide ^{81}Kr in meteorites are typically of the order of several 10^5 atoms per g of sample, but are considerably lower in large iron meteorites. This limitation required the development of a new mass spectrometer with very high sensitivity. Such a facility has been developed at CNAB (University of Bordeaux), making use of resonant laser ionization (RIS) for Kr at 216.4 nm, using cryogenic sample concentrator and a time-of-flight mass analyzer (Lavielle *et al.*, 2007). This technique is about 50 times more sensitive than a conventional mass spectrometer and achieves a mass resolution of about 400 and a detection limit for ^{81}Kr $\leqslant$ 1,000 atoms (Lavielle *et al.*, 2007); first results have been obtained for iron meteorite Old Woman.

2. Flux monitors during the past 10 Ma

The pairs ^{10}Be $- ^{21}$Ne ($t_{1/2} = 1.6$ Ma) and ^{53}Mn $- ^{53}$Cr ($t_{1/2} = 3.7$ Ma) are largely self-correcting for shielding variability and are considered to represent suitable monitors for the last 10 Ma. The radionuclides (by AMS) and integrating stable spallation components (^{53}Cr the decay product) need to be measured.

3. Flux monitor during the past ~100 Ma

The chronometer ^{129}I $- ^{129}$Xe$_c$ ($\lambda_{129} = 4.6 \times 10^{-7}$ a^{-1}) with ^{129}I (t) = P$_{129}/\lambda_{129}$ [1 $-$ exp ($-\lambda_{129}t$)] and a ratio [^{129}Xe$_n$]/[^{129}I] $= \lambda_{129}$ T$_{CRE}$ / [1 $-$ exp ($-\lambda_{129}$ T$_{CRE}$)]$^{-1}$ is suitable to study a possible recent (<100 Ma) flux increase.

This change is assessed by the nuclide of appropriate half-life, ^{129}I which is produced by neutron reactions on Te. This $t_{1/2} = 16$ Ma is ideal for monitoring changes in the GCR flux over the last 100 Ma. The ^{129}I-^{129}Xe$_n$ chronometer (Marti, 1986) which is based on the pair ^{129}I and its integrating stable decay product ^{129}Xe was used in a study of troilite in Cape York (Murty and Marti, 1987). However, the cosmic-ray-produced ^{129}Xe$_n$ needs to be resolved from products of "extinct" ^{129}I. The presence of a GCR spallation component is observed by elevated ^{124}Xe/^{130}Xe and ^{126}Xe/^{130}Xe ratios, compared to corresponding ratios in the trapped Xe component. In iron meteorites products due to low energy neutron capture reactions on ^{127}I, ^{128}Te and ^{130}Te are found as excesses ^{128}Xe$_n$, ^{129}Xe$_n$ and ^{131}Xe$_n$. A ^{129}I-^{129}Xe$_n$ chronometer appears to be especially suitable in Te-rich minerals since low-energy secondary neutrons are predominant and the chronometer may provide CRE ages in cases where production rates of commonly used nuclides are not known due to heavy shielding in large meteorites. Both reactions ^{128}Te(n, γ)^{129}Te, $\beta^- \rightarrow ^{129}$I, and ^{130}Te(n, 2n) ^{129}Te, $\beta^- \rightarrow ^{129}$I are reaction channels for the production of ^{129}I from Te nuclides. ^{129}Xe$_n$ is produced via decay of precursor ^{129}I; this presents the parent-daughter system that is used for the chronometry. GCR secondary neutron reactions on Te provide a system which is independent of shielding, as long as the exposure geometry remains constant.

Mathew and Marti (2008) improved the experimental techniques required for the Xe isotopic measurements in Te-rich troilite of iron meteorites. They showed that the neutron-produced excesses ^{129}Xe$_n$ and ^{131}Xe$_n$ show a linear correlation, and the slope identifies GCR epithermal neutrons as the prevalent source of particle reactions in Cape York troilite. Therefore the reaction excess ^{131}Xe$_n$ serves as a suitable monitor of GCR reactions and permits the calculation of the total excess ^{129}Xe$_n$.

4. Conclusions and Outlook

Although the GCR flux in the inner solar system is observed to be variable over a variety of time-scales because of solar modulation effects, longer time variations reflect changes in the local interstellar medium and in the sources of GCR. We discussed potentially useful nuclide pairs, which represent suitable GCR flux monitors for the time-scales of 1 Ma, 10 Ma and 100 Ma. Two of those methods were developed to the point that they can be used for flux calibrations. Other chronometers need to be further developed. For example, a very sensitive technique already exists for ^{53}Mn (3.7 Ma halflife) measurements by accelerator mass spectrometry (Korschinek *et al.*, 1987), while GCR-produced excesses on stable nuclide ^{53}Cr need to be assessed, but can be measured on multi-collector instruments with high precision. Calibrated average GCR fluxes over the one, ten and hundred million year time-scales are expected to provide the data required to assess the flux environment during recent crossings of star-forming and of inter-arm regions of the solar system. For average flux data for the entire circular motion in the galaxy an integration over longer time-scales is required. Therefore, it is desirable to improve isotopic abundance measurements of K in iron meteorites with documented one-stage exposure to GCR. This could provide average flux data over the time-scale of ^{40}K which corresponds to several revolutions of the solar system in the galaxy.

Among possible applications, the currently hypothetical connection between GCR flux variations and the terrestrial cloud cover is of interest. The implied temperature reductions and the appearance of ice ages need to be assessed (e.g. Calogovic *et al.*, this volume), once documentations of flux variations and of the time of crossing star-forming regions in the galaxy are known.

References

Axford, W. I. 1981, *Proc. 17th Intl. Cosmic-Ray Conference (Paris)*, 12, 155

Begelman, M. C. & Rees, M. J. 1976, *Nature*, 261, 298

Calogovic, J., Arnold F., Desorgher L., Flueckiger E. O., & Beer, J. 2008, Forbush Decreases: No change of global cloud cover. *Universal Heliophysical Processes*, IAU Symposium 257 Abstracts, 30

De Zeeuw, P. T., Hoogerwerf, R., De Bruijne, J. H. J., Brown, A. G. A., & Blaauw, A. 1999, A Hipparcos Census of the Nearby OB Associations. *Astrophys. J.*, 117, 354–399

Dragicevich, P. M., Blaie, D. G., & Burman, R. R. 1999, *MNRAS*, 302, 693.

Higdon, J. C. & Lingenfelter, R. E. 2003, The myriad-source model of cosmic rays: I. Steady state age and path length distributions. *Astrophys. J.*, 582, 330–341

Korschinek, G., Morinaga, H., Nolte, E., Preisenberger, E., Ratzinger, U., Urban, A., Dragov-itsch, P., & Vogt, S. 1987, Accelerator mass spectrometry with completely stripped 41-Ca and 53-Mn ions at the Munich tandem laboratory. *Nucl. Instrum. Methods Phys. Res.*, B29, 67

Knie, K., Korschinek, G., Faestermann, T., Dorfi, E. A., Rugel, G., & Wallner, A. 2004, ^{60}Fe Anomaly in a Deep-Sea Manganese Crust and Implications for a Nearby Supernova Source. *Phys. Rev. Letters*, 93(17), 171103–1

Lavielle, B., Marti, K., Jeannot, J.-P., Nishiizumi, K., & Caffee, M. W. 1999, The 36Cl-36Ar-40K-41K records and cosmic ray production rates in iron meteorites. *Earth Planet. Sci. Lett.*, 170, 93–104

Lavielle, B., Gilabert, E., & Thomas, B. 2007, A new facility for the determination of cosmic ray exposure ages in small extraterrestrial samples using 81Kr-Kr dating method. *70th Met. Soc. Mtg.*, A92 (abstract)

Marti, K. 1967, Mass-spectrometric detection of cosmic-ray-produced ^{81}Kr in meteorites and the possibility of Kr-Kr dating. *Phys. Rev. Lett.* 18(7), 264–266

Marti, K. 1986, Live ^{129}I-^{129}Xe dating. In *Workshop on Cosmogenic Nuclides* (ed. P. A. J. Englert and R. C. Reedy), pp. 49-51. LPI Tech. Rpt. 86-06. Lunar and Planetary Institute

Mathew, K. J. & Marti, K. 2008, Galactic cosmic-ray-produced 129Xe and 131Xe excesses in troilites of the Cape York iron meteorite. *Met. & Planet. Sci.*, accepted for publication.

Murty, S. V. S. & Marti, K. 1987, Nucleogenic noble gas components in the Cape York iron meteorite. *Geochim. Cosmochim. Acta*, 51(1), 163–172

NASA: www.spitzer.caltech.edu/Media/index.shtml
www.spitzer.caltech.edu/features/articles/20070103.shtml
www.spitzer.caltech.edu/Media/releases/ssc2008-10/ssc2008-10a.shtml

Shaviv, N. J. 2002, Cosmic ray diffusion from the galactic spiral arms, iron meteorites, and a possible climatic connection. *Phys. Rev. Lett/*, 89(5)

Shaviv, N. J. 2003, *New Astronomy*, 8, 2003, 39–77

Voshage, H. 1962, Eisenmeteorite als Raumsonden fur die Untersuchung des Intensitatsverlaufes der komischen Strahlung während der lezten Milliarden Jahre. *Z. Naturforsch.*, 17a, 422–432

Discussion

ANONYMOUS: When did the increase in cosmic ray flux by 38% occur?

MARTI: Less than 150 My ago, based on the 38% flux increase calibration.

ANONYMOUS: Is there any evidence on Earth of life extinctions through ozone loss, coincident with the penetration of GCR due to supernova explosions as might be expected when the solar system crosses a star forming region?

MARTI: The chronometers discussed in the talk aim at establishing the time of the last GCR-flux increase due to passage through a star-forming region. Results are not available at present. Ozone loss may be a possible result, but is not established.

Universal Heliophysical Processes
Proceedings IAU Symposium No. 257, 2008
Nat Gopalswamy & Dave Webb, eds.

© 2009 International Astronomical Union
doi:10.1017/S1743921309029718

Cosmic ray spectra in planetary atmospheres

M. Buchvarova[1] and P. Velinov[2]

[1]Space Research Institute, Bulgarian Academy of Sciences,
6 Moskovska Str., Sofia 1000, Bulgaria
email: marusjab@yahoo.com

[2]Solar-Terrestrial Influences Laboratory, Bulgarian Academy of Sciences
Acad. G. Bonchev, bl.3, Sofia 1113, Bulgaria

Abstract. Our model generalizes the differential $D(E)$ and integral $D(>E)$ spectra of cosmic rays (CR) during the 11-year solar cycle. The empirical model takes into account galactic (GCR) and anomalous cosmic rays (ACR) heliospheric modulation by four coefficients. The calculated integral spectra in the outer planets are on the basis of mean gradients: for GCR – 3%/AU and 7%/AU for anomalous protons. The obtained integral proton spectra are compared with experimental data, the CRÈME96 model for the Earth and theoretical results of 2D stochastic model. The proposed analytical model gives practical possibility for investigation of experimental data from measurements of galactic cosmic rays and their anomalous component.

Keywords. Cosmic ray spectra, modelling, planet atmospheres

1. Introduction

The observed CR spectrum can be distributed into the following five energy intervals (Dorman 1977): I ($E = 3 \times 10^6$–10^{11} GeV/n), II ($E = 3 \times 10^2$–3×10^6 GeV/n), III ($E = 30$ MeV/n – 3×10^2 GeV/n), IV ($E = 1$–30 MeV/n), V ($E = 10$ KeV/n – 1 MeV/n), where E is the kinetic energy of the particles. The modulation above 100 GeV is negligible due to the magnetic field of the heliosphere, and the energy spectrum of cosmic rays is described by power law. Particles with energy below 100 GeV are subject to solar modulation due to the presence of the heliosphere. In this paper it is proposed an empirical model for the calculation of the cosmic ray proton and helium spectra on the basis of balloon and satellite measurements in the energy intervals III and IV.

2. Cosmic ray differential spectrum

The empirical model for the differential spectrum (energy range E from about 30 MeV to 100 GeV) of protons and other groups of cosmic ray nuclei on account of the anomalous cosmic rays (energy range E from 1 MeV to about 30 MeV) is given by Velinov (2000) and Buchvarova (2006):

$$D(E) = K(0.939 + E)^{-\gamma} \left(1 + \frac{\alpha}{E}\right)^{-\beta} \left\{\frac{1}{2}\left[1 + \tanh\left(\lambda\left(E - \mu\right)\right)\right]\right\}$$

$$+ \frac{x}{E^y}\left\{\frac{1}{2}\left[1 - \tanh\left(\lambda(E - \mu)\right)\right]\right\} \quad (2.1)$$

The coefficients α, β, x and y are related to modulation levels in corresponding energy intervals. The terms with tanh are smoothing functions (Velinov 2002). The dimensionless parameter $\lambda = 100$ is inversely proportional to the length of the smoothing interval between the two addends. The physical meaning of μ (GeV) is the energy at which the differential spectrum of GCR crosses the differential spectrum of ACR (Buchvarova 2006).

Experimental data (E_i, D_i) for protons and helium nuclei were gotten from Hillas (1972) for solar cycle 20. The power index $\gamma = 2.63$ (Hillas 1972). The normalization constants $Kp = 14.55671$ GeV$^{2.63}$/(s.m^2 ster. MeV) for protons and $K\alpha = 0.90985$ GeV$^{2.63}$/(s.m^2 ster. MeV/n) for alpha particles are chosen to match the modulated data near to 100 GeV/n, where the modulation effect is negligible. The coefficient values for the protons with these model parameters are: $\alpha = 9.848438$, $\beta = 0.881461$, $x = 0.000012$, $y = 1.825847$, $\mu = 0.043094$ at solar minimum and $\alpha = 1.108842$, $\beta = 1.029644$, $x = 0.000189$, $y = 1.79179$, $\mu = 0.012681$ at solar maximum. We obtain the following values for alpha particles: $\alpha = 0.575525$; $\beta = 0.622310$, $x = 0.000193$, $y = 1.641234$; $\mu = 0.022171$ at solar minimum and the values $\alpha = 1.325028$, $\beta = 1.027078$, $x = 0.000059$, $y = 1.432015$ and $\mu = 0.034635$ at solar maximum.

The calculation of coefficients α, β, x, y and μ is performed by Levenberg-Marquardt algorithm (Press *et al.* 1991), applied to the special case of a least squares. The described programme is realized in algorithmic language C^{++}.

3. Application of the model to planetary atmospheres

The solar EUV radiation is weaker than the galactic cosmic ray intensity for Jupiter, Saturn, Uranus and Neptune. This shows the importance of the galactic cosmic rays in the formation of outer-planet ionospheres. The integral spectra of the outer planets are obtained using mean integral gradient of GCR – 3%/AU (Fuji et al. 1997; Heber *et al.* 1995) and 7%/AU for anomalous protons (McKibben *et al.* 1979; Christian *et al.* 1999) and the obtained values of coefficients from Eq. (2.1) for the Earth. For integration we use the equation:

$$D(> E) = \int\limits_{E}^{\infty} D(E)dE \qquad (3.1)$$

$D(E)$ – differential spectrum of galactic and anomalous CR, $D(>E)$ – integral spectrum, expressed by the number of particles per unit solid angle, square centimeter, and second, with total energies at least E. Integration is performed by Simpson rule (Press *et al.* 1991).

In Figure 1 is presented the modelled integral spectrum for protons at minimum and maximum of solar activity for Earth and Jupiter. The data for CR protons by Shopper (1967) are noted with crosses and with full squares for the CRÈME96 model (CREME96). The computations are compared with satellite data: ▲ – Voyager 2 (5 July 1979 – near to solar maximum) for Jupiter and theoretical results of 2D stochastic model built from Bobik *et al.* (2006).

It is seen from Figure 1 that in Bobik's model, the integral spectrum of the Earth almost tally with the Jupiter's integral spectrum at solar minimum. It is due to lower average values of integral radial gradient in this model. Actually at high values of the diffusion coefficient radial gradient has lower values in the inner heliosphere. Above few GeV the difference between modulated spectra for different planets becomes negligible in Bobik's model. In our model integral spectra are computed in first approximation. We assume mean integral gradient of GCR as 3%/AU for all rigidities, irrespectively of the distance in the heliosphere or solar activity level.

The Bobik's model with the noted parameters (from Figure 2) reproduces comparatively average level of the solar activity.

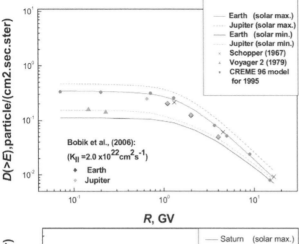

Figure 1. The modelled integral spectra $D(>E)$ of CR protons for maximum and minimum levels of solar activity for Earth and Jupiter. The computations are compared with the experimental data: × – Schopper (1967) for Earth, ▲ – Voyager2 (5 July 1976 – near to solar minimum) for Jupiter and theoretical results of 2D stochastic model built from Bobik *et al.* (2006) and ● – CRÈME model (CREME96) for Earth.

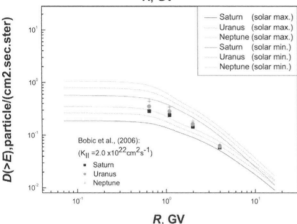

Figure 2. The modelled integral spectra $D(>E)$ of CR protons for maximum and minimum levels of solar activity for Saturn, Uranus and Neptune. The results are compared with computations of Bobik *et al.* (2006) for 2D transport model with drift.

The calculated integral proton spectra for solar minimum and maximum for Jupiter, Saturn, Uranus and Neptune are show in Figure 2.

References

Bobik, P. *et al.* 2006, in: *Proc. of 20 ECRS*, (Lisbon, Portugal)

Buchvarova, M. 2006, PhD Thesis, Sofia

Christian, E. *et al.* 1999, in: B. L. Dingus, D. B. Kieda, & M. H. Salamon (eds.), *Proc. of 26 ICRC., Salt Lake City, Utah* (USA, New York: American Institute of Physics Press)

CREME96, Cosmic ray ion differential and integral spectra. Available from: http://tec-ees.esa.int/ProjectSupport/ISO/CREME96.html.

Dorman, L. I. 1977, in: *Proc. of 15 ICRC*, (Sofia: Bulg. Acad. Sci. Publ. House), p. 405

Fuji, Z. & McDonald, F. B. 1997, *J. Geophys. Res.*, 102, 24201

Heber, B. *et al.* 1995, *Space Sci. Rev.*, 72, 391

Hillas, A. M. 1972, *Cosmic Rays* (Oxford; Pergamon Press)

McKibben, R. B, Pyle, K. R., & Simpson, J. A. 1979, *ApJ.* (Letters), 227, L147

Velinov, P. I. Y. 2000, *Compt. Rend. Acad. Bulg. Sci.*, 53, 37.

Press, W. H., Flannery, B. P., Teukolsky, S. A., & Vetterling, W. T. 1991, *Numerical Recipes in C – the Art of Scientific Computing* (Cambridge University Press)

Schopper, E. 1967, in: K. Sitte (ed.), *Encyclopedia of Physica, v. XL* (Berlin: Springer), p. 372

Universal Heliophysical Processes
Proceedings IAU Symposium No. 257, 2008
N. Gopalswamy & D.F. Webb, eds.

© 2009 International Astronomical Union
doi:10.1017/S174392130902972X

SEPs and CMEs during cycle 23

Pertti Mäkelä[1], Nat Gopalswamy[2], Seiji Yashiro[3], Sachiko Akiyama[1], Hong Xie[1] and Eino Valtonen[4]

[1] Dept. of Physics, Catholic University of America,
200 Hannan Hall, Washington, DC 20064, USA
email: `pertti.makela,sachiko.akiyama,hong.xie@nasa.gov`

[2] Nasa Goddard Space Flight Center,
Code 695, Greenbelt, MD 20771, USA
email: `nat.gopalswamy@nasa.gov`

[3] Interferometrics, Inc,
13454 Sunrise Valley Drive, Herndon, VA 20171, USA
email: `seiji.yashiro@nasa.gov`

[4] Dept. of Physics & Astronomy, University of Turku,
Vesilinnantie 5, FI-20014 Turku University, Finland
email: `eino.valtonen@utu.fi`

Abstract. We present a study of solar energetic particles (SEPs) in association with coronal mass ejections (CMEs) and type II radio bursts. The particle and CME observations cover the years 1996–2007. We find that heavy-ion events in association with type II bursts and proton events are produced in more western and most energetic CMEs. In addition, the source distribution of type II associated proton events with heavy ions reminds the source distribution expected for events with flare particles. Therefore, the estimation of relative contributions by flares and shocks in SEP events and separation of suggested different particle acceleration models is complicated.

Keywords. Acceleration of particles, Sun: coronal mass ejections (CMEs), Sun: particle emission

1. Introduction

Large Solar Energetic Particle (SEP) events are closely associated with Coronal Mass Ejections (CMEs). It is believed that solar wind particles are accelerated by CME-driven coronal and interplanetary (IP) shock fronts (Reames 1999). However, based on heavy-ion observations Cane *et al.* (2003) have suggested that higher-energy SEPs could originate from solar flares. Instead, Tylka *et al.* (2005) attribute variations in high-energy heavy-ion abundances to different shock geometries and seed particle populations. Coronal and IP shocks may also generate type II radio bursts. Type II bursts occur either continuously or intermittently in metric, decameter-hectometric (DH) and kilometric wavelengths. The emission wavelength depends on the density of the ambient plasma, which decreases with distance from the Sun. The metric type II bursts are emitted by shocks in dense coronal plasma close to the Sun ($\lesssim 2.5$ R$_S$) and kilometric type II bursts by IP shocks in less dense plasma further out from the Sun ($\gtrsim 10$ R$_S$). Previously Gopalswamy (2003) has found that large SEP events are all associated with DH type II bursts. The association of CMEs, proton events and type II burst has also been studied recently by Gopalswamy *et al.* (2008). In this study we concentrate on CMEs associated heavy-ion events and DH type II radio bursts. We also use proton events for comparison.

Table 1. Average properties of CMEs.

Heavy-ion Enhancement	CME properties Speed [km/s]	Width [deg]	Halo-%
Yes	1518±160	176±35	72
No	1111±130	166±28	56

2. Data analysis

Our analysis focus on CMEs associated with proton events and DH type II radio bursts. Particle and CME measurements by the Energetic and Relativistic Nuclei and Electron (ERNE; Torsti *et al.* 1995) instrument and by Large Angle and Spectrometric Coronagraph (LASCO; Brueckner *et al.* 1995) cover the period 1996–2007. Both instruments are onboard the Solar and Heliospheric Observatory (SOHO) spacecraft. We search for proton intensity increases in the 1.8–4.1 MeV, 4.1–12.7 MeV, 13.8–28.0 MeV and 25.9–50.8 MeV energy channels. Plasma measurements by Proton Monitor (Hovestadt *et al.* 1995), also onboard SOHO, are used to identify events associated with IP shocks, i.e. Energetic Storm Particle (ESP) events, and Corotating Interaction Regions (CIRs). We inspect the O and Fe intensities in 4–15 MeV/n and 43–70 MeV/n energy channels for concurrent heavy-ion enhancements. Proton event onsets are compared with the LASCO CME catalogue (http://cdaw.gsfc.nasa.gov/CME_list/) in order to identify a possible CME. We search for an associated DH type II radio burst observed by Wind/WAVES (Bougeret *et al.* (1995); http://lep694.gsfc.nasa.gov/waves.waves.html; http://cdaw.gsfc.nasa.gov/CME_list/radio/waves_type2.html) and a solar source location based on a list of GOES X-ray flares (Solar Geophysical Data).

3. Results and discussion

In Table 1 we give the average properties with respective statistical errors of CMEs associated with DH type II bursts and proton events with and without a heavy-ion enhancement. CMEs with heavy ions are faster and possibly wider than CMEs without heavy ions. In addition, larger fraction of CMEs with heavy ions is halo CMEs. The properties of CMEs without heavy-ions resemble the general population CMEs with type II bursts studied by Gopalswamy *et al.* (2005). They found that the average speed of type II associated CMEs is 1115 km/s, the average width is 139°, and 42.5% are halo CMEs. CMEs associated with SEPs and type II bursts are wider and faster than an average CME (Gopalswamy *et al.* 2008). Therefore, it appears that heavy-ion events are produced in the most energetic CMEs.

The distributions of solar sources (flares) of proton events with and without heavy-ion enhancement are plotted in Fig. 1. Proton events with type II bursts and heavy-ion events originate from the western hemisphere, mostly near the western limb (left panel). Events without heavy-ion enhancements are evenly distributed over solar longitudes (right panel). Therefore, even in proton events were type II radio emission indicate the existence of a shock, a good connection to the acceleration site in the western hemisphere is essential for the detection of heavy ions at 1 AU. The western location suggest that the heavy ions are accelerated near the nose of the shock where it is strongest. Furthermore, the source distribution of type II associated proton events with heavy ions reminds the source distribution expected for SEP events with particles accelerated in flares. Therefore, the estimation of relative contributions by flares and shocks in SEP events is complicated. The two models suggested by Cane *et al.* (2003) and Tylka *et al.* (2005) cannot be separated based on source locations alone.

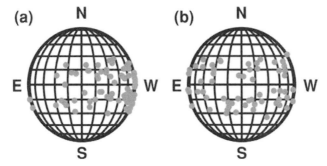

Figure 1. Solar source (flare) locations of type II associated proton events (a) with and (b) without a heavy-ion increase.

Acknowledgements

SOHO is an international cooperation project between ESA and NASA. This research was supported by NASA grant NNX08AD60A.

References

Bougeret, J.-L., Kaiser, M. L., Kellogg, P. J., Manning, R., Goetz, K., Monson, S. J., Monge, N., Friel, L., Meetre, C. A., Perche, C., Sitruk, L., & Hoang, S. 1995, *Space Sci. Revs*, 71, 231

Brueckner, G. E., Howard, R. A., Koomen, M. J., Korendyke, C. M., Michels, D. J., Moses, J. D., Socker, D. G., Dere, K. P., Lamy, P. L., Llebaria, A., Bout, M. V., Schwenn, R., Simnett, G. M., Bedford, D. K., & Eyles, C. J. 1995, *Solar Phys.*, 162, 357

Cane, H. V., von Rosenvinge, T. T., Cohen, C. M. S., & Mewaldt, R. A. 2003 *Geophys. Res. Lett.*, 30, 8017

Gopalswamy, N. 2003, *Geophys. Res. Lett.*, 30, 8013

Gopalswamy, N., Aguilar-Rodriguez, E., Yashiro, S., Nunes, S., Kaiser, M. L., & Howard, R. A. 2005, *J. Geophys. Res.*, 110, A12S07

Gopalswamy, N., Yashiro, S., Akiyama, S., Mäkelä, P., Xie, H., Kaiser, M. L., Howard, R. A., & Bougeret, J.-L. 2008, *Ann. Geophys.*, 26, 3033

Hovestadt, D., Hilchenbach, M., Bürgi, A., Klecker, B., Laeverenz, P., Scholer, M., Grünwaldt, H., Axford, W. I., Livi, S., Marsch, E., Wilken, B., Winterhoff, H. P., Ipavich, F. M., Bedini, P., Coplan, M. A.,Galvin, A. B., Gloeckler, G., Bochsler, P., Balsiger, H., Fischer, J., Geiss, J., Kallenbach, R., Wurz, P., Reiche, K.-U., Gliem, F., Judge, D. L., Ogawa, H. S., Hsieh, K. C., Möbius, E., Lee, M. A., Managadze, G. G.,Verigin, M. I., & Neugebauer, M. 1995, *Solar Phys.*, 162, 441

Reames, D. V. 1999, *Space Sci. Revs*, 90, 413

Torsti, J., Valtonen, E., Lumme, M., Peltonen, P., Eronen, T., Louhola, M., Riihonen, E., Schultz, G., Teittinen, M., Ahola, K., Holmlund, C., Kelhä, V., Leppälä, K., Ruuska, P., & Strömmer, E. 1995, *Solar Phys.*, 162, 505

Tylka, A. J., Cohen, C. M. S., Dietrich, W. F., Lee, M. A., Maclennan, C. G., Mewaldt, R. A., Ng, C. K., & Reames, D. V. 2005, *ApJ*, 625, 474

Session VIII

Heliosphere Boundaries, Interfaces and Shocks

Universal Heliophysical Processes
Proceedings IAU Symposium No. 257, 2008
N. Gopalswamy & D.F. Webb, eds.

Multi-spacecraft observations to study the shock extension in the inner heliosphere

Aline de Lucas[1,2], Rainer Schwenn[2], Eckart Marsch[2], Alisson Dal Lago[1], Alicia L. Clúa de Gonzalez[1], Ezequiel Echer[1], Walter D. Gonzalez[1] and Marlos R. da Silva[1]

[1] Instituto Nacional de Pesquisas Espaciais,
Av. dos Astronautas, 1758, Jardim da Granja, 12227010, São José dos Campos, São Paulo,
Brazil
email: delucas@dge.inpe.br

[2] Max-Planck-Institut für Sonnensystemforschung – MPS,
Max-Planck-Str. 2, 37191, Katlenburg-Lindau, Germany
email: schwenn@mps.mpg.de

Abstract. The two Helios probes traveled at variable longitudinal and radial separations through the inner heliosphere. They collected most valuable high resolution plasma and magnetic field data for an entire solar cycle. The mission is still so successful that no other missions will collect the same kind of data in the next 20 years. One of the subjects studied after the success of the Helios mission was the identification of more than 390 shock waves driven by Interplanetary Coronal Mass Ejections (ICMEs). Combining the data from both probes, we make a statistical study for the extension of the shock waves in the interplanetary medium. For longitudinal separations of 90° we found a cutoff value at this angular separation. A shock has 50% of chance to be observed by both probes and the same probability for not being observed by two spacecrafts at the same time, when the angle between them is around 90°. We describe the dependence of the probability for shocks to be observed by both probes with decreasing spacecraft separation. Including plasma data from the ISEE-3 and IMP-8 spacecrafts improves our statistical evaluation substantially.

Keywords. shock waves, longitudinal extension, inner heliosphere, interplanetary coronal mass ejections.

1. Introduction

Interplanetary shock waves are the strongest abrupt perturbation in the solar wind, playing an important role in the solar-terrestrial environment variability. They are large-scale phenomena resultant of the propagation of interplanetary structures, such as ICMEs – the interplanetary counterparts of the coronal mass ejections (CMEs)(see terminology discussion by Burlaga 2001 and Russell 2001). The reason why they are formed is the fact that the relative speed between a fast stream (the ICME, in this case) and the background solar wind is often greater than the characteristic speed of the medium-magnetosonic speed. In the inner heliosphere – inside 1 *AU* – shocks driven by ICMEs are well formed, and the ICME-shock association has been observed since the first images from Solwind coronagraph were available (Sheeley *et al.* 1985). Sheeley *et al.* (1985) were the first to confirm that fast ICMEs were related to shock formation.

The purpose of this work is to study the shock longitudinal extension in the inner heliosphere using Helios, IMP-8, and ISEE-3 observations for the entire solar cycle 21. Section 2 will present the event selection and data analysis, Section 3 will discuss the results, and, finally, section 4 will summarize the conclusions.

2. Event Selection and Data Analysis

Helios was a mission composed by two twin probes, Helios 1 (H1) and Helios 2 (H2), that operated at the same time from 1976 until the beginning of 1981 (Porsche 1984). Due to the long life of H1 (1974–1986), it has become possible to collect one of the most complet sets of plasma data over the time span of a full solar cycle for studying the solar wind evolution and variation into the inner heliosphere (Schwenn & Rosenbauer 1984). Among the total set of shock waves detected by the instruments onboard Helios, 395 were classified as those driven by ICMEs. Corotating Interaction Regions (CIRs) were not included in the present work since they are normally related to shocks at distances further than 1 *AU* (Hundhausen & Gosling 1976; Smith & Wolfe 1976).

Each of the shocks from the full set of events was analyzed separately. Solar wind and magnetic field data from three different positions (H1, H2, and ISEE-3/or IMP-8) contributed to the comparison of the shock signatures in these reference points. This provided the opportunity to estimate the total angular distance in longitude which one could expect a shock to expand to.

By using the list of shock/ICME events studied by Sheeley *et al.* (1985), we had the possible flare locations associated to the limb CMEs when H1 was located close to $\pm 90°$ from the Sun-Earth line. This enabled to correlate the shocks observed by H1 with the ones observed at Earth by IMP-8 and/or ISEE-3, once the flare location was giving further information about the possible direction the shock wave was being driven.

Among the large set of events, there were periods without observations from the solar wind and/or magnetic field instruments onboard the three missions. Sometimes gaps filled the period when the shock was expected to arrive. These cases were not included in the statistical analysis carried out in this study. Only safe events, i.e., the events with visible signatures of shocks, made part of the considered sample. However, for some of these cases with gaps, we could see a level enhancement in all solar wind parameters and magnetic field strength before and after the gap. For those cases, we saw that there was a shock, even though we could not say exactly its time of occurrence.

Another difficulty was to determine the periods when we expected IMP-8 to observe the shock arriving at Earth. The periods when IMP-8 was in the solar wind were not many, but contributed to improving the estimate, until ISEE-3 appeared in the scenario at the second half of the year 1978 with an orbit around point L1, constantly in the solar wind. Another aspect that have influenced a lot in our sample is the fact that H2 did not operate during the full solar cycle like H1 did. For many years we just had a constellation (a pair of probes) as an input to our statistical analysis.

The inspection is basically observational, based in the comparison among the different points of references and their observations. The left plot on Figure 1 is one example of a strong shock observed by Helios 1 at 0.952 *AU* driven by a magnetic cloud in the interplanetary medium. As it is shown in the right plot of Figure 1, the same shock was seen by H2 at 0.978 AU. However, the signatures of the "ejecta" are not visible on the solar wind and magnetic field parameters. Since the variation in $\vec{B}$ is very smooth and plasma β is going down after almost one day, counting from the day of the shock, we may conclude that the probe was crossing only the shock wave and not the ICME structure itself. Near Earth, IMP-8 was the only spacecraft operating during this time (Figure 2). In this day, IMP-8 was outside the magnetospheric cavity, and the solar wind parameter profiles observed by IMP-8 were similar to the ones seen by H2. This was already expected since H2 and IMP-8 were separated by only 9°. At the end of DOY 29/1977, the shock was detected by H2, and some hours later, by IMP-8.

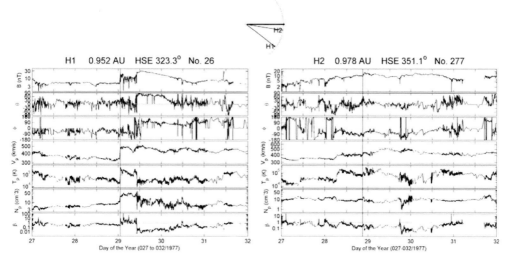

Figure 1. H1 (left plot) and H2 (right plot) magnetic field and plasma data for the shock detected on DOY 29/1977, at 01:03 UT by H1, and on DOY 28/1977, at 21:07 UT by H2. From top to bottom, one can see the magnetic field strength and angular components, followed by the solar wind proton speed, density, temperature, and the plasma beta. A magnetic cloud drives the shock, observed by H1 at 0.952 AU, 37° away from Earth. At the top of the plot the position of the two probes H1 and H2 is shown, as well the radial distance and longitude (in the counterclockwise direction in relation to the Sun-Earth line) of H1/H2. Earth is schematically represented on the upper plot, as well the Sun, and H1 and H2 positions at the period of the shock. The Sun is the central point of the circumference sector from where the location lines of H1 and H2, for the period of the shock, originate. The thicker solid line connects the center (Sun) with Earth.

Based on the observations at the considered points we proceeded with the estimate of the shock extension in the interplanetary medium. We separated the three points of observation in three constellations of two spacecrafts each: H1 and H2, H1 and IMP-8/ISEE-3, and H2 and IMP-8/ISEE-3. In Figure 1 we associated the shock occurrence on H1 with the one at H2, which means that we considered this shock as being the same in each probe. Based on this association we say that the minimum longitudinal angular distance reached by the shock was the separation between H1 and H2 – in this case the shock extended at least to 28°. When IMP-8 is included in the statistical analysis a larger angle is considered: H1 and the Sun-Earth line are about 37° of longitude away from each other. Again the minimum distance in longitude the shock reached was the one separating IMP-8 and H1, since the shock was crossed by these two spacecrafts when traveling in the interplanetary medium outwardly from the Sun. The angular separation between H2 and IMP-8 makes also part of the estimate once we are considering the three constelations independently.

We wanted to make sure we were seeing the same shock rather than to increase the number of cases studied without certainty. For this reason, the rate of shocks does not correspond to the total number of shocks registered during the mission. The histograms showed in the next figures are a result of two different classes: shocks observed by a pair of probes (first pannel), and shocks observed by a single probe (lower pannel).

Since our results depend on the orbit of the probes, that probably did not contribute to observe all the shocks for the period and the probes might have crossed the shocks in only one part, we might expect that there are larger angles than the ones we found. This guides us to the estimate of the margin of error.

A. de Lucas *et al.*

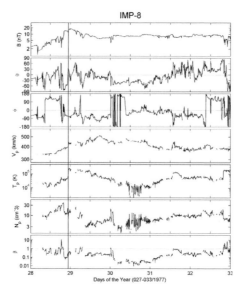

Figure 2. Interplanetary shock observed on DOY 28/1977 by IMP-8. This is the same shock previously observed by H2 and later on by H1. From top to bottom, one can see the profiles of the magnetic field strength and angular componentes, the solar wind protons speed, temperature, density, and plasma beta. IMP-8 was in the solar wind near Earth during the period of observation of this shock.

3. Results

For each case the angular separation between a pair of spacecrafts represented the minimum separation we could expect a shock to extend into the interplanetary medium. As we separated the observations according to the three different contellations, three different estimates were obtained for the whole period of observation. From the group H1 and H2, smaller angles separated the two probes for most of the time of operation, so our estimate was limited to the angles they formed during their orbits. From H1 and H2 observations we could say primarily which shocks were observed in both spacecrafts, then look for other observations near Earth – first with IMP-8 and then with ISEE-3.

When near Earth observations by IMP-8/ISEE-3 were included on the estimate, larger angles started to appear and new shock extensions were revealed. The full longitudinal range of the inner heliosphere was covered by the new points included on the observations and a new scenario for the shocks extension estimate took place. The left side of Figure 3 shows the rate of shocks for each longitudinal separation considered. In that figure, the values vary from 10 to 170°, and each column centered in a given Φ represents the sum of all events in the interval I ($\Phi \leqslant I < (\Phi + 10°)$). This means that the number of cases centered in 20° are a result of the number of cases in our sample where the angles were bigger than or equal to 20° and smaller than 30°, and consecutively for the other angles in the x-axis.

As it is shown in Figure 3, there are bars that are in the right side up and others that are upside down for both the plots. The former ones correspond to those events where one of the constellations (two different points in the space) had seen the same shock. So from the different constellations separated by colors – H1 and H2 (black); H1 and IMP-8 (gray); and H2 and IMP-8 (white) -, we have the total number of cases in each angular separation considered from the set of shocks under study. And the later ones represent those shock waves observed by only one of the three points of reference. As it is shown in the left side of Figure 3, increasing the angular distance between two different observational points diminishes the number of events observed by each of the constellations.

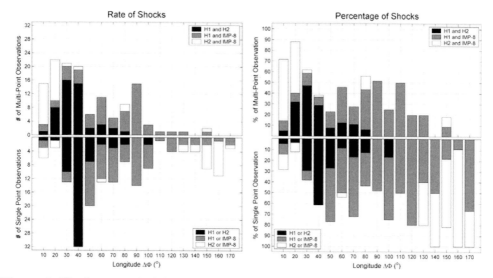

Figure 3. The figure on the left shows the number of shock waves observed from 1974 to 1985 by a pair of spacecrafts/probes (upper panel), or only by one of the spacecrafts (lower panel) as a function of the longitudinal separation ($\Delta\Phi$) between the probes. The constellations are divided in three groups according to each pair of probes: Helios 1 and 2 (black), Helios 1 and IMP-8/ISEE-3 (gray), and Helios 2 and IMP-8/ISEE-3 (white). The figure on the right shows the same results, but as percentages.

In percentage, the distribution of our sample shows a clear trend that is illustrated in the right side of Figure 3. As we go to bigger angular separations, the percentage of shocks seen by both spacecrafts decreases, following a quasi-exponential decrease. Even though we have some special cases with large angles, like those events at 120°, 130°, and 150°, we need to investigate them in details in order to certify that they really correspond to the same event seen in different points. According to the percentage we found in the right side of Figure 3, at $\Delta\Phi = 90°$ one has 50% of chance of seeing a shock or not seeing the same shock in two different points of observation.

The critical interval for the percentage of shocks (right side of Figure 3) was determined by using the test of proportions analysis (for details, see Kalbfleisch 1979). Figure 4 shows the error bars that represent a 95% confidence intervals for each angular separation. The estimated value is more accurate as we have a larger number of cases from the sample, like it is shown in Figure 4. A critical value at $\Delta\Phi = 110°$ is found as we have just two cases inside this angular distance.

4. Conclusions

We have studied shock angular extension in the inner heliosphere using observations from H1, H2, and IMP-8/ISEE-3 spacecrafts. By using a pair of these probes each time, we found that shock extension decreases as the probes angular separation increases. When a CME is observed at the solar limb, for example, there is 50% of probability of seeing

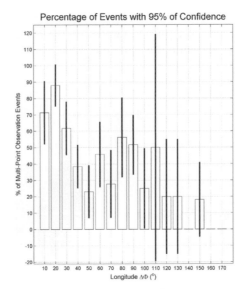

Figure 4. This is the same plot as shown before in the percentage of shocks (upper panel of Figure 3). The error margin for the percentage of shock observed by multi-points into each longitudinal separation as seen by Helios-1,2 and IMP-8/ISEE-3. As one goes further in degrees, the uncertainty for observing a shock in the angular separation ($\Delta\Phi$) increases. Observe that in $\Delta\Phi = 110°$ the biggest error for our estimate is found. That is because only two events (left side in Figure 3) were registered for that angle: one was detected by a pair of probes, and the other, by a single probe.

the shock driven by the ICME at Earth. Further investigation is needed to evaluate those cases with large ($> 110°$) separation.

5. Acknowledgements

The authors acknowledge the Instituto Nacional de Pesquisas Espaciais/INPE-MCT and the Brazilian government agency CNPQ for doctorate fellowship and projects (142012/2005-0, 302523/2005-7, and 472031/2007-4). The authors also thank the Max-Planck-Institut für Sonnensystemforschung for Helios mission data, and to NSSDC for the IMP-8 and the ISEE-3 data.

References

Burlaga, L. F. 2001, *Eos Trans. AGU*, 82, 433

Hundhausen, A. J. & Gosling, J. T. 1976, *J. Geophys. Res.*, 81, 1436

Kalbfleisch, J. G. 1979, *Probability and Statistical Inference II*, New York: Springer-Verlag, p. 303

Porsche, H. 1984, *10 Years HELIOS*, Munich: Hirmer-Verlag

Russell, C. T. 2001, *Eos Trans. AGU*, 82, 433

Schwenn, R. & Rosenbauer, H. 1984, in: Porsche, H. (ed.) *10 Years HELIOS*, Munich: Hirmer-Verlag, p. 66

Sheeley, Jr., N. R., Howard, R. A., Michels, D. J., Koomen, M. J., Schwenn, R., Mühlhäuser, K. H., & Rosenbauer, H. 1985, textit*J. Geophys. Res.*, 90, 163

Smith, E. J. & Wolfe, J. H. 1976, *Geophys. Res. Let.*, 3, 137

Discussion

THEJAPPA: Type II radio bursts are observed as very patchy, which is interpreted in terms of ripples in the shock front. Is there any evidence for such ripple-like structure in the shock front from the two spacecraft observations, i.e., from two points of measurements from Helios?

DeLUCAS: Yes, there is. In many of the cases we studied we believe that the shock shape is represented by a ripple-line. This shape is the only explanation for these shocks we observed, when we expected the shock to arrive first at one point and it didn't. This point in general was closer to the Sun in relation to the other one with similar shock speeds and the further one observed before.

HOWARD: What was the average/most probable width of shocks that you found? I believe that one of the events on the Sheeley *et al.* (1985) list was a streamer blow-out event observed on one limb and the associated shock was observed when Helios was off the opposite limb, implying a longitudinal separation of more than 90 degrees from the "nose".

DeLUCAS: The average width of shocks we found is mostly concentrated around 90 degrees, I would say that 90 degrees is a cutoff value for our distribution. We also have many cases observed around 30, 40, 50 degrees, but since we are limited in our angular distribution, that is the separation between the probes, we can have an even greater number of multi-point observation at other angles.

Universal Heliophysical Processes
Proceedings IAU Symposium No. 257, 2008
N. Gopalswamy & D.F. Webb, eds.

© 2009 International Astronomical Union
doi:10.1017/S1743921309029755

Modeling and prediction of fast CME/shocks associated with type II bursts

H. Xie[1,2], N. Gopalswamy[2] and O. C. St. Cyr[2]

[1] Catholic University of America, Washington DC, 20064, USA
email: hong.xie@nasa.gov

[2] NASA Goddard Space Flight Center, Greenbelt, MD 20771, USA

Abstract. A numerical simulation with ENLIL+Cone model was carried out to study the propagation of the shock driven by the 2005 May 13 CME. We then conducted a statistical analysis on a subset of similar events, where a decameter-hectometric (DH) type II radio burst and a counterpart kilometric type II have been observed to be associated with each CME (DHkm CME). The simulation results show that fast CME-driven shocks experienced a rapid deceleration as they propagated through the corona and then kept a nearly constant speed traveling out into the heliosphere. Two improved methods are proposed to predict the fast CME-driven shock arrival time, which give the prediction errors of 3.43 and 6.83 hrs, respectively.

Keywords. Coronal mass ejections (CMEs), shock waves, solar-terrestrial relations.

1. Introduction

Many efforts have been made to study the CME and shock propagation using theoretical models (e.g., Chen 1996; Borgazzi *et al.* 2008), numerical simulations (e.g., Cargill 2004; Odstrcil & Pizzo 1999; Odstrcil *et al.* 2005), and empirical models (e.g., Gopalswamy *et al.* 2001a; Gopalswamy *et al.* 2005; Cremades *et al.* 2007; Vršnak & Žic 2007). In this work, we use numerical simulations with ENLIL+Cone model (Odstrcil *et al.* 2005) to conduct a case study of the 2005 May 13 CME. We then study a subset of similar CMEs, where a decameter-hectometric (DH) type II radio burst (14–1 MHz) and a counterpart kilometric (< 1000 kHz) type II were observed to be associated with each event (DHkm CME). Two new methods are proposed to improve the prediction of the shock arrival of such CMEs.

2. Numerical simulation

The ENLIL+Cone model we use is a three-dimensional MHD model of the heliosphere that forecasts CME/shock propagation from the ENLIL inner boundary ($21.5\ R_s$) to the point of interest. The simulation is done in two stages: 1) setup of the steady state ambient solar wind background and 2) insert the transient disturbances (CMEs) propagating in that background. The input CME is specified as a conical plasma cloud with the location, angular width, and velocity corresponding to the cone model fit (Xie *et al.* 2004) to the coronagraph observations. Recent positive results using this method have been reported by Taktakishvili (2008).

The ENLIL+Cone model has well simulated the kinematic evolution of the 2005 May 13 CME and its driven shock, with a prediction error of the shock arrival of ∼ 4 hrs. The CME is a fast halo CME with a M8.0 flare and the CME onset time is 16:47 UT. The observed sky-plane speed and space speed from the CME cone model are 1689 km/s and 2171 km/s, respectively. Figure 1 (left panel) shows that the CME-driven shock

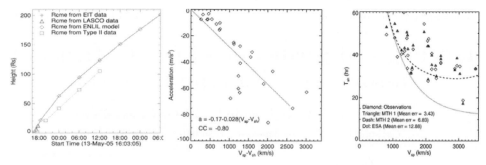

Figure 1. (left) Height-time profile of CME and its driven shock propagation determined from the cone model fit to coronagraph observations and ENLIL simulation.(middle) Shock acceleration vs. $(V_{sp} - V_{sh})$ for 25 DHkm CMEs. (right) Comparison of observations and model predictions of T_{sh} vs. V_{sp}. Rcme is the CME radial distance, V_{sp} is the shock speed near the Sun, V_{sh} is the shock speed obtained from kmII dynamic spectrum data, a is the shock acceleration, and CC is the correlation coeffient between a and $(V_{sp} - V_{sh})$.

propagates at a nearly constant speed of ~ 1000 km/s after a rapid deceleration within 0.3 AU with an average acceleration of ~ 50 m/s^2. The linear speed obtained from the simulation is in agreement with the shock speed of 1016 km/s derived from kilometric type II data.

3. Statistical analysis

3.1. Data selection

We study the shock propagation for a subset of CMEs similar to the 2005 May 13 event. The primary criterion we use is that there must be a decameter-hectometric (DH) type II radio burst occurring at frequencies between 14 and 1 MHz and a counterpart kilometric (km) type II occurring at frequencies less than 100 KHz associated with each CME (DHkm CME). DHkm CMEs are faster and wider than the general population of CMEs (e.g., Gopalswamy *et al.* 2001b). Since there are both DH and km type II bursts associated with each event, we are able to identify a unique CME-shock pair. In addition, we have excluded limb events (solar source within $25°$ from limb).

3.2. Shock propagation

We assume that shocks driven by fast CMEs decelerate through the outer corona and inner heliosphere until the CMEs reach the kilometric type II (kmII) domain at constant acceleration. The shock acceleration is given as $a = (V_{sp} - V_{sh})/t_1$, where V_{sp} is the shock speed near the Sun, V_{sh} is the shock speed derived from the frequency draft of kmII, and t_1 is acceleration cessation time. We approximate the CME space speed obtained from the cone model fit to LASCO observations as the shock speed near the Sun (since the CME leading edges observed in white-light coronagraphs might be in fact compressed shock fronts ahead of CMEs).

Figure 1 (middle panel) shows the scatter plot of shock acceleration vs. $(V_{sp} - V_{sh})$ for 25 DHkm CMEs from 2000–2006. The linear fitting gives:

$$a = -0.17 - 0.028(V_{sp} - V_{sh}) \tag{3.1}$$

The obtained acceleration magnitudes are consistent with the accelerations from Gopalswamy *et al.* (2001b), which obtained by a second-order polynomical fitting to the height-time plots of a subset of limb CMEs (associated with DH type II bursts) within LASCO C2 and C3 field of view:

$$a = -30. + 0.0666u - 4.5 \times 10^{-5}u^2 \tag{3.2}$$

where u denotes the CME initial speed.

3.3. *Improved shock prediction methods*

Method 1 uses V_{sp} and V_{sh} to obtain the shock acceleration, a, acceleration cessation time, t_1, and acceleration cessation distance, d_1, then computes the shock travel time using (3.1) and the following equations:

$$\begin{aligned} t_1 &= (V_{sp} - V_{sh})/a, \ d_1 = V_{sp}t_1 + 1/2at_1^2 \\ t_2 &= (1AU - d_1)/V_{sh}, \ T_{sh} = t_1 + t_2 \end{aligned} \tag{3.3}$$

Method 2 uses the formula (3.2) to obtain the CME speed at t_1:

$$V_{sp1} = V_{sp} + at_1, \qquad t_1 = \frac{-V_{sp} + \sqrt{V_{sp}^2 + 2ad_1}}{a} \tag{3.4}$$

where d_1 is set to be 0.1 AU (21.0 R_s), and then uses V_{sp1} as input to the Empirical Shock Arrival (ESA) model (Gopalswamy *et al.* 2005) to compute T_{sh}.

Figure 1 (right panel) compares observations and model predictions of T_{sh} vs. V_{sp}, where diamonds denote observed T_{sh} vs. V_{sp}, triangles denote T_{sh} vs. V_{sp} from method 1, and dotted and dashed curve represent the prediction from the ESA model and the improved ESA model (method 2), respectively.

4. Summary

1) The ENLIL+Cone model can well simulate the kinematic evolution of the 2005 May 13 event. 2) Method 1 combined with kilometric type II data gives a mean prediction error of 3.43 hrs, implying that CME-driven shocks travel at near constant speed after a rapid deceleration phase (< 0.5 AU). 3) Method 2 improves the ESA model by taking into consideration the fast deceleration phase within the field of view of C3, decreasing the mean prediction error from 12.88 to 6.83 hrs. 4) The ESA model underestimates the deceleration of fast CMEs in the outer corona.

HX was partly supported by NASA (NNX08AD60A).

References

Borgazzi, A., Lara, A., Romero-Salazar, L., & Ventura, A. 2008, *Geofisica International*, 47, 301
Cargill, P. J. 2004, *Sol. Phys.*, 221, 135
Chen, J. 1996, *J. Geophys. Res.*, 101, 27499
Cremades, H., St. Cyr, O. C., & Kaise, M. L. 2007, *Space Weather*, 5, S08001
Gopalswamy *et al.* 2001a, *J. Geophys. Res.*, 106, 29207
Gopalswamy *et al.* 2001b, *J. Geophys. Res.*, 106, 29219
Gopalswamy *et al.* 2005, *Adv. Space Res.*, 36, 2289
Odstrcil, D. & Pizzo, V. J. 1999, *J. Geophys. Res.*, 104, 483
Odstrcil, D., Pizzo, V. J., & Arge, C. N. 2005, *J. Geophys. Res.*, 110, A02106
Taktakishvili *et al.* 2008, submitted to *Space Weather*
Vršnak, B. & Žic T. 2007, *A&A*, 423, 717
Xie, H., Ofman, L., & Lawrence, G. 2004, *J. Geophys. Res.*, 109, A08103

Universal Heliophysical Processes
Proceedings IAU Symposium No. 257, 2008
N. Gopalswamy & D.F. Webb, eds.

© 2009 International Astronomical Union
doi:10.1017/S1743921309029767

Simulations of shock structures of a flare/CME event in the low corona

Jens Pomoell[1], Rami Vainio[1] and Silja Pohjolainen[2]

[1] Department of Physics, University of Helsinki
P.O. Box 64, 00014 University of Helsinki, Finland
email: `jens.pomoell@helsinki.fi`, `rami.vainio@helsinki.fi`

[2] Department of Physics and Astronomy, University of Turku,
Tuorla Observatory, 21500 Piikkiö, Finland
email: `silpoh@utu.fi`

Abstract. We study the MHD processes related to a flare/CME event in the lower solar corona using numerical simulations. Our initial state is an isothermal gravitationally stratified corona with an embedded flux rope magnetic field structure. The eruption is driven by applying an artificial force to the flux rope. The results show that as the flux rope rises, a shock structure is formed, reaching from ahead of the flux rope all the way to the solar surface. The speed of the shock quickly exceeds that of the driving flux rope, and the shock escapes from the driver. Thus, the shock exhibits characteristics both of the driven and blast wave type. In addition, the temperature distribution behind the shock is loop-like, implying that erupting loop-like structures observed in soft X-ray images might be shocks. Finally, we note that care must be taken when performing correlation analysis of the speed and location of type II bursts and ejecta.

Keywords. Sun: coronal mass ejections (CMEs); flares; radio radiation, shock waves, plasmas

1. Introduction

Large-amplitude waves and shocks launched by coronal mass ejections (CMEs) and flares are believed to play an important role in the generation of a number of solar transient phenomena such as type II radio bursts, Moreton waves, EIT waves and SEP events (see, e.g., Warmuth 2007 for an overview). However, the exact mechanisms linking the eruptions with the observed disturbances continue to be elusive. For instance, for type II radio bursts and Moreton waves, the debate continues whether the shock responsible for the disturbance is a flare-generated blast wave or instead driven by mass motions related to CME lift-off (see Vršnak & Cliver 2008 for a recent review).

In this study, we employ MHD simulations of CME lift-off to study the shock structures induced by the eruption. We focus especially on the shock formation process, and point out features that are of importance when interpreting observations.

2. Model

We perform ideal-MHD simulations of an erupting CME in a local model of the low corona. The coronal plasma is assumed to be isothermal with an exponentially decreasing density profile balancing gravity, while the potential background magnetic field is quadrupolar. Thus, the Alfvén speed of the model corona increases as a function of height. In addition, we superpose a flux rope with increased density on top of the background plasma. Fig. 1 shows the density and Alfvén speed of the initial state. The filament-like structure is then made to erupt by invoking an artificial force that acts on the filament plasma during the simulation. For details of the model, consult Pomoell *et al.* (2008).

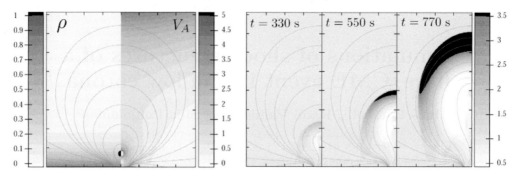

Figure 1. Left: The initial state of the simulation: density (left half) and Alfvén speed (right half). The lines depict magnetic field lines. The tick marks are drawn at intervals of 5×10^4 km. The units in the color bars are for the density (left bar) 1.67×10^{-12} kg m^{-3} and for the Alfvén speed (right bar) 91 km s^{-1}. Right: Temperature at three different times of the eruption. The unit in the color bar is 0.636×10^6 K. Note the clipping of the color bar; a black (white) color indicates values larger (smaller) than the color bar maximum (minimum).

3. Results

The dynamics of the eruption is as follows:

(a) A perturbation surrounding the flux rope is quickly formed as the flux rope starts to rise under the influence of the artificial force.

(b) As the driving flux rope picks up speed, the outward propagating wave surrounding the flux rope develops to a shock ahead of the flux rope. The speed and strength of the shock are highest at the leading edge, and decrease towards the flanks, degenerating to a fast-mode wave close to the solar surface.

(c) The speed of the shock quickly exceeds that of the driving flux rope, and the shock escapes from the flux rope as its speed continues to increase due to the increasing Alfvén speed of the ambient corona. However, the shock starts to lose strength once it escapes from the driver.

(d) At the end of the simulation, when the shock approaches the upper boundary, the eruption has evolved into a large global structure. However, the driving flux rope has remained roughly the same size during the eruption.

4. Discussion

4.1. Shock formation: a driven blast wave

The flux rope motion launches a coronal shock wave, which propagates in all directions from the driver. It nevertheless remains strongest near the leading edge of the shock, which emphasizes the role of the driver. However, due to the increasing Alfvén speed of the corona, the shock starts to escape from the driver. In this sense, the shock expands more like a freely propagating than a driven wave. Thus, caution must be practised in labeling shocks as being either driven waves or blast waves, as low cadence observations could lead one to make an erroneous conclusion about the mechanism responsible for generating the shock.

4.2. Shocks and soft X-ray observations

The temperature plot (Fig. 1) reveals an arc of extremely hot plasma (downstream of the shock front), which could easily be interpreted as a hot erupting coronal loop. However, the feature is not a loop but a wave. Thus, one must be careful when interpreting propagating loop-like features in coronal soft X-ray images, some of them might actually be

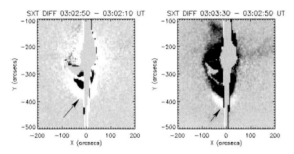

Figure 2. SXT difference images (AlMg filters) at 03:02:50 UT and 03:03:30 UT, 13 May 2001, showing a loop-like eruption front in soft X-rays. See Pohjolainen *et al.* (2008) for details.

shock waves. For instance, in an event studied by Pohjolainen *et al.* (2008), SXT difference images (Fig. 2) show a loop-like eruption front, which could in fact be the signature of a shock wave. Similar structures have been identified as shocks in Yohkoh SXT images in conjunction with Moreton waves, see Khan & Aurass (2002) and Narukage *et al.* (2002).

4.3. *Ejecta and type II burst correlations*

Recently, Shanmugaraju *et al.* (2006) analysed 18 events of X-ray plasma ejections associated with coronal shocks inferred from metric type II bursts, and concluded that the absence of correlation between the speeds of ejecta and type IIs as well as the sub-Alfvénic speeds of the ejections are factors not in favor of the ejecta to be the main driver of all coronal shocks. Our results suggest a number of important points to note when performing such an analysis of observations. First, a sub-Alfvénic ejection is capable of launching a shock, since a wave can steepen to a shock due to nonlinear evolution of the wave profile. Also, if a wave enters a region with low Alfvén speed, the wave can quickly steepen to a shock. Such behaviour of the shock has actually been proposed to cause fragmented high-frequency type II emission (Pohjolainen *et al.* 2008).

Furthermore, depending on the variations of the Alfvén speed in the corona, the ejection can at times act as the driver, while at other times the shock may propagate freely. If we assume that type II bursts are generated at the leading edge of the shock, where the shock is strongest, the speeds and locations of the ejecta and burst may not be correlated in any simple way. Thus, we conclude that observations in conjunction with modeling are needed in order to resolve such correlation issues.

References

Khan, J. I. & Aurass, H. 2002, *A&A*, 383, 1018

Narukage, N., Hudson, H. S., Morimoto, T., Akiyama, S., Kitai, R., Kurokawa, H., & Shibata, K. 2002, *ApJ*, 572, L109

Pohjolainen, S., Pomoell, J., & Vainio, R. 2008, *A&A*, 490, 357

Pomoell, J., Vainio, R., & Kissmann, R. 2008, *Solar Phys.*, 253, 249

Shanmugaraju, A., Moon, Y.-J., Kim, Y.-H., Cho, K.-S., Dryer, M., & Umapathy, S. 2006, *A&A*, 458, 653

Vršnak, B. & Cliver, E. W. 2008, *Solar Phys.*, 253, 215

Warmuth, A. 2007, in: Klein, K.-L. & MacKinnon, A. L. (eds.), *The High Energy Solar Corona: Waves, Eruptions, Particles*, Lecture Notes in Physics, vol. 725, p. 107

Session IX

Planetary Atmospheres, Ionospheres, Magnetospheres

Universal Heliophysical Processes
Proceedings IAU Symposium No. 257, 2008
N. Gopalswamy & D.F. Webb, eds.

Planetary ionospheres – sources and dynamic drivers

Joseph M. Grebowsky[1] and Arthur C. Aikin[2]

[1]NASA Goddard Space Flight Center,
Greenbelt, Maryland, USA
email: `joseph.m.grebowsky@nasa.gov`

[2]Institute for Astrophysics and Computational Sciences,
Department of Physics, Catholic University of America,
Washington, DC, USA
email: `aaikin1@verizon.net`

Abstract. External energy inputs into all planetary upper atmospheres (including more than a half dozen moons with atmospheres) are comprised of combinations of solar EUV, soft x-rays, solar energetic particles, solar wind charged particles, magnetospherically accelerated particles, solar wind electric field, interplanetary dust particles as well as propagating lower atmosphere disturbances. Each input has analogous physical interactions with all planetary ionospheres and upper atmospheres, but the integrated consequences of the multiple energy inputs vary from planet to planet. The Earth forms the framework for most fundamental processes because of extensive measurements of the effects of each of the inputs. However the conditions at Earth are far different from those at the carbon dioxide atmosphere of magnetic field-free, slow-rotating Venus, the carbon dioxide atmosphere of Mars with patchy remnant magnetic fields, while the outer planets have hydrogen atmospheres, are fast rotating with intrinsic magnetic fields, and encompass moons that interact with the magnetospheres and have exotic atmospheres. Although the physical processes are known, our understanding of our solar system's ionospheres diminishes with increasing distance from the Sun.

Keywords. earth, meteors, meteoroids, planets and satellites: general,magnetic fields

1. Introduction

The uppermost atmospheric layers of all atmospheric-laden planets and moons comprise typically less than one part in a million of the mass of the total atmosphere but these are the planetary boundary regions where energy inputs from interplanetary space and planetary magnetospheres are deposited. The solar wind/magnetosphere energy inputs in all cases produce distinct ionospheric layers and drive motions of the neutral atmosphere that both drag, and are dragged by, the ionospheric ions. Although ionospheric layers are prominent structures often studied by themselves, separate from the atmosphere, their charged particle maximum concentrations are generally much less that those of the neutral atmospheres in the same region. Hence in the main layer of the ionosphere the atmospheric gases and ionization are inseparable parts of the physics. The neutral atmosphere comprises the parent neutrals for the ionized particles responsible for the ionospheric ion composition and the location of the peak ionization. The global topology and motions of the atmosphere and ionosphere are inextricably tied to one another by collisions. All the principle processes controlling the ionization layers throughout the solar system are in general known, lacking still, however, is an understanding of the details of the energy inputs and how different combinations of these inputs produce uniquely different ionospheric variability from one atmosphere to atmosphere. Our knowledge of the ionospheric complexities decrease as one moves away from Earth and is perhaps weakest for the atmospheric-laden moons of the outer planets, but even the terrestrial ionosphere is incompletely understood. The intent of this paper is to give a review of the fundamental sources of ionospheric layers and delineate the unique changes in energy inputs into the upper atmosphere-ionosphere and their consequences for the different atmosphere-laden solar system bodies. This will encompass a survey of seven planets (Venus, Earth, Mars, Jupiter, Saturn, Uranus, Neptune) and 5 moons with substantial ionospheres (the moons of Jupiter Io, Europa, Ganymede, and Callisto, as well as Saturn's moon Titan.).

2. Basic energy inputs and ionization layer production

The ionizing sources that prevail throughout the solar system are: solar EUV radiation, solar wind ionized particles, interplanetary meteoroids, and energetic charged particles produced by solar wind and/or magnetosphere induced electric fields. The solar radiation and interplanetary dust particles deposit their energy and react directly with the upper atmospheres of all atmosphere-surrounded bodies. However the solar wind can have a direct atmospheric interaction only for magnetic field-free planets like Venus and some regions of Mars. For magnetized planets the solar wind and its imbedded magnetic field has an indirect impact through the production of strong magnetospheric electric fields induced by the penetration of the solar wind dynamo electric field (i.e., the solar wind being a conducting medium moving across the interplanetary magnetic field generates a dynamo electric field across solar system magnetospheres) and by connection of the planetary magnetic field with the interplanetary magnetic field. For the magnetic planets, these processes produce energetic charged particles that in turn impact the planetary atmospheres. The fast rotation of the outer planets, in particular Jupiter and Saturn, itself also produces energized plasma from the ionization of the planetary ionosphere production and/or gases emitted from their moons.

The diverse energy inputs lead not only to the production of ionospheres, but spatially varying atmospheric and ionospheric heating that drive ionospheric motions and upper atmospheric winds causing redistributions of the ionospheric morphology. The sources of ionization vary in magnitude and spatial extent from atmosphere to atmosphere and the resulting ionospheres are unique to each planetary and lunar atmosphere in the solar system. The results are further dependent on the atmospheric composition which varies from the dominance of N_2 and O_2 at Earth, through the CO_2 atmospheres of Venus and Mars, the hydrogen atmospheres of the giant planets and the varied atmospheric complexities of planetary moons arising from gaseous emissions of water or volcanic gases.

In all cases the ionosphere and upper atmosphere comprise one integral system. A comprehensive list of processes that can control an ionosphere consists of: Solar radiation; Neutral atmosphere composition topology; Photochemistry; Meteoroid ionization (dependent on ablation of interplanetary micro-particles, ambient atmosphere and ionosphere and photochemistry), Magnetospheric coupling (through energetic particle production, joule heating, magnetosphere reservoir of ionospheric particles); Solar wind coupling (precipitation of SW ionization, induced magnetic fields and electric field accelerations), Atmospheric drag (winds, tides, waves).

3. Nominal background ionospheres

The nominal ionospheric layers, corresponding to the electron density (or total ion density) resulting from the photoionization by solar EUV are well established throughout the solar system from theory. This understanding evolved from comprehensive global measurements of ionospheric ion composition of the layers at Earth, Pioneer Venus Orbiter measurements in the topside of the Venus ionosphere, from the descent of two Viking lander probes at Mars and from radio occultation measurements of vertical electron density profiles made on all planetary spacecraft missions.

The general origin of ionospheres is best described by considering a typical vertical profile of the terrestrial ionosphere (Figure 1, adapted from Rich, 1985) and the effects of photochemistry. With decreasing altitude, ionization is produced by the more energetic wavelengths of the solar radiation with the ionization peaks occurring near the altitude with maximum atmospheric absorption of the shorter wavelength bands. It leads to ionization density peaks corresponding to the D and E layers. Molecular ions, produced by photoionization of atmospheric species, dominate the ion composition in these layers. The ions are in photochemical equilibrium with the atmosphere, as is also the case for the higher layer, F1, where atomic ions begin to appear as the neutral molecular species drop off with increasing altitude. In addition, from the E region peak to below 80 km metal ion species are also present (as seen in every ion composition measurement from sounding rockets, e.g., Grebowsky and Aikin 2002) as are layers of neutral metal species (seen from ground based LIDAR measurements). These are the products of ablated metal atoms from incoming meteoroids. The metal ions are produced by photoionization and charge exchange of the ablated neutral atoms with ambient molecular ions. The metal ions have low ionization potentials and long lifetimes.

The main terrestrial ionospheric layer, the F2 peak, is no longer in photochemical equilibrium. The atmosphere is no longer dense enough to prohibit vertical diffusion of the ionization. Diffusion plays a role in the formation of the layer structure. At night, where solar EUV photoionization is absent and only backscattered radiation is available for photoproduction, F region ionization is supplied from transport of ionization from the dayside or downward transport of ionization that was transported up during the day (see Figure 1). Metal ion species are still measured at night at the lower altitudes because of their long chemical lifetime.

This simple picture of the ionosphere, however, is altered with motions induced in the neutral atmosphere and ionosphere by the energy inputs from the solar wind and magnetosphere, particularly in the form of auroral energetic particle precipitations and heating effects and transport of the ionospheric plasma by the solar wind electric field penetrating to low altitudes. These motions, that vary in response to changes in solar activity, lead to a very complex ionospheric morphology (see Figure 2). We are beginning to model successfully many of the complexities for Earth, but for other solar system bodies we do not yet have the measurements to even begin to understand the complexities. The Earth's

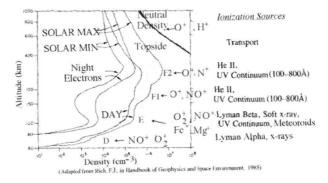

Figure 1. Description of ionospheric layers on Earth, showing the composition of major ions.

internal magnetic field, observed auroral heating effects, and atmospheric/ionospheric variations puts it in a classification with the outer giant planets that have intrinsic magnetic fields. On the other hand, magnetic field-free planets Mars and Venus have novel ionospheric processes operating. We have only a few spacecraft that sampled the outer planets and these are mostly from remote observations, with the exception of Cassini, which is currently sampling directly the ionosphere of Titan at Saturn. The Mars ionosphere has and is being explored remotely by Mars Explorer (MEX) and Mars Global Surveyor (MGS). However, there are extensive in situ ionospheric measurements of Venus from Pioneer Venus Orbiter (PVO), and also a few in situ ionospheric measurements at Mars from the Viking landers that provide glimpses of the types of processes operating in magnetic field-free atmospheres.

4. Magnetic field-free atmospheres: Mars and Venus

The first upper atmosphere/ionosphere difference in moving from Earth to the other terrestrial planets of Mars and Venus is the change of atmospheric composition from nitrogen/oxygen to carbon dioxide dominances. Another distinguishing factor from Earth is the absence of a global magnetic field, although Mars has localized pockets of remnant magnetic fields (Connerney *et al.* 2001) near its surface. Also Mars and Venus have different solar radiation influxes than Earth. The incident solar EUV flux is ∼4 times lower for Mars, resulting in a much weaker ionosphere at Mars compared to Venus. (Mendillo *et al.* 2003, showed that on average the dayside peak electron density of all planetary ionospheres decreases with increasing distance from the Sun as expected for the EUV flux drop off). One consequences of a weakened ionosphere for Mars and the increased scale height of the atmosphere due to Mars's low gravity is that its topside ionospheric pressure is not sufficient to protect the topside ionospheric plasma from being stripped off by the solar wind. On the other hand, Pioneer Venus observations have shown that during solar maximum the ionospheric pressure is high enough to form a barrier to the solar wind that is far above the collision-dominated, photochemical ionosphere region. Figure 3 shows dayside observations on both Venus and Mars (the

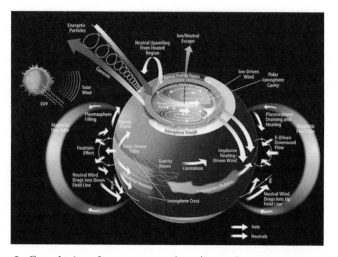

Figure 2. Complexity of upper atmosphere/ionosphere structures at Earth.

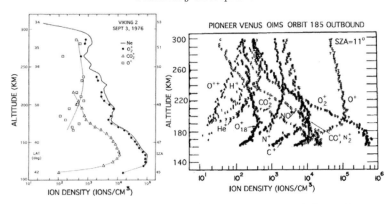

Figure 3. Measurements of ion composition (Hanson *et al.* 1977) from the entrance of one of the two Viking landers at Mars and one Pioneer Venus Orbit from the Orbital Ion Mass Spectrometer (Taylor *et al.* 1980).

former obtained from the Planetary Data System Pioneer Venus observations, the later from Hanson *et al.* 1977). The main O_2^+ layers on both planets are chemical equilibrium layers as on Earth, but there is a transition into an O^+ dominant ionosphere on Venus. The solar wind influence on Mars readily strips the ions produced by photoionization on the topside ionosphere. At Venus however, the boundary of penetration of the shocked solar wind influence (the ionopause) reaches as high as 1000 km at solar maximum. During solar minimum, when solar EUV and photoproduction is reduced, the Venus ionosphere becomes analogous to the Mars situation in that the solar wind electrodynamics results in stripping off of ionization deep in the ionosphere – the ionopause moves in to 200–300 km (Kliore and Luhmann 1991).

Earth has a substantial nightside ionosphere because of transport of photoionized-produced ions, that corotate with the Earth, from the dayside. However, Venus has a very slow rotation rate, so it was somewhat surprising when a substantial nightside ionosphere was observed. The clue to its origin was confirmed by the pioneer Venus Orbiter Retarding Potential Analyzer observations of horizontal ion drifts, which reached supersonic speeds, from the day into the night (Knudsen *et al.* 1980) – the nightside ionosphere was being supplied from the day, the motions driven by the pressure gradient between the nightside chemical sink and the dayside photoproduction source. There were periods, however, during the PVO mission when the solar wind dynamic pressure was high and the nightside ionosphere became vanishingly weak. Indeed observations near solar minimum, just before the end of the mission, found the nightside ionosphere to be in a persistently low density state similar to the solar maximum, high solar wind dynamic pressure events. The topside ionosphere often vanished. On Mars on the other hand radio occultation measurements have indicated in general a very weak, variable, or vanishing nightside ionosphere. These measurements are consistent with the interpretation that there is competition between photoproduction of the dayside ionosphere and stripping away of the ionization by the solar wind. The magnitude of the solar wind dynamic pressure relative to the ionospheric pressure controls the nightside ionosphere. Mars is always in the state where the solar wind dominates. Venus is in that state only during solar minimum or under enhanced solar wind conditions.

Current understanding of the Venus nightside is shown in Figure 4 (adapted from Brace and Kliore, 1991). The nightside ionosphere is formed by the transport of O^+ from the day that then falls under the force of gravity to chemically produce an O_2^+ layer. There are also distinctive holes in the ionosphere that are related to the draping of the

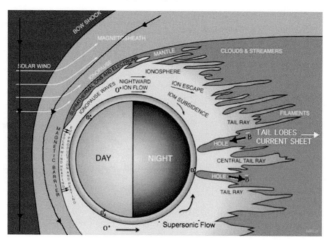

Figure 4. Model of the Venus plasma environment (Adapted from Brace and Kliore 1991).

interplanetary magnetic field in the ionosphere. Although there were a large number of PVO passages through these regions, their origin is still not fully understood. During solar minimum or periods of high solar wind dynamic pressure all the topside ionosphere during the day is ripped off by the solar wind electric field and transported energetically down the Venus tail eliminating the nightside ionosphere transport source. However, even under these conditions an O_2^+ layer is still often present at night – this is attributed to the precipitation of solar wind electrons, energetic enough to ionize the atmosphere.

The depleted nightside ionosphere of Venus appears to be a close analogy to the Martian nightside state. Overall the solar wind-forcing accelerates most of the topside ionosphere production on the dayside down the tail, rather than allow it to subside at night due to gravity, which is less than at Venus. The exceptions may occur in the magnetic anomaly regions, since these magnetic fields when closed, act as retainers for ionospherically produced plasma that will rotate with these field lines. Observations from Mars Express have seen UV auroral emissions (Bertaux *et al.* 2005) in the atmosphere and kev energetic ions and precipitating particles in the cusp regions of the magnetic fields where the planetary field connects with the interplanetary magnetic field. This is a source for ionospheric electrons and auroral emissions (Brain *et al.* 2006) as occurs in the Earth's polar cap.

Although we have a general concept of the global variations of the ionospheres at Mars and Venus, we are missing vital measurements. The observations thus far have not enabled us to develop global empirical models of the ionosphere, and measurements of the atmospheric dynamical control of the ionosphere are totally lacking as is the global consequences of the magnetic anomalies on Mars.

5. The giant planets

5.1. *Jupiter*

Jupiter is the largest planet in the solar system with a radius 11.2 times that of Earth. The planet possesses a strong magnetic field that results in a significant magnetosphere. There is interaction between many of Jupiters 63 moons and its magnetosphere. Jupiter's atmosphere is 89.8% H_2 and 10.2% He, with trace amounts of CH_4 and NH_3. Dissociation of methane leads to the production of hydrocarbons including ethane and acetylene.

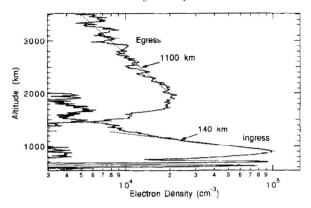

Figure 5. Jupiter ionosphere ingress and egress electron density profiles (Hinson *et al.* 1997).

Hydrocarbons are restricted to the lower and middle atmosphere. Helium falls off with altitude so that H_2 and H are the chief component of the thermosphere and exosphere.

The ionosphere has been measured using radio occultation experiments on both the Pioneer and Voyager spacecraft. Figure 5 shows electron density profiles from Voyager for both an ingress and egress measurement (Hinson *et al.* 1997). The ingress measurement was taken when the solar zenith angle was 89 degrees in the evening as opposed to the egress measurement taken for a solar zenith angle of 91 degrees in the morning. Peak electron densities are 10^5 cm^{-3} and 2×10^4 cm^{-3} for the two cases. Altitudes for the peaks are 900 and 2000 km. Ingress measurements are consistent with daytime photoionization equilibrium conditions both from the altitude and magnitude standpoint. The egress profile is more difficult to explain in both altitude and magnitude. It is important to keep in mind that conditions are sunrise and that the Jupiter ionosphere is strongly influenced by energetic electrons and other charged particles from Jupiters magnetosphere. Between 600 and 800 km altitude on the ingress measurement there are several narrow ionization regions. One possible explanation is that the origin of these layers is the focusing of long-lived metallic ions, whose origin is meteoric debris, by wind shears moving across Jupiters magnetic field. The possibility that metal ions are an important constituent of the Jupiter ionosphere has been considered by Kim *et al.* (2001).

5.2. *Saturn*

With a similar atmosphere and size as Jupiter one would expect the ionosphere of Saturn to have an electron density and ion composition as Jupiter. In Figure 6 are presented a series of electron density profiles taken with the radio occultation experiment on the Cassini spacecraft (Moore *et al.* 2006). The peak electron densities are in the 10^3 to

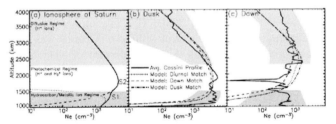

Figure 6. Saturn average electron density profiles and profiles for dusk and dawn. Results from a model are also shown for comparison. (Adapted from Moore *et al.* 2006.)

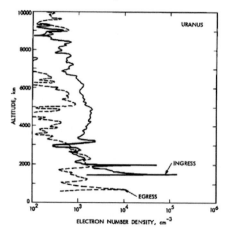

Figure 7. Electron density ingress and egress profiles for Uranus (Lindal *et al.* 1987).

10^4 cm^{-3}. In contrast the Jupiter maximum electron densities are in the 10^4 to 10^5 cm^{-3}. This order of magnitude difference in number density cannot be explained simply as the difference in solar radiation between Jupiter and Saturn. The explanation for the difference is that the Saturn ionosphere has a predominance of molecular ions compared with Jupiter where H$^+$ is the dominant ion. Atomic ions recombine with electrons by radiative recombination as opposed to molecular ions that recombine by dissociative recombination. Radiative recombination is slower than dissociative recombination by several orders of magnitude. The added neutral constituent that leads to an increase in recombination rate is water, which is moved from Saturn's rings. Reactions such as

$$H^+ + H_2O \rightarrow H_2O^+ + H \qquad (5.1)$$

and

$$H_2O^+ + H_2O \rightarrow H_3O^+ + OH \qquad (5.2)$$

lead to the molecular ion H_3O^+. Examination of the profiles in Figure 6 shows regions of electron density depletion. These regions can be explained by sudden water influxes from the rings. Saturn possess a magnetosphere and exhibits aurora at both poles.

5.3. *Uranus*

Unlike Jupiter and Saturn, Uranus is covered by a deep ocean of methane, water and ammonia. These compounds are important components of the lower atmosphere. Photodissociation of methane and subsequent chemical reactions leads to a variety of hydrocarbon compounds including ethane, and acetylene. Both hydrogen and helium are the principal constituents of the thermosphere with helium falling off by diffusive separation so that the bulk of the thermosphere is H and H$_2$. The planet has an atmosphere that extends more than 2 planetary radii. The ionosphere of Uranus was measured using the radio occultation experiment on Voyager 2. A sample pair of electron density profiles is shown in Figure 7 (Lindal *et al.* 1987). For the most part the electron densities are in the 10^3 to 10^4 cm^{-3} range. This is consistent with results from Saturn, but the densities are less than Jupiter. This result is consistent with the requirement for an ion composition with molecular ions as the principal constituent so that the chief electron loss process is dissociative recombination leading to lower electron densities. This implies that water is a component of the atmosphere.

5.4. *Neptune*

The atmosphere of Neptune is similar to that of Uranus with He, H_2, and H principal constituents of the upper atmosphere. Like Uranus, Neptune possess a magnetic field, giving rise to a magnetosphere. The ionosphere has been measured using the radio occultation experiment on Voyager 2 (Tyler *et al.* 1989). The electron density profile between 1000 and 5000 km decreases from 2×10^3 cm^{-3} at 1000 km to less than 100 cm^{-3} at 5000 km.

6. Jupiter's moons with ionospheres

6.1. *Io*

Io is the closest moon to Jupiter. With a radius of 0.286 that of Earth, close to the radius of the Earth's moon, and a surface gravity only 0.183g of Earths value, the moon would not be expected to possess an atmosphere. However, because of the moons proximity to Jupiter and the composition of the moon, a rocky core with a silicate mantle and surface, strong tidal forces give rise to volcanism. The large number of active volcanoes produce a large variety of sulfur compounds. Some eruptions reach 500 km above the surface. As a result Io has an atmosphere of mostly SO_2 with a surface pressure of 10^{-12} bar. There is an ionosphere with a peak electron density of 3×10^5 cm^{-3}. Solar radiation is only 0.04 that of the amount striking the Earth. Precipitation of energetic charged particles from Jupiters magnetosphere is an important source of ionization. The effect of the Jupiter magnetosphere is evident in the existence of an aurora zone on Io. The zone is centered near Io's equator. The Jupiter magnetic field lines passing through the moon are a source of ionizing particles for the atmosphere, and aid in the escape of sulfur compounds into the Jovian magnetosphere.

6.2. *Europa*

Jupiter's moon Europa has a radius that is 0.245 that of Earth. Compared to Earth its surface gravity is 0.134g. It has been found that Europa possesses an atmosphere composed of molecular oxygen (Hall *et al.* 1995). The origin of the oxygen atmosphere is the fact that the surface of Europa is covered with ice and also subject to the bombardment by charged energetic particles and electrons from Jupiters magnetosphere. The impact of these particles causes sputtering of water from the surface. Further interaction of the particles with the freed water dissociates the water to produce free oxygen, which is less reactive than the hydrogen produced by the water dissociation reaction. The surface pressure is 1×10^{-11} bar.

The measurements of the Europa electron density are based on radio occultation data from the Galileo spacecraft (Kliore *et al.* 1997). The entry point for the occultation occurs on the dayside of the moon. This is also on the ram side for the entry of magnetospheric particles. On this side there is a significant ionosphere with a surface electron density of 2×10^4 cm^{-3}. The plasma scale height is 240 ± 40 km. The exit occultation for this measurement occurs on the nightside of the moon. For this situation there is no measurable ionosphere. The comparison of these electron density profiles indicates that the ionosphere is the result of photoionization of the atmospheric O_2 and the entry of magnetospheric particles. Based on the neutral atmospheric measurements of O_2 as the major constituent the principal ion is O_2^+.

6.3. *Ganymede*

Ganymede is the largest of Jupiters moons with a radius that is 0.413 of Earth. The acceleration of gravity is 0.146g that of Earth. Ganymede is very similar to Europa in its covering of water ice and being subject to bombardment by energetic charged particles and electrons from the Jovian magnetosphere. This bombardment gives rise to a molecular oxygen atmosphere with a surface pressure of 1×10^{-14} bar.

There is an ionosphere with the peak density between 400 and 2500 cm^{-3}. Ganyede is unique because it possesses its own magnetic field and a weak magnetosphere. Ions and electrons from the surface are swept along the field lines to form the magnetosphere together with input from Jupiter's magnetosphere. In addition to its intrinsic magnetic field, Ganymede possesses an induced magnetic field.

6.4. Callisto

The Jovian moon Callisto is the fourth in distance from the planet. Its size is approximately the same as Mercury, with a radius of 0.378 that of Earth and an acceleration due to gravity of 0.126g. The moon is not subject to orbital induced tidal forces and is covered with carbon dioxide ice. Sublimation of CO_2 from the surface has given rise to a CO_2 atmosphere with a surface pressure of 7.5×10^{-12} bar and a surface number density of 4×10^8 cm^{-3}.

As a result of photoionization of the CO_2 atmosphere Callisto possesses an ionosphere. The maximum electron density is 1.74×10^4 cm^{-3} at an altitude of 47.6 km. The plasma scale height is 49 km. The ionosphere corotates with the planet, which always presents the same face to Jupiter.

7. Titan

Titan is the largest moon of the Saturn system and the only one to possess an atmosphere. The radius of the moon is 0.4 that of the Earth with a surface gravity of 0.14g. Although the surface pressure is 1.4 bar, the surface temperature is only 94 K. The composition of the atmosphere is 98.4% N_2 and 1.6% methane with trace amounts of more complex hydrocarbons. The duration of the diurnal cycle is 30 years.

The daytime ionosphere has a peak electron density of 10^4 cm^{-3} in spite of the fact that the solar ionizing radiation is only 1% of the flux reaching Earth. In addition there is the ionization by electrons and ions associated with Saturns magnetosphere. The importance of this ionization source is demonstrated by the substantial nighttime ionosphere with a peak ionization of 10^3 cm^{-3} over an altitude range of 1000 to 1400 km. Based on comparison with the daytime ionosphere the magnetosphere ionization source is about 1% of the daytime photoionizatioin source. Cosmic radiation is a small additional ionization source operative at altitudes below the daytime and nighttime ionization peaks. The principal positive ion of the ionosphere is $HCNH^+$. Secondary ions include $C_2H_5^+$, $C_3H_5^+$, and CH_5^+. Negative ions have been measured at 960 km. The negative ions can be arranged by mass range: 10 to 30 amu, candidate ion species CN^-; 30 to 50 amu candidate ion species NCN^-, $HNCN^-$, C_3H^-; 50 to 80 amu candidate ion species $C_5H_5^-$, C_6H^-, $C_6H_5^-$; 80 to 110 amu candidate ion species polynes, nitriles, PAH; 110 to 200 amu candidate ion species polynes, nitriles, PAH; > 200 amu candidate ion species polynes, nitriles, PAH. Positive ion composition measurements (Waite et $al.$ 2007) have shown that the ion $C_6H_7^+$ is present. This ion can be correlated with the presence of large concentrations of benzene, C_6H_6, in the thermosphere. Dissociative electron-ion recombination of $C_6H_7^+$ by the reaction

$$C_6H_7^+ + e^- \rightarrow C_6H_6 + H \qquad (7.1)$$

yields benzene.

8. Conclusions

All bodies in the solar system are subject to solar radiation, which includes ultraviolet and x radiation capable of ionizing atmospheric gases of those bodies possessing atmospheres, forming ionospheres. Those planets and moons not possessing an intrinsic magnetic field, including Venus and Mars, have their atmospheres subject to the ionizing radiation of the charged energetic particles that comprise the solar wind. The interaction of the solar wind with the intrinsic magnetic field of planets such as Earth and Jupiter produces a magnetosphere for each of these bodies. Electrons in magnetospheres are accelerated to higher energies. Precipitation of these electrons into a planet's atmosphere gives rise to an auroral zone. Moons of magnetospheric planets are subject to energetic particles, which can modify the moons' ionospheres. Moons in this category are the Jovian moons of Io, Europa, Ganymede, and Callisto as well as Saturn's moon Titan. Although we now understand which bodies have ionospheres and how they are formed, the detailed behavior of these ionospheres is unknown. Future ionospheric missions will not only fill in details, but may uncover new phenomena.

References

Bertaux, J.-L., Leblanc, F., Witasse, O., Quemerais, E., Lilensten, J., Stern, S. A., Sandel, B., & Korablev, O. 2005, *Nature*, 435, 790

Brace, L. H. & Kliore, A. J. 1991, *Space Science Reviews*, 55, 81

Brain, D. A., *et al.* 2006, *Geophys. Res. Lett.*, 33, 1201

Connerney, J. E. P., Acuña, M. H., Wasilewski, P. J., Kletetschka, G., Ness, N. F., Rème, H., Lin, R. P., & Mitchell, D. L. 2001, *Geophys. Res. Lett.*, 28, 4015

Grebowsky, J. M. & Aikin, A. C. 2002, Meteors in the Earth's atmosphere. Edited by Edmond Murad and Iwan P. Williams. Publisher: Cambridge, UK: Cambridge University Press, 2002., p.189, 189

Hall, D. T., Strobel, D. F., Feldman, P. D., McGrath, M. A., & Weaver, H. A. 1995, *Nature*, 373, 677

Hanson, W. B., Sanatani, S., & Zuccaro, D. R. 1977, *J. Geophys. Res.*, 82, 4351

Hinson, D. P., Flasar, F. M., Kliore, A. J., Schinder, P. J., Twicken, J. D., & Herrera, R. G. 1997, *Geophys. Res. Lett.*, 24, 2107

Kliore, A. J. & Luhmann, J. G. 1991, *J. Geophys. Res.* , 96, 21281

Kliore, A. J., Hinson, D. P., Flasar, F. M., Nagy, A. F., & Cravens, T. E. 1997, *Science*, 277, 355

Kim, Y. H., Pesnell, W. D., Grebowsky, J. M., & Fox, J. L. 2001, *Icarus*, 150, 261

Knudsen, W. C., Spenner, K., Miller, K. L., & Novak, V. 1980, *J. Geophys. Res.*, 85, 7803

Lindal, G. F., Lyons, J. R., Sweetnam, D. N., Eshleman, V. R., & Hinson, D. P. 1987, *J. Geophys. Res.*, 92, 14987

Mendillo, M., Smith, S., Wroten, J., Rishbeth, H., & Hinson, D. 2003, *J. Geophys. Res.*, 108, 1432

Moore, L., Nagy, A. F., Kliore, A. J., Müller-Wodarg, I., Richardson, J. D., & Mendillo, M. 2006, *Geophys. Res. Lett.*, 33, 22202

Rich, F. J. 1985, Ionospheric Physics: Chapter 9 in the Handbook of Geophysics and the Space Environment. Edited by A. S. Jursa. Publisher: Air Force Geophysics Laboratory: United States Air Force, 1985

Taylor, H. A., Brinton, H. C., Bauer, S. J., Hartle, R. E., Cloutier, P. A., & Daniell, R. E. 1980, *J. Geophys. Res.*, 85, 7765

Tyler, G. L., *et al.* 1989, *Science*, 246, 1466

Waite, J. H., Young, D. T., Cravens, T. E., Coates, A. J., Crary, F. J., Magee, B., & Westlake, J. 2007, *Science*, 316, 870

Discussion

BOCHSLER: Mars Express has recently observed an increase by 1 order of magnitude in the ion outflow from the Martian ionosphere (Futaana et al., PSS, 2008) during an SEP event in December 2006. Would this affect the communications in a similar way to the Earth?

GREBOWSKY: Ion outflow would have no effect on communications unless SEP was associated with the production of an ionospheric density enhancement.

GIRISH: 1. The galactic cosmic ray effects in night-side ionospheres of Mars and Venus? 2. The effect of very slow rotation (243 days) of Venus on its atmospheric dynamics?

GREBOWSKY: 1. Measurements of cosmic ray effects on ionospheres of Mars and Venus have not been made, but models have included ionization by cosmic rays. 2. Venus' atmosphere actually rotates faster than planet-i.e. it super rotates.

JARDINE: Can you say a little about observations of Enceladus?

GREBOWSKY: Enceladus sporadically has intense emissions of water and ice. These ongassed products are a significant source of Saturn's magnetospheric plasma.

Universal Heliophysical Processes
Proceedings IAU Symposium No. 257, 2008
N. Gopalswamy & D.F. Webb, eds.

© 2009 International Astronomical Union
doi:10.1017/S1743921309029792

On the nature of the longitudinal anomaly of the southern hemisphere lower stratospheric temperature

Luis Eduardo Antunes Vieira[1] and Ligia Alves da Silva[2]

[1] Max-Planck-Institut für Sonnensystemforschung, 37191 Katlenburg-Lindau, Germany
email: `vieira@mps.mpg.de`

[2] Instituto Nacional de Pesquisas Espaciais, 12227-010 Sao Jose dos Campos, Brazil
email: `dasilva@mps.mpg.de`

Abstract. The effects of changes in the solar radiative emission on ozone levels in the strato-sphere have been considered as a candidate to explain the link between solar activity and its effects on the climate. As ozone absorbs electromagnetic radiation, changes in ozone concentrations alter Earth's radiative balance by modifying both incoming solar radiation and outgoing radiation. In this way, ozone controls solar energy deposition in the stratosphere and its variations alter the thermal structure of the stratosphere. These changes are assumed to propagate downward through a chain of feedbacks involving thermal and dynamical processes. The effects of high energy particle precipitation on mesospheric and stratospheric ozone have also been investigated. However, while the effects of high energy particle precipitation on ozone distribution in the auroral region has been investigated during the last decades, little is known about the role of the high energy particle precipitation on the stratospheric composition and thermal structure in the tropical/subtropical region. Here we show that the spatial distribution of the lower stratosphere temperature is affected by the presence of the southern hemisphere magnetic anomaly. We found that during the austral winter and spring, in the subtropical region (below 30 deg S), the reduction of the lower stratosphere temperature occurs systematically in the magnetic anomaly area. This result is consistent with the observations that in the southern hemisphere subtropical region the energy of precipitating particles is deposited lower in altitude in regions with weaker magnetic field intensity.

Keywords. Climate Change, Lower Stratosphere Temperature, Space Weather

1. Introduction

The increase of the surface temperature during the last century is the most important evidence that the Earths climate is changing. However, the increase of the surface temperature is not uniform (see Fig. 1). A large increase of the surface temperature is observed in the North Hemisphere over the continents. Regions without changes or reductions in the temperature are observed in the Pacific and Atlantic Oceans. Most of the changes of the surface temperature are closely related to changes of the regional atmospheric and oceanic circulation patterns.

Several authors argue that the changes of the atmospheric and oceanic patterns are related to the increase of the emission of greenhouse gases due to the anthropogenic activity. One strong argument is that the solar activity during the last 30 years is not increasing while a rapid increase of the surface temperature is observed. However, there are large uncertainties of natural and anthropogenic forcings of climate change. For example, the long-term variability of the total and spectral solar electromagnetic emission are not known.

Recently, a link between changes of the Walker circulation in the South Pacific and the westward drift of the Southern Hemisphere Magnetic Anomaly (SHMA), a region of low magnetic field intensity over South America and adjacent oceans, was proposed. Many interesting and significant atmospheric circulation features, such as the ENSO phenomena, are observed in the equatorial and southern low-latitude regions of the Pacific Ocean (e.g., Vincent 1994, and references therein). Two important large-scale features that occur in the Pacific SHMA region are: the Intertropical Convergence Zone (ITCZ), and the South Pacific Convergence Zone (SPCZ). In this paper, we use the effects of long-wavelength (LW) cloud effects, on the radiative flux in the atmosphere to indicate tropical convection and rainfall. In order to study the effects on the radiative flux in the SHMA region, we used the ISCCP D2 data (Rossow & Schiffer 1991). Figure 2 shows the distribution of LW cloud effects for December and June, averaged from 1983–2004. Positive values indicate energy input and negative values indicating energy loss. The convective precipitation is observed in the heating regions. The superimposed black lines in Figure 3 show the iso-intensity contours of the geomagnetic field at 10 km in the region of the magnetic anomaly for year 1990. The geomagnetic field was estimated using the International Geomagnetic Reference Field (IGRF) model [IAGA, Division V, Working Group VMOD, 2005]. We used in this paper the magnetic field parameters in order to identify the magnetic anomaly region. In December, the axis of heating stretches eastward, along $5° - 10°$ S, from the eastern Indian Ocean to $180°$ W. It then extends southeastward toward higher latitudes, to about $45°$ S, $130°$ W. West of the $180°$ meridian, the near-zonal band of convection is commensurate with the ITCZ. East of the $180°$ meridian, the diagonal cloud band is referred to as the SPCZ. In December, it is most intense and extends farther poleward. The SPCZ cloud band is confined between the 0.35 to 0.45 Gauss iso-intensity contours of the geomagnetic field. In June, the ITCZ cloud band has shifted into the Northern Hemisphere; the axis of minimum values of LW cloud effects stretches from Southeast Asia to $10°$ N, near $140°$ E, and then generally eastward across the entire Pacific Ocean. The cloud band associated to the SPCZ extends to approximately $10°$ S also confined in the 0.35 to 0.42 Gauss iso-intensity contours of the geomagnetic field.

The authors speculated that the physical mechanism should involve effects of particle precipitation in the magnetic anomaly region on the thermal structure of the

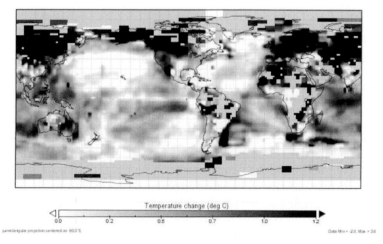

Figure 1. Surface temperature trend from 1951 to 2007. Source: Goddard Institute for Space Studies (GISS) Surface Temperature Analisis.

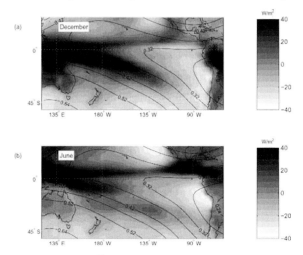

Figure 2. Long-wavelength cloud effects on the net radiative flux in the atmosphere for (a) December and (b) June, averaged for the years 1983–2001. The superimposed black lines show the iso-intensity contours of the geomagnetic field near the surface (10 km) for year 1990.

stratosphere. The effects of changes in the solar radiative emission and particle precipitation on ozone levels in the stratosphere have been considered as a candidate to explain the link between solar activity and its effects on the Earths atmospheric composition and thermal structure (e.g., Cubasch & Voss 2000, Randal *et al.* 2006). Pinto *et al.* (1990) estimated that the ozone depletion due to electron precipitation at 70-80 km in the Southern Hemisphere Magnetic Anomaly (SHMA) region during large geomagnetic storms can be as much as 30 %. Estimatives of the ionization rate due to electron precipitation in middle latitude suggests a peak near 75–90 km altitude due to primary electron energy deposition, whereas the secondary peaks near 35–45 km are due to Bremsstrahlung X ray penetration (Vampola & Gorney 1983). In the SHMA region the precipitation penetrates deeper in the atmosphere and can lead to more enhanced ionization from Bremsstrahlung at low altitudes. From this estimative, the ionization rate from the Bremsstrahlung X ray penetration is about four orders of magnitude smaller than the primary electrons. Balloon-born X-ray measurements have detected energetic electron precipitation effects at stratospheric heights during intense magnetic disturbances (e.g., Pinto & Gonzalez 1986, Pinto & Gonzalez 1989).

The effects of the zonal asymmetry in the ozone on climate have been examined in the Northern and Southern Hemispheres. These analyses suggest that changes in the zonal asymmetry of ozone have had important impacts on Southern Hemisphere Climate (Crook *et al.* 2008, Vieira *et al.* 2008). In the next section we compare the patterns of the lower stratosphere temperature with the configuration of the Earth's magnetic fiels in the South Hemisphere. The results presented are discussed in detail by Da Silva *et al.* (2008).

2. Southern hemisphere lower stratospheric temperature

In order to estimate the monthly lower stratosphere temperature climatology, we used monthly distributions of the lower stratosphere temperature (TLS channels) obtained by Microwave Sounding Units (MSU) operating on nine NOAA polar-orbiting platforms from 1979 to 2007. The weighting function for the TLS channel peaks between 15 and

20 km of altitude. Warming events caused by the eruptions of El Chichon (1982) and Mt Pinatubo (1991) were observed during the analyzed period.

Figure 3 presents the southern hemisphere lower stratosphere brightness temperature climatology for June to November, estimated from 1979 to 2005. The atmospheric temperature varies with altitude according to chemical and physical processes taking place. In the lower stratosphere air is transported from the tropics towards the poles by the Brewer-Dobson circulation. In this region, the upper tropospheric convective processes affect the tropics so that the temperature has a minimum near the equator and maxima at the summer pole and in winter mid-latitudes. In the upper stratosphere and mesosphere there is a solstitial circulation with upward motion in the summer hemisphere, a summer-to-winter transport in the mesosphere and descent near the winter pole. Following Grytsai et al. (2007), we focus on these months, as this is generally when both the chemical and dynamical contributions to the zonal ozone asymmetry act, and transport processes associated with the break-down of the polar vortex are less important. The superimposed black lines in Figure 3 show the iso-intensity contours of the geomagnetic

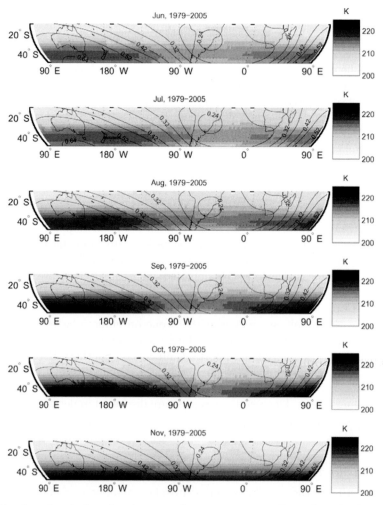

Figure 3. Southern hemisphere lower stratosphere temperature climatology for June to November. The superimposed black lines show the iso-intensity contours of the geomagnetic field at 10 km for year 1990. After Da Silva et al. 2008

field at 10 km in the region of the magnetic anomaly for year 1990. For the purpose of this analysis the boundaries of the magnetic anomaly were not fixed.

We note a large reduction of the temperature in the region of the magnetic anomaly in the belt between 60° S and 30° S during the austral winter and spring, while during the austral summer (not shown here) the reduction in the temperature is not noticeable. This reduction of the temperature coincides quite well with the region which presents higher electron flux. The maximum difference between the temperatures inside the magnetic anomaly (60° W) and outside the anomaly (150° E) for the latitude of 42.5° S is approximately 5.9 K and occurs in October during the austral spring.

3. Conclusions

During the austral winter and spring, in the subtropical region (below 30° S), the reduction of the lower stratosphere temperature occurs systematically in the magnetic anomaly area. This result is consistent with the assumption that in the subtropical region the energy of precipitating particles is deposited lower in altitude in regions with weaker magnetic field intensity. However, from this analysis it is not possible to distinguish if the effects of particle precipitation occur in the lower stratosphere or, most probably, in the lower mesosphere/upper stratosphere propagating downward.

References

Crook, J. A., Gillett, N. P., & Keeley, S. P. E. 2008, *Geophys. Res. Lett.*, 35, L07806, doi:10.1029/2007GL032698.

Cubasch, U. & Voss, R. 2000, *Space Science Reviews*, 94(1-2), 185–198.

Da Silva, L. A., Vieira, L. E. A., Echer, E., & Satyamurty, P. 2008, *JASTP*, submitted.

Gledhill, J. A. 1976, *Reviews of Geophysics*, 14 (2), 173–187.

Grytsai, A. V., Evtushevsky, O. M., Agapitov, O. V., Klekociuk, A. R., & Milinevsky, G. P. 2007, *Ann. Geophys.*, 25, 361–374.

Pinto, O. & Gonzalez, W. D. 1986, *Journal of Geophysical Research*, 91(A6), 7072–7078.

Pinto, O. & Gonzalez, W. D. 1989, *Journal of Atmospheric and Terrestrial Physics*, 51(5), 351–365.

Pinto, O., Kirchhoff, V., & Gonzalez, W. D. 1990, *Annales Geophysicae-Atmospheres Hydrospheres and Space Sciences*, 8 (5), 365–367.

Randall, C. E., *et al.* 2006, *Geophysical Research Letters*, 33.

Rossow, W. B. & Schiffer, R. A. 1991, *Bulletin of the American Meteorological Society*, 72, 2–20.

Vampola, A. & Gorney, D. 1983, *J. Geophys. Res.*, 88 (A8), 6267–6274.

Vieira, L. E. A. & da Silva, L. A. 2006, *Geophys. Res. Lett.*, 33, L14802, doi:10.1029/2006GL026389.

Vieira, L. E. A., da Silva, L. A., & Guarnieri, F. L. 2008, *J. Geophys. Res.*, 113, A08226, doi:10.1029/2008JA013052.

Vincent, D. G. 1994, *Monthly Weather Review*, 122, 1949–1970.

Discussion

DAVILA: Is there a physical mechanism to explain the magnetic field correlations you showed?

VIEIRA: The physical mechanism seems to be related to changes of the atmosphere composition and thermal structure due to particle precipitation in the magnetic anomaly region.

Universal Heliophysical Processes
Proceedings IAU Symposium No. 257, 2008
N. Gopalswamy & D.F. Webb, eds.

© 2009 International Astronomical Union
doi:10.1017/S1743921309029809

Characteristic signatures of energetic ions upstream from the Kronian magnetosphere as revealed by Cassini/MIMI

Olga E. Malandraki[1], S. M. Krimigis[2,3], E. T. Sarris[4], N. Sergis[2],
K. Dialynas[2], D. G. Mitchell[3], D. C. Hamilton[5] and A. Geranios[6]

[1]Institute for Astronomy and Astrophysics, National Observatory of Athens,
Pedeli, Athens, Greece
email: omaland@astro.noa.gr

[2]Office for Space Research and Technology, Academy of Athens, Athens, Greece

[3]Applied Physics Laboratory, Johns Hopkins University, Laurel, Maryland, USA

[4]Democritus University of Thrace, Xanthi, Greece

[5]University of Maryland, Department of Physics and Astronomy, MD, USA

[6]Nuclear and Particle Physics Department, University of Athens, Greece

Abstract. We present unique observations obtained by the Magnetospheric Imaging Instrument (MIMI) on the Cassini spacecraft, of the energetic ion population in the environment upstream from the dawn-to-noon sector of the Kronian magnetosphere during the approach phase and subsequent several orbits of the Cassini spacecraft around the planet. High sensitivity observations of energetic ion directional intensities, energy spectra, and ion composition were obtained by the Ion and Neutral Camera (INCA) of the MIMI instrument complement with a geometry factor of $\sim 2.5\ cm^2 sr$. Charge state information was provided by the Charge-Energy-Mass-Spectrometer (CHEMS) over the range ~ 3 to 220 keV per charge. The observations revealed the presence of distinct upstream bursts of energetic hydrogen and oxygen ions up to distances of $\sim 135 R_S$. The observations are presented and their theoretical implications are addressed.

Keywords. planets and satellites: general, acceleration of particles, plasmas

1. Introduction

Energetic charged particle events upstream from planetary bow shocks have been observed at Earth, Jupiter, Saturn and Uranus (e.g., Krimigis 1992). Over the last 3 decades there have been several attempts to interpret the observations in the context of one or more theoretical concepts. Generally the models are divided into those favouring solar wind ions undergoing first-order Fermi acceleration at the bow shock (e.g., Lee 1982; Scholer 1985; Trattner *et al.* 2003) and those suggesting leakage of pre-accelerated magnetospheric ions into the interplanetary medium (e.g., Krimigis *et al.* 1985; Anderson 1981; Krupp *et al.* 2002; Anagnostopoulos *et al.* 2005).

O^{+1} has been perceived as an ideal tracer ion of ionospheric origin accelerated to keV energies in the magnetosphere (Krimigis *et al.* 1986). Recently, ion composition measurements have provided clear evidence for low-charge-state ions of magnetospheric origin (e.g. O^{+1}, N^{+1}, O^{+2}) upstream of the Earth's bow shock (Christon *et al.* 2000, Posner *et al.* 2002; Keika *et al.* 2004). Analysis of the Voyager observations at Jupiter showed that the ions consisted primarily of oxygen and sulphur (Zwickl *et al.* 1981; Krimigis *et al.* 1985). Jupiter's magnetosphere with a plasma composition dominated by oxygen and sulphur (Hamilton *et al.* 1981) emitted from Io's volcanos offered a unique opportunity to utilize "tracer ions" to determine the origin of the upstream ions. This composition is clearly different from that of the solar wind.

The existence of energetic ions upstream and downstream of Saturn's bow shock was established during the Voyager 1 and 2 encounters with Saturn in 1980 and 1981, respectively. The energetic ions were found to exist in the interplanetary medium up to distances of $\sim 200 R_S$ upstream (Krimigis *et al.* 1983; Krimigis 1986) from the planet. Analysis of the Voyager 2 observations showed the ion increases extended in energy to ~ 500 keV. Definitive separation of a magnetospheric from a solar wind source has been more difficult than in the case of Jupiter since no unique compositional signature was apparently associated with the magnetosphere of Saturn. In this work, we present unique observations obtained by the Magnetospheric Imaging Instrument (MIMI) on the Cassini spacecraft of the energetic ion population in the environment upstream from the dawn-to-noon sector of the Kronian magnetosphere during the approach phase and subsequent several orbits of the Cassini spacecraft around the planet. These measurements by Cassini/MIMI, which offers unprecedented observational capabilities, are utilized to investigate characteristic signatures of the ion events in terms of the time history of intensities, spectra, composition and charge state of energetic ions upstream from the Kronian Magnetosphere as revealed in the Cassini era. The theoretical implications of the observations are also addressed.

2. Instrumentation

High sensitivity observations of energetic ion directional intensities, energy spectra, and ion composition are obtained by the Ion and Neutral Camera (INCA) which is part of the Magnetospheric IMaging Instrument (MIMI) complement on the Cassini spacecraft (Krimigis *et al.* 2004). MIMI/INCA (which has a 90° by 120° field of view) with its large geometry factor $\sim 2.5 cm^2$ sr in the ion mode is ideal for studies of upstream ion activity, also providing some capability of separating light (H, He) and heavier (C, N, O) ion groups (henceforth referred to as "hydrogen" and "oxygen" respectively). The intensities are integrated over the entire image obtained by INCA. Detailed composition measurements are provided by a second MIMI sensor, the Charge-Energy-Mass-Spectrometer (CHEMS), which is capable of determining independently the charge state, mass, and energy of ions over the range ~ 3 to ~ 220 keV per charge.

3. Observations and Data Analysis

The Cassini spacecraft was injected into orbit around Saturn (SOI) on July 1, 2004. Figure 1 (left) presents the orbit of Cassini in an equatorial projection in Saturn Solar Orbit (SSO) coordinates for the period from June 12 (DOY 164) till October 27 (DOY 301) in 2004, during its approach phase and subsequent orbits around Saturn. As denoted by the arrow the Sun is to the left of the diagram. Model bow shock and magnetopause curves are presented. For reference the orbits of Titan and Rhea are also shown. The highlighted parts of the Cassini trajectory denote periods during which individual upstream events are analyzed in detail in this work. (For color versions of all figures see online electronic version of the paper).

In Figure 1 (upper right), from top to bottom energetic ion differential intensities versus time are shown in the energy range 3 - > 220 keV as measured by the MIMI/INCA hydrogen channels (H7-H0) for the period DOY 175-180 in 2004, during the approach phase of Cassini, prior to its entry into the magnetosphere of Saturn. This period corresponds to the first period highlighted in Figure 1 (left). On DOY 175, Cassini was at $\sim 90 R_S$ radial distance from Saturn. As Cassini approaches Saturn, distinct ions bursts

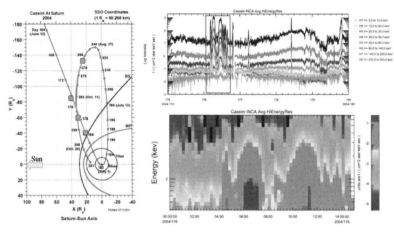

Figure 1. Left: Orbit of Cassini for period DOY 164-301 in 2004. Upper right: Energetic ion intensities (3 - >220 keV) for DOY 175-180 in 2004. Bottom right: Dynamic spectrogram of intensities measured in the Oxygen channel on DOY 176.

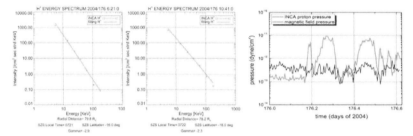

Figure 2. Left: Hydrogen spectra for the events observed on DOY 176. Right: Comparison of the particle and magnetic and field pressures.

are observed by MIMI/INCA. On DOY 179 multiple bow shock crossings started to be observed (Masters *et al.* 2008). We next focus on a series of upstream ion events delimited by a box in Figure 1 (left), observed by Cassini/MIMI on DOY 176 in 2004 at $\sim 80R_S$.

Figure 1 (bottom right) shows a dynamic spectrogram (Energy vs Time) of intensities measured by the Oxygen channel of INCA from 0000 UT to 1500 UT on DOY 176. The measurements reveal the upstream events near Saturn are also observed in the Oxygen channel up to ~ 500 keV energy. The Pulse Height Analyzed (PHA) observations available by INCA (not shown) also reveal significant oxygen presence in the upstream events observed on this day. We have also analyzed the magnetic field measurements during this period (not shown) and found evidence for wave structures consistent with ion cyclotron waves for both H^+ and O^+.

In Figure 2 (left panels), energy spectra (assuming protons) at the peak of the events on DOY 176 are shown. The spectra appear to fit well in energy with spectral indices $\gamma \sim 2.9$, and 2.3. An event previously detected near Saturn by Voyager 2 that was examined in detail exhibited a power law of ~ 2.1 (Krimigis 1986). Cassini, thus observed somewhat softer spectra during the events on DOY 176. Figure 2 (right panel) shows a comparison of the magnetic field and energetic particle (as derived from the hydrogen channel in the energy range ~ 3-200 keV) pressures on DOY 176. During the ion events the particle pressure by far exceeds the magnetic field pressure, with beta ratio values (energetic particle pressure to magnetic field pressure) of ~ 4-10.

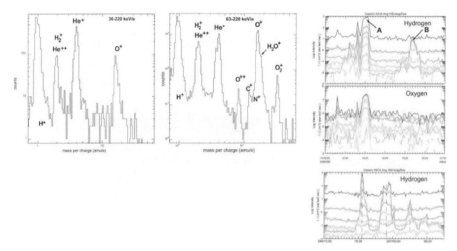

Figure 3. (Left panels) MIMI/CHEMS observations for the periods DOY 242-296 in 2004 (first panel) and DOY 29-41 in 2005 (second panel). (Right, two top panels) INCA/MIMI observations of the upstream events $\sim 134R_S$ on DOY 265, (Right, bottom panel) INCA/MIMI observations of the upstream events on DOY 286/287.

Figure 3 (left first panel) presents a histogram of counts vs Mass/Charge ratios from the MIMI/CHEMS sensor summed over the period DOY 242-296 in 2004 when Cassini was upstream from the Saturnian bow shock (Masters *et al.* 2008). The observations reveal that oxygen in the energy range 36 - 220 keV/charge is present throughout Rev A, from ~150 to ~50 Rs upstream. Figure 3 (left second panel) presents composition and charge states of ions as measured by MIMI/CHEMS during Rev C (DOY 29, $\sim 56R_S$, to DOY 41, $\sim 43R_S$ in 2005). During this period, Cassini was upstream (Masters *et al.* 2008) but close to the bow shock. The measurements reveal significant fluxes of $63 < E < 220$ keV/e Water group ions (W^+) are present.

The two panels on the right in Figure 3 present differential intensities as measured by the INCA Hydrogen and Oxygen channels from 1000-2200 UT on DOY 265 during upstream events observed by Cassini at $\sim 134R_S$. The PHA observations also showed significant oxygen presence during the first event of this period. The spectrum at the peak of this event (A) appears to fit well a power law in energy with spectral index $\gamma \sim 2.5$. The spectrum of the second event (B) showed a deficiency of low energy ions. A noteworthy feature is that at the end of the first event the particle intensities were observed to decrease abruptly and simultaneously at all energies, in association with an abrupt change in the direction of the magnetic field (not shown), no longer connecting the spacecraft to the planetary bow shock.

The bottom panel on the right in Figure 3 presents differential intensities as measured by the INCA Hydrogen channel from 12:00 UT on DOY 286 to 09:00 UT on DOY 287 in 2004 during upstream events observed by Cassini at ~ 86 Rs. PHA observations showed significant oxygen presence during the events. The events are symmetrical around their peak intensities. The start and stop times of high-energy and low-energy ions do not match, with the higher energies evidently observed earlier than the low-energies. The reverse is observed at the end of the event. The observations provide evidence of filaments filled with energetic particles being convected over the spacecraft, with the higher energy particles, which have larger gyroradii, starting to be detected earlier than the lower energy particles which have smaller gyroradii. This is also consistent with the reverse observation as the spacecraft exits the IMF filament.

4. Summary and Conclusions

We have presented preliminary results of the characteristic signatures of energetic ions observed upstream from the Kronian magnetosphere by Cassini/MIMI. Hydrogen ion bursts are observed in the energy range 3 - 220 keV (and occasionally to E > 220 keV) and Oxygen ion bursts in the energy range 32 to $\sim$ 500 keV. The duration of the ion bursts is several minutes up to 4 hrs. Some ion bursts are accompanied by distinct diamagnetic field depressions with plasma $\beta \sim 4$ to 10. There is evidence for wave structures consistent with ion cyclotron waves for both H^+ and O^+. The bursts have a filamentary structure with some exhibiting distinct signatures of 'velocity-filtering effects' at the edges of convecting IMF filaments. MIMI has provided for the first time charge state measurements of upstream ion events at Saturn. The events are of varying composition, with some exhibiting significant fluxes of oxygen. Time-averaged detailed composition measurements upstream show all elements/charge states identified within Saturns MSPH including O^+, O^{++}, O_2^+ etc. Given that energetic ions trapped within the magnetosphere of Saturn are mostly H^+, O^+, and other water group ions W^+ (Krimigis *et al.* 2005) we conclude that O^+ -rich upstream events must be particles leaking from Saturns magnetosphere under favourable IMF conditions. We have thus measured and identified for the first time unique tracer ions at Saturn to discriminate between solar wind and magnetospheric sources. Recent results by the CAPS plasma instrument onboard Cassini (Thomsen *et al.* 2008) have shown no detectable contribution from magnetospheric W^+ ions to upstream suprathermal ions between 3–50 keV/q. This also supports our results on the origin of the particles and strongly suggests the energetic W^+ group ions observed by Cassini/MIMI leak directly from the magnetosphere and are not produced by energization of lower-energy particles by bow shock-related processes.

References

Anagnostopoulos, G. C., Efthymiadis, D., Sarris, E. T., & Krimigis, S. M. 2005, *J. Geophys. Res.*, 110, A10203

Anderson, K. A. 1981, *J. Geophys. Res.*, 86, 4445

Christon, S. P., Desai, M., Eastman, T. E., *et al.* 2000, *Geophys. Res. Lett.*, 27(16), 2433

Hamilton, D. C., Gloeckler, G., Krimigis, S. M., & Lanzerotti L. J. 1981, *J. Geophys. Res.*, 86, 8301

Keika, K., Nose, M., Christon, S. P., & McEntire, R. W. 2004, *J. Geophys. Res.*, 109, A11104

Krimigis, S. M. 1986, in: CNES & CEPADUES (eds.), *Comparative Study of Magnetospheric Systems* (France), p. 99

Krimigis, S. M. 1992, *Space Sci. Revs*, 59, 167

Krimigis, S. M., Zwickl, R. D., & Baker, D. N. 1985, *J. Geophys. Res.*, 90, 3947

Krimigis, S. M., Carbary, J. F., Keath, E. P. *et al.* 1983, *J. Geophys. Res.*, 88, 8871

Krimigis, S. M., Sibeck, D. G., & McEntire, R. W. 1986, *Geophys. Res. Lett.*, 13, 1376

Krimigis, S. M., *et al.* 2004, *Space Sci. Rev.*, 114, 233

Krimigis, S. M., *et al.* 2005, *Science*, 307, 1270

Krupp, N., Woch, J., Lagg, A., *et al.* 2002, *Geophys. Res. Lett.*, 29, 1736

Lee, M. A. 1982, *J. Geophys. Res.*, 87, 5063

Masters, A., Achilleos, N., Dougherty, M. K., *et al.* 2008, *J. Geophys. Res.*, 113, A10210

Posner, A., *et al.* 2002, *Geophys. Res. Lett.*, 29(7), 1099

Scholer, M. A. 1985, in: B. Tsurutani & R. G. Stone (eds.), *Collisionless Shocks in the Heliosphere: A Tutorial Review* (Washington, D.C.: AGU), p. 287

Thomsen, M. F., *et al.* 2008, *J. Geophys. Res.*, 112, A05220

Trattner, K. J., Fuselier, S. A., Peterson, W. K., Chang, S. W., Friedel, R., & Aellig, M. R. 2003, *J. Geophys. Res.*, 108, 1303

Zwickl, R. D., Krimigis, S. M., Carbary, J. F. *et al.* 1981, *J. Geophys. Res.*, 86, 8125

Discussion

SPANGLER: What processes are occurring which allow ions to escape from the Saturnian magnetosphere into the upstream region?

MALANDRAKI: Processes by which the energetic magnetospheric particles escape from the outer magnetosphere into the Magnetosheath have been discussed for the Earth's case. The particles reach the magnetopause by following magnetospheric drift paths, enter into the magnetosheath and fail to re-enter the magnetosphere (Sibeck *et al.*, JGR 92 (11), 12, 097, 1987; Sarafopoulos *et al.*, JGR, 105 (7), 15, 729, 2000). The loss of magnetospheric particles is caused by finite gyroradius effects. Alternately the energetic magnetospheric particles might escape into the magnetosheath along interconnected magnetosphere – magnetosheath magnetic field lines that result from merging (e.g. Speiser and Williams; JGR, 87(4), 2177, 1982)

It has been shown that lower-energy magnetoshealth ions have ready access to the upstream interplanetary medium. The quasi-perpendicular bow shock transition is characterized by 2 downstream ion populations including high-energy gyrating ions in addition to the directly transmitted anisotropic ions. Tanaka *et al.*, JGR, 88(4), 3046, 1983 have shown by particle simulations that this highly anisotropic downstream ion distribution can excite electromagnetic ion cyclotron waves, which, in turn, pitch-angle scatter the gyrating ions in a few ion gyro-periods. As a result, some ions acquire large parallel velocities and move fast enough along the connecting downstream magnetic field to escape back across the bow shock, into the upstream region. Within the model of Edmiston *et al.*, GRL, 9(5), 571, 1982, significant upstream ion fluxes occur for the quasi-parallel portion of the bow shock. The ions are heated and thermalized in a thin layer at the shock front. They calculated the ions returning upstream from a hot Maxwellian distribution at this layer.

Similar processes are apparently taking place in the vicinity of Saturn's magnetosphere.

DAVILA: Is the time variability observed really a time varying process, or are they spatial structures that the spacecraft is passing through?

MALANDRAKI: The upstream ion events exhibit a filamentary structure which strongly suggests we are dealing with spatial structures that the spacecraft is passing through. In some of the events, as presented, signatures of 'velocity filtering effects' at the edges of connecting IMF filaments are observed.

Universal Heliophysical Processes
Proceedings IAU Symposium No. 257, 2008
N. Gopalswamy & D.F. Webb, eds.

© 2009 International Astronomical Union
doi:10.1017/S1743921309029810

MHD Kamchatnov-Hopf soliton in the model of primordial solar nebula

Vladimir V. Salmin

Siberian Federal University,
Svobodny ave., 79, Krasnoyarsk, 660041, Russia
email: vsalmin@gmail.com

Abstract. Stereographic projection of Hopf field on the 3-sphere into Euclidean 3-space is used as a model of 3D steady flow of ideal compressible fluid in MHD. In such case, flow lines are Villarceau circles lying on tori corresponding to the levels of Bernoulli function. Existence of an optimal torus with minimal relative surface free energy is shown. Beat of oscillations with wave numbers corresponding to structural radii of optimal torus leads to scaling of optimal tori. Spatial intersection of homothetic tori within one torus result in formation of cluster with the size depending on scaling factor. Optimal tori are considered as precursors of planetary orbits.

Keywords. MHD, solar system: formation

1. Introduction

Solution of the equations of ideal magnetohydrodynamics describes a localized topological soliton with use of Hopf mapping shown by Kamchatnov (1982). Example of introducing the Euler's potential into a topological MHD soliton which has non-trivial helicity called MHD Kamchatnov-Hopf soliton was described by Semenov *et al.* (2001). Hopf field on S^3 has minimal energy among all the fields diffeomorphic to it (Arnold & Khesin 1998). Stability of Hopf field on S^3 has been proved (Gil-Merdano & Llinares-Fuster 2001, Yampolsky 2003). Hopf fibration of S^3 with stereographic projection induces the toroidal coordinates on E^3 (Gibbons 2006).

2. Overview

On a torus $\nu = \operatorname{arcsinh}(k)/k$ contravariant metrical tensor in toroidal coordinates is $g^{ij}|_T = \delta_i^j \left(\sqrt{k^2 + 1} - \cos(k\alpha)\right)^2 /R^2 k^2$. For steady flow of ideal compressible fluid, stress tensor is determined by metrical tensor only $p^{ij}|_T = -p\delta_i^j \left(\sqrt{k^2 + 1} - \cos(k\alpha)\right)^2 /R^2 k^2$. Since $\left(\sqrt{k^2 + 1} - \cos(k\alpha)\right)^2 /R^2 = r^{-2}$, where r-distance from of torus axis to the point on the torus, then the pressure on the surface $P \propto r^{-2}$. The volume of the solid torus with main radius $c = 1$ is $V(g) = 2\pi^2 g^2$ where $g = a/c$ - form factor, and a-tube radius. Notice that $g = \sin(\theta)$ where θ - inclination of Villarceau circles to the main plane of torus symmetry, we call this angle as "stream inclination". The force of pressure $\Sigma(g) = 4\pi g A \int_0^\pi (1 - g\cos(\phi))^{-1} d\phi$. Let's find the form factor g when the ratio of the solid torus volume to the force of pressure on it's boundary has a maximum: maximize $(V(g)/\Sigma(g), g = 0..1, location)$, $g = (\sqrt{2})^{-1} \approx 0.7071\ldots$. We will call the torus meeting form factor as "optimal torus". Physically, it corresponds to the minimal relative surface free energy.

Torus in the principal symmetry plane perpendicular to its axis is characterized with two structural radii r_m, r_o. The corresponding wave numbers are: $k_m = 2\pi/c(1-g)$, $k_o =$

$2\pi/c(1+g)$. Indices m-massive, and o-outer are used as mnemonic. Beat of oscillations is calculated as: $\cos(k_m r) + \cos(k_o r) = 2\cos\left(r(k_m + k_o)/2\right)\cos\left(r(k_m - k_o)/2\right)$. If half-sum or half-difference of wave numbers would be equal to wave numbers of structural radii of tori homothetic to original one, we would have the following four expressions for possible scaling factor $K_1 = 1 + g$, $K_2 = 1 - g$, $K_3 = (1 + g)/g$, $K_4 = (1 - g)/g$. If we suggest that the orbits of at least two neighbor planets have been forming due to interactions between two tori formed in nebula, the scaling relation should be taken into consideration when their interactions are described. Scaling factor K would have the value similar to the ratio of the semi-major axes of neighbor planets. We call it as an orbital scaling factor. For any value of the form factor g, $|\ln(K_1)| < |\ln(K_2)|$, $|\ln(K_1)| < |\ln(K_3)|$, and $|\ln(K_1)| < |\ln(K_4)|$ if $g < \sqrt{2} - 1$ or $g > (\sqrt{5} - 1)/2$ thereby it corresponds to the larger volume of tori intersection, and results in maximal interactions for K_1.

Arrangement of tori group results in their intersection in space, therefore, it should affect the proximity of physical characteristics of neighbor planets in a case of planetary system formation. Actually, we would like to know what is the maximal number C of the planet group members able to have similar physical parameters. Let's call the indicated group as cluster, and determine the cluster's size having intersection within one torus $r_m K^{(C-1)} < r_o$. The cluster size $C = 2 - \ln(1 - g)/\ln(1 + g)$. Since $K = 1 + g$, the latter would be re-written as: $C = 2 - \ln(2 - K)/\ln(K)$. First threshold corresponding to transition from the cluster size 3 to the size 4 at orbital scaling factor $K \approx 1.3894$, being in agreement with stream inclination $\theta \approx 22.917°$.

3. Implications

Solar system. Orbital scaling factor K for solar system $K = 1.6995 \pm 0.0224$. Difference between the theoretical value and the real one does not exceed the value of standard deviation: $\Delta K = 1.7071 - 1.6995 = 0.0076$. Cluster size $C = 4$. Stream inclination to ecliptic $\theta_\varepsilon = 44.4 \pm 2°$.

Galilean moons. Orbital scaling factor for Galilean moons $K = 1.6414 \pm 0.028$. Value of orbital scaling factor is less than theoretical one on a magnitude exceeding the standard deviation: $\Delta K = 0.066$. Cluster size is just as in the Solar system $C = 4$. Stream inclination $\theta = 39.9 \pm 2.3°$.

Saturn's satellites. The distinction of the satellite system of Saturn is the lower value of the orbital scaling factor $K = 1.2949 \pm 0.00377$, and geometrical progression is clearly recognizable up to Rhea. Further analysis is possible if for the next satellite Titan we omit 2 orbits. The same situation is seen with Iapetus. The size of cluster $C = 3$. Stream inclination $\theta = 17.15 \pm 0.23°$.

What is the reason for decreasing the orbital scaling factor in satellite systems of above-mentioned planets comparatively to Solar system or theoretical value? The orbital planes lie close to the equatorial plane of the central planet, and the latter has some axial tilt, thus stream inclination on the tori forming satellite system differs from $\pi/4$ on a value of the axial tilt. When the axial tilt does not exceed $\pi/4$, expression for orbital scaling factor is $K_\varepsilon = 1 + \sin(\pi/4 - \varepsilon)$, where ε - axial tilt. Axial tilt of Jupiter $\varepsilon = 3.08°$, thus corrected values of orbital scaling factor for Galilean moons $K_\varepsilon = 1.668$. Axial tilt of Saturn $\varepsilon = 26.7°$, therefore corrected values of orbital scaling factor for regular moons of Saturn $K_\varepsilon = 1.314$. Corrected values well agree with actual ones. Values of stream inclination to ecliptic for Galilean moons $\theta_\varepsilon = \theta + \varepsilon = 43 \pm 2.3°$ and Saturn's moons - $\theta_\varepsilon = 43.88 \pm 0.23°$. Deviation from the optimal $\pi/4$ does not exceed $2°$.

Neptune's satellites. Orbital scaling factor for Neptune's regular satellites $K = 1.176 \pm 0.002$. Therefore, the size of cluster, like in the Saturn's system, $C = 3$. Also, like in the

Saturn's system, two orbits between Larissa and Proteus are omitted. Omitting two orbits might reflect regularization when the size of cluster C is 3. Stream inclination to the equatorial plane $\theta = 10.15 \pm 0.12°$. Since the Neptune's axial tilt $\varepsilon = 28.32°$, the calculated value of orbital scaling factor is $K_\varepsilon = 1.287$. Significant difference from the actual value is evident. For its analyzing, suppose that stream inclination on the torus forming Neptune's orbit differs from the optimal value $\theta_\varepsilon < \pi/4$. Value of stream inclination on the torus forming the Neptune's orbit: $\theta_\varepsilon = \theta + \varepsilon = 38.47°$. Thereafter, ratio of Neptune's to Uranus' semi-major axes should be less than 1.707: $K = 1 + \sin\theta_\varepsilon = 1.622$. Indeed, ratio of Neptune's to Uranus' semi-major axes is lowest in the family of giant planets. Taking into consideration the eccentricity of Uranus' and Neptune's orbits, the actual value of ratio of semi-major axes is $a_N/a_U = 1.57 \pm 0.045$. Thus, the estimation fits well the actual value.

Uranus' satellites. Orbital scaling factor $K = 1.455 \pm 0.0146$. The cluster's size $C = 4$. Stream inclination to the Uranus' equatorial plane $\theta = 27.07 \pm 0.95°$. Ratio of Uranus' to Saturn's semi-major axes is largest not only in the group of giant planets, but in the Solar system in general. Subject to eccentricity, $a_U/a_S = 2.01 \pm 0.1$. Orbital scaling factor is higher than maximal value 2 allowed by the model, however, there is an area of values $K \in [1.91..2]$ due to eccentricity of Saturn's and Uranus' orbits allowed by the model. The corresponding interval of the stream inclination to the ecliptic is $\theta_\varepsilon \in [65.6°..90°]$. Since Uranus' axial tilt $\varepsilon = 97.77°$, stream inclination on torus forming Uranus' orbit in relation to ecliptic is $\theta_\varepsilon = \varepsilon - \theta = 70.70°$. Corresponding value of orbital scaling factor for Uranus vs Saturn would be $K = 1 + \sin\theta_\varepsilon = 1.944$, well fitting the actual value.

Pluto's satellites. Significant gap exists between the Charon's and Nix's orbits. Presumably, this gap is caused by regularization in a cluster sized $C = 3$, and two orbits are omitted. Then orbital scaling factor $K = 1.351 \pm 0.003$. Stream inclination to equatorial plane is $\theta = 20.55 \pm 0.18°$. Pluto's orbit is characterized by significant inclination $i = 17.14175°$, and it should be taken into consideration while calculating the stream inclination to the ecliptic plane. Since Pluto's axial tilt $\varepsilon = 119.591°$, stream inclination in the area of Pluto's orbit $\theta_\varepsilon = \pi - (i + \varepsilon + \theta) = 22.7°$. Thus, ratio of semi-major axes of Pluto and Neptune $K = 1.386$. Actual value is $a_P/a_N = 1.313$.

Phobos and Deimos. Ratio of semi-major axes $a_{Deimos}/a_{Phobos} = 2.50$. Let's imagine that there is lack of two orbits as it is usual for the cluster with size $C = 3$. In such case, orbital scaling factor $K = (a_{Deimos}/a_{Phobos})^{1/3} = 1.357$, and the corresponding value of stream inclination to the equatorial plane $\theta = 20.95°$. Taking into consideration Mars' axial tilt $\varepsilon = 25.19°$, we found flow inclination to the ecliptic plane $\theta_\varepsilon = \theta + \varepsilon = 46.14°$ being close to the optimal value $\pi/4$.

References

Kamchatnov, A. M. 1982, *Sov. JETP*, 82, No1, 117

Semenov, V. S., Korovinski, D. B., & Biernat, H. K. 2001, preprint (arXiv: physics/0111212v1)

Arnold, V. I. & Khesin, B. A. 1998, in *Topological Methods in Hydrodynamics*, (Springer-Verlag, New York)

Gil-Merdano, O. & Llinares-Fuster, E. 2001, *Math. Ann.*, 320, 531

Yampolsky, A. 2003, *Acta Math. Hungar.*, 101, 73

Gibbons, G. W. 2006, Applications of Differential Geometry to Physics Part III (Cambridge University, Great Britain)

Session X

Waves and Turbulence in Heliospace

Universal Heliophysical Processes
Proceedings IAU Symposium No. 257, 2008
N. Gopalswamy & D.F. Webb, eds.

© 2009 International Astronomical Union
doi:10.1017/S1743921309029834

Radio remote sensing of the corona and the solar wind

Steven R. Spangler and Catherine A. Whiting

Department of Physics and Astronomy, University of Iowa, Iowa City, Iowa, 52242, USA
email: `steven-spangler@uiowa.edu`

Abstract. Modern radio telescopes are extremely sensitive to plasma on the line of sight from a radio source to the antenna. Plasmas in the corona and solar wind produce measurable changes in the radio wave amplitude and phase, and the phase difference between wave fields of opposite circular polarization. Such measurements can be made of radio waves from spacecraft transmitters and extragalactic radio sources, using radio telescopes and spacecraft tracking antennas. Data have been taken at frequencies from about 80 MHz to 8000 MHz. Lower frequencies probe plasma at greater heliocentric distances. Analysis of these data yields information on the plasma density, density fluctuations, and plasma flow speeds in the corona and solar wind, and on the magnetic field in the solar corona. This paper will concentrate on the information that can be obtained from measurements of Faraday rotation through the corona and inner solar wind. The magnitude of Faraday rotation is proportional to the line of sight integral of the plasma density and the line-of-sight component of the magnetic field. Faraday rotation provides an almost unique means of estimating the magnetic field in this part of space. This technique has contributed to measurement of the large scale coronal magnetic field, the properties of electromagnetic turbulence in the corona, possible detection of electrical currents in the corona, and probing of the internal structure of coronal mass ejections (CMEs). This paper concentrates on the search for small-scale coronal turbulence and remote sensing of the structure of CMEs. Future investigations with the Expanded Very Large Array (EVLA) or Murchison Widefield Array (MWA) could provide unique observational input on the astrophysics of CMEs.

Keywords. Sun: corona,plasmas,turbulence,waves; Sun: coronal mass ejections

1. Introduction

Understanding the physics of the solar corona and its transition to the solar wind requires specifying the plasma physics properties in the region from the coronal base to heliocentric distances of tens of solar radii. We require knowledge of parameters such as the plasma density, vector magnetic field, plasma flow velocity, and ion and electron temperatures. Properties of turbulence are also important, so that we may test various theories which invoke turbulence as an important agent.

It is in this part of space that radioastronomical propagation measurements can play an important role. The basic idea is to observe a radio source which is viewed through the corona. This radio source can be an extragalactic radio source like a radio galaxy or a quasar, or the transmitter of a spacecraft. A cartoon illustrating the geometry is shown in Figure 1 of Spangler (2002). The radio waves propagate through the corona or the inner solar wind, and are modified in their amplitude and phase. A radio telescope on Earth can measure these amplitude or phase changes, and infer properties of the intervening plasma. Further details are discussed in the literature; illustrative papers are those of Bird and Edenhofer (1990) and Spangler (2002).

This paper will emphasize observation of Faraday rotation in the plasma of the outer corona and inner solar wind. A linearly polarized radio wave propagating through a magnetized plasma will undergo a rotation in the plane of linear polarization, with the change in polarization position angle $\Delta\chi$ given by

$$\Delta\chi = \left[\left(\frac{e^3}{2\pi m_e^2 c^4}\right)\int_L n_e \vec{B}\cdot d\vec{z}\right]\lambda^2 \qquad (1.1)$$

Equation (1) is in cgs units. The term in square brackets on the right hand side is referred to as the *rotation measure (RM)*. It is conventionally reported in SI units of radians/m^2; the cgs value of the rotation measure is converted to SI by multiplying by a factor of 10^4.

The geometry of coronal Faraday rotation is illustrated in Figure 1 of Spangler (2005). Radio waves from a linearly polarized radio source, in this case a radio galaxy or quasar, propagate towards the Earth and pass close to the Sun. Because of the higher plasma density and magnetic field strength close to the Sun, the measured RM is mainly determined by plasma close to the "proximate point", i.e. the point on the line of sight which is closest to the Sun. The heliocentric distance of the proximate point is called the "impact parameter".

Our investigations (e.g. Sakurai and Spangler (1994), Mancuso and Spangler (1999), Mancuso and Spangler (2000), Spangler and Mancuso (2000), Spangler (2005), Ingleby *et al.* (2007), Spangler (2007)) have used the Very Large Array (VLA) of the National Radio Astronomy Observatory to make such measurements. With the VLA, there are many radio galaxies and quasars which are sufficiently strong to make these measurements. An observation of one or more such sources juxtaposed to the Sun can be made almost every day. Radio galaxies and quasars radiate by synchrotron radiation, and are linearly polarized at the level of several percent or more.

2. The magnitude of coronal Faraday rotation

Independent estimates of the plasma density in the inner heliosphere exist and allow us to make rough estimates of the Faraday rotation as a function of impact parameter and frequency of observation. These estimates obviously require knowledge of the strength and form of the coronal magnetic field as well. For the most part, these magnetic field estimates result from prior Faraday rotation measurements.

We assume that the plasma density (strictly speaking, electron density) is a function only of the heliocentric distance r,

$$n(r) = N_0 \left(\frac{r}{R_\odot}\right)^{-2.5} \qquad (2.1)$$

We also assume that the magnitude of the magnetic field is similarly a function only of heliocentric distance, and is radial

$$\vec{B}(r) = B_0 \left(\frac{r}{R_\odot}\right)^{-2} \hat{e}_r \qquad (2.2)$$

where $\hat{e}_r$ is the unit vector in the radial direction. The direction of the magnetic field will change with location in the corona. We introduce a dimensionless function of heliographic coordinates, m, to describe the polarity of the coronal field. In the simplest case, $m = \pm 1$, depending on location in the corona, but in the more general case $-1 \leqslant m \leqslant 1$.

We substitute equations (2.1) and (2.2) into (1.1). As pointed out by Pätzold *et al.* (1987), the resulting integral is simplified if we make a change of variable from z, the spatial coordinate along the line of sight, to β, an angle which is defined by the point along the line of sight, the center of the Sun, and the proximate point (see definition of β in Figure 1 of Ingleby *et al.* (2007)). The resultant expression for the Faraday rotation is

$$\Delta\chi = 2.63 \times 10^{-17} R_\odot \left[N_0 B_0 \right] \frac{\lambda^2}{R_0^{3.5}} \int_{-\pi/2}^{\pi/2} d\beta \cos^{5/2}\beta \sin\beta m(\beta) \qquad (2.3)$$

In equation (2.3), $R_\odot$ is the radius of the Sun, 2.63×10^{-17} is the cgs value of the set of fundamental constants in parentheses in equation (1.1), and R_0 is the (dimensionless) value of the impact parameter in units of the solar radius. The wavelength of observation is λ. The modulation function $m(\beta)$ is now expressed as a function of the angle β.

The integral in equation (2.3) is very important in determining the magnitude of the coronal Faraday rotation. As pointed out in Ingleby *et al.* (2007), if the magnetic field is of constant polarity along the line of sight, and only a function of r, $\Delta\chi$ is exactly zero. This is obvious from the form of the integral. The maximum value of the integral occurs if

$$m(\beta) = -1, \beta < 0 \qquad (2.4)$$
$$m(\beta) = +1, \beta \geqslant 0$$

In this case, the integral has a value of 4/7. In what follows, and to parameterize nonoptimum coronal conditions, we introduce a parameter ϵ to give the magnitude of the integral in equation (2.3). The parameter ϵ is defined as $\epsilon \equiv \int_{-\pi/2}^{\pi/2} d\beta \cos^{5/2}\beta \sin\beta m(\beta)/(4/7)$. $|\epsilon| \leqslant 1$, and can be zero.

We adopt representative values of the constants $N_0 = 1.83 \times 10^6$ cm^{-3} and $B_0 = 1.01$ G (Ingleby *et al.* (2007)).Our observations with the VLA have been made at wavelengths of about 20 cm, and at impact parameters of $5R_\odot$ or more. To obtain a useful empirical formula, we then let $\lambda = 20\lambda_{20}$, and $R_0 = 5R_5$.

With all of these normalizations and substitutions, we have the following handy, empirical formula for the magnitude of coronal Faraday rotation.

$$\Delta\chi = 158° \left[\frac{\lambda_{20}^2}{R_5^{3.5}} \right] \epsilon \qquad (2.5)$$

It is again worth emphasizing that $|\epsilon| \leqslant 1$, and in most cases it will be much less than unity. However, the observations of Spangler (2005) showed coronal Faraday rotation about equal to the prediction of equation (2.5) with $\epsilon = 1$.

For some applications, it is more useful to express the rotation measure rather than the position angle rotation. For the same set of parameters as above, the coronal rotation measure is

$$RM = \frac{69\epsilon}{R_5^{3.5}} \text{ rad/m}^2 \qquad (2.6)$$

An appealing feature of coronal Faraday rotation (in contrast to the case for Faraday rotation in interstellar and ionospheric plasmas) is that the reference observations in the absence of the coronal plasma are easily made. Observations of the source are made when the line of sight passes through the corona, and then later when the Sun has moved away from that part of the sky. With relatively rare exceptions, extragalactic radio sources do not change their polarization properties on timescales of a few weeks. The maps of polarization position angle can then be differenced to yield the coronal Faraday rotation. The process is illustrated in Figure 3 of Ingleby *et al.* (2007).

3. Information obtainable from coronal Faraday rotation

Observations of coronal Faraday rotation provide valuable, and often unique constraints on properties of the coronal plasma. A partial list of coronal properties which may be inferred is as follows.

• **The Large Scale Coronal Magnetic Field.** Measurements on a large number of lines of sight over a range in the impact parameter R_0 produce our best model for the coronal magnetic field at heliocentric distances of $\sim 5 - 10 R_\odot$. Results in this area are given in Pätzold *et al.* (1987), Mancuso and Spangler (2000), and Ingleby *et al.* (2007).

• **MHD Turbulence in the Corona.** Like all astrophysical fluids, the corona should be turbulent. This turbulence may play an important role in the thermodynamics of the corona. The stochastic fluctuations in plasma density and magnetic field which occur in plasma turbulence generate corresponding fluctuations in the coronal Faraday rotation. Measurements of, or limits to these fluctuations have been reported by Hollweg *et al.* (1982), Sakurai and Spangler (1994), Efimov *et al.* (1993), Mancuso and Spangler (1999), and Spangler and Mancuso (2000), among others. Although Faraday rotation fluctuations have been measured which are probably due to magnetohydrodynamic turbulence, it is not certain at this point whether this turbulence has a sufficient energy density and dissipation rate to make a significant contribution to coronal heating.

• **Detection of Electrical Currents in the Corona.** Electrical currents must flow in the solar corona to produce the intricate structure seen there. They may also play a role in coronal heating via Joule Heating. Nonetheless, remote measurement of these currents is very difficult. Spangler (2007) pointed out that VLA observations of an extended radio source through the corona (so that simultanous lines of sight on two or more paths can be measured) can yield information on these currents. The observational signature of currents is *differential Faraday rotation*, or a difference in the rotation measure on two closely juxtaposed lines of sight. Spangler (2007) reported just such observations during VLA measurements of the source 3C227 in August, 2003. The inferred electric current between the two lines of sight was 2.5×10^9 Amperes in the clearest case, with smaller values for, or upper limits to the current in the remainder of two days of observation.

• **Internal Structure of Coronal Mass Ejections.** If a coronal mass ejection (CME) crosses the line of sight to a polarized radio source, the resulting Faraday rotation can be used to extract information on the magnetic field and plasma density in the interior of the CME. This can provide unique information about the structure of these objects close to the Sun and in interplanetary space. Further discussion of this topic is given in Section 6 below.

4. The spectrum of turbulence in the solar corona

Central to the question of turbulent heating of the corona is the spatial power spectrum of the turbulence. It is usually assumed that this spectrum is a power law extending from the outer scale of several tenths of a solar radius to a few solar radii, to an inner scale comparable to plasma microscales such as the ion inertial length. Depending on location in the corona, the inner scale should be a few kilometers.

This assumption about the coronal spectrum is based on observed astrophysical power spectra, such as the turbulence spectrum in the solar wind at the location of the Earth, or the inferred spectrum for the interstellar medium. They possess this power law nature, with an inertial range of a few to many decades. A good illustration of the power law spectrum of magnetic field fluctuations in the solar wind is given in Figure 1 of Bavassano *et al.* (1982).

However, it is not certain that the turbulence power spectrum in the corona must be of this power law form. There could be an enhancement of power closer to the plasma microscales, generated by plasma kinetic processes, and occurring in a part of wavenumber space where dissipation processes are more active. Such turbulence would have a larger energy density on small scales where the dissipation rate is higher, and thus have a larger volumetric heating rate.

A well-known example of turbulence with these properties is the wave environment upstream of the quasi-parallel portion of the Earth's bow shock. Power spectra of the magnetic field components showing substantial enhancements on small spatial scales over the solar wind background are shown, for example, in Figure 4 of Spangler *et al.* (1997). These quasi-monochromatic fluctuations correspond to obliquely-propagating fast mode magnetosonic waves generated by an ion streaming instability. There have been theoretical suggestions that enhanced turbulence on short spatial scales also exists in the corona, the dissipation of which could be responsible for heating of the corona (McKenzie *et al.* (1995), Marsch and Tu (1997))

5. Faraday screen depolarization: A diagnostic for small scale coronal turbulence

A radioastronomical effect called *Faraday Screen Depolarization* can detect the presence of small-scale, intense turbulence in the solar corona (Spangler and Mancuso (2000)). The physical concept is illustrated in cartoon form in Figure 1 of that paper. Briefly stated, if intense, small-scale turbulence is present in a plasma medium between an imaging radio telescope and an extended, polarized radio source, the turbulence will produce small scale randomization of the polarization position angle. Averaging over the beam of the radio telescope will then cause a drop in the degree of linear polarization which is measured. This phenomenon is probably responsible for the well-established observational fact that radio sources are less strongly linearly polarized at low radio frequencies than high.

Spangler and Mancuso (2000) presented formulas relating the drop in the degree of linear polarization to properties of the turbulent screen in the corona. The analysis identified two parameters which governed the depolarization, or decrease in the degree of polarization. The first is $x \equiv l_0/L_f$, where l_0 is the outer scale of the turbulence, and L_f is the "footprint" of the radiotelescope beam. The radiotelescope beam subtends an angle which corresponds to a physical size in the corona; this is L_f. The second parameter is the asymptotic value of the rotation measure structure function, D_∞. This is the value of the rotation measure structure function which would be measured for spatial (or angular) lags greater than the outer scale. It is twice the variance in the rotation measure introduced by the turbulent screen.

The depolarization D is defined and given by (Spangler and Mancuso (2000), equation (8))

$$D^2 \equiv \frac{m^2}{m_0^2} = \frac{K}{2} \left[\frac{x^2}{\xi} \left(1 - e^{-\xi} \right) + \frac{2}{K} e^{-\xi} \right] \tag{5.1}$$

In equation (5.1) m is the observed degree of linear polarization of a radio source (or part of a radio source) when viewed through the turbulent screen, and m_0 is the corresponding intrinsic degree of linear polarization, i.e. that which would be measured in the absence

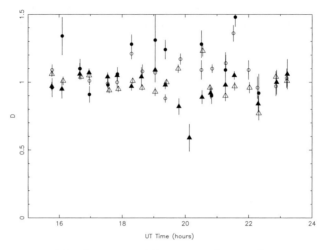

Figure 1. Measurements of depolarization of the source 3C228 on August 16, 2003. Each plotted data point represents a measurement of the depolarization D (see equation (5.1) for definition) resulting from a single, 15 minute scan. Circles and triangles distinguish measurements for the two hot spots in the source. Filled symbols represent measurements at 1465 MHz, open circles are those at 1665 MHz.

of the screen. The variable ξ is defined as

$$\xi \equiv \frac{Kx^2}{2} + 2\lambda^2 D_\infty \tag{5.2}$$

where $K \equiv 4\log 2$ and λ is the wavelength of observation.

Spangler and Mancuso (2000) analysed existing observations for evidence of screen depolarization, and found none. The above formulas were used to place constraints on the properties of small scale turbulence in the corona. The results of those calculations are given in Sections 4 and 5 of Spangler and Mancuso (2000).

Subsequently, better observations were made with this type of analysis in mind. The radio source 3C228 was observed on August 16 and August 18, 2003. Results from these observations have been presented in Spangler (2005), which shows a radio image of the source and describes its polarization characteristics (Figure 2 of Spangler (2005)). 3C228 is a radio galaxy with a double structure, consisting of two highly polarized "hot spots" separated by about 46 arcseconds, corresponding to a linear separation of 33,000 km in the corona. The footprint L_f for these observations was determined by a convolution of the interferometer beam and the angular broadening function due to coronal scattering (itself a consequence of small-scale coronal turbulence). The net footprint was about 4000 km. Simultaneous observations were made at frequencies of 1465 and 1665 MHz. The source was observed for a period of 8 hours, with the basic observational unit consisting of a scan of 15 minutes duration. A depolarization measurement was made for each scan at both frequencies of observation.

The results of our observations on August 16, 2003 are shown in Figure 1. The different symbols indicate measurements of the two source components at the two frequencies of observation. The data from Figure 1 are consistent with $D = 1$ throughout the observation, i.e. there is no evidence for screen depolarization. There is a single low point at 20 UT. The scan from which this measurement was made was of marginal quality, and the depolarization is not confirmed by adjacent scans at 18cm. The large scatter in the measurements results from the noise-sensitivity of the depolarization measurement; both m and m_0 are obtained from a ratio of two measurements (that of the polarized intensity and the total intensity), each of which has its characteristic measurement errors. There is no indication of a *systematic*, session-averaged offset of D from 1.0. During the observing session, the rotation measure to the source increased (Spangler (2005)) and the line of sight passed deeper into a coronal streamer. It is therefore significant that there is no trend towards smaller D as the session progressed. Similar results (i.e no evidence for depolarization) were obtained on August 18, 2003, when the impact parameter of the observations was even smaller.

An analysis similar to that in Spangler and Mancuso (2000) has been made on these new data, and new limits obtained for the properties of small scale turbulence in the corona. These results are shown in Figure 2. The model curves were calculated in a way similar to those in Spangler and Mancuso (2000), and employ models for the background coronal plasma properties similar to those written in equations 2.1 and 2.2.

Figure 2 shows that a wide range of possible models for small scale turbulence are compatible with our depolarization measurements. However, there are constraints which can be imposed. Turbulence with an outer scale of order the beam footprint $l_0 = 0.5 - 1.0L_f$ is compatible with the observations, but the turbulent amplitude $\frac{\delta b}{B_0}$ in that case must be of order $\frac{\delta b}{B_0} \leqslant 0.2 - 0.3$. Similarly, large turbulent amplitudes could be present in the corona ($\frac{\delta b}{B_0} \geqslant 0.5$), but the outer scale would have to be quite small ($l_0 \leqslant 0.1 - 0.2L_f$).

To summarize, our basic observational result is that the corona does not depolarize radio sources, at least at the level probed in the observations to date. A number of mathematical models for small scale coronal turbulence are still compatible (i.e "not inconsistent") with the data. However, it is interesting, within the context of a general exploration of turbulence in the universe, that there is no observational indication of coronal turbulence with a spectrum like that seen upstream of the Earth's bow shock.

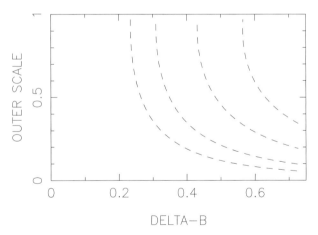

Figure 2. Range of parameter space allowed by the depolarization measurements of 3C228 on August 16 and 18, 2003. The abscissa of the parameter space is $\frac{\delta b}{B_0}$, where δb is the rms value of the fluctuating magnetic field in the turbulence, and B_0 is the mean magnetic field. The ordinate is $\frac{l_0}{L_f}$ (ratio of outer scale of turbulence to beam footprint). The dashed curves correspond to different values of the reduced χ_ν^2. The curves correspond to reduced chi-square values of 1.90, 2.30, 3.30, and 5.00, progressing from left to right. The first two curves correspond to chi-squared probability of occurrence of 5 % and 2 %, respectively. The part of parameter space below and to the left of a curve represents parameters which are acceptable at that level.

6. Faraday rotation as a remote diagnostic of coronal mass ejections

Coronal Mass Ejections (CMEs) are among the most interesting objects in heliospheric plasma physics. A static structure like a solar prominence loses its stability and expands at supersonic speeds into the corona and interplanetary medium. CMEs are the most important agents in space weather, since their interaction with Earth is the cause of pronounced terrestrial response to solar activity. We would like a better knowledge of the internal plasma properties of CMEs at all points between liftoff and the orbit of the Earth, so that tests may be made of the numerous theoretical descriptions of these objects. Faraday rotation is an ideal diagnostic for CMEs, since it provides information on both the density and magnetic field. Some of the questions we would like to ask about CMEs are as follows.

- What is the internal plasma structure of a typical CME?
- Are CMEs approximately describable as force-free flux ropes?
- What is the physical significance of the "three part structure", i.e. an outer loop, a middle void, and an inner loop or core, which is generally observed for these objects?

In a recent study, Liu et al. (2007) calculated the expected Faraday rotation signatures of various models for CMEs. The simplest such model is that of a force-free flux rope. The calculated RM for this model is shown in Figure 6 of Liu et al. (2007). It shows an RM perturbation with an amplitude of about 9 rad/m^2 for a structure at a heliocentric distance of 10 $R_\odot$, comparable to what is observed. In this model, regions of the sky with opposite sign of the RM are closely juxtaposed. As discussed by Liu et al. (2007), this is a consequence of the helical magnetic field structure in a force-free flux rope. Liu et al. (2007) note that the helicity of the flux rope and the orientation of the flux rope axis to the line of sight could be determined by RM observations.

To date, the only high quality set of observations of RM anomalies due to a CME are those of Bird et al. (1985), which were of occultations of the Helios spacecraft transmitter by CMEs. Liu et al. (2007) describe qualitative agreement between their models and the observations of Bird et al. (1985).

7. CME observations with the VLA

The Very Large Array is well-suited for observations of CMEs. Among the desirable features of the VLA for this type of observation are

- the capability of making simultaneous observations at 1465 and 1665 MHz, and thus confirming the λ^2 dependence of the position angle rotation,
- relative easy measurement of RMs as small as ~ 1 rad/m^2,
- imaging of extended radio sources, which allows measurements of differential Faraday rotation through the CME,
- the option of observing a number of sources on a given day which are available for occultation by a CME.

Ingleby et al. (2007) briefly describe a CME observation with the VLA, and we now consider those results in more detail. Some of the observations of Ingleby et al. (2007) were made on March 12, 2005. On that day, there were 7 sources arranged in a particularly good "constellation" around the Sun (see Figure 4 of Ingleby et al. (2007)). A coronal mass ejection occurred on the east limb of the Sun, and moved in the direction of two of the sources, 2335-015 and 2337-025. The outer loop of the CME was in the vicinity of the two sources near the end of the observing session.

Faraday rotation data for these two sources are shown in Figures 3 and 4. Plotted are the residual polarization position angles as a function of time. By residual position angles,

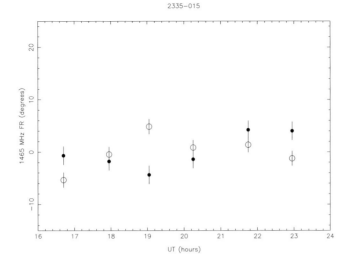

2335−015

Figure 3. Faraday rotation observations of radio source 2335-015 on March 12, 2005. Plotted is the residual polarization position angle (position angle minus the mean for the whole eight-hour session) as a function of UT. Solid symbols are the measurements at 1465 MHz, open symbols are measurements at 1665 MHz, with the 1665 MHz residual multiplied by the square of the ratio of the observing wavelengths. No definite evidence is seen of a Faraday rotation event associated with the approach of the CME.

we mean that the mean χ for the session has been subtracted from the position angle time series. This permits us to study variations in the polarization position angle as a function of time. Measurements at both 1465 MHz (solid symbols) and 1665 MHz (open symbols) are shown. The residuals at 1665 MHz have been multiplied by the square of the ratio of the observing wavelengths, $(\lambda_{1465}/\lambda_{1665})^2$, where λ_{1465} is the wavelength corresponding to 1465 MHz, and λ_{1665} is that corresponding to 1665 MHz. Since Faraday rotation is proportional to the square of the wavelength, after this adjustment the measurements at the two frequencies should agree.

The data for 2335-015 show no credible variation during this period. There is no significant departure or trend from zero that appears in both the 1465 and 1665 MHz data. However, the observations of 2337-025 do show a temporal variation in the Faraday rotation. A progressive change is seen in the last two scans, and the 1465 and 1665 MHz data are in agreement. The total change in position angle (defined as the difference between the position angle measured in the last scan, and the average position angle from 16 - 21h UT) is about $28°$, corresponding to a rotation measure of about 12 rad/m^2. This is comparable in magnitude to the model value of Liu *et al.* (2007) of 9 rad/m^2 at a somewhat larger impact parameter (10 $R_\odot$ for the model calculations, compared to 6.6 $R_\odot$ for the observations).

A curious result of these observations is that it appears the main part of the CME had not occulted the line of sight to the source when the RM variation began. The data shown in Figure 4 indicate that a plasma structure with detectable RM first occulted the line of sight around 21h, and certainly before 22h. In the remaining one to two hours of the session, the line of sight appears to have penetrated deeper in the structure. However, coronagraph observations from the LASCO C2 coronagraph seem to indicate that the "outer loop" of the CME crossed the line of sight to the source after the end of the session. An observation with the LASCO C2 coronagraph at 23h12m UT appears to show that

2337−025

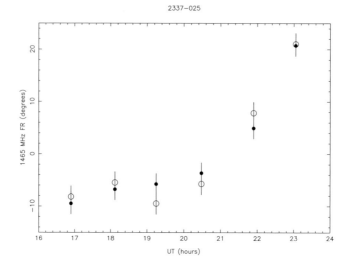

Figure 4. Faraday rotation observations of radio source 2337-025 on March 12, 2005. Format is the same as Figure 4. For this source, a change in the Faraday rotation is seen in the last 2 scans, which is plausibly attributed to the approaching CME.

the outer loop of the CME had not quite occulted the line of sight. However, as may be seen in Figure 4, the RM transient began long before this time.

If the simple astrometry used in the preceding analysis is correct, we can conclude that there is a magnetohydrodynamic precursor associated with the CME, ahead of the outer loop, and capable of producing an easily measurable variation in the RM.

8. Future Faraday rotation studies of the heliosphere

The scientific results described here as well as in the papers we have referenced illustrate the uniqueness and utility of Faraday rotation observations for studies of the plasma physics of the solar corona and inner solar wind. Furthermore, enhanced capability in this field will be available in the very near future due to the availability of the Expanded Very Large Array (EVLA) of the US National Radio Astronomy Observatory and the Murchison Widefield Array, developed by a consortium of the Massachusetts Institute of Technology and several Australian universities. This enhanced capability will benefit general studies of the heliosphere and its turbulence, and should be of particular interest in new studies of coronal mass ejections. These future studies offer the potential of an improvement on the study of Bird *et al.* (1985) in two ways. First, better coronagraphs are now available in space in the form of SOHO/LASCO and STEREO/SECCHI. These coronagraphs provide a better view of the way in which the line of sight to the radio source passes through the CME structure, as well as information on the line-of-sight integral of the electron density. Second, Faraday rotation observations with the EVLA will permit more and better simultaneous observations along multiple lines of sight. They will also allow differential Faraday rotation measurements which will diagnose the internal structure of the CME. Finally, future CME measurements may be possible at greater heliocentric distances than in the past. This would permit a study of how the CME structure evolves with distance in the interplanetary medium.

8.1. *The expanded Very Large Array*

The Very Large Array is the premier radio synthesis telescope in the world. It is presently in the process of a major upgrade in the antennas, receivers, and electronics. When completed in roughly 2010, it will have greater sensitivity and frequency agility. This will benefit coronal Faraday rotation measurements. Since our research program utilizes radio galaxies and quasars for background sources, increased instrumental sensitivity is as important for this type of work as for studies of these objects *pro suis*. Improved sensitivity will permit accurate rotation measure measurements on weaker sources than is currently possible. Since the number of weaker sources is larger than stronger sources, it will be possible to probe more lines of sight through the corona, as well as more interesting lines of sight (e.g. through coronal streamers, etc), and pairs of lines of sight with a wide range of angular separations. Continuous frequency coverage between 1 and 50 GHz will permit observations over a wide range of heliocentric distances, most importantly deeper in the corona.

8.2. *The Murchison Widefield Array*

The Murchison Widefield Array (MWA) is in the process of being constructed in Western Australia. It will operate at low radio frequencies, 80 - 300 MHz. The appealing feature of such frequencies in the present context is that they should permit Faraday rotation measurements at greater heliocentric distances than are possible with the VLA, an instrument which mainly operates at microwave frequencies. This is illustrated by the discussion in Section 2, particularly equation (2.5). At a wavelength of 300 cm (corresponding to the low end of the MWA frequency range), a given Faraday rotation $\Delta\chi$ can be measured at a heliocentric distance 4.7 times greater than the distance at which this rotation would be measured at 20 cm. Faraday rotation measurements can be made with the VLA at 20 cm at impact parameters out to about 10 $R_{\odot}$; the same amount of rotation might be measurable with MWA out to $\sim 50 R_{\odot}$. This would permit a new class of Faraday rotation measurements, and the extension of the Faraday rotation technique to new parts of the heliosphere.

The MWA will face two clear challenges in heliospheric Faraday rotation observations. The first is the empirical fact that extragalactic radio sources become progressively less linearly polarized with decreasing frequency below 1 GHz. This behavior has been known for decades (e.g. Kronberg and Conway (1970)). Simply stated, at frequencies of a few hundred megahertz and lower, radio galaxies and quasars are not very effective radiators of polarized radiation.

The second issue to arise in MWA Faraday rotation observations is that heliospheric physicists will, perforce, become ionospheric physicists as well. The Earth's ionosphere produces Faraday rotation which varies with direction in the sky, time of day, and date. For observations at large distances from the Sun, the ionospheric contribution will be much larger than that due to the interplanetary medium. The problem can be illustrated with equation (2.6). At an impact parameter of 30 $R_{\odot}$, the expected solar wind RM should be of order 0.13ϵ rad/m^2. Again, recall that ϵ will usually be less than 1. The ionospheric rotation measure is typically of order $0.5 - 1.0$ rad/m^2 and can be much higher. As a not-atypical case, Ingleby *et al.* (2007) encountered ionospheric rotation measures of $2.5 - 3.8$ rad/m^2 on one of the four days of observation in their investigation. Detection of interplanetary RMs of a few tenths of a rad/m^2 in the presence of ionospheric contributions which are larger (sometimes much larger), as well as spatially and temporally variable, will take better diagnosing of the ionosphere than has been previously attempted by radio astronomers.

9. Summary and conclusions

Faraday rotation measurements on radio waves which have propagated through the solar corona and inner solar wind provide a unique plasma diagnostic on that region of space. Measurements to date have provided information on the coronal magnetic field, the amplitude and spatial power spectrum of MHD turbulence in the corona, estimates of electrical currents, and the plasma structure of coronal mass ejections. Two new instruments which will become available in the next few years, the Expanded Very Large Array (EVLA) and the Murchison Widefield Array (MWA), will permit much better measurements, as well as measurements of a different type.

Acknowledgements

We thank the US National Science Foundation, Division of Atmospheric Sciences, for supporting this research via grant ATM03-54782. We also thank the National Radio Astronomy Observatory for allocating significant amounts of VLA observing time to this research program. The National Radio Astronomy Observatory is a facility of the National Science Foundation, operated under cooperative agreement with Associated Universities, Inc.

References

Bavassano, B., Dobrowolny, M., Mariani, F., & Ness, N. F. 1982, *J. Geophys. Res.*, 87, 3616

Bird, M. K., Volland, H., Howard, R. A., Koomen, M. J., Michels, D. J., Sheeley, N. R., Armstrong, J. W., Seidel, B. L., Stelzried, C. T., & Woo, R. 1985, *Solar Phys.*, 98, 341

Bird, M. K. & Edenhofer, P. 1990, in Physics of the Inner Heliosphere II, R. Schwenn and E. Marsch, ed., (Springer-Verlag:Berlin), p13

Efimov, A. I., Chashei, I. V., Shishov, V. I., & Bird, M. K. 1993, *Astron. Lett.*, 19, 57

Gudiksen, B. V. & Nordlund, Å. 2005, *ApJ*, 618, 1020,; *Erratum: ApJ*, 623, 600

Hollweg, J. V., Bird, M. K., Volland, H., Edenhofer, P., Stelzried, C. T., & Seidel, B. L. 1982, *J. Geophys. Res.*, 87, 1

Ingleby, L. D., Spangler, S. R., & Whiting, C. A. 2007, *ApJ*, 668, 520

Kronberg, P. P., & Conway, R. G. 1970, *MNRAS*, 147, 149

Liu, Y., Manchester, W. B. IV, Kasper, J. C., Richardson, J. D., & Belcher, J. W. 2007, *ApJ*, 665, 1439

Mancuso, S. and Spangler, S. R. 2000, *ApJ*, 525, 195

Mancuso, S. and Spangler, S. R. 2000, *ApJ*, 539, 480

Mancuso, S. & Garzelli, M. V. 2006, *A & A*, 466, 5

Marsch, E. & Tu, C. Y. 1997, *A & A*, 319, L17

McKenzie, J. F., Banaszkiewicz, M., & Axford, W. I. 1995, *A & A*, 303, L45

Pätzold, M., Bird, M. K., Volland, H., Levy, G. S., Seidel, B. L., & Stelzried, C. T. 1987, *Solar Phys.*, 109, 91

Sakurai, T. and Spangler, S. R. 1994 *ApJ*, 434, 773

Spangler, S. R., Leckband, J. A., & Cairns, I. H. 2000 *Phys. Plasm.*, 4, 846

Spangler, S. R. & Mancuso, S. 2000 *ApJ*, 530, 491

Spangler, S. R., Kavars, D. W., Kortenkamp, P. S., Bondi, M., Mantovani, F., & Alef, W. 2002, *A & A*, 384, 654

Spangler, S. R. 2005 *Space Sci. Revs*, 121, 189

Spangler, S. R. 2007 *ApJ*, 670, 841

Discussion

WEBB: What is the problem of the polarization degree in galaxies with the MWA vs. your VLA measurements of Faraday rotation?

SPANGLER: The VLA observations are made at frequencies of 1-2 GHz, at which extra-galactic radio sources have degrees of linear polarization a few percent to several percent. The MWA will operate at frequencies well below 1 GHz, where degrees of polarization are much smaller. See section 8.2 of the paper. There are significantly polarized sources at 300 MHz, but they are much rarer than at 1.5 GHz.

ARGE: What techniques are used to separate net B from $\int nBds$?

SPANGLER: To obtain information on the magnetic field, which is generally the point of Faraday rotation measurements, one needs independent information on the plasma density. In some of our analyses, we have used radio propagation information from the dispersion of radio waves in a plasma (M. Biral and coworkers). Another source of information we have used is coronagraph observations. Given the advances in coronagraphs, this latter technique holds great promise for the future.

DAVILA: How does the turbulence result that you presented reconcile with measurements by Bill Coles or Richard Woo?

SPANGLER: It is well-established from radio oscillation observations of Coles, Woo, and others that turbulence with a large inertial sub-range exists in the corona. These observations are sensitive in density fluctuations in the turbulence. The point of the screen depolarization measurements is to see if, in addition, there is an intense narrow-band turbulence component at high wave numbers close to the ion inertial scale.

Universal Heliophysical Processes
Proceedings IAU Symposium No. 257, 2008
N. Gopalswamy & D.F. Webb, eds.
© 2009 International Astronomical Union
doi:10.1017/S1743921309029846

Energy balance and cascade in MHD turbulence in the solar corona

Francesco Malara, Giuseppina Nigro and Pierluigi Veltri

Dipartimento di Fisica, Università della Calabria,
via P. Bucci, I-87036, Rende (CS), Italy
email: malara@fis.unical.it, nigro@fis.unical.it, veltri@fis.unical.it

Abstract. The dynamics of fluctuations in a closed coronal structure is regulated both by resonance with motions at bases that stores energy in the structure in form of discrete eigenmodes, and by nonlinear couplings that move this energy along the spectrum to smaller scales. The energy balance is evaluated both analytically and, numerically, using an hybrid shell model. The input energy flux is independent of nonlinear effects and is determined by slow (DC) perturbations. Coherent eigenmode couplings determine the nonlinear energy flux and, consequently, the level of fluctuations at large scales. The estimated velocity fluctuation level is in agreement with measures of nonthermal velocity in corona. The resulting turbulence spectrum contains both a pre-inertial range where coherent interactions dominate, and a standard inertial range where the turbulence behaves as in an unbounded system.

Keywords. Sun: corona, (magnetohydrodynamics:) MHD, turbulence, waves

1. Introduction

Within the theoretical modelling of coronal heating, there are two important questions that are currently debated: first, in which form and at which time scale the energy of photospheric motions reaches the corona; second, which are the mechanisms that carry this energy to the very small dissipative scales. Concerning the former point, models of coronal heating have been traditionally classified in two groups: 1) DC models consider slow photospheric motions at time scales $t_{ph} \gg T_A$, T_A being the Alfvén crossing time. The coronal structures continuously re-arrange through a sequence of quasi-equilibrium states, while the magnetic free energy increases in time; energy enters the corona mainly as magnetic energy. 2) AC heating models consider perturbations at $t_{ph} \lesssim T_A$; kinetic energy is comparable with magnetic energy. Due to the finite length of closed coronal structures, exciting such perturbations is similar to put in resonance an elastic string of given length by an external driver (Ionson 1982). Typically, the Alfvén crossing time $T_A \sim 10-15$ s, while the energy of photospheric motions is concentrated around $t_{ph} \sim 300$ s; this suggests that photospheric motions are expected to energize the corona mainly by DC perturbations. However, the distinction between AC and DC mechanisms is quite artificial (Milano *et al.* 1997) because, in the context of resonant systems a DC mechanism can be simply considered as corresponding to a resonance at zero frequency.

The second question concerns mechanisms that move energy towards small enough scales (probably of the order of the proton Larmor radius), where it can be efficiently dissipated. Linear mechanisms related to the interaction between perturbations and an inhomogeneous background are: (i) *resonant absorption* (e.g., Davila 1987), in which the perturbation energy concentrates around magnetic fieldlines where the wave frequency matches the local Alfvén frequency; (ii) *phase-mixing* (e.g., Heyvaerts & Priest 1983) where differences in the local phase velocity bend wavefronts progressively decreasing the transverse wavelength; (iii) *3D fast dissipation* (Petkaki *et al.* 1998), where exponential separation of nearby magnetic lines builds up small scales faster than in ordinary phase-mixing. In contrast, turbulence models assume a uniform background and study how nonlinear interactions build up an energy cascade from large to small dissipative scales (Einaudi et al. 1996, Dmitruk & Gomez 1997, Nigro et al. 2004). The injection range of the turbulence corresponds to the length of the photospheric velocity pattern: $\sim (1.5 \times 10^4 - 1.5 \times 10^3)$ km, the dissipative length is of the order of the proton Larmor radius ($\sim (10-1)$ m); the inertial range of the turbulence, formed by several decades, where nonlinear interactions play the major role, is located in between.

A model of coronal heating should include both the aspect of energization of the corona and that of the internal dynamics which leads to dissipation. Milano et al. (1997) considered a model of a coronal loop with a spectrum of driving frequencies ranging from low (DC) to higher (AC) values. Resonances play a key role allowing for the fluctuating energy to enter the loop. The resulting turbulence, described by an EDQNM technique, moves energy to microscales and it has the net effect of an enhanced dissipation on macroscales.

Direct numerical simulations give a detailed representation of spatial structures, but cannot adequately describe the wide range of spatial scales of the coronal turbulence. This is more easily represented in "reduced" models, like shell models (Giuliani & Carbone 1998, Boffetta et al. 1999), in which the dynamics of turbulence is represented in a simplified Fourier space. A drawback of shell models is that phenomena related to wave propagation, like resonance, cannot be described due to the lack of spatial information. A compromise is represented by the *hybrid shell model* (Nigro et al. 2004, Buchlin & Velli 2007), based on the reduced MHD (RMHD) equations, that includes both nonlinear effects using a shell technique, and linear propagation along the magnetic field. This gives an adequate description of a wide turbulent range as well as of the resonance phenomenon. In the present paper we focus on the energy input due to motions at the loop basis at large scales and on the energy flux towards small scales due to nonlinear effects, deriving, by an analytical treatment, some properties of fluctuations and of nonlinear interactions at large scales. The results are compared with those derived from the hybrid shell model (Nigro et al. 2004).

2. RMHD and energy balance equations

A coronal magnetic structure is characterized by a strong longitudinal magnetic field, by a large aspect ratio $R = L/L_\perp$ (L and $L_\perp$ being the longitudinal and the transverse lengths, respectively), and by a low value of the plasma β ($\sim 10^{-2}$). In these conditions the large scale dynamics can be described by the RMHD equations (Strauss 1976), that are written in the following form, after a Fourier expansion in the transverse directions:

$$\rho_0 \frac{\partial v_n(\mathbf{k}_\perp)}{\partial t} - \frac{B_0}{4\pi} \frac{\partial B_n(\mathbf{k}_\perp)}{\partial z}$$
$$= -ik_{\perp n} P(\mathbf{k}_\perp) + \sum_{\mathbf{p}_\perp, \mathbf{q}_\perp} iq_{\perp j} \left[\frac{1}{4\pi} B_j(\mathbf{p}_\perp) B_n(\mathbf{q}_\perp) - \rho_0 v_j(\mathbf{p}_\perp) v_n(\mathbf{q}_\perp) \right] \delta_{\mathbf{p}_\perp + \mathbf{q}_\perp, \mathbf{k}_\perp}$$

(2.1)

$$\frac{\partial B_n(\mathbf{k}_\perp)}{\partial t} - B_0 \frac{\partial v_n(\mathbf{k}_\perp)}{\partial z} = \sum_{\mathbf{p}_\perp, \mathbf{q}_\perp} iq_{\perp j} \left[B_j(\mathbf{p}_\perp) v_n(\mathbf{q}_\perp) - v_j(\mathbf{p}_\perp) B_n(\mathbf{q}_\perp) \right] \delta_{\mathbf{p}_\perp + \mathbf{q}_\perp, \mathbf{k}_\perp}$$

(2.2)

where $\mathbf{v}$ and $\mathbf{B}$ are the transverse velocity and magnetic field, P is the total (fluid + magnetic) pressure, z is the longitudinal coordinate. These equations indicate that the dynamics of perturbations is determined by two mechanisms: (i) linear propagation along

$\pm \mathbf{B}_0$ at the Alfvén velocity $c_{A0} = B_0/(4\pi\rho_0)^{1/2}$ (LHS terms); (ii) nonlinear couplings between Fourier modes at different transverse wavevectors (RHS terms). The kinetic and magnetic energy at a given transverse wavevector $\mathbf{k}_\perp$ are:

$$E_{kin}(\mathbf{k}_\perp, t) = \frac{1}{2}\rho_0 L_\perp^2 \int_0^L dz\, |v_n(\mathbf{k}_\perp, z, t)|^2; \quad E_{mag}(\mathbf{k}_\perp, t) = \frac{L_\perp^2}{8\pi}\int_0^L dz\, |B_n(\mathbf{k}_\perp, z, t)|^2$$
(2.3)

whose time evolution satisfies the energy balance equations:

$$\frac{dE_{kin}(\mathbf{k}_\perp)}{dt} = L_\perp^2\left[\Phi_{kin}^{in}(\mathbf{k}_\perp) + \Phi_{kin}^{nl}(\mathbf{k}_\perp)\right]; \quad \frac{dE_{mag}(\mathbf{k}_\perp)}{dt} = L_\perp^2\left[\Phi_{mag}^{in}(\mathbf{k}_\perp) + \Phi_{mag}^{nl}(\mathbf{k}_\perp)\right]$$
(2.4)

where Φ_{kin}^{nl} and Φ_{mag}^{nl} are the spectral energy fluxes due to nonlinear effects, defined by

$$\Phi_{kin}^{nl}(\mathbf{k}_\perp) = \frac{1}{2}\int_0^L dz \sum_{\mathbf{p}_\perp, \mathbf{q}_\perp} iq_{\perp j}\left[\rho_0\left[v_n(\mathbf{k}_\perp)v_j^*(\mathbf{p}_\perp)v_n^*(\mathbf{q}_\perp) - v_n^*(\mathbf{k}_\perp)v_j(\mathbf{p}_\perp)v_n(\mathbf{q}_\perp)\right]\right.$$
$$\left. + \frac{1}{4\pi}\left[v_n^*(\mathbf{k}_\perp)B_j(\mathbf{p}_\perp)B_n(\mathbf{q}_\perp) - v_n(\mathbf{k}_\perp)B_j^*(\mathbf{p}_\perp)B_n^*(\mathbf{q}_\perp)\right]\right]\delta_{\mathbf{p}_\perp + \mathbf{q}_\perp, \mathbf{k}_\perp}$$
(2.5)

and

$$\Phi_{mag}^{nl}(\mathbf{k}_\perp) = \frac{1}{8\pi}\int_0^L dz \sum_{\mathbf{p}_\perp, \mathbf{q}_\perp} iq_{\perp j}\left[B_n(\mathbf{k}_\perp)v_j^*(\mathbf{p}_\perp)B_n^*(\mathbf{q}_\perp) - B_n^*(\mathbf{k}_\perp)v_j(\mathbf{p}_\perp)B_n(\mathbf{q}_\perp)\right.$$
$$\left. + B_n^*(\mathbf{k}_\perp)B_j(\mathbf{p}_\perp)v_n(\mathbf{q}_\perp) - B_n(\mathbf{k}_\perp)B_j^*(\mathbf{p}_\perp)v_n^*(\mathbf{q}_\perp)\right]\delta_{\mathbf{p}_\perp + \mathbf{q}_\perp, \mathbf{k}_\perp}$$
(2.6)

while the input energy flux through the loop bases is

$$\Phi_{tot}^{in}(\mathbf{k}_\perp) = \Phi_{kin}^{in}(\mathbf{k}_\perp) + \Phi_{mag}^{in}(\mathbf{k}_\perp) = \frac{B_0}{4\pi}\left[\Re\left[v_n(\mathbf{k}_\perp)B_n^*(\mathbf{k}_\perp)\right]_{z=0}^{z=L}\right]$$
(2.7)

Since only one variable (v or B) can be specified at each boundary, the input flux is determined both by the boundary conditions (representing the external forcing), and by the internal dynamics of the system.

3. Linear dissipative model

We consider the velocity and the magnetic field at a given transverse wavevector $\mathbf{k}_\perp$ within the energy injection range $k_\perp \sim 2\pi/l_\perp^{in}$. Nonlinear couplings represented by nonlinear terms in equations (2.1), (2.2) put away energy from this mode towards smaller transverse scales. We assume that in the injection range these nonlinear terms are small with respect to the linear ones (weak nonlinearity). Thus, a simplified model can be derived (Milano *et al.* 1997), in which nonlinear terms in equations (2.1) and (2.2) are replaced by fictitious linear dissipative terms:

$$\frac{\partial v_\perp}{\partial t} = \frac{B_0}{4\pi\rho_0}\frac{\partial B_\perp}{\partial z} - \nu k_\perp^2 v_\perp; \quad \frac{\partial B_\perp}{\partial t} = B_0\frac{\partial v_\perp}{\partial z} - \lambda k_\perp^2 B_\perp$$
(3.1)

$v_\perp(z, t)$ and $B_\perp(z, t)$ are real quantities representing the real or the imaginary part of perturbations, while the fictive dissipative coefficients ν and λ are small quantities and represent free parameters. Boundary conditions give the velocity perturbation at the loop ends:

$$v_\perp(z = 0, t) = u(t); \quad v_\perp(z = L, t) = 0$$
(3.2)

where $u(t)$ is a random gaussian-distributed signal with a given auto-correlation time T_c.

We assume the following ordering among characteristic times

$$T_A \ll T_c \ll T_d, T_{d0} \qquad (3.3)$$

where $T_A = L/c_{A0}$ is the Alfvén crossing time and $T_d = 1/\omega_d = 2/(\nu + \lambda)k_\perp^2$, $T_{d0} = 1/\omega_d^{(0)} = 1/(\lambda k_\perp^2)$ are the dissipative times giving a measure of the nonlinear time T_{nl}).

The results derived from the linear model can be summarized as follows. In the weak nonlinearity limit perturbations are a superposition of eigenmodes, each at frequency $\omega_n = (c_{A0}\pi/L)n$, representing Alfvén standing waves; at $\omega_n = 0$ the velocity is a linear function of z while magnetic field is constant. Eigenmodes have a well-defined parity in space and time and their amplitude is determined by dissipative coefficients, i.e., by nonlinear effects (Nigro $et\ al.$ 2008).

The input energy flux $\langle \Phi_{tot}^{in} \rangle_t$ is a convolution in the frequency space between the spectrum $G(\omega) = 2u_0 T_c/(1 + \omega^2 T_c^2)$ of velocity at the loop base (u_0 being the velocity amplitude) and a function $H_{uB}(\omega)$ describing the response of the loop at each frequency ω_n. The main contribution to $\langle \Phi_{tot}^{in} \rangle_t$ is due to the resonance at $\omega_n = 0$ (DC motions), which gives (Nigro $et\ al.$ 2008):

$$\langle \Phi_{tot}^{in} \rangle_t \simeq -\frac{B_0^2}{4\pi}\frac{T_c}{L}u_0^2 \qquad (3.4)$$

This expression is independent of T_d and T_{d0}. Thus, the input flux does not depend on nonlinear effects but only on the loop resonance. The scaling law (3.4) is the same as in "stochastic buildup models" (Sturrock & Uchida 1981), where random photospheric motions increase the magnetic energy by twisting flux tubes. Assuming $B_0 = 100$ G, $L = 3 \times 10^9$ cm, $T_c = 300$ s and $u_0 = 10^5$ cm s^{-1}, we get $\langle \Phi_{tot}^{in} \rangle_t = 8 \times 10^5$ erg cm^{-2} s^{-1}, which is of the order of the energy flux required to sustain the quiet-Sun corona (Withbroe 1988).

The main contribution to fluctuating kinetic energy comes from the $n = 1$ eigenmode (AC), which gives an estimation for the velocity fluctuation:

$$\delta v \simeq u_0 \left(\frac{T_d}{6T_c}\right)^{1/2} \qquad (3.5)$$

We note that $\delta v \gg u_0$, in accordance with measures of nonthermal velocities in corona much larger than photospheric velocities. Magnetic fluctuations are mainly due to the $n = 0$ eigenmode (DC), giving an estimation for the magnetic field fluctuation:

$$\delta B \simeq \frac{B_0}{c_A}u_0 \left(\frac{T_c T_{d0}}{2T_A^2}\right)^{1/2} \qquad (3.6)$$

Since $\delta B/B_0 \gg \delta v/c_A$ the turbulence injection range is magnetically dominated. Note that both δv and δB depend on dissipative coefficients, i.e., are determined by the amplitude of nonlinear effects.

4. Nonlinear effects: spectral energy fluxes

In a statistically stationary situation, the input energy flux must be balanced by the spectral flux moving energy from the whole injection range to smaller scales along the spectrum. The spectral flux can be estimated using the expressions (2.5), (2.6), as well as the form of eigenmodes derived from the linear analysis. Nonlinear interactions take place among triads of wavevectors $(\mathbf{k}_\perp, \mathbf{p}_\perp, \mathbf{q}_\perp)$. The perturbation at a given wavevector is a superposition of infinite eigenmodes. However, due to the symmetry properties of eigenmode profiles, nonlinear interactions are dominated by *coherence effects*. This fact has some consequences that make this situation peculiar with respect to a "standard" MHD turbulence (Nigro *et al.* 2008): (i) in each interacting triad, two AC eigenmodes ($n \neq 0$) interact with the velocity perturbation of a DC ($n = 0$) eigenmode; (ii) only eigenmodes with the same parallel wavelength (at $n \neq 0$) can efficiently interact; in the other cases the energy flux is vanishing; (iii) only interactions where two wavevectors are in the injection range and the third in the inertial range contribute to the energy flux; (iv) the kinetic and magnetic spectral energy fluxes are equal; (v) the above properties makes the spectral energy flux much smaller than in a standard MHD turbulence; this justifies the hypothesis of weak nonlinearity (equation (3.3)).

As a consequence, the turbulence spectrum is characterized by the presence of a *pre-inertial* range at scales just smaller than those of the injection range. In the pre-inertial range nonlinear interactions are strongly influenced by coherence effects, due to the presence of linear eigenmodes. In particular, in the pre-inertial range magnetic energy dominates kinetic energy, as is the injection range. With increasing $k_\perp$ the nonlinear time T_{nl} decreases, while T_A keeps unchanged; this implies that at smaller scales resonance lines becomes wider and wider. When T_{nl} becomes $\lesssim T_A$ nonlinear effects dominate linear ones: resonant eigenmodes disappear and the turbulence is determined by incoherent mode interactions; this corresponds to a standard inertial range, where $E_{kin} \sim E_{mag}$ and the two spectra both follow a Kolmogorov law. The transition between pre-inertial and inertial range is characterized by $T_{nl} \sim T_A$ giving the condition

$$k_\perp^{inc} \delta v(k_\perp^{inc}) \sim \frac{L}{c_{A0}} \tag{4.1}$$

where $k_\perp^{inc}$ is the lower limit of the pre-inertial range. An estimation of the spectral energy flux from the injection range to smaller scales is (Nigro *et al.* 2008):

$$\langle \Phi_{tot}^{nl} \rangle_t \simeq \frac{2\pi}{\sqrt{3}} \frac{L}{l_\perp^{in}} \rho_0 u_0 \delta v^2 (l_\perp^{in}) \tag{4.2}$$

ρ_0 being the background density. Energy conservation requires $\langle \Phi_{tot}^{in} \rangle_t + \langle \Phi_{tot}^{nl} \rangle_t = 0$. Then, from the expressions (3.4) and (4.2) we obtain a scaling law for the velocity fluctuation in the injection range:

$$\delta v(l_\perp^{in}) \simeq \left(\frac{\sqrt{3}}{2\pi} \frac{l_\perp^{in}}{L} \frac{T_c}{T_A} c_{A0} u_0 \right)^{1/2} \tag{4.3}$$

Using the values $l_\perp^{in} = 3 \times 10^8$ cm, $L = 3 \times 10^9$ cm, $T_c = 300$ s, $T_A = 15s$, $c_{A0} = 2 \times 10^8$ cm s^{-1}, $u_0 = 10^5$ cm s^{-1} we obtain $\delta v(l_\perp^{in}) = 3 \times 10^6$ cm s^{-1}, which is consistent with typical nonthermal velocities measured in the corona (e.g., Warren *et al.* 1997). Since the spectral flux is independent of the amplitude of DC magnetic perturbation, this method does not allow us to obtain a scaling law for δB.

5. A numerical approach: the hybrid shell model

The hybrid shell model (Nigro *et al.* 2004, Buchlin & Velli 2007) is based on RMHD equations (2.1), (2.2); propagation terms in the parallel (z) direction are explicitly included, while the perpendicular spectral space is divided into concentric shell of exponentially increasing width. At each shell a value $k_{\perp n} = k_0 2^n$ ($k_0 = 2\pi/l_\perp^{in}$, $n = 0, 1, \ldots, n_{max}$) of the wavevector is assigned, while velocity and magnetic fluctuations (normalized to c_{A0} and to B_0, respectively) are represented by complex scalar quantities $v_{\perp n}(z, t)$, $b_{\perp n}(z, t)$. Nonlinear couplings are represented by quadratic nonlinear terms

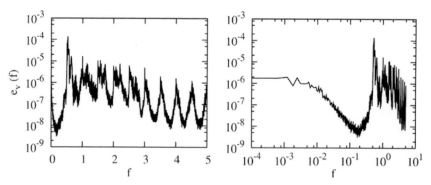

Figure 1. Velocity spectrum e_v as a function of the normalized frequency $f = \omega L/2\pi a_{A0}$ in semilogarithmic scale (left) and in logarithmic scale (right).

whose coefficients are chosen in order to have conservation of 2D quadratic invariants. The equations of the hybrid shell model are (Nigro *et al.* 2004):

$$\left(\frac{\partial}{\partial t} \mp \frac{\partial}{\partial z}\right) Z_n^\pm = ik_{\perp n}\left(\frac{13}{24}Z_{n+2}^\pm Z_{n+1}^\mp + \frac{11}{24}Z_{n+2}^\mp Z_{n+1}^\pm - \frac{19}{48}Z_{n+1}^\pm Z_{n-1}^\mp - \frac{11}{48}Z_{n+1}^\mp Z_{n-1}^\pm \right.$$

$$\left. + \frac{19}{96}Z_{n-1}^\pm Z_{n-2}^\mp + \frac{13}{96}Z_{n-1}^\mp Z_{n-2}^\pm\right) + \chi k_{\perp n}^2 Z_n^\pm \quad (5.1)$$

where $Z_n^\pm = v_{\perp n} \pm b_{\perp n}$ are the Elsässer variables, χ is a dissipative coefficient which determine the dissipative scale, z is normalized to L and t to L/c_{A0}. Boundary conditions are the same as in the analytical treatment (equation 3.2) and are imposed on the first three shells; then, these shells represent the energy injection range of the turbulence.

The hybrid shell model is able to reproduce both nonlinear effects and linear phenomena, like resonance, that are naturally included into the model. Then, results derived from it can be compared with those of the analytical treatment previously described; in particular, the hypothesis of weak nonlinearity and the related dominance of resonant eigenmodes at large scales. Moreover, a wide spectral range can be described with a limited numerical effort; this allows to get a detailed description of the spectra.

The evolution equations (5.1) have been numerically solved, using the following values for the parameters: $L = 3 \times 10^9$ cm, $L_\perp = \pi \times 10^8$ cm, $c_{A0} = 2 \times 10^8$ cm s^{-1}, $\rho_0 = 1.67 \times 10^{-16}$ g cm^{-3}, $u_0 = 10^5$ cm s^{-1}, $T_c = 300$ s, $\chi = 2 \times 10^{-9}$. This value of χ gives a dissipative length $l_d = 1.2 \times 10^3$ cm, which is a factor $\simeq 4 \times 10^{-6}$ smaller than the injection scale. Such a wide range of scales for the turbulence (6 decades) is inaccessible to direct MHD numerical simulations.

From the space-time dependence of fluctuations $v_{\perp n}(z, t)$, $b_{\perp n}(z, t)$ in each shell, we calculated the frequency spectra $e_v(\omega)$ of velocity and $e_b(\omega)$ of magnetic field, integrated in space. We select the shell $n = 1$ which is at the middle of the energy injection range:

$$e_v(\omega) = \int_0^1 dz|\hat{v}_{\perp 1}(z, \omega)|^2; \quad e_b(\omega) = \int_0^1 dz|\hat{b}_{\perp 1}(z, \omega)|^2 \quad (5.2)$$

where $\hat{v}_{\perp 1}$ and $\hat{b}_{\perp 1}$ are the Fourier time transform of velocity and magnetic field fluctuations:

$$v_{\perp 1}(z, t) = \sum_\omega \hat{v}_{\perp 1}(z, \omega)e^{-i\omega t}; \quad b_{\perp 1}(z, t) = \sum_\omega \hat{b}_{\perp 1}(z, \omega)e^{-i\omega t} \quad (5.3)$$

The velocity frequency spectrum e_v is plotted in Figure 1, as a function of the normalized frequency $f = \omega L/2\pi a_{A0}$. The spectrum displays a sequence of peaks approximately

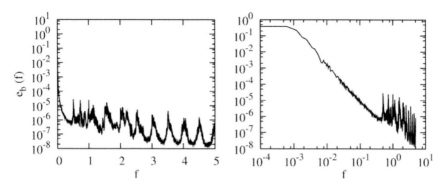

Figure 2. The same as in Figure 1, for the magnetic field spectrum e_b.

located at semi-integer values of f, that corresponds to the resonance frequencies ω_n. This clearly show the presence of resonant eigenmodes that dominate large scale fluctuations. The most energetic eigenmode corresponds to the 1st-order resonance , while much less energy is present in the 0th-order resonance. Thus, in accordance with the analytical model, AC perturbations (in particular, the $n = 1$ resonance) dominate the velocity fluctuations. The width of the dominant peak is much less than the separation T_A^{-1} between peaks; this indicates that at large scales the nonlinear time is much longer than T_A, thus supporting the assumption of weak nonlinearity made in the analytical treatment.

The magnetic field frequency spectrum e_b is plotted in Figure 2, as a function of f. Also this spectrum is formed by a sequence of peaks localized at the resonance frequencies (semi-integer f). In contrast with the velocity spectrum, magnetic fluctuations are dominated by the resonant eigenmode $n = 0$, corresponding to slow DC perturbations. Comparing the two spectra we see that, at resonance frequencies $n \neq 0$, $e_v \sim e_b$ (except for $n = 1$ where e_v is slightly larger than e_b), this indicates that resonant AC perturbations are stationary Alfvén waves trapped into the closed magnetic structure. At the resonance $n = 0$ we find $e_b \gg e_v$, as expected. All these features are in accordance with the results of the analytical model.

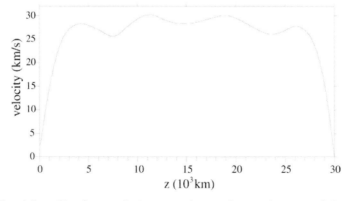

Figure 3. Spatial profile of rms velocity perturbation δv as a function of the longitudinal coordinate z.

In Figure 3 we show the spatial profile of the rms velocity perturbation, defined by

$$\delta v(z) = \left(\int_0^T \frac{dt}{T} \sum_n |v_{\perp n}(z,t)|^2 \right)^{1/2} \tag{5.4}$$

The contribution of the resonant eigenmode profiles at $n \lesssim 4$ can be recognized in Figure 3. The typical value for the velocity perturbation found in the run is $\delta v \simeq 3 \times 10^6$ cm s^{-1}, which is in accordance with the estimation (4.3) derived from the analytical treatment. As expected, this value of δv is much larger than at the loop basis, which is of the order of 10^5 cm s^{-1}.

The rms magnetic field perturbation found numerically is $\delta B/B_0 \simeq 0.2 \gg \delta v/c_{A0}$. This value of δB is mainly due to the $n = 0$ resonance; for this reason, such a value cannot be predicted using energy flux conservation, as we did for δv.

The total energy E, the net incoming power F and the dissipated power W are shown in Figure 4 as functions of time. During a transient, which lasts about 15 hours, the energy increases. For subsequent times the energy does not stabilize but displays strong irregular variations. These variations are due both to energy input/output at the lower boundary and to dissipation. The energy input depends on how the forcing at the base couples with perturbations which are present inside the system. This coupling contributes to determine the sign of the incoming power F which continuously changes sign on a short time scale (Figure 4). Excluding the initial transient, the time average of net incoming and dissipated powers are $\langle F \rangle_t \sim \langle W \rangle_t \simeq 2.6 \times 10^{22}$ erg s^{-1}. We compare this value with the predictions of the analytical model; from equation (3.4) we derive

$$\langle F \rangle_t \sim \frac{B_0^2}{4\pi c_{A0}} \frac{T_c}{T_A} \delta u^2 L_\perp^2 \tag{5.5}$$

Using the above values for the parameters, this equation gives $\langle F \rangle_t \sim 2 \times 10^{22}$ erg s^{-1}, which is quite close to the value of the input power in the hybrid shell model. Then, the analytical estimation (3.4) of the input energy flux is in agreement with the results of the numerical model.

Motions at the loop basis inject energy at scales $\sim l_\perp^{in}$ and nonlinear couplings transfer this energy to smaller transverse scales forming a spectrum which extends down to dissipative scales. The average spectra of kinetic and magnetic fluctuations are defined by

$$\langle e_n^{(v)} \rangle_t = \int_0^1 dz \int_0^T \frac{dt}{T} |v_{\perp n}(z,t)|^2; \quad \langle e_n^{(b)} \rangle_t = \int_0^1 dz \int_0^T \frac{dt}{T} |b_{\perp n}(z,t)|^2 \tag{5.6}$$

These perpendicular spectra are shown in Figure 5. Note that a linear scale in the index n corresponds to a logarithmic scale in $k_{\perp n}$. In the injection range ($0 \leqslant n \leqslant 2$) magnetic fluctuations dominate velocity fluctuations, but the average slope of $\langle e_n^{(b)} \rangle_t$ is much larger than that of $\langle e_n^{(v)} \rangle_t$. Thus, for increasing n there is a tendency to an equipartition between $\langle e_n^{(b)} \rangle_t$ and $\langle e_n^{(v)} \rangle_t$, which is approximately verified in the range $5 \lesssim n \lesssim 10$. Within such a range both the kinetic and the magnetic energy spectra approximately follow a Kolmogorov law, in accordance with what expected from the analytical treatment. The kinetic energy spectrum has a similar slope also in the injection range. In the considered case we have $l_\perp^{(nr)} \sim 2^{-5} L_{\perp 0} \simeq 2 \times 10^7$ cm. Then, from equation (4.1) we find $\delta v(l_\perp^{(nr)}) \sim 1.3 \times 10^6$ cm s^{-1}, which is compatible with the results of the numerical model.

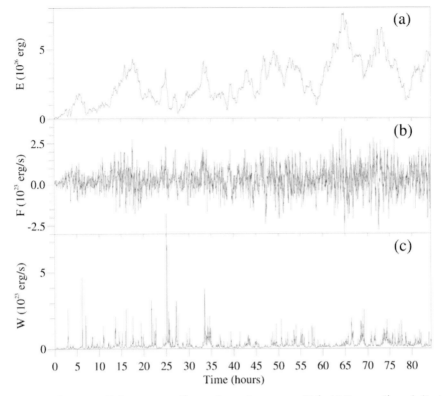

Figure 4. Total energy E (upper panel), net incoming power F (middle panel) and dissipated power W (lower panel) as functions of time t.

6. Conclusions

We have examined and discussed some features of the energy input in a coronal loop due to photospheric motions at large transverse scales and of the subsequent generation of a turbulence towards smaller scales, where this energy can be dissipated. We focused on the energy injection range which is dominated by coherent dynamics. Perturbations are essentially formed by resonant eigenmodes at discrete frequency multiples of the fundamental Alfvén frequency, including the eigenmode at $\omega = 0$ corresponding to DC perturbations. The latter is mainly responsible for the input energy flux, that essentially depends on the coupling between such an eigenmode and the external driver. The coherence properties of eigenmodes imply that at large scales nonlinear effects are much smaller than linear ones (weak nonlinearity). In this limit, the input flux is independent of nonlinear effects. The spectral flux is determined by interactions between AC perturbations at times $\sim T_A$ and the velocity perturbation of DC modes: then, both AC and DC mechanisms regulate the energy balance of the loop. Moreover, the weak nonlinearity allows for the formation of a pre-inertial range of the turbulence, where nonlinear interactions are influenced by coherence effects: a dominance of magnetic on kinetic energy and a reduced spectral energy flux. This theory has given scaling laws for the input energy flux $\langle \Phi_{tot}^{in} \rangle_t$, the spectral flux $\langle \Phi_{tot}^{nl} \rangle_t$ from large to small scales, and the velocity perturbation level δv. The estimation for $\langle \Phi_{tot}^{in} \rangle_t$ is in agreement with the energy flux required to sustain the quiet-Sun corona against radiative losses (Withbroe 1988), while δv is of the order of nonthermal velocities deduced from nonthermal broadening (e.g.,

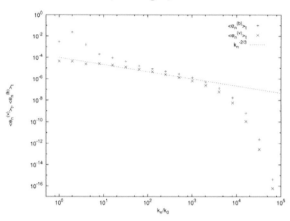

Figure 5. Spectra $\langle e_n^{(v)} \rangle_t$ of velocity perturbation ("×" symbols) and $\langle e_n^{(b)} \rangle_t$ of magnetic field perturbation ("+" symbols) as functions of the transverse wavevector $k_{\perp n}$. The line corresponds to a Kolmogorov spectrum.

Warren *et al.* 1997). All the above properties have been checked on a numerical model (the hybrid shell model) finding a good agreement.

References

Boffetta, G., Carbone, V., Giuliani, P., Veltri, P., & Vulpiani, A., 1999, *Phys. Rev. Lett.*, 83, 4662

Buchlin, E. & Velli, M., 2007, *ApJ*, 662, 701

Davila, J. M. 1987, *ApJ*, 317, 514

Dmitruk, P. & Gomez, D. O., 1997, *ApJ*, 484, L83

Einaudi, G., Velli, M., Politano, H., & Pouquet, A., 1996, *ApJ*, 457, L113

Giuliani, P. & Carbone, V., 1998, *Europhys. Lett.*, 43, 527

Heyvaerts, J. & Priest, E. R., 1983, *Astron. Astrophys.*, 117, 220

Ionson, J. A., 1982, *ApJ*, 254, 318

Milano, L., Gomez, D. O., & Martens, P. C. H., 1997, *ApJ*, 490, 442

Nigro, G., Malara, F., Carbone, V., Veltri, P. 2004, *Phys. Rev. Lett.*, 92, 194501, 1.

Nigro, G., Malara, F., & Veltri, P. 2008, *ApJ*, 685, 606

Petkaki, P., Malara, F., & Veltri P. 1998, *ApJ*, 500, 483

Strauss, H. 1976, *Phys. Fluids*, 19, 134

Sturrock, P. A. & Uchida, Y. 1981, *ApJ*, 246, 331

Withbroe, G. L., 1988, *ApJ*, 325, 442

Warren, H. P., Mariska, J. T., Wilhelm, K., & Lamaire, P. 1997, *ApJ*, 484, L91

Discussion

DAVILA: I wonder if the Reduced MHD solutions violate the initial assumptions used to derive the equations originally.

MALARA: Reduced MHD is based on the hypothesis that perpendicular gradients are made larger than parallel gradients. When a current sheet forms (at large scales) the associated gradient is quasi-perpendicular to B, so the above assumption is not violated. Moreover, perturbations responsible for the current sheet formation are included in these approaches as DC perturbations ($\omega = 0$ resonant eigenmodes).

TSAP: Why do you think about the fact that the crossing Alfven time in the solar atmosphere is about 20 s? Its value is significantly greater in the photosphere.

MALARA: I tried to estimate how much the first-order resonance frequency decreases when chromosphere is included. For this I used a simple exponential dependence of the Alfven speed on the altitude. I found that this frequency is decreased by a factor of 3, while the input energy flux is only slightly increased.

Universal Heliophysical Processes
Proceedings IAU Symposium No. 257, 2008
N. Gopalswamy & D.F. Webb, eds.

© 2009 International Astronomical Union
doi:10.1017/S1743921309029858

Generation and propagation of Alfvén waves in solar atmosphere

Yuri T. Tsap[1,2], Alexander V. Stepanov[2], and Yulia. G. Kopylova[2]

[1]Crimean Astrophysical Observatory, Nauchny, Crimea, Ukraine
email: yur@crao.crimea.ua

[2]Central Astronomical Observatory at Pulkovo, Russia
email: stepanov@gao.spb.ru; yulia00@mail.ru

Abstract. The propagation of Alfvén waves from the photosphere into the corona with regard to the fine structure of the magnetic field is considered. The energy flux of Alfvén–type waves generated in the photosphere by convective motions does not depend on the ionization ratio. The reflection coefficient continuously decreases with a decrease of wave period. Influence of the external magnetic field on the Spruit cutoff frequency for transverse (kink) modes excited in the thin magnetic flux tubes is analyzed. Torsional modes can penetrate into the upper atmosphere most effectively since their amplitudes does not increase with height in the photosphere while kink ones can be transformed into shock waves in the lower chromosphere because of a significant increase of amplitudes. In spite of stratification the linearity of Alfvén–type modes in the chromosphere is conserved due to violation of the WKB approximation. The important role of the magnetic canopy is discussed. Alfvén waves generated by convective motions in the photosphere can contribute significantly to the heating of the coronal plasma in quite regions of the Sun.

Keywords. Sun: magnetohydrodynamics, waves, atmosphere, magnetic fields, corona

1. Introduction

It is currently believed that magnetohydrodynamic (MHD) waves (AC models), quasi–stationary electric currents or current sheets (DC models) can be responsible for the coronal heating (e.g., Narain & Ulmshneider 1996). Since the role of nanoflares and microflares in the coronal heating is not clear the AC models proposed more than sixty years ago (e.g., Alfven 1947) remain relevant.

Alfvén waves, as distinguished from other MHD modes, do not compress plasma and they are considered as the main carrier of the energy of convective motions into the corona by many authors (e.g., Narain & Ulmshneider 1996; Noble *et al.* 2003; Cranmer & Ballegooijen 2005; Ofman 2005). However, the region of their generation is unknown; Alfvén waves can be excited in the convective zone, photosphere, chromosphere or corona.

Recently Vranjes *et al.* (2008) based on the MHD equations concluded that the generally accepted expression for estimates of energy fluxes of Alfvén waves $F = \rho \delta v_0^2 / 2 v_A$, where δv_0 is the amplitude of the velocity perturbation, ρ is the plasma density, $v_A = B/\sqrt{4\pi\rho}$ is the Alfvén speed, is unsuitable for the solar photosphere. They argued that due to the small ionization ($n_i/n_a \sim 10^{-4}$) "the ion collisions do not feel the effects of the magnetic field". To our opinion, this inference is not valid since Vranjes *et al.* (2008) did not take into account electromagnetic forces in momentum equations (see next section).

Results, concerning the reflection of Alfvén waves, are very contradictory. For example, Thomas (1978) came to conclusion that these waves can not penetrate into the corona. Bel & Leroy (1981) argued that energy flux, reaching the corona, is less than 10^{-5} of the input energy flux density for waves with periods $T_p = 100-500$ s. On the other hand, according to Geronicolas (1977), reflection of Alfvén waves is negligible in the upper atmosphere of the Sun. Cranmer & Ballegooijen (2005) concluded that waves are strongly reflected at the transition region and only about 5% of the wave energy can penetrate into the corona.

The urgency of these studies still increases in the light of results obtained by Noble et al. (2003). According to them, the propagation Alfvén–type waves (torsional and transverse) with $T_p \gtrsim 10$ s in the isolated thin magnetic flux tubes is impossible in the solar chromosphere because of restrictions connected with the Spruit cutoff frequency. Though Musielak et al. (2007) have shown that Noble et al. (2003) made a mistake and the frequency cutoff for torsional modes can be neglected, nevertheless, as follows from some estimates (Noble et al. 2003), transverse modes are generated by convective motions more productively than torsional ones.

The coronal plasma density is tens and hundreds of millions of times lower than the photosphere one. As a result Alfvén modes should be transformed into the strongly dissipated shock waves (e.g., Hollweg 1982; Kudoh & Shibata 1999) due to a sharp increase of amplitudes with height. Consequently, a question arises: can Alfvén waves penetrate from the photosphere into the corona of the Sun without significant energy losses?

2. On the energy flux of Alfvén waves in the weakly ionized plasma

Using the standard notation, the simplified momentum equations for electrons, ions, and neutrals can be written as

$$-en_e\delta\mathbf{E} - \frac{en_e}{c}\delta\mathbf{v}_e \times \mathbf{B} = 0; \tag{1}$$

$$n_i M\frac{\partial\delta\mathbf{v}_i}{\partial t} = en_i\delta\mathbf{E} + \frac{en_i}{c}\delta\mathbf{v}_i \times \mathbf{B} + n_i M\nu_{ia}(\delta\mathbf{v}_a - \delta\mathbf{v}_i); \tag{2}$$

$$n_a M\frac{\partial\delta\mathbf{v}_a}{\partial t} = n_a M\nu_{ai}(\delta\mathbf{v}_i - \delta\mathbf{v}_a). \tag{3}$$

It should be emphasized that Vranjes et al. (2008) suggested that the role of Lorentz's force $f_L = |\delta\mathbf{v}_i \times \mathbf{\Omega}_i|$ is negligible on the right–hand side of equation (2) since $\Omega_i \ll \nu_{ia}$. However, the force caused by the ion drag $f_{ia} = \nu_{ia}|\delta\mathbf{v}_i - \delta\mathbf{v}_a| \to 0$ at $\delta\mathbf{v}_a \to \delta\mathbf{v}_i$ and the frequency ν_{ia} does not characterize the value of f_{ia}.

Taking into account that $n_i = n_e$ and current density $\mathbf{j} = en_i(\delta\mathbf{v}_i - \delta\mathbf{v}_e)$, equations (1) and (2) give

$$n_i M\frac{\partial\delta\mathbf{v}_i}{\partial t} = \frac{\delta\mathbf{j} \times \mathbf{B}}{c} + n_i M\nu_{ia}(\delta\mathbf{v}_a - \delta\mathbf{v}_i). \tag{4}$$

If the magnetic field $\mathbf{B} \parallel \mathbf{Z}$, the wave vector $\mathbf{k} \parallel \mathbf{B}$, and $\delta\mathbf{B} = \delta\mathbf{B}_0\exp(-i\omega t + ikz)$, then, adopting for the sake of simplicity $z = 0$, equation (4), in view of Ampere's Law, $\delta\mathbf{j} = c/4\pi\nabla \times \delta\mathbf{B}$, takes the form

$$\frac{\partial\delta v_i}{\partial t} = ibe^{-i\omega t} + \nu_{ia}(\delta v_a - \delta v_i), \quad b = \frac{kB\delta B_0}{4\pi n_i M}, \tag{5}$$

where δv_i and δv_a are the transverse with respect to the magnetic field direction $\mathbf{B}$ components of the disturbed velocity.

Assuming $x = \delta v_i - \delta v_a$ and combing (3) and (5), we find

$$\frac{\partial x}{\partial t} + (\nu_{ia} + \nu_{ai})x = ibe^{-i\omega t}. \tag{6}$$

Solution of equation (6) at $x(t = 0) = x_0$ can be represented as

$$x = \frac{ibe^{-i\omega t}}{\nu_{ia} + \nu_{ai} - i\omega} + \left(x_0 - \frac{ib}{\nu_{ia} + \nu_{ai} - i\omega}\right)e^{-(\nu_{ia} + \nu_{ai})t}. \tag{7}$$

On the other hand, from equations (3) and (5), taking into account that

$$n_i\nu_{ia} = n_a\nu_{ai}, \tag{8}$$

we have

$$\frac{\partial\delta v_i}{\partial t} + \frac{\nu_{ia}}{\nu_{ai}}\frac{\partial\delta v_a}{\partial t} = ibe^{-i\omega t}.$$

Whence, assuming $y = \delta v_i + \nu_{ia}/\nu_{ai}\delta v_a$, we obtain

$$\frac{\partial y}{\partial t} = ibe^{-i\omega t}.$$

Solution of last differential equation at $y(t = 0) = y_0$ is reduced to the form

$$y = y_0 + \frac{b}{\omega}\left(1 - e^{-i\omega t}\right). \tag{9}$$

Thus, suggesting $\delta v_{0a} = 0$, $\delta v_{0i} = -b/\omega$, $\omega \ll \nu_{ai}$, $t \gg 1/\nu_{ia}$, and $n_a \gg n_i$, equations (7)–(9) give

$$\delta v_i \approx \delta v_a \approx -\frac{n_i}{n_a}\frac{b}{\omega}e^{-i\omega t}. \tag{10}$$

At $\Omega_i \gg \omega$, taking $\delta\mathbf{v} = \delta\mathbf{v}_i$, from equations (2) and (10) we obtain the frozen-in condition, i.e., $\delta\mathbf{E} \approx -\delta\mathbf{v} \times \mathbf{B}/c$. Since the averaging over a period the energy flux $\mathbf{F} = c/8\pi\delta\mathbf{E}^* \times \delta\mathbf{B}$, in view of (10) and the dispersion equation for Alfvén waves, $\omega = kv_A$, we get

$$F_z \approx \frac{\delta B_0^2}{8\pi}v_A.$$

It means, for instance, that in the solar photosphere, for which $\nu_{ai} \approx 10^5$ s^{-1} (Vranjes *et al.* 2008), the energy flux of waves F_z with periods $T_p \gg 2\pi/\nu_{ai} \approx 10^{-4}$ s does not depend on the degree of ionization. Equation (10) with the help of the energy conservation law, $\rho\delta v_0^2/2 \approx \delta B_0^2/8\pi$, may be rewritten as $F_z \approx \rho\delta v_0^2/2v_A$, while as follows from Vranjes *et al.* (2008) the energy flux $F_z \approx \rho\delta v_{0i}^2/2(n_i/n_a)v_A$.

3. On the reflection and propagation of Alfvén waves

Model of the solar atmosphere. In terms of the structure of the magnetic field, we can divide the solar atmosphere into two parts: the region of the thin isolated magnetic flux tubes (photosphere–chromosphere) and the region of the quasi–homogeneous magnetic field (chromosphere–corona). The boundary between these regions is called the magnetic canopy. Based on this simple model we shall consider the propagation of Alfvén waves from the photosphere into the corona.

Reflection of Alfvén waves. The wave equation for Alfvén waves in the stratified atmosphere at $B = $ const can be represented as (e.g., Tsap 2006)

$$\frac{\partial^2\delta v_\perp}{\partial t^2} = v_A^2(z)\frac{\partial^2\delta v_\perp}{\partial z^2}. \tag{11}$$

At $v_A \propto \exp(z/2H)$ and $\delta v, \delta B \propto \exp(-i\omega t)$ the solution of equation (11) is

$$\delta v_\perp = [C_1 H_0^{(1)}(\eta) + C_2 H_0^{(2)}(\eta)]e^{i\omega t}, \tag{12}$$

where C_1 and C_2 are arbitrary constants, $\eta = 2H\omega/v_A$, $H_0^{(1)}$ and $H_0^{(2)}$ are the Hankel functions, which describe waves propagating in different directions.

Ferraro & Plumpton (1958), taking into account that Macdonald function $N_0(\eta) \to -\infty$ at $\eta \to 0$, wrote the solution of equation (11) in the upper solar atmosphere as

$$\delta v_\perp = C_3 J_0(\eta).$$

Last equation describes oscillations, which cannot transfer the wave energy since the energy flux is equal to zero in this case. As a result, for example, Thomas (1978) came to conclusion about the total reflection of Alfvén waves in the upper atmosphere of the Sun. An *et al.* (1989) for justification of this approach used a "paradox" connected with the value of Alfvén velocity ($v_A \to \infty$ at $\rho \to 0$). However, if we take into account the displacement current, the dispersion relation of Alfvén waves at $\omega \ll \Omega_i$ takes a form: $\omega/k = c/\sqrt{1 + 4\pi\rho c^2/B^2}$. Therefore $\omega/k \to c$ at $\rho \to 0$, i.e., Alfvén waves are transformed into electromagnetic ones. Moreover, this suggests that the idea of the continues reflection of Alfvén waves in the stratified atmosphere (Bel & Leroy 1981; An *et al.* 1990; Musielak & Moore 1995) is not sufficiently correct.

There are many indications that the transmission of Alfvfen waves into the corona is determined by the reflection in the transition region (see, e.g., Schwartz *et al.* 1984; Cranmer & Ballegooijen 2005). According to estimates, obtained by Tsap (2006), under conditions of the solar atmosphere the transmission coefficient for waves with periods less than few tens of seconds is about 0.3. It means that Alfvén modes with $T_p = 10 - 40$ s (waves with $T_p < 10$ s are strongly damped in the solar chromosphere due to ion–neutral collisions) can effectively penetrate into the corona.

Equilibrium condition for a thin magnetic flux tube. If a vertical magnetic flux tube is thin, the condition of equilibrium along the vertical axis Z at the temperature $T(z) = $ const is

$$p(z) = p(0)e^{-z/H}. \tag{13}$$

The balance of pressures at the tube boundary reduces to equality

$$p_i(z) + \frac{B_i^2(z)}{8\pi} = p_e(z) + \frac{B_e^2(z)}{8\pi}. \tag{14}$$

In particular, for $T_i(z) = T_e(z) = $ const and $B_i(z) \propto B_e(z)$ or $B_e(z) = 0$, we have from (13) and (14)

$$e^{-z/H} = \frac{B_i^2(z)}{B_i^2(0)} = \frac{B_e^2(z)}{B_e^2(0)}, \tag{15}$$

i.e., $v_A^2(z) = $ const. Consequently, Alfvén velocity both inside and outside of a magnetic flux tube must be constant.

Transverse waves and the generalized Spruit cutoff frequency. As distinguished from Spruit (1981) we consider transverse oscillations of the thin magnetic flux tubes surrounded by the magnetic field B_e. The force related to the response of the external medium we write as follows

$$\mathbf{f}_\perp = -\rho_e \frac{\partial \delta \mathbf{v}_\perp}{\partial t} + \frac{B_e^2}{4\pi}\mathbf{c},$$

where $\mathbf{c} = \partial \mathbf{e}_l/\partial l$ is the vector of curvature of perturbed magnetic field lines, $\mathbf{e}_l$ is the unit vector directed along the magnetic field. If $\delta \mathbf{v}_\perp = \delta v_x$, $\mathbf{g} = -g\mathbf{e}_z$, the equation, describing transverse Alfvén modes of a thin magnetic flux tube in the stratified atmosphere, can be represented as follows

$$(\rho_i + \rho_e)\frac{\partial^2 \delta v_x}{\partial t^2} - \frac{B_i^2 + B_e^2}{4\pi}\frac{\partial^2 \delta v_x}{\partial z^2} + (\rho_i - \rho_e)g\frac{\partial \delta v_x}{\partial z} = 0. \tag{16}$$

Note that equation (16) takes the form of the wave equation of torsion waves at $\delta v_x = \delta v_\varphi$, $B_e = 0$, and $\rho_e = \rho_i$ (see, e.g., Noble *et al.* 2003)

$$\frac{\partial^2 \delta v_\varphi}{\partial t^2} = v_{Ai}^2 \frac{\partial^2 \delta v_\varphi}{\partial z^2}. \tag{17}$$

Using (13)–(15), for transverse waves, instead of (16), we have

$$v_k^2 \frac{\partial^2 \delta v_x}{\partial z^2} - \frac{\Delta_B}{2H}\frac{\partial \delta v_x}{\partial z} + \omega^2 \delta v_x = 0, \tag{18}$$

where

$$v_k^2 = \frac{B_i^2 + B_e^2}{4\pi(\rho_i + \rho_e)} = \text{const}, \quad \Delta_B = \frac{B_i^2 - B_e^2}{4\pi(\rho_i + \rho_e)} = \text{const}.$$

Solution of equation (18) is

$$\delta v_x = e^{\Delta_B z/(4v_k^2 H)}(C_1 e^{i\kappa z} + C_2 e^{-i\kappa z}), \quad \kappa = \frac{1}{2}\sqrt{\left(\frac{2\omega}{v_k}\right)^2 - \left(\frac{\Delta_B}{2v_k^2 H}\right)^2}, \tag{19}$$

i.e, the generalized Spruit cutoff frequency is equal to $\Omega_A = \Delta_B/(4v_k H)$. When the plasma parameter $\beta < 1$, in view of (14), we can accept $B_i \approx B_e$, therefore we obtain the amplitude $|\delta v_x(z)| = \text{const}$ and frequency $\Omega_A \approx 0$ from (19). Meanwhile, if a magnetic tube is isolated ($B_e = 0$), then, in contrast to the previous case, the wave amplitude δv_x must increase significantly with height. It means that the external magnetic field not only suppresses the growth of amplitudes of transverse waves in the isothermal atmosphere but also considerably reduces the Spruit cutoff frequency Ω_A.

It is interesting to note that for $\mathbf{B} = \text{const}$, $\Delta_B = 0$, and $\rho_i = \rho_e$ equation (16) is transformed to the expression, which coincides with the wave equation (11), describing Alfvén waves in the stratified atmosphere with the quasi–homogeneous magnetic field.

Amplitudes of Alfvén–type modes in the stratified atmosphere. Let us consider the variation of amplitudes of transverse and torsional modes above and below the magnetic canopy, suggesting that the wave generation takes place in the photosphere of the Sun.

As follows from equation (19), the expression connected amplitudes of transverse waves of the isolated magnetic flux tubes $|\delta v_p|$ and $|\delta v_h|$ at the photospheric level $z = 0$ and some given height z, respectively, takes the form

$$\frac{|\delta v(z)|}{|\delta v(0)|} = \left(\frac{n(0)}{n(z)}\right)^{1/4}. \tag{20}$$

If we use for the solar atmosphere the VAL C model and assume number densities $n(0) \approx 10^{17}$ cm^{-3} and $n(z) = 10^{14}$ cm^{-3} (≈ 700 km), from (20) we obtain $|\delta v(z)|/|\delta v(0)| \approx 6$. Thus, nonlinear effects can play an important role even in the lower chromosphere for transverse waves. In contrast, according to (17), amplitudes of torsional waves do not depend on the height z since the internal Alfvén velocity $v_{Ai} = \text{const}$.

Let us now estimate the growth of amplitudes in the region above the magnetic canopy, where magnetic field is quasi–homogeneous and the waves are described by equation (11). Adopting $v_A = v_{A0} \exp[(z - z_0)/2H]$, where z_0 is the initial height, using (12), we derive

$$\frac{|\delta v(z)|}{v_A(z)} = \frac{|\delta v(z_0)|}{v_A(z_0)}\sqrt{\frac{J_0^2(\eta) + N_0^2(\eta)}{J_0^2(\eta_0) + N_0^2(\eta_0)}}e^{-(z-z_0)/2H}. \tag{21}$$

Numerical analysis of equation (21) shows that the relative amplitudes $|\delta v(z)|/v_A(z)|$ decrease with height z. In turn, since the disturbed magnetic field $\delta B \propto \partial \delta v/\partial z$, it is easy to make sure that the relative amplitudes $|\delta B(z)|/B(z_0)$ decrease with height as well.

4. Conclusions

In this work we have obtained the following results:

(a) The energy flux of Alfvén waves with periods $T_p \gg 10^{-4}$ s does not depend on the degree of ionization in the solar photosphere.

(b) The energetic losses of linear waves with periods less than several tens of seconds, propagating from the solar photosphere to the corona, are about 70% due to reflection.

(c) The external magnetic field of the thin magnetic flux tubes slows down with height the growth of amplitudes of transverse waves as well as decreases the Spruit frequency.

(d) The capability of transverse waves to heat coronal plasma depends on the height of the formation of the magnetic canopy.

(e) Torsional waves transfer the energy of convective plasma motions on the Sun and stars from the photosphere to the corona more efficiently than transverse ones since their amplitudes in the thin magnetic flux tubes are not changed.

Acknowledgements

We would like to thank the anonymous referee for helpful comments and suggestions. This work was partially supported by the Russian Foundation for Basic Research (projects No 06-02-16859-a, 09-02-00624-a), the Program of Presidium of Russian Academy of Sciences "Solar Activity and Physical Processes in the Sun-Earth System", the Program of the Department of Physical Sciences of RAS "Study of Solar-Wind Disturbance Sources" and the Program for Support of Leading Scientific Schools (NSh-6110.2008.2).

References

Alfven, H. 1947, *MNRAS*, 107, 211.
An, C.-H., Musielak, Z. E., Moore, R. L., & Suess, S. T. 1989, *ApJ*, 345, 597.
An, C.-H., Suess, S. T., Moore, R. L., Musielak, Z. E. 1990, *ApJ*, 350, 309.
Bel, N. & Leroy, B. 1981, *Astron. Astrophys.*, 104, 203.
Cranmer, S. R. & van Ballegooijen, A. A. 2005, *ApJS*, 156, 265.
Ferraro, V. C. A. & Plumpton, C. 1958, *ApJ*, 127, 459.
Geronicolas, E. A. 1977, *ApJ*, 211, 966.
Hollweg, J. V. 1982, *ApJ*, 254, 806.
Kudoh, T. & Shibata, K. 1999, *ApJ*, 514, 493.
Musielak, Z. E. & Moore, R. L. 1995, *ApJ*, 452, 434.
Musielak, Z. E., Routh, S., & Hammer R. 2007, *ApJ*, 659, 650.
Narain, U. & Ulmshneider, P. 1996, *Space Sci. Revs*, 75, 453.
Noble, M. W., Musielak, Z. E., & Ulmschneider P. U. 2003, *Astron. Astrophys.*, 409, 1085.
Ofman, L. 2005, *Space Sci. Revs*, 120, 67.
Schwartz, S. J., Cally, P. S., & Bel, N. 1984, *Solar Phys.*, 92, 81.
Spruit, H. C. 1981, *Astron. Astrophys.*, 98, 155.
Thomas, J. H. 1978, *ApJ*, 225, 275.
Tsap, Y. T. 2006, in Proccedings IAU Symposium No. 233, p.253.
Vranjes, J., Poedts, S., Pandey, B. P., & de Pontieu, B. 2008, *Astron. Astrophys.*, 478, 553.

Discussion

GABRIEL: Your estimate of 10% wave reflection should still allow sufficient Alfven wave energy flux to penetrate and heat the corona, since it is estimated that there is large wave energy flux in the chromosphere and below (for example, see DePontieu *et al.*, Science, 2007).

TSAP: You are right. However, according to our calculations the period of Alfven waves must be less than 40 s. This follows from the expression for the reflection coefficient, which is different from the formula obtained by Hollweg (1984).

Universal Heliophysical Processes
Proceedings IAU Symposium No. 257, 2008
N. Gopalswamy & D.F. Webb, eds.

© 2009 International Astronomical Union
doi:10.1017/S174392130902986X

Alfvén waves in a gravitational field with flows

A. Satya Narayanan, C. Kathiravan & R. Ramesh

Indian Institute of Astrophysics
Koramangala II BLock, Bangalore - 560 034, India
email: satya@iiap.res.in

Abstract. The gravitational stratification effect on magnetohydrodynamic waves at a single interface in the solar atmosphere has been studied in the penumbral region of the sunspot recently. The existence of slow and fast magneto acoustic gravity waves and their characteristics has been discussed. The effect of flows on magneto acoustic gravity surface waves leads to modes called flow modes or v-modes. The present geometry is that of a plasma slab moving with uniform velocity surrounded by a plasma of different density. As is applicable to the corona, we assume that the plasma β to be small. The dispersion characteristics change significantly with a change in the value of G (gravity) and uniform flow.

Keywords. Sun: Oscillations

1. Introduction

It is clear from the literature that gravity is one of the dominant forces in the solar atmosphere. It plays an important role in the dispersion characteristics of Alfven and magneto acoustic waves and several authors (Erdelyi *et al.*, 1999, Miles & Roberts 1989, Miles & Roberts 1992, Miles *et al.* 1992, Roberts 1991) have studied these waves in different contexts. Varga & Erdelyi (2001) have extended the work of Miles *et al.* (1992) by assuming the uniform flow of plasma over a field free medium and found that the flow causes some modes to appear and others to disappear. Satya Narayanan (2000) and Sengottuvel & Somasundaram (2001) have also discussed the effect of equilibrium flow on Magneto Acoustic Gravity Waves in the solar atmosphere and suggested the running penumbral waves from penumbra to be the gravity influenced magneto acoustic surface waves. Recently, McEvan & Diaz (2007) have investigated the effect of gravity on a horizontal coronal slab by extending the work of Edwin & Roberts (1982) and reported that the presence of gravity modifies the oscillatory frequencies of the slab and resulting in the possible transition between surface and body modes.

It is also well known that gravity waves play an important role in studying the coupling of lower and upper solar atmospheric regions and are therefore of tremendous inter- disciplinary interest. The gravity waves in the Sun may be divided into two types, namely, (i) the internal gravity waves, which are confined to the solar interior, and (ii) the atmospheric gravity waves, which are related to the photosphere and chromosphere, and may be further beyond. In general, the observation of gravity mode oscillations of the Sun would provide a wealth of information about the energy- generating regions, which is poorly probed by the p-mode oscillations. It has been suggested that the turbulent convection below the photosphere will generate the high order, non-radial g-mode oscillations (internal gravity waves). There are theoretical studies earlier on solar-atmospheric gravity waves by several authors (Lighthill (1967), Stein (1967), Schmieder (1977). Traces of internal gravity waves are present in the ω - k diagrams of Frazier (1968). One of the earliest observations of internal gravity waves from individual granules is from the work of Deubner (1974). Cram (1978) has investigated the evidence of low, but significant, power at frequencies relevant to internal gravity waves. He also concluded that from the studies of the phase lag between successive layers, there was an upward energy flux. In addition, Brown & Harrison (1980) observed indications of the possible existence of trapped gravity waves by analyzing the brightness fluctuations of the visible continuum. These internal gravity waves, by-products of the granulation, are expected to be fairly common and may not be negligible in the energy balance of the lower chromosphere. Deubner, (1981), has pointed out that gravity waves are extremely difficult to observe because these are local, small-scale features requiring very high spatial resolution observations. High temporal and spatial resolution data will reveal that these gravity waves are small-scale phenomena. There have been lots of claims on the detection of g-modes in the Sun, but, so far, there is not much of a convincing evidence from the observational side. However, Straus & Bonaccini (1997), have presented observationally, the strongest evidence of gravity wave present in the middle photosphere using the wavenumber and frequency resolved phase-difference spectra and horizontal propagation diagram. Some recent works on the effects of gravitational stratification on Alfven waves are found in (Mullan & Khabibrakhmanov 1999, McKenzie & Axford 2000, DeMoortel & Hood 2004). Rathinavelu *et al.* (2007) studied the effect of flow on Alfven Gravity Surface Waves in a plasma slab surrounded by plasma and neutral gas in the context of magnetosphere.

There are some observational investigations to show that there is a signature of atmospheric gravity waves at the chromospheric level using the time sequence of filtergrams and spectra obtained in CaII H & K and Mg b2 lines (Dame *et al.* 1984, Kneer & von Uexkull 1993, Rutten & Krijger 2003).

In the last few years, progress in the spatial and time resolution of solar coronal instruments have given us enough evidence on the presence of Magnetohydrodynamic activity in the corona (Nakariakov & Verwichte 2005). In uniform plasmas, it is well known that there are three types of waves, namely the Alfven wave, the slow and fast magnetosonic waves. In addition to the above, there are possibly additional modes such as the kink, sausage, longitudinal ones, since it is well known by now that the corona is highly structured due to magnetic fields. The magnetoacoustic modes have been identified in the corona. However, there is still no direct evidence of Alfven modes. Recently, Tomczyk *et al.* (2007), using the Coronal Multi-Channel Polarimeter (CoMP) have identified waves in the corona. The revelation that waves were observed in the corona is an important development which is certainly worth probing into. Also the observation has important implication in the context of coronal heating and coronal seismology. They interpreted their observations in terms of Alfven waves, based on the facts that 1) the observed phase speeds are much larger than the sound speed, 2) the waves propagate along the field lines, and 3) the waves were incompressible. However, in the work of Van Doorsselaere *et al.* (2007), the observations were interpreted in terms of fast kink waves rather than Alfven waves. A very interesting article on the existence of Alfven waves in the Solar Atmosphere is found in Erdelyi & Fedun (2007).

2. The dispersion relation

In the present study, we consider a very simple three layer model whose densities and magnetic fields are different, though uniform in each of them. This model may be thought of a simple extension to a current sheet that one comes across in many physical situations. We also assume that the thickness of the middle layer is not very large, while the layers above and below this layer is semi-infinite in extent. We are not introducing any rigid boundaries in this model. Also we assume that the magnetic fields are plane parallel in each of the layer with the strength differing. The middle layer is assumed to move with a uniform speed and gravity is the external force in addition to the magnetic field and pressure forces acting in the field. We invoke MHD approximation, which is valid to a large extent as most of the scales (length and time) are large compared to the mean free path of the particles. That is, we use the continuum hypothesis and also assume that the role of electric fields are negligible in the present scenario. The plasma β which is the ratio of the gas pressure to the magnetic pressure is assumed to be very small which is also a valid approximation, say in the solar corona.

Before discussing the dispersion relation of the waves that is under consideration, we give briefly below the standard dispersion relation for Alfven, Magneto Acoustic and Magneto Acoustic Gravity waves in an infinite fluid with constant magnetic field and compressibility.

It is well known that the effects of tension in an elastic string is to allow transverse waves to propagate along the string. In analogy, it is reasonable to expect the magnetic tension to produce transverse waves that propagate along the magnetic field B_0 with a speed

$$v_A = \frac{B_0}{(\mu\rho_0)^{1/2}} \tag{2.1}$$

In the case of shear Alfven waves, the dispersion relation is given by

$$\omega = kv_A cos\theta_B \tag{2.2}$$

These waves propagate at a certain angle to the direction of the magnetic field. When compressibility is taken into account, in addition to the Alfven mode, two additional modes appear, namely, the fast and slow magneto acoustic modes, with a dispersion relation given by

$$\omega^4 - \omega^2 k^2 (c_s^2 + v_A^2) + c_s^2 v_A^2 k^4 Cos^2\theta = 0 \tag{2.3}$$

When gravity is included, the dispersion relation is modified to yield

$$\omega^4 - \omega^2 (N^2 + k'^2) + N^2 sin^2\theta k'^2 c_s^2 = 0 \tag{2.4}$$

Here N^2 is the Brunt-Vaisala frequency which is a very important parameter for the static stability of a stratified fluid. More details about the nature of these modes can be found in Priest (1982).

For the model under consideration, the basic equations of motion are linearized by assuming perturbations of the form

$$f(x, z, t) = f(x)exp(i(kz - \omega t)) \tag{2.5}$$

The approach is well known in the literature and we skip the details for the sake of brevity. The dispersion relation after algebraic substitution reduces to

$$[\epsilon_1\epsilon_2 + \epsilon_1\epsilon_g + g\epsilon_1 \frac{(\rho_g - \rho_2)}{\omega}] + [\epsilon_1^2 + \epsilon_2\epsilon_g + \epsilon_2 g \frac{(\rho_g - \rho_1)}{\Omega} + \epsilon_g g \frac{(\rho_1 - \rho_2)}{\Omega}]tanh(2ka) = 0 \tag{2.6}$$

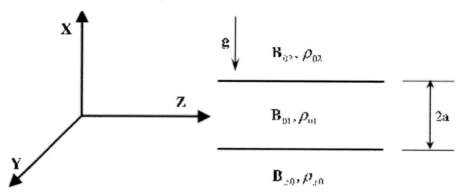

Figure 1. The Geometry.

The Epsilons and Omega are defined as

$$\Omega = \omega - kU_0 \tag{2.7}$$

$$\epsilon_1 = \frac{(\rho_1 \Omega)}{k(v_{ph} - \hat{U})^2}[1 - (v_{ph} - \hat{U})^2] \tag{2.8}$$

$$\epsilon_2 = \frac{(\rho_1 \omega)}{kv_{ph}^2}[\beta_1^2 - v_{ph}^2 \eta_1] \tag{2.9}$$

$$\epsilon_g = \frac{(\rho_1 g)}{kv_{ph}^2}[\beta^2 - v_{ph}^2 \eta] \tag{2.10}$$

Introducing the non-dimensional quantities

$$v_{ph} = \frac{\omega}{kv_{A1}}, \ \hat{U} = \frac{U_0}{v_{A1}}, \ \beta = \frac{B_{0g}}{B_{01}}, \ \beta_1 = \frac{B_{02}}{B_{01}}, \ \eta = \frac{\rho_{0g}}{\rho_{01}}, \ \eta_1 = \frac{\rho_{02}}{\rho_{01}} \ G = \frac{g}{kv_{A1}^2}$$

The normalized dispersion relation can be simplified to yield

$$[1 - (v_{ph} - \hat{U})^2][\beta_1^2 - v_{ph}^2 \eta_1) + (\beta^2 - v_{ph}^2 \eta) + G(\eta - \eta_1)]$$

$$\left[\frac{v_{ph}}{(v_{ph} - \hat{U})}(1 - (v_{ph} - \hat{U})^2)^2 + \frac{(v_{ph} - \hat{U})}{v_{ph}}(\beta_1^2 - v_{ph}^2 \eta_1)(\beta^2 - v_{ph}^2)\right.$$

$$\left. +(\beta_1^2 - v_{ph}^2 \eta_1)G(\eta - 1) + (\beta^2 - v_{ph}^2 \eta)G(1 - \eta_1)\right]tanh(2ka) = 0 \tag{2.11}$$

3. Discussion of the results

The normalized dispersion relation presented in equation (2.16) is much more complicated than the ones presented for the uniform fluid without density, magnetic field discontinuities, gravity and flows as is expected. The geometry presented in Figure 1, being a three layer model introduces more parameters, characterizing the geometry, say, the ratios of densities, magnetic fields, non-dimensional flow and gravity. The number of parameters characterizing the model is large and so in the present work, we have restricted our discussion for a chosen (rather arbitrary) set of values, in order to discuss the effects of flow and gravity.

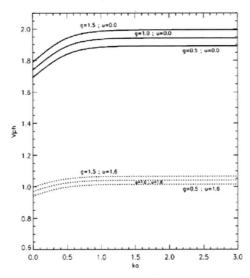

Figure 2. The Normalized Phase Velocity as a function of
non-dimensional wavenumber.

The dispersion relation is solved numerically and the phase speed is plotted as a function of dimensionless wave number for various values of the interface parameters $\beta^2 = 0.5$, $\beta_1^2 = 1.5$, $\eta = 1.8$, $\eta_1 = 1.2$ and different values of non-dimensionalised g and u as shown in Figure 2. For the compressible case, we will have both the fast Alfven- gravity surface wave and slow Alfven-gravity surface wave. However, in the present study, since the plasma beta is small, the slow mode does not appear. This is a well known result in MHD. It is interesting to note that for increasing values of g, say 0.5, 1.0, 1.5, the normalized phase speed of the fast Alfven gravity surface mode increases as a function of ka. This is the case for $u = 0$. However, when flow is introduced ($u > 0$), it is interesting to note that the phase speed of the mode is significantly reduced. This means that the effect of flow has a dampening effect on the phase speed of the fast Alfven gravity surface mode. Another interesting observation from the above Figure is that the phase speed has significant variation only for $ka \approx 1$ while for $ka > 1$, the phase speed remains the same and asymptotically approaches the phase speed of the body wave.

Figure 3 presents the normalized phase speed as a function of ka as a three dimensional plot. It is evident that the dispersion is more pronounced for small values of ka only. More computations need to be done to get a full picture of the nature of these waves. The present model can have interesting applications in coronal streamers and solar wind. Computations for other sets of parameters are being worked out and applications to coronal streamers and solar wind is being considered. The detailed study will be reported later.

Acknowledgements

I wish to thank the members of the SOC of IAU 257, Dr. N. Gopalswamy, Dr. David Webb, Dr. Alexander Nindos for giving me an opportunity to participate in this meeting. I wish to thank the authorities of IAU for providing partial support to attend the meeting. I also thank Prof. S. S. Hasan, the Director of IIA, for encouraging me to attend the meeting and providing travel support.

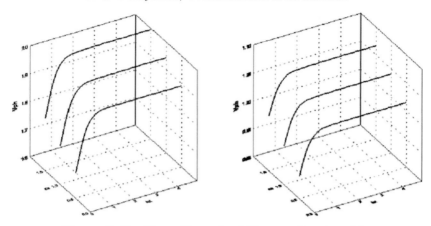

Figure 3. The Three Dimensional Plot as a function of ka

References

Brown, T. M. & Harrison, R. L., 1980, *Ap.J.*,236, L169

Cram, L. E., 1978, *Astron & Astrophys.*, 70, 345

Dame, L., Gouttebroze, P., & Malherbe, M. 1984 *Astron & Astrophys.*, 130, 331

De Moortel, I. & Hood, A. W. 2004 *Astron & Astrophys.*, 415, 705

Deubner, F. L. 1974 *Solar Phys.* 39, 31

Deubner, F. L. 1981 *NASA - SP* 450, 65

Edwin, P. M. & Roberts, B. 1982 *Solar Phys.*, 76, 239

Erdelyi, R., Varga, E., & Zetenyi, M. 1999 *ESA - SP*, 448, 269

Erdelyi, R. & Fedun, V. 2007 *Science*, 318, 1572

Frazier, E. N. 1968 *Z. Astrophys.*, 68, 345

Kneer, F. & von Uexkull, M. 1993 *Astron & Astrophys.*, 274, 584

Lighthill, M. J. 1967 *IAU Symp.*, 28, 429

McEvan, M. P. & Diaz, A. J. 2007 *Solar Phys.*, 246, 243

McKenzie, J. F. & Axford, W. I. 2000 *Solar Phys.*, 193, 153

Miles, J. W. & Roberts, B. 1989 *Plasma Phenomena Solar Atmosphere*, 77

Miles, J. W. & Roberts, B. 1992 *Solar Phys.*, 141, 205

Miles, J. W., Allen, H. R., & Roberts, B. 1992 *Solar Phys.*, 141, 235

Mullan, D. J. & Khabibrakhmanov, I. K. 1999 *ESA - SP*, 446, 503

Nakariakov, V. M. & Verwichte, E. 2005 *Living Reviews in Solar Phys.*, 2, 3

Priest, E. 1982 *Solar Magnetohydrodynamics*,Reidel Publishing Company

Rathinavelu, D. G., Sivaraman, M., & Satya Narayanan, A. 2007 *Plasma 2007*, 189

Rutten, R. J. & Krijger, J. M. 2003 *Astron & Astrophys.*, 407, 735

Satya Narayanan, A. 2000 *BASI*, 28, 85

Schmieder, B. 1977 *Solar Phys.*,54, 269

Sengottuvel, M. P. & Somasundaram, K. 2001 *Solar Phys.*, 198, 79

Stein, R. F. 1967 *Solar Phys.*, 2, 385

Strauss, T. & Bonaccini, D. 1997 *Astron & Astrophys.*, 324, 704

Tomczyk *et al.* 2007 *Science*, 317, 1192

Van Doorsselaete *et al.* 2007 *Astron & Astrophys.* 471, 311

Varga, E. & Erdelyi, R. 2001 *ESA - SP*, 464, 255

Universal Heliophysical Processes
Proceedings IAU Symposium No. 257, 2008
N. Gopalswamy & D.F. Webb, eds.

© 2009 International Astronomical Union
doi:10.1017/S1743921309029871

Properties of lower hybrid waves

Alix L. Verdon[1], I. H. Cairns[1], D. B. Melrose[1] and P. A. Robinson[1]

[1]School of Physics A28,
University of Sydney, NSW 2006,
Australia
email: a.nulsen@physics.usyd.edu.au

Abstract. Most treatments of lower hybrid waves include either electromagnetic or warm-plasma effects, but not both. Here we compare numerical dispersion curves for lower hybrid waves with a new analytic dispersion relation that includes both warm and electromagnetic effects. Very good agreement is obtained over significant ranges in wavenumber and plasma parameters, except where ion magnetization effects become important.

Keywords. waves, plasmas, acceleration of particles, solar system: general

1. Introduction

Waves in heliospheric plasmas are involved in many universal heliophysical processes, including plasma heating, particle acceleration, and emission processes that are signatures of energy releases. Lower hybrid (LH) waves are of particular interest in contexts that involve both electrons and ions; these waves can transfer energy between parallel motions of electrons and perpendicular motions of ions (Omel'chenko *et al.* 1989; Melrose 1986; Cairns 2001; Cairns & Zank 2002).

LH waves are nearly electrostatic with wavevectors $\mathbf{k}$ nearly perpendicular to the magnetic field, and involve oscillations of both the ions and electrons. In a cold plasma LH waves occur at a frequency which depends on the angle θ between the magnetic field $\mathbf{B}$ and $\mathbf{k}$ and is given by

$$\omega^2 = \omega_{LH}^2 \left(1 + \frac{m_i}{m_e} \cos^2 \theta \right). \tag{1.1}$$

Here m_e and m_i are, respectively, the masses of electrons and ions in the plasma, and the LH frequency is

$$\omega_{LH} \approx \frac{1}{\sqrt{1/\omega_{pi}^2 + 1/\Omega_e \Omega_i}} , \tag{1.2}$$

where ω_{pi} is the ion plasma frequency and Ω_e and Ω_i are the electron and ion gyrofrequencies, respectively.

At the LH frequency the ions are unmagnetized and free to move across $\mathbf{B}$, but the electrons are magnetized and may only move along $\mathbf{B}$. If the wave electric field is nearly perpendicular to $\mathbf{B}$ then the electron response time is greatly increased. The LH resonance occurs only when the ion response time is less than or comparable to the electron response time; i.e. when the following is satisfied,

$$\cos^2 \theta \lesssim \frac{m_e}{m_i}. \tag{1.3}$$

When warm plasma or electromagnetic effects are included in the plasma response the LH resonance becomes a propagating mode which must still satisfy equation (1.3). Hence, the components of $\mathbf{k}$ parallel and perpendicular to $\mathbf{B}$ are of such different magnitude, with

$k_{||}/k_{\perp} \lesssim m_e/m_i \ll 1$, that LH waves at the same frequency ω can satisfy the resonant conditions for interacting with the unmagnetized ions ($\omega = \mathbf{k} \cdot \mathbf{v}_i$) and the magnetized electrons ($\omega = k_{||}v_{e||}$). Thus LH waves may transfer energy from the perpendicular motions of ions to the parallel motions of electrons or vice versa, either accelerating particles or heating them.

Whenever there is a non-thermal perpendicular distribution of ions or parallel distribution of electrons, LH waves may play a role in redistributing the energy. Such distributions occur in many regions of the heliosphere, including in the outer heliosheath where pick-up ions form a ring beam distribution (Cairns & Zank 2002), and near magnetic reconnection sites where bulk ions flows across $\mathbf{B}$ and electrons accelerated along $\mathbf{B}$ are observed (Cairns 2001), such as occur in the Earth's magnetotail and possibly in the solar corona. However, to determine where this mechanism is efficient and could be relevant we must calculate the wave growth and particle diffusion rates. To do so, we must first find the LH dispersion relation, so that we can calculate the phase and group speeds.

Section 2 introduces the existing analytic dispersion relations for LH waves, which are compared with numerical dispersion relations in Section 3. In Section 4 we compare the numerical results with a new analytic dispersion relation we have derived. Section 5 contains a discussion and summary of our results.

2. Existing dispersion relations

The dispersion relations most commonly used for LH waves either include electromagnetic (EM) effects and assume the plasma is cold, or include warm plasma effects and assume that the waves are electrostatic. In general when deriving a dispersion relation for LH waves it is assumed that equation (1.3) is satisfied and $\Omega_i \ll \omega \ll \Omega_e \ll \omega_{pe}$, where ω_{pe} is the electron plasma frequency, so the ions may be treated as unmagnetized and the electrons as magnetized.

Assuming the plasma is cold but including EM effects gives (Omel'chenko *et al.* 1989)

$$\omega^2/\omega_{LH}^2 = \frac{1}{1 + \omega_{pe}^2/k^2c^2}\left(1 + \frac{m_i}{m_e}\frac{\cos^2\theta}{1 + \omega_{pe}^2/k^2c^2}\right), \qquad (2.1)$$

where c is the speed of light. This expression makes no assumption on the size of ω_{pe}^2/k^2c^2.

Including warm plasma effects and assuming the waves are longitudinal leads to the dispersion relation (Melrose 1986)

$$\omega^2/\omega_{LH}^2 = \left(1 + \frac{m_i}{m_e}\cos^2\theta + \left(3\frac{T_i}{T_e} + \frac{3}{4}\right)\frac{k^2V_e^2}{\Omega_e^2}\right), \qquad (2.2)$$

where T_e and T_i are the electron and ion temperatures, respectively, and V_e is the electron thermal speed. This expression includes terms only to first order in $k^2V_e^2/\Omega_e^2$.

A third dispersion relation for LH waves was used by Bingham *et al.* (2002). This dispersion relation includes warm plasma effects to first order in $k^2V_e^2/\Omega_e^2$ and EM effects to first order in ω_{pe}^2/k^2c^2.

3. Comparison of numerical and existing dispersion relations

The deviation of the LH mode frequency from the cold plasma resonance frequency [equation (1.1)] is greatest when $k^2V_e^2/\Omega_e^2$ is largest in the warm electrostatic approximation, and when ω_{pe}^2/k^2c^2 is greatest in the EM cold plasma approximation. The dispersion relation used by Bingham *et al.* (2002) differs from equation (1.1) at both large and small

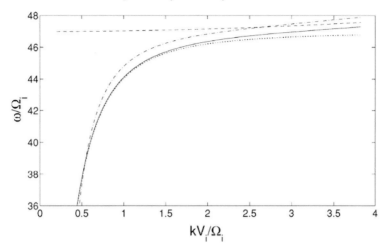

Figure 1. Warm electrostatic (dashes), cold EM (dots), Bingham *et al.* (2002) (dash-dot) and numerical (solid) dispersion relations for $\pi/2-\theta = 0.011$ rad, $T_i = T_e = 4000$ K and $\omega_{pe} = 10\Omega_e$.

k. In order to determine whether any of these were appropriate we compared them with a numerical dispersion relation.

The numerical LH dispersion relation was found for some parameters using the code of Willes & Cairns (2000). The code calculates the fully electromagnetic response tensor for the specified plasma parameters, including both electron and ion magnetization effects, and finds $\omega(\mathbf{k})$ as a root of its determinant in the complex plane. The code follows a single mode as k is changed and θ is kept constant, extrapolating from previous $\omega(\mathbf{k})$.

The three existing dispersion relations introduced above are compared with the real part of the numerical dispersion relation in Fig. 1. For low k the cold EM dispersion relation matches best with the numerical result. As k increases the slopes differ, with the cold EM result asymptotically approaching the cold plasma resonance frequency while the numerical frequency continues to increase. Consequently the group speed for LH waves calculated in the cold EM approximation is much lower than the actual value. At higher k the slope and magnitude of the warm electrostatic frequency best matches the numerical dispersion relation. Neither of these dispersion relations is adequate to describe LH waves over the range of k that interact with particles with speeds around the thermal speed and slightly faster.

The Bingham *et al.* (2002) dispersion relation gives a better overall fit than the other two dispersion relations. Ideally, one needs a dispersion relation that allows more accurate calculations of the phase and group speed of LH waves.

4. New results

We derived a new analytic dispersion relation including both EM and warm effects, with terms to all orders in ω_{pe}^2/k^2c^2 but only to first order in $k^2V_e^2/\Omega_e^2$. This dispersion relation will be presented elsewhere. As for the other, existing, dispersion relations we assume that equation (1.3) is satisfied and the electrons are magnetized but that the ions are not, and that there is no thermal damping of the waves. This dispersion relation reduces to equation (2.2), the warm electrostatic dispersion relation, and equation (2.1), the cold EM dispersion relation, in the appropriate limits.

A comparison of this new dispersion relation with the numerical and Bingham *et al.* (2002) dispersion relations is shown in Fig. 2 for a low temperature plasma. The new

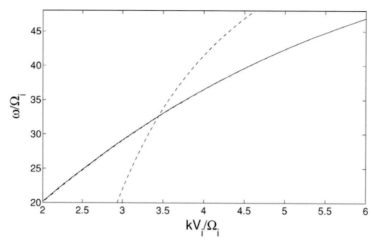

Figure 2. Bingham *et al.* (2002) (dashed), new analytic (dot-dashed) and numerical (solid) dispersion relations for $\pi/2 - \theta = 0.024$ rad, $T_i = T_e$, $T_e = 6000$ K and $\omega_{pe} = 100\Omega_e$.

analytic and numerical results are almost indistinguishable, especially for large k, for these plasma parameters. Our new dispersion relation matches the numerical results far more closely than the Bingham *et al.* (2002) dispersion relation.

A comparison of the new analytic dispersion relation with the numerical results at higher temperatures is shown in Fig. 3. In this case the agreement is not as good. At both high and low k no numerical solution to the dispersion relation can be found at or near multiples of Ω_i. As the frequency approaches a multiple of Ω_i the LH mode appears to break up into a series of ion Bernstein modes that make a transition between harmonics of Ω_i along a locus that is effectively the dispersion curve for LH waves. The analytic dispersion relation and the numerical results agree except near multiples of Ω_i.

Increasing the ion temperature by even a small factor, with or without changing the electron temperature, decreases the range of k over which numerical solutions can be found at multiple of Ω_i. Increasing θ so that the waves are even closer to perpendicular propagation also dramatically decreases the range of k over which numerical solutions at multiples of Ω_i can be found. There are also weaker dependencies on other parameters of the plasma at which a numerical solution for the LH wave can no longer be found at frequencies very close to multiples of Ω_i. Similar results were found numerically, for example by Feng *et al.* (1992); the ion acoustic mode becomes a series of perturbations to the ion Bernstein modes for low k and nearly perpendicular propagation.

5. Discussion and Summary

Ion magnetization effects, which are not included in any explicit analytic dispersion relation for LH waves, are strongest at multiples of Ω_i. As these are included in the numerical results it is not surprising that the greatest difference between our numerical and analytic results occur at multiples of Ω_i, and may be attributed to ion magnetization effects.

To lowest order, ion magnetization effects contribute only to the imaginary part of ω, $\mathrm{Im}(\omega)$, and not to the real part, $\mathrm{Re}(\omega)$. All the numerical results shown here are only for the real part of the complex frequency. In all cases $|\mathrm{Im}(\omega)| \ll \mathrm{Re}(\omega)$, so the mode is not heavily damped and the weak damping approximation is valid. Using this approximation to include the ion magnetization effects analytically to lowest order we reproduce the same

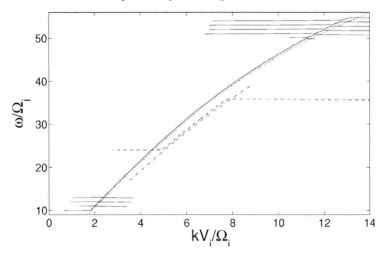

Figure 3. New analytic and numerical dispersion relations for $\omega_{pe} = 100\Omega_e$, $\pi/2 - \theta = 0.024$ rad, $T_e = 24000$ K and $T_e = T_i$ (dotted – analytic, and solid – numerical) and $T_e = 0.8T_i$ (dash-dot – analytic, and dashed – numerical).

features as the numerical results, but there are significant quantitative differences. Both numerical and analytic results show that $|\mathrm{Im}(\omega)|$ is greatest when $\mathrm{Re}(\omega)$ is a multiple of Ω_i and that $|\mathrm{Im}(\omega)|$ at local maxima increases from a minimum at intermediate k to greater values at large and small k. Ranges of k where the peaks of $|\mathrm{Im}(\omega)/\mathrm{Re}(\omega)|$ exceed some threshold (which is $\ll 1$) correspond very closely with ranges of k where the numerical LH mode breaks up into a series of ion Bernstein modes.

Relativistic effects are not included in either analytic or numerical work and we need to check that they are not important. Although the speeds involved are low, for waves very close to perpendicular propagation weakly relativistic effects are necessarily important even for what would normally be considered non-relativistic speeds (Robinson 1987).

In summary, our analytic dispersion relation for LH waves agrees far better with our numerical results than any previous dispersion relation over a large range of plasma parameters and wavenumbers. The agreement is extremely good when ion magnetization effects are small. However, when ion magnetization effects become larger, the numerical and analytic results differ qualitatively, with no numerical solution being found at frequencies very close to multiples of Ω_i where the ion magnetization effects are greatest.

References

Bingham, R., Dawson, J. M., & Shapiro, V. D. 2002, *J. Plasma Phys.*, 68, 161

Cairns, I. H. 2001, *Publ. Astron. Soc. Aust.*, 18, 336

Cairns, I. H. & Zank, G. P. 2002, *Geophys. Res. Lett.*, 29, 1143

Feng, W., Gurnett, D. A., & Cairns, I. H. 1992 *J. Geophys. Res.*, 97, 17005

Melrose, D. B. 1986 *Instabilities in space and laboratory plasmas* (Cambridge: Cambridge University Press)

Omel'chenko, Yu. A., Sagdeev, R. Z., Shapiro, V. D., & Shevchenko, V. I. 1989, *Sov. J. Plasma Phys.*, 15, 427

Robinson, P. A. 1987, *Phys. Fluids*, 31, 107

Willes, A. J. & Cairns, I. H. 2000, *Phys. Plasmas*, 70, 3167

Session XI

Flows, Obstacles, Circulation

Universal Heliophysical Processes
Proceedings IAU Symposium No. 257, 2008
N. Gopalswamy & D.F. Webb, eds.

© 2009 International Astronomical Union
doi:10.1017/S1743921309029895

Flows and obstacles in the heliosphere

John D. Richardson

Kavli Center for Astrophysics and Space Science,
Room 37-655, M.I.T., Cambridge, MA USA 02139
email: jdr@space.mit.edu

Abstract. The supersonic solar wind is highly variable on all time scales near the Sun but fluctuations are moderated by self-interaction as this plasma moves outward. The solar wind runs into many obstacles on its way out. The neutrals from the interstellar medium slow it down. Magnetospheres and interplanetary coronal mass ejections (ICMEs) cause shocks to form so that the flow can divert around these obstacles. Finally the solar wind is stopped by the circum-heliospheric interstellar medium (CHISM); it slows at the termination shock and then turns down the heliotail. The shocks and sheaths formed by these interactions cover scales which vary by orders of magnitude; some aspects of these shocks and sheaths look very similar and some very different. We discuss solar wind evolution, interaction with the neutrals from the CHISM, foreshocks, shock structure, shock heating, asymmetries, and sheath variability in different sheath regions.

Keywords. solar wind, shock waves, plasmas

1. Introduction

The space age started 50 years ago with the launch of the first spacecraft above Earth's atmosphere. In those 50 years we have learned much about the planets, comets, asteroids, and the interplanetary medium. The Voyager spacecraft are now approaching the local interstellar medium and may cross into this region in roughly 10 years; they have already sampled a great deal of the solar wind/CHISM boundary region. One common feature in all the regions we have encountered are plasma flows. In particular, the heliosphere is filled with the supersonic magnetized plasma called the solar wind and the region outside the heliosphere is filled with the combination of neutrals and magnetized plasma called the CHISM. The heliosphere is filled with obstacles which affect the solar wind flow which include neutral atoms from the CHISM, planets, moons, comets, asteroids, and dust. The Sun is very dynamic which causes large variations in the solar wind and causes complex interactions within the solar wind itself. The final and largest interaction of the solar wind is that with the CHISM. The physical processes in these flows are often similar. The magnetic field and plasma are tied together since plasma can only move along, not across, the magnetic field. Neutrals are not affected by the magnetic field, so neutrals can move between plasmas and link different plasma regimes. When the plasma flows encounter a non-conducting obstacle, such as the moon or asteroids, the flow slams into the surface and is absorbed. When these flows encounter a conducting obstacle, such as a magnetosphere or another magnetized flow, the flow must divert around the obstacle since the magnetic fields cannot pass through it. If the flow is supersonic like the solar wind, a reverse shock forms upstream of the obstacle causing the flow to compress, heat, become subsonic, and divert around the obstacle. This paper discusses the evolution of the solar wind flow and compares the interactions of the solar wind with the multitude of obstacles in its path. We compare the shocks and sheath formed in front of planets, solar wind transient structures, and the CHISM. All these involve the diversion of the solar wind but at very different scales and with different flow and obstacle parameters. We will discuss the interaction of the solar wind with interstellar neutrals, compare the shocks and sheaths at these various obstacles, discuss the heating of plasma at these shocks, and discuss the asymmetries in these interaction regions.

2. Overview

Fig. 1 shows a schematic diagram of a variety of flows inside and outside the heliosphere. The top right panel shows the heliosphere, the bottom right shows a planetary magnetosphere, the bottom left shows a coronal mass ejection (CME) leaving the Sun, and the top left shows an astrosphere. In the heliosphere the solar wind flows radially outward from the Sun and the CHISM flows from left to right. The heliopause, the boundary between the solar wind and the CHISM, forms where the solar wind and CHISM pressures balance. The CHISM flows around the outside of the heliopause and the solar wind turns and flows downstream inside the heliopause. If the CHISM were supersonic, a bow shock would form in the CHISM flow upstream of the termination shock (TS) which would make the CHISM flow subsonic and allow it to divert around the heliosphere. This region of shocked CHISM flow is called the outer heliosheath; we do not know the CHISM magnetic field strength so we do not know if the CHISM is supersonic. If it were not, the CHISM flow would not need to be shocked to divert around the heliosphere. The solar wind becomes subsonic at the TS; the shocked solar wind plasma makes up the heliosheath and moves down the heliotail. Voyager 1 and 2 have both crossed the TS and in 2008 are in the heliosheath.

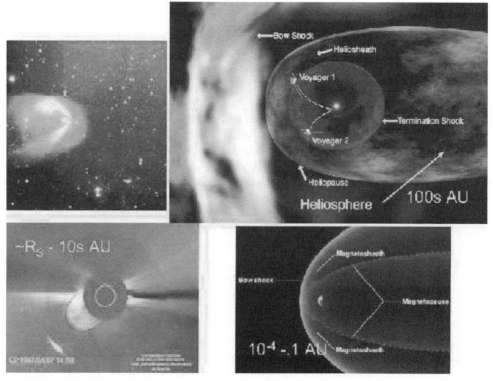

Figure 1. Examples of shocks and sheaths in the heliosphere and beyond. From top right clockwise are the heliosphere, a magnetosphere, a CME driven shock, and an astrosphere.

Magnetospheres can be either intrinsic or induced. An intrinsic magnetosphere is one where the planet has a magnetic field strong enough that a magnetopause, the pressure balance point between the solar wind dynamic pressure and the planet's magnetic pressure, forms outside the planet. The solar wind must flow around this conducting obstacle, the magnetosphere; a bow shock forms in from the the planet and the shocked solar wind plasma in the magnetosheath is diverted around the planet. Planets without strong fields which have ionospheres have induced magnetospheres; the solar wind magnetic field cannot flow through the conducting ionosphere, but compresses the ionosphere until pressure balance is reached between the ionospheric thermal pressure and the solar wind pressure. This boundary is called the ionopause, and acts like a magnetopause in that the solar wind slows at a bow shock and then flows around the planet. CMEs are explosive events on the Sun which eject large amounts of plasma from the Sun, sometimes at very high speeds. Called interplanetary CMEs (ICMEs) in the solar wind, they are large magnetic flux ropes which maintain their connection to the Sun well past 1 astronomical unit (AU). Since they move faster than the ambient solar wind, a shock forms ahead of the ICME which allows the faster ICME to flow through the ambient solar wind. This region of shocked solar wind plasma is called the ICME sheath. The astrosphere in the upper left was observed by the Hubble Space Telescope and gives us confidence that our ideas about the heliosphere are basically correct. The CHISM is coming from the right, gets heated at the bow shock, and diverts around the astrosphere. The stellar wind comes outward, then becomes shocked and hot at the termination shock. Many similar astrospheres have been observed.

3. Solar wind evolution

The Sun is dynamic on scales from minutes to centuries. A solar cycle (11 years) of Wind data at 1 AU shows that solar wind speeds vary from 260–2140 km/s with an average of 440 km/s, densities vary from 0.1–135 cm^{-3} with an average of 7.3 cm^{-3}, and temperatures vary from 6,000 to 3,700,000 K. Over the solar cycle the configuration changes from a nearly uniform slow dense flow at solar maximum to a bi-model flow with hot tenuous solar wind at heliolatitudes above 20–30 ° and slow dense solar wind nearer the equator. The solar wind pressure changes by about a factor of 2 over the solar cycle, with minimum pressure near solar maximum, a sharp increase in the 2-3 years after solar maximum, then a slow decline until the next solar maximum. ICMEs are more prevalent near solar maximum and in the declining phase of the solar cycle. As the solar wind moves outward in interacts with itself and this variability relaxes. Fast solar wind streams run into slow streams and transfer momentum reducing variability. Fig. 2 shows daily averages and 1 AU average speeds and relative standard deviations of the speed and density versus distance. The amount of small scale variation decreases with distance as is apparent in both the daily average plots and the relative standard deviations. By 20 AU most of the daily change in speed is gone leaving only larger structures. The density relative standard deviation continues to decline with distance. The termination shock crossing is at 84 AU where the speed decreases and density increases; the relative standard deviation of density stays the same while that of the speed increases.

4. Flows and neutrals

Neutral atoms and molecules are not affected by magnetic fields and can move through the plasma. The plasma and neutrals are coupled when collisions lead to charge exchange, in which an ion gains an electron from a neutral. The ion is now a neutral and continues moving in the same direction since it is not bound by the magnetic field. The new ion (the former neutral) is bound by the magnetic field and is accelerated to the speed of the plasma. It has an initial thermal energy equal to that of the plasma flow energy. Charge exchange is an important process in both the solar wind which interacts with interstellar neutrals and in planetary magnetospheres where neutrals from a planet's moons, rings, and ionosphere interact with the plasma. In magnetospheres the magnetic field lines are connected to the planet; when a new ion is formed the energy to accelerate it comes from the planet's rotation. The accelerating force is a $J \times B$ force with the current closing along field lines and through the planet's ionosphere. If mass-loading were large enough that sufficient current can't flow through the ionosphere to accelerate the plasma to the corotation speed, then the plasma will subcorotate. Subcorotation occurs in both the magnetospheres of Jupiter and Saturn, where moons and rings produce large numbers of neutrals, many of which are ionized through charge exchange. The hot ions formed in this manner are important energy sources for these magnetospheres. The solar wind has no energy source after it leaves the Sun, so the energy to accelerate pickup ions comes from the bulk flow energy of the solar wind and the pickup ions cause the solar wind to slow down and heat up. Fig. 3 shows 101-day averages of the speeds and temperatures observed by Voyager 2 (V2) and the speeds observed at 1 AU by IMP 8. The comparison between V2 and IMP 8 speeds shows clearly that the solar wind is slowing down. In the inner heliosphere the V2 and IMP 8 speeds are very similar. The exceptions are solar minimum in 1986-87 when V2 is at a lower average latitude than IMP 8 and sees slower flow and in the 1995-97 solar minimum when V2 is at higher heliolatitudes and sees slower flow. After 1998 V2 observes consistently lower speeds than IMP 8; by 2005 this slowdown is about 60 km/s which implies the pickup ion density is about 16% of the total solar wind density. The temperature profile shows that the solar wind cools until 1986-87

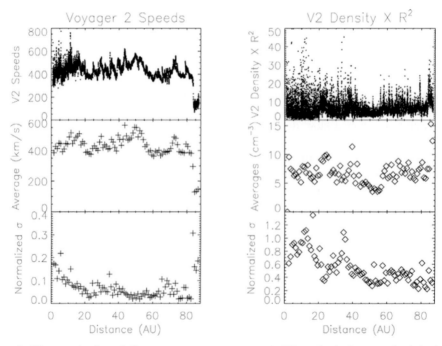

Figure 2. The panels show daily averages, averages over 1 AU, and relative standard deviations (σ/average) of the solar wind speed (left) and normalized density NR^2 (right) observed by Voyager 2.

at about 20 AU, then the temperature increases (Richardson & Smith, 2003). The pickup ions are not measured directly so this plot shows only the thermal proton temperature. The pickup ions are formed with unstable ring distributions; magnetic fluctuations are generated when these distributions become spherical and these fluctuations heat the thermal plasma (Smith *et al.*, 2006; Isenberg *et al.*, 2005). The CHISM neutrals remove about 30% of the flow energy from the solar wind before it reaches the TS and provide almost all the energy in the thermal plasma in the outer heliosphere.

5. Shocks

Shocks form upstream of obstacles in the solar wind: magnetospheres, ICMEs, and the CHISM. These shocks all make the downstream flow subsonic so the solar wind can divert around the obstacle, but have major differences. Fig. 4 superposes plasma data from Neptune's bow shock, the TS, and an interplanetary shock. The time scales are not adjusted, but the upstream parameters are normalized to those observed at the TS. The ICME shock is from about 13 AU; ICME shocks vary depending on the speed of the CME and weaken with distance. The ICME shock is a fast forward shock; the speed, density, and temperature all increase. Speed and density jumps at ICME shocks range from very

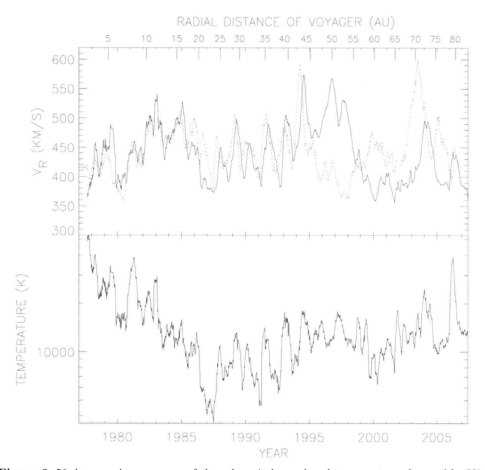

Figure 3. 51-day running averages of the solar wind speed and temperature observed by V2. The speeds measured by IMP 8 at 1 AU are superposed.

small to a factor of 4 and the proton temperature in the strongest shocks reaches a few hundred thousand degrees. The ICME shocks propagate outward, with shock speeds up to 200 km/s faster than the solar wind flow. Planetary bow shocks and the TS are both stationary reverse shocks, so the speed drops and the density and temperature increase. Planetary bow shocks are usually supercritical, quasi-perpendicular shocks at which the speed decreases and density increases by a factor of about 4. The temperature jumps by roughly 3 orders of magnitude to a few million degrees K. Most of the flow energy ends

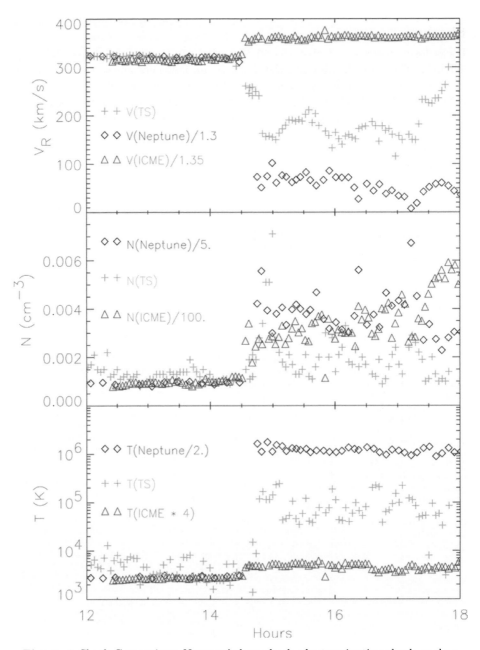

Figure 4. Shock Comparison: Neptune's bow shock, the termination shock, and an interplanetary shock. The upstream parameters are normalized to those at the TS.

up in the thermal plasma. The TS differs from the planetary bow shocks in that the shock is weaker; the speed and temperature change by about a factor of 2, not 4, and the heating of the thermal plasma is much less at the TS. The flow in the heliosheath remains supersonic with respect to the thermal plasma. Another difference between these shocks is the downstream proton distributions. Fig. 5 shows an example of spectra observed in the Jovian magnetosheath, the heliosheath, and in an ICME sheath. At planetary bow shocks, a percentage of the ions encountering a quasi-perpendicular shock are reflected, then convected back to the shock and further heated, forming a hot proton component (Sckopke *et al.*, 1983). At Earth, the percentage of hot ions depends on the Mach number; theory predicts that the percentage of hot ions should reach an asymptotic value of 20-25% at high Mach numbers (Fuselier and Schmidt, 1994). This prediction seems valid at Earth, but Table 1 shows that Jupiter, Saturn, and Neptune often have much larger percentages of ions in the hot component, 30-60% (Richardson, 1987; 2002).

The heliosheath does not have a reflected ion component even though the TS is a perpendicular shock. The temperature of the heliosheath is much lower than that of planetary magnetosheaths, 100,000K as compared to a few million K. The available data support a picture in which the pickup ions gain most of the plasma flow energy at the TS. Zank *et al.* (1996) showed that pickup ions, not thermal ions, were likely to be reflected and heated at the TS. Gloeckler *et al.* (2005) present Voyager 1 energetic particle spectra for energies greater than 30 keV. They showed that extrapolation of these spectra to lower energies suggests that 80% of the plasma flow energy ends up in the pickup ions. Only about 15% of the plasma flow energy ends up heating the thermal plasma (Richardson *et al.*, 2008a). At the TS shock about 20% of the protons are pickup ions (Richardson *et al.*, 2008b). If the flow energy were all going into the pickup ions then these ions should have a temperature of order 5 keV. Wang *et al.* (2008) reported STEREO observations of energetic neutral atoms with energies of 6-10 keV which appear to originate near the nose of the TS (the direction from which the CHISM is flowing); these ENAs form from charge exchange of pickup ions in the heliosheath. These 6-10 keV pickup ions have thermal speeds much faster than the average plasma flow speed of 150 km/s in the heliosheath, thus the flow is subsonic with respect to the pickup ions. Thus these observations seem to form a consistent picture if the energy from the plasma flow is transferred at the TS not to the thermal plasma but to the pickup ions. ICME sheath spectra are now being investigated. Fig. 5 shows an example of an ICME sheath with two

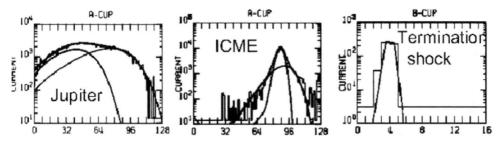

Figure 5. Spectra observed in Jovian magnetosheath, and ICME sheath, and the heliosheath The histograms are the data plotted versus a log energy scale and the curves are fits of convected isotropic proton Maxwellians to the individual components and the sum of these currents. The Jovian and ICME sheaths have hot components which comprise about 40% of the total density. The temperature of the hot component is about 9 times that of the cold component. The heliosheath spectra is well fit by a single Maxwellian distribution.Spectra observed in Jupiter's magnetosphere, an ICME sheath, and in the heliosheath. The histogram shows the measured currents and the curves the fit of convected isotropic proton Maxwellians to the data.

proton components formed at the interplanetary shock (the proton distributions in the solar wind upstream of the shock were well fit by single Maxwellians). The temperature of the hot component is about 9 times that of the cold component. The hot particles in this sheath carry the bulk of the thermal pressure in the ICME sheath, change wave propagation speeds, and could provide seed particles for further acceleration. But this hot component is not always present in the ICME sheaths we have studied; 3 of the 10 cases we have looked at show only a single proton Maxwellian. These spectra are similar to the heliosheath spectra, which is cold and fit well by a single Maxwellian. Why some ICME sheaths have reflected ions and some don't is under investigation; possibly this difference depends on the shock parameters. As the percentage of pickup ion densities increases with distance the amount of reflected thermal ions in ICME sheaths may decrease. The one outer planet which does not show evidence of hot ions in its magnetosheath is Uranus, which is surrounded by a corona of H. This result leads us to hypothesize that there are sufficient pickup ions from interstellar gas plus the corona that thermal ions are not reflected at Uranus's bow shock.

6. Sheath size

An ICME sheath differs from that of planetary magnetosheaths in two important respects: 1) since the ICME is a long flux rope, the flow can be approximated by a 2-D flow around a cylindrical obstacle (unlike the 3-D semi-spherical obstacle posed by a magnetosphere) and 2) the ICME is expanding so interaction is not in a quasi-steady-state (like a magnetosheath or the heliosheath). The thickness of the sheath increases as the shock propagates through the solar wind ahead of the ICME. MHD models predict that the thickness of the sheath is smaller than that of a planetary magnetosphere, about 0.1 (as opposed to 0.2) times the radius of curvature of the obstacle (Siscoe et al., 2007); the sheath layer is smaller because of the expansion of the ICME. The flow speeds around the obstacle are smaller than the expansion speed of the ICME, so the sheath accretes material as its size increases. ICME expansion stops at 10-15 AU (Liu and Richardson, 2004); perhaps at this distance the thickness of ICME sheaths may increase and become comparable in width to magnetosheaths.

7. Foreshocks

The particles heated at the shock can stream along the solar wind magnetic field lines into the heliosphere; this region of streaming particles is called the foreshock. At planetary magnetospheres the foreshock region is dominated by waves generated by the shock and waves generated by particles with energies of a few keV reflected and heated at the shock which move upstream. For planets only the lower energy particles encounter the bow shock again; most particles stream along field lines and remain in the solar wind. The Voyager spacecraft observed a large foreshock region upstream of the TS; this region was filled with streaming tens of keV to MeV particles (Krimigis et al., 2003; MacDonald et al., 2003). Both V1 and V2 first entered this region about 2.5 years before they crossed the TS. Essentially all of these particles will encounter the shock again. The foreshock region of the TS is not filled by waves as for planetary magnetospheres, perhaps because of the much higher particle energies. At planetary bow shocks the solar wind slows in the foreshock region as the plasma, waves, and streaming particles interact. At the TS, a decrease in speed is observed upstream of the shock (Richardson et al., 2008b) but whether this decrease is associated with foreshock particles is not known.

Table 1. Shock and Sheath Parameters

	Jupiter	Saturn	Uranus	Neptune	ICME	TS
Distance	55-99 R_J	24-26R_S	24 R_U	35 R_N	0.4-82 AU	84-94 AU
Mach No	4-19	8-14	17.	9		8
Beta	0.4-7.	0.5-1.4	3.6	0.2		2?
V_R (km/s)	60	85	42	115		140
T (10^6 K)	5.1	4.8	2.6	3.6		0.18
N_H/N_C	0.34	0.45	0	0.48	0 - 0.5	0
T_H/T_C	6.3	7.5	0	13.0	0-10	0
$\sigma(V_R)/V_R$	0.48	0.41	2.2	0.14		0.19
$\sigma(N)/N$	0.53	0.54	0.41	0.13		0.52

8. Asymmetries

The sheaths of planets, ICMEs, and the heliosheath may all be asymmetric. Studies of Earth's magnetosheath show higher densities in the dawn than dusk in both data and MHD results (Paularena *et al.*, 2001) and asymmetries in the magnetopause locations (Dmitriev *et al.*, 2004). MHD simulations predict similar asymmetries in ICME sheaths driven by the angle between the solar wind magnetic field and the solar wind flow (Siscoe *et al.*, 2007). The tilted field drapes around the ICME asymmetrically, giving an east-west asymmetry in the sheath radius and thickness. The Voyager spacecraft showed that the TS is 10 AU closer in the V2 than the V1 direction (Stone *et al.*, 2008; Richardson *et al.*, 2008). Models show that CHISM field tilted with respect to the CHISM flow can produce such an asymmetry (Opher *et al.*, 2007; Pogorolev *et al.*, 2007).

9. Sheath plasmas

Sheaths are turbulent regions with variable plasma and magnetic fields. Table 1 compares the shocks and sheath variabilities at the outer planets, ICMEs, and the TS. The TS has a smaller Mach number and larger plasma beta than most of the bow shock crossings. The relative standard deviation of V in the heliosheath is smaller than in all the magnetosheaths except Neptune, which was apparently encountered during a time of very steady solar wind. Planetary bow shocks and magnetopauses, and thus the magnetosheaths, move in and out with changes in solar wind pressure causing large motions and speed variability in the magnetosheaths. The heliosheath is a much larger region which responds more slowly to solar wind changes. Large merged interaction regions, high density and magnetic field regions formed when ICMES merge, may be able to move the TS outward several AU (Zank and Mueller, 2003) and V1 Low-energy Charged Particle Experiment data suggest that the heliosheath plasma was moving inward following the TS crossing (Decker *et al.*, 2005), but variability would be expected to be much less, and is observed to be less, away from the TS. The density variation, however, is very similar in the magnetosheaths and in the heliosheath. This similarity suggests similar changes occur in the planetary bow shocks and TS, either motions of the shock surface or temporal variability in shock structure. The TS speeds calculated for the two complete TS crossings observed TS speeds are 90 km/s when the TS moved out past V2 and 70 km/s when the TS moved inward past V2; these speeds are comparable to the speeds of planetary bow shocks; the average speed of Earth's bow shock, for example, is 85 km/s (Formisano *et al.*, 1973). Flow angles are much more variable in magnetosheaths than in the heliosheath data measured to date. The EW flow angle has remained fairly constant in the heliosheath. Part of this lesser variability results from V2 traversing only the inner edge of the heliosheath. The flow is expected to rotate as the heliopause is approached and this rotation has been observed on V1 (Decker *et al.*, 2007).

10. Summary

Shocks and sheaths are ubiquitous features of the Universe. Comparison of these features at ICMEs, planets, and the heliosphere boundary show both similarities and differences. The planetary and TS shocks are stationary reverse shocks whereas the ICME shock is a fast forward shock. The magnetosheath and heliosheath are in quasi-steady state while the ICME sheath accretes material and are thinner than the other sheaths. These are high-Mach number shocks; the planetary bow shocks (except Uranus) and most ICME sheaths have two populations of thermal ions, including a hot population presumably reflected at the shock. At the TS, most of the flow energy goes into the existing hot component, the pickup ions. All of these sheaths are asymmetric with asymmetries driven by a non-zero angle between the shock normal and the flow velocity which leads to asymmetric draping of the field around the obstacle. The heliosheath speeds and flow angles are much less variable than those in magnetosheaths, but the density variations are similar.

11. Acknowledgments

This work was supported under NASA contract 959203 from the Jet Propulsion Laboratory to the Massachusetts Institute of Technology and NASA grants NAG5-8947 and NNX08AC04G.

References

Burlaga, L. F., Ness, N. F., Acuna, M. H., Lepping, R. P., Connerney, J. E. P., & Richardson, J. D. 2008 Observations of magnetic fields at the termination shock by Voyager 2, *Nature*, 454, 75-77.

Decker, R. B., Krimigis, S. M., Roelof, E. C., Hill, M. E., Armstrong T. P., Gloeckler, G., Hamilton, D. C., & Lanzerotti, L. J. 2005 Voyager 1 in the foreshock, termination shock, and heliosheath. *Science 309*, 2020 - 2024 DOI: 10.1126/science.1117569.

Decker, R. B., Krimigis, S. M., Roelof, E. C., & Hill, M. E. 2006, Low-energy ions near the termination shock. In *Physics of the Inner Heliosheath: Voyager Observations, Theory, and Future Prospects*, AIP Conference Proceedings 258, pp. 73-78.

Decker, R. B., Krimigis, S. M., Roelof, E. C., & Hill, M. E. 2007 *Eos Trans. AGU*, 88(52), Fall Meet. Suppl., Abstract SH11A-05,.

Decker, R. B. *et al.* 2008 Shock that terminates the solar wind is mediated by non-thermal ions. *Nature*, 454, 67-70.

Dmitriev, A. V., Suvorova, A. V., Chao, J. K., & Yang, Y.-H. 2004, Dawn-dusk asymmetry of geosynchronous magnetopause crossings. *J. Geophys. Res., 109*, A05203.

Formisano, V., Hedgecock, P. C., Moreno, G., Palmiotto, F., & Chao, J. K. 1973, Solar wind interaction with the Earth's magnetic field, 2. Magnetohydrodynamic bow shock, *J. Geophys. Res. 78* 3731.

Fuselier, S. A. & Schmidt, W. K. H. 1994 *J. Geophys. Res.*, 99, 11539-11546.

Gloeckler, G., Fisk, L. A., & Lanzerotti, L. J. 2005 Acceleration of Solar Wind and Pickup Ions by Shocks. In *Solar Wind 11/SOHO 16 Programme and Abstract Book (pdf file)*, European Space Agency, 52.

Isenberg, P. A, Smith, C. W., Matthaeus, W. H., & Richardson, J. D. 2005 Turbulent heating of the distant solar wind by interstellar pickup protons with a variable solar wind speed. In *Proceedings of Solar Wind 11: Connecting Sun and Heliosphere, ESA SP-592* (B. Fleck & T. H. Zurbuchen, eds.), European Space Agency, The Netherlands, 347-350.

Krimigis, S. M., Decker, R. B., Hill, M. E., Armstrong, T. P., Gloeckler, G., Hamilton, D. C., Lanzerotti, L. J., & Roelof, E. C. 2003 Voyager 1 exited the solar wind at a distance of 85 AU from the Sun. *Nature, 426*, 45-48, 10.1038/nature02068.

Linde, T. J., Gombosi, T. I., Roe, P. L., Powell, K. G., & DeZeeuw, D. L. 1998 Heliosphere in the Magnetized Local Interstellar Medium: Results of a Three-Dimensional MHD Simulation, *J. Geophys. Res., 103*, 1889-1904.

McDonald, F. B. *et al.* 2003 Enhancements of energetic particles near the heliospheric termination shock *Nature, 426*, 48-51.

McComas, D. J. & Schwadron, N. A. 2006 An explanation of the Voyager paradox: particle acceleration at a blunt termination shock. *Geophys. Res. Lett., 33* L04102.

McComas, D. J., Ebert, R. W., Elliot, H. A., Goldstein, B. E. & Gosling, J. T. 2008. Weaker solar wind from the polar coronal holes and the whole Sun, submitted to *Geophys. Res. Lett.*

Opher, M., Stone, E. C., & Gombosi, T. I. 2007, The orientation of the local interstellar magnetic field, *Science, 316*, 875-878 DOI: 10.1126/science.1139480.

Pogorelov, N. V., Stone, E. C., Florinski, V., & Zank, G. P. 2007 Termination shock asymmetries as seen by the Voyager spacecraft: The role of the interstellar magnetic field and neutral hydrogen. *Astrophys. J.* 668, 624.

Richardson, J. D. & Smith, C. W. 2003 The radial temperature profile of the solar wind. *Geophys. Res. Lett., 30*, 1206-1209, 10.1029/2002GL016551.

Richardson, J. D., Liu, Y., Wang, C., & McComas, D. J. 2008 Determining the LIC H density from the solar wind slowdown. *Astron. Astrophys.*, in press.

Richardson, J. D., Kasper, J. C., Wang, C., Belcher, J. W., & Lazarus, A. J. 2008 Termination shock decelerates upstream solar wind but heliosheath plasma is cool, *Nature*, 454, 63-66.

Sckopke, N., Paschmann, G., Bame, S. J., Gosling, J. T., & Russell, C. T. 1983 *J. Geophys. Res.*, 88, 6121-6136.

Siscoe G., MacNeice, P. J., & Odstrcil, D. 2007, East-west asymmetry in coronal mass ejection geoeffectiveness, *Space Weather*, 5, S04002, doi:10.1029/2006SW000286.

Smith, C. W., Isenberg, P. A., Matthaeus, W. H., & Richardson, J. D. 2006 Turbulent Heating of the Solar Wind by Newborn Interstellar Pickup Protons. *Astrophys J.*, 638, 508-517.

Stone, E. C. *et al.* 2008. Voyager 2 finds an asymmetric termination shock & explores the heliosheath beyond. *Nature*, 454, 71-74.

Stone, E. C., Cummings, A. C., McDonald, F. B., Heikkila, B., Lal, N., & Webber, W. R. 2005 Voyager 1 explores the termination shock region and the heliosheath beyond. *Science, 309*, 2017-2020.

Wang, L., Lin, R. P., Larson, D. E., & Luhmann, J. G. 2008 Domination of heliosheath pressure by shock-accelerated pickup ions from observations of neutral atoms, *Nature* 454, 81-83.

Zank, G., Pauls, H., Cairns, I., & Webb, G. 1996 Interstellar pickup ions and quasi-perpendicular shocks: Implications for the termination shock and interplanetary shocks. *J. Geophys. Res. 101*, 457.

Zank, G. P. & Meuller, H.-R. 2003, The dynamical heliosphere, *J. Geophys. Res.*, 108, 1240.

Universal Heliophysical Processes
Proceedings IAU Symposium No. 257, 2008
N. Gopalswamy & D.F. Webb, eds.

© 2009 International Astronomical Union
doi:10.1017/S1743921309029901

Evolution of stellar winds
from the Sun to red giants

Takeru K. Suzuki

School of Arts and Sciences, University of Tokyo, Komaba, Meguro, Tokyo, Japan 153-8902
email: `stakeru@ea.c.u-tokyo.ac.jp`

Abstract. By performing global 1D MHD simulations, we investigate the heating and acceleration of solar and stellar winds in open magnetic field regions. Our simulation covers from photosphere to 20-60 stellar radii, and takes into account radiative cooling and thermal conduction. We do not adopt ad hoc heating function; heating is automatically calculated from the solutions of Riemann problem at the cell boundaries. In the solar wind case we impose transverse photospheric motions with velocity ~1 km/s and period between 20 seconds and 30 minutes, which generate outgoing Alfvén waves. We have found that the dissipation of Alfvén waves through compressive wave generation by decay instability is quite effective owing to the density stratification, which leads to the sufficient heating and acceleration of the coronal plasma. Next, we study the evolution of stellar winds from main sequence to red giant phases. When the stellar radius becomes ~10 times of the Sun, the steady hot corona with temperature 10^6 K, suddenly disappears. Instead, many hot and warm ($10^5 - 10^6$ K) bubbles are formed in cool ($T < 2 \times 10^4$ K) chromospheric winds because of the thermal instability of the radiative cooling function; the red giant wind is not a steady stream but structured outflow.

Keywords. waves, MHD, Sun: photosphere, Sun: chromosphere, Sun: Corona, solar wind, stars: chromospheres, stars: coronae, stars: magnetic fields, stars: mass loss, stars: late-type

1. Introduction

Research on heliophysical processes covers broad fields from basic plasma physics to various applications such as space weather forecasts, which are widely discussed in this conference. One direction of heliophysics research is to understand the evolution of the solar system, or more generally, star(s)-planets systems.

For example, UV/X-ray radiations and winds from young stars affect the evolution of protoplanetary disks and the formation of planets. Recent observations have shown that younger stars are more active than main sequence stars, giving considerably larger X-ray and UV flux luminosity (Güdel 2004) and higher wind mass flux (Wood *et al.* 2005). We can also infer the properties of the early sun from observations of young solar-like stars (Güdel 2008). Current planet formation models are constructed by incorporating these observational facts.

Asterospheres of stars on red giant branch or later stages are also expected to play a significant role in the evolution of planets. The radius of a star expands after the end of main sequence phase. Close planets would be eventually engulfed. Before the engulfment, properties of atmosphere and wind change through the evolution; as inferred from observations of red giants, the wind mass flux increases considerably (Judge & Stencel 1991), which could affect the evolution of planets. In this talk, we mainly introduce our own work on the evolution of stellar winds from the Sun to red giant.

In this proceeding, we firstly summarize our self-consistent MHD simulations of solar winds from the photosphere to sufficiently outer region (Suzuki & Inutsuka 2005; 2006; hereafter AI05 & SI06). We introduce a scaling of the solar wind speeds near 1AU based on our simulations (Suzuki 2006). Then, we discuss the evolution of stellar winds from the Sun to red giant stars (Suzuki 2007).

2. Solar wind

2.1. *Simulation set-up*

We consider 1D open flux tubes which are super-radially open, measured by heliocentric distance, r. The simulation regions are from the photosphere ($r = 1R_\odot$) with density, $\rho = 10^{-7}$g cm^{-3} to $65R_\odot$ (0.3AU). Radial field strength, B_r, is given by conservation of magnetic flux as $B_r r^2 f(r) = $ const., where $f(r)$ is a super-radial expansion factor. For the fast solar wind case, we adopt $B_{r,0} = 161$ G at the photosphere and the total expansion factor $= 75$ (see SI05 and SI06 for detail).

We input the transverse fluctuations of the field line by the granulations at the photosphere, which excite Alfvén waves. In this paper we only show results of linearly polarized perturbations with power spectrum proportional to $1/\nu$, where ν is frequency (for circularly polarized fluctuations with different spectra, see SI06). Amplitude, $\langle dv_{\perp,0} \rangle$, at the photosphere is chosen to be compatible with the observed photospheric velocity amplitude $= 0.7$km s^{-1} (Holweger *et al.* 1978). At the outer boundaries, outgoing condition is imposed for all the MHD waves, which enables us to carry out simulations for a long time until quasi-steady state solutions are obtained without unphysical wave reflection.

We dynamically treat the propagation and dissipation of the waves and the heating and acceleration of the plasma by solving ideal MHD equations. In the energy equation we take into account radiative cooling and Spitzer thermal conduction (SI06). We adopt the second-order MHD-Godunov-MOCCT scheme (Sano & Inutsuka 2008 in preparation) to update the physical quantities. We initially set static atmosphere with a temperature $T = 10^4$K to see whether the atmosphere is heated up to coronal temperature and accelerated to accomplish the transonic flow. At $t = 0$ we start the inject of the transverse fluctuations from the photosphere and continue the simulations until the quasi-steady states are achieved.

2.2. *Results of simulation*

Figure 1 plots the initial condition (dashed lines) and the results after the quasi-steady state condition is achieved at $t = 2573$ minutes (solid lines), compared with recent observations of fast solar winds. Figure 1 shows that the initially cool and static atmosphere is effectively heated and accelerated by the dissipation of the Alfvén waves. The sharp transition region which divides the cool chromosphere with $T \sim 10^4$K and the hot corona with $T \sim 10^6$K is formed owing to a thermally unstable region around $T \sim 10^5$K in the radiative cooling function (Landini & Monsignori-Fossi 1990). The hot corona expands outward as the transonic solar wind. The simulation naturally explains the observed trend quite well.

The heating and acceleration of the solar wind plasma in inner heliosphere is done by the dissipation of Alfvén waves. Here we inspect waves in more detail. Figure 2 presents contours of amplitude of v_r, ρ, $v_\perp$, and $B_\perp/B_r$ in $R_\odot \leqslant r \leqslant 15R_\odot$ from $t = 2570$ min. to 2600 min. Blue (gray) shaded regions denote positive (negative) amplitude. Above the panels, we indicate the directions of the local 5 characteristics, two Alfvén, two slow, and one entropy waves at the respective positions. Note that the fast MHD and Alfvén modes degenerate in our case (wave vector and underlying magnetic field are in the same direction), so we simply call the innermost and outermost waves Alfvén modes. In our simple 1D geometry, v_r and ρ trace the slow modes which have longitudinal character, while $v_\perp$ and $B_\perp$ trace the Alfvén modes which are transverse.

Alfvén waves areseen in in $v_\perp$ and $B_\perp/B_r$ diagrams, which have the same slopes with the Alfvén characteristics shown above. One can also find the incoming waves propagating from lower-right to upper-left as well as the outgoing modes generated from the surface. These incoming waves are generated by the reflection at the 'density mirrors' of the slow modes. At intersection points of the outgoing and incoming characteristics the non-linear wave-wave interactions take place, which play a role in the wave dissipation.

The slow modes appears in v_r and ρ diagrams. After inspecting the diagrams we find that the most of the patterns are due to the outgoing slow modes† which are generated from the

---

† The phase correlation of the longitudinal slow waves is opposite to that of the transverse

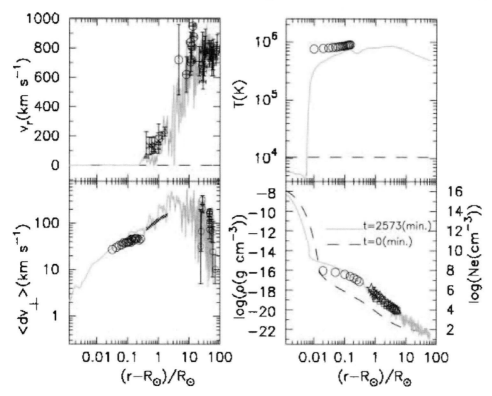

Figure 1. Results of fast solar wind mode with observations in polar regions. Outflow speed, v_r(km s^{-1}) (top-left), temperature, T(K) (top-right), density in logarithmic scale, $\log(\rho(\text{g cm}^{-3}))$ (bottom-right), and rms transverse amplitude, $\langle dv_\perp\rangle$(km s^{-1}) (bottom-left) are plotted. Observational data in the third panel are electron density, $\log(N_e(\text{cm}^{-3}))$ which is to be referred to the right axis. Dashed lines indicate the initial conditions and solid lines are the results at $t = 2573$ minutes. In the bottom panel, the initial value ($\langle dv_\perp\rangle = 0$) dose not appear. The observational data in the inner region ($< 6R_\odot$) are from SOHO (Teriaca *et al.* 2003; Zangrilli *et al.* 2002; Fludra *et al.* 1999; Wilhelm *et al.* 1998; Lamy *et al.* 1997; Banergee *et al.* 1998; Esser *et al.* 1999) and those in the outer region are from interplanetary scintillation measurements (Grall *et al.* 1996; Habbal *et al.* 1994; Kojima *et al.* 2004; Canals *et al.* 2002).

perturbations of the Alfvén wave pressure, $B_\perp^2/8\pi$ (Kudoh & Shibata 1998& Tsurutani *et al.* 2002). These slow waves steepen eventually result in the shock dissipation.

Figure 3 presents the dissipation of the waves more quantitatively. We plot so-called wave actions,

$$S_c = \rho \delta v^2 \frac{(v_r + v_{\text{ph}})^2}{v_{\text{ph}}} \frac{r^2 f(r)}{r_c^2 f(r_c)}, \tag{2.1}$$

of outgoing Alfvén , incoming Alfvén , and outgoing slow MHD (sound) waves, where δv and v_{ph} are amplitude and phase speed of each wave mode. S_c is an adiabatic constant derived from wave action (Jacques 1977) in unit of energy flux. For the incoming Alfvén wave, we plot the opposite sign of S_c so that it becomes positive in the sub-Alfvénic region. The outgoing and incoming Alfvén waves are decomposed by correlation between $v_\perp$ and $B_\perp$. Extraction of the slow wave is also from fluctuating components of v_r and ρ.

Figure 3 exhibits that the outgoing Alfvén waves dissipate quite effectively; S_c becomes only $\sim 10^{-3}$ of the initial value at the outer boundary. A sizable amount is reflected back downward

Alfvén waves. The outgoing slow modes have the positive correlation between amplitudes of v_r and ρ, ($\delta v_r \delta\rho > 0$), while the incoming modes have the negative correlation ($\delta v_r \delta\rho < 0$).

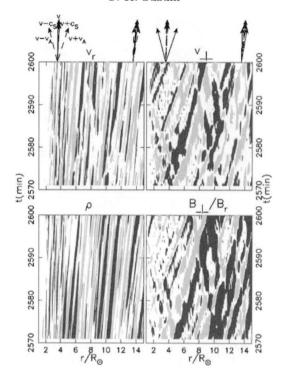

Figure 2. $r-t$ diagrams for v_r (upper-left), ρ (lower-left), $v_\perp$ (upper-right), and $B_\perp/B_r$ (lower-right.) The horizontal axises cover from $R_\odot$ to $15R_\odot$, and the vertical axises cover from $t = 2570$ minutes to 2600 minutes. Dark(blue) and light shaded regions indicate positive and negative amplitudes which exceed certain thresholds. The thresholds are $dv_r = \pm96$km/s for v_r, $d\rho/\rho = \pm0.25$ for ρ, $v_\perp = \pm180$km/s for $v_\perp$, and $B_\perp/B_r = \pm0.16$ for $B_\perp/B_r$, where $d\rho$ and dv_r are differences from the averaged ρ and v_r. Arrows on the top panels indicate characteristics of Alfvén , slow MHD and entropy waves at the respective locations.

below the coronal base ($r - R_S < 0.01R_S$), which is known from the incoming Alfvén wave following the outgoing component with slightly smaller level. This is because the wave shape is considerably deformed owing to the steep density gradient; a typical variation scale ($< 10^5$km) of the Alfvén speed becomes comparable to or even shorter than the wavelength ($= 10^4 - 10^6$km). Although the energy flux, $\simeq 5 \times 10^5$erg cm^{-2}s^{-1}, of the outgoing Alfvén waves (S_c in the static region is equivalent with the energy flux) which penetrates into the corona is only $\simeq 15\%$ of the input value, it is sufficient for the energy budget in the coronal holes (Withbroe & Noyes 1977).

The processes discussed here are the combination of the direct mode conversion to the compressive waves and the parametric decay instability due to three-wave (outgoing Alfvén , incoming Alfvén , and outgoing slow waves) interactions (Goldstein 1978; Terasawa *et al.* 1986) of the Alfvén waves. These processes, which are not generally efficient in homogeneous background, become effective by amplification of velocity amplitude in the density decreasing atmosphere. The Alfvén speed also changes even within one wavelength of Alfvén waves with periods of minutes. This leads to both variation of the wave pressure in one wavelength and partial reflection through the deformation of the wave shape (Moore *et al.* 1991). The density stratification plays a key role in the propagation and dissipation of the Alfvén waves.

2.3. *Solar wind speed*

It is widely believed that properties of open flux tubes are important parameters that control the solar wind speed. Wang & Sheeley (1990; 1991) showed that the solar wind speed at $\sim$1AU is anti-correlated with a super-radial expansion factor, $f_{\rm tot}$, from their long-term observations as well as by a simple theoretical model. Ofman & Davila (1998) showed this tendency by

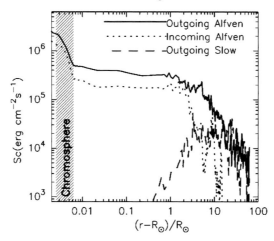

Figure 3. S_c of outgoing Alfvén mode (solid), incoming Alfvén mode (dotted), and outgoing MHD slow mode (dashed) at $t = 2573$ mins. Hatched region indicates the chromosphere and low transition region with $T < 4 \times 10^4$ K.

time-dependent simulations as well. Fisk *et al.* (1999) claimed that the wind speed should be positively correlated with photospheric field strength, $B_{r,0}$, by a simple energetics consideration.

Kojima *et al.* (2005) have found that the solar wind velocity is better correlated with the combination of these two parameters, $B_{r,0}/f_{tot}$, than $1/f_{tot}$ or $B_{r,0}$ from the comparison of the outflow speeds obtained by their interplanetary scintillation measurements with the corresponding flux tubes (Figure 4). Suzuki (2004; 2006) and SI06 also pointed out that $B_{r,0}/f_{tot}$ should be the best control parameter provided that the Alfvén waves play a dominate role in the coronal heating and the solar wind acceleration. This is because the nonlinearity of the Alfvén waves, $\langle \delta v_{A,+} \rangle / v_A$ is controlled by $v_A \propto B_r \propto B_{r,0}/f_{tot}$ in the outer region where the flux tube is already super-radially open. Wave energy does not effectively dissipate in the larger $B_{r,0}/f_{tot}$ case in the subsonic region because of relatively small nonlinearity and more energy remains in the supersonic region. In general, energy and momentum inputs in the supersonic region gives higher wind speed, while those in the subsonic region raises the mass flux (ρv_r) of the wind by an increase of the density (Lamers & Cassinelli 1999). This indicates that the solar wind speed is positively correlated with $B_{r,0}/f_{tot}$.

Suzuki (2006) further derived a relation between the solar wind speed, v_{1AU}, at 1AU and surface properties from a simple energetics argument:

$$
\begin{aligned}
v_{1AU} &= \left[2 \times \left(-\frac{R_\odot^2}{4\pi (\rho v r^2)_{1AU}} \frac{B_{r,\odot}}{f_{tot}} \langle \delta B_\perp \delta v_\perp \rangle_\odot + \frac{\gamma}{\gamma - 1} R T_C - \frac{G M_\odot}{R_\odot} \right) \right]^{1/2} \\
&= 300 (\text{km/s}) \left[5.9 \left(\frac{-\langle \delta B_\perp \delta v_\perp \rangle_\odot}{8.3 \times 10^5 \, (\text{cm s}^{-1}\text{G})} \right) \left(\frac{B_{r,\odot}(\text{G})}{f_{tot}} \right) + 3.4 \left(\frac{\gamma}{1.1} \right) \left(\frac{0.1}{\gamma - 1} \right) \left(\frac{T_C}{10^6 \,(\text{K})} \right) - 4.2 \right]^{1/2},
\end{aligned}
\tag{2.2}
$$

where we have the three free parameters, surface amplitude, $\langle \delta B_\perp \delta v_\perp \rangle_\odot$, *effective* coronal temperature, T_C, and ratio of specific heats, γ. In the equation, these parameters are evaluated by the standard values, which should be used for actual prediction of wind speed. The first term is the contribution from Alfvén waves, the second term represents net heating minus cooling due to radiation loss and thermal conduction, and the third term is the gravitational loss. The first term, due to Alfvén waves, exhibits the dependence on $B_{r,0}/f_{tot}$, which reflects the Alfvén waves in expanding flux tubes. In previous works this term is neglected (Fisk *et al.* 1999) or more simplified (Schwadron & McComas 2003) because this is a nonlinear term. However, on account of the dissipationless character of Alfvén waves it plays a role in the solar wind in spite of the nonlinear term. Figure 4 shows the prediction of Equation (2.2), in comparison with the observation using interplanetary scintillation measurements by Kojima *et al.* (2005). The observed data are nicely explained by the relation based on the simple energetics.

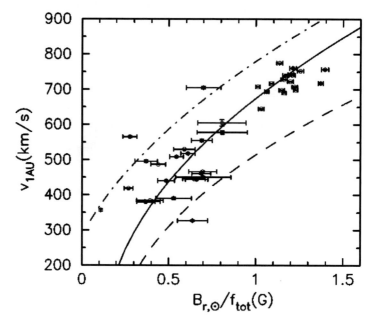

Figure 4. Relations between $v_{1\mathrm{AU}}$ and $B_{r,\odot}/f_{\mathrm{tot}}$. Lines are theoretical prediction from Equation (2.2). Solid line indicates the fiducial case ($(\langle \delta B_\perp \delta v_\perp \rangle = 8.3 \times 10^5 \mathrm{G}$ cm s^{-1} and $T_{\mathrm{C}} = 10^6$K). Dot-dashed line adopt higher coronal temperature ($T_{\mathrm{C}} = 1.5 \times 10^6$K) with the fiducial $\langle \delta B_\perp \delta v_\perp \rangle$. Dashed line adopt smaller $\langle \delta B_\perp \delta v_\perp \rangle$ ($= 5.3 \times 10^5 \mathrm{G}$ cm s^{-1}) with the fiducial temperature. Observed data are from Kojima et al. (2005). Coronal magnetic fields are extrapolated from $B_{r,\odot}$ by the potential field-source surface method Hakamada & Kojima (1999). f_{tot} is derived from comparison between the areas of open coronal holes at the photosphere and at the source surface ($r = 2.5 R_\odot$). $v_{1\mathrm{AU}}$ is obtained by interplanetary scintillation measurements. $v_{1\mathrm{AU}}$, $B_{r,\odot}$, and f_{tot} are averaged over the area of each coronal hole and the data points correspond to individual coronal holes.

3. Evolution to red giants

We investigate the evolution of stellar winds with stellar evolution to red giants. Red giant stars, similarly to the Sun, generally possess surface convective layers. Also, red giants are slow rotators as a result of angular momentum loss (magnetic braking) through stellar evolution. Then, our solar wind simulation is directly applicable to red giant winds. We would also like to note that our simulation can be applied to winds from young stars, such as proto-stars and T Tauri stars, because they are expected to be driven mainly by surface turbulence as well, whereas the effects of accretion (Cranmer 2008) and rotation might have to be taken into account in these objects.

We consider winds from $1M_\odot$ and $3M_\odot$ stars in various evolutionary stages from the main sequence to the red giant branch. The properties of surface fluctuations (e.g. amplitude and spectrum) can be estimated from conditions of surface convection which is negatively correlated with surface gravity and positively correlated with temperature (e.g. Renzini et al. 1977; Stein et al. 2004). Because there are very few observations of the magnetic fields, we simply use the same surface strength $B_{r,0} = 240$ G and super-radial expansion factor, $f_{\mathrm{tot}} = 240$, for all the cases. Then, we carry out the simulations of the red giant winds in a similar manner to the solar wind simulations.

Figure 5 presents the evolution of stellar winds of a $1M_\odot$ star from main sequence to red giant stages. The middle panel shows that the average temperature drops suddenly from $T \simeq 7 \times 10^5$K in the sub-giant star (dash-dotted) to $T \leqslant 10^5$K in the red giant stars, which is consistent with the observed "dividing line"(Linsky & Haisch 1979). The main reason of the disappearance of the steady hot coronae is that the sound speed (≈ 150 km s^{-1}) of $\approx 10^6$ K plasma exceeds the escape speed, $v_{\mathrm{esc}}(r) = \sqrt{2GM_\star/r}$, at $r \gtrsim$ a few $R_\star$ in the red giant stars; the hot corona cannot be confined by the gravity any more in the atmospheres of the red giant stars. Therefore, the

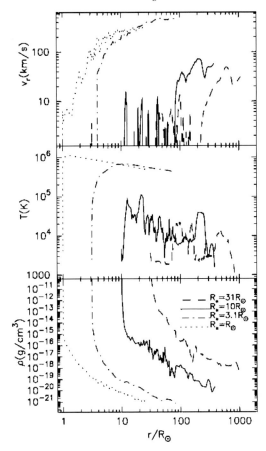

Figure 5. Time-averaged stellar wind structure of the $1M_\odot$ stars. From the top to the bottom, radial outflow velocity, v_r (km s^{-1}), temperature, T (K), and density, ρ(g cm^{-3}), are plotted. The dotted, dash-dotted, solid, and dashed lines are the results of stellar radii, $R = R_\odot$ (the present Sun), $3.1R_\odot$ (sub-giant), $10R_\odot$ (red giant), and $31R_\odot$ (red giant), respectively.

material flows out before heated up to coronal temperature. In addition, the thermal instability of the radiative cooling function (Landini & Monsignori-Fossi 1990) plays a role in the sudden decrease of temperature; since the gas with $T = 10^5 - \lesssim 10^6$ K is unstable, the temperature quickly decreases from the subgiant to red giants.

The densities of the winds increase with stellar evolution due to the decrease of the surface gravity. On the other hand, the velocities of the winds decrease because slower winds can escape from the stellar gravity in low-gravity evolved stars. The mass loss rate increases due to the increase of the stellar surface ($\propto R^2$) and the increase of the density. From the sun to the star with $R_\star = 31R_\odot$, the mass loss rate increases 10^5–10^6 times. The physical properties at 1AU greatly change for the red giant Sun: for the $R_\star = 31R_\odot$ star the density is 10^8 (1/cm^3), which is $\sim 10^7$ times larger than the current value and the wind velocity is very slow ~ 10 km/s and still in the acceleration region. The wind temperature mostly $\lesssim 10^4$ K in chromospheric winds, but hot bubbles with $\gtrsim 10^5$–10^6 K sometimes pass through, which we discuss below.

The thermal instability also results in structured and time-dependent stellar winds from red giant stars. Figure 6 shows the snap-shot wind structure of a 3 $M_\odot$ red giant star at $t = 6909$(hr) (solid) in comparison with the time-averaged structure (dashed). Figure 6 illustrates that the simple picture of layered atmosphere, photosphere – chromosphere – transition region – corona – wind from the stellar surface, does not hold in red giant stars. A characteristic feature is that a number of hot bubbles with low densities are distributed in cool background material. For example, in an inner region ($r - R_\star < R_\star$) the two bubbles with the peak temperatures $> 10^6$ K are formed; in $R_\star < r - R_\star \lesssim 10\,R_\star$ a couple of warm bubbles with $T \gtrsim 10^5$ K are formed.

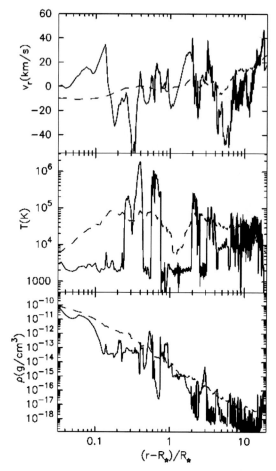

Figure 6. Snap-shot wind structure (solid) of a $3 M_\odot$ red giant star with surface gravity $\log g = 1.4$ at $t = 6909 (\mathrm{hr})$, compared with the time-averaged structure (dashed). From the top to the bottom, v_r (km s^{-1}), T (K), and ρ (g cm^{-3}) are plotted on $(r - R_*)/R_*$.

The hot and warm bubbles and the cool background materials are connected by the transition regions, at which the temperature drastically changes because of the thermal instability. These hot bubbles are also seen in 1 $M_\odot$ red giant stars.

   The densities of the hot bubbles are lower than the ambient media to satisfy the pressure balance. As seen in the bottom panel of Figure 6 the density fluctuates typically by 2-3 orders of magnitude in the wind, which is related to the fact that the gas mainly consists of hot ($> 10^6$ K) and cool ($< 10^4$ K) components†. The outflow speed fluctuates as well to fulfill the mass conservation relation. The red giant wind is not a steady outward stream but an outflow consisting of many small-scale structures. Our simulations show that both hot plasma and cool chromospheric wind coexist in red giant stars, which is consistent with reported hybrid activities (Hartmann et al. 1980, Harper et al. 1995, Ayres et al. 1998).

4. Summary & discussions

   We have studied the evolution of stellar wind from the Sun to red giants by performing 1D MHD numerical simulations. The Alfvén waves are generated by the footpoint fluctuations of the magnetic field lines. We have treated the wave propagation and dissipation, and the heating

   † Strictly speaking, however, magnetic pressure also needs to be taken into account for the force balance (Suzuki 2007).

and acceleration of the plasma in a self-consistent manner. Our simulation is the first simulation which treats the wind from the photosphere to the (inner) heliosphere with the relevant physical processes.

We have shown that the dissipation of the Alfvén waves through the generation of the compressive waves (decay instability) and shocks (nonlinear steepening) is one of the solutions for the heating and acceleration of the plasma in the coronal holes. However, we should cautiously examine the validity of the 1-D MHD approximation we adopt. There are other dissipation mechanisms due to the multidimensionality (e.g. Ofman 2004), such as turbulent cascade into the transverse direction (Goldreich & Sridhar 1995; Oughton *et al.* 2001) and phase mixing (Heyvaerts & Priest 1983). If Alfvén waves cascade to higher frequency, kinetic effects (e.g. Axford & McKenzie 1997; Nariyuki & Hada 2006) becomes important.

We have also extended the solar wind simulations to red giant winds. With stellar evolution, the steady hot corona with temperature, $T \approx 10^6$ K, suddenly disappears because the surface gravity becomes small; hot plasma cannot be confined by the gravity. Thermal instability also generate intermittent magnetized hot bubbles in cool chromospheric winds. When the solar radius expands $\sim$30 times, the earth orbit is still in the wind acceleration region and the wind density is 10^7 times larger than the present value there. In the red giant simulations, we assume the same photospheric field strength and super-radial expansion factor because very little information is known from observations at the moment. It is important to study the evolution of magnetic fields with stellar evolution from both observation and modelling.

Acknowledgements

The author thanks the organizers of IAU257 for the nice conference. This work is supported in part by Inamori Foundation and a Grant-in-Aid for Scientific Research (19015004 & 20740100) from the Ministry of Education, Culture, Sports, Science, and Technology of Japan.

References

Axford, W. I. & McKenzie, J. F. 1997 The Solar Wind in "Cosmic Winds and the Heliosphere", Eds. Jokipii, J. R., Sonnet, C. P., and Giampapa, M. S., University of Arizona Press, 31

Ayres, T. R., Simon, T., Stern, R. A., Drake, S. A., Wood, B. E., & Brown, A. 1998 ApJ, 496, 428

Banerjee, D., Teriaca, L., Doyle, J. G., & Wilhelm, K. 1998 A&A, 339, 208

Canals, A., Breen, A. R., Ofman, L., Moran, P. J., & Fallows, R. A., 2002 Ann. Geophys., 20, 1265

Cranmer, S. R. 2008, ApJ in press (arxiv0808.2250)

Esser, R., Fineschi, S., Dobrzycka, D., Habbal, S. R., Edgar, R. J., Raymond, J. C., & Kohl, J. L., 1999 ApJ, 510, L63

Fisk, L. A., Schwadron, N. A., & Zurbuchen, T. H. J. Geophys. Res., 104, A4, 19765

Fludra, A., Del Zanna, G., & Bromage, B. J. I., 1999 Spa. Sci. Rev., 87, 185

Goldreich, P. & Sridhar, S., 1995 ApJ, 438, 763

Goldstein, M. L., 1978, ApJ, 219, 700

Grall, R. R., Coles, W. A., Klinglesmith, M. T., Breen, A. R., Williams, P. J. S., Markkanen, J., & Esser, R., 1996 Nature, 379, 429

Güdel, M. 2004, ARA&A, 12, 71

Güdel, M. 2007, Living Rev. in Sol. Phys., 4, 3

Habbal, S. R., Esser, R., Guhathakura, M., & Fisher, R. R., 1994 Gephys. Res. Lett., 22, 1465

Hakamada, K. & Kojima, M. 1999, Sol. Phys., 187, 115

Harper, G. M., Wood, B. E., Linsky, J. L., Bennett, P. D., Ayres, T. R., & Brown, A. 1995, ApJ, 452, 407

Hartmann, L., Dupree, A. K., & Raymond, J. C. 1980, ApJ, 236, L143

Heyvaerts, J. & Priest, E. R., 1983 A&A, 117, 220

Holweger, H., Gehlsen, M., & Ruland, F., 1978 A&A, 70, 537

Jacques, S. A. 1977, ApJ, 215, 942

Judge, P. G. & Stencel, R. E. 1991, ApJ, 371, 357

Kojima, M., Breen, A. R., Fujiki, K., Hayashi, K., Ohmi, T., & Tokumaru, M., 2004 J. Geophys. Res., 109, A04103

Kudoh, T. & Shibata, K., 1999 ApJ, 514, 493

Lamers, H. J. G. L. M. & Cassinelli, J. P. (1999), 'Introduction to Stellar Wind', Cambridge

Lamy, P., Quemerais, E., Liebaria, A., Bout, M., Howard, R., Schwenn, R., & Simnett, G., 1997 Fifth SOHO Worshop, The Corona and Solar Wind near Minimum Activity, ed A. Wilson (ESA-SP 404; Noordwijk:ESA), 491

Landini, M. & Monsignori-Fossi, B. C., 1990 A&AS, 82, 229

Linsky, J. L. & Haisch, B. M., 1979 ApJ, 229. L27

Moore, R. L., Suess, S. T., Musielak, Z. E., & An, A.-H., 1991 ApJ, 378, 347

Nariyuki, Y. & Hada, T., 2006 Phys. Plasma, 13, 124501

Ofman, L. 2004, J. Geophys. Res., 109, A07102

Ofman, L. & Davila, J. M. 1998, J. Geophys. Res., 103, 23677

Oughton, S., Matthaeus, W. H., Dmitruk, P., Milano, L. J., Zank, G. P., & Mullan, D. J., 2001 ApJ 551, 565

Renzini, A., Cacciari, C., Ulmschneider, P., & Schmitz, F., 1977 A&A, 61, 39

Schwadron, N. A. & McComas, D. J. 2003 ApJ, 599, 1395

Stein, R. F., Georgobiani, D., Trampedach, R., Ludwig, H.-G., & Nortlund, Å., 2004, Sol. Phys., 220, 229

Suzuki, T. K., 2004 MNRAS, 349, 1227

Suzuki, T. K. 2006, ApJ 640, L75

Suzuki, T. K., 2007 ApJ, 659, 1592

Suzuki, T. K. & Inutsuka, S., 2005 ApJ, 632, L49

Suzuki, T. K. & Inutsuka, S., 2006 J. Geophys. Res., 111, A6, A06101

Terasawa, T., Hoshino, M., Sakai, J. I., & Hada, T., 1986, J. Geophys. Res., 91, 4171

Teriaca, L., Poletto, G., Romoli, M., & Biesecker, D. A., 2003 ApJ, 588, 566

Tsurutani, B. T. et al., 2002 Geophys. Res. Lett., 29, 23-1

Wang, Y.-M. & Sheeley, Jr, N. R.: Solar wind speed and coronal flux-tube expansion, Astrophys. J., 355, 726–732, 1990

Wang, Y.-M. & Sheeley, Jr, N. R.: Why fast solar wind originates from slowly expanding coronal flux tubes, Astrophys. J. Lett., 372, L45 - L48, 1991

Wilhelm, K., Marsch, E., Dwivedi, B. N., Hassler, D. M., Lemaire, P., Gabriel, A. H., & Huber, M. C. E., 1998 ApJ, 500, 1023

Withbroe, G. L. & Noyes, R. W., 1977, ARAA, 15, 363

Wood, B. E., Müller, H.-R., Zank, G. P., Linsky, J. L., & Redfield, S. 2005, ApJ, 628, L143

Zangrilli, L., Poletto, G., Nicolosi, P., Noci, G., & Romoli, M., 2002 ApJ, 574, 477

Discussion

SHIBATA: Very nice work! I have two questions: 1) Sometimes very low density solar wind is observed. Can you explain such low density solar wind? 2) As for red giant wind, how did you assume magnetic field distribution and strength?

SUZUKI: 1) Non linear effects should be employed (Suzuki and Inutsuku 2006), but because it is complicated, I will discuss this issue with you later 2) I'm using the same B strength (240 G) in the photosphere and same super-radial expansion factor ($\sim$240 G). This is just an assumption.

ARGE: Please elaborate on your comment regarding that Alfven wave term not properly considered in previous work.

SUZUKI: The new point of my work is that I include the Alfven wave term. This is a nonlinear term, but important in solar wind region because Alfven waves propagate through a long distance.

BOCHSLER: How does stellar rotation influence winds of red giants?

SUZUKI: The stellar rotation of red giants is negligible in terms of wind acceleration because red giants are slow rotators. But if we consider younger stars, the rotation may be important.

JARDINE: Just a comment. While I agree that centrifugal effects may not be important in slow rotators, the effect of rotation rate on the strength of the magnetic field that is generated is a more serious concern.

SUZUKI: I agree with the comment. Thank you.

OFMAN: 1. Have you considered the effect of ion-neutral dissipation on Alfven wave driven wind in red giants? 2) How can you justify the large expansion factor you used in Red Giants?

SUZUKI: 1) The collision frequency between ions and neutrals is shorter than the frequency of the Alfven waves I'm considering. So we can treat as 1-fluid for the ion frequency Alfven waves. But for high frequency waves, that effect is important. 2) In my opinion the large expansion is a consequence of the pressure balance between magnetic pressure and gas pressure. It is our working hypothesis.

Universal Heliophysical Processes
Proceedings IAU Symposium No. 257, 2008
N. Gopalswamy & D.F. Webb, eds.
© 2009 International Astronomical Union
doi:10.1017/S1743921309029913

3D evolution of solar magnetic fields and high-speed solar wind streams near the minimum of solar cycle 23

O. S. Yakovchouk[1], I. S. Veselovsky[1,2], K. Mursula[3], Yu. S. Shugai[1]

[1]Skobeltsyn Institute of Nuclear Physics, Moscow State University,
Moscow 119992, Russia
email: olesya@dec1.sinp.msu.ru

[2]Space Research Institute (IKI),
Russian Academy of Sciences, Moscow 117997, Russia
[3]Department of Physical Sciences, University of Oulu,
Oulu, 90014 Finland

Abstract. The numerical method developed by Veselovsky & Ivanov (2006), together with magnetograms of the Sun obtained at the photospheric level were used to calculate the coronal magnetic field with open, closed and intermittent topology during March-December 2007. The results of the modelling are compared with stereoscopic images and movies of the corona observed by EUV telescopes onboard STEREO and SOHO spacecraft. The sources of the permanent and transient high speed solar wind streams as well as the sector structure and the heliospheric plasma sheet observed at the Earth's orbit by the ACE and STEREO spacecraft are discussed.

Keywords. Magnetic field, Solar wind, Solar cycle, Coronal holes.

1. Introduction

The goal of this paper is to present and discuss new observational data during the minimum of solar cycle 23. Available space born telescopic and coronagraph observations onboard the twin STEREO, SOHO and ACE spacecraft helped to find the origins of the high-speed solar wind streams and sector structure of the magnetic field and investigate their dynamics during March–December 2007. In our study we use magnetic field data from Wilcox Solar Observatory and SOLIS for visualization of the three-dimensional structure of the magnetic field within the solar corona. The problem of visualization of three-dimensional magnetic-line patterns in the solar corona is an important part of the more general problem of development of graphical tools for adequate visualization of the magnetic field and more complete information about the state of the near-Sun environment. It is difficult to measure the coronal magnetic field. The lack of such information is usually compensated to some extent by computations. There are many models that can be applied to calculate the magnetic field vector in the solar corona and near-Sun environment from photospheric observations. Models based on potential and force-free approximations are most widely used. There also exist more complete and complicated self-consistent numerical models that describe the joint dynamics of the field and plasma on the basis of magnetohydrodynamic equations. The modern state of the investigations in this field and the difficulties arising during computations are partially reviewed by Aschwanden (2004), Obridko *et al.* (2006), Veselovsky (2002).

2. A method for visualization of the solar magnetic field

In the present study, we use the coefficients of the spherical harmonic expansion of the solar magnetic field in the potential approximation, which were calculated by G.V. Rudenko from magnetic field data observed at the photospheric level. The method used for extrapolation is described by Rudenko (2001) and the results of the current computations of the harmonic coefficients for certain dates can be found at http://bdm.iszf.irk.ru/. We developed a program that can visualize the field lines derived from these coefficients. This is a convenient tool for visualization of the three-dimensional structure of the magnetic field within the solar corona from its lowest layers to the surface of the solar-wind source. The projection of the field lines onto the skyplane gives important information, which is difficult or even impossible to obtain from the original data in another way. The specific feature of our visualization method is that the magnetic flux density is directly introduced into the computations as an important parameter. The spatial density of the field lines is proportional to the field strength. Thus, in addition to the global pattern of open and closed field lines in the corona, we obtain information about the magnitude and direction of the magnetic field. In Figure 1 closed field lines incoming and outgoing from the photospheric level, are shown in yellow. Open field lines outgoing from the solar surface, that is, lines for which the vector of the magnetic field strength is directed away from the Sun at R = 1, are given in red. Finally, the blue lines correspond to the incoming open field lines. Visualization of the coronal magnetic field works well during the minimum of solar activity in the absence of sporadic perturbations. The results obtained with this method of visualization agree with the observations. The closed magnetic field lines are associated with active regions, the open field lines with coronal holes and high-speed solar wind streams observed onboard ACE.

3. High-speed solar wind streams and magnetic sector observations

High-speed solar wind streams (HSSWSs) and sector structure of the magnetic field observed simultaneously onboard the STEREO, SOHO and ACE spacecraft were used for better understanding of the three-dimensional and time dependent properties of the heliospheric plasma and magnetic field parameters at the Earth's orbit around the minimum of solar activity. We have considered 11 Carrington Rotations (March–December, 2007) during the minimum of solar cycle 23. Sources of HSSWSs with velocities of about 600-700 km/s were the coronal holes which had variable angular spans, geometry and time histories (Veselovsky & Shugay 2008). The four-sector structure of the

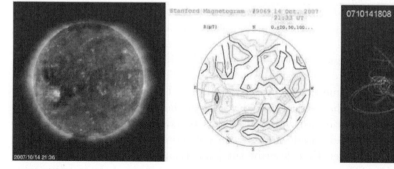

Figure 1. Left panel: the Sun observed by the SOHO/EIT in the Fe XII bandpass (195 Å) on 14 October 2007. Middle panel: Wilcox solar magnetogram for the same day. Right panel: The magnetic field lines in the solar corona according to our calculations.

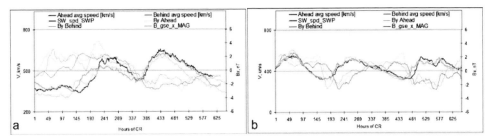

Figure 2. The average value of the interplanetary magnetic field and HSSWSs from three-point observations, ABE (Stereo-A, Stereo-B, ACE) a) The period from March to August 2007 (CR 2054-2058) b) The period from September 2007 to February 2008 (CR2060-CR2064).

interplanetary magnetic field and three HSSWSs were observed during the period from March to August 2007 (CR 2054-2058). Three (or four) coronal holes with different polarity (-+ - +) were the sources of HSSWSs (Fig. 2a). In July-August, 2007 (CR2059) a gradual change from four-sector to two-sector structures of the interplanetary magnetic field took place (−+ +). The two-sector structure of the interplanetary magnetic field (-+) was observed from September 2007 to December 2007 (CR2060-CR2064) (Fig. 2b).Two coronal holes (CH) were the sources of HSSWSs. One of them was a recurrent negative polarity equatorial CH with an active region near/inside it (complex AR+CH) and the other was a recurrent positive polarity middle latitude CH located in the south hemisphere.

4. Summary

We have considered 11 Carrington Rotations (March-December, 2007) during the minimum of solar cycle 23. The method of visualization of the coronal magnetic field works well in a minimum of solar activity in the absence of sporadic perturbations. The global space-time pattern of HSSWSs (600-700 km/s) and their sources can be only partially reconstructed from three-point observations using Stereo-A, Stereo-B, SOHO/EIT and ACE spacecraft. Inclination of the heliospheric current sheet, its shape as well as its position, and the shape and the size of coronal holes determine this pattern in the heliosphere. The geometry of streams appreciably changes from one rotation to another, but preserves the overall zonal and sector structure manifested in the solar wind and the interplanetary magnetic field at the Earth 's orbit.

This study was supported by the RFBR grants 07-02-00147. The authors thank SOLIS, STEREO, SOHO/EIT, WSO and ACE instrument for making the data available. We are grateful to Dr. G.V. Rudenko for permission to use his data.

References

Veselovsky, I. S. & Ivanov, A. V. 2006, *Solar System Research*, 40, 5, 432

Aschwanden, M. J. 2004, *Physics of the Solar Corona, Chichester, UK: Springer, Praxis*

Obridko, V. N., Shelting, B. D., & Kharshiladze, A. F. 2006, *Geomagn. Aeron.*, 46, 3, 294

Veselovsky, I. S. 2002, *Adv. Space Res.*, 29, 10, 1532

Rudenko, G. V. 2001, *Sol. Phys.*, 198, 1, 5

Veselovsky, I. S. & Shugay, Yu. 2008, *37th COSPAR 2008*. http://www.cospar-assembly.org/abstractcd/COSPAR-2008

Author Index

605

Subject Index